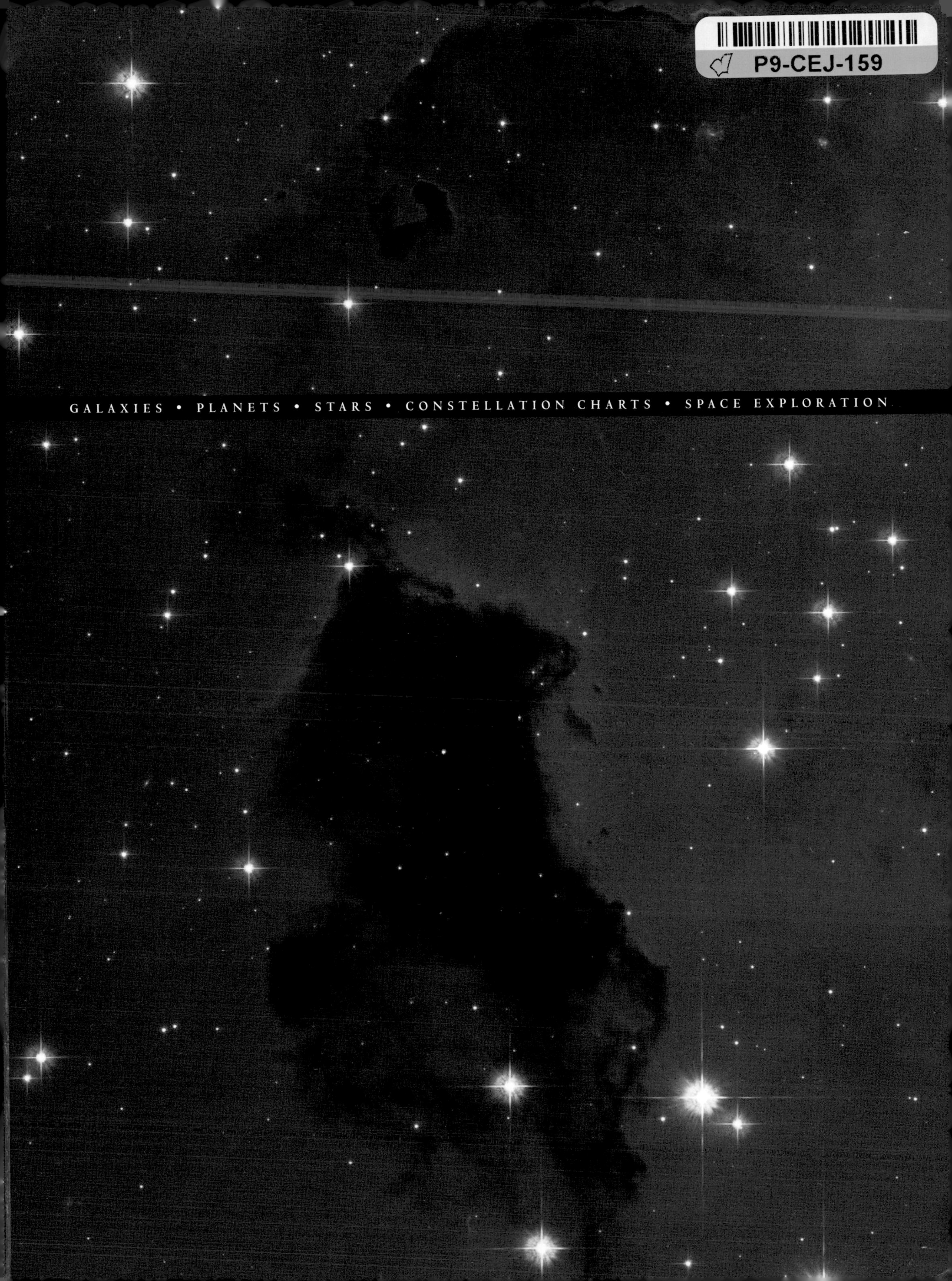
GALAXIES • PLANETS • STARS • CONSTELLATION CHARTS • SPACE EXPLORATION

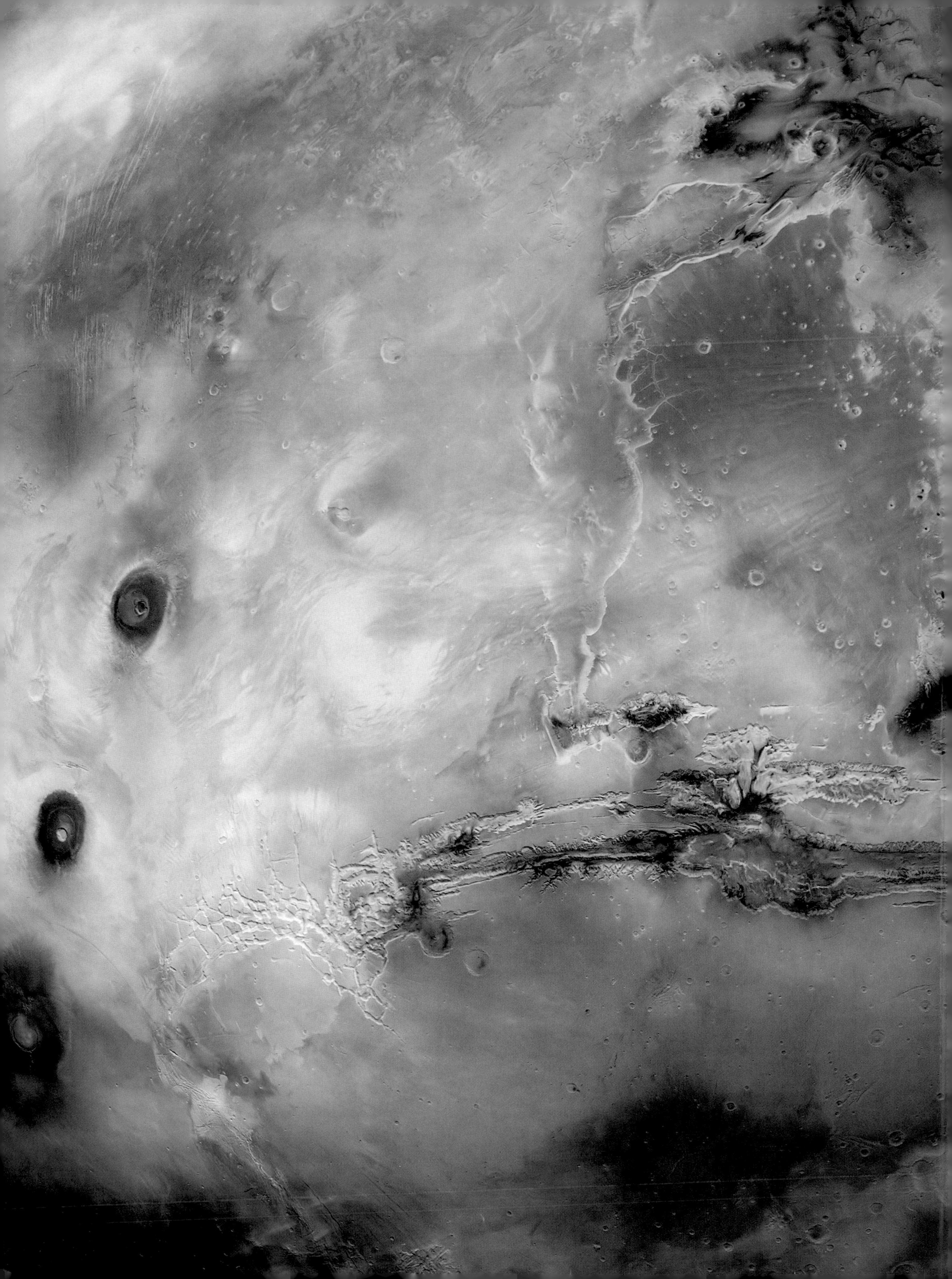

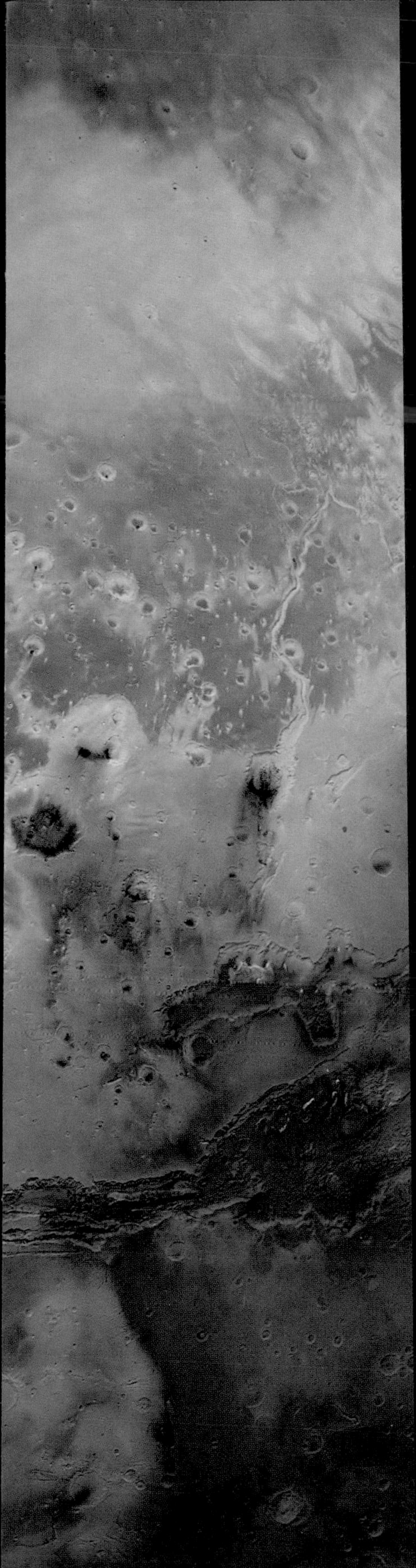

GALAXIES • PLANETS
STARS • CONSTELLATION CHA
SPACE EXPLORATION

foreword by
Sir PATRICK MOORE

chief consultant
Professor FRED WATSON

First published in 2007 as *Astronomica* by
Millennium House Pty Ltd
52 Bolwarra Rd, Elanora Heights, NSW, 2101, Australia
Ph: 612 9970 6850
Fax: 612 9913 3500
Email: editor@millenniumhouse.com.au

Reprinted 2009, 2010

ISBN: 978-1-921209-73-4

Authors:
Millennium House would be pleased to receive submissions from authors. Please send brief submissions to editor@millenniumhouse.com.au

Photographers and Illustrators:
Millennium House would be pleased to receive submissions from photographers or illustrators. Please send submissions to editor@millenniumhouse.com.au

Color separation by Pica Digital Pte Ltd, Singapore
Printed in China

Photographs on cover and preliminary pages:

Front Cover Hubble Space Telescope image of the Orion Nebula.

Back Cover Hubble Space Telescope image of the Pinwheel Galaxy (M101) in the constellation Ursa Major.

Page 1 The dark areas seen here in NGC 281 are known as Bok globules. These molecular clouds attract gas and dust and may sometimes form stars at their core.

Pages 2–3 This photo mosaic of the Valles Marineris region of Mars focuses on the canyon system of the area, which stretches for more than 1,860 miles (3,000 km).

Pages 4–5 *Apollo 12* crew member Alan Bean descends the ladder of the Lunar Module *Intrepid* to become one of the select few who have walked on the lunar surface.

Pages 10–11 Launched in October 1997, the *Cassini* probe has provided a wealth of images of Saturn. This image taken by *Cassini* shows the planet from an edge-on view.

Publisher	Gordon Cheers
Project manager	Janet Parker
Art director	Stan Lamond
Managing editors	Janet Healey Carol Jacobson
Editors	Loretta Barnard Helen Cooney Heather Jackson Melody Lord Anne Savage Marie-Louise Taylor
Cover design	Stan Lamond
Designers	Warwick Jacobson Lena Lowe Avril Makula Ingo Voss
Picture research	Jane Cozens Carol Jacobson Oliver Laing Melody Lord Debi Wager
Illustrators	Susan Cadzow Andrew Davies Paula Kelly Stephen Pollitt Glen Vause
Cartography	Andrew Davies David Hosking Samantha Hosking Alan Smith
Index	Jon Jermey
Production	Simone Russell
International Sales	Kanagasabai Suppiah
Publishing assistants	Bernard Roberts Liam Wilcox

WARNING: To avoid blindness, never look directly at the Sun at any time, including an eclipse. Neither the authors nor the publisher may be held responsible for any type of damage or harm if this advice is ignored.

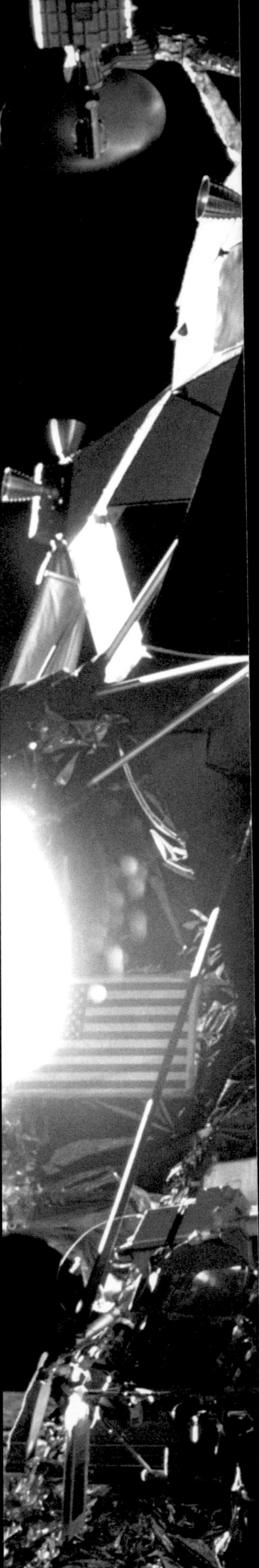

Contributors

CHIEF CONSULTANT

Professor Fred Watson
Professor Fred Watson says he has spent so many years working in large telescope domes that he has started to look like one. He is Astronomer in Charge of the Anglo-Australian Observatory at Coonabarabran, where his main scientific interest is gathering information on very large numbers of stars and galaxies. He is also an adjunct professor at the Queensland University of Technology and the University of Southern Queensland. Fred is the author of *Stargazer—the life and times of the telescope*, and is a regular broadcaster on ABC radio. In 2003, he received the David Allen Prize for communicating astronomy to the public, and in 2006 was the winner of the Australian Government Eureka Prize for Promoting Understanding of Science. Fred has an asteroid named after him (5691 Fredwatson), but says that if it hits the Earth it won't be his fault.

CONTRIBUTORS

Dr Martin Anderson
Dr Martin Anderson is an astrophysicist with a keen interest in the education and history of astronomy. His main areas of expertise are in radio and x-ray astronomy, with a particular interest in the development of 3-D stereo multimedia to promote a better understanding of astronomical concepts. He has over 15 years experience as an astronomy educator, working at the Nepean Observatory and currently at Sydney Observatory. He has taught at the University of Western Sydney and Flinders University, and has spent the last decade actively promoting the need to establish a large planetarium in Sydney.

Colin Burgess
Sydney-born space historian Colin Burgess is the author of several non-fiction books on military and spaceflight subjects, as well as children's books and a humorous history of Qantas Airways, with whom he worked as a flight attendant and crew manager for 32 years before retiring in 2002. His interest in human spaceflight began with the successful 1962 launch of Mercury astronaut John Glenn, and today he is recognized as the author of several popular biographical books on spaceflight topics. These include *Teacher in Space: Christa McAuliffe and the Challenger Legacy*; *Oceans to Orbit* (an authorized biography of oceanographer/astronaut Paul Scully-Power); *Fallen Astronauts*, his tribute to the eight NASA astronauts who died prior to the first manned lunar landing; and the self-explanatory *Animals in Space*. He is currently working with an American university press as series editor and author for a ten-volume set of books detailing the complete history of space exploration.

Les Dalrymple
Les Dalrymple is one of the most experienced amateur astronomers in Australia and has been observing the night sky for over 36 years. He boasts of never having purchased a telescope—preferring instead to build his own from a mix of home-made and commercially manufactured components. He currently observes with 12-inch (30 cm) and 18-inch (46 cm) Newtonian reflectors. An avid observer, his chief interest is deep-sky observing—particularly galaxies, planetary nebulae, and globular star clusters. He is a past president of the Sutherland Astronomical Society in Sydney, Australia, and currently works as an evening guide and lecturer at the Sydney Observatory, where he also conducts practical astronomy courses. Les is in considerable demand as a speaker at all manner of amateur astronomical gatherings, is a contributing editor to *Sky & Telescope* magazine and *Australian Sky & Telescope* magazine, a former consulting editor to *Sky & Space* magazine, and has contributed to several other astronomical publications.

Paul Deans
For many years Paul Deans wrote, produced, and directed multimedia shows for planetarium theaters in Edmonton, Toronto, and Vancouver (in Canada). Until recently he worked as the book editor at Sky Publishing (Cambridge, MA), where he was also the editor of *SkyWatch* magazine, and an associate editor for *Sky & Telescope* and *Night Sky* magazines. Now based in Edmonton, Paul is currently a freelance science writer and editor who also specializes in science-themed travel.

Stefan Dieters
Stefan Dieters started his interest in astronomy as an eight-year-old with the *Apollo 8* mission round the Moon and a fabulous look at Saturn through a old 8-inch (20 cm) refractor in Nelson, New Zealand. He became an active amateur, concentrating on the observation of variable stars. As an undergraduate, Stefan was editor of Australia's first space news magazine. He graduated from Monash University with honors in physics in 1983, after studying spotted stars; PhD with the x-ray and optical astronomy group at the University of Tasmania, studying neutron stars and supernova 1987A. In a series of postdoctoral positions with the University of Amsterdam, Max Planck Institute for Astrophysics in Germany, and a joint position with University of Alabama Huntsville and the NASA Marshall Space Flight Center, he studied the time and spectral variability of black hole and neutron interacting binaries. This included the discovery of the first magnetar—a neutron star with an ultra-strong magnetic field. Stefan is currently an Honorary Research Associate with the University of Tasmania, with the international PLANET consortium looking for extra solar planets using gravitational microlensing. In 2006, the group announced two new planets—a Jupiter analog and a frozen super-Earth.

Kerrie Dougherty
Kerrie Dougherty is Curator of Space Technology at the Powerhouse Museum, Sydney, Australia. She developed Australia's first major museum space technology display, *Space—Beyond This World*, at the Powerhouse and co-authored *Space Australia*, the first overview history of Australia's involvement in space activities. Kerrie is also a member of the Faculty of the International Space University, Strasbourg, France, where she lectures in space and society studies, and is active on international committees in space education and outreach and space history. She is the author of a number of scholarly and popular articles on space history and actively communicates space to the public through adult education and community programs. A lifelong science fiction fan, Kerrie has also co-authored two official Star Wars guides, *Star Wars: Complete Locations* and *Star Wars: Complete Cross-Sections*, and a Dr. Who reference book *Dr. Who: the Visual Dictionary*.

Professor Anthony Fairall
Anthony Fairall is Professor of Astronomy at the University of Cape Town and Planetarium Astronomer at Iziko Museums of Cape Town. He completed a BSc Hons at the University of Cape Town and a PhD at the University of Texas. His research career spans 40 years; his best known discovery is the extreme Seyfert galaxy known as "Fairall 9." He has produced some 200 research papers, 25 planetarium scripts, countless popular articles, and four books. He is an Associate of the Royal Astronomical Society (UK) and a Fellow of the University of Cape Town. He is also a Fellow of the International Planetarium Society.

Francis French
Francis French is originally from Manchester, England, although he now lives and works in southern California. He has been working for over a decade in the field of science education, particularly in making science and technology accessible and understandable to family audiences in informal learning settings such as museums. This has included positions at the Museum of Science and Industry in Manchester, the San Bernardino County Museum, and the Reuben H. Fleet Science Center in San Diego, California, where he also served as the Spaceflight and Astronomy Spokesperson, briefing media on space-related stories. His work has included regular collaborations with NASA, retired astronauts, notable astronomers, and astronomical observatories around the world, and a banner he designed was flown on the space shuttle *Columbia*'s last successful mission. He currently serves as Director of Events with Sally Ride Science, working for America's first woman in space. Francis is the co-author of two spaceflight history books.

James Inglis
James Inglis specializes in reviews, interviews, and opinion pieces, and has been published in various Australian national and state newspapers and periodicals. His interests include adult learning, workplace training, design and presentation, and analysis of language.

Nick Lomb
Nick Lomb studied at Sydney University, Australia, obtaining a BSc (Hons) and a PhD. He has been at Sydney Observatory for 28 years, mostly as curator of astronomy. At the Observatory he is closely involved in the planning and mounting of exhibitions, such as the current *By the Light of the Southern Stars* exhibition. He is often called upon by the media to provide information to the public whenever interesting things happen in the sky. Each year Nick prepares the *Australian Sky Guide*, which is now sold around the country. He is the author of the *Transit of Venus: the scientific event that led Captain Cook to Australia*, Powerhouse Publishing 2004, and the co-author of two books: a children's book, *Astronomy for the Southern Sky*, Nelson, 1986, and a pictorial history of Observatory Hill and Sydney Observatory, *Observer and Observed*, Powerhouse Publishing, 2001.

Sir Patrick Moore
Sir Patrick Moore, CBE FRS, has his private observatory at Selsey, South England, and has been concerned mainly with observations of the Moon. He has written a number of astronomical books, and since 1957 has presented a monthly program, *The Sky at Night*, on BBC Television. He is a member of the International Astronomical Union, and an Honorary Member of the Royal Astronomical Society of New Zealand.

Jonathan Nally
Jonathan Nally is an award-winning Australian author who specializes in the fields of astronomy and space exploration. In 1987 he was the co-founder of Australia's first astronomy magazine, *Southern Astronomy* (the name was later changed to *Sky & Space*), which he led for 16 years, developing it into one of the world's leading astronomical publications. In 2004 he became the founding editor of *Australian Sky & Telescope* magazine. Jonathan has also been a regular on Australian radio and television since the early 1990s, appearing on everything from children's television to live coverage of space events. In 2000, he was awarded the inaugural David Allen Prize by the Astronomical Society of Australia (the body of professional astronomers) for excellence in communicating the science of astronomy to the public. He has served on astronomical advisory bodies, and these days concentrates full time on writing for radio, print, and multimedia productions.

Christophe Rothmund
Born in Offenburg, Germany, Christophe Rothmund obtained degrees in mechanical engineering degrees and aerospace engineering, and has been working in the French space industry ever since. Christophe was first involved in the manned European shuttle plane project *Hermes*, before switching to Ariane 4 and the development of Ariane 5. He was then assigned to advanced projects such as the French hypersonics research program PREPHA and some research activities for the European Space Agency. Besides his engineering duties, Christophe is also a recognized space historian. Since 1991 he has participated actively in the Space History Symposia of the International Academy of Astronautics. He has written over 20 papers on the history of French rocket propulsion systems. He was also a member of the team that authored the book *Spaceflight* and its subsequent versions in German and Italian languages.

Alan Whitman
Retired from the Canadian Meteorological Service, Alan Whitman is a full-time amateur astronomer. His British Columbia backyard is dark enough that the Gegenschein is sometimes visible. He uses 16-inch and 8-inch Newtonians, an 80 mm refractor, and binoculars. His club's observatory has a 24-inch Cassegrain. Four magazines have published his articles, including 13 articles in *Sky & Telescope*, and many astronomy clubs and star parties have invited him to be a speaker. In 1984, he founded the Mount Kobau Star Party, western Canada's major annual observing event. Alan has observing notes on every significant deep-sky object, having spent 26 nights using large Dobsonians at Australian dark sites. He was the volunteer director at New Mexico's Chaco Observatory for a year and used its 25-inch Obsession. He also observes the planets, asteroid occultations, and grazing lunar occultations. An avid eclipse-chaser, he has traveled the world to see five total solar eclipses and one annular eclipse.

EARTH is an epic publishing feat never to be repeated, proudly created by Millennium House

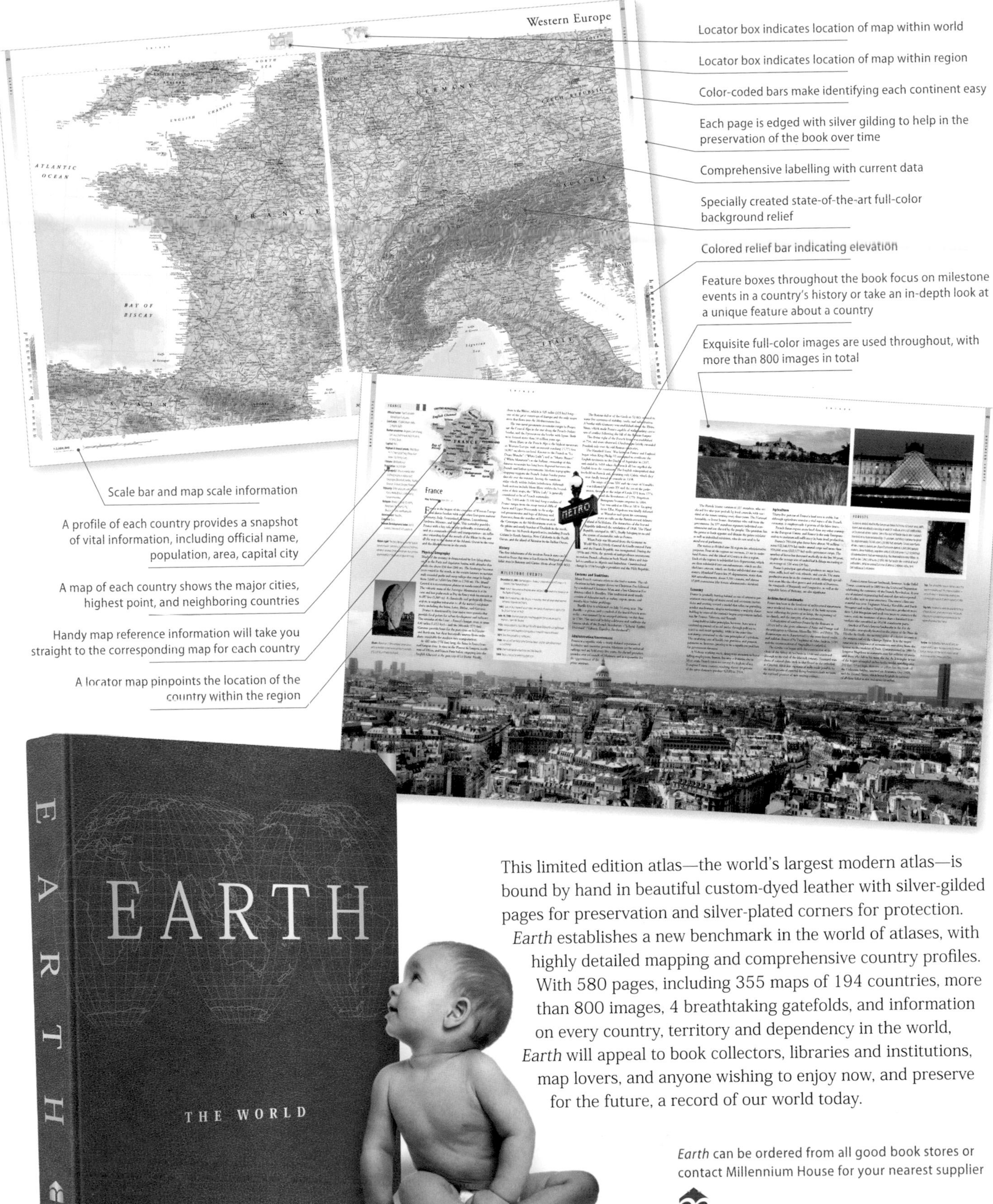

Western Europe
ATLANTIC OCEAN
FRANCE
BAY OF BISCAY
Locator box indicates location of map within world
Locator box indicates location of map within region
Color-coded bars make identifying each continent easy
Each page is edged with silver gilding to help in the preservation of the book over time
Comprehensive labelling with current data
Specially created state-of-the-art full-color background relief
Colored relief bar indicating elevation
Feature boxes throughout the book focus on milestone events in a country's history or take an in-depth look at a unique feature about a country
Exquisite full-color images are used throughout, with more than 800 images in total
Scale bar and map scale information
A profile of each country provides a snapshot of vital information, including official name, population, area, capital city
A map of each country shows the major cities, highest point, and neighboring countries
Handy map reference information will take you straight to the corresponding map for each country
A locator map pinpoints the location of the country within the region
France
METRO
EARTH
EARTH
THE WORLD
This limited edition atlas—the world's largest modern atlas—is bound by hand in beautiful custom-dyed leather with silver-gilded pages for preservation and silver-plated corners for protection. *Earth* establishes a new benchmark in the world of atlases, with highly detailed mapping and comprehensive country profiles. With 580 pages, including 355 maps of 194 countries, more than 800 images, 4 breathtaking gatefolds, and information on every country, territory and dependency in the world, *Earth* will appeal to book collectors, libraries and institutions, map lovers, and anyone wishing to enjoy now, and preserve for the future, a record of our world today.
Earth can be ordered from all good book stores or contact Millennium House for your nearest supplier
www.millenniumhouse.com.au

C o n t

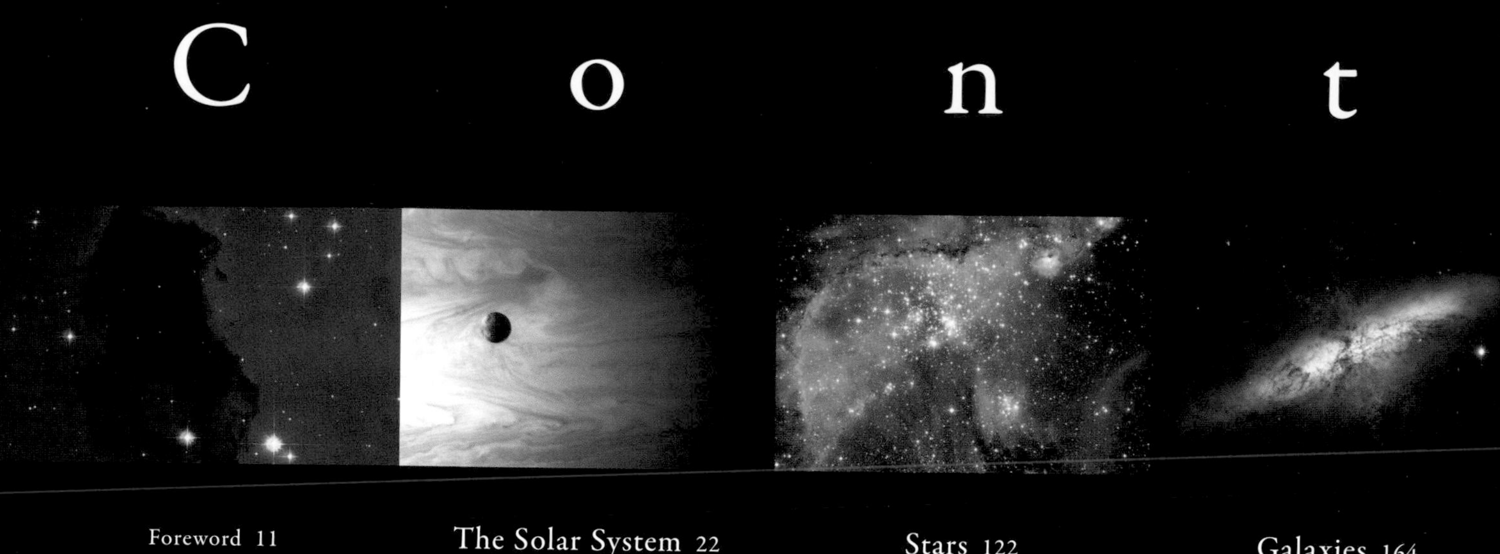

e n t s

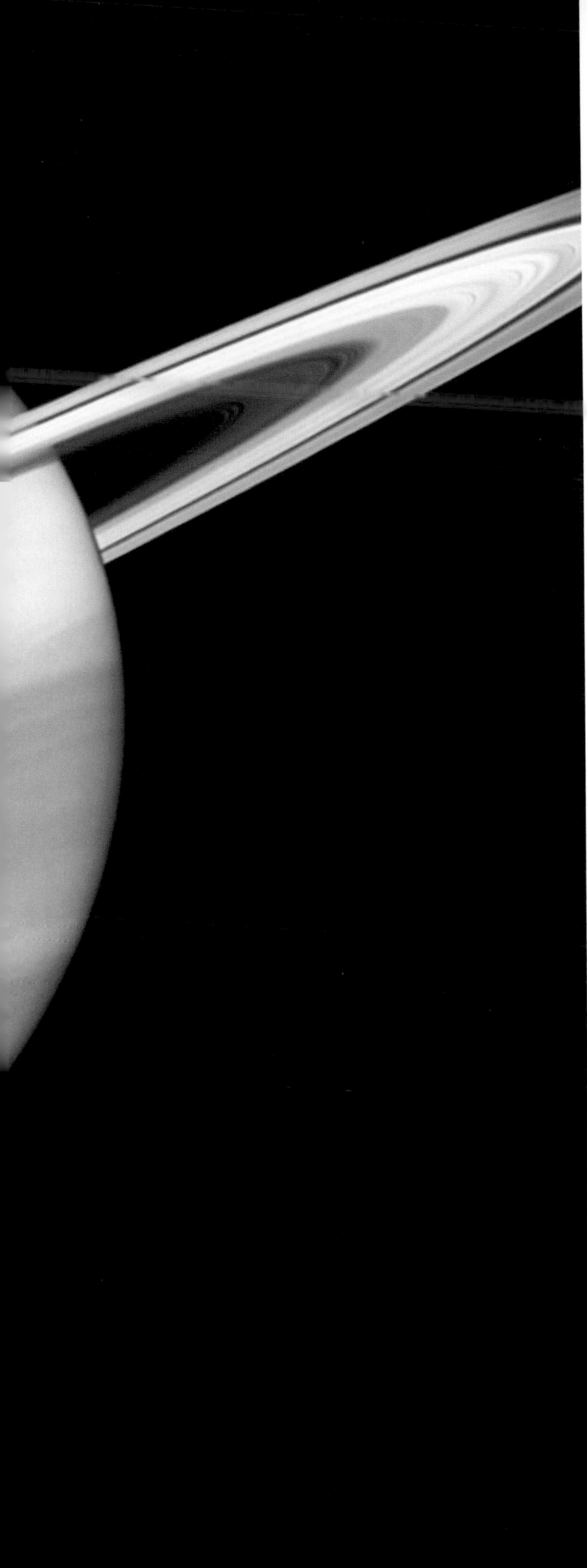

Foreword

Many books on astronomy are now published each year, and this is natural enough—everyone must surely take at least a passing interest in the sky. Some of the books are written for experts, and will be beyond the grasp of readers who have no prior knowledge; others cater for complete beginners, and are purely elementary. This present book, due to leading astronomers around the world, strikes a happy mean. It is written in a way that makes the text comprehensible to the novice, and it also contains a great deal of information that will be of real use to the serious student.

Astronomy is one of the best of all hobbies. Anyone can play a part, and there is no need to invest in a large and expensive telescope, or indeed in any advanced equipment. Amateurs can do work of importance, and they are welcomed by professional astronomers—one need think only of men such as the Rev. Robert Evans, an Australian clergyman who has earned himself an international reputation as an observer, and is known and respected all over the world. Remember, too, that your beginner of today is your researcher of tomorrow.

If you decide to become an astronomer, either as an amateur or (ultimately) a professional, you will not regret it. You will make many new friends—and there is always something new to see in the sky. Whichever path you intend to follow, you will find that this book is of the utmost value to you.

Sir Patrick Moore

Asking the Big Questions

Most of us think of astronomy as the science of the night sky and, indeed, that's a good starting point. Literally, astronomy means "arranging the stars." But even the most cursory glance through this book will reveal that today's astronomy encompasses far more than mere stargazing. Painstaking detective work, using information from every nook and cranny of the universe, has enabled us to construct a picture of things "out there" that is extraordinarily cogent and complete.

Right The Cassini spacecraft sent back this image of Saturn's A ring taken at ultraviolet wavelengths. Turquoise rings are icier than the "dirty" red rings that contain more dust and smaller particles.

Our knowledge now encompasses the entire spread of space and time, from the Earth itself to the horizon of the universe, and from the Big Bang to the grand finale—whatever that might be. If astronomy is defined as the study of the universe and all it contains, then truly it is the science of everything.

The popular conception of an astronomer is likewise misleading. Almost everyone imagines a slightly crazed individual looking through an old-fashioned spyglass in a hopeful search for new celestial wonders. The reality is quite different, and many people are disappointed to know that it is well over a century since professional astronomers routinely made their discoveries peering through an eyepiece. Today, they invariably stare at a computer screen.

Above The Earth and Moon were imaged by *Mariner 10* in 1973 as it departed Earth for the inner Solar System. The images have been combined to illustrate the relative sizes of the two bodies.

If that suggests—as it does to some—that the romance has gone out of astronomy, then consider the kinds of questions that astronomers have in mind when they are planning and executing their research. Where did we come from? What is our destiny? And—perhaps most significantly—are we alone in this wild and wonderful universe? Those "big questions" go to the heart of our most profound understanding of who and what we are. Truly, romance abounds in the ultimate aims of astronomy.

The process by which astronomers address these "big questions" is the same as that used by all scientists in carrying out their research. It often involves making a succession of relatively small discoveries that are gradually pieced together to build up the bigger picture. And there is usually a step-wise process of refinement, in which theoretical ideas suggest targeted observations—the results of which are, in turn, used to improve the theory. Eventually, with further cycles of iteration, a self-consistent picture is built up in which observation and theory are in good agreement, leading to a consensus of opinion among the scientific community.

Unlike most other scientists, however, astronomers are almost always limited to remote observations of their subjects rather than hands-on experimentation. And no-one would be surprised to hear that the main tool at their disposal is the telescope. Until 1932, when Karl Jansky of the Bell Telephone Laboratories discovered radio waves coming from space, there was only one kind of astronomical telescope. It collected and focused ordinary visible light, using the science of optics to allow information to be retrieved—originally by human eyes, but, from the 1880s, by photographic plates.

Today, however, there are as many different kinds of telescope as there are varieties of natural radiation traversing the universe. A plethora of names identifies these ghostly emissions. Gamma-rays, X-rays, ultraviolet rays, infrared rays, millimeter- and radio-waves—as well as visible light. Together, these emissions comprise what is known as the electromagnetic spectrum.

It has often seemed odd that the various rays of the electromagnetic spectrum are bereft of a convenient mnemonic to indicate their correct order. Generations of English-speaking schoolchildren learned "Richard of York gained battles in vain" to remember the spectrum colors of the rainbow—red, orange, yellow, green, blue, indigo, and violet. Indigo was never really there, but it made the mnemonic easier. If kids could recall the ill-fated duke's lifespan (1411–1460), so much the better—the phrase had served a dual purpose.

But gamma-rays, X-rays, ultraviolet, visible, infrared, millimeter, and radio is a list that doesn't easily lend itself to such useful trivia. How about "great xylophonist uttered valediction in masterly recital?" While that is perfectly true (of a certain Joseph Gusikof, who died while giving a virtuoso performance in 1837), it is rather easier to remember the individual radiations by name.

Today, we recognize these various rays as different manifestations of the same thing—undulations in electric and magnetic field-strength, or so-called electromagnetic waves. They are best imagined as ripples on the surface of a pond, except that they are transmitted through three-dimensional space rather than over a two-dimensional surface. The details are unimportant; the essential aspect is that all the varieties of electromagnetic rays are distinguished from one another only by the characteristic distance between one wave crest and the next—the wavelength.

The electromagnetic spectrum runs all the way from gamma-rays—with vanishingly small wavelengths measured in billionths of a millimeter—to radio waves, which span whole meters. Somewhere near the middle is visible light, whose wavelength increases with color from violet to red. But, in a curious paradox, physics tells us that these waves can equally well be thought of as packets of energy—or "particles" of radiation—which we normally call photons. Contrary to what might be expected, the energy content of the photons falls as the wavelength of the radiation gets longer.

One of the heartwarming facts that astronomy has brought to us is that the Earth is constantly bathed in radiation—covering the whole electromagnetic spectrum—from sources everywhere in the universe. Some of it comes from the Big Bang itself, a faint whisper of the time 13.7 billion years ago when the universe was still filled with the fires of its own creation. In fact, you can see that particular radiation in your own home if you tune your TV slightly off-station. About one in ten of those flickering white dots comes from the Big Bang.

All radiation carries information about both its source and the space through which it has traveled, and detailed analysis of the signal to reveal the information hidden within it is the main aim of observational astronomy. But most of the radiation never reaches the surface of the planet—it is absorbed by our stalwart and trusty atmosphere, the protector and sustainer of the biosphere.

Visible light, of course, reaches the ground, and so do some categories of radio and infrared radiation. Telescopes designed to probe the universe using these types of radiation can therefore be built on *terra firma*. Among practitioners, they are accurately but unimaginatively known as ground-based telescopes. Explorations using any other type of radiation have to be carried out from above the atmosphere, and that usually means the use of spacecraft—although occasionally high-flying aircraft or balloons can be used. Either way, it puts observing into a different financial and operational league from ground-based endeavors. It also places limits on the physical sizes of the telescopes that can be used.

Astronomy based on observation with visible light is known as optical astronomy, and by analogy, visible-light telescopes are called optical telescopes. Today they always contain a large dished mirror as their main light-collecting element. It is the diameter of the mirror that determines the effectiveness of the telescope and, in contrast with most other aspects of

Below Many ground-based telescopes provide images of the night sky. This image of the star-forming region NGC 281 in the constellation Cassiopeia was taken with the WIYN 0.9 m telescope at Kitt Peak National Observatory, Arizona, USA.

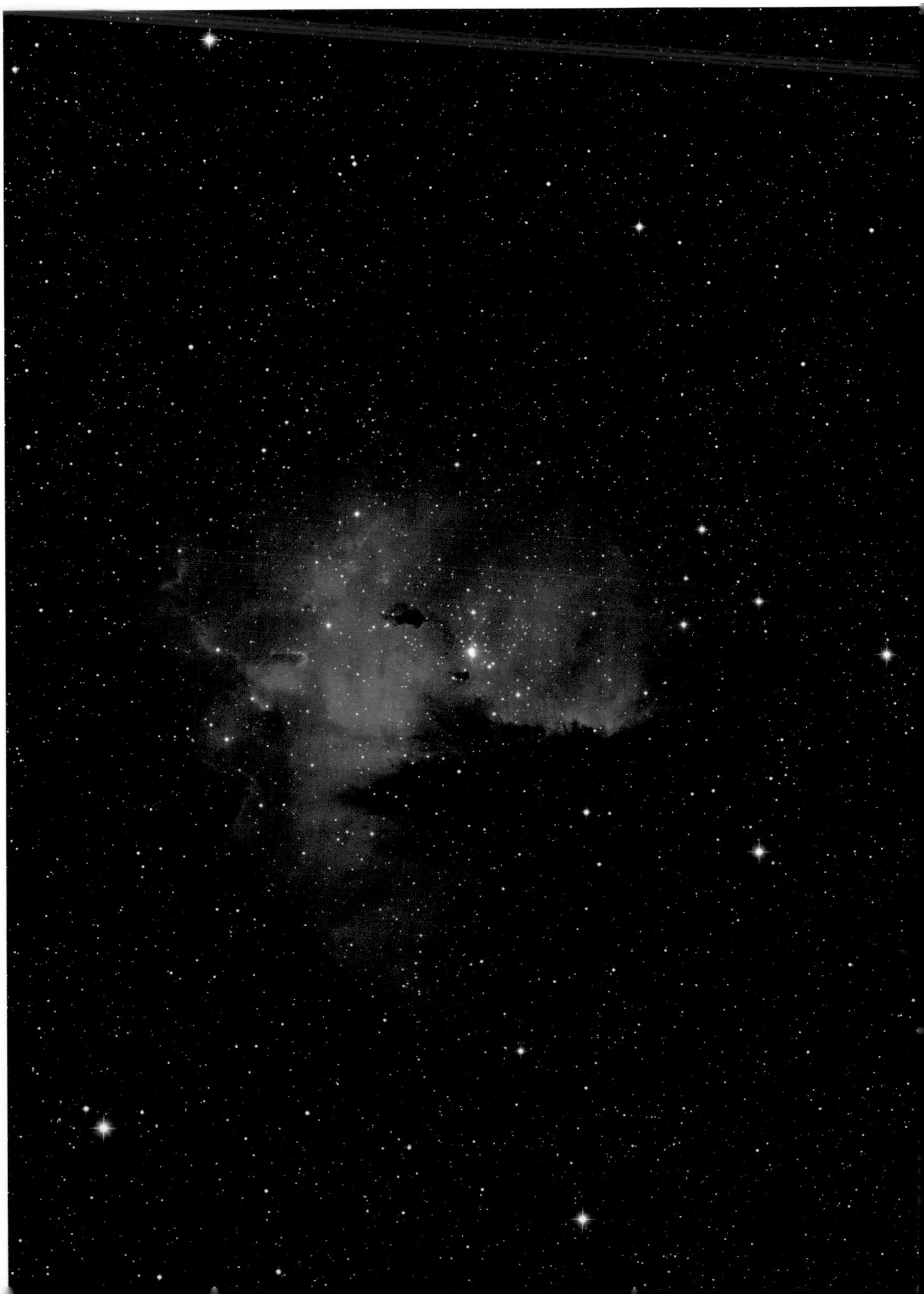

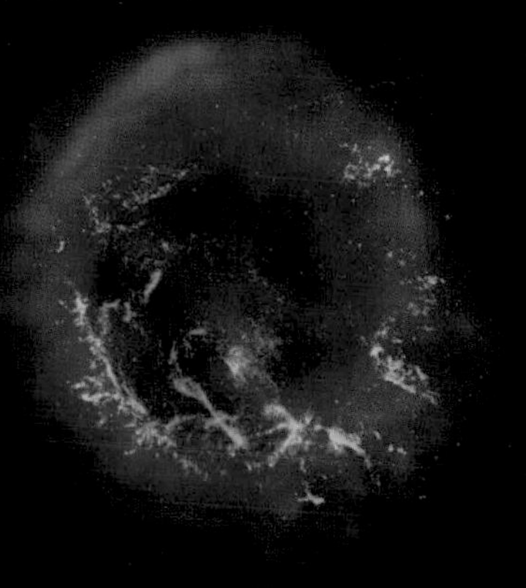

Below The supernova remnant E0102-72 was imaged in X-rays (blue), optical (green), and radio (red). The star exploded outward at speeds in excess of 12 million mph (20 million km/h) and collided with surrounding gas. This collision produced two shock waves, or cosmic sonic booms, one traveling outward, and the other rebounding back into the material ejected by the explosion.

modern technology, size is everything. The bigger the better. Today, we live in an era in which the largest optical telescopes have mirrors 8 to 10 meters in diameter, but a new age of 20-, 30-, and even 40-meter monsters is already on the horizon.

Unlike their more exotic cousins operating in other wavebands, optical telescopes require darkness to function, so observing with them is always night work. They also require clear skies. But no matter what sorts of telescope are used, whether they be gamma-ray, optical, millimeter-wave or whatever, the observations made with them are mutually complementary. They allow astronomers to "see" celestial objects with widely differing perspectives.

There is another form of radiation permeating the universe, and that is the stream of subatomic particles known as cosmic rays. These emissions originate in high energy processes taking place deep in the universe—exploding stars, for example—and their detection requires telescopes with quite different characteristics from conventional electromagnetic instruments. A further "non-electromagnetic" technique is likely to emerge over the next few years as gravitational waves begin to be observed. These elusive oscillations of the fabric of spacetime itself are known to be emitted by objects undergoing gravitational disturbance—massive stars collapsing into black holes at the end of their short lives, or superdense neutron stars in orbit around one another.

While gravitational waves have not yet been seen at first hand, their existence has been confirmed by indirect observations. When the finely tuned "telescopes" that are needed to observe them have reached the required level of sensitivity, a new observational window on the universe will be opened up. It is one that will allow us to probe regions of space from which electromagnetic rays can never emerge. Perhaps even the strange interiors of black holes will reveal their secrets.

With all this sophisticated hardware at their disposal, what are astronomers working on in these early years of the twenty-first century? There are several hot topics, but perhaps the hottest of all is the search for life beyond the Earth. It remains the case that we know of nowhere in the universe where living organisms exist other than here on Earth. And there is absolutely no doubt that humankind as a whole is fascinated with the possibility that we might share the universe with other life-forms.

That curiosity has given rise to a new scientific discipline, which has emerged only in the past 25 or so years. Indeed, its origins can be traced back to a meeting of the International Astronomical Union that took place back in 1982 in Patras, Greece. At that conference, a commission was formed to explore the possibilities of finding life beyond the Earth. The new commission was—and still is—known as the "bioastronomy commission," but the subject itself tends to be known today as astrobiology.

Astrobiology brings together a wide range of sciences, all of which have some bearing on the possible existence of life beyond the Earth. Thus biology, astronomy, and planetary science are key components, but other relevant disciplines like chemistry, geophysics, paleontology, and climatology are also encompassed. It is under the umbrella of astrobiology that the search for life beyond Earth is progressing across a wide range of fronts.

Above Arecibo Observatory in northern Puerto Rico, is one of the most powerful radio telescopes in the world. The reflectors are configured to receive radio information from space with minimal interference from man-made transmissions.

Perhaps the most spectacular is the direct exploration of the Solar System—space missions like the two robotic NASA rovers currently operating on Mars, and the NASA/ESA/ASI Cassini mission to explore Saturn and its moons. Thus far, such enterprises have been concerned with general exploration, with a particular emphasis on the search for places that might harbor life. Some missions, however (such as the two Viking Mars landers of 1976), have also carried out targeted biological experimentation in an attempt to detect the presence of living organisms. No conclusive evidence has yet emerged, although the Viking results are still controversial.

It is quite possible that during the next few years, we will discover rudimentary life-forms elsewhere in the Solar System by this process of direct exploration. The reasons for optimism center around the discovery that water seems to be abundant, locked up, for example, in subsoil permafrost on Mars, or in putative oceans under the icy surfaces of some of the moons of Jupiter and Saturn. Water is still thought to be a crucial prerequisite to the emergence of life, at least as we know it. Another reason to be hopeful comes from the discovery that many earthly bacteria can survive extremely hostile conditions. Even the vacuum of space—seared as it is with lethal radiation—presents no problems for some hardy microbes.

It seems very unlikely that we will find advanced extraterrestrial life within the confines of the Solar System. To find E.T., we will have to look further afield, and it is here that astronomy's arsenal of strategic hardware will come to the fore. The search for intelligent life elsewhere in the Sun's vicinity requires the exploration of planets revolving around other stars. That these are relatively commonplace has been understood since the first one was discovered back in 1995. Well over 200 "extra-solar planets" are now known, betraying their presence by the wobble they impart to the motion of their parent stars. Unlike the planets of the Solar System, their distances are measured in light-years rather than mere millions of kilometers, so it is impossible to contemplate sending space missions to observe them directly.

Left The *Stardust* spacecraft was sent to collect interstellar dust and carbon-based samples during a close encounter with Comet Wild 2. The *Stardust* sample capsule reentered Earth's atmosphere in the spectacular fashion pictured, in January 2006. Analysis of the sampled dust could yield important insights into the evolution of the Solar System and possibly even the origin of life itself.

Opposite Large radio telescopes gather information that occurs in the form of radio waves, allowing constant updates to current knowledge. Significant discoveries are made as a result of the data retrieved by these huge structures.

THE SCALE OF THE UNIVERSE

The Universe is so vast that it is difficult to picture the enormous distances and sizes of the structures it contains. This diagram uses a logical progression of scale to show just how big it really is. Beginning with Earth at the top left and moving clockwise, each cube represents an area of space one hundred times (two orders of magnitude) larger than the one before. For example, the diameter of Earth is approximately 12,750 kilometers, so you can see the planet taking up virtually the whole volume of a cube of space with sides measuring ten thousand kilometers (10^4]). In the next cube, with its sides measuring one million (10^6) kilometers, you can see how the 768,800-kilometer diameter of the moon's orbit fits nicely inside.

Earth
10^4 km

10^{22} km
Virgo Supercluster

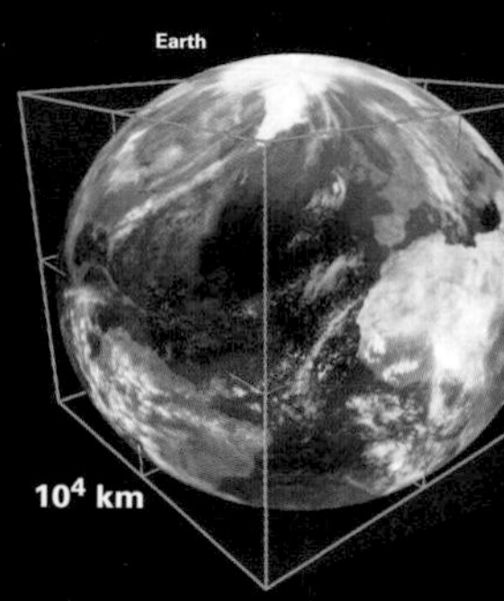

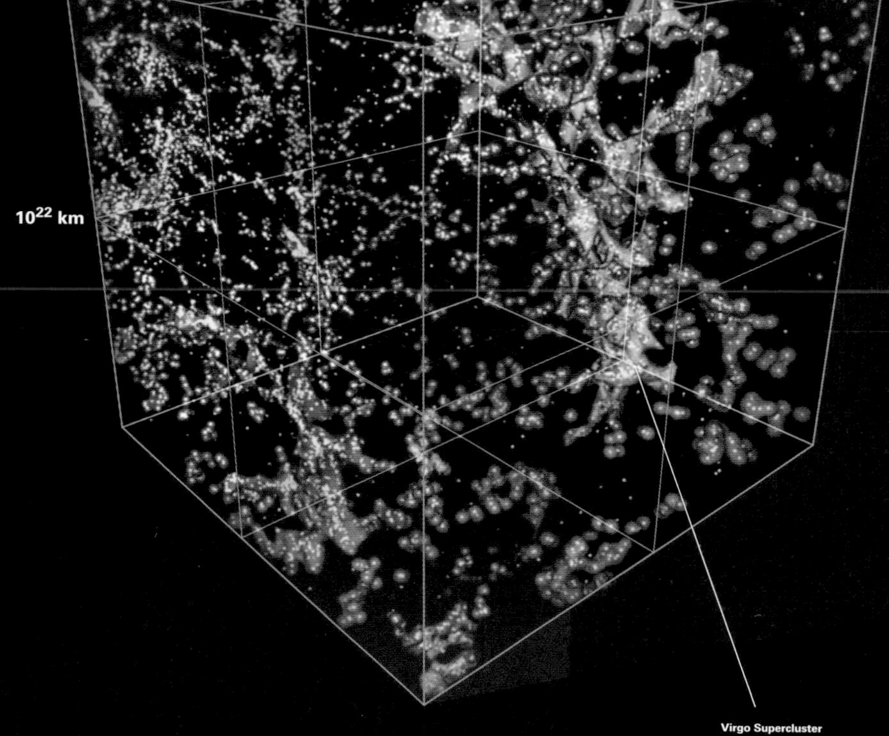

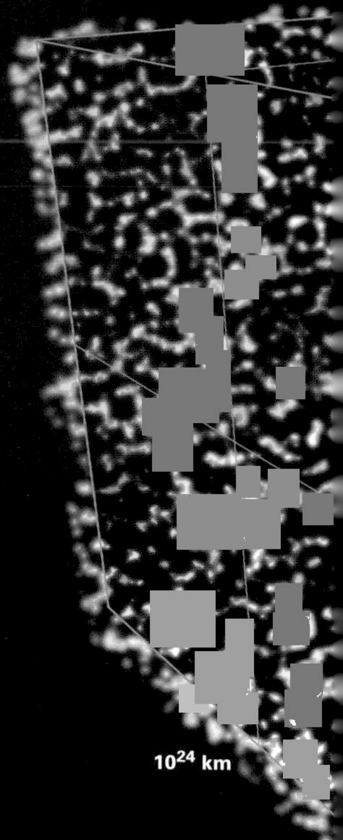

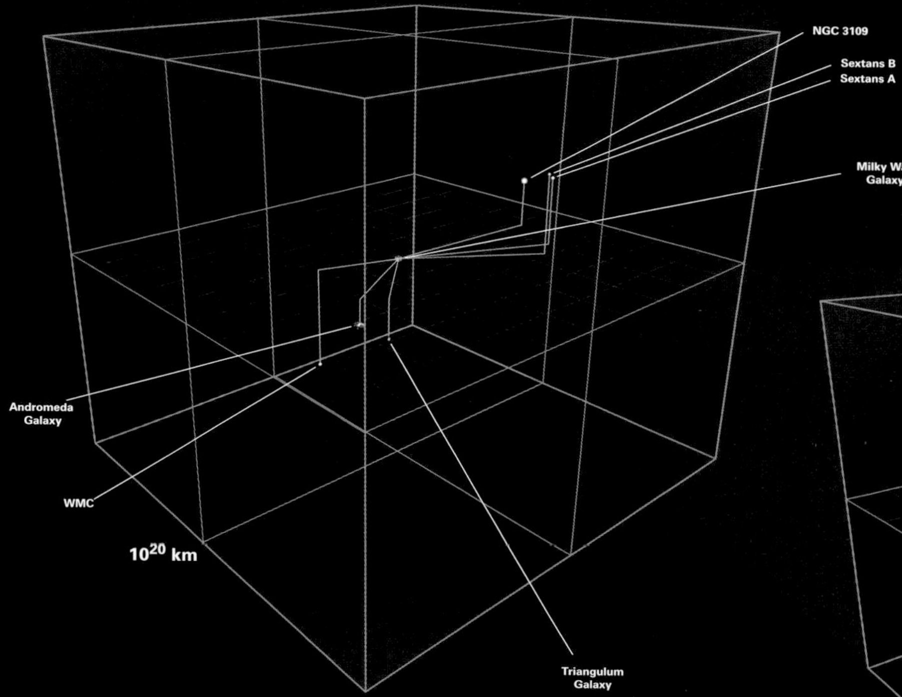

10^{24} km

NGC 3109
Sextans B
Sextans A
Milky Way Galaxy
Andromeda Galaxy
WMC
10^{20} km
Triangulum Galaxy

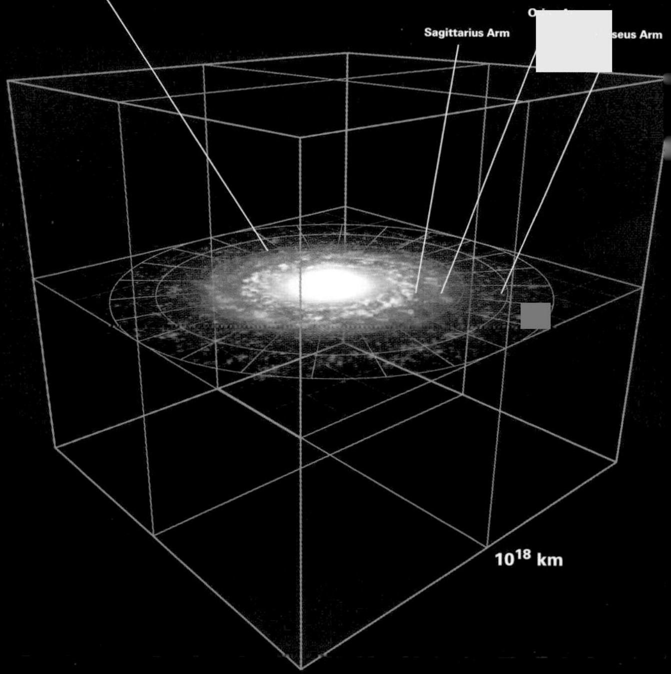

Sagittarius Arm
10^{18} km

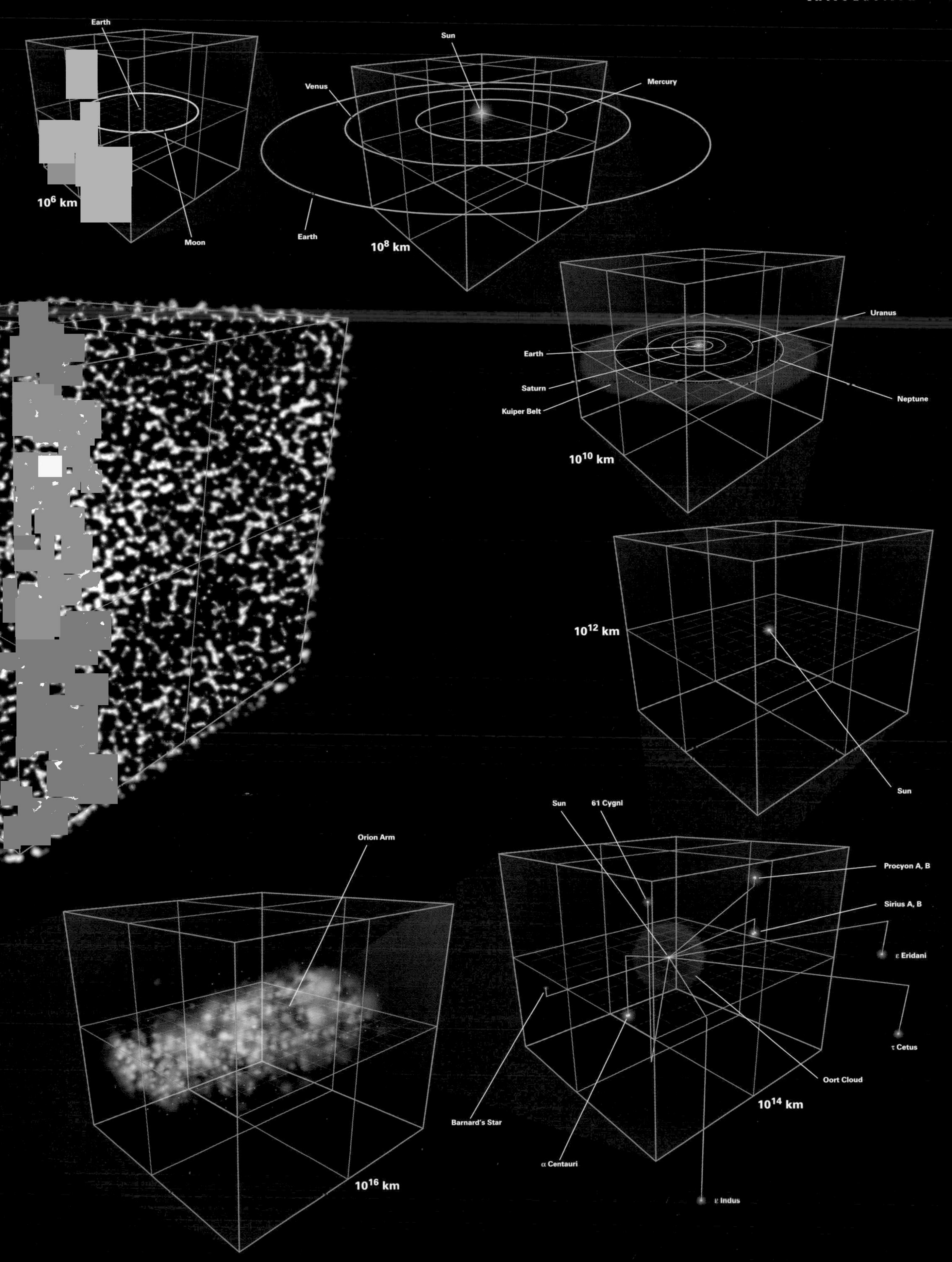
Earth
Moon
10^6 km
Sun
Venus
Mercury
Earth
10^8 km
Uranus
Earth
Saturn
Kuiper Belt
Neptune
10^{10} km
10^{12} km
Sun
Sun
61 Cygni
Procyon A, B
Sirius A, B
ε Eridani
τ Cetus
Oort Cloud
10^{14} km
Barnard's Star
α Centauri
ε Indus
Orion Arm
10^{16} km

We are thus thrown back onto large telescopes as our only means to explore these extra-solar planets. It's one thing to infer the presence of a planet by detailed observations of a star, but quite another to see it and analyze its atmosphere and surface for signs of life. The former technique is well within the capabilities of modern-day optical telescopes. This instrument has been very successful in the quest to find extra-solar planets. The technique of atmospheric analysis, however, requires direct observations of the planets themselves, and that can only be done with the coming generation of 20-meter-plus telescopes.

Below Since its discovery by Art Hoag in 1950—for whom this unusual galaxy is named—Hoag's Object Galaxy has intrigued astronomers. Space telescope images may aid in resolving the questions that surround this galaxy.

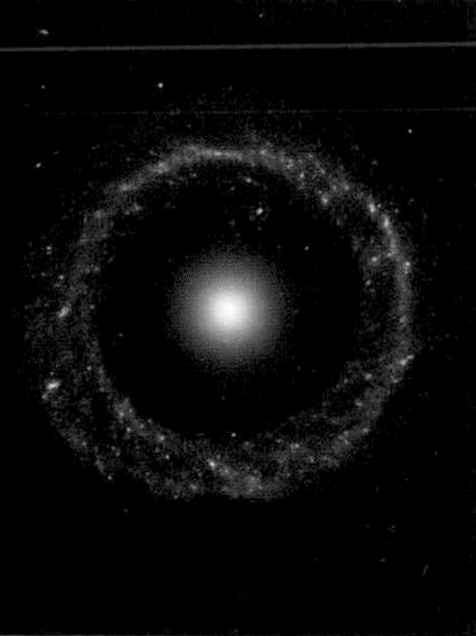

What these instruments will do is disentangle the faint image of a planet from the glare of its parent star, which is billions of times brighter. The light can then be decomposed into its rainbow spectrum colors using the magical instrument known as a spectrograph, and the hidden "bar-code" of telltale signatures examined. Astrobiologists will look for biomarkers that indicate the presence of living organisms—spectral features like oxygen and the so-called "vegetation red edge." It is even possible that industrial pollutants like carbon dioxide and heavier chemical compounds might be detected, hinting at the presence of intelligent life. The fact that those life-forms are trashing their planet with similar recklessness to the way we treat our own will surely do little to diminish the excitement of such discoveries, made across the abyss of space.

Other hot topics in modern astronomy raise some even bigger questions—and, indeed, they are among the biggest we can possibly ask. They are also among the most embarrassing, since astronomers have had to admit that they don't know what most of the universe is made of. We scientists hide our embarrassment behind two quite independent entities known as dark matter and dark energy. Those apparently simple ideas conceal two of the most puzzling problems in contemporary astrophysics. The mystery is deepened by the fact that between them, these invisible components of our cosmic environment make up an overwhelming 95 percent of the mass-energy budget of the universe.

Above Astronomers can only detect dark matter by observing how gravity affects light. By comparing this X-ray image and optical images, it has been determined that the galaxy clusters seen here (in blue) are located close to dark matter.

The quest to discover the nature of dark matter and dark energy presents challenges that demand all the ingenuity of astronomers and physicists. Moreover, it requires a formidable array of scientific hardware of the kind we have already described, particularly in the use of large ground-based optical and infrared telescopes. This mission to understand nature at its most fundamental level also places strict limits on sky quality at observatory sites, since the measurement of exceedingly faint objects is required. Thus, the darkest secrets of astronomers also include their quest for dark skies unpolluted by stray light from human activity.

Taking dark matter first, how do we know that it is present in the universe? It reveals itself only by one thing—the effect of its gravitational attraction on matter that we *can* see. Other than that, we have no way of detecting it. We can, however, sense this gravitational "smoking gun" by a number of methods. While many scientists imagine that the investigation of dark matter is a new topic with a short history, it was as long ago as 1933 that the US–Swiss astronomer Fritz Zwicky (1898–1974) first noticed that something didn't add up. He was investigating clusters of galaxies—the largest concentrations of matter in the universe.

Most people know that galaxies are aggregations of stars—hundreds of billions at a time—and that they often occur in the shape of a flattened disc with spectacular spiral arms. We live in such a galaxy—usually known as the Milky Way Galaxy, since we see its disc as a milky band of starlight stretching all the way around the sky. Such spiral galaxies also contain large amounts of gas (mostly hydrogen) and dust embedded in their discs. Other types of galaxies occur too, however, including the large football-shaped objects we call elliptical galaxies.

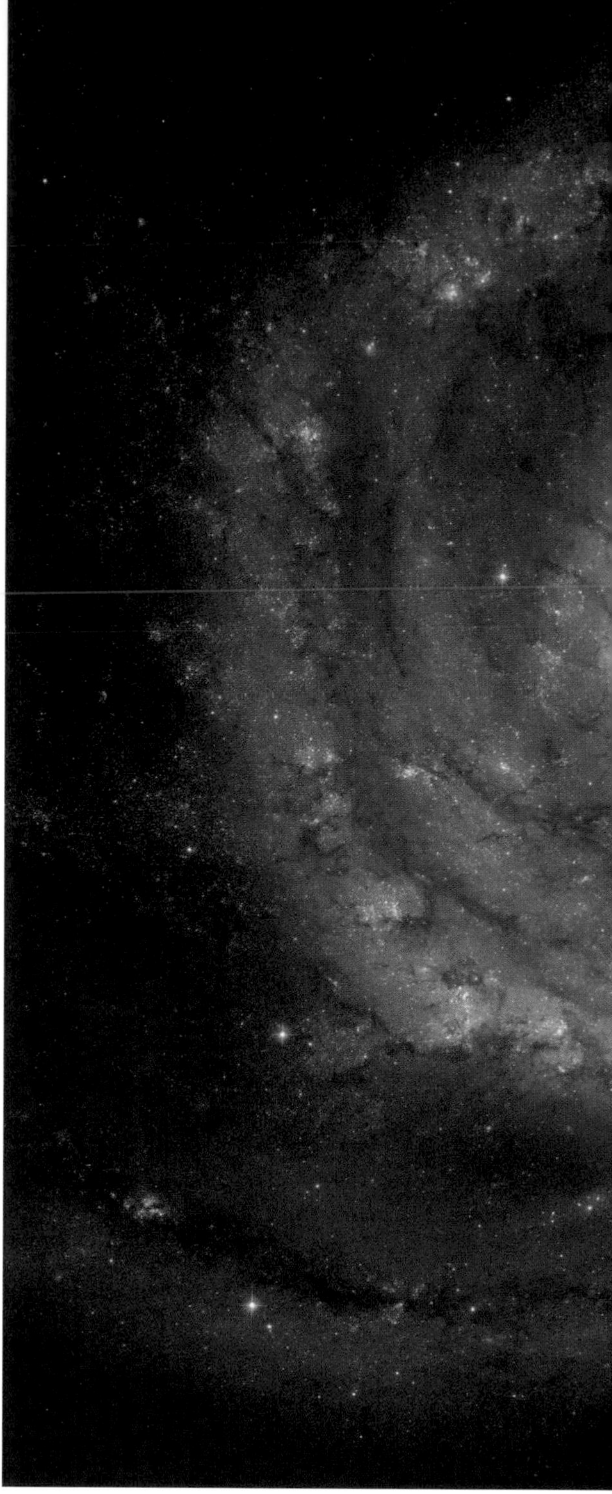

Zwicky was particularly interested in a very rich cluster of galaxies in the constellation of Coma Berenices in the northern hemisphere sky. Using a spectrograph, he was able to measure the speeds of several members of the cluster. He was amazed to find that these galaxies seemed to be moving too fast for the cluster to hold onto them, and quickly established that the gravitational attraction of the visible matter in

the cluster (the stars, gas, and dust in the constituent galaxies) was not enough to bind the whole thing together. Given their velocities, the galaxies he was observing should have escaped from the cluster long ago.

Zwicky inferred from this that something else was present—an invisible component that neither emitted light nor absorbed it from the radiation of background objects. Whatever it was, it exerted a gravitational pull on the members of the cluster. He was spot-on with that deduction, but little attention was paid to it at the time.

Forty-five years later, the US astronomer Vera Rubin made careful studies of the rotation of galaxies, and discovered new evidence for the existence of dark matter. We know that individual galaxies rotate, and, by selecting spiral galaxies that we see almost edge-on, can measure the way the rotational velocity changes with distance from the center of the galaxy. Rubin was surprised to find that far from falling off with distance as expected, the rotational velocity stayed almost constant out to the galaxy's extremities. This could only be explained if each galaxy was enveloped in a gigantic halo of dark matter. Thus was born the modern era of dark matter studies, culminating today in new sky surveys designed to explore its properties in fine detail.

The best tracers of dark matter are moving objects such as stars, and by measuring large numbers of these in our own and other galaxies, we can hope to obtain sufficient information

Above This Hubble Space Telescope image shows the striking structure of the two swirling spiral arms—comprised of stars and dust—of the Whirlpool Galaxy (M51).

Below By studying the distorted light created by gravitational lensing, astronomers have concluded that a ring of dark matter surrounds this galaxy cluster—Cl 0024+17. The gravitational lensing is determined by the light being bent and magnified by the powerful gravity of the cluster.

that the true nature of this mysterious substance can be identified. Such large-scale velocity surveys are currently being carried out in our own Milky Way Galaxy. These and other studies have already revealed much about dark matter, including its tendency to concentrate where normal visible matter does. In fact, we believe it was such concentrations of dark matter in the early universe that led to the formation of the first galaxies. The detailed exploration of dark matter in galaxies beyond our own remains in the future, requiring giant telescopes in the 20-meter class.

Dark energy has a similarly long pedigree, although the observational evidence only began to emerge more recently. It was in Einstein's adoption of his so-called cosmological constant—introduced belatedly into his equations of general relativity early in 1917—that the idea first appeared. General relativity is a theory of the way gravity acts by distorting space, and it remains the best theory of gravity we have. It has been tested many times, and our observations fit the theory with an accuracy that is now better than one part in a hundred thousand.

In 1917, however, Einstein thought his new theory was in big trouble. If his equations were applied to the universe as a whole, they became unstable, producing a universe that would have to expand or contract. But to the best of his knowledge, the universe was static. So he did something very clever. He introduced a mathematical entity that he called the

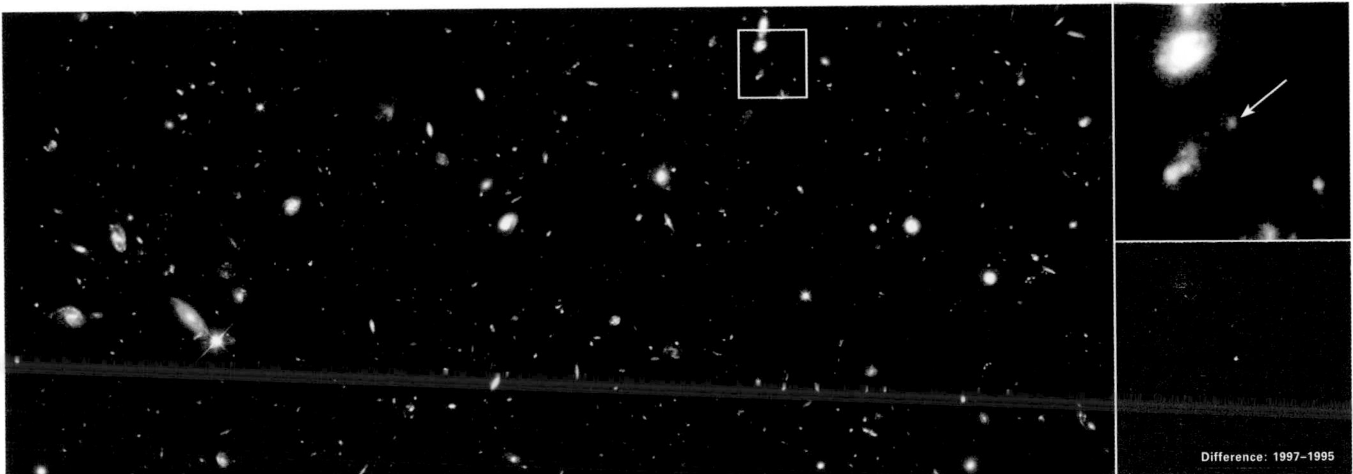

cosmological constant. This was supposed to represent an inbuilt repulsive (or attractive) force in the fabric of space that would counteract any tendency for it to expand or contract. When the real expansion of the universe was discovered 12 years later, however, Einstein quickly withdrew the idea in embarrassment, famously describing it as his "biggest blunder."

Except for a very few scientists, most people then simply assumed that the cosmological constant was zero, and that space had no inbuilt force field. But in 1998, two separate groups of astronomers produced hard evidence that far from slowing down as expected, the universe is expanding more rapidly today than it was seven or eight billion years ago.

This acceleration is attributed to an inherent springiness of space—or dark energy—that overcomes the tendency of the universe to decelerate because of the mutual gravitational attraction of everything within it. Its action may be the same as Einstein's cosmological constant, but there might also be subtle differences. A number of different hypotheses could give rise to a universe whose expansion accelerates, and so, in the best scientific tradition, tests are required to differentiate between them.

Exploring the properties of dark energy to try to understand exactly what causes it ranks with the dark matter question as the most pressing problem facing today's astrophysicists. In response, they are contemplating ambitious strategies to enable the research, proposing experiments that demand instrumentation at the cutting edge of astronomical technology. Once again, large-scale surveys are required, but this time they are surveys of millions of faint galaxies at distances measured in billions of light years. Within the next decade or so, it is likely that such strategies will have paid off, yielding a better understanding of the true nature of dark energy.

Big new telescopes. The search for extraterrestrial life. The quest to discover dark matter. The hunt for dark energy. These are just some of the exciting ventures being undertaken by today's astronomers in response to the "big questions." They epitomize the way this most far-reaching of all the sciences uses the resources at its disposal to confront some of the deepest issues facing humankind.

It is important to remember, however, that astronomers themselves are human, and have human foibles and fallibilities. Generally speaking, it would be wrong to imagine them as star-struck romantics, mystic souls coveting an intimacy with the universe for its own sake. True, a minority of astronomers might point to inspiration from the night sky as the wellspring of their calling, but most pursue their science as a hard-nosed intellectual enterprise—a means to advance humanity's understanding of its wider environment.

Individuals in this position belong to a small and rather privileged group. Worldwide, they total only a few thousand. They are paid—chiefly by universities and government institutions—to be part of the workforce gathering a harvest of understanding. There are many reasons why nations elect to invest a small fraction of their hard-earned gross domestic product in pure sciences like astronomy. They include education, national pride, the support of industry, and even political gain. And perhaps, too, the well-being of their people, for knowledge brings its own benefits in satisfying deep-seated needs. *Homo sapiens* is, after all, gifted above all species with curiosity. A culture that prizes knowledge and learning is likely to be one that is civilized.

Astronomy represents a pinnacle of human achievement comparable with music and art in its capacity to make the spirit soar. Perhaps alone among the sciences, it is a field in which knowledge is valued principally for the inspiration it brings to its recipients. If that is the case, then this book is a key that will unlock a whole universe of insight and understanding. Turn it, and you will embark on the biggest learning adventure it is possible to have.

Fred Watson

Above Using NASA's Hubble Space Telescope, astronomers pinpointed a blaze of light from the farthest supernova ever seen, a dying star that exploded 10 billion years ago. The detection is greatly bolstering the case for the existence of dark energy pervading the cosmos. This panel of images shows the supernova's cosmic neighborhood; its home galaxy; and the dying star itself.

Left Albert Einstein's contribution to astronomy has been immense. He is seen here in 1930, on a visit to the observatory at Mount Wilson in the United States. At the time, this was the world's largest telescope.

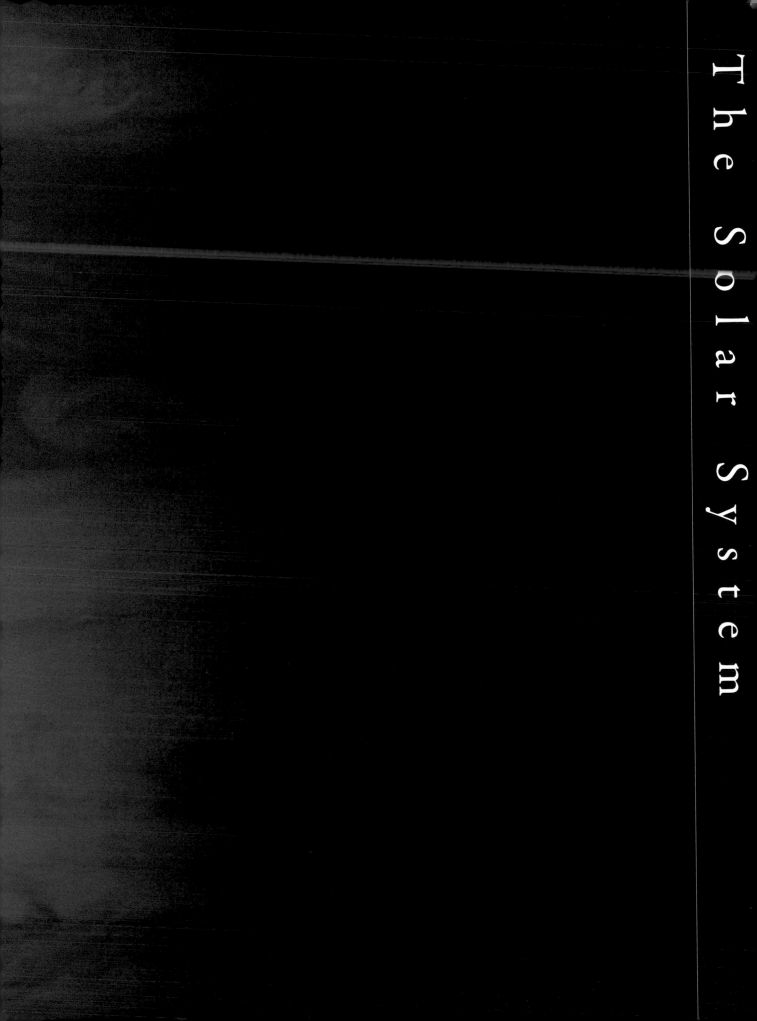

The Solar System

Orbiting the Sun

If planet Earth is our home, then the Solar System is our neighborhood.

Right This beautifully preserved early fifteenth-century manuscript is part of an illuminated treatise on astronomy, showing Earth encircled by the planets as they were known then.

Below The space shuttle, *Discovery,* gives a sublime view of Earth's horizon during sunrise, with Mars and Venus rising. Venus is usually visible to the naked eye, and can even be seen by keen observers in daylight in a clear sky.

OUR CELESTIAL NEIGHBORS

Humans have always looked up in wonder at the night sky. Filled with thousands of sparkling lights in countless patterns, it has been a source of inspiration and curiosity for countless millennia.

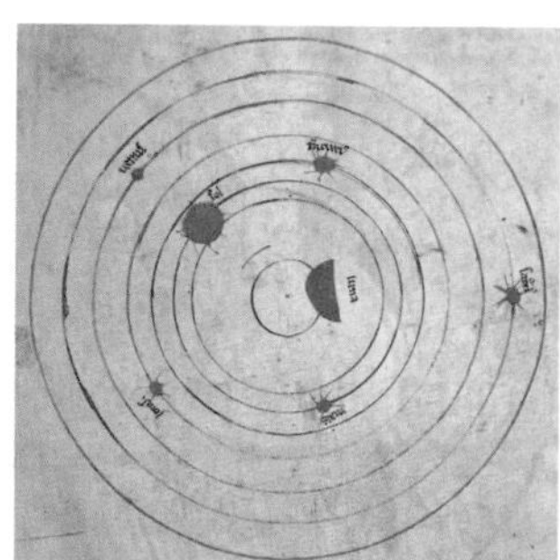

For almost all of that time, people had no real idea of what those lights—stars—were. So they wove them into their mythologies and superstitions, connecting the sky and Earth in the only way they knew how. Although the positions of the stars shift slightly over tens of thousands of years and more—as the stars move through space and as Earth's wobbly axis alters our perspective—the changes take place so slowly that the stars seem to be effectively fixed in place. But it has been known since antiquity that a handful of those twinkling lights do move, and quite dramatically. They wander slowly from one star grouping to another, sometimes reversing direction, and often disappearing for many weeks at a time. Why are these "stars" different, and where do they fit into our carefully organized schemes for the heavens?

These moving stars are, of course, the planets. The name comes from a Greek word meaning "wanderers," aptly describing their antics in the sky.

For a long time it was assumed that these non-fixed stars were circling Earth, a notion that fit in nicely with many of the religious and superstitious ideas of the ancients. But gradually superstition gave way to fact and evidence, and we came to see the wanderers for what they really are—other worlds, orbiting the Sun like Earth does, and each with its own unique characteristics and mysteries. Planets come in many different sizes, and their location in the Solar System has a lot to do with how they came to be as big or as small as they are, and how they came to made of rock, gas, or ice, or a combination of the three.

In this chapter, we'll look at each planet in turn, revealing what we know about its history, its structure, and its atmosphere, moons, rings, and much more. We'll come to see how incredible each different world is, and we will also learn to appreciate why Earth is so special.

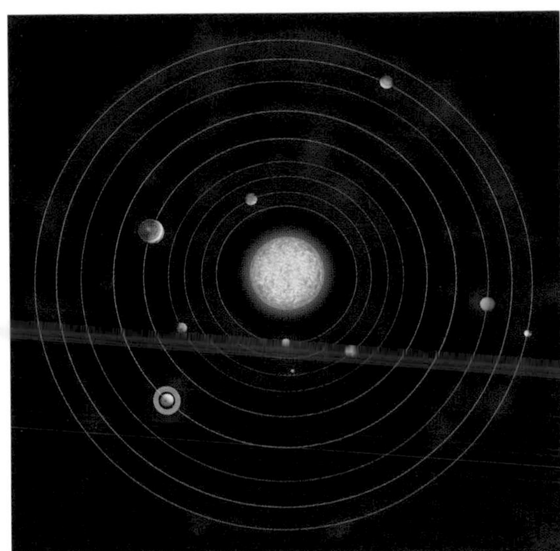

Left Digital image of the Solar System, showing the orbits of the planets around the Sun. The inner planets are called terrestrial—Mercury, Venus, Earth, and Mars. The outer planets, the gas giants, are Jupiter, Saturn, Uranus, and Neptune. Pluto is now classified as a dwarf planet.

We'll also examine the other bodies that populate the Solar System—comets, asteroids, and the new and quite controversial category of dwarf planets. We'll also learn where each of these families of objects lives, and what scientists are doing to find out more about them.

Planets and their smaller siblings are special. They are the only places beyond our Earth that we are able to reach with our spacecraft—the stars are too far away. Planets are our laboratories for examining different geological and climatological processes, and for comparing those processes with what occurs on Earth. Can other planets help us face our challenges here? And what clues can they give us about why Earth is so different...and especially, why our planet is filled with abundant life yet our neighbors, seemingly, are not?

And in this modern age of exploration, the search for life on other planets—if it exists—is one of the prime driving forces behind space science. As research has progressed, it has become more and more apparent that the conditions for simple life may, just may, exist in certain niches throughout the Solar System. Whether it is in any underground water deposits on Mars, or a subsurface ocean on Jupiter's moon Europa, the more we learn, the more we realize the potential for life to exist elsewhere—if not in our Solar System, then perhaps in planetary systems around the myriad other stars.

We've only made a small start on the road to discovery in our Solar System, and there is so much still to learn. Why are Venus and Mars so different to Earth? What role did comets play in developing life on our planet? How did Saturn get its rings? And how many icy worlds like Pluto are there still to be found in the dark distant reaches beyond Neptune? The questions are endless.

The Solar System is our neighborhood, and we should take the time to get to know it and find our way around. On the pages that follow we'll take a look at each member of the Solar System and try to uncover its secrets.

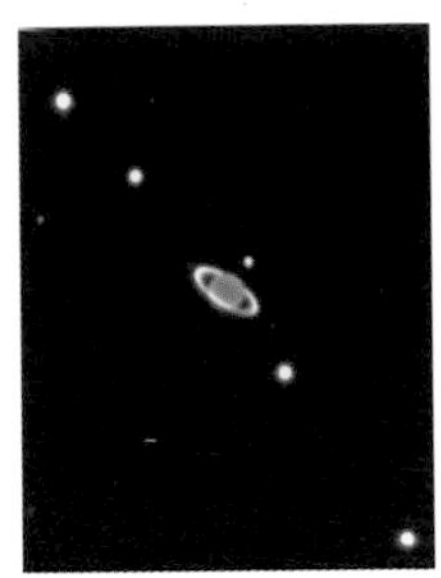

Above Uranus surrounded by six of its moons in near infra-red. From the top, the moons are Titania, Umbriel, Miranda, Ariel, and Oberon. The tiny round object to the left is a background star.

Below Artist's impression of spacecraft *New Horizons* on its flyby of Jupiter. The dim crescent shape at the upper right of the Sun is Callisto, one of Jupiter's four largest moons.

The Solar System

The Solar System is our celestial backyard, comprising the Sun (our nearest star) and its retinue of planets large and small, and lots of other smaller bodies, all held in the Sun's gravitational thrall. It takes its name from *Sol*, the Latin word for the Sun. There are eight known major planets, three dwarf planets, and countless Small Solar System Bodies (SSSBs), such as asteroids and comets.

Below This rich allegory of the *Creation of the Cosmos* is an oil on canvas and shows the influence of religion on early cosmic thinking. It resides in the Pavlovsk Palace, St Petersburg, Russia.

IMPERFECT ORBITS

The planets orbit the Sun more or less within a plane called the ecliptic plane, which is an imaginary plane joining the Sun and Earth's orbit. They don't have perfectly circular orbits—all of them are slightly elliptical, to a lesser or greater extent. Some, like Earth and Venus, are not too far removed from a circle; Mercury, though, has a very elliptical orbit—so does the dwarf planet Pluto.

The farthest point from the Sun in a planet's orbit is called its aphelion; the closest point is called its perihelion. Many of the SSSBs, such as asteroids and comets, have highly elliptical orbits that are also inclined at significant angles to the ecliptic plane. All the planets, and most of the other bodies, orbit the Sun in the same direction as the Sun's rotation—counterclockwise to an observer looking "down" on the Solar System from above the Sun's north pole. This almost common direction of revolution is a direct result of the original rotation direction of the pre-solar nebula from which the Solar System formed.

The Solar System is dominated, of course, by the Sun. It resides in the center of the system, with the major and dwarf planets all orbiting it at various distances. The

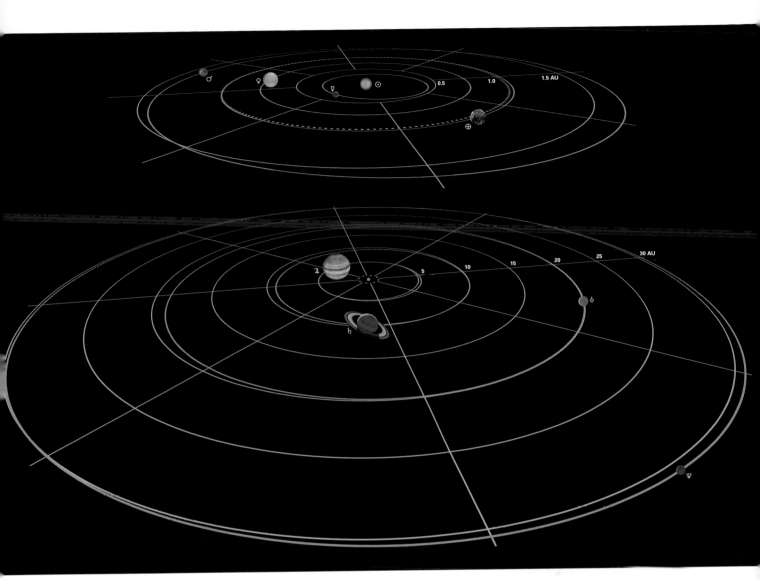

myriad minor planets (asteroids) and comets are mostly contained in numerous different populations, at different locations, within the overall Solar System.

The Sun contains over 99 percent of all the mass in the Solar System, with Jupiter and Saturn making up most of the rest and all the other planets and bodies comprising the tiny remainder.

SOLAR SYSTEM FORMATION

Many theories for the origin and formation of the Solar System have been advanced over the years, but the one that is currently accepted, and which has the strongest evidence, is called the "nebular hypothesis," and was proposed independently by two famous scientists, Immanuel Kant and Pierre-Simon Laplace. In this scenario, everything in the Solar System originated in a huge interstellar cloud or nebula of gas, known as a molecular cloud.

This cloud was comprised of material ejected into space during the death throes of earlier generations of stars. It contained not only large amounts of hydrogen and helium—the most abundant elements in the universe—but also heavier elements produced by that earlier stellar generation. Slowly, local regions within this cloud began to collapse in on themselves—perhaps triggered by a shockwave from a nearby supernova (exploding star)—and began to spin as they did so. One such region was destined to become our Solar System, and is referred to as the pre-solar nebula. Gradually, as it contracted and began to spin faster—held together by its own overall gravity and magnetic fields—it began to flatten out into a huge spinning disk of gas, thousands of times wider than our Solar System is now.

Above The top illustration shows the orbits of the terrestrial planets of Mercury, Venus, Earth, and Mars around the Sun. The lower illustration, on a smaller scale, shows the orbits of the more distant gas planets of Jupiter, Saturn, and Neptune, with the dwarf planet Pluto on the farthest orbit. (not to scale)

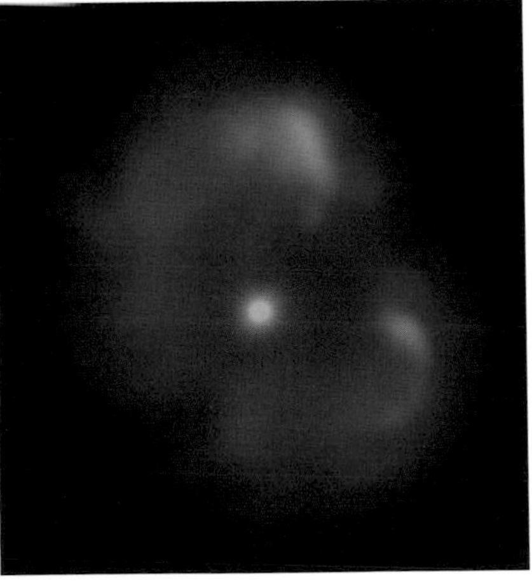

Right Called a proto star, this is a digital illustration of plasma forming into a planetary body. The Sun is actually a big ball of plasma.

Right This digital composite shows a hypothetical Solar System. After the formation of the proto-Sun, the gas molecules combine to become larger and larger particles that clump together over millions of years, eventually growing in size until large enough to become planets.

As it continued to contract, most of the material gravitated (literally!) into the center of the nebula, becoming hotter and under increasing pressure—this central portion would become the proto-Sun.

As the rest of the material whirled around in the disk, the gas molecules combined to form larger particles called "dust grains," and those grains stuck together to form larger and larger grains.

Gradually, over millions of years, the spinning disk of material became "clumpy." Small clumps stuck together to form larger clumps, and larger and still larger, until eventually some of them grew large enough to become planets.

The inner parts of the disk became too hot for ices to form or gases to liquefy, so the only materials able to remain solid were those with high melting points, i.e. rocky substances and metals. Consequently, the inner planets Mercury, Venus, Earth, and Mars, are rocky worlds.

Further out the temperatures were cooler, so ices were able to form. The gas planets, Jupiter and Saturn, became the dominant bodies in the middle reaches of the Solar System, with Uranus and Neptune further out still, and somewhat smaller.

Jupiter's massive bulk gave it a large gravitational influence, preventing any other rocky planet from coalescing any further in toward the Sun in the region known today as the asteroid belt.

By this time, nuclear fusion had begun in the infant Sun's core, and it began to release copious amounts of energy. A super-solar wind began to blow through the Solar System, sweeping it clean of the remaining gases and preventing any more major planets from forming.

Although this whole scenario was worked out on a theoretical basis, it is no longer all speculation—strong evidence has been found that indicates this very process is occurring right now around many infant stars elsewhere in our galaxy.

Above Digital composition of microscopic crystals in a dusty disk surrounding a brown dwarf, or "failed star." The crystals, made up of a green mineral found on Earth called olivine, are thought to help seed the formation of planets.

Left It is thought that the presence of dusty disks slows down the acclerated rotation of developing stars, as illustrated here in this artist's impression, preventing them from spinning out of control. Dusty disks are also believed to be involved in planet formation.

VOYAGER'S VIEW

Where is the edge of the Solar System? Is it at the orbit of the last planet, Neptune? Is it at the edge of the Kuiper Belt, the family of icy bodies beyond Neptune that includes Pluto? Or is it in the Oort Cloud, the hypothetical cloud of comets that surrounds our Solar System at enormous distances from the Sun?

One way to define the "edge" is the place at which the Sun's solar wind finally peters out and gives way to the near-emptiness of interstellar space. The solar wind blows a "bubble" in space out to over three times the distance of Pluto, called the heliosphere. Beyond this is a point called the termination shock, where the wind begins to mix with the thin gases of interstellar space in the heliosheath. Almost half as far again is where the solar wind finally gives up—the heliopause—and true interstellar space begins.

NASA's *Voyager* spacecraft are heading in different directions toward this zone. Running on the small amount of residual electricity produced by their atomic power units, it is hoped one or both of them will last long enough to reach interstellar space. Already they have sent back indications that they are getting close.

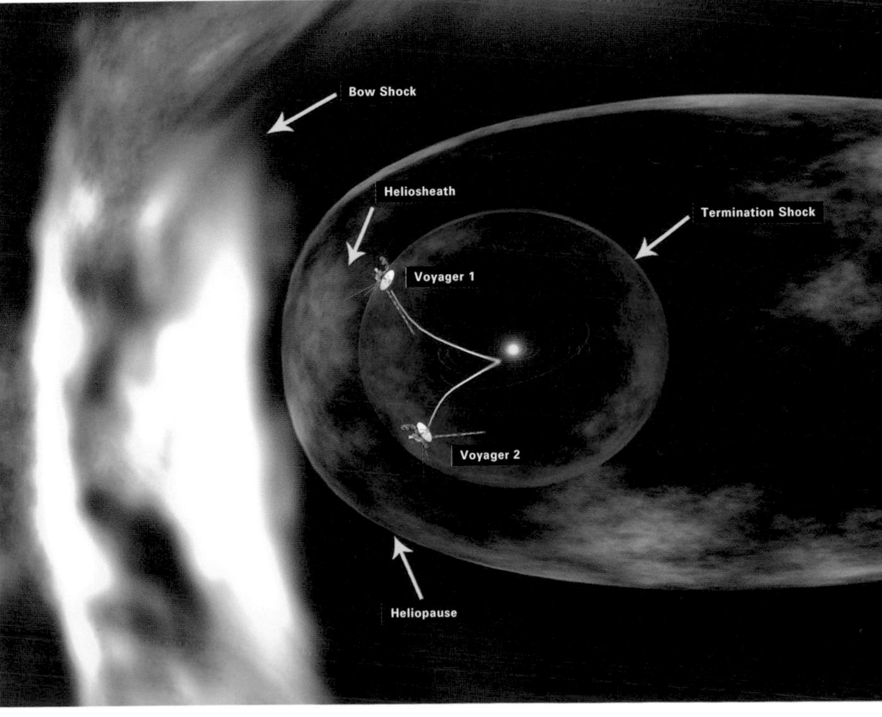

Right Artist's impression showing the areas around which the solar wind forms around our Sun. The approximate positions of the twin *Voyagers* are shown about to enter the heliosheath.

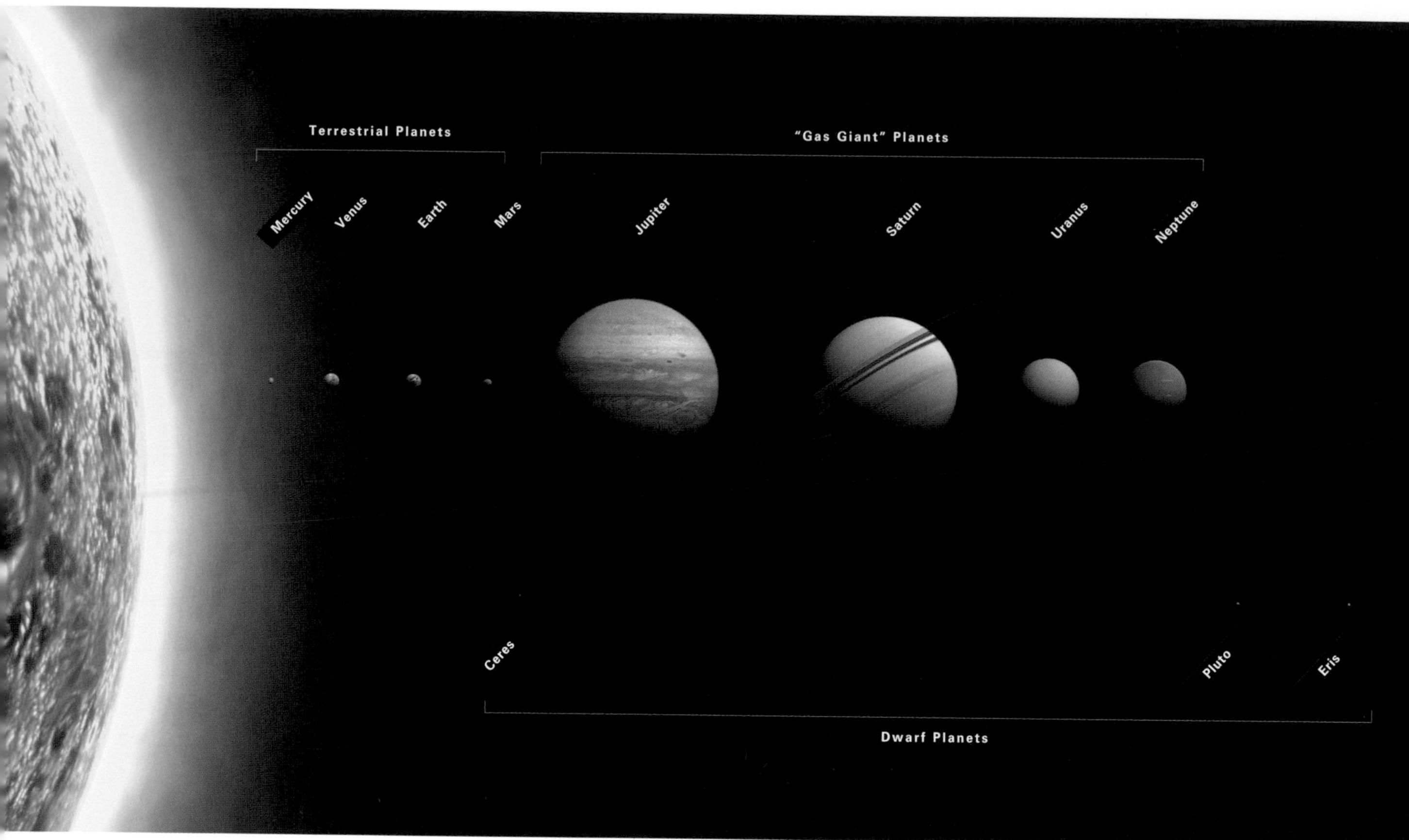

Above The Solar System as it is today with eight planets recognized by the IAU: Mercury, Venus, Earth, Mars, Jupiter, Saturn, Uranus, and Neptune. Pluto has been demoted to a dwarf planet, joining Ceres and Eris.

PARADE OF PLANETS

How do astronomers define a planet? Until recently, the answer was obvious—a planet was any really large body that goes around the Sun. Most people alive today have grown up with the certain knowledge that our Solar System has nine planets—Mercury, Venus, Earth, Mars, Jupiter, Saturn, Uranus, Neptune, and Pluto—and for generations schoolchildren learned simple rhymes to remember their order.

But there was always a problem with Pluto. Ever since its discovery in 1930, astronomers have known that there was something odd about it. It was much smaller than the other planets for a start; it was made of ice instead of rock or gas; and it had a highly elliptical orbit inclined at a steep angle to the orbits of the other planets. In some ways it was very similar to a comet, except that it seemed far too big. It was an anomaly.

But in the closing years of the twentieth century and the early years of the twenty-first century, astronomers began to discover lots of other Pluto-like objects out beyond the orbit of Neptune—one of them (called Eris) is actually slightly bigger than Pluto.

It seemed that Pluto was not alone, and so the question had to be asked: If Pluto is considered a planet, shouldn't these extra objects be called planets too? Or, if they do not satisfy the criteria for classification as planets, then what does that make Pluto?

A decision had to be made, and in 2006 many of the world's astronomers gathered at a meeting of the International Astronomical Union (IAU) to thrash out the question once and for all. What they came up with was, for the first time, a technical definition of what constitutes a planet, and a whole new category for objects that don't fit the mold. It was very controversial, and stirred public interest in astronomy as few things have done for many many years.

This is what they came up with. A planet, then, is any body that is in an independent orbit around the Sun; is big enough to have formed itself into a spherical shape; and which has cleaned up, collected in, or swept away, any competing bodies in its orbital zone. Under this new official definition the Solar System has only eight planets, from Mercury out to Neptune.

Pluto has been placed into the newly created category of dwarf planet. This category has basically the same definition as a full planet, with two important exceptions: That the object has not cleared its orbital zone of competing bodies, and it is not in itself a satellite or moon of another body. By this definition the Solar System has three dwarf planets: Pluto, Ceres (the largest body in the asteroid belt), and Eris.

At the same time as the IAU made its decision on Pluto, it also tried to clarify the status of the smaller bodies in the Solar System—asteroids and comets—by categorizing them overall as Small Solar System Bodies (SSSBs).

THE INNER PLANETS

Long before the IAU made its 2006 decision, astronomers recognized that the four innermost planets were quite distinct from the others. Mercury, Venus, Earth, and Mars are known as the terrestrial planets. Because they formed close in to the very hot young Sun, they are made of materials that could solidify at high temperatures—rock and metals. They do not have extensive retinues of moons like the gas giant planets have, nor do they have ring systems. Of the four, only two—Earth and Mars—have natural satellites (moons). Earth has the Moon—an unusually large body in comparison to its parent planet—and Mars has two very small natural satellites, Phobos and Deimos, which are almost certainly captured asteroids. Venus and Earth have dense atmospheres, Mars has a thin atmosphere (possibly far more dense in the past), and Mercury has a barely perceptible atmosphere.

Earth is located in a special position in the Solar System. It is right in the middle of what is called the habitable zone; that is, it is at just the right distance from the Sun to be neither too hot nor too cold for liquid water to exist in abundance, and for water-based life to have evolved and flourished. Lucky for us.

Next outward from the Sun is the Asteroid Belt. Some scientists like to include it in the same zone as the terrestrial planets, since asteroids, too, are rocky worlds, albeit small ones, that probably formed under the same sort of conditions in the infant Solar System. Popular culture, such as movies, often represents this region as a swarming tumbling minefield of millions of mountain-sized rocks. In reality, the asteroids are spread very far apart, and space probes have passed through this region with no trouble at all.

OUR PLACE IN THE GALAXY

Our Solar System is situated in the Orion Arm of our Galaxy, a spiral galaxy known as the Milky Way. Located around 25,000–28,000 light-years from the galactic center, the Solar System is on a circular orbit around that center over a period of around 220–250 million years. Our Sun and its family are heading in the direction of the constellation Hercules.

In our local neighborhood, the Solar System is embedded in a region called the Local Interstellar Cloud, itself part of a larger volume known as the Local Bubble—a 300-light-year-wide void in the surrounding space.

It was long believed that our Solar System was the only planetary system in the universe. We now know this is not the case. Astronomers have found dozens of planetary systems in our Galaxy, with more of these "exoplanets" being discovered every month. Limitations of technologies have meant that almost all of the planets detected have, by necessity, been very big—Neptune-sized and larger. But in April 2007, European scientists announced the detection of a planet not much bigger than Earth, orbiting a nearby red dwarf star in that star's habitable zone. It was quickly dubbed the first habitable Earth-like planet to be found outside our Solar System. No doubt there will be many more to come.

Below Artist's impression of our Milky Way as seen from above. The core (which has a peanut-shaped center) has been shown to exist by infra-red images, and is made up of old stars. The Sun sits on the inner edge of the Local Arm, frequently called the Orion Arm.

THE GIANTS

Next out from the Sun are four much larger bodies with very different compositions to those of the terrestrial planets. Jupiter, Saturn, Uranus, and Neptune are often called "gas giants," although lately Uranus and Neptune have increasingly been referred to as "ice giants." These four bodies are comprised almost entirely of gas. They have complex weather systems, with several of them sporting enormous long-lived storms. They each have systems of rings—spectacularly so in the case of Saturn—and many moons. Some of those moons are bigger than the planet Mercury; some have volcanic activity; and a few are considered potential habitats for simple life forms. They have each been visited by spacecraft, and more missions are on the drawing boards.

THE OUTER LIMITS

Finally, much further out are several different populations of smaller bodies. First is the Kuiper Belt, beyond the orbit of Neptune. The leading example of this population is Pluto, and there are dozens more Pluto-like bodies in this region. Next is what is called the "scattered disk," a region of similar icy bodies that are believed to have formed closer in toward Neptune but have been flung into more distant orbits through gravitational encounters with that planet. Eris is a member of this group. Much further out is a theoretical swarm of cometary bodies that surrounds the entire Solar System, known as the Oort Cloud. It is too far away to have been detected directly, and its presence can only be inferred indirectly from the trajectories of some long-period comets.

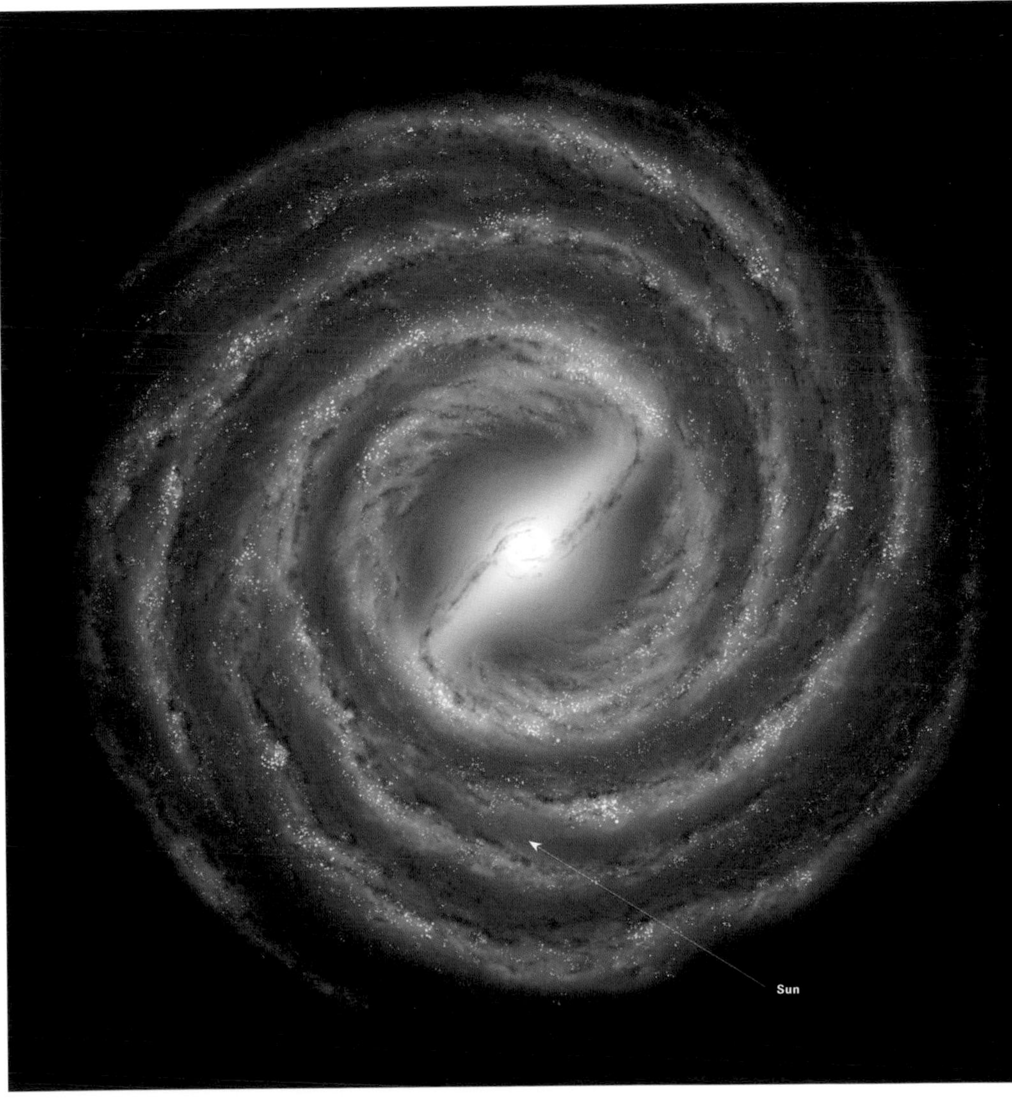

The Sun

FACT FILE

Average Distance from Earth
1 AU/92,955,820 miles/ 149,597,870 km

Age
4.6 billion years

Length of day
25.38 Earth days

Equatorial radius
432,200 miles/695,500 km (109 times that of Earth)

Mass
4.385 x 10^{30} lbs/ 1.98892 x 10^{30} kg
(332,900 times that of Earth)

Average Temperature
10,000°F/5,500°C

Composition by mass
74 percent hydrogen, 25 percent helium

Notable features
The Sun is a G2 star. Energy generated in the Sun's core takes a million years to reach its surface.

It holds the planets in their orbits, it provides life-giving energy, and it dominates our sky. The Sun truly is the center of our local universe.

The Sun is located at the center of the Solar System. All the planets, minor planets (asteroids), Kuiper Belt Objects, and most comets are gravitationally bound to it in orbits. To the ancient Greeks it was Helios; to the Romans, Sol; to the Egyptians it was associated with one of their leading gods, Ra.

The Sun contains 99.8 percent of all the mass in the Solar System, with the eight planets, the dwarf planets, asteroids, comets, and meteoroids make up the remaining two-tenths of a percent. It dominates everything in its region of space.

The Sun is a massive 865,000 miles (1,392,000 km) in diameter; by comparison, Earth measures only 7,920 miles (12,746 km) wide. In fact, the Sun is so massive that you could fit Earth 1.3 million times inside, and it has about 330,000 times more mass than Earth!

Because the Sun is not a solid body, like a rocky planet is, but rather is a huge globe of intensely hot gas, it does not rotate as a solid mass. It rotates faster at the equator—around 25 days for a full rotation—than it does at its polar regions where it takes a much longer 35 days. The Sun, along with its retinue of planets and other bodies, orbits around the center of the Galaxy, taking around 220–260 million years to complete one circuit at a speed through space of around 135 miles (217 km) per second. It, and everything else in the Solar System, is made from the material ejected during the explosive destruction of earlier generations of stars. In fact, scientists think our star probably belongs to the third generation of stars born after the Big Bang. Although it looks big and bright in our daytime skies, the Sun is, in fact, a fairly ordinary star as stars go. Its surface temperature is a white-hot 9,932°F (5,500°C), though to our eyes it looks yellow, as part of its blue light is scattered in the atmosphere of Earth.

The Sun is what is known as a main sequence star, which means that it is in its middle age—roughly 4.6 billion years old. Scientists estimate it has probably another five billion years of life before it peters out and becomes a white dwarf star—a dying stellar ember.

Above right An ancient Roman carving, from Tyre, of the Sun god, Helios. Helios was imagined as a handsome god, crowned with the shining aureole of the Sun, who drove a chariot across the sky.

GREAT BALLS OF FIRE

The Sun is composed mainly of hydrogen (74 percent by mass) and helium (25 percent), with small amounts of some other elements. Its internal structure is broken into three zones: The core, the radiation zone, and the convection zone. In the core, the pressure is so great that hydrogen atoms get squashed together (fused) to form helium, releasing prodigious amounts of energy in the process. The temperature

Right Digital impression of solar flares swirling off the surface of the Sun. This region is where the solar wind originates.

Opposite Amazing snapshot of the solar corona taken by SOHO (Solar and Heliospheric Observatory). The corona is a plasma and can reach temperatures greater than 1 million°K. The North and South poles clearly show up as "cooler" regions.

c. 4.567 Billion BCE	5000-3500 BCE	3000 BCE	*c.* 2700 BCE	1223 BCE	*c.* 200 BCE	965-1039 CE	1543	1610	*c.* 1660	1687	1800
The Sun, a star, is born (give or take a few millennia).	The first device for indicating the time of day is built. It consists of a vertical stick casting a shadow by the Sun. The length of the shadow gives an indication of the time of day. (The first sundial.)	Known as "the cave of the sun," Newgrange, Ireland, is built. On winter solstice, the sunlight perfectly aligns with an opening in the structure to illuminate the inner chamber.	Stonehenge, in England, is built. It is a giant circle of huge stones that are aligned to the position of the Sun.	The oldest eclipse is recorded on a clay tablet uncovered in the ancient city of Ugarit, (in what is now Syria).	Aristarchus of Samos, a Greek mathematician and astronomer, announces a theory of a Sun-centered universe. He also attempts to mathematically calculate the sizes and distances of the Sun and Moon.	Muslim scholar, Abu Ali Al-Hasan, invents the camera obscura and becomes the first known person to use a device to observe the Sun.	Copernicus publishes his theory that Earth travels around the Sun. This contradicts the Church teachings.	Galileo Galilei describes spots on the Sun that he views with his early telescope.	Isaac Newton shows that sunlight can be divided into separate chromatic components via refraction through a glass prism.	Sir Isaac Newton publishes his finding, *Principia Mathematica*, establishing the theory of gravitation and laws of motion. This allows astronomers to understand the interacting forces among the Sun, the planets, and their moons.	William Herschel extends Newton's experiment by demonstrating that invisible "rays" exist beyond the red end of the solar spectrum.

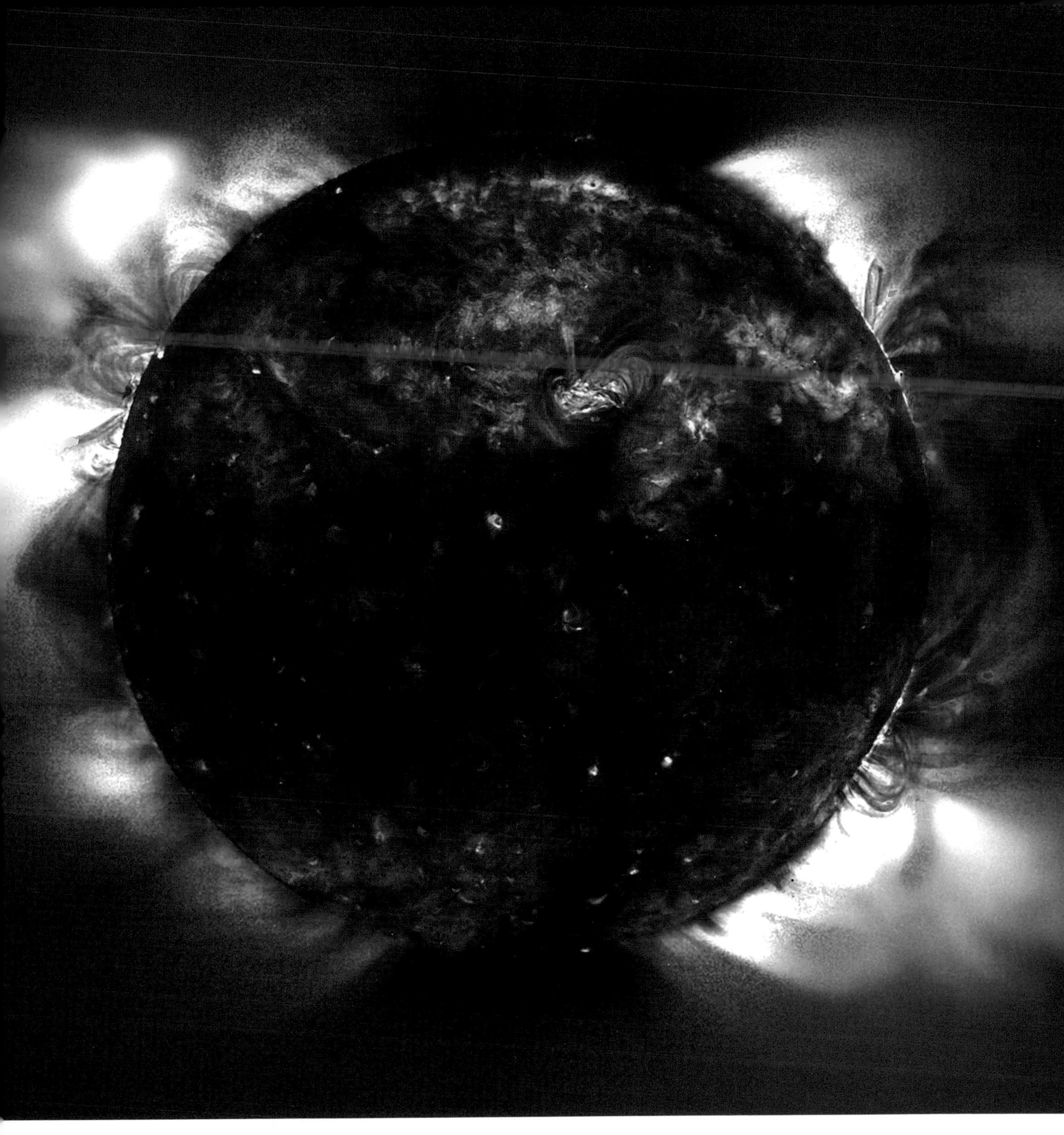

1814
Joseph von Fraunhofer builds the first accurate spectrometer, and uses it to study the spectrum of the Sun's light.

1843
German amateur astronomer, Heinrich Schwabe, who has studied the Sun for 17 years, announces his discovery that the number and positions of sunspots vary over an 11-year period.

1845
First solar photograph is made on April 2.

1860
The total solar eclipse of July 18, 1860, is probably the most thoroughly observed eclipse up to this time.

1868
During an eclipse, astronomers observe a new bright emission line in the spectrum of the Sun's atmosphere. As a result of observations, British astronomer, Norman Lockyer, identifies and names helium.

1908
American astronomer, George Ellery Hale, shows that sunspots contain magnetic fields that are thousands of times stronger than Earth's magnetic field.

1938
German physicist Hans A. Bethe and American Charles L. Critchfield show how a sequence of nuclear reactions called the proton-proton chain make the Sun shine.

1982
Helios 1, a joint German and US deep space mission, sends back the last of its data indicating the presence of fifteen times more micrometeorites close to the Sun than near Earth.

1990
Ulysses, an interplanetary spacecraft, is launched with the mission to measure the solar wind and magnetic field over the Sun's poles during periods of both high and low solar activity.

1991
Launch of the YOHKOH spacecraft, to photograph the Sun in X-ray emission over a full solar cycle, (11 years).

1995
The Solar and Heliospheric Observatory (SOHO), reaches a point where the Sun's gravitational pull balances Earth's. The satellite orbits the Sun with Earth studying the Sun from its core to the outer corona, and the solar wind.

2006
NASA's two Solar TErrestrial RElations Observatory (STEREO) satellites take the first three-dimensional images of the Sun.

in the core is a staggering 27 million°F (15 million°C), and every second 4.4 million tons (4 million tonnes) of matter is converted into energy.

The radiation zone extends about two-thirds of the way out from the core. The energy released in the core's fusion reactions fights its way through the thick radiation zone, being continually absorbed and re-emitted. Once the energy has reached the edge of the radiation zone, it heats the gases at the bottom of the convection zone. These gases form into huge currents that flow up to the Sun's surface—where the energy is finally released into space as electromagnetic radiation, e.g. light, infrared, ultraviolet—and then the currents sink back down again to pick up more energy from the top of the radiation zone, and start the whole process over again. The journey of energy from fusion in the core to emission at the surface can take anywhere up to millions of years.

The visible surface of the Sun is called the photosphere. Close inspection using professional telescopes shows the surface to have a speckled appearance, which is known as granulation. These "granules" are actually the tops of the convection currents.

Extending out from the photosphere is the Sun's atmosphere, and it, too, is divided into different regions. The first region is called the chromosphere. It is about 1,250 miles (2,000 km) deep, and temperatures in the gas here can reach up to 180,000°F (100,000°C). Above this is a transition region, where temperatures steadily climb to around 1.8 million°F (1 million°C). Finally, the outer region or corona is reached, where temperatures peak at up to several million degrees more. The corona extends a long way out from the Sun, and can be seen from Earth during total solar eclipses. Why the temperature increases so markedly from

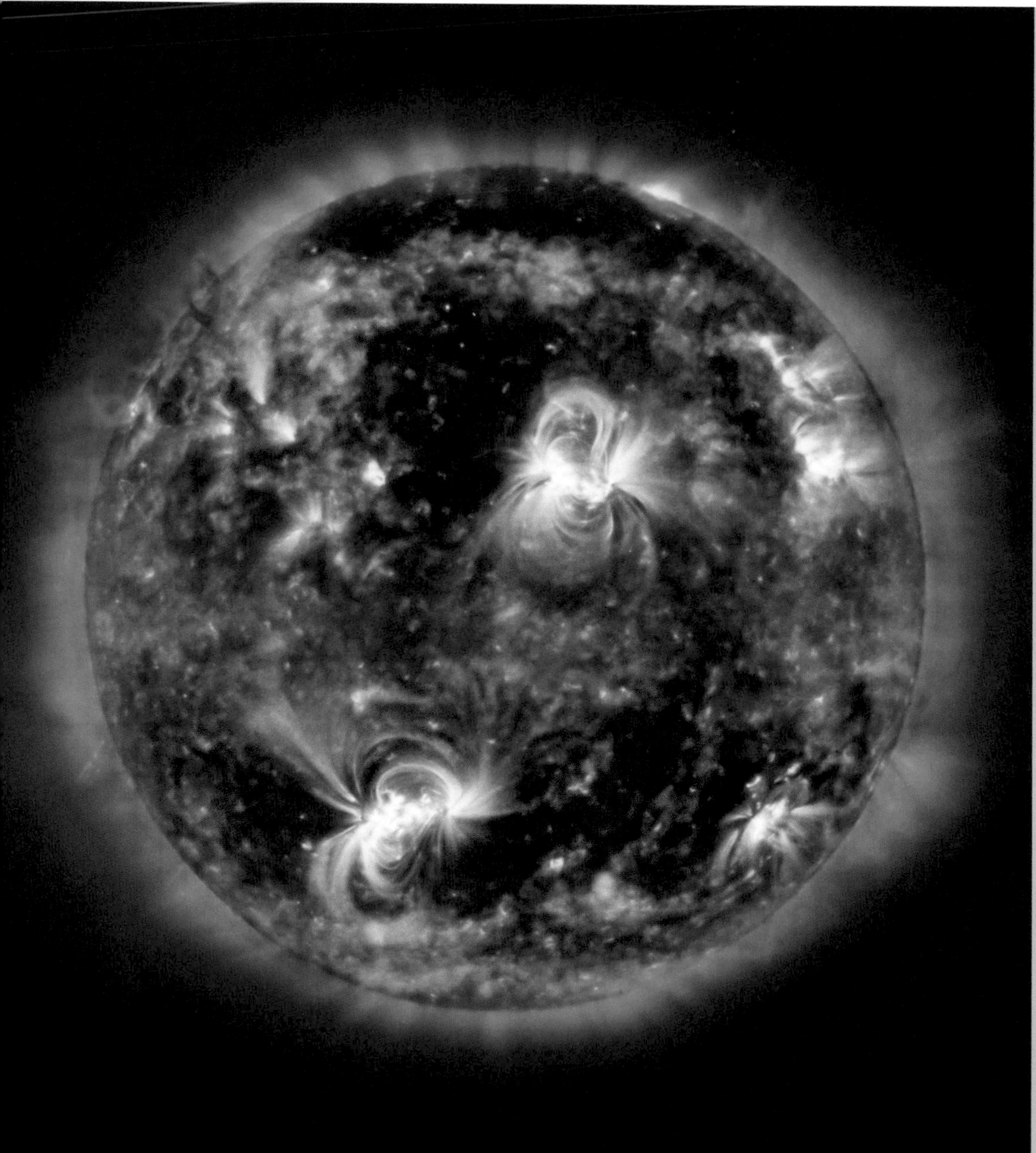

Below SOHO-EIT image in the extreme ultraviolet, showing the solar corona at a temperature of about 1 million°K. It is dominated by two large active region systems, composed of numerous magnetic loops.

SUN SCANNERS

The human eye is too sensitive to be used for studying the Sun directly. Telescopes and film or electronic detectors can be used, but ground-based facilities have the disadvantage of having to endure the vagaries of the weather, not to mention the episodes of darkness we call nighttime. But observatories positioned in space can be set to watch the Sun essentially uninterrupted for years at a time. And indeed, a whole flotilla of scientific spacecraft is doing just that.

Perhaps the most famous and successful of these is the joint European-USA craft known as SOHO (Solar and Heliospheric Observatory). Launched in 1995, it was intended to operate for only a few years. However, it has proved far more resilient than its makers had hoped for, and is still going strong in 2007. From its vantage point 1 million miles (1.5 million km) sunward of Earth—where the Sun and Earth's gravity balance out—it has kept its electronic eyes trained on the solar disk almost continuously during this time.

Other spacecraft include the unique *Ulysses* probe, which was put on a trajectory that has taken it out of the plane of the ecliptic and into an orbit that takes it over the poles of the Sun. And launched in 2006 was *Stereo*, twin spacecraft that circle the Sun inside and outside Earth's orbit, providing views which, when combined, will give a unique stereo perspective on the Sun's activity.

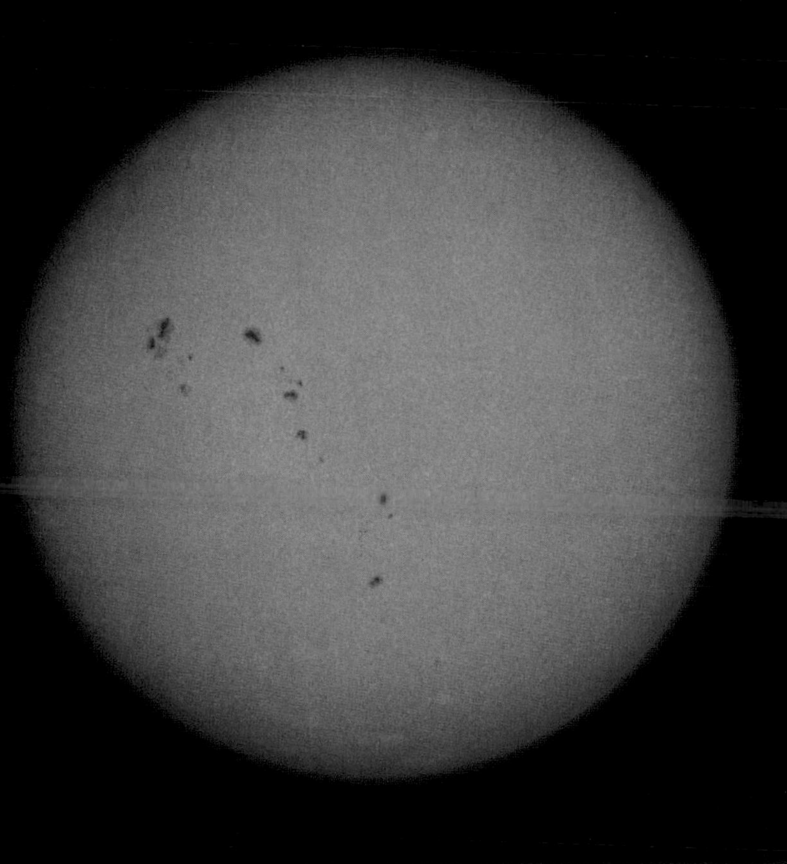

It has been known for a long time that sunspot numbers rise and fall in an 11-year cycle that has a definite minimum and maximum. At the time of minimum, very few sunspots are seen, and those that are present are usually found at high solar latitudes (i.e. toward the poles). As the sunspot cycle moves toward maximum, the number of spots increases, and they appear closer and closer to the Sun's equator. During the period of each sunspot cycle, the Sun's magnetic field completely reverses, so after 22 years the magnetic field returns to its original orientation.

From time to time the Sun lets off violent explosions, such as solar flares and coronal mass ejections (CMEs). Both involve huge amounts of energy and can send streams of particle heading out into the Solar System.

If Earth, or a spacecraft, happens to get in the way, the effects can be pronounced. Damage can be done to the electronics aboard spacecraft, and on Earth the effects can range from disruption to radio communications, to the wonderful formation of the beautiful Aurora Borealis and Aurora Australis (Northern and Southern Lights.)

Left Sunspots—caused by intense magnetic fields emerging from the interior, a sunspot appears to be dark only when contrasted against the rest of the solar surface, because it is slightly cooler than the unmarked regions.

the Sun's surface out to the corona is not fully understood, but it is tied in with the strong solar magnetic field. As well as electromagnetic radiation, particles are also emitted from the Sun's surface. The bulk of this is known as the solar wind, which "blows" through the Solar System way out to beyond Pluto. In fact, the Sun blows a huge "bubble" in space, known as the heliosphere.

STORMY WEATHER

The Sun has an extremely strong magnetic field that reaches far out beyond Pluto, gradually weakening the further out it goes (although the sparse gas in the almost-vacuum between the planets actually helps maintain the solar magnetic field at higher strengths than it otherwise would.)

Being largely made of plasma, the Sun creates its own magnetic field. Its rotation causes the field to become twisted and distorted. This leads to phenomena like sunspots and prominences. Sunspots are small regions of cooler gas on the Sun's surface. They reach temperatures around 7,000–8,000°F (4,000–4,500°C)—it's only the comparison with the hotter adjacent gas that makes them look dark. Sunspots are tied in with complexities in the Sun's magnetic field and it is thought that the field reduces the amount of convection from the convection zone, which lowers the energy reaching the surface in the local region of the sunspot.

ECLIPSES

While it is extremely unwise for people to look directly at the Sun, as serious eye damage or blindness can occur, there is one time when it is possible to see the corona with the naked eye—during a total solar eclipse.

It just so happens that the Moon is about 400 times smaller than the Sun, but also 400 times closer. As the Moon orbits Earth, it sometimes gets in the way of the Sun and blocks its light, causing an eclipse.

When this occurs, the photosphere is blocked from view and the beautiful, wispy corona stands out. Enthusiasts travel to remote corners of the world just to witness a few precious seconds of this incredible spectacle.

Above Solar eclipse (montage). Total solar eclipses are of special interest to astronomers because it is the only time the Sun's corona can be seen from Earth's surface. Observers can detect and measure properties of the Sun's outer atmosphere, such as temperature, density, and chemical composition, when the light of the disk is completely blocked by the Moon.

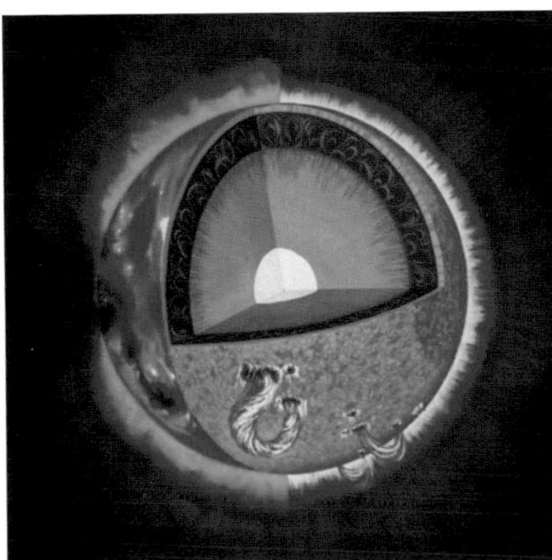

SUN CORE FACTS

The Sun is a ball of hot gas, mainly hydrogen and helium. It has three zones: core, radiation zone, and convection zone. The massive core pressure causes hydrogen atoms to fuse and form helium, releasing energy. Energy from the core bounces around the radiation zone, taking about 170,000 years to get to the convection zone where the temperature drops, as large bubbles of hot plasma move upward. The Sun's "surface"—photosphere—is a 300-mile (500-km) thick region, from which most of the Sun's radiation escapes as the sunlight we observe on Earth about eight minutes later.

Mercury

FACT FILE

Planetary Order
1st planet from Sun

Average Distance from Sun
0.39AU/35,983,095 miles/
57,909,175 km

1 Mercury Day
59 Earth days

1 Mercury Year
0.241 Earth years

Equatorial radius (Earth = 1)
0.383

Mass (Earth = 1)
0.055

Mean Surface Temperature
353°F (178.9°C)

Atmosphere
31.7 percent potassium, 24.9 percent sodium, 9.5 percent atomic oxygen, 7.0 percent argon, 5.9 percent helium, 5.6 percent molecular oxygen, 5.2 percent nitrogen, 3.6 percent carbon dioxide, 3.4 percent water, 3.2 percent hydrogen

Moons/satellites
No moons, no rings

Notable features
Crater, Caloris Basin, 812 miles (1,300 km) in diameter.

Above right Marble statue of *Mercury* by Danish sculptor Berthel (Albert) Thorwaldsen. Mercury was a major god of trade, profit and commerce, and the son of Jupiter. Mercury has influenced the name of a number of things in a variety of scientific fields. The word mercurial is commonly used to refer to something or someone erratic, volatile, or unstable, derived from Mercury's swift flights.

Mercury is the closest planet to the Sun, but it's not the hottest. It looks just like the Moon, yet it's smaller than some moons. Mercury is a mystery.

Mercury is the innermost planet of our Solar System. Its orbit is not as circular as most of the other planets; in fact it is quite elliptical. The closest it comes to the Sun is around 28.5 million miles (46 million km), while the farthest point of its orbit is 43.3 million miles (69.8 million km).

The planet has been given many different names from different cultures and languages. Mercury was the Roman god of commerce, and also a messenger for other gods. His speediness as a messenger is reflected in Mercury's rapid appearance and disappearance in the morning and evening skies.

The planet has been connected in German legends with the god Wodin, and in Norse legends with Odin. The Greek name for Mercury translates as "the gleaming," and its Hebrew name translates as "the star of the hot one"—the "hot one" is the Sun, in reference to the fact the planet never strays far from the Sun in the sky.

EXTREME CONDITIONS

It used to be thought that Mercury was in "tidal lock" with the Sun, meaning that it kept the same face turned toward the Sun at all times—just like the Moon does with Earth. But in the 1960s, scientists used huge radio dishes to bounce radio waves off Mercury's surface (i.e. radar). By analyzing the reflected waves, they could determine its rotation speed. They were astonished to find that the planet rotates three times for every two orbits around the Sun. Its slow rotation on its axis —one Mercury "day"—takes 58.7 Earth days, while it takes 88 Earth days to complete one orbit—one Mercury year. (A "day" can also be defined as the rotation time needed for the Sun to appear in the same spot in the sky (on the meridian), called a "solar day"—due to Mercury's long day and shortish year, this means that one solar day on Mercury is 176 Earth days long, twice as long as its year.)

Mercury's orbit is inclined at an angle of just on 7° to the ecliptic—the plane of Earth's orbit, and it has almost no axial tilt. Its proximity to the Sun means that parts of its surface get very hot, with a maximum in the equatorial regions of around 554°F (290°C). By contrast, in the deep shadows of craters near the poles, the temperature is a frigid −292° F (−180°C), giving the planet the greatest temperature range in the Solar System. This range is mainly due to the lack of an atmosphere, as atmospheres work to smooth out heat distribution. (Venus actually is slightly hotter, even though it is further from the Sun, due to the runaway greenhouse effect in its atmosphere.)

ROCKY WORLD

Mercury is one of the Solar System's four "terrestrial planets," a category that also includes Venus, Earth, and Mars. The planet is thought to be composed of around 30 percent rocky material (in its crust and mantle), and around 70 percent metallic material (in the form of a very large iron core). It has the second highest density in the Solar System, after Earth. (In fact, if compression of Earth's bulk due to its gravity is removed from the equation, Mercury actually has a higher density.) The core is believed to comprise around 40 percent of Mercury's total volume, compared to around 17 percent for Earth. The mantle is around 372 miles (600 km) thick, and on top of that is the crust, around 60–125 miles (100–200 km) in depth. Scientists think that Mercury was struck by another small planet billions of years ago, which smashed away much of the rocky mantle and crust, leaving only the thin remnants we see today.

Although definitely considered a planet under both the old and new classifications, Mercury is so small that it is actually smaller than two of the Solar System's moons—Titan (Saturn's largest moon) and Ganymede (Jupiter's largest moon). In fact, Mercury closely resembles Earth's Moon, with smooth plains and countless craters. The large crater count indicates that its surface is very old; that is, it hasn't been geologically active for a very long time (as any such activity would have wiped away many of the craters).

Like Earth's Moon, Mercury shows abundant evidence of a period of intense bombardment by asteroids and comets around 3.8 billion years ago. There was also a lot of volcanic activity on the planet's surface, which flooded low-lying areas with magma, forming the vast plains.

4.5 billion years ago
An asteroid collides with the still-forming Mercury sending chunks of the planet hurtling through space.

3rd millennium BCE
Mercury is known to the Sumerians, who call it Ubu-idim-gud-ud.

Up to 6th century BCE
The planet Mercury has two names, as it was not realized it could alternately appear on one side of the Sun and then the other. It was called Hermes when in the evening sky, but was known as Apollo in honor of the Roman god of the Sun when it appeared in the morning.

5th century BCE
Pythagoras is credited for pointing out that Hermes and Apollo were one and the same planet.

4th century BCE
Heraclitus of Ephesus (*ca.* 535–475 BCE) believed that Mercury and Venus orbited the Sun, not Earth.

265 BCE
Greek scientists begin studies of Mercury in the morning and evening skies.

12th century
A transit is expected but not seen, leading the Moroccan astronomer Alpetragius to conclude that Mercury emits its own light.

807 CE
At the time of Charlemagne, Annales Loiselianos states, "the star Mercury was seen in the Sun like a small black spot, a little above the center of that very body, and it was seen by us for 8 days." (However it was clearly a sunspot due to its visible size.)

1610
Galileo Galilei, the Italian astronomer, makes the first telescopic observation of Mercury.

1644
Johannes Hevelius discovers the phases of Mercury.

MERCURY CORE FACTS

Mercury's molten core is believed to be composed mostly of iron, taking up about 70 percent of the planet's radius. This is covered by a mantle up to 400 miles (600 km) thick, and topped with a surface crust rather like Earth's Moon. The crust is scarred by impact craters resulting from collisions with asteroids and comets. The outer crust has continued to contract over the last half-billion years and grown strong enough to prevent magma from reaching the surface, ending the planet's period of geologic activity. Mercury's high average density is second only to Earth.

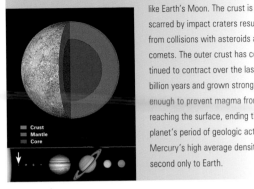

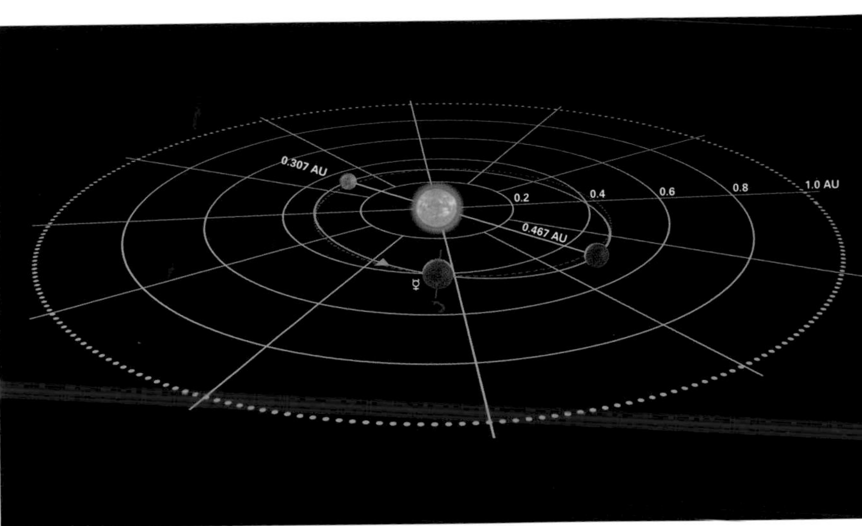

Above Mercury takes 88 Earth days to complete an orbit of the Sun, with an inclination of just 7° to the ecliptic. Its axial tilt is only 0.01°, the smallest of all the planets, and causing no variation in seasons. Mercury's orbit is prograde—counterclockwise around the Sun when viewed from above. The dash line indicates the degree of tilt of Mercury's orbit from the plane of the Solar System. (not to scale)

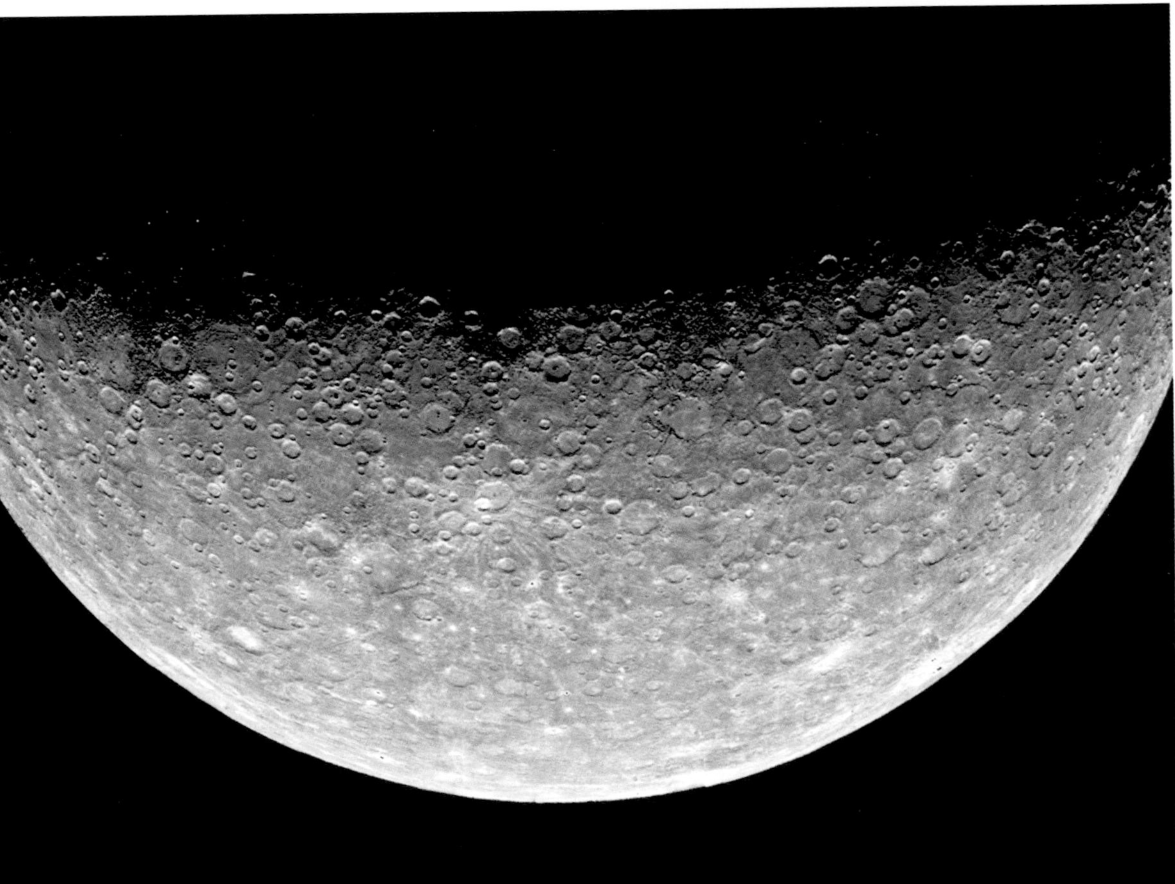

Left Close-up of the surface of Mercury taken by *Mariner 10*. Rather than an atmosphere, Mercury possesses a thin exosphere made up of atoms blasted off its surface by the solar wind and striking micrometeoroids. With such a thin exosphere, there has been no wind erosion of the surface and meteorites do not burn up due to friction.

1676
Sir Edmund Halley travels to an island in the South Atlantic to create an atlas of southern stars. While there, he attempts to observe the transit of Mercury in 1677 but bad weather made the sightings virtually impossible

1800s
Astronomers realize that the physics of the day could not correctly predict Mercury's orbital path.

1915
Albert Einstein uses his new theory of gravity, General Relativity, to correctly predict Mercury's orbit—and explains the reason for previous errors: Mercury is so close to the Sun that its orbit is affected by the "warp" in space created by the Sun's powerful gravitational field.

1965
After believing for centuries that the same side of Mercury always faced the Sun, astronomers discover the planet rotates three times for every two orbits.

1974/1975
NASA's *Mariner 10* spacecraft photographs about half of Mercury's surface. The spacecraft makes three flybys. *Mariner 10* takes first photograph of Mercury in detail on March 29.

1991
Scientists, using Earth-based radar to study Mercury, find signs of ice tucked in permanently shadowed areas of craters in the planet's polar regions.

2003
Mercury makes a rare pass—a transit—across the face of the Sun as seen from Earth. On average, there are 13 transits of Mercury each century.

2004
NASA's MESSENGER spacecraft is launched on a long circuitous trek to Mercury, and should enter orbit in 2011 for a one-year mission to this least-explored of the Solar System's inner planets.

The future—2011
BepiColumbo is a future mission that will be a collaboration of Japan and the European Space Agency (ESA). It will consist of two probes, one to study Mercury's surface and the other to study its magnetosphere. The launch is planned for 2013.

Opposite Digital image of the planet Mercury. While there are areas of smooth terrain, there are also lobe-shaped scarps or cliffs, some hundreds of miles long and soaring up to a mile high, no doubt formed by early contraction of the crust.

Above A simulation of NASA's *Mariner* spacecraft arriving at the planet Mercury. The craft passed by Venus on its way to Mercury, and conducted investigations and observations of both the planets.

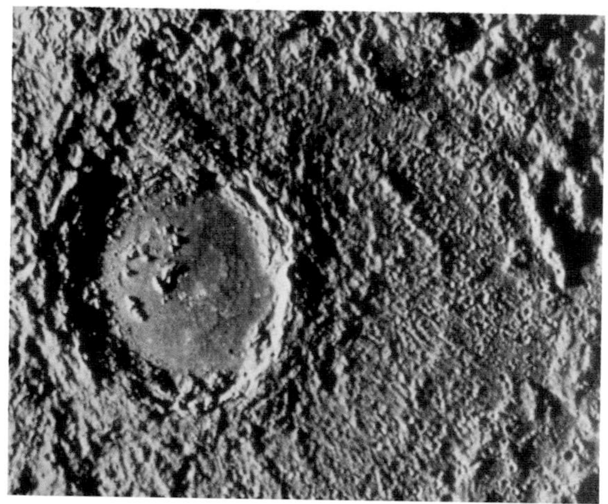

Right A close-up of the cratered surface of the planet Mercury, photographed by the *Mariner 10* spacecraft. Mercury's surface resembles that of Earth's Moon, scarred by impact craters resulting from collisions with comets and meteoroids.

Mercury's largest crater is the Caloris Basin, a huge depression that spans 807 miles (1,300 km) in diameter. The impact that caused it was so great that the effects are still apparent on the opposite side of the planet today, where a region called the "Weird Terrain" was formed—either by shockwaves traveling around the planet and meeting up at that point, or by debris thrown up by the impact being flung to a final landing point there. Mercury also has a series of ridges or folds, which are thought to have been formed as the planet cooled and shrank slightly.

The planet does not have a significant atmosphere. Traces of gases are present—hydrogen, helium, oxygen, calcium, potassium, sodium, and even water vapor—some perhaps from outgassing from within the planet, others released by interaction of the solar wind with the surface. The solar wind also has probably caused a gradual darkening of the whole surface over the aeons.

Mercury has a magnetic field, but it is only one percent as strong as Earth's. It is unknown whether the field is generated by currents in the still molten iron core or whether it is simply "frozen" into the rocks, a leftover from an earlier period in the planet's history.

Intriguingly, it appears that Mercury might even have regions of ice cover! Deep within the permanent shadows of craters at the poles, a thin layer of ice has possibly built up either from water vapor outgassed from within the planet, or from ices brought there by comet impacts.

FUTURE FLIGHTS

Only one spacecraft has so far visited planet Mercury, the US *Mariner 10* probe. It made three flybys of the planet in 1974–75. At the time of each close approach, the same side of the planet was illuminated by the Sun, which meant that *Mariner 10* was able to photograph only 45 percent of the total surface. The images returned are still the best we have of the planet's surface.

Sending a spacecraft to Mercury is difficult, as it will pick up a lot of speed as it gets closer in to the Sun. A large amount of fuel must be carried to slow the spacecraft down at its destination. But, by using gravitational slingshots around Venus and Earth, mission planners can alter a craft's orbit and velocity so that it sneaks up on Mercury rather than rushing by. This is what was done with *Mariner 10*, and also what is being done with two new Mercury spacecraft. A US space probe, MESSENGER (MErcury Surface, Space ENvironment, GEochemistry, and Ranging) was launched in 2004 and will go into orbit around Mercury in 2011. It carries numerous instruments to image and analyze the planet's geology. Another Mercury probe, the joint Japanese-European craft *Bepi-Colombo*, is due for launch in 2013, with an arrival date of 2019. It will comprise two orbital craft; a proposal for a lander has been shelved.

EINSTEIN'S TRIUMPH

A planet's orbit is quite easy to model using mathematics, but for a long while Mercury's orbit posed a problem. The point of perihelion—closest approach to the Sun—was shifting by a greater-than-expected amount each year. It was only a tiny amount, but nevertheless significant. Newton's laws of motion could not explain it. Some astronomers proposed that there might be another planet, which they dubbed Vulcan, orbiting closer to the Sun, and giving a slight gravitational tug on Mercury. It wasn't until Albert Einstein came along with his *Theory of Relativity* that the real explanation was forthcoming.

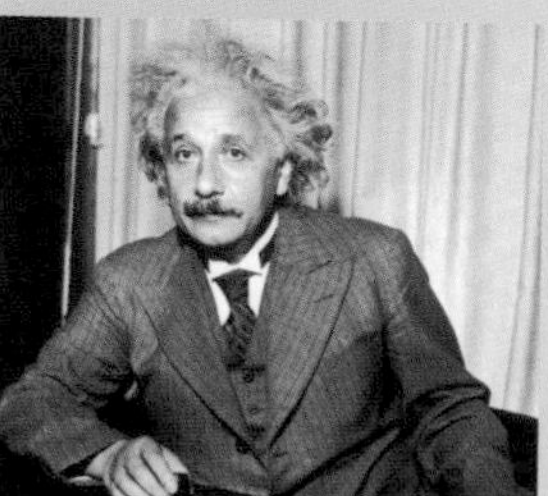

Mercury's orbital speed was increasing slightly around perihelion due to an effect of relativity, and this meant that the perihelion point was then shifted further along its orbit each time.

NEVER FAR FROM THE SUN

Mercury orbits much closer to the Sun than does Earth, and this means that to observe it in the sky you need to look very close to the Sun. During the day the Sun's light is too intense for Mercury to stand out in the blue sky, but after the Sun has set (or, in the morning, before it rises), Mercury can often be seen low down near the horizon. It looks just like a bright star. Because of the speed of its orbital motion, it doesn't stay in the same position for very long, but moves up into the sky and then back down to the horizon again over a period of usually only weeks.

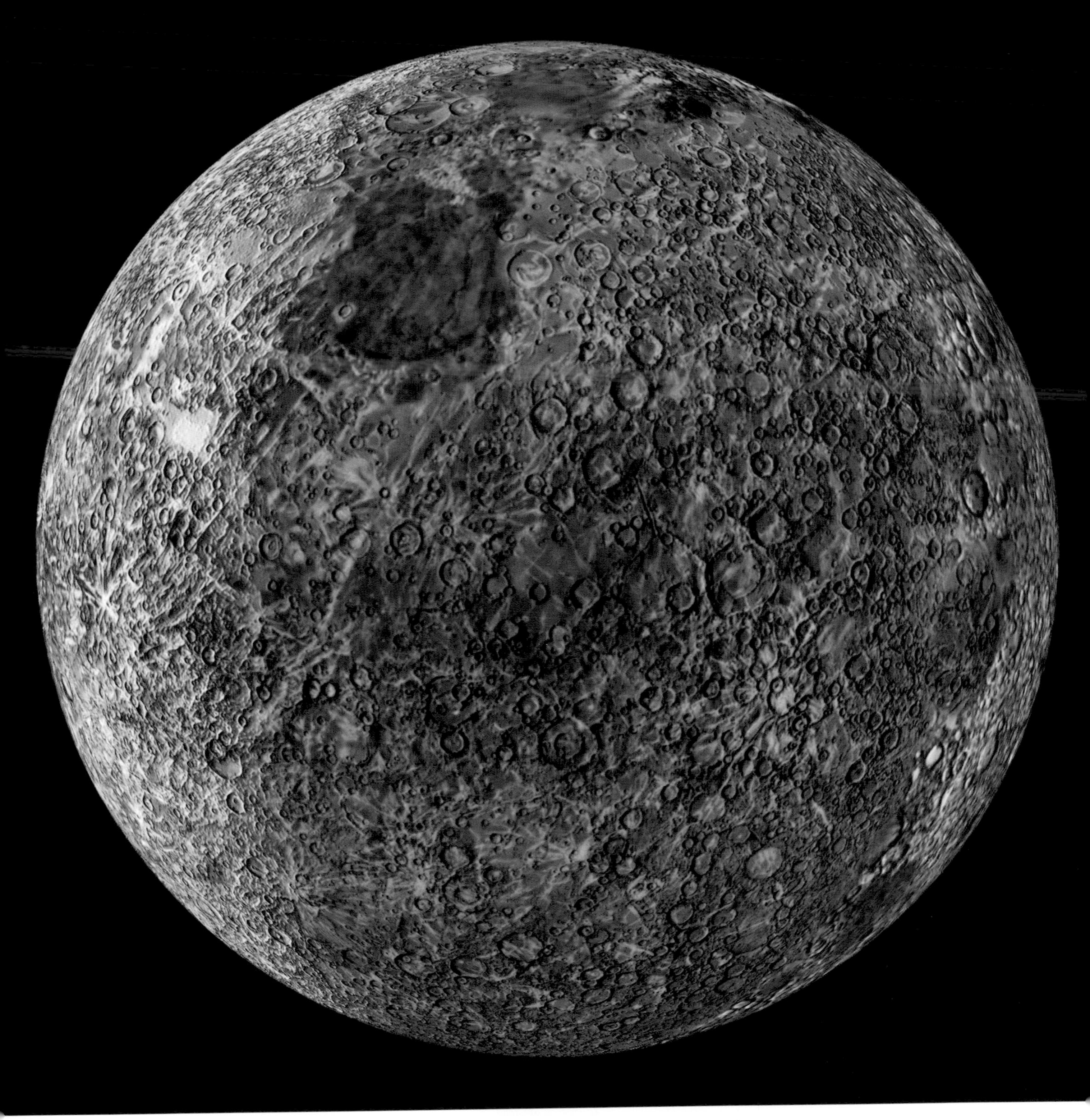

Being an inner planet, Mercury—like Venus—shows phases like the Moon. A telescope will show those phases, as well as the change in apparent size of its disk as the planet moves closer to us and then further away as it orbits the Sun.

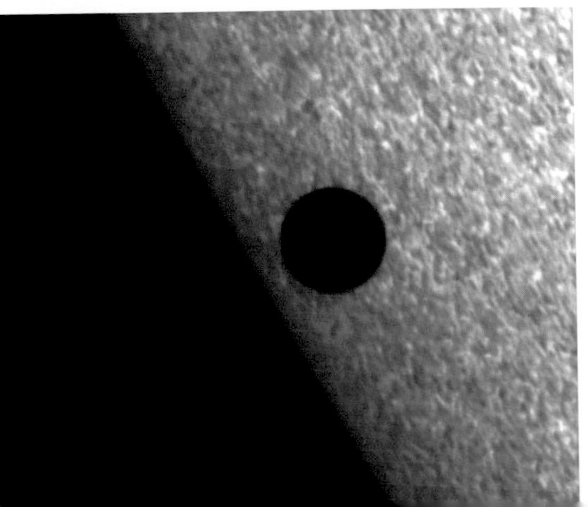

Left This image of Mercury passing in front of the Sun, a transit, was captured by the Solar Optical Telescope, one of three primary instruments on *Hinode*, a collaboration between the space agencies of Japan, USA, UK, and Europe to investigate the interaction between the Sun's magnetic field and its corona.

Venus

FACT FILE

Planetary Order
2nd planet from Sun

Average Distance from Sun
0.72AU/67,237,910 miles/ 108,208,930 km

1 Venus Day
–243 Earth days (retrograde)

1 Venus Year
0.615 Earth years

Equatorial radius (Earth = 1)
0.95

Mass (Earth = 1)
0.815

Surface Temperature
864°F/462°C

Atmosphere
96.5 percent carbon dioxide, 3.5 percent nitrogen, traces of sulfur dioxide, argon, water vapor, carbon monoxide, helium, neon

Moons/satellites
No moons, no rings

Notable features
More than 1,000 volcanoes and a crushing atmospheric pressure 90 times that of Earth.

In many ways Venus is nearly Earth's twin, but in other ways it couldn't be more different. It appears to be the brightest planet in the sky, but Venus holds its secrets tight.

SISTER WORLD

Venus is a fascinating planet. For millennia it was known only as a bright light in the sky—third brightest, in fact, after the Sun and the Moon. Even with the advent of the telescope little more was learned about the planet, for it quickly became apparent that Venus is completely covered by clouds that prevent the surface from being seen. It wasn't until space probes began to make the journey to Venus in the last decades of the twentieth century that we began to get an idea of what makes this planet tick. Venus is named for the Roman goddess of love, and in keeping with astronomical tradition of naming Solar System objects according to themes, almost every feature on Venus is named after women, real or mythological. The planet Venus, itself, is the only planet in the Solar System with a female name.

Above right Detail from *The Birth of Venus*, c.1485, by Filipepi Sandro Botticelli, the Italian painter. The original resides in the famous Galleria degli Uffizi, Florence, Italy. Venus was a major Roman goddess principally associated with love, beauty, and fertility. She was the consort of Vulcan, and played a key role in many religious festivals and myths.

MORNING AND EVENING STAR VISIBILITY

Being the brightest planet in the sky, Venus has naturally been known to mankind for as long as people have admired the night sky. Being an inner planet (like Mercury), however, set it apart from the known outer planets, Jupiter and Saturn, as Venus never gets very far from the Sun in the sky—it is visible for up to a few hours before dawn in the east or for up to a few hours after sunset in the west.

This led to it being called the "morning star" and "evening star," although at first it was not realized that these two "stars" were the same object. It was Pythagoras who first suggested they were one and the same. Venus is so bright that, if you know exactly where to look, it can even be seen with the naked eye in the daylight sky.

Often it will be near the Moon in the sky, making its identification easy. In fact, it is so bright that some people sometimes mistake it for a UFO—long before he became the US President, Jimmy Carter once reported Venus as such!

Venus sometimes comes between Earth and the Sun, in an event known as a transit. These come in pairs, eight years apart, and every 243 years. The last one was in June 2004, and the next will occur in June 2012.

COOK'S TRANSIT

One of the most famous sea voyages of all time was undertaken to (among other things) make a very important observation of Venus. Captain James Cook's first voyage to the South Pacific in 1769 was primarily intended to witness a rare transit of Venus, where the planet comes between Sun and Earth, and is seen to move or "transit" across the Sun's face. Scientists were eager not to miss this event, as they could use it to help sort out one of the thorniest astronomical questions then facing them—how big is the Solar System?

Cook traveled to Tahiti in 1769 and set up a small observatory there in anticipation of the transit, which was to occur on June 3. The transit was seen, but unfortunately the accuracy of the observations—although the best Cook and his team could do with the resources of the time—was not as good as had been hoped. Nevertheless, they were combined with other data to refine the knowledge of Venus's distance from the Sun. This done, the distances to the other planets could be calculated, leading to a better appreciation of the size of the Solar System.

Cook, of course, then went on to map the coastlines of New Zealand and Australia. The shuttle *Endeavour* was named after Cook's ship.

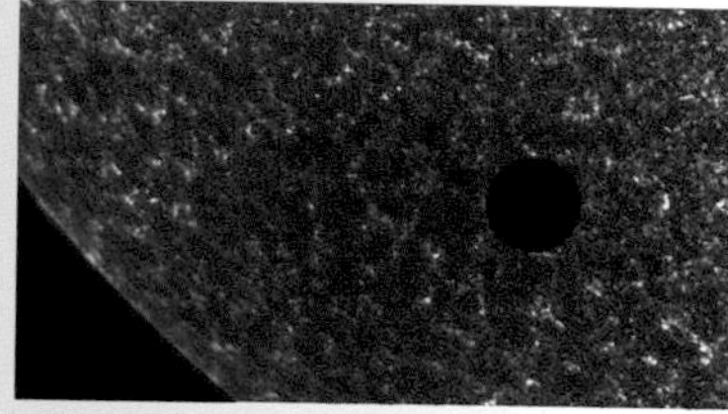

NASA's TRACE satellite captured this image of Venus crossing the face of the Sun as seen from Earth orbit. The next Venus transit will be visible in 2012.

Prehistory
Venus is known as the brightest object in the sky except for the Sun and the Moon.

3rd century BCE
Venus is thought to be two separate bodies: Eosphorus as the morning star and Hesperus as the evening star. Venus's apparition as the morning star is also called Lucifer.

1610
Galileo Galilei is the first person to actually see Venus as more than just a bright point of light in the sky. He observed Venus through different phases.

1631
Johannes Kepler, using laborious hand calculations, predicts that Venus will pass in front of the Sun on December 6, 1631, but the transit will not be visible from Europe.

1663
Rev. James Gregory, mathematician, suggests that a more accurate measurement of the Earth-Sun distance could be made using the transit of Venus.

1677
Sir Edmund Halley (1656-1742), makes the same suggestion 14 years later in 1677 and publishes a paper on the details of this technique in 1716.

1680
Halley suggests using Venus' transits to measure the Astronomical Unit (Astronomical Unit is approximately the mean distance between Earth and the Sun.)

1761
During the transit of June 5, 1761, which was observed by over 176 scientists from more than 117 stations all over the world, the Russian astronomer, Mikhail V. Lomonosov (1711-1765), discovers a beautiful halo of light all around Venus's dark edge, indicating that it has an atmosphere.

1768
On August 12, 1768, His Majesty's bark *Endeavour* captained by Lt. James Cook, departs England bound for Tahiti, to observe a transit of Venus on June 3, 1769.

1874
Hundreds of photographs are taken of the next transit on December 8, 1874. This was the first use of the new technology. Few photographic plates turned out to be scientifically useful, and only a few are actually preserved for us to look at today.

VENUS CORE FACTS

Venus is believed to have a very similar internal structure to Earth, with a core, mantle, and crust. Although the liquid iron core is large enough, it does not produce a magnetic field like Earth's (probably because it is too hot, preventing convection currents that generate magnetism). The mantle is some 1,860 miles (3,000 km) thick, above which the crust extends around 6–18 miles (10–30 km) to the surface. About 90 percent of the surface of the planet appears to be solidified basalt lava, no doubt remaining from volcanic activity 300 to 500 million years ago.

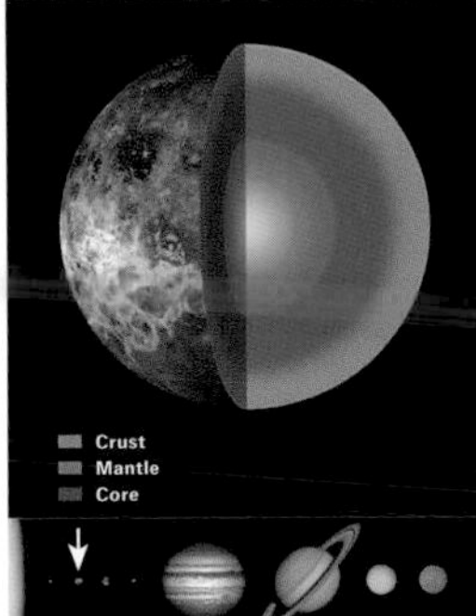

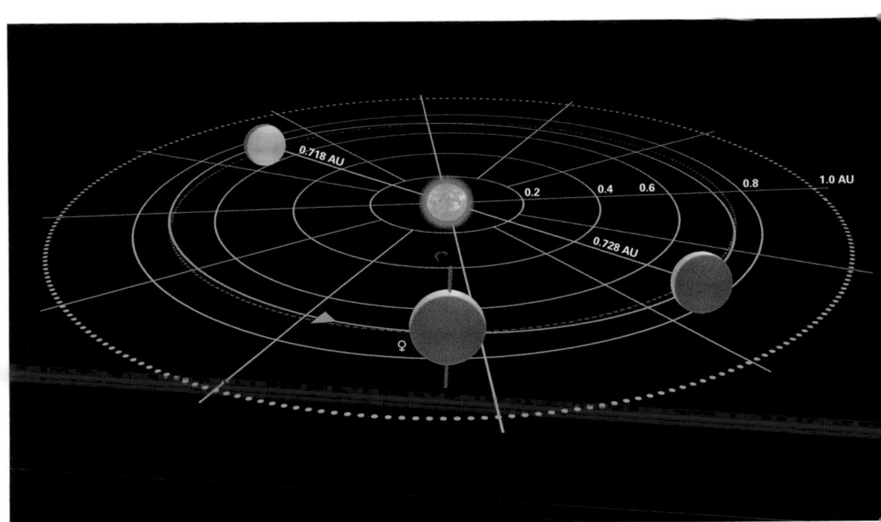

Above Venus takes 224 Earth days to complete an orbit of the Sun that is just off circular by less than one percent, making it the most circular of all the planets' orbits. Venus's orbit is inclined at 3.39° to the ecliptic. Its axial tilt is 2.64°. Its orbit is prograde—counterclockwise around the Sun when viewed from above. However, Venus's rotation is retrograde or clockwise. The dash line indicates the degree of tilt of Venus's orbit from the plane of the Solar System. (not to scale)

Left Venus and the Moon rising, taken April 22, 2004, in Germany. This is one of the many shots taken every year by amateur astronomers around the world.

1882
There is enormous public interest in the next transit on December 6, 1882. It makes the front pages of every national and international newspaper of the time, with lavish full-length articles covering several pages.

1891
Simon Newcomb, who was in charge of the US Venus Transit Commission, publishes his best estimate of the solar parallax angle in 1891, based on all of the assembled data from several transits.

1932
Astronomers Walter Adams and Theodore Dunham use refined spectroscopic instruments to detect carbon dioxide in the atmosphere of Venus.

1961
Russian craft *Sputnik 7* attempts to launch a rocket, with the *Venera* probe aboard, toward a landing on Venus after one Earth orbit, but ignition fails.

1962
The first spacecraft to flyby Venus is *Mariner 2*, backup for the *Mariner 1* mission that failed shortly after launch. US astronomer, Carl Sagan calculates the effect of the atmosphere on the temperature of Venus.

1967
Mariner 5 does a flyby of Venus, revealing new information about its atmosphere, including its composition of almost 97 percent carbon dioxide.

1970
Venera 7 enters the atmosphere of Venus on December 15, 1970, and a landing capsule is jettisoned. Signals are returned for 35 minutes. then another 23 minutes of very weak signals are received after landing. The capsule is the first man-made object to return data after landing on another planet.

1978
Pioneer 13, or *Pioneer–Venus 2*, carries four smaller probes to be dropped into the atmosphere of Venus.

1989
The interplanetary space probe *Magellan* leaves Earth and falls into orbit around Venus on August 10, 1990. It sends back spectacular radar images for a new, more-detailed map of Earth's cloud-shrouded sister planet.

2005
The European Space Agency sends *Venus Express* to probe the second planet. The spacecraft explores the Venusian atmosphere where scientists wish to understand the origin of violent winds that blow around the planet.

Above Surface of Venus from the *Venera 10* Lander, which touched down in October, 1975, and then returned this image during the 65 minutes of its operation on the surface.

Right Digital composite image provided by the *Venus Express* space probe on April 12, 2006. The South pole of the planet Venus is seen in a false-color view, and shows the planet's day side at the left and night side at the right.

ODD ROTATION

Venus has the most circular orbit of all the planets, having an eccentricity of just slightly less than one percent. (For comparison, Earth's orbit has an eccentricity of 1.7 percent, Jupiter has 4.8 percent, and Mercury a whopping 20.5 percent.)

Venus orbits the Sun at an average distance of around 67 million miles (108 million km), which means it is the planet that comes closest to Earth, around 25 million miles (40 million km); Mars can approach to within about 35 million miles (56 million km), although usually it is a lot more.

Galileo's early observations of Venus showed it having a crescent shape at certain times in its orbit, just like the Moon. Mathematically, there was only one way this could be explained—Venus had to be in orbit around the Sun, not Earth. This revelation helped pave the way for the change from an Earth-centered view of the heavens to a Sun-centered one. Observations in the late eighteenth century first gave hints that Venus had a thick atmosphere. The length of the Venusian year is just under 225 Earth days, but for a long time scientists were unable to gauge the length of the Venusian day.

For many other planets it is easy; just look through a telescope, pick a prominent feature on the planet's surface and measure the length of time it takes to go right around. But with Venus' thick cloud cover this would not work. As a planet (or any other body for that matter) rotates, one edge of it is moving toward the observer while the other edge is heading away. Scientists can measure the difference using the Doppler effect, and can use this technique to estimate a planet's rotation speed, and hence the length of its day. But when they first tried to do this with Venus it just wouldn't work. Eventually they had to conclude that the planet's rotation must be too slow to be picked up by their Doppler observations.

Observations of the clouds in the middle of the twentieth century gave hints that Venus rotated backward compared to all the other planets.

It wasn't until scientists turned powerful radars upon the planet, capable of penetrating the clouds and bouncing off the ground below, that they were able to accurately measure the rotation—and sure enough, it was backward, or retrograde. And the Venusian day was incredibly long—it takes a whole 243 Earth days for Venus to rotate once on its axis.

There are two main theories to account for the odd rotation. One theory is that Venus was hit by a wandering small planet at some stage of its life, stopping its normal prograde rotation, and making it turn backward. The other theory is that tidal forces acting on Venus's thick atmosphere could have operated like friction to slowly reverse the rotation.

The early radar observations showed little detail of the surface, but they could make out two broad areas that were different to their surroundings. These were named simply Alpha Regio and Beta Regio. A smaller, but more prominent, area was named Maxwell Montes, after James Clerk Maxwell, a physicist who laid the foundation for our understanding of electricity and magnetism. These three geological features are the only ones on Venus not named after female figures.

TWINS

Venus and Earth are quite alike in some ways. They are similar in size—7,500 miles (12,100 km) in diameter at the equator for Venus; Earth is 7,928 miles (12,759 km).

Also, Venus has around 82 percent of Earth's mass. This, plus the fact that Venus is covered with clouds—early observers more or less assumed they were made of water, like Earth—and the fact that Venus is closer to the Sun,

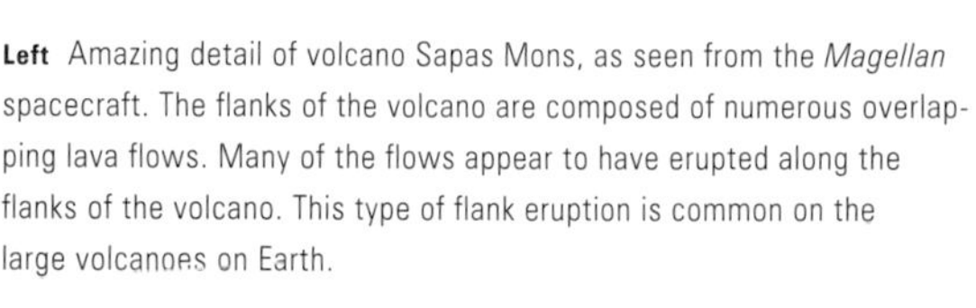

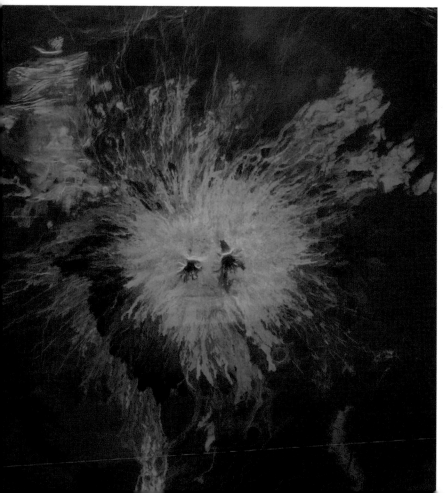

Left Amazing detail of volcano Sapas Mons, as seen from the *Magellan* spacecraft. The flanks of the volcano are composed of numerous overlapping lava flows. Many of the flows appear to have erupted along the flanks of the volcano. This type of flank eruption is common on the large volcanoes on Earth.

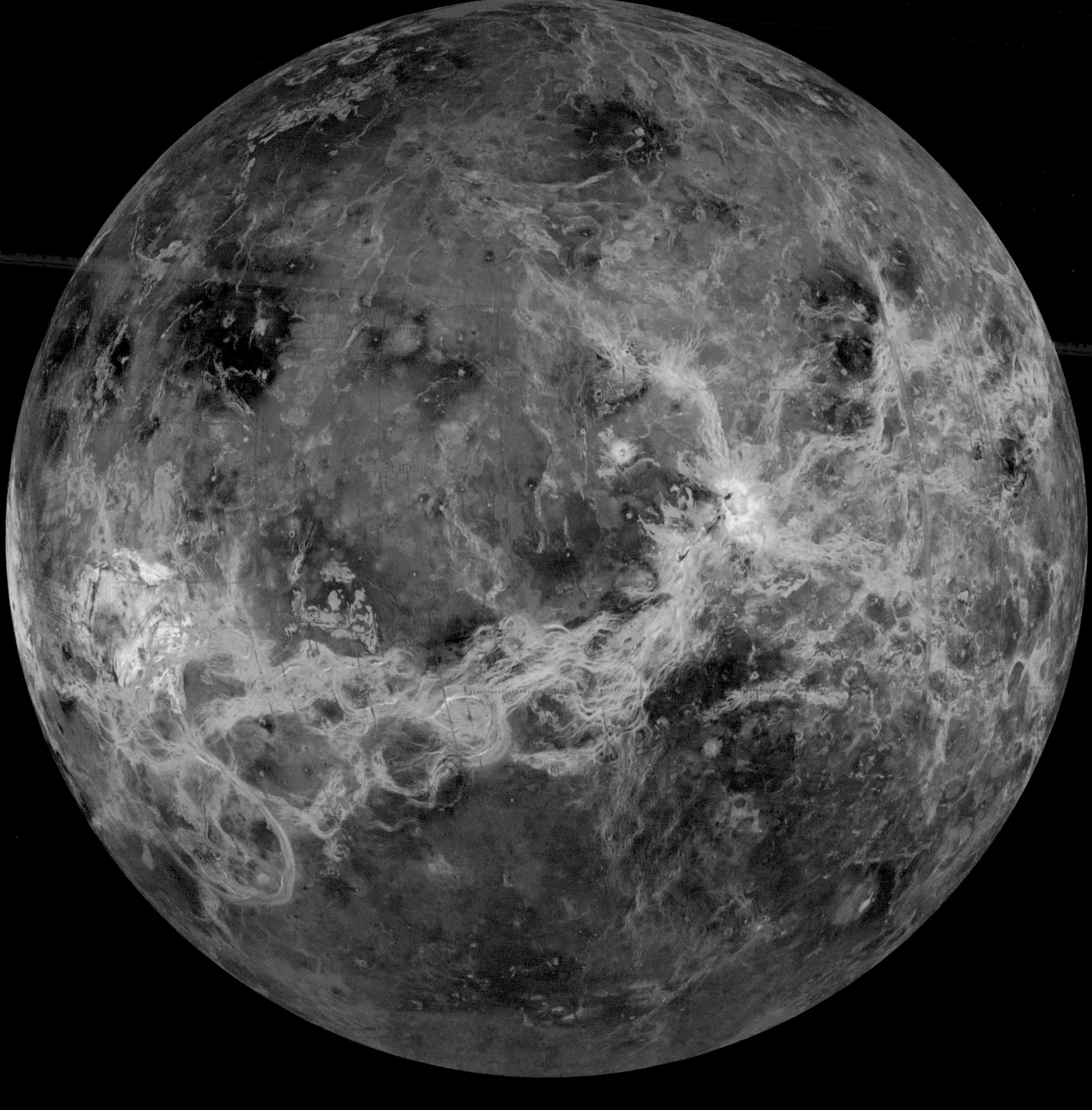

and hence should be warmer, led some to conclude that it was probably a jungle world. Like Earth's tropics, scientists thought Venus would have abundant planet and animal life, maybe even dinosaurs roaming the countryside. They couldn't have been more wrong.

Venus is believed to have a very similar internal structure to that of Earth, with a core, mantle, and crust. The core is intriguing, as it is thought to be technically large enough to produce a magnetic field like Earth's, yet it seems that no such field exists. To produce such a field, the core needs to have convection currents from the hotter inner part to the outer cooler part. Venus doesn't appear to have this convection, either because the core has solidified, or else it is all at much the same temperature. The latter situation could come about because Venus is not able to get rid of its internal heat like Earth does, because the planet does not have the energy "relief valve" of plate tectonics.

The planet does, however, have a small weak magnetic field produced by the interaction of the solar wind with the upper levels of the planet's atmosphere.

Above Global view of the surface of Venus centered at 180° E longitude. Scientists used radar to peer through the clouds to map the surface. These clouds reflect sunlight and trap heat, usually making Venus the brightest planet in the sky.

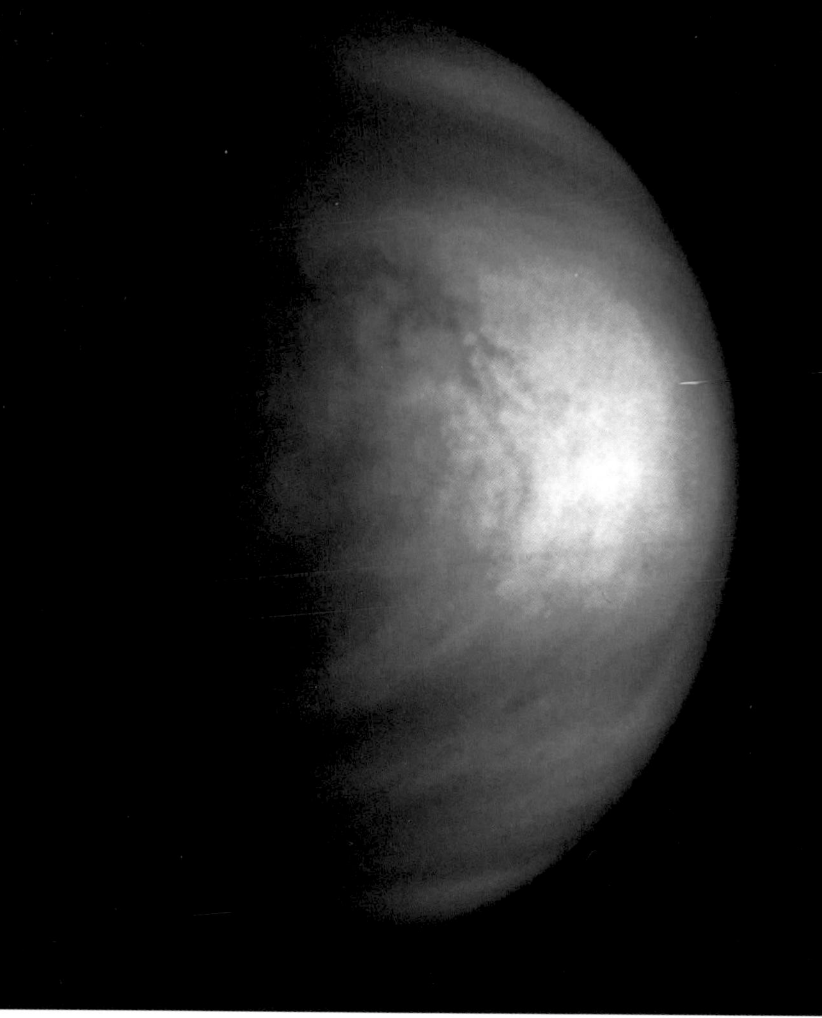

Above *Mariner 2* was the world's first successful interplanetary spacecraft. Launched August 27, 1962, it passed within about 21,000 miles (34,000 km) of Venus, sending back valuable new information about interplanetary space and the Venusian atmosphere.

Above Colorized picture of Venus taken about six days after *Galileo*'s closest approach. The image shows cloud banding caused by winds that flow from east to west at about 230 mph (370 km/h).

SPACE MISSIONS

The coming of the space age at the beginning of the 1960s brought with it the possibility of learning more about Venus by sending space probes there—and Venus became a target of some of the earliest planetary spacecraft. Almost everything we now know about the planet is as a result of sending spacecraft to investigate.

The first attempts to send probes—the Soviet *Venera 1* and the US *Mariner 1*—both failed. The first successful mission was the USA's *Mariner 2*, launched on July 22, 1962, which flew by Venus at a distance of just over 21,600 miles (34,800 km) on December 14, 1962. *Mariner 2*'s instruments revealed that the planet's clouds were cold but that the surface was very hot, over 750°F (400°C)—not quite the comfortable abode of life it was once thought to be. More Soviet and US missions followed, some successful, some not. *Venera 4* entered Venus' atmosphere in October 1967, and determined that it was more than 90 percent carbon dioxide, and a lot thicker than expected. The spacecraft's parachute slowed it down more than anticipated, with the result that it didn't reach the surface before *Venera*'s batteries gave out. Parachutes on following missions were made smaller to increase the descent rate.

Right *Venera 1* was the first spacecraft to attempt a flyby of Venus. On February 19, 1961, seven days after the launch, contact with *Venera 1* was lost.

Venera 5 and *6* followed in 1969. Both were built stronger than *Venera 4*—the planet's surface pressure had been calculated to up to 100 times that of Earth—but still both were crushed at around the 12.5-mile (20 km) altitude level by the increasing atmospheric pressure.

The US probe *Mariner 4* flew past the planet in October 1967, just one day after the arrival of *Venera 4*. It carried instruments to study the atmosphere.

The Soviets then concentrated on landing probes on the surface, sending *Venera 7*, *8*, *9*, and *10* between 1970 and 1975, all of them strengthened to withstand the surface pressures. *Venera 7* made a descent through the atmosphere and had a hard landing (probably due to a torn parachute). It radioed data back to Earth for over 20 minutes; this was the first data ever received from the surface of another planet (not counting the Moon). *Venera 8* was similar to *Venera 7*, and sent back data for just under an hour.

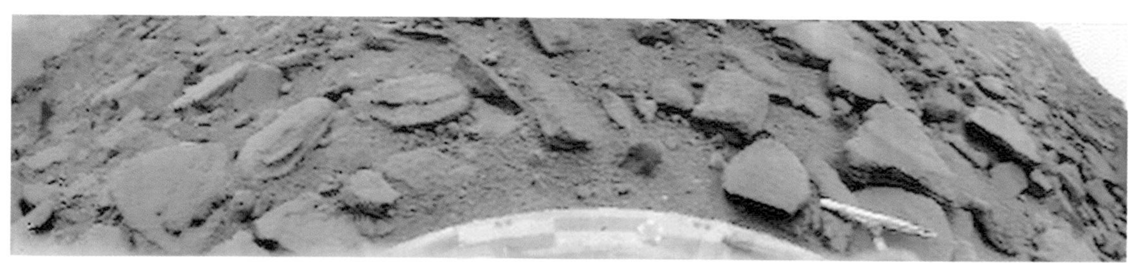

Right *Venera 9* lander image of the surface of Venus. The lander touched down with the Sun near zenith on October 22, 1975, and operated for 53 minutes, allowing the return of this single image.

Then came *Venera 9* and *10*, both equipped with cameras in addition to their other instruments, so that images of the surface could be beamed back. They were launched six days apart in June 1975 and landed three days apart in October 1975, over 1,250 miles (2,000 km) away from each other in the planet's Northern hemisphere. The images sent back showed a baked and desolate rocky landscape.

Buoyed by their success, the Soviets sent four more landers to Venus during the period 1978 to 1981. *Venera 11*, *12*, *13*, and *14* all landed successfully, although several instruments failed to function properly or at all—the cameras aboard *Venera 11* and *12* did not work as the lens caps failed to eject; the *Venera 11* soil analyzer did not work; and the *Venera 14* impact probe could not operate after an ejected camera lens cap got in the way. Nevertheless, the missions were highly successful. Between them they returned data on the atmosphere and its constituent gases, analyzed the planet's surface, and detected lightning. *Venera 13* even recorded loud thunder, which was the first sound ever to be detected on another planet.

Meanwhile, the USA had continued its investigations of Venus, although not to the same extent as the Soviets.

The American space program was very busy with its *Apollo* manned lunar missions and also with a fleet of Mars probes. The USA did not land on Venus, but concentrated on flybys and orbiters.

The *Mariner 10* spacecraft was sent on a trajectory that took it past Venus on its way to Mercury, its real destination. It passed within 3,728 miles (6,000 km) of the Venusian surface, its ultraviolet camera revealing details in the clouds that were not visible in ordinary light.

NASA's *Pioneer Venus Orbiter* and *Multiprobe* missions were launched in May and August 1978, and arrived in December 1978. The *Orbiter* went into orbit around the planet and operated for over 14 years. It was equipped with a large suite of instruments to study the planet's atmosphere, clouds, surface (using a radar mapper), magnetic field, and gravitational field. This was a highly successful mission, brought to a close in 1992 only when its thruster fuel had been exhausted. The probe descended into the atmosphere where it was destroyed.

The *Multiprobe* mission comprised one large and three small probes that directly entered the atmosphere—the large probe had a parachute, but the smaller probes did not carry them.

Each probe sent back data on the atmosphere and its constituent gases, as well as wind measurements and pressure readings.

The probes were placed on separate entry trajectories that saw them descend over different regions of the planet. None were intended to survive impact, but one of the smaller probes did, and continued to send back data for over an hour.

The Soviets, meanwhile, had not given up on Venus. In June 1983 they launched *Venera 15* and *Venera 16*, both orbiters, which arrived at the planet within days of each other in October 1983. These space probes were designed to study the surface from orbit, using a radar system to bounce signals off the rocky terrain. They successfully mapped the Venusian surface from about 30° North latitude up to the planet's North pole. They also carried altimeters and instruments to detect radiation in space.

Below Computer-enhanced three-dimensional view of the surface of the planet Venus, showing volcanoes and lava flows.

Right The *Venus In-situ Explorer* (VISE) is a New Frontiers Class mission that will study the composition and surface properties of Venus. It is expected to launch in 2013.

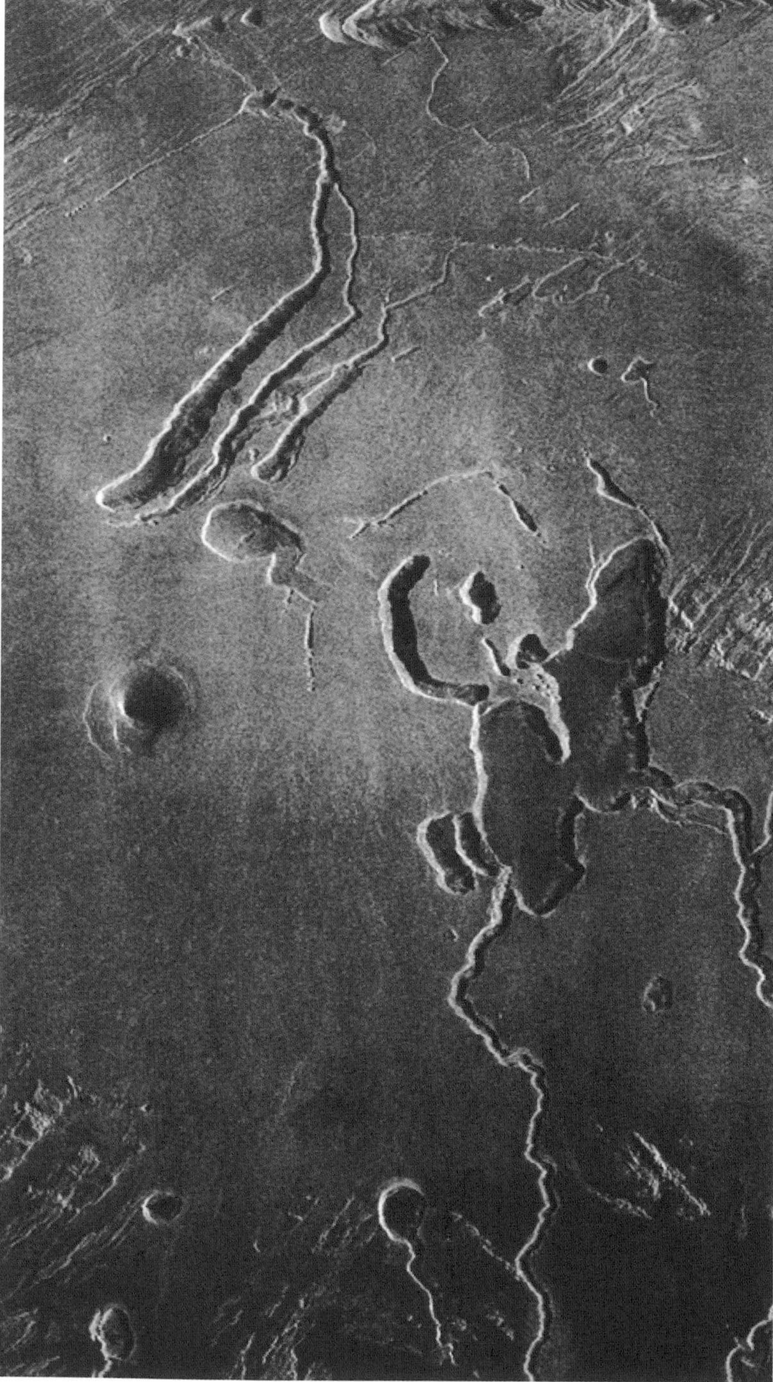

In 1986, a small flotilla of spacecraft was sent to investigate Comet Halley during its pass through the inner Solar System. The Soviets took the opportunity of launching two dual-use spacecraft, *Vega 1* and *Vega 2*, which would first fly past Venus before continuing on to the comet. The two spacecraft carried balloon probes and landers to be dropped into the planet's atmosphere and onto its surface. Launched in December 1984, the *Vega*s reached Venus in June 1985 and released their balloons and landers.

Vega 1's lander began its descent but activated its surface experiments too early—a jolt from the Venusian winds

Left *Magellan* radar image of the lava channels north of Ovda Regio on the surface of Venus. This image shows the Lo Shen Valles, a system of deep channels and large collapsed source areas caused by volcanic eruptions on the planet.

Left Sapas Mons is in the center of this computer-generated three-dimensional perspective view of the surface of Venus. Lava flows extend across the fractured plains shown in the foreground to the base of Sapas Mons.

Below Color-enhanced, three-dimensional perspective (achieved by a process called radar-clinometry) of a portion of the eastern edge of Alpha Regio on a hilly region on the planet Venus. Taken by the spacecraft *Magellan*, it shows several places where the ridges are flooded by radar-dark (smooth) lava flows, indicating that volcanic activity postdates the formation of the ridges.

tricked the lander into thinking it had touched down—and consequently it didn't return any data. However, *Vega 2*'s lander did successfully reach the surface and sent back data for just under an hour.

Meanwhile, the balloons had been released into the atmosphere. Inflated with helium to a diameter of 12 ft (3.5 m), they each carried a tether 43 ft (13 m) long with an instrument package attached.

The balloons descended until they reached an altitude at which the pressure was sufficient for them to float, a little over 30 miles (50 km) above the surface. The instrument package studied the atmospheric gases and measured wind speeds, temperatures, and pressures.

Powered by batteries, they each lasted almost two Earth days, sending their data back to a network of radio dishes on Earth. That was the end of the Soviet investigations of Venus but the US had more to come.

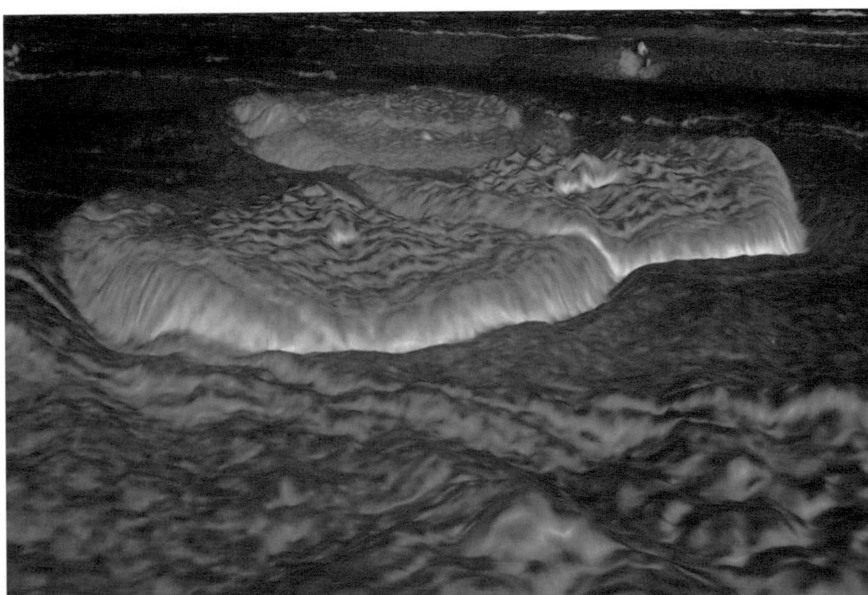

Above Artist's impression of *Venus Express* in orbit around the planet Venus. The simulation shows cloud wave structure being clearly pushed by winds across the surface.

In May 1989, the US launched the sophisticated *Magellan* probe from the space shuttle *Atlantis* from Earth orbit—the first interplanetary probe to be launched from the shuttle. *Magellan* was essentially a giant radar system, designed to penetrate the Venusian clouds with its powerful radar beams and map the planet in high detail.

Reaching Venus in August 1990, it went into a series of looping orbits, which, over the next four years, took it over almost the entire surface of the planet. Aiming its radar downwards, it scanned long strips of terrain with around 328-feet (100-m) resolution.

Magellan also spent time mapping the gravitational field of the planet—as it flew over different terrain, it would slightly speed up or slow down depending upon how much mass was below it.

By measuring the Doppler shift in its radio waves produced by these velocity changes, scientists on Earth could refine their understanding of the geology of the planet.

EXPRESS TO VENUS

There's now a new space mission studying Venus. The European Space Agency's *Venus Express* spacecraft went into orbit around the planet on April 11, 2006. Its primary mission will last for two Earth years and there are enough fuel resources onboard to extend the mission even further.

Venus Express will comprehensively study the planet. The instruments will analyze the various aspects of the atmosphere, the magnetic field, clouds, and surface temperatures. Within about six months of arriving, the spacecraft had already sent back intriguing images of a strange double vortex storm formation at the South pole, as well as a temperature map of parts of the Southern hemisphere.

Scientists hope that *Venus Express* will help answer many questions regarding the ongoing chemical balances in the planet's atmosphere and clouds, and whether or not volcanic activity is still occurring on Venus today.

WHAT'S IN A NAME?

With the exception of three features discovered many years ago, all geological features and structures on Venus are named after women, either real or mythological, according to rules laid down by the International Astronomical Union. For instance, round features called *coronae* are named after goddesses of fertility and Earth. Elongated features called *linea* are named after goddesses of war. Ridges are named after sky goddesses. Plateaus are named after goddesses of prosperity. Large regions are named for giantesses and titanesses.

But lest you think that only mythological women are getting all the recognition, real women are honored too. Craters under 12 miles (20 km) in diameter are given common female first names, while those over that diameter are given the names of deceased women who have made important contributions to their fields.

Marble statue of the Greek goddess Athena. Athena Tessera, a large area of terrain in the Northern hemisphere of Venus is named after her.

And the geology revealed by *Magellan* was fascinating. Scientists found no evidence for the kind of continental plate tectonics that we have on Earth. But the radar maps did show that the surface of Venus is dominated by volcanic structures and their aftermath—large flat volcanoes, ancient lava flows thousands of miles long, volcanic domes, and huge plains of solidified lava. There were relatively few impact craters evident, which is a good indication that the planet's overall surface is "young" geologically speaking—that is, volcanism has erased the signs of older impacts. On Earth, wind and water erosion, and tectonic plate movements do the same job. A massive planet-wide "resurfacing event" is believed to have taken place on Venus around 800–500 million years ago.

Straddling the equator are two continent-sized upland regions, known as Ishtar Terra and Aphrodite Terra. Ishtar Terra has Venus' highest mountain peaks, including the monstrous Maxwell Montes, which stands 6.8 miles (11 km) above the surface of the planet.

The *Magellan* mission eventually came to an end, but even at its conclusion the spacecraft was able to perform useful science. Mission controllers used the technique of aerobraking—when a spacecraft dips slightly into the upper levels of a planet's atmosphere, slowing it down—to lower *Magellan*'s orbit bit by bit. Eventually, it got so low that it could not keep orbiting, and descended into the atmosphere and destruction. But even as it was doing so, it was radioing back data, that was used to refine the knowledge of Venus's atmosphere.

After a long gap, there's now a new spacecraft investigating Venus. The European *Venus Express* spacecraft has begun a multi-year study of the planet, and just months after arrival had already shown its potential by producing a partial temperature map of the Southern hemisphere of the planet. This is a very important and useful development, as it will help scientists to pinpoint any potential "hot spots" that could show active volcanism, if any.

CLOUDS AND ATMOSPHERE

Venus has an incredibly thick atmosphere, far thicker than Earth's, composed mainly of carbon dioxide with traces of other gases. The pressure at the surface is a crushing 90 times the surface pressure on Earth—early Soviet spacecraft were squashed by the atmosphere even before they reached the ground. Carbon dioxide is a very potent "greenhouse" gas, as the inhabitants of planet Earth are coming to realize. Venus has so much of it that the planet has a runaway greenhouse effect, leading to surface temperatures over 750°F (400°C)—that's hot enough to melt lead. Venus's surface is, in fact, hotter than that of Mercury, even though Mercury is much closer to the Sun and receives much more solar energy.

Some scientists believe that Venus, like Mars, might once have had a more earth-like atmosphere and surface conditions, with lots of liquid water. If that water had evaporated—water is an even more potent greenhouse gas than carbon dioxide—it could have started the deadly spiral of rising temperatures that led Venus to the state it is in today. Venus's clouds might look bland and uninteresting at visible light wavelengths, but they hide a complex climate system that is of great interest to scientists. The clouds themselves are made of sulfuric acid droplets and sulfur dioxide. The clouds reflect a lot of sunlight back into space—this means that less heat gets through to the surface, making it plain that the greenhouse effect is responsible for the incredible surface temperatures.

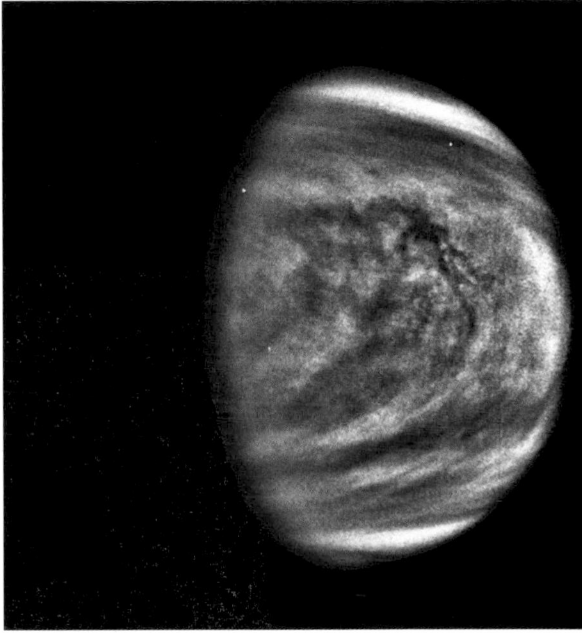

Above This photograph of Venus taken by the *Galileo* spacecraft has been colored to a bluish hue to emphasize the subtle contrasts in the cloud markings. The sulfuric acid clouds indicate considerable convective activity in the equatorial regions of the planet to the left, and downwind of the subsolar point (afternoon on Venus). They are analogous to "fair weather clouds" on Earth.

Winds at ground level are slow, but the density of the atmosphere means that they pack a fair punch. This is the opposite to the situation on Mars, where wind surface speeds are high, but the air is so thin that gale force winds would seem like a gentle breeze.

High up at the top of the clouds however, the wind speeds are very fast—around 185 mph (300 km/h). This is fast enough to make the clouds circle the entire planet in only four or five Earth days.

Observations from the *Venus Express* spacecraft have shown that the clouds near the equator are more unevenly shaped than in the temperate regions, where they tend to be elongated and follow the strong winds.

At the poles are strange vortex circulations. On its first data-taking orbit, *Venus Express* revealed the south polar vortex to be like a huge cyclone with two "eyes."

Spacecraft analyzes showed that atmospheric levels of sulfur dioxide dropped significantly between the late 1970s and mid 1980s. This lead scientists to speculate that a large volcanic eruption may have taken place sometime earlier on the planet, causing the drop.

Measurements such as these are indirect indications that volcanism still shapes the face of Venus.

Earth and Moon

FACT FILE

EARTH

Planetary Order
3rd planet from Sun

Average Distance from Sun
1 AU/92,955,820 miles/149,597,870 km

Equatorial radius
3,963.19 miles/6,378.137 km

Mass
5.879x10^{21} tons/5.9737x10^{24} kg

Surface Temperature
–126 to136°F/–88 to 58°C

Atmosphere
78.08 percent nitrogen, 20.95 percent oxygen, 0.03 percent argon, 0.038 percent carbon dioxide, trace water vapor (varies with climate)

Moons/satellites
One Moon

Notable features
Only known inhabited planet in the Solar System.

Oceans at least 2.5 miles (4 km) deep cover nearly 70 percent of Earth's surface.

Earth's land surfaces are in motion through tectonic plates.

MOON

Average distance from Earth
238,855 miles /384,400 km

1 Moon Day
27.32 Earth days

1 Moon Year
0.075 Earth years

Equatorial radius (Earth = 1)
0.2724 x Earth

Mass (Earth = 1)
0.0123

Average Surface Temperature
–387 to 253°F/ –233 to 123°C

Atmosphere
Atmosphere is so thin that it is almost negligible.

Notable features
Only satellite of Earth.

Earth is unique. As far as we are aware, it is the only planet in the entire universe that supports life. And now life has begun to reach out and explore the universe around it, starting with its nearest neighbor, the Moon.

EARTH

For almost the entire history of our species, we humans had it completely wrong. We thought that our planet, Earth, was the center of the universe, and that everything else revolved around us. One can quite understand why that view would have arisen. Simple observation of the changing seasons, and the nightly spectacle of the stars and planets, all conspire to give the impression that we live in the middle of a cosmic orchestra, with Earth as the band leader. But we now know that that is not the case. Earth revolves around the Sun in concert with the seven other major planets (and myriad smaller bodies) that together make up our Solar System. And our Solar System is only one of probably countless billion planetary systems throughout the wider universe.

Earth is just one name for our planet, of course. Every language and culture has its own name, often woven in with mythologies and creation legends. To the ancient Greeks it was *Gaia*, the Earth goddess; to the Romans, *Terra Mater* or *Tellus Mater* (both of which mean Mother Earth). Science fiction authors have often used Tellus and Terra in place of the more mundane Earth; they have even used Sol 3, referring to the fact that Earth is the third planet of our particular star, *Sol*—Sun in Latin.

Yet despite being just one planet of many, Earth is exceptional. It is the only one known to support life...life that has evolved to a point where it can examine itself and the cosmos around it. And all this is possible only because of the special confluence of conditions in which Earth finds itself.

JUST RIGHT

The Earth is around 4.57 billion years old, having formed from the pre-solar nebula just like the other bodies in the Solar System. It is the third of the terrestrial planets, and is actually located just beyond the outer edge of the Sun's habitable or "Goldilocks zone"—the region where the temperature would be neither too hot nor too cold, but just right for liquid water to exist. That Earth, nevertheless, does have vast quantities of water is testament to the heat-retaining properties of the atmosphere and oceans, as well as the protective effects of the planet's magnetic field, and the contribution made by heat sources such as volcanism. If it didn't have these, Earth would be an icy world with lots of frozen water—and it is liquid water, of course, that is one of the essentials for life on our planet.

Earth goes around the Sun in an almost circular orbit at an average distance of 93 million miles (150 million km). One complete orbit is one year. One rotation of Earth so that the Sun returns to the meridian—an imaginary line

Above right Ancient Sumerian terracotta and lime plaque of the Earth Goddess, dated *c.* 2000 BCE. It resides in the Archeological Museum, Aleppo, Syria.

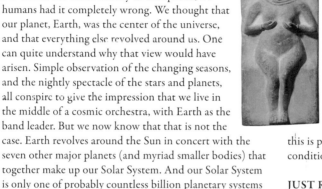

Left This view of the rising Earth greeted the *Apollo 8* astronauts as they came from behind the Moon after the lunar orbit insertion burn. The photo is displayed here in its original orientation.

4.57 billion years ago	Earth forms out of the solar nebula, along with the other planets in our Solar System. The Moon forms soon after.
5 million years ago	Volcanoes erupt and create the area of land that joins North and South America. Mammals from North America move south and cause extinction of mammals there. Human ancestors speciate from the ancestors of the chimpanzees.
1.8 million years ago	*Homo erectus* evolves in Africa and migrates to other continents.
***c.* 400 BCE**	Aristotle deduces that the Earth is round from the shape of Earth's shadow on the Moon during a lunar eclipse.
***c.* 350s BCE**	Heraclides proposes that the apparent daily motion of the stars is created by the rotation of Earth on its axis once a day.
***c.* 200 BCE**	Hipparcos carries out a calculation of the dimensions of the Earth–Moon system.
3rd century BCE	Eratosthenes, Greek mathematician, astronomer, and geographer, devises a map of the world. He also estimates the circumference of Earth, and the distance to the Moon and the Sun, and constructs a method for finding prime numbers.
1 CE	Human population reaches 150 million.
1543	Copernicus states that Earth revolves around the Sun.
1610	Galileo makes telescopic observations of planets. He is jailed for his theory that the Sun is fixed, contrary to Holy Scriptures. Released on house arrest, he later goes blind, presumably from looking directly at the Sun.
1835	Human population reaches 1 billion.
1915	Albert Einstein publishes his *General Theory of Relativity.*

EARTH AND MOON CORE FACTS

Earth is the densest planet in the Solar System. It has a core mostly made up of nickel and iron. This hard core is covered with a mantle consisting of rock containing silicon, iron, magnesium, aluminum, oxygen, and other minerals. The crust, or rocky surface of Earth, is mostly oxygen, silicon, aluminum, iron, calcium, potassium, sodium, and magnesium.

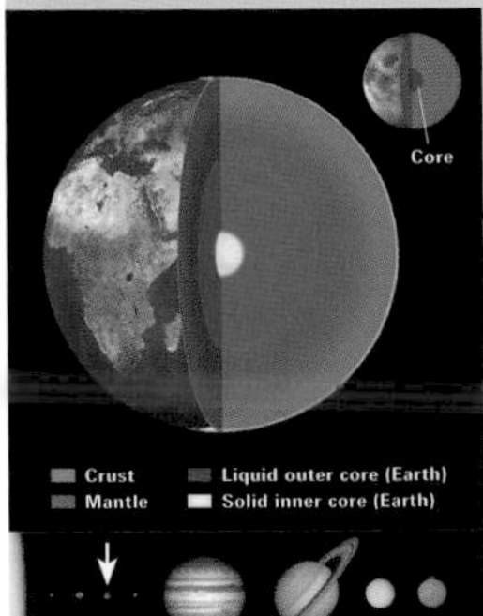

The Moon has a small iron core, the outer part of which is thought to be liquid. It has a thick mantle and a thin crust. The very cratered surface is covered by a thin layer of a dusty substance called regolith.

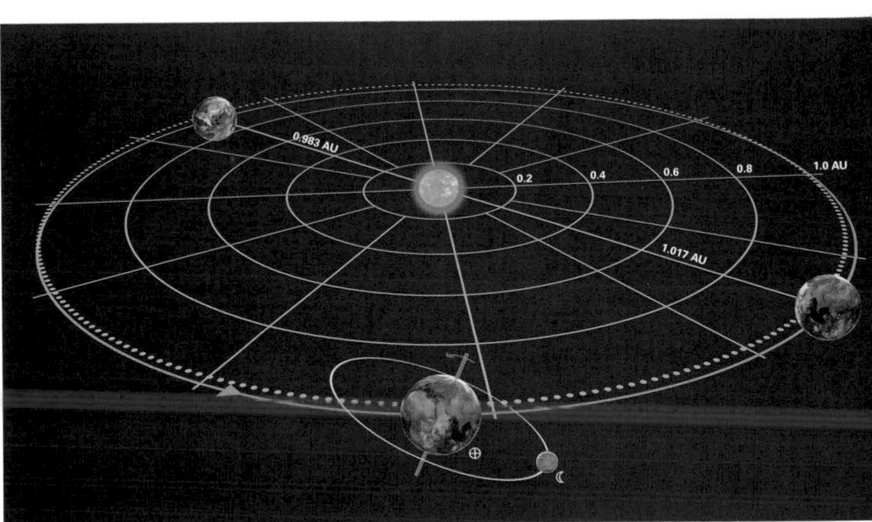

Above Earth takes 365.25 Earth days to orbit the Sun with an inclination of 7.25° to the ecliptic. Earth's axial tilt is about 23.5°, causing the change of seasons. Its orbit is prograde—counter-clockwise around the Sun when viewed from above. The dash line indicates the degree of tilt of Earth's orbit from the plane of the Solar System. The Moon orbits Earth every 29.53 days in an elliptic orbit. The axial tilt between Earth and the Moon is about 5°. The Moon's orbit is prograde—counterclockwise around Earth when viewed from above. (not to scale)

Left True-color satellite mosaic image of Earth: Africa is on the left, Asia along the top, with India centered under the clouds. The body of water is the Indian Ocean.

1927 Georges Lemaître presents his theory of the "Big Bang" and later publishes it in the pages of *Nature,* in 1931. Einstein believes in an eternal universe and expresses his skepticism.

1957 *Sputnik* is launched by the former Soviet Union. It orbits Earth, sending back a beeping signal for 23 days.

1958 The National Aeronautics and Space Administration—NASA—is formed in the USA.

1959 *Luna 1*, an unmanned Russian spacecraft, makes the first flyby of the Moon.

1960 On May 22, the largest earthquake in the world, with a magnitude of 9.5, strikes Chile. Approximately 1,655 are killed, with thousands of injuries.

1961 On April 12, the first human, Russian cosmonaut Yuri Gagarin, travels beyond the atmosphere of Earth.

1969 *Apollo 11* is the first manned mission to land on the Moon. The first steps made on another planetary body are taken by Neil Armstrong and Buzz Aldrin on July 20. They also return the first samples from another planetary body.

1981 Early on April 12, the world's first reusable spacecraft, *Columbia*, known as the shuttle, is launched. It carries two astronauts and completes 36 orbits, then glides to a perfect touchdown at Edwards Air Force Base, California, USA.

1990 On April 24, the Space Shuttle *Discovery* lifts off from Earth carrying the Hubble Space Telescope. The following day, Hubble is released into space where it continues to offer mankind unique glimpses of our Solar System.

1995 Space shuttle *Discovery* maneuvers to within 37 feet (11 m) of Russian space station *Mir*, in preparation for a shuttle-*Mir* docking (STS-63). This is the first shuttle mission to be flown by a female pilot.

2004 On December 26, an earthquake centered off Sumatra triggers a devastating tsunami that kills over 275,000 people across Asia. Measuring 9.3 on the Richter scale, it is the deadliest in recorded history.

2007 Human population approaches 6.6 billion, with China being the most populated country on Earth, with close to 1.5 billion people.

Above Incredible thematic map of the Himalayas, which are home to Mt Everest, the highest mountain on Earth. The Himalayas, on the border of Nepal and China, were formed when a large land mass broke away from Africa and was pushed up against Asia. The small ocean that formed was eventually drained as the mass continued to raise up, forming the Himalayan Mountains. Marine fossils can be found at the top of these great peaks.

Opposite Awe-inspiring images from *Galileo* were combined to give this view of Earth and the Moon. The large continent on Earth is South America with Panama above. The Moon's surface is pockmarked with craters.

in the sky joining North and South—is one day, which our system of timekeeping breaks down into 24 hours. (This is almost four minutes more than it actually takes Earth to spin once on its axis. The difference comes because Earth is moving through space, and it takes a smidge more than one full rotation in order for the Sun to once again be on the meridian.) One full orbit around the Sun takes about 365.26 days, which is why we have to add an extra day every four years—a leap year—to keep our calendar balanced.

Earth's axis of rotation is tilted with respect to the ecliptic by an angle of 23.4°, and it is this tilt that gives us our seasons. The spinning mass of Earth acts as a giant gyroscope, keeping its axis fixed in space. From this, it can be seen that when Earth is on one side of the Sun, its Northern hemisphere will be facing the Sun, while the Southern hemisphere will not—this is summer in the north and winter in the south. Six months later, the Earth's Northern hemisphere will be facing away from the Sun, while the Southern hemisphere now faces toward it—winter in the north and summer in the south. The autumn and spring seasons are mid-way between these two extremes. The axial tilt does vary a little over time, but that doesn't have much effect on our seasons.

If Earth didn't have an axial tilt—if the rotation axis were straight up and down—we would not have seasons. The equatorial regions would be directly beneath the Sun, and the polar regions would get very little warmth indeed.

THIRD ROCK FROM THE SUN

Earth's diameter at the equator is 7,926 miles (12,756 km). The distance between the poles from north to south is marginally less, 7,900 miles (12,713 km), making our planet only slightly less than perfectly round. Earth is bigger than the other terrestrial planets, the dwarf planets, and all the Small Solar System bodies. Only the four gas giant planets and the Sun are larger than Earth.

In terms of internal structure, our planet has a core, a mantle, and an outer crust. The core is in two parts: A central portion about 1,500 miles (2,440 km) in diameter of solid iron surrounded by a zone of liquid iron that extends for about 1,360 miles (2,200 km). The core is believed to also contain small amounts of nickel and sulfur. The core generates a substantial magnetic field that creates a magnetosphere—a magnetic "bubble"—around the planet, which serves to protect it from much of the radiation from the Sun and interstellar sources. The North and South poles of this field are located very close to the planet's geographic North and South poles.

The mantle is semi-solid rock, and extends from the outer boundary of the core almost all of the way up to the surface. The upper, more solid part of the mantle, and the solid crust above it, "float" on the mantle below. The crustal thickness is least at the bottom of the oceans (around 3.7 miles [6 km] in thickness) and greatest where the continents lie (up to 31 miles [50 km] in thickness).

Earth's solid surface is actually broken up into a number of segments called "plates," which comprise the hard upper layer of the mantle, as well as the crustal rocks. These plates move relative to each other—where they meet, they sometimes submerge one under the other, or crash head on. The movement of the plates and collisions between them create volcanic activity, earthquakes, and raise mountain ranges.

The highest point above mean sea level is the peak of Mt Everest, at 29,000 feet (8,848 m) elevation, on the border between China and Nepal. The lowest point on land is the Dead Sea, which is 1,371 feet (418 m) below mean sea level in the Jordan Rift Valley. In the oceans, the lowest point is the Mariana Trench, a deep fissure that extends to an incredible 35,761 feet (10,900 m) below mean sea level, in the Pacific Ocean, near the island of Guam.

Most of Earth—around 70 percent—is covered with saltwater seas and oceans. The land area is comprised of several large continental masses and innumerable islands. The southern polar region is permanently frozen, with an enormous ice sheet overlying a large continental landmass, Antarctica. Much of Earth's fresh water is locked up in this ice sheet

THE LIVING PLANET

Planet Earth has both brought forth life, and been substantially modified over the aeons by that very life. Earth's combination of suitable temperature (because of its distance from the Sun), axial tilt (which gives the seasons), Moon (which provides tides), large supply of liquid water, appropriate atmosphere, and protective magnetic field, all conspired to help create the conditions needed to bring forth self-replicating molecules. Those molecules have grown in complexity over the billions of years to produce the variety of life forms that exist here today.

At the same time, that life has changed the way the world works—interacting with the land, oceans, and atmosphere. For instance, were it not for life, Earth would not have the large ratio of oxygen that is present in its atmosphere. Oxygen is a very reactive substance, and soon becomes locked up in chemical bonds with other elements if it is not continually replenished.

Below Jettisoned external fuel tank from space shuttle *Atlantis*, falls toward Earth's atmosphere during the completion of the launch phase. It will burn up during re-entry.

Below High clouds lit by sunglint over the Indian Ocean, captured by space shuttle *Discovery* during the STS-96 mission. The blue haze of Earth's atmosphere can be seen over the horizon.

ATMOSPHERE AND CLIMATE

Earth has a dynamic atmospheric and climatological system that is mostly driven by the energy received from the Sun, and the cycle of water from liquid to gas and back again.

The planet's present atmosphere is around 78 percent nitrogen and 21 percent oxygen, with small amounts of other gases such as carbon dioxide and water vapor. The gaseous mix has not always been this way. In the geological past, the oxygen level has risen and fallen on many occasions, at one time being perhaps only a few percent, but reaching a peak perhaps as high as 35 percent within the last half billion years.

Some of the gases—prime among them water vapor, carbon dioxide, and methane—prevent some surface heat from radiating back out into space, keeping the surface of the planet warmer than it would otherwise be. This is the greenhouse effect. In addition, in the upper reaches of the atmosphere there exists a thin layer of ozone—a molecule comprising three oxygen atoms—that helps to shield the planet from harmful ultraviolet radiation from the Sun.

Air density falls with increasing altitude, with most of the atmosphere being constrained within the first 6–7 miles (10–11 km) above mean sea level. The atmosphere is loosely divided into several layers. The troposphere is first—this is where almost all of our weather takes place—followed by the stratosphere, mesosphere, thermosphere, and finally the exosphere. Temperatures change at different rates in the different layers, ironically reaching a maximum high above the ground in the thermosphere.

The reason for this is that even though there are very few atoms and molecules in the rarefied air at that altitude, the energy of motion of the few that are there is very high, and this is how their temperature is measured.

Weather patterns around the globe are roughly separated into bands. There is a tropical or equatorial band that spans the region around the equator, adjacent to which is the subtropical zone. Then there is the temperate zone and finally the polar region. These bands run north and south of the equator.

Right Satellite image of Hurricane Lili as it continues to gain strength and move northwest toward the Gulf Coast of North America in 2002. It moved forward at 15 mph (24 km/h) with wind gusts of 120 mph (193 km/h). Hurricane Lili was a Category 3 hurricane on the Saffir-Simpson scale.

THE OCEANS

Oceans cover most of the surface of our planet, in enormous depressions or basins where the crust is thin and there are no continents above. The oceans hold a huge amount of water—more than 312 million cubic miles (1,300 million^3 km)—with much of them having a depth of over 1.8 miles (3 km).

If the solid surface of Earth could be leveled and smoothed out, the water in the oceans would cover the entire globe to a depth of over 1.5 miles (2.5 km).

The oceans are home to an enormous variety of animal and plant life. They also act as a huge heat sink, absorbing warmth from the Sun, and currents within these vast oceans circulate the warmer and cooler flows of water to all the different parts of the globe.

EYES IN THE SKY

Mankind's view of its home planet was changed forever with the coming of the Space Age in the late 1950s. An International Geophysical Year was declared from July 1957 to December 1958, and an all-out effort was made by the USA, and USSR, in particular, to develop Earth-orbiting satellites to learn more about the planet's upper atmosphere, interaction with near-Earth space, and also to provide a perspective on Earth from above.

Since those early days, many dozens of Earth observation satellites have been placed into orbit, and used to monitor everything from weather and climate, to ocean currents, atmospheric changes and man-made pollution, land-use patterns, verification of international treaties, and more. These eyes in the sky have become essential not only for the modern way of life, but also for understanding and managing the changes that are taking place on our planet, many of which are a direct result of mankind's own activities.

THE MOON

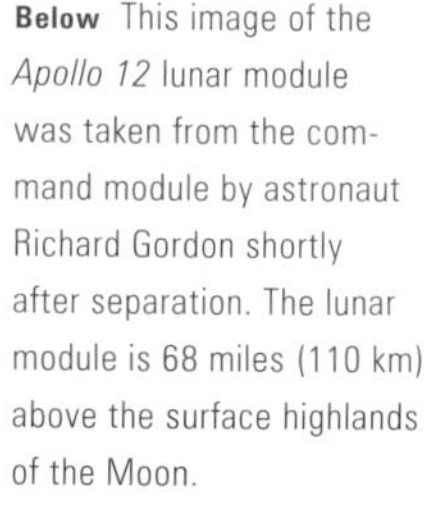

Below This image of the *Apollo 12* lunar module was taken from the command module by astronaut Richard Gordon shortly after separation. The lunar module is 68 miles (110 km) above the surface highlands of the Moon.

The Moon is Earth's only permanent natural satellite, which is the formal term for a body that orbits another. With regard to the natural satellites of the other planets, technically speaking they should not really be called moons, but common usage over the years has made the term stick. However, there will always be only one Moon, and that is ours.

At 2,159 miles (3,475 km) in diameter, the Moon is a little over one quarter the width of Earth. It's only one hundredth the mass of Earth, and its surface gravity is about one sixth that of Earth's. Although these numbers might make our neighbor appear a great deal smaller than Earth, it is in fact rather large as natural satellites go. The moons of the other planets in our Solar System are much much smaller than their parent planets —with the exception of Pluto, and its moon Charon. This has led some people to suggest that Earth and the Moon (and Pluto and Charon) should really be considered as a double planet system.

For a long time there were a number of competing hypotheses as to the origin of the Moon. One proposed that it was a wandering body that became captured into orbit around Earth. Another speculated that it formed along with Earth as a home-grown (not captured) satellite. Another suggested that it formed from a gigantic lump of material that broke off from Earth's crust, leaving behind the Pacific Ocean basin. But all of these ideas had serious problems, chief among them being their inability to explain the high angular momentum

Above First picture of the Moon by a US spacecraft, *Ranger 7*, on July 31, 1964, about 17 minutes before impacting the lunar surface. The large crater at center right is the 67-mile (108 km) diameter Alphonsus.

Above *Galileo* spacecraft took this image of the Moon in 1992, on its way to explore Jupiter. The distinct bright crater at the bottom is the Tycho Impact Basin. The dark areas are lava rock-filled impact basins.

of the combined Earth–Moon system—which results from the mass or density of the two bodies and their velocities.

The concept that holds consensus today is called the giant impact hypothesis. It postulates a body about the size of Mars crashing into the young Earth only 50 million or so years after it first formed. In those early days of the Solar System, there would have been quite a lot of large bodies zooming around, and the other planets were probably hit as well. The collision would have shattered the incoming body, and much of its heavy iron would have merged into Earth. The debris of the impactor—along with shattered portions of Earth—would have been blasted back into space. Some of that material went into orbit around the young Earth and eventually coalesced to form the Moon. However, there are still some problems with this hypothesis and scientists are working to answer them.

SURFACE FEATURES

The Moon is the second densest natural satellite in the Solar System, but is believed to have only a small iron core less than 250 miles (400 km) in diameter, the outer part of which is thought to be liquid. Like Earth, it has a thick mantle and a thin crust, but it does not have—and shows no sign of ever having—the plate tectonics that Earth has.

The Moon has a feeble magnetic field, 100 times weaker than Earth's. It comes from the crustal rocks rather than the core. Perhaps it is all that remains of magnetism produced by the core in younger days before it solidified.

The lunar farside and nearside surfaces are quite different in appearance. The nearside has large regions called *mare*, which, from Earth, appear dark and flat. These are huge basins gouged out of the Moon by impacts with asteroids and comets, and which then filled with lava, long since solidified. The brighter regions we see are mostly highlands—the mountains and raised areas. The farside has almost no mare.

The dominant features of the lunar surface are, of course, craters. Large and small, there are countless numbers of them. Almost all of them are the result of asteroid, meteoroid, or comet impacts; only a very few have a volcanic origin. Many have been given names. The surface is covered by a thin layer of a dusty substance called regolith. It is often mistakenly called lunar "soil," but soil is found exclusively on Earth, and is characterized by the inclusion of living organisms. The regolith is the powdered surface rock that has broken up and been pulverized over the aeons by the never-ending "rain" of micrometeorites from space.

MOON ROCKS

Moon rocks are among the most precious substances on Earth. Specimens were brought back to Earth by the *Apollo* astronauts, by unmanned Soviet probes, and there are a small number of meteorites believed to have come from the Moon. The nearly 880 pounds (400 kg) of rocks brought back during the *Apollo* missions, in particular, were essential for developing a comprehensive understanding of the Moon's formation

Being so precious, the *Apollo* rocks have been handed out sparingly to scientists, usually only fractions of a gram per time. Around 660 pounds (300 kg) of the rocks remain untouched in NASA vaults. Small samples were given as gifts to countries that helped the US Moon effort. Over the years, a handful of lunar samples have been stolen; the thieves attempted to sell them, but in each case they were recovered.

Right Close-up of *Apollo 16* lunar sample no. 60017—a black matrix breccia of extremely small grain size.

Right *Apollo 16* astronaut, John Young, jumps off the ground and salutes for this superb tourist picture. He was off the ground about 1.45 seconds which, in the lunar gravity field, means that he launched himself at a velocity of about 3.8 feet (1.17 m) per second and reached a maximum height of 1.3 feet (42 cm).

THE MOON ILLUSION

The full Moon is an amazing sight. Over the years it has inspired songwriters and poets, painters, and priests. The Moon is full when it is on the opposite side of the Earth from the Sun. This being so, it rises in the east around the time that the Sun sets in the West, forming a magnificent spectacle in the eastern sky. And it is around this time that a curious illusion takes place—to many people's eyes, the Moon appears much bigger near the horizon than it does when it is high overhead.

During the years there have been many explanations given for this apparent size increase, but the truth is rather prosaic—it's just an illusion. When the Moon is near the horizon the eye has other things to contrast it with, such as trees, buildings, and so on. But when it is high in the sky there is nothing to compare it with, and it seems smaller. It's easy to test this. Grab a coin or some other small object and, holding it at arms length, place it next to the Moon when it is near the horizon to gauge the Moon's size. Then do it again later when the Moon is high overhead. You'll find you won't be able to tell the difference.

ORBIT AND PHASES

The Moon circles our planet in an elliptical orbit that brings it as close as 225,622 miles (363,104 km) and takes it as far away as 252,087 miles (405,696 km). It doesn't orbit around our equator—as most other moons do with their planets, but instead has an orbit inclined at 5° to the plane of Earth's orbit around the Sun—the ecliptic. One lunar sidereal orbit (i.e. one based on the background stars) takes 27.3 Earth days. But because the Earth is moving along in its own orbit during this time, the cycle of lunar phases (which depends on the changing Sun–Earth–Moon angle) actually takes 29.5 days.

The Moon has become tidally locked with Earth, giving it a synchronous rotation; that is, it rotates on its own axis in the same time it takes to orbit Earth, which means that it keeps the same face—the lunar "nearside"—toward our planet at all times, with the "farside" facing away from us.

This is why we get to see only one side of the Moon; only some robotic spacecraft and 18 *Apollo* astronauts have ever seen the farside.

Strictly speaking, we do see a little more than half the Moon from Earth. A phenomenon called libration, which results from the varying speed of the Moon in its orbit, and another phenomenon called parallax, allows us to look "around the edge" of the Moon, and over time see an extra 9 percent of the lunar surface.

The Moon, and to a much lesser extent, the Sun, is responsible for the tides on Earth. Although the tides are most noticeable in large bodies of water, the tidal effect also affects Earth's landmass, for instance, causing Earth's solid surface to rise and fall by about 0.3 miles (0.5 km) at the equator. These tidal interactions cause Earth's daily rotation rate to slow ever so slightly every day, which is why a leap second has to be added to our timekeeping system every year or two. The decrease in Earth's angular momentum is transferred to the Moon, with the result that the Moon's orbital speed is increasing by a tiny amount and causing it to move away from Earth by just under 2 inches (4 cm) per year.

The Moon's presence also acts to stabilize Earth's spin axis, helping to keep it at its present angle and thereby contributing to the phenomenon of seasons.

The Moon shows phases because of the changing angle between the Sun, Earth, and Moon. When the Moon is on the same side of Earth as the Sun, its sunlit side is facing away from us and we see only the thinnest of crescents or no crescent at all—new Moon. When the Moon is on the opposite side of us from the Sun, we look straight on at its fully sunlit face—full Moon. In between are various degrees of crescent shape, including first quarter and last quarter.

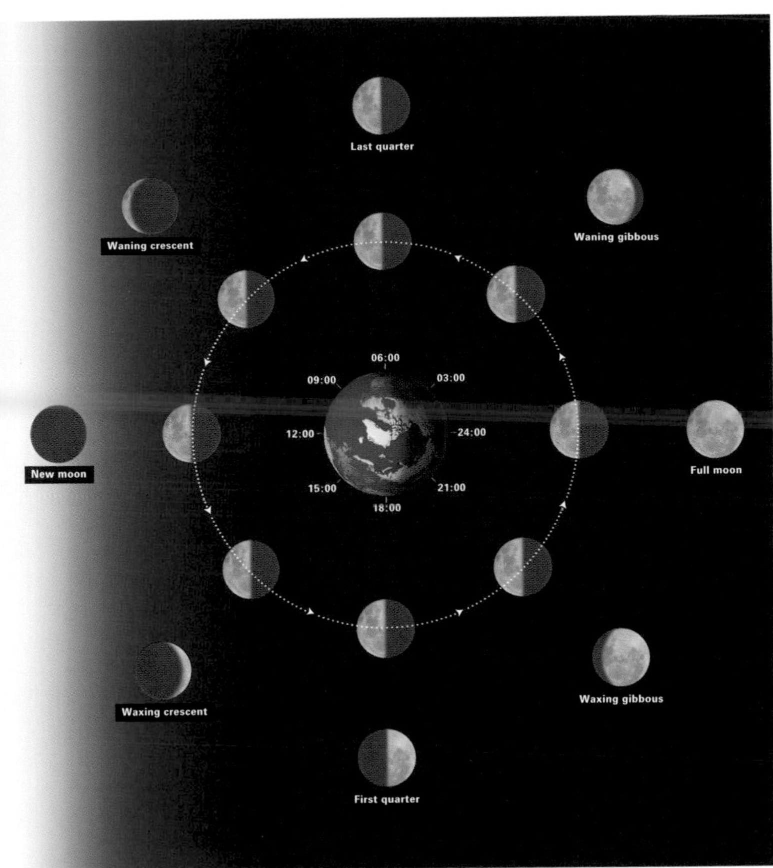

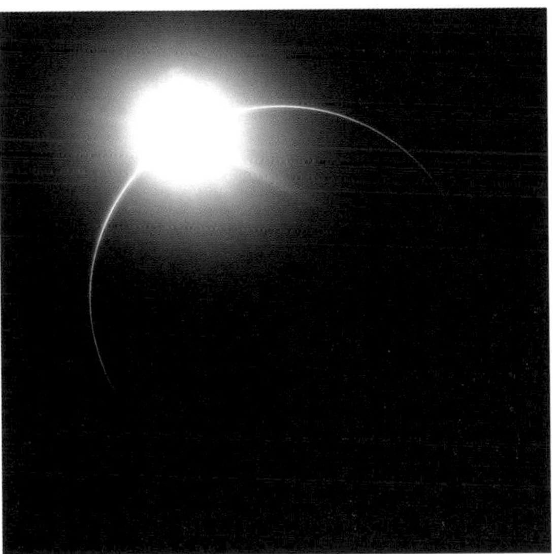

Above The inner ring shows the Moon as it would appear from above. From Earth we only see the part that is facing toward us (the near side). This is the lit part of the Moon shown inside the white circle. What is outside the circle is never visible to us on Earth (the far side). The phases of the Moon, or the shape of the lit part that we can see, are in the outer ring. The moon phases shown are as seen from the Northern hemisphere—if the left side is dark then the light part is waxing or growing. The opposite happens when viewed from the Southern hemisphere—if the left side is dark then the light part is waning or shrinking.

Above left View from Earth of a lunar eclipse. The shadow of the Earth is just moving away, allowing the Sun to once again hit the surface of the Moon.

ECLIPSES

The Moon is 400 times smaller than the Sun, but coincidentally it is 400 times closer. This means that at those times when the Moon's orbit makes it cross the line joining the Sun and Earth, it blocks out the Sun and we get a total solar eclipse. If the alignment is not exact, we get a partial solar eclipse. If the alignment occurs when the Moon is at the furthest point in its orbit, it won't cover the Sun completely; it will leave a thin ring of sunlight around it, which is called an annular solar eclipse.

If the point of crossing is in line with the Sun and Earth, but on the side of Earth opposite to the Sun, the Moon will pass through Earth's shadow and we will see a lunar eclipse. And like partial solar eclipses, if the alignment is not exact, the Moon will only partly pass through the darkest part of the shadow and we will see a partial lunar eclipse.

ATMOSPHERE

The Moon's atmosphere is so thin that it is almost not there. Small amounts of gas, such as radon, escape from below the surface. Other particles of gas are liberated directly from the surface due to a constant "rain" of micrometeorites striking the regolith, as well as sunlight and Sun particles. Much of the atmosphere is blown away into space by the solar wind.

Mars

FACT FILE

Planetary Order
4th planet from Sun

Average Distance from Sun
1.52AU/141,633,260 miles/ 227,936,640 km

1 Mars Day
1.026 Earth days

1 Mars Year
1.8807 Earth years

Equatorial radius (Earth = 1)
0.5326

Mass (Earth = 1)
0.10744

Average Temperature
-125 to 23°F/-87 to -5°C

Atmosphere
95.72 percent carbon dioxide, 2.7 percent nitrogen,1.6 percent argon, traces of oxygen, carbon monoxide, water vapor

Moons/satellites
Two moons—Phobos and Deimos

Notable features
Largest volcano in the Solar System—Olympus Mons, 15 miles (24 km) above the surrounding plain and three times higher than Mt Everest.

Throughout history people have been fascinated by the red planet. Identified with ancient Rome's god of war, Mars caused great excitement when some astronomers claimed that they could see canals built by intelligent beings on its surface. Today, scientists are using a whole fleet of spacecraft to scrutinize the planet, as it is the most likely abode of life in the Solar System.

THE GOD OF WAR

In ancient Rome, Mars was initially a god of fertility and agriculture, but later he also became the god of war. As such, he was widely worshipped, for battles and war were important components of life as the city-state expanded its empire. It is likely that the fourth planet from the Sun became associated with this war-like god because of its red color—the color of blood.

Military commanders took part in ceremonies to Mars before setting out for war, and soldiers believed that the god, together with a female companion, appeared on the battlefields. The importance of the god was also boosted by the legend that Mars, the god of war, was the father of the twins Romulus and Remus, who were the mythical founders of Rome.

Above right Marble bust of Mars, from the temple at the forum of Augustus. Honored on coins and revered as the mythical father of all Romans, the name is still in use in words such as martial law and martial arts.

THE CANALS OF MARS

Mars is normally so far away that even through a large telescope it appears just a tiny featureless red dot. Fortunately, every two years or so, there is an opportunity to observe the planet close up when it is relatively close to Earth at opposition. This occurs when Earth, in its yearly circuit around the Sun, catches up with slower-moving Mars, placing it exactly on the opposite side of Earth to the Sun. As Mars has a somewhat elongated path around the Sun, on some of those occasions it is closer to Earth and brighter, and hence the opposition is regarded as more favorable. A most favorable opposition took place in August 2003, while the next similar one will be in July 2018.

A number of astronomers observed Mars during the favorable opposition of 1877. Among them was the Italian astronomer Giovanni Schiaparelli, who thought he saw dark lines criss-crossing the planet. In publications, Schiaparelli described the lines as "canali," which in English means "channels." Of course, the inevitable happened and "canali" was mistranslated as "canals," a word meaning waterway, but with the implication that they are artificial and constructed by intelligent creatures.

Right This view of Mars was taken on the last day of Martian spring in the Northern hemisphere. The annual north polar carbon dioxide frost (dry ice) cap is rapidly evaporating from solid to gas, revealing the much smaller permanent water ice cap, along with a few nearby detached regions of surface frost.

PERCIVAL LOWELL

American astronomer Percival Lowell was inspired by Schiaparelli's work to set up an observatory at Flagstaff, Arizona, in 1894, specifically to look for life on Mars. Lowell theorized that to maintain crops, Martians built a series of great waterways to carry water from the ice-covered poles to the rest of their parched planet. This great work could only be achieved with cooperation among the creatures of the whole planet, and Lowell considered that provided a good model for nations on Earth.

The Lowell Observatory published maps of Mars with increasingly greater complexity. On those maps, not only were there a large number of lines depicting canals, but also some of them were actually double—like railway lines. Where two canals crossed, small round spots were seen which were sometimes referred to as "oases." Observers at Flagstaff were

c. 1570 BCE to 1293 BCE
Egyptians have knowledge of Mars, speaking of the planet traveling backward, a reference to its retrograde motion.

c. 300 BCE
Aristotle concludes that Mars is higher up in the heavens than the Moon.

c. 1600
Tycho Brahe plots Mars's position in the sky. In 1604, Johannes Kepler calculates that Mars has an elliptical orbit.

1609
Galileo Galilei, (1564-1643), often called the "father of modern astronomy," makes the first telescopic observation of Mars and goes on to make notes of the different phases of the planet. He uses telescopes he makes himself, with approximately 32x magnification.

1619
Kepler publishes his *3rd Law of Planetary Motion*, in which he makes notes of Mars's movement through the universe.

1659
Huygens estimates the size of Mars and arrives at an approximate 24-hour rotational period.

1666
Gian Cassini, the famous Italian-born astronomer, observes the polar cap on Mars. He also calculates and measures the length of the Martian day to be 24 hours and 40 minutes.

1671
Gian Cassini measures the distance from Earth to Mars.

1672
Huygens observes a white spot on the South pole of Mars.

1698
Huygens publishes *Cosmotheoros*, a book about whether or not there is life on Mars.

1704
Giacomo Filippo Maraldi (the nephew of the famous astronomer Gian Cassini) observes white spots on the North and South poles of Mars. In 1719, he suggests that the white spots could be ice caps on the polar regions.

MARS CORE FACTS

Mars has a partially liquid core that is almost one-half the radius of the planet, and made mostly of iron along with perhaps 15 percent sulfur, which helps to keep it molten. The once active, but now solidified mantle comprises iron oxide-rich rocky silicates, topped by a crust that is proportionately much thicker than Earth's. Surface features include giant extinct volcanoes and huge canyon systems, and lots of evidence of a watery past. That water has now largely disappeared. Great dust storms often engulf the planet leaving giant dunes, wind streaks, and wind-carved features.

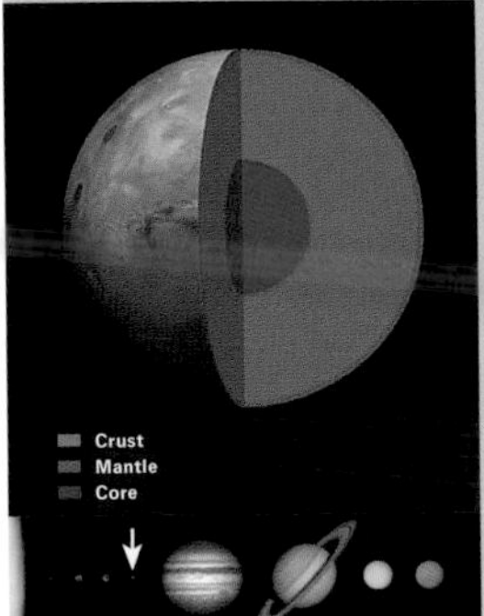

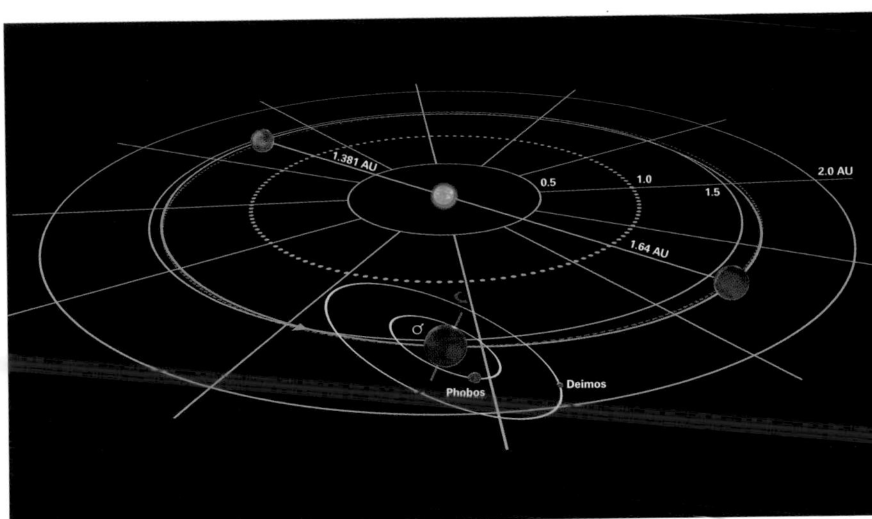

Above Mars takes 687 Earth days to complete an orbit of the Sun, with an inclination of 5.65° to the ecliptic. Its axial tilt is 25.19°, making it similar to Earth, and causing it to have seasons. Mars' orbit is prograde—counterclockwise around the Sun when viewed from above. The dash line indicates the degree of tilt of Mars's orbit from the plane of the Solar System. (not to scale)

Left Digital composition of Earth, the Sun, and a red planet that could be Mars. In fact, the planet Mars is merely a tiny dot when viewed from Earth.

not the only ones to see the canals, as astronomers from around the world reported similar observations. Sadly, modern observations of Mars from Earth, and from spacecraft circling the planet, have shown that the canals do not exist. It seems that the observers, in trying to distinguish undiscernible details on the tiny disk of the planet, were at the limit of human vision and saw more than what was there.

ATMOSPHERE

The atmosphere of Mars is so thin that average atmospheric pressure on the surface is less than one-hundredth of that on Earth. The composition is unsuitable for humans as it is mainly carbon dioxide with small amounts of oxygen and argon, as well as traces of oxygen and water vapor. Despite the thinness of the atmosphere, strong winds can blow that, on occasion, cause major dust storms covering the whole planet.

Mars is also much cooler than Earth, since its orbit around the Sun is 50 percent further away, and hence receives less than half the amount of solar energy. However, the planet's path around the Sun is somewhat elongated, so there are large variations in the solar energy it receives. Surface temperatures can vary from −207°F (−133°C) at the poles during winter, to a pleasant 81°F (27°C) on a summer's day.

During oppositions, North and South polar caps are easily visible on the disk of the planet through a telescope. These visible polar caps are largely solid carbon dioxide or dry ice. When it is summer at one of the poles, the dry ice turns into gas and shoots into the atmosphere in jets with speeds of around 100 mph (160 km/h), leaving behind a layer of water ice. In winter, the process reverses and a layer of solid carbon dioxide is again deposited to cover the pole.

1781 Sir William Herschel discovers that the inclination of Mars's axis of rotation is approximately 24°.

1784 Sir William Herschel observes the seasonal changes on the polar caps of Mars and raises the possibility that they might be composed of snow and ice just as Maraldi had done some 80 years earlier.

1894 Lowell Observatory is established in the USA for the sole purpose of studying Mars.

1905 C.O. Lampland of the Lowell Observatory takes a photograph of Mars that shows 38 canals.

1953 On the initiative of the Lowell Observatory, an International Mars Committee is organized to coordinate continuous observation of Mars during the favorable opposition of 1954.

1962 NASA builds a series of spacecraft to explore the inner Solar System, including Mars, Venus, and Mercury. As a result, *Mariner 4* sends first close-up pictures of Mars during its flyby.

1971 *Mariner 9* becomes the first satellite orbiting Mars for almost a year.

1976 *Viking 1* and *2* land on Mars after first imaging the surface to enable landing spots to be selected. These are the first landings of spacecraft on another planet. They both return images of the entire planet in what was then high resolution.

1997 *Mars Global Surveyor* begins orbiting Mars. *Mars Pathfinder* also lands on Mars and begins to send back excellent data.

2004 NASA successfully lands Mars twin rovers, *Spirit* and *Opportunity*, on Mars, watched by millions on live television feeds back on Earth. The rovers were seen to roll off their landing gear and begin exploration.

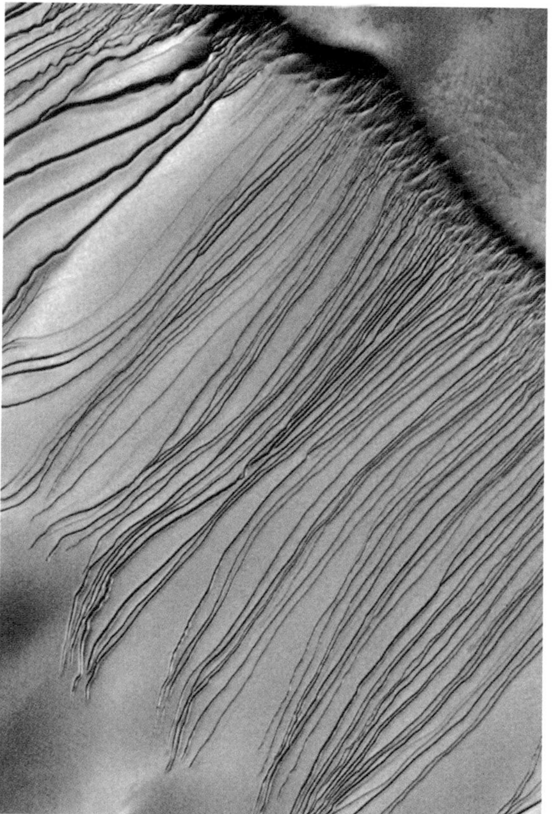

Above NASA's *Mars Reconnaissance Orbiter* shows an impact crater at Meridiani Planum, near the equator of Mars. It has a distinctive scalloped shape to its rim, caused by erosion and downhill movement of the crater wall.

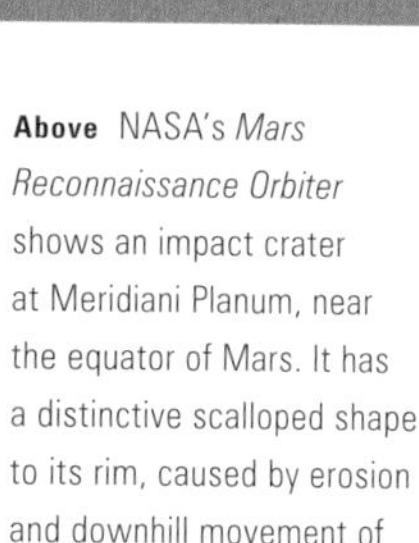

Right Mysterious dune gullies of Russell Crater are of unknown origin, although it is known that they tend to occur only on slopes facing southward. Some scientists have noted that these resemble mud-flows, suggesting liquid water may be present.

Top right Although resembling a lone footprint in Gusev Crater, this impression is actually soil that has been disturbed by the left front wheel of the *Spirit* rover while on Mars.

THE SURFACE

Mars has a surface that is generally desert-like with everything covered by red dust. Due to past impacts, numerous craters of various sizes dot the surface. Surprisingly, there are major differences between the Southern and Northern hemispheres. The Southern hemisphere is much more heavily cratered than the Northern and is higher by about 3 miles (5 km). Scientists consider that the heavily cratered Southern highlands are very ancient. They do not fully understand the differences between the two hemispheres, but most consider that, in the North, younger material has buried an ancient surface similar to that in the south.

THE LARGEST CRATER

Hellas Planitia is a giant impact crater in the Southern hemisphere of Mars. It is over 1,250 miles (2,000 km) across and is 6 miles (9 km) deep as measured from its rim. Scientists consider that it was created by an impact during the early days of the Solar System when huge numbers of large rocks were moving around the Solar System and bumping into Earth, the Moon, and other bodies. Much material was thrown up by the impact and piled up to form a very wide elevated rim around the crater.

The giant crater is large enough to be seen from Earth during oppositions. It was Giovanni Schiaparelli of canals fame who named it Hellas—meaning Greece—on his 1877 map of Mars. Through a telescope, observers can easily mistake Hellas Planitia for the South Pole of the planet, as it appears white in contrast with its surroundings and can be fairly bright.

WHAT DO WE KNOW OF MARS?

Mars is a small planet that is half the width of Earth. Gravity on its surface is correspondingly less—so if you were to move to Mars, your bathroom scales would show your weight as one third of what it is on Earth. You would also find it easy to jump higher than any athlete at the Olympics. However, you would be handicapped by having to wear spacesuit and life support equipment, for Mars has only a thin atmosphere.

Opposite Hellas Planitia, (Hellas Basin), on Mars, compiled from 50 *Viking Orbiter* images in 1980. The Hellas Basin is an impact crater with a diameter of over 1,250 miles (2,000 km).

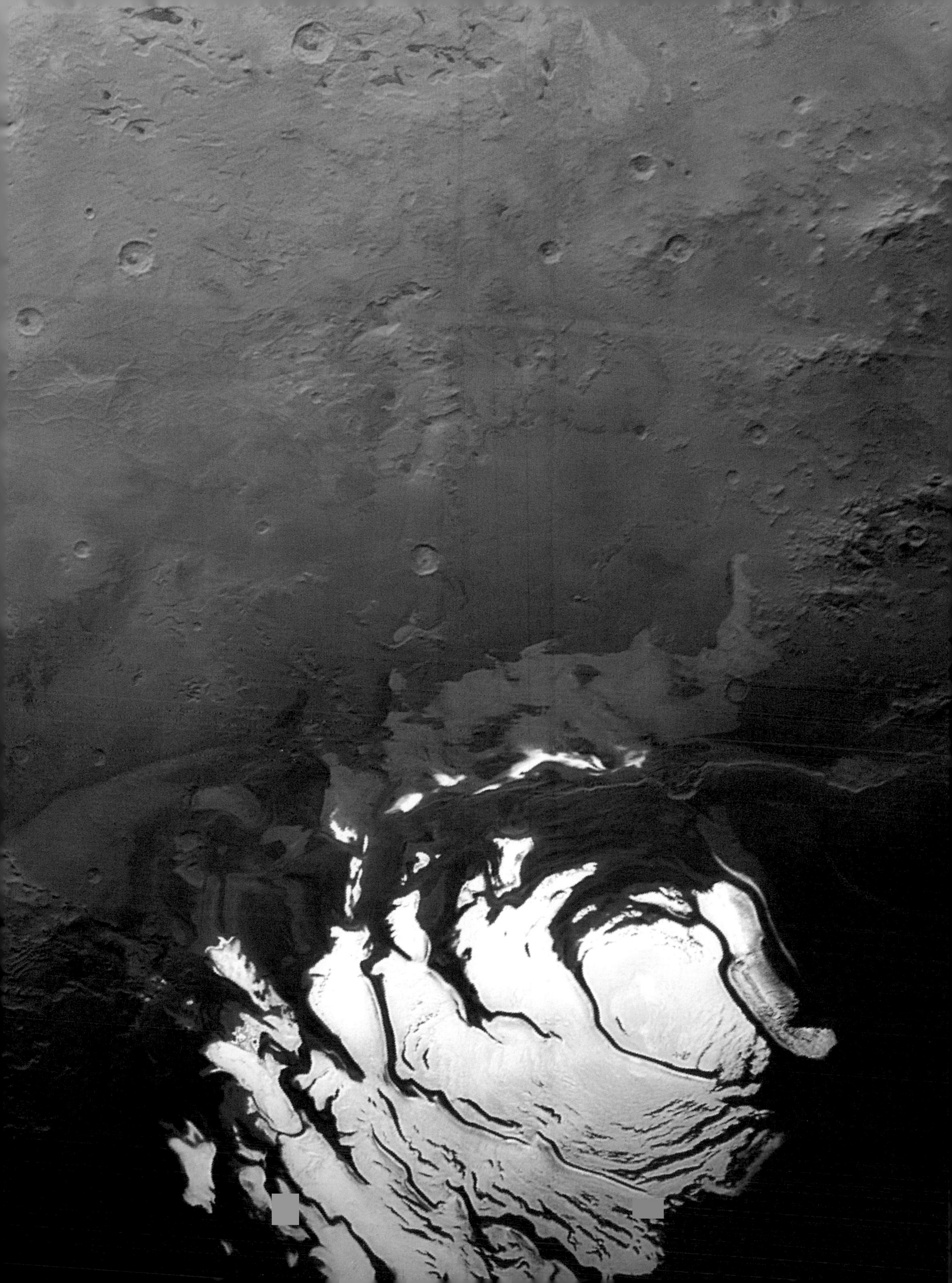

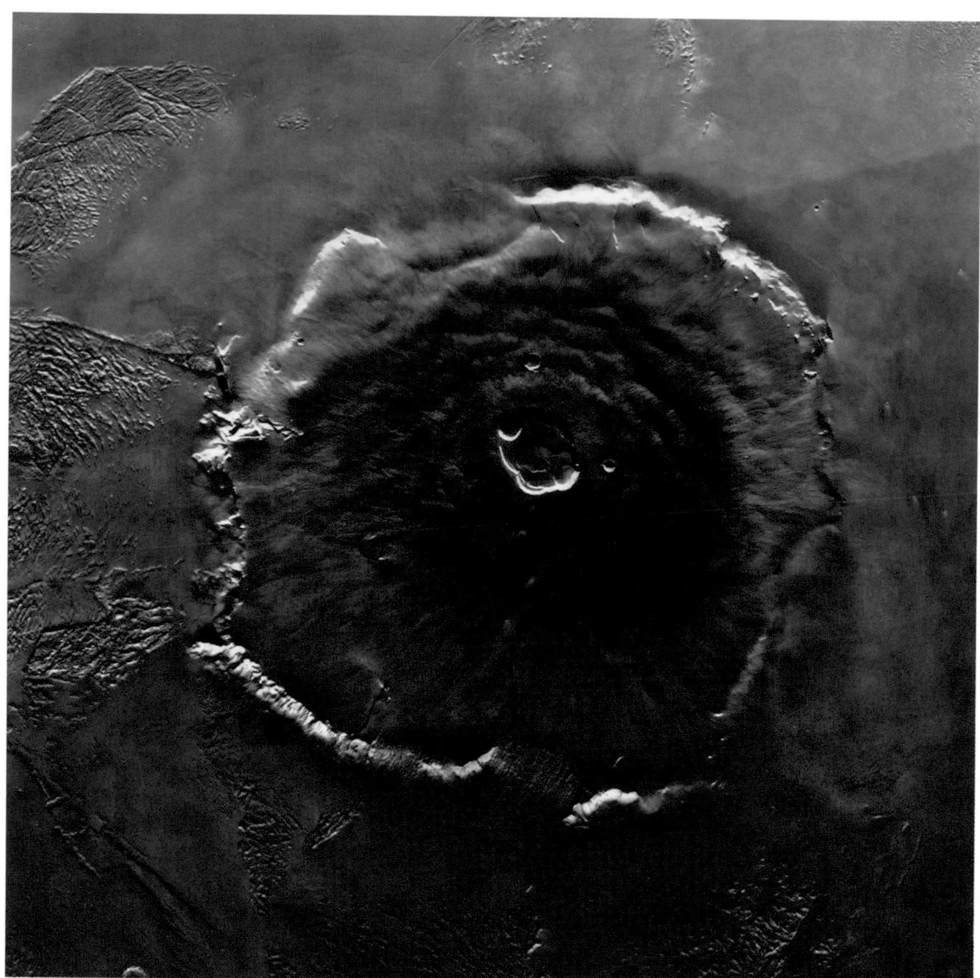

Above Satellite image of Olympus Mons volcano on Mars, the largest volcano in the Solar System. It is 310 miles (500 km) across, with a multiple collapse pit at its summit.

VOLCANOES ON MARS

Mars has an array of fascinating features on its surface including the largest volcano in the Solar System and a huge system of canyons, valleys, and deserts, as well as polar ice caps.

Olympus Mons, which is Latin for Mount Olympus, is the largest volcano in the Solar System. It rises 15 miles (24 km) above the surrounding plain and is nearly three times higher than the tallest mountain on Earth, Mt Everest. At that height, the atmospheric pressure is less than one-tenth of that on the surface, but it still can be covered by clouds of carbon dioxide ice. The volcano stands on a base 340 miles (550 km) across, which has steep cliffs 4 miles (6 km) high at its edge. The base is so huge that it spreads over an area similar to some American states, such as Arizona and Nevada. Since the base is so large, there are only gentle slopes from the cliffs at the edge all the way to the top of the volcano.

Why does a relatively small planet like Mars have a volcano so much larger in scale than any on Earth? The answer seems to lie in the different structure of the crust of the two planets. Earth's crust is made up of slowly moving pieces called tectonic plates that move over hot spots that form volcanoes. As Mars does not have the same system of tectonic plates, the crust remained stationary while lava or molten rock kept flowing from a hot spot below. As the lower layers of lava cooled and solidified, the fresh lava kept building up the volcano until it reached its current giant size.

Is Olympus Mons still active? Is there lava still flowing from vents at its top? A group of scientists using the *Mars Express* spacecraft circling Mars have recently studied the ages of solidified lava flows on the sides of Olympus Mons. They found the ages of the lava ranged from 115 million years to just two million years. This lower age is so recent in a geological time scale that the scientists speculate that the volcano may still be active today.

Tharsis is a continent-sized dome, measuring approximately 2,500 miles (4,000 km) across and uplifted by about 6 miles (10 km) above the surrounding terrain. It contains the four largest volcanoes on Mars including, at its edge, Olympus Mons.

The other three are Ascraeus Mons, Pavonis Mons, and Arsia Mons. Though these three are smaller than Olympus Mons, reaching upward only about twice as high as Mt Everest on Earth, they stand on a high part of the Tharsis dome. Hence the summits of the smaller volcanoes are at about the same level as that of Olympus Mons.

The exact processes that formed the Tharsis bulge are still controversial among scientists, but most likely it was due to hot rock rising from deep inside the planet. As the hot rock rose, it pushed out the surface into a bulge, while at the edges cold rock sank into the interior. This kind of material flow, called mantle convection, is of great importance in terrestrial geology.

Valles Marineris, Latin for Mariner Valley, is named after the *Mariner 9* spacecraft that circled the planet for a year

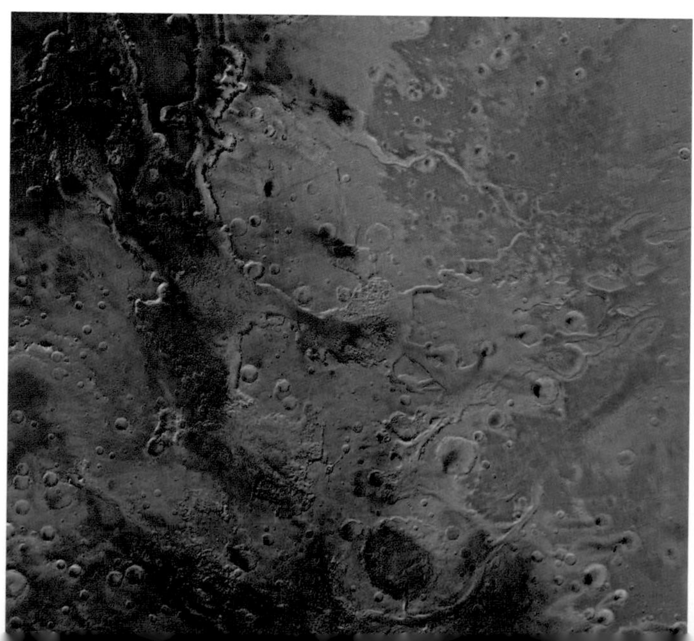

Right Satellite image of Valles Marineris, Chryse Basin, part of the vast equatorial canyon system on Mars. It is surrounded by now dry channels, which were carved out by running water in a period when Mars was capable of sustaining liquid water on its surface.

from late 1971. The *Mariner* spacecraft discovered the vast system of canyons that stretch along the planet's equator for 2,500 miles (4,000 km). The largest known such structure in the Solar System, it is up to 5 miles (8 km) deep and up to 125 miles (200 km) wide. If placed on Earth, it would stretch all the way across the United States. The Grand Canyon—with a length of 275 miles (440 km), a maximum depth of just over 1 miles (1.6 km) deep, and a maximum width of 15 miles (24 km)—appears tiny in comparison. Scientists currently think that Valles Marineris originally formed as a rift in the crust during the cooling of the planet after its formation. Over the long geological periods since that time, erosion by water, carbon dioxide, or wind, enlarged the valley, as did collapses of the canyon walls.

Above Digital composition of a satellite view of planet Mars showing details of its ice clouds above volcanos, and the disturbances around the North polar region.

Right Deimos, one of the moons of Mars. This image was obtained by the *Viking Orbiter* spacecraft in 1977.

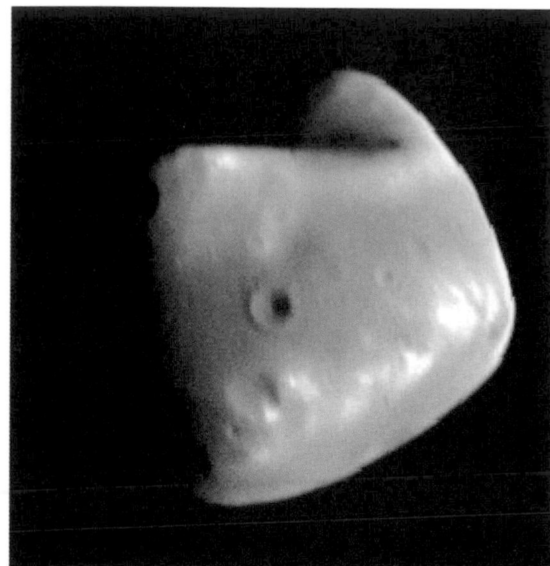

Below This image, taken by ESA's *Mars Express* spacecraft, is of the Martian moon Phobos.The image shows the Mars-facing side of the moon, taken from a distance of less than 125 miles (200 km).

THE MOONS OF MARS

Mars has two tiny moons—Phobos and Deimos. The moons are named after the two attendants of the Greek god of war, Ares, and mean panic and fear respectively. American astronomer Asaph Hall discovered them in August 1877, during the same opposition that Giovanni Schiaparelli saw his "canali" on the disk of the planet. Hall used what was then the world's largest lens telescope, the 26-inch (66 cm) lens telescope at the US Naval Observatory in Washington DC. Both moons are so tiny and faint that even today they are difficult to see through ground-based telescopes.

Phobos is the larger of the two moons, with a jagged irregular shape that has an average width of about 14 miles (22 km). It is also the closer of the two to Mars, circling above the surface at a distance of only 3,750 miles (6,000 km), which is a little more than the distance between New York City, USA, and London, England. Being so close to Mars, Phobos has to move so fast that when seen from the surface of the planet, it rises and sets twice a day. Another unusual aspect is that it rises in the west and sets in the east.

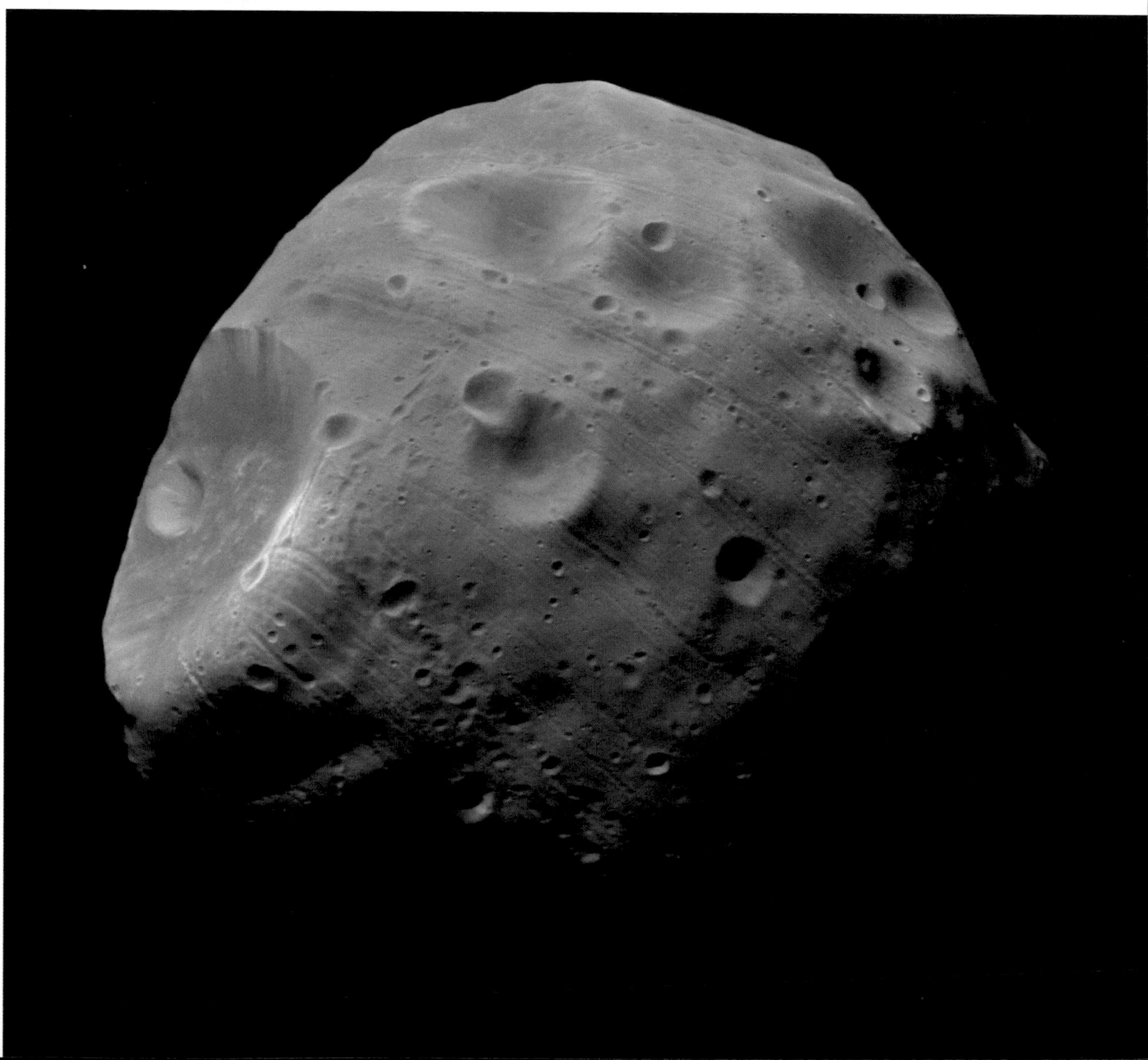

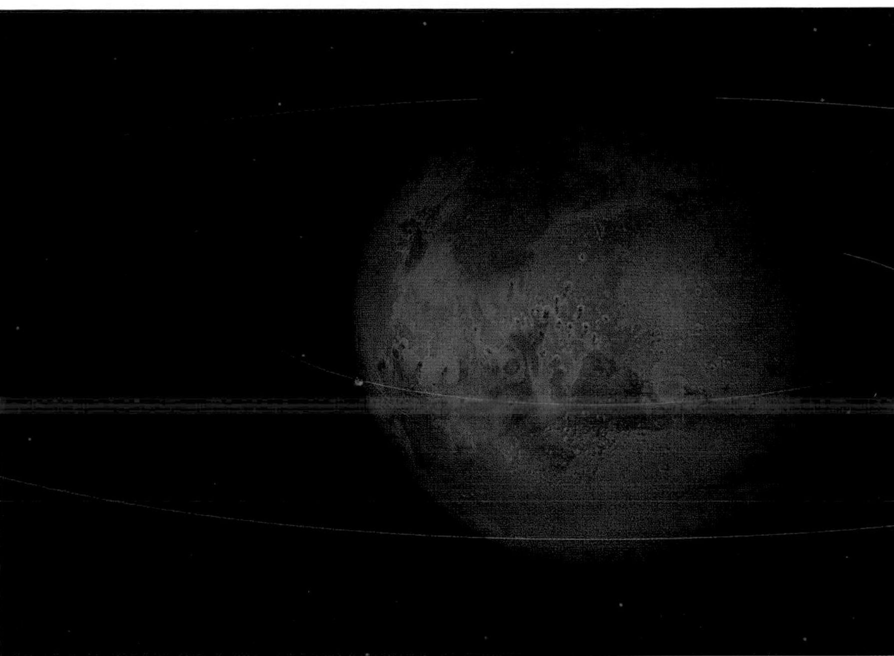

Right Digital illustration of Mars and its moons, Phobos and Deimos. Both moons are small and resemble Earth-like rocks. Phobos orbits so fast that it rises and sets twice a day when seen from the surface of Mars.

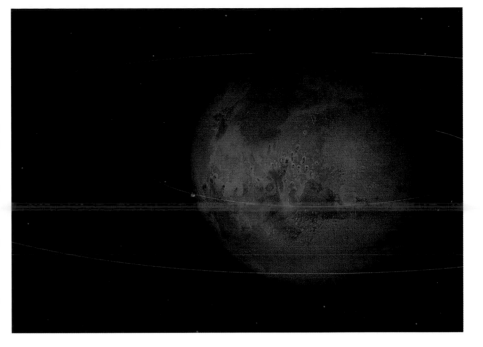

Spacecraft images show that Phobos is covered by craters from impacts with small and large rocks in the past. Its main feature is a large crater named Stickney, which is named after the maiden name of Asaph Hall's wife.

Deimos is smaller than Phobos and further from Mars. Its average width is 8 miles (13 km), and it circles Mars just over 12,500 miles (20,000 km) above its surface. Like Phobos, it is covered in craters. Scientists believe that the two moons are both asteroids that ventured too close to Mars and were captured by the planet.

METEORITES FROM MARS

On June 28, 1911, at about 9 A.M. in the morning, a rain of stones fell from the sky onto the small village of El Nakhla el Bahariya in Egypt. According to an eyewitness, one of the stones killed a dog, a story that has neither been proved nor disproved since that time. If the story is true, then the dog was one of the unluckiest dogs in Earth's history. Not only would it be the only living creature known to have been killed by a rock from space, or meteorite, but it would have been killed by a very rare meteorite from Mars.

Of the 24,000 meteorites that have been found on Earth, scientists have so far classified 34 as coming from the planet Mars. These have provided invaluable information on the geological history of the planet. They reached Earth because at various times in the past they were ejected into space by impacts on the surface of Mars. Each then circled the Sun until millions of years later it happened to hit Earth's atmosphere and land on the planet's surface.

The Martian meteorites are all igneous rocks, that is, rocks formed by solidified lava. There are other types of rocks on the planet, but presumably they are not strong enough to survive being thrown into space by an impact. Most of the meteorites probably come from the flat volcanic plains of the planet's Northern hemisphere. This suggestion fits in well with their estimated age of 1.3 billion years, a young age compared with the age of the formation of the Solar System, 4.6 billion years ago.

The proof of Martian origin of these meteorites is from the study of meteorite EETA79001, which was found in the Elephant Moraine Icefield of Antarctica in 1979. This meteorite has small dark lumps of a glassy substance containing trapped gas. The impact on Mars that ejected the rock on its long path to reach Earth also heated it, and the glass lumps formed by melting. When scientists analyzed the trapped gas, they found it to be a perfect match to the unique Martian atmosphere that the two *Voyager* landers had measured in 1976. Subsequently, at least four other meteorites have been examined and been found to contain samples of the Martian atmosphere.

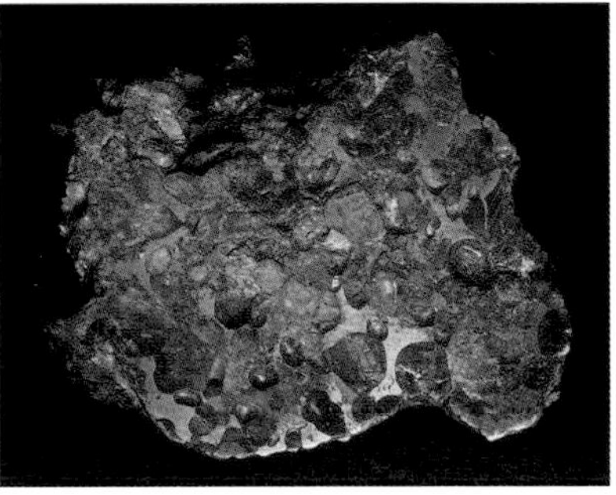

Above Close-up of a meteoritic pallasite fragment found in the Haviland Crater, Kansas, USA. The fragment is believed to be from Mars.

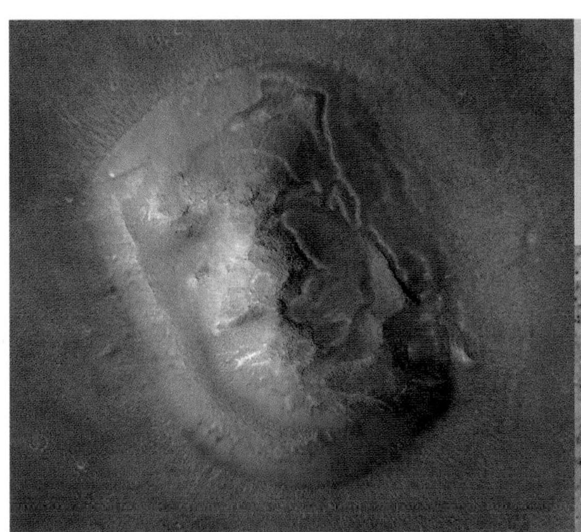

FACE ON MARS

A hill in the Cydonia region of Mars acquired temporary notoriety when, in July 1976, it was imaged by the *Voyager 1* orbiting spacecraft. In one image, scientists, using the spacecraft to photograph potential landing sites on the planet, saw a hill with shadows in just the right places to make it resemble a human face. They thought that it would make a light-hearted story and released the image to the public, noting this interesting—but completely accidental—resemblance. There was a much greater public fascination with the image than they had expected, with a number of authors writing sensational articles and books claiming the hill to have been shaped by non-natural forces. Disappointingly for these authors, a subsequent spacecraft produced high-resolution images of the flat-topped hill that removed the resemblance and ended the speculation.

Original press release image from *Voyager 1* in 1976 (right), and a more recent image of the same hill taken from *Mars Global Surveyor* in 2001, (loft). Tho now imago dobunkc tho "Faco on Marc" thoory.

LIFE ON MARS

If there were any form of life on Mars, it would be in the form of microbes and not in the form of intelligent creatures visualized by science fiction writers like H. G. Wells. The microbes would have to be underground, as the thin atmosphere provides little protection from the wind of charged particles from the Sun, or its ultraviolet radiation, and hence the surface is likely to be sterilized. It is also possible that there were life forms that became extinct when conditions changed on the planet.

MICROBES IN A METEORITE?

The most famous of all Martian meteorites, ALH84001, is much older than the others that have been found. It has an estimated age of 4.5 billion years and is likely to have originated from the Southern hemisphere highlands.

The code name of the meteorite just means that it was the first meteorite collected from the Allan Hills in Antarctica in 1984. The meteorite is famous because in August 1996, a team of NASA scientists announced that it shows evidence of microscopic life forms.

Using sophisticated tools such as an electron microscope, the team found a variety of indicators in the meteorite, such as tiny grains of the mineral magnetite in a form similar to that deposited by Earth-based bacteria. They also found features that resembled terrestrial bacteria, though they were considerably smaller. If correct, these findings would have great significance as the first serious suggestion of life beyond Earth. However, most scientists are sceptical, as they consider that the features of the meteorite can be explained by non-biological events, and by contamination during the the time the meteorite spent on Earth.

WATER

Mariner 9—the first spacecraft to circle Mars—revolutionized our knowledge of the red planet. The images it sent back to Earth in 1972 showed no canals, but instead showed ancient riverbeds, suggesting that water once flowed on the planet's surface. These images excited scientists, since water is a necessary prerequisite for life, and the presence of water anytime in Mars' history could have led to the development of simple microscopic life forms. Though the surface of the planet may be dry today, except for the water ice at the polar caps, there seems to have been flowing water on the surface in the distant past. Scientists set themselves the task of finding out what happened to that water.

Since January 2004, two robotic rovers, *Spirit* and *Opportunity*, have obtained ground-based evidence for past water.

Below right Close-up of "Yogi" rock, investigated by the *Sojourner* rover on Mars. It was named because it resembled rocks found in the former quarry where "Yogi" Berra baseball stadium was constructed on the Montclair State University, NJ, USA.

Right In January 2004, two robotic geologists named *Spirit* and *Opportunity* landed on opposite sides of Mars. Since then, the twin rovers have sent back thousands of spectacular images of Martian terrain.

Below Panorama of the surface of Mars dubbed a "Presidential Panorama" by the Mars *Pathfinder* team. It shows in colorful detail the surroundings of the Sagan Memorial Station. The big rock midway through the picture is called "Yogi."

SCIENCE FICTION

Writers were not slow to grasp the public interest in Mars created by Percival Lowell's discoveries. Among them was the British writer H. G. Wells, whose book *The War of the Worlds*, published in 1898, is by far the most famous work of fiction relating to Mars. In this book, Wells tells the story of a fictitious Martian attack on Earth that devastates the English countryside and only ends when the invaders succumb to terrestrial bacteria. Forty years later, American actor Orson Welles produced a radio adaptation of the book that was so highly realistic that it created panic among many of its listeners.

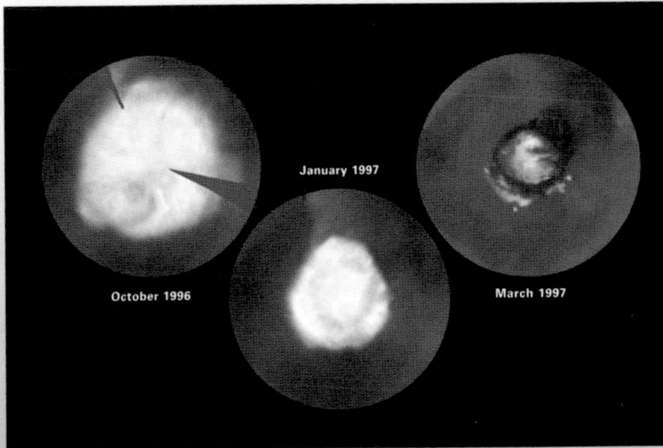

Opportunity found large numbers of small round pebbles of the mineral haematite, which is an iron mineral normally deposited by water. It also examined rock layers exposed in the walls of craters and found indications that they had once been soaked by water containing minerals such as chlorine and bromine. *Spirit*, operating on the other side of the planet inside a large crater called Gusev Crater, had difficulties finding exposed rock, as lava flows had covered the original floor of the crater. When it finally reached exposed rocks, it also found minerals inside them indicating that sometime in the past the rocks had been in a wet environment.

Not only do scientists think that there has been water on Mars in the distant past, it seems that on some occasions and at some locations on Mars, liquid water emerges from the ground today. This crucial discovery is from the *Mars Global Surveyor* spacecraft that circled and studied the planet for nine years until falling silent in late 2006. Scientists operating the spacecraft had previously noticed numerous ditches on crater walls, and similar places, that resembled gullies carved out by water on Earth. They monitored these gullies for changes and found two where new deposits, probably carried by flowing water, have appeared over a period of a few years.

The surface of Mars is presently completely dry. The prospect of a reservoir of underground water gives rise to the exciting possibility that it may provide a suitable habitat for simple bacterial life forms. As yet, we do not know the extent and depth of this underground water nor whether it is usually frozen or in a liquid state. Future explorers of Mars have the challenge of finding the water and studying its contents. Obviously, they need to do this without contamination by bacteria from Earth, for if contamination occurs we may never know if Mars' microbes had ever existed.

Above Enhanced-color view of an unnamed crater in the Terra Sirenum region of Mars. The gullies in the crater appear to have once carried liquid water.

Above left These images by Hubble Space Telescope were created by assembling mosaics of three sets of images taken during late 1996 and early 1997. They capture the seasonal retreat of Mars's north polar cap.

Jupiter

FACT FILE

Planetary Order
5th planet from Sun

Distance from Sun
5.2AU/483,682,792 miles/ 778,412,000 km

1 Jupiter Day
9.92 Earth hours

1 Jupiter Year
11.9 Earth years

Equatorial radius (Earth = 1)
11.2

Mass (Earth = 1)
318

Effective Temperature
About -234°F/-148°C

Atmosphere
90 percent hydrogen, 10 percent helium, traces of other gases

Moons/satellites
Galilean moons—Callisto, Europa, Ganymede, Io, plus 59 others

Notable features
At least 2,000 Trojans (asteroids that follow either 60° ahead or behind a planet in its orbit)

After the Sun, Jupiter is the most dominant body in the Solar System. And like any king, it has its own retinue of attendants, in orbit around the planet.

KING OF THE PLANETS

Above right *Jupiter and Semele* by Gustave Moreau. Portrayed as a powerful and often ruthless character, Jupiter was the king of the gods in Roman mythology.

Jupiter is the largest planet in the Solar System. It is truly huge, with an equatorial diameter of 88,846 miles (142,984 km)—eleven times wider than Earth.

It's also one of the best-known sights in the sky for stargazers and amateur astronomers, being bright and prominent. Only the Sun, the Moon, Venus, (and sometimes Mars), are brighter. Through a telescope, at least four of its moons can be seen as tiny pinpricks of light, and the banding of its cloud patterns is easily visible. Its prominence in the night sky means that it has been known to mankind since antiquity, and it has played a role in the mythologies and legends of many peoples across Earth. It is named for the chief god of Roman mythology.

Like the other giant planets of the Solar System—Saturn, Uranus, and Neptune—when we look at Jupiter we do not see a solid surface, such as we do with the four inner planets. Rather, we see the tops of clouds.

Jupiter is a "gas giant," composed of roughly 75 percent hydrogen (by mass) and about 25 percent helium, with trace amounts of other gases. It has not been confirmed, but it is suspected, that there is a rocky core right in the middle of the planet. On top of that is hydrogen that has become so squashed by the intense pressures, that it has turned into liquid metallic hydrogen. Above this is a region of liquid hydrogen and helium that gradually gives way to hydrogen and helium gas as it gets closer to the visible surface. At the very top of the atmosphere are cloud layers composed of ammonia ice crystals, and possibly some ammonium hydrosulfide and water/water ice. It was hoped that the *Galileo* atmospheric probe would detect some significant amounts of water in one of the cloud layers during its descent into the atmosphere, but in the end it didn't. Scientists think the probe just happened to strike a "dry" region.

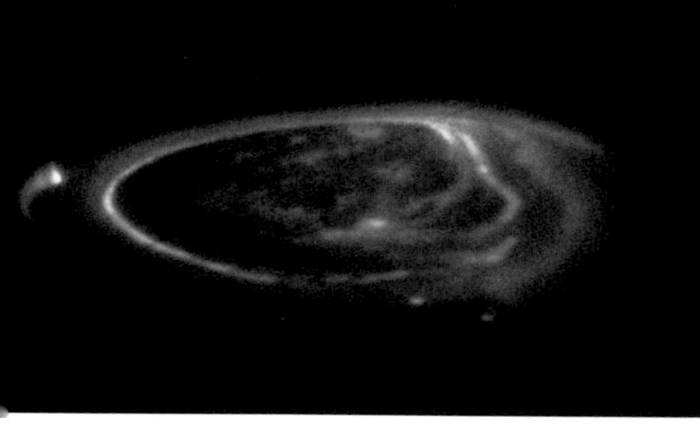

Left This is a spectacular Hubble Space Telescope close-up image of an electric-blue aurora that can be seen eerily glowing on the giant gas planet, Jupiter.

BIG BROTHER

Jupiter rotates faster than any other planet in the Solar System, completing a rotation in just over 9 hours 55 minutes. This rapid rotation has made the planet "flatten out" into an oblate spheroid, with a sideways bulge at the equator. Being a non-solid planet, the atmosphere actually rotates at slightly different rates depending on the latitude—for instance, the poles rotate about five minutes slower than the equator.

Jupiter orbits the Sun in a roughly circular orbit which, at its closest point, is just under five times—460 million miles (740.5 million km)—further from the Sun than Earth is, and at its furthest point is just under 5.5 times the distance between Earth and the Sun, approximately 507.4 million miles (816.6 million km). Its orbit is inclined only slightly—1.3° to the ecliptic plane. Jupiter's year—the time it takes to complete one orbit—is 11.9 Earth years.

Its position in the middle of the Solar System, and its large gravitational field, make it a dominant player in the movement of small Solar System bodies, such as asteroids and comets. It is responsible for keeping some of them in their orbits, while others are flung onto different trajectories if they get too close to the giant planet.

Jupiter has often been called a "failed star." It is less than one percent as massive as the Sun. If it had been only about a dozen times more massive it would have become a "brown dwarf," a feeble kind of star that emits a small amount of heat. If it had been around 80 times as massive, it would have become a true (though small) star.

An intriguing fact is that Jupiter actually puts out about as much heat as it receives from the Sun. All that energy is then slowly released as the planet continues to slowly contract over its life, making it now half the diameter it was on formation.

c. 3300 BCE Babylonian Empire shows records of their knowledge of Jupiter's existence.

c.3000 BCE Ancient Greek astronomy records knowledge of the planet Jupiter.

c.2000 BCE Chinese astronomers keep records of Jupiter's orbit.

1200 BCE–476 CE Roman records of Jupiter are kept. In 476 CE, Chinese astronomer, Gan De, makes observations of a body which is now believed to be Ganymede, a moon of Jupiter.

1610 Galileo discovers Jupiter's four largest moons: Callisto, Europa, Ganymede, and Io. They have since been known as the Galilean moons.

1664 A British chemist and physicist, Robert Hooke, discovers Jupiter's Great Red Spot.

1892 Amalthea, another satellite of Jupiter, is discovered by Edward Barnard.

1904 Himalia, a large moon, is discovered by Charles Perrine, the US astronomer. Himalia has a diameter of 105 miles (170 km) and is the largest in the group that now bears its name. Perrine goes on to discover Elara the following year.

1908 Pasiphae, a moon of Jupiter, is discovered by astronomer, Philibert Melotte.

1914 Sinope, a retrograde irregular moon of Jupiter, is discovered by Seth Barnes Nicholson. It is named after Sinope of Greek mythology.

1938 Lysithea and Carme are also discovered by Seth Barnes Nicholson.

1951 Yet another moon, Ananke, is once again discovered by Nicholson.

JUPITER CORE FACTS

Jupiter is a giant gas planet, and is the largest of the Solar System's planets. Scientists are not sure if it has a solid core or whether gas extends all the way down to the center, gradually becoming denser.

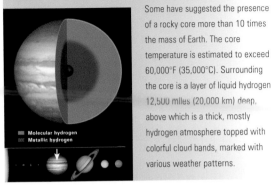

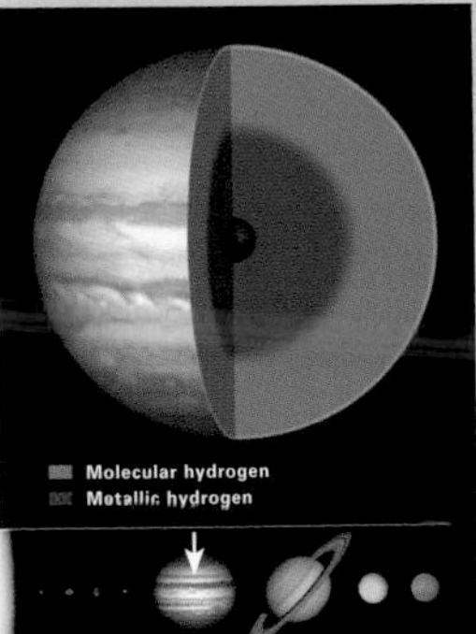

Some have suggested the presence of a rocky core more than 10 times the mass of Earth. The core temperature is estimated to exceed 60,000°F (35,000°C). Surrounding the core is a layer of liquid hydrogen 12,500 miles (20,000 km) deep, above which is a thick, mostly hydrogen atmosphere topped with colorful cloud bands, marked with various weather patterns.

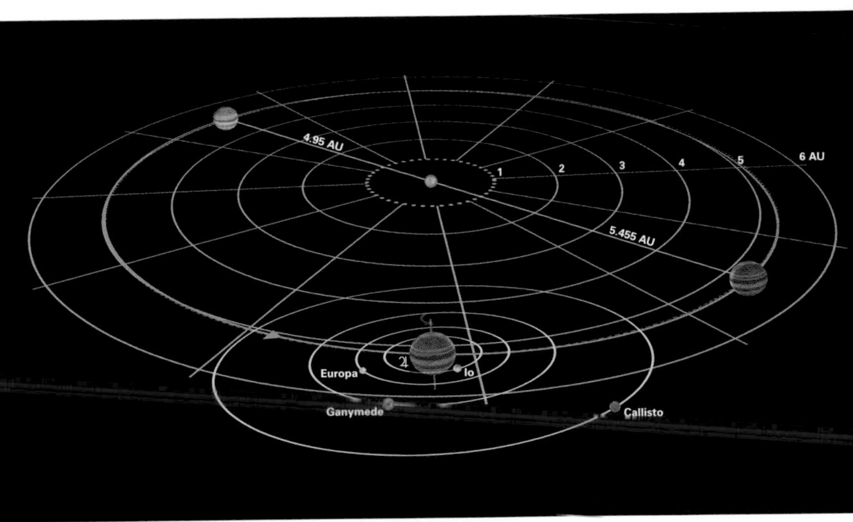

Above Jupiter takes almost 12 Earth years to complete an orbit around the Sun, with an inclination of 1.3° to the ecliptic. Its axial tilt is 3.13°. Jupiter's orbit is prograde—counterclockwise around the Sun when viewed from above. The dash line indicates the degree of tilt of Jupiter's orbit from the plane of the Solar System. (not to scale)

Left This Hubble Space Telescope image gives astronomers the first view of the birth of a new red spot on the giant planet. Actually a storm, it is roughly one-half the diameter of the already well-known Great Red Spot.

1955 Kenneth Franklin, US astronomer, detects some radio transmissions from the planet.

1973 Jupiter is visited by *Pioneer 10*, the first spacecraft to travel though the Asteroid Belt and reach the outer Solar System.

1974 Leda discovered by Charles Kowa. Space probe *Pioneer 11* reaches Jupiter and uses the planet's gravitational force to send it on past Saturn and onto other regions of the Solar System. It lost contact with Earth in 1996.

1975 Themisto, another moon, is discovered by astronomers Elizabeth Roemer and Charles Kowal.

1979 Thebe and Metis are discovered by Stephen Synnott. *Voyager 1* discovers Jupiter's rings. *Voyager 2* also visits Jupiter.

1991 Hubble Space Telescope sends first images of Jupiter, revealing details never imagined before. Hubble continues to return data helping to improve our knowledge of Jupiter and the Solar System.

1992 Space probe *Ulysses* passes Jupiter, using its gravity to hurl it onto the correct trajectory for its examination of the Sun. During its Jupiter flybys, scientists use its instruments to discover more of the planet.

1994 Fragments of comet Shoemaker-Levy 9 collide with Jupiter, leaving large "scars" on the southern surface. It is observed by scientists around the world.

1999 Another new moon, S/1999 J1, is discovered.

2000 *Cassini-Huygens* probe passes Jupiter on a flyby. It's on its way to Saturn but is deliberately sent via Jupiter to uncover new information. It reveals a vast whirling bubble of charged particles surrounding Jupiter.

2001 Due to the rapid increase in technology, 11 new moons are discovered.

2003 *Galileo* probe plunges into Jupiter's atmosphere and is destroyed.

Right Extraordinary view of Comet Shoemaker-Levy 9 on its path to impact Jupiter. The "string of pearls" shows the path of the comet.

CLOUDY WEATHER

Through even a small telescope, it can be seen that Jupiter's visible surface is divided into distinct bands or stripes. These are cloud patterns that circle the planet—dark colored bands are known as belts, while the lighter colored bands are called zones.

Each hemisphere is divided into three main belts and three main zones, with an extra zone straddling the equator. There are also two polar regions, and several other belts and zones that come and go. The Great Red Spot is located within the south equatorial belt and south tropical zone.

The different colorations of the zones and belts reflect the gases that make up the clouds, and the clouds themselves are in three layers. The top white layer is made of ammonia; the middle layer is ammonium hydrosulfide, which gives it its reddish-brown color; and below that is a layer of water clouds, but these are not usually visible underneath the two top layers.

Like Earth, Jupiter has circulation patterns that give it its complex climate system. Warmer air near the equator moves towards the poles, and cold air moves away from the poles, which combined with the strong Coriolis effect caused by Jupiter's rapid rotation, leads to the creation of the strong atmospheric currents.

SHOEMAKER-LEVY 9

In 1994, a once-in-a-lifetime event occurred when fragments of a comet known as Shoemaker-Levy 9 ended up on a collision course with Jupiter. Around 24 pieces of the broken comet slammed into the planet's atmosphere with such velocity that they exploded upon impact, causing dark scars in Jupiter's clouds that were easily visible from Earth. This was the first time that a comet impact on a planet had ever been witnessed, and astronomers made the most of the opportunity—telescopes on the ground, in the air, and in space were trained on the impact points.The data collected from the event helped to refine scientists's understanding of both Jupiter's atmosphere, and also the composition of comets.

GALILEO'S MOONS

Jupiter has the largest system of moons in the Solar System, with 63 known at present. Some were discovered shortly after the telescope was invented almost 400 years ago; others had to wait until the advent of spacecraft that could make the long journey to the planet.

The Jovian satellite system comprises four large moons—the Galilean moons—plus numerous smaller bodies at various distances from the planet. Many are believed to have formed

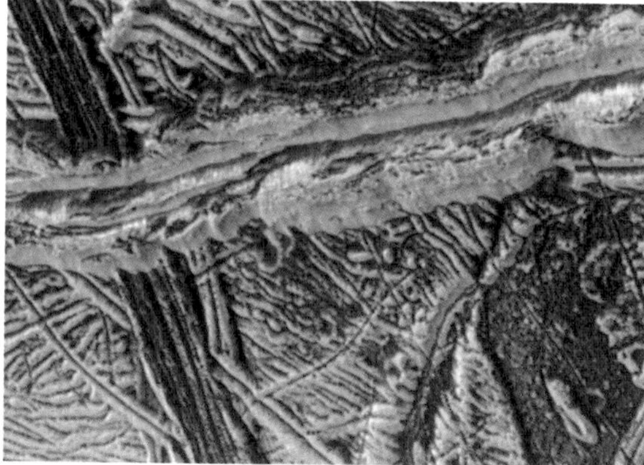

Above Close-up of the surface of Europa taken by *Galileo,* the Jupiter probe spacecraft. The frozen surface shows crisscross lines of dark old cracks, while the lighter cracks appear to be much newer.

Left On its way to Saturn, *Cassini* took some extraordinary shots of Jupiter's South pole. Combined, the maps show colorful cloud features, including reddish-brown and white bands, and the Great Red Spot.

Opposite True-color mosaic of Jupiter constructed from images taken by *Cassini* in 2000. The energetic, small, bright clouds to the left of the Great Red Spot and in the northern half of the planet grow and disappear over a few days and generate lightning. Streaks form as clouds and are sheared apart by the intense jet streams that run parallel to the colored bands.

Above View of eruptions of sulfur compounds from the volcano, Pele, on one of Jupiter's Galilean moons, Io, taken by spacecraft *Voyager 2*.

along with the planet, while others are considered to be independent bodies that became caught by Jupiter's immense gravity, and swept into orbit.

The first four moons to be found were discovered by Galileo Galilei in January 1610. Using his new telescope—one of the first ever built—he spotted four tiny dots surrounding the planet. He continued to watch them over the course of a few nights and it became obvious to him that they were circling the planet.

This observation had far-reaching consequences, as it showed that there were at least some heavenly bodies that were not in orbit around Earth—at the time, the prevailing view (largely based on religious beliefs) was that Earth was the center of the Universe and everything revolved around it.

Right Digital composite photo of Europa, one of Jupiter's Galilean moons. This image shows the surface covered by dark brown areas marking rocky material derived from the interior of the planet, implanted by impact, or from a combination of interior and exterior sources.

Galileo's discovery eventually led to the abandonment of that view in favor of one based on reason and hard evidence.

After their discovery, the four moons were known simply as Jupiter I, II, III, and IV.

However, in the middle of last century, the moons were officially named according to a proposal originally put forward by Simon Marius, a contemporary of Galileo who had always claimed that he discovered them first. The Galilean moons were given the names Io, Europa, Ganymede, and Callisto, and they each have their own distinct features and peculiarities.

VOLCANO WORLD

Io is the innermost Galilean moon of Jupiter, and one of the most intriguing bodies in the Solar System. It is a large moon—just over 2,236 miles (3,600 km) in diameter. There are only three moons larger than it in the entire Solar System, and it is the densest moon in the Solar System. It is also the Solar System's most volcanically active body.

The *Voyager 1* probe was first to spot an eruption from Io's surface, and ended up seeing nine such eruptions. The volcanoes do not eject lava as we know it here on Earth; rather, they project sulfur compounds high above the surface. Some of that sulfur gets caught in Io's orbit, creating a continuous torus—a donut-shaped ring—that stretches all the way around Jupiter. Io's volcanism is responsible for the geologically young appearance of the surface; hardly any craters are visible—any craters that do form from impacts are quickly covered by volcanic activity. Io get its energy from the constant gravitational tug of war that takes place between it and Jupiter on the one hand, and the other Galilean moons on the other. Each of these bodies pulls at Io from a different direction, causing the moon to flex and stretch. This creates the internal heat that drives the volcanism. As well as its dramatic volcanoes, Io also has lakes of liquid sulfur, mountains, and lava flows. Sulfur compounds are different colors at different temperatures, and this leads to Io's amazingly varied color scheme, which has been likened to a moldy orange. The moon has essentially no water, unlike its siblings further out.

It is thought that the young Jupiter's heat must have driven off any water present. Clearly sulfur and its compounds play a major role on the moon, and initially it was thought that the lava flows were liquid sulfur. But studies of the temperatures of these flows show that they are too hot for sulfur—it would have boiled away—so they might instead be made of molten silicate rock. The *Galileo* spacecraft found that Io has an iron core of at least 560 miles (900 km) diameter.

LIFE UNDER THE ICE?

The next of the Galilean moons outward from Jupiter is Europa, a body that holds a lot of interest for scientists. Over the past decade or so, there has been increasing speculation on the possibility of simple life existing elsewhere in the Solar System. This speculation has arisen through a better understanding of the hardiness of life of Earth—microorganisms, for example, have been found to thrive in places once thought completely unsuitable for life—along with new information on the conditions extant on some other Solar System bodies.

One of the latter is Europa, which is believed by many to have an ocean of liquid water perhaps 60 miles (100 km) deep caught between its rocky mantle and icy crust. Europa's surface is very smooth, indicating that it is geologically "young." It also is covered in intriguing colored patterns, the origin of which is unknown. Some scientists have proposed that they are the result of liquid material breaking through the ice from below, possibly in the form of geysers (as with Neptune's moon Triton). It has even been speculated that microorganisms could be the cause of the reddish-brown coloration. Europa's possible subsurface ocean could potentially support simple life, but the question of how life might have evolved on such a frozen world in the first place remains unanswered. Space missions have been proposed that would see an automated craft drill or melt through the ice to reach the ocean below. Equipped with sensitive instruments, it could analyze the water and possibly do direct tests for signs of life.

The moon's surface is crisscrossed with lines and markings. Imagery reveals them to be areas where the ice surface has cracked and moved apart. Most of them are thought to have been formed by the tidal stresses induced by Jupiter, but some may be a result of eruptions of geysers from below. There are smaller markings on the surface too, which could be places where material has welled up from below.

Data from the *Galileo* mission has shown that Europa interacts with Jupiter's magnetic field to produce one of its own—indirect evidence, perhaps, of the presence of a salty subsurface ocean. Europa's tenuous atmosphere is composed of oxygen liberated from water ice molecules by the effect of sunlight.

Below This montage (not to scale) shows Jupiter accompanied by the Galilean moons in their relative positions. Io, Europa, Ganymede, and Callisto are known as the Galilean moons in honor of their discoverer, Galileo.

Above Jupiter's largest moon, Ganymede, as depicted by *Voyager 1* spacecraft. Ganymede is the largest moon in the whole Solar System and is the third moon out from Jupiter.

THE BIGGEST MOON

Jupiter's largest moon, Ganymede, is also the largest moon in the Solar System—it is 3,268 miles (5,260 km) in diameter; Earth's Moon, by comparison, is much less at 2,159 miles (3,475 km) wide. It is the third Galilean moon out from Jupiter.

Scientists think Ganymede has three layers: A small core of iron (perhaps with sulfur mixed in), then a thick layer of rock, and finally an icy crust. The surface has three distinct types of terrain. The first are younger, bright colored regions covered with ridges and linear patterns, which are believed to be created as tectonic forces stretch the crust. The second darker areas are covered with impact craters, evidence that these areas are far older geologically than the lighter areas. The third dark areas are thought to be quite ancient, around four billion years old.

Ganymede is the only moon to have its own magnetosphere, probably generated by currents in the core. It also has some magnetism induced by Jupiter's magnetic field, probably as a result of a layer of salty water deep under the surface.Ganymede has a very thin atmosphere of oxygen; like Europa's it is produced by the effect of sunlight on the icy surface, which breaks the water molecules into its constituent hydrogen and oxygen.

CALLISTO

The fourth Galilean moon is Callisto, another one that's larger than our Moon and about the same size as Mercury.

Like the other moons, Callisto is believed to be a mixture of ice and rock. Its icy surface—perhaps up to 60 miles (100 km) thick—is absolutely covered with craters, making it one of the most cratered bodies in the Solar System. Under the crust could be an ocean of salty water—again, inferred through measurements of the moon's magnetic field as induced by Jupiter's magnetosphere—that could be anywhere between 6 and 180 miles (10 and 300 km) thick, depending upon its exact chemical makeup. The rest of the interior of the moon is thought to be rock, gradually increasing in density toward the center.

The surface is actually a mixture of rock and ice, and it has several types of terrain. By far the most common terrain is the cratered plain, but there are also small areas of much smoother ground, thought to result from outpourings of icy volcanoes.There are two standout features on Callisto: Valhalla and Asgard, both of them classified as multi-ring basins. These areas of concentric circles are thought to have originated through symmetric cracking of the crust. Callisto's atmosphere is extremely thin, and consists mainly of carbon dioxide.

Below Taken from *Voyager*, this image shows one of Jupiter's four major moons, Callisto, as a hailstone scarred by meteorite craters. Callisto is named after one of Zeus's many love interests.

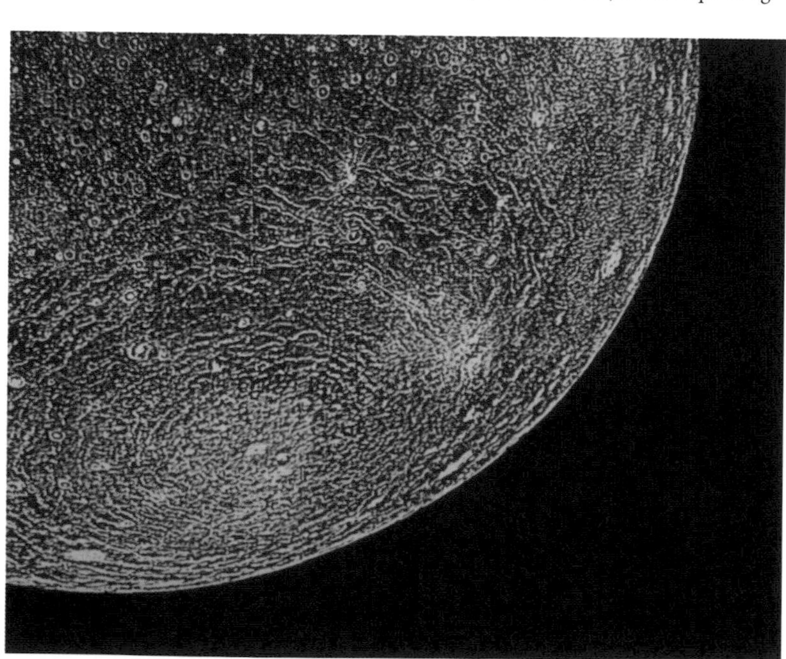

OTHER MOONS

The Galilean moons grab most of the limelight when discussing Jupiter's moons, but there are many more. Scientists have classified them into a number of groups based on similar orbital characteristics. The Galilean moons are one such group.

The first group is a collection of four small moons that orbit closer to Jupiter than their Galilean siblings—Metis, Adrastea, Amalthea, and Thebe. Each of these four is small—around 120 miles (200 km) diameter. It is thought they formed at the same time as the Galilean moons. One moon on its own is Themisto, which is located between the Galilean four and the next group out. It is only about 5 miles (8 km) across.

The Himalia group is next, comprising of Leda, Himalia, Lysithea, and Elara—there might even be a fifth member, yet to be confirmed. These all orbit Jupiter at an angle of around 27°; this, plus the fact that they are all similar chemically, suggest they were once a single body broken into pieces.

Next out is Carpo, another small single moon approximately 1.8 miles (3 km) wide. And again, there is another recently discovered small moon that may belong with this one.

THE GREAT RED SPOT

Imagine a planetary storm so big you could comfortably fit three Earths inside. Well, space scientists don't have to imagine; all they have to do is take a look at the most prominent feature on Jupiter. The Great Red Spot is an enormous anti-cyclone that has been in existence for over 175 years, and is probably the spot seen by the astronomer Giovanni Cassini in 1665.

With dimensions of around 7,500–8,700 miles (12,000–14,000 km) by 15,000–25,000 miles (24,000–40,000 km), it is by far the largest storm in the Solar System—other planets, like Saturn and Neptune, have smaller versions.

The Spot is located in Jupiter's Southern hemisphere at around latitude 22° South. The oval storm rotates in an anticlockwise direction on the boundary between two equatorial cloud belts.

Over the years it has changed color from pale pink to a darker reddish color. Measurements show that the Spot is actually higher than the surrounding cloud tops.

All the moons discussed so far are prograde; that is, they circle Jupiter in the same direction as it rotates. All the remainder of the moons are retrograde, meaning that they go around the planet in the opposite direction. This almost confirms that they were wandering bodies that were captured into orbit around the planet.

The first of these groups has seven main members: Ananke, Praxidike, Iocaste, Harpalyke, Thyone, Euanthe, and Euporie. Ananke is by far the largest at around 18 miles (30 km) in diameter, while the rest are each somewhere between 1.2 and 4.3 miles (2 and 7 km) wide. The next group has 13 main members, including Carme, which is about 30 miles (45 km) in diameter. The outermost group has seven main members, including Pasiphae, which is around 37 miles (60 km) wide. Many of the bodies in these latter groups are suspected of originating from single, larger bodies that were broken apart by Jupiter's gravity.

Several extra moons were discovered in the early years of this century. Little is known about them, and their membership of individual groups is under study.

Below One of Jupiter's best known features is the Great Red Spot. This impressive element, easily viewed through an 8-inch (20 cm) telescope, is a region of turbulence and storm activity that extends from the equator to the southern polar latitudes.

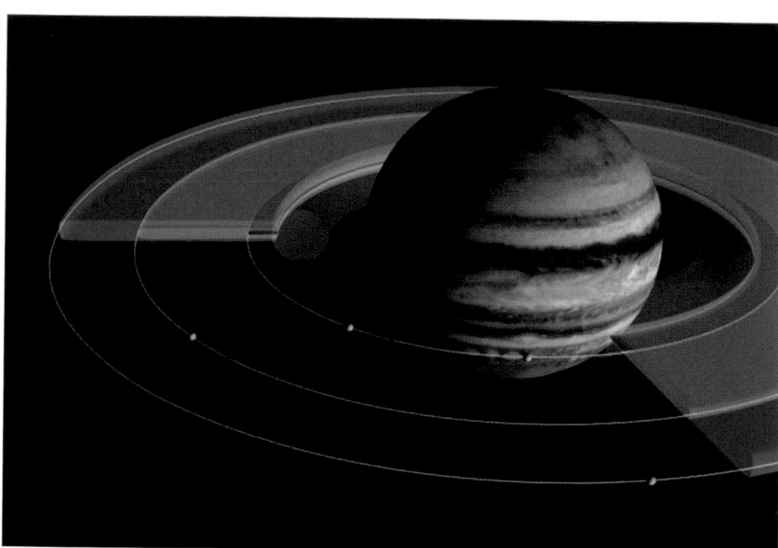

Above Mosaic of Jupiter's ring system taken by *Galileo* when the spacecraft was in Jupiter's shadow looking back toward the Sun. Jupiter's ring system is composed of three parts: An outermost gossamer ring, a flat main ring, and an innermost donut-shaped halo. These rings are made up of dust-sized particles that are blasted off the nearby inner satellites by small impacts.

RINGS

Like Saturn, Jupiter, too, has a system of rings, although nowhere near as spectacular as Saturn's. Jupiter's ring system is made up of what are believed to be dust particles, unlike Saturn's, which are made of chunks of ice.

Hardly anyone suspected that Jupiter might have rings, but some researchers with the *Voyager 1* mission convinced mission controllers to get the spacecraft to look for them during its flyby in March 1979. It found three rings: The inner halo torus (that is, it is shaped like a donut, with the thickest part of its cross-section in the middle), a brighter main ring, and a three-part gossamer ring.

The rings have small moons associated with them, and scientists think the rings are composed of material blasted from the surface of these moons by meteorites.

Metis and Adrastea are the two moons responsible for the main ring, while the gossamer ring is most likely to be associated with Amalthea and Thebe.

The halo torus extends out from about 55,300–76,400 miles (89,000–123,000 km) measured from Jupiter's center; the Main ring extends from 76,400–80,000 miles (123,000–129,000 km); and the gossamer ring out from about 80,000–174,000 miles (129,000–280,000 km).

MAGNETOSPHERE

Jupiter has a magnetic field that is both very strong and very large, extending far out into space—so large that the Sun and its corona could easily fit within it. If it could be seen from Earth, it would be several times the size of the Moon. Jupiter's magnetosphere is, in fact, the largest thing in the Solar System. It is believed to be generated by electrical currents in the metallic hydrogen core of the planet. The magnetosphere of Jupiter reaches out about 3 million miles (5 million km) in front of the planet in the direction of the Sun, but can stretch out to an amazing 400 million miles (650 million km) behind the planet in a long tail—that's out beyond the orbit of Saturn in fact!

Contained with the magnetic field is a mixture of hydrogen, helium, oxygen, and sulfur ions. The sulfur comes from the moon Io, which has volcanoes that spew large quantities of sulfur into space. Some of this sulfur is swept up by the magnetic field on Jupiter into a torus (donut-shaped ring) that surrounds Jupiter at Io's orbital distance.

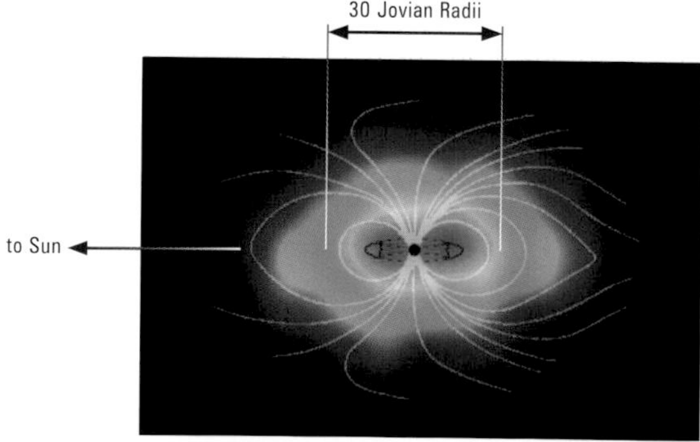

Right The vast magnetosphere of charged particles whirling around Jupiter. The black circle shows the size of Jupiter, lines show Jupiter's magnetic field, and a cross-section of the Io torus, a doughnut-shaped ring that originates from eruptions on Jupiter's moon and circles the planet at about the orbit of Io. The largest area in the Solar System, the magnetosphere would appear two to three times the size of the Sun or Moon to observers on Earth.

SPACE MISSIONS

Eight spacecraft have made it as far as Jupiter, but only one has spent an extended time at the planet. The first to encounter the giant planet were NASA's twin *Pioneer 10* and *Pioneer 11* craft, which reached Jupiter in December 1973 and December 1974, respectively. They were followed in March and July 1979 by the twin *Voyager 1* and *Voyager 2* spacecraft.

All four missions were responsible for giving us our first close-up look at the gas giant planet, revealing new moons, the rings, details in the atmosphere, and revelations about Jupiter's huge magnetic field.

The *Pioneer* and *Voyager* missions were also taking advantage of the "gravitational slingshot" method, whereby a spacecraft can pick up speed by making a close flyby of a planet.

In this case the increase in velocity made it possible for each of the spacecraft to continue on to Saturn and, in the case of *Voyager 2*, further out to Uranus and Neptune as well.

Three other spacecraft have used Jupiter for a slingshot: The solar probe *Ulysses*, which swung around Jupiter in 1992 and onto an orbit that would take it over the Sun's poles;

Right Eight impact sites from fragments of Comet Shoemaker-Levy 9 are visible as dark reddish spots in this image from NASA's Hubble Space Telescope's Planetary Camera. While some of Jupiter's features, such as the Great Red Spot, have remained over time, many of the features seen here are rapidly evolving in the turbulent atmosphere.

Cassini, which flew past in 2000 on its way to Saturn; and *New Horizons*, a Pluto-bound probe that passed the planet in February 2007. While *Ulysses* did not carry cameras that could record its passage through the Jovian system, *Cassini* and *New Horizons* did, sending back valuable images of Jupiter and its moons. But the main space probe player at Jupiter was *Galileo*, which was launched in October 1989, and arrived at Jupiter in December 1995. It then spent eight years orbiting the planet and making close flybys of its fascinating moons. It carried a special probe that was sent into Jupiter's atmosphere to take the first direct measurements of its chemical constituents, wind speeds, pressures, and so on. As the probe descended through 95 miles (152 km) of the top layers of the atmosphere, it collected 58 minutes of data. *Galileo* also spent a lot of its time focusing on Jupiter's moons, about many of which little had been known.

Below left
Digital composition of the spacecraft *Pioneer 11* orbiting the Jovian surface. *Pioneer 11* was the second mission to investigate Jupiter and the outer Solar System.

TAKING THE PLUNGE

While seven space missions have flown past Jupiter, only one has ever taken up residence at the planet—the *Galileo* mission, which arrived at the giant planet in December 1995. Just prior to arrival, *Galileo* released a probe that was directed into Jupiter's atmosphere. Protected from the heat of entry by a heatshield and then suspended by parachutes, the probe spent just under an hour radioing back data as it descended into the clouds. Eventually the increasing pressure in the depths of the atmosphere would have crushed the probe.

Eight years after *Galileo* arrived at the planet, and with it reaching the end of its life, mission controllers directed it to also burn up in Jupiter's atmosphere. This was done to avoid it drifting uncontrolled through the Jovian system and possibly impacting on Europa, potentially contaminating the moon with Earth organisms.

Below Artist's impression of the *Galileo* probe as it was directed into Jupiter's atmosphere. The actual probe, protected by heatshields, transmitted for almost an hour before being crushed by the atmosphere.

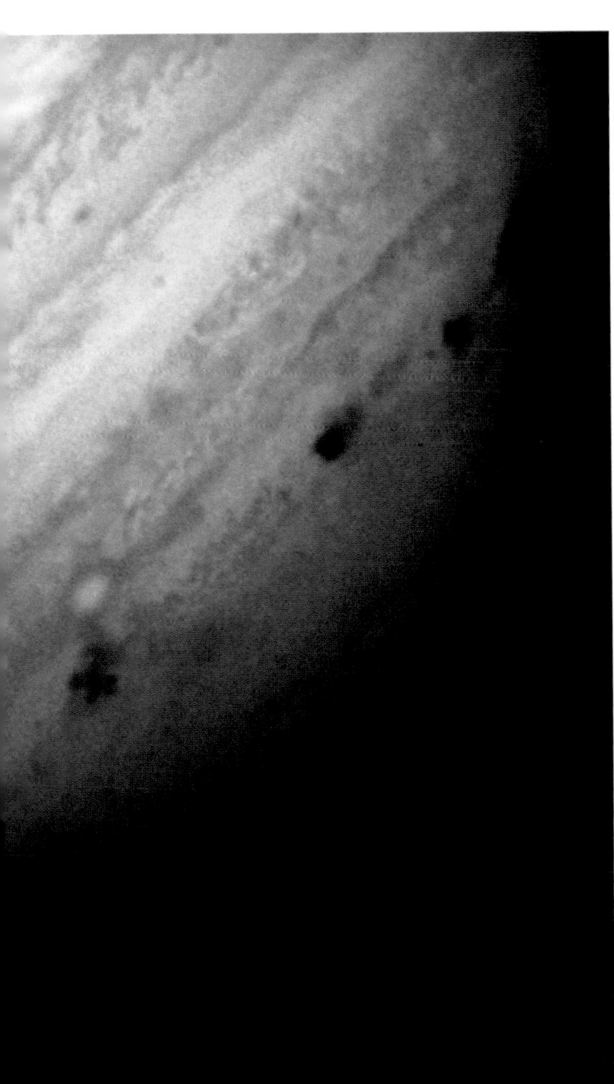

Saturn

Planetary Order
6th planet from Sun

Average Distance from Sun
9.53AU/885,904,700 miles/ 1,426725.400 km

1 Saturn Day
10.656 Earth hours

1 Saturn Year
29.4 Earth years

Equatorial radius (Earth = 1)
9.449

Mass (Earth = 1)
95.16

Effective Temperature
-288ºF/-178ºC

Atmosphere
93 percent hydrogen, 5 percent helium, trace of methane, water vapor, ammonia, ethane, phosphine

Moons/satellites
Titan—second-largest moon in Solar System—and 55 others

Notable features
Thousands of rings made of up billions of particles of ice and rock, believed to be pieces of comets, asteroids, or shattered moons.

Above right Detail from a fresco of Saturn, father of Jupiter, on the ceiling of college hall, Collegio del Cambio, Perugia, Italy.

It is the one planet that is instantly recognizable by a layperson and experienced astronomer alike. With its spectacular system of rings, Saturn really is the jewel of the Solar System.

Saturn is the sixth planet from the Sun, and probably the most beautiful and identifiable of all the planets. It gets its name from the Roman god, Saturnus, the deity of agriculture, and it has also been associated with the Greek figure, Cronus.

Saturn is easily visible to the naked eye, and, as such, has been known to skywatchers since people first turned their eyes to the heavens. Orbiting the Sun in a slight ellipse that ranges from nine to ten times further out than Earth, the planet takes a little under 30 Earth years to complete one orbit.

Saturn's day, the time it takes to rotate once on its axis, is thought to be about 10 hours and 45 minutes of Earth time, which is rapid as planetary rotations go. This is one reason why Saturn bulges out at its equator and is slightly flat at the poles. It's an oblate spheroid—the equatorial diameter is 74,893 miles (120,530 km), while the distance between the poles is 67,561 miles (108,730 km).

Saturn is around 95 times the mass of Earth, but you could fit more than 760 Earths within its volume. These figures give Saturn a low overall density. In fact, its density is the lowest of all the planets, around 70 percent that of water. If you had an ocean wide enough and deep enough, Saturn would be able to float in it!

Left This image taken by *Cassini* shows bright clouds swimming in the giant banks and shoals in Saturn's dream-like atmosphere. The streaks of cloud rise and cast shadows on their surroundings.

INTERNAL STRUCTURE AND COMPOSITION

Like the other gas giants—Jupiter, Uranus, and Neptune—Saturn does not have a visible solid surface. Rather, what we see are the tops of its clouds. Like Jupiter, Saturn is composed mainly of hydrogen (around 93 percent) and helium (just under 7 percent) with small traces of other gases such as methane, ethane, and ammonia. The upper clouds are composed of ammonia ice crystals; lower levels are probably ammonium hydrosulfide and/or water—a lot like Jupiter.

Saturn and Jupiter are twins in many respects, despite their difference in size. Like Jupiter, Saturn is thought to have a rocky core hidden deep beneath its cloudy exterior. Temperatures there would be very high, around 21,500°F (12,000°C). In fact, Saturn actually gives off more energy than it receives from the Sun. This is believed to be caused by heat produced as the planet continues to slowly contract under its own gravity. It has also been proposed that some heat might be given off by droplets of liquid helium that form deep within the planet and fall to lower levels.

Surrounding the core is a thick layer of metallic hydrogen (which is hydrogen compressed under huge pressure), and then, finally, there is the outer layer of hydrogen gas supporting the cloud layers swirling around the planet.

CLOUDY SKIES

Like Jupiter, Saturn too has cloud bands that circle the planet, although they aren't as dramatic and colorful as those of its larger sibling. Also like Jupiter, Saturn can have stormy weather, and oval storms are often spotted. In the years between the visits by the *Voyager* spacecraft (1981) and observations by the Hubble Space Telescope (1990), a large white cloud formation had developed, and another smaller storm was witnessed a few years later.

Saturn seems prone to developing a Great White Spot in its Northern hemisphere that in many respects resembles Jupiter's Great Red Spot (GRS). But unlike the GRS, the Great White Spot seems to be a short-lived phenomenon that comes and goes, reappearing every 30 years or so—once per Saturnian orbit—when the planet's Northern hemisphere is turned toward the Sun. It has been observed before—in 1876, 1903, 1933, and 1960. The 1933 spot was particularly impressive. Scientists will have to wait until around 2020 to see if it returns as expected.

700 BCE
The oldest written records documenting Saturn are attributed to the Assyrians. They describe the ringed planet as a sparkle in the night and named it "Star of Ninib."

400 BCE
Ancient Greeks name Saturn, which they think is a wandering star, in honor of Cronos, the god of agriculture.

c. 150 BCE
Romans, who adapted much of their culture from the Greeks, change the name of the ringed planet from Cronos to Saturnus, the root of Saturn's English name.

1610
Galileo Galilei is the first to view Saturn's system of rings, but mistakenly interprets them to be two moons similar to those he had discovered near the planet Jupiter.

1655
Dutch astronomer, Christiaan Huygens discovers Saturn's moon, Titan. He also makes detailed examination (at the time) of Saturn's planetary rings, and decides it is one large flat solid ring around the planet but not touching it anywhere.

1659
On further observation, Huygens determines that Saturn's rings are separate from the planet.

1660
Jean Chapelain suggests that Saturn's rings may be made up of small moons orbiting the planet.

1671
Cassini discovers a second moon of Saturn, Iapetus, and observes that the moon has light and dark sides. He also discovers that Saturn has more than one ring, by observing the distance between the rings. The largest of these gaps is known today as the Cassini Division.

1789
Sir William Herschel discovers two more moons around Saturn. They are named Tethys and Dione. He also determines the rotation rate of Saturn to be 10 hours and 32 minutes.

SATURN CORE FACTS

Saturn is one of the giant gas planets and is the second largest in the Solar System. It is believed to have a small rocky core where the temperature could be up to 21,500°F (12,000°C). Saturn actually radiates more heat than it receives from the Sun and this is most likely due to the process of gravitational compression, which provides the planet with heat. The core is surrounded with a layer of metallic hydrogen. The atmosphere consists of a layer of hydrogen gas supporting the cloud layers that cover the planet. It is the only planet that is less dense than water.

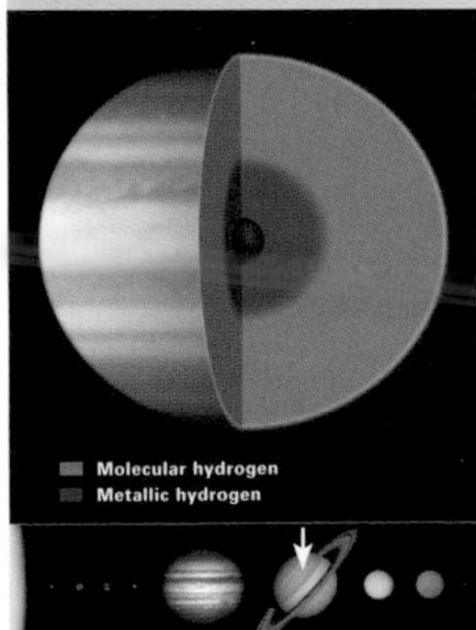

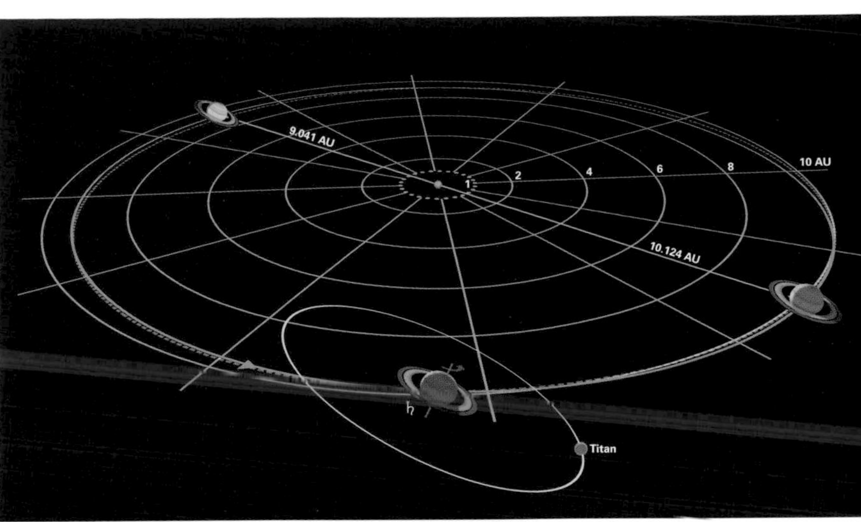

Above Saturn takes almost 30 Earth years to orbit the Sun, with an inclination of 5.51° to the ecliptic. Saturn does not orbit at a uniform rate. Its axial tilt is 26.73°. Saturn's orbit is prograde—counterclockwise around the Sun when viewed from above. The dash line indicates the degree of tilt of Saturn's orbit from the plane of the Solar System. (not to scale)

Left Image of Saturn and two of its moons, Tethys (above) and Dione, photographed by *Voyager 1* in 1980, from a distance of 8 million miles (13 million km). The shadows of Saturn's three bright rings and Tethys are cast onto the cloud tops.

1837
German astronomer, Johann Encke observes a dark band in the middle of the A ring around Saturn. This is later determined to be a division between rings. This has since been known as the Encke Division.

1883
British astronomer takes the first images of Saturn's rings.

1967
Walter Feibelman discovers Saturn's E ring.

1979
Pioneer 11 approaches Saturn in September of 1979 for a flyby. It takes close-up pictures and is sent in a highly dangerous maneuver through the plane of rings sending back new images and data. The craft survives and goes off to probe the outer Solar System and beyond.

1981
Voyager 1 does flyby and photographs Saturn, returning stunning images of its rings, showing that they are made mostly of water ice. They also found "braided" rings, ringlets, and "spokes."

1995
Hubble Space Telescope shows four new moons orbiting Saturn.

1997
NASA launches the *Cassini* spacecraft to orbit Saturn. It carries the *Huygens* probe bolted to its side, designed to investigate the moon, Titan.

2005
Cassini frees its probe *Huygens* on a further voyage to discover more about Titan. It lands on its surface making Titan the only moon other than Earth's to have a spacecraft land on it.

2007
Cassini continues its orbit of Saturn returning data and images of Saturn as well as relaying data and images from *Huygens*.

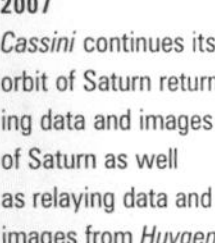

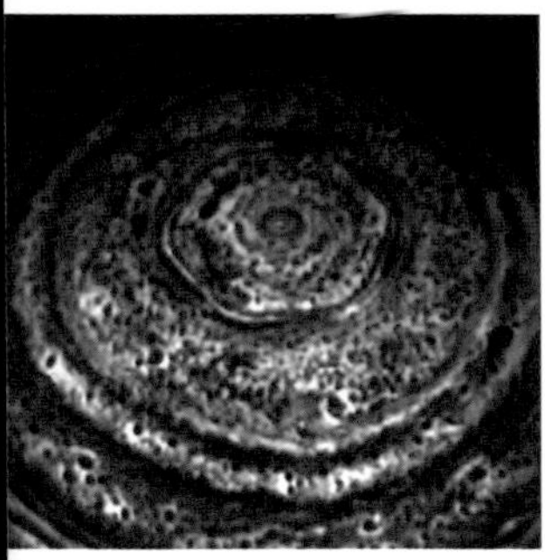

Above A bizarre six-sided feature encircling the North pole of Saturn near 78°N latitude has been seen by the visual and infrared mapping spectrometer on NASA's *Cassini* spacecraft.

Above top Dramatic close-up of Saturn's South pole shows the hurricane-like vortex that resides there. The entire polar region is dotted with bright clouds, including one that appears to be inside the central ring of the polar storm.

Saturn's polar regions are very interesting, with the North and South poles displaying different atmospheric features. The planet's North pole has an intriguing hexagonal cloud formation, with each side of the hexagon being over 8,300 miles (13,500 km) long—that's longer than the Earth is wide! Not only that, but the hexagon clouds extend about 62 miles (100 km) down into the atmosphere.

This feature was first spotted by the *Voyager* probes almost 30 years ago, but still no one really knows why it has formed. Many planetary scientists think it could be the result of "standing waves," which are a type of wave current that remains in the same position.

Interestingly, radio emission that emanates from Saturn varies with the same period as the hexagon takes to rotate. Some scientists had speculated that the emission and the hexagon are connected somehow, but this is not now considered to be very likely.

The South pole is different. It has a more traditional round-shaped storm of winds and clouds. Observations by the *Cassini* spacecraft have shown this region to have an "eye," such as a cyclone or hurricane does on Earth.

Saturn also has some of the fastest winds in the Solar System, with speeds of around 1,120 miles (1,800 km) per hour being recorded regularly.

MAGNETOSPHERE

Although similar to the other gas giants, Saturn's magnetic field is smaller and simpler than Jupiter's, and more centered and closely aligned to its spin axis than Uranus.

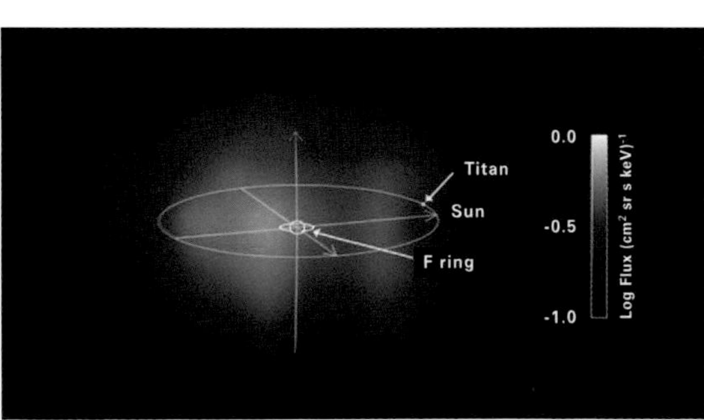

Right The magnetosphere of Saturn is captured in this image taken by the *Cassini* spacecraft in 2004. The magnetosphere is a magnetic envelope of charged particles that surrounds some planets, including Earth.

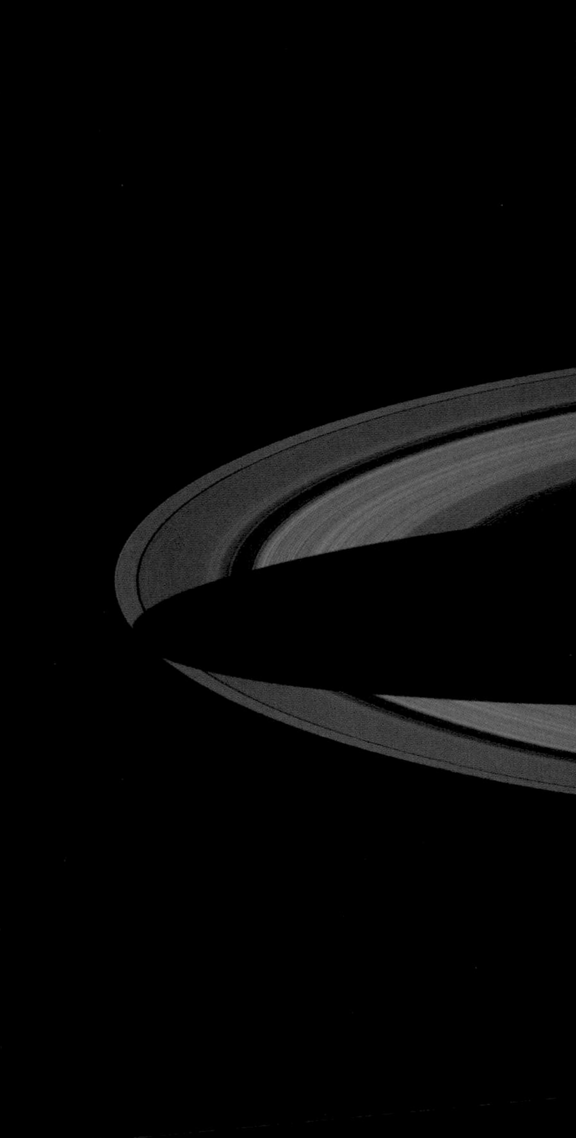

In fact, Saturn's magnetic field is more similar to Earth's. The field forms a huge magnetosphere that surrounds the planet and encompasses within it, the rings, Titan, and many moons. The magnetosphere traps gas molecules and particles within it, some sourced from the top of Titan's atmosphere, while others come from Saturn's rings and some of the other moons. The interaction between these molecules, particles, and sunlight, is similar to that which happens with comets.

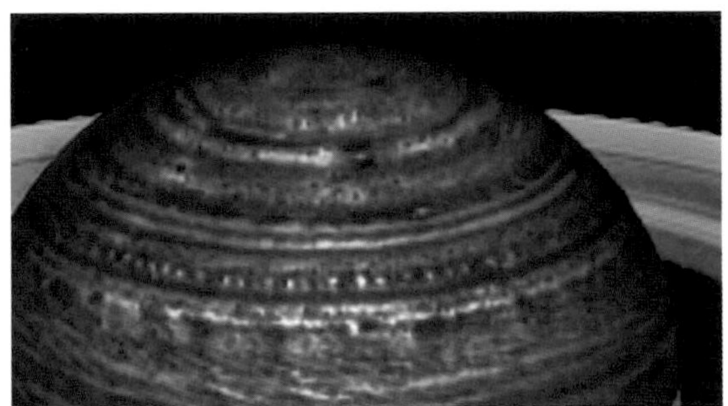

Left Seen in images acquired by *Cassini*'s mapping spectrometer, and lit from below by Saturn's internal thermal glow, the bright "pearls" seen around the planet are actually clearings in Saturn's very deep cloud system.

GOING 'ROUND IN CIRCLES

Because Saturn does not have a solid visible surface, scientists have had great difficulty in determining the planet's rotation rate. Like Jupiter, different bands of cloudy atmospheric circle the planet at different rates, so they can't be used as yardsticks to measure physical rotation of Saturn on its axis. The equatorial region completes one rotation in around 10 hours and 14 minutes (Earth time), while the rest of the atmosphere completes a rotation in just over 10 hours and 39 minutes.

Interestingly, this latter time has also been close to the time of variation of the natural radio emissions that emanate from the planet. For a long time, scientists had made an assumption that the radio emission period was connected with the physical rotation of the planet's core, but recent studies have shown that the radio emission is affected by the amount of gaseous material held in check by Saturn's magnetic field, causing the field's rotation to be slowed down. The scientists now have to go back to the drawing board and try to develop a new way to gauge the planet's true rotation rate.

Below *Cassini* mosaic image shows the subtle color variations across the rings. Ring shadows are cast against the blue Northern hemisphere where the planet's shadow makes its way across the rings.

Left Strikingly beautiful, Saturn is embraced by the shadows of its stately rings. *Cassini*'s image shows the subtle northward gradation of color, possibly due to seasonal influences in the Northern hemisphere.

Below The ridges confirm that a small moon is orbiting within the narrow Keeler Gap. The body is named S/2005 S1. The Keeler Gap is located about 155 miles (250 km) inside the outer edge of the A ring.

RINGS

By far the most impressive, intriguing, and beautiful aspect of Saturn is its large and complex series of rings, which are the reason it is so easily identified.

When Galileo turned his telescope onto the planet in 1610, the rings were slightly tilted with respect to an observer on Earth. Because his telescope's optics were not very good, he couldn't make out their true nature; rather, what he saw were two blobs or "ears" on either side of the planet, and seeming to almost touch it. It looked like three planets, one large in the middle and one small on either side, not moving with respect to each other, and that's how he reported it.

Because Saturn's rotation axis is tilted with respect to its orbital plane, and because its orbital plane is inclined slightly to the ecliptic, for an Earthbound observer the rings (which lie above the planet's equator) sometimes seem to be tilted more to the line of sight, and sometimes less, and sometimes they are edge-on and seem to disappear! And indeed, when Galileo looked at Saturn in 1612 the rings were edge on to Earth and could not be seen, causing him great consternation. They turned up again the next year as the tilt increased once more, and he became even more confused.

It wasn't until 1655 that Christiaan Huygens, using a better telescope, determined the true nature of the "ears." He saw that it was actually a broad but flat ring surrounding the planet without touching it. Twenty years later, Giovanni Cassini could make out gaps, proving that it wasn't just one ring but a series of rings. The largest of the gaps he saw is now named after him.

A little over 80 years later, in 1859, the famous scientist James Clerk Maxwell worked out that the rings couldn't possibly be solid and remain intact, so they therefore had to be made of particles or fragments, all orbiting Saturn in unison. Spectroscopic studies later showed his suggestion to be correct.

The true nature of the rings was further refined over the following century and more, and today scientists have a good idea of their size and composition, largely as a result of images and data sent back by spacecraft. The rings are made mainly of chunks of ice, ranging in size from tiny particles up to pieces the size of a car. There are also probably lumps of rock and lots of dusty particles.

The rings begin about 4,350 miles (7,000 km) from Saturn's cloud tops and extend outward for up to another 248,550 miles (400,000 km). Despite their width, they are actually very thin, ranging from only yards in thickness up to just over half a mile thick—about one kilometer.

There are several main rings and several smaller ones, with a number of gaps in between. The rings were named with letters of the alphabet in order of discovery. The main rings—going outward from Saturn—are the D ring, C ring, B ring, A ring, F ring, G ring, and E ring. There are also some smaller rings. The main gaps in the rings are the Cassini Division and the Encke Division—the former separates the B and A rings, while the latter divides the A ring and F rings. (The Cassini Division can easily be seen with a backyard telescope if the observing conditions are clear and steady.) Some of the gaps in the rings are caused by small moons sweeping material out

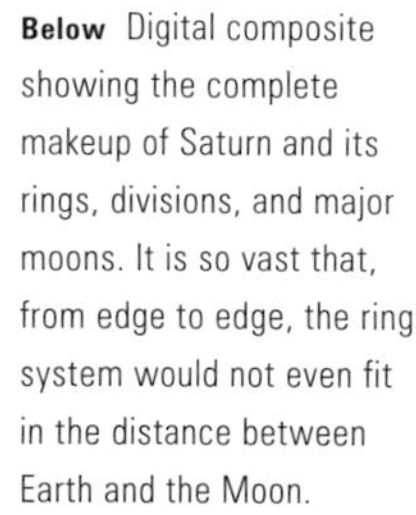

Below Digital composite showing the complete makeup of Saturn and its rings, divisions, and major moons. It is so vast that, from edge to edge, the ring system would not even fit in the distance between Earth and the Moon.

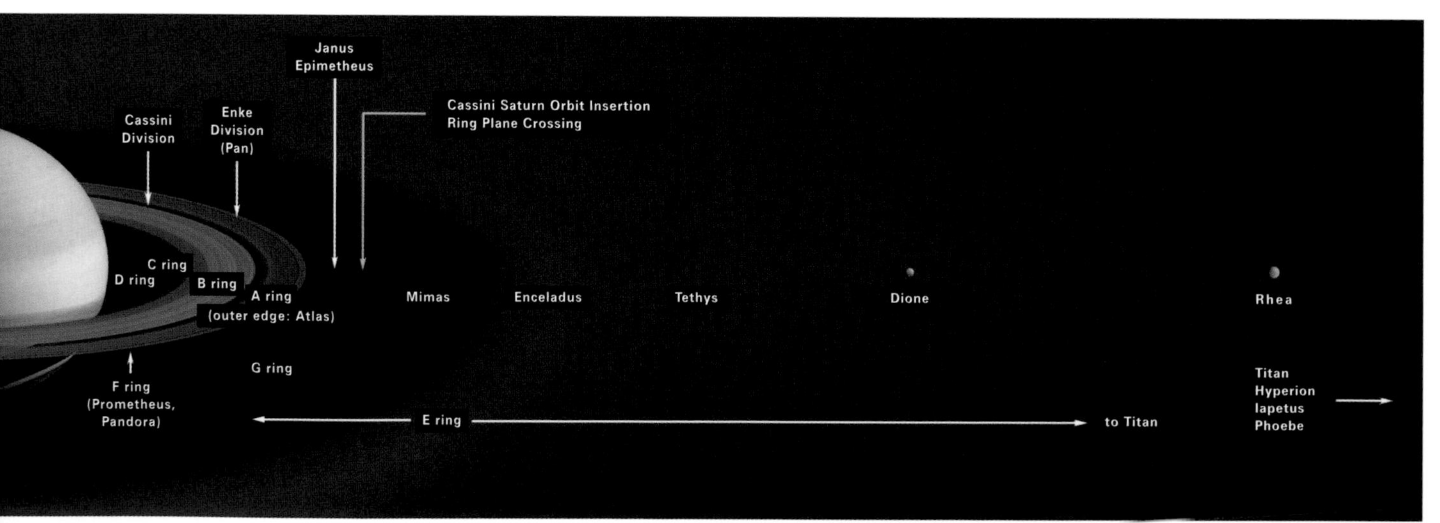

of the way. Others are formed by the gravitational tug of other moons keeping the particles and chunks in check—this is known as "shepherding." The gravitational influence of the moon Mimas is responsible for keeping the wide Cassini Division cleared.

The F ring is intriguing. Spacecraft imagery revealed what seemed to be a "braided" structure, as if three thin rings had been wound around each other. The small moons, Pan and Prometheus, orbit just inside and outside these rings respectively, so it was suspected that they played a role not only in keeping the F ring together, but also perhaps contributing to its strange nature. The latest images from the *Cassini* mission have clarified the F ring situation—it has one central ring and a wavy ring surrounding it. There are also kinks and bumps, all of which explain the initial impression of a braided ring.

Ever since their real nature was revealed, astronomers had puzzled over the origin of the rings. The two main ideas were: They came from a moon that got too close to Saturn and was torn apart by gravitational tidal forces (or, alternatively, was hit by another object and fragmented); or, it was rubbly material left over from the when Saturn first formed, and which was unable to form itself into a moon. Recent computer simulations show that the latter option cannot be correct, as the rings became unstable over periods of hundreds of millions of years. The rings we see now could not have been there a few hundred million years ago. Another ring puzzle is the nature and origin of dark markings found spanning the B ring. Known as "spokes," they are dark regions that seem to change shape and come and go.

There's no way these apparent changes can be explained by physical movement of particles within the ring, so the suspicion is that they are somehow connected with Saturn's magnetic field, as they do seem to move and change in concert with changes in the field.

RING WORLD

Saturn's rings are broad but flat, and are comprised of countless millions of chunks of ice, ranging from small particles up to car-sized lumps. There's also a component of rocky rubble.

Even the smallest backyard telescope will show Saturn's rings, although the larger the 'scope the better the view. The orientation of the rings appears to change from year to year as the alignment of Earth's orbit, Saturn's orbit, and Saturn's axial tilt all conspire to line up in different ways. Occasionally the alignment will be such that the rings appear to be edge-on from Earth.

For a long time it was thought that only Saturn had rings, but it is now known that Jupiter, Uranus, and Neptune also have ring systems; however, they are nowhere near as impressive as Saturn's.

Above Digital impression of Saturn looking through one of its rings, which are made up of ice chunks, along with dusty particles. They extend outward hundreds of miles from the planet and are divided into bands by the orbits of Saturn's moons.

TITAN

Titan was discovered by Dutch astronomer Christiaan Huygens in 1655, for whom the space probe that landed on its surface in January 2005 was named.

It is a moon like no other. It's bigger than Earth's Moon, and it's bigger than Mercury, and it's the only moon in our Solar System to have a thick atmosphere (at its surface the pressure is about 1.5 times that on Earth's surface). In fact, it has been said that because Titan's atmosphere is so thick, and its gravity is so low, a person in a spacesuit with wings attached to their arms could actually fly in it!

The atmosphere is mostly nitrogen gas (98.4 percent) with the remainder being methane plus a few trace gases. Its surface has hydrocarbon lakes of methane and/or ethane, making it the only body other than Earth to have large bodies of stable surface liquids.

Some scientists consider Titan's atmosphere to be very similar to that of early Earth, hence the great interest in studying this distant world. The moon is covered with a layer of haze, which not only makes it difficult to see the surface, it also reflects a lot of sunlight and heat back into space, helping to keep the planet very cold. In fact, surface temperature is a very chilly -290°F (-179°C). It is not thought likely that life could exist there.

Below Hubble Space Telescope image of Titan chasing its own shadow across the face of Saturn. Taking almost 30 years to orbit the Sun, these edge-on alignments of the rings occur roughly once every 15 years.

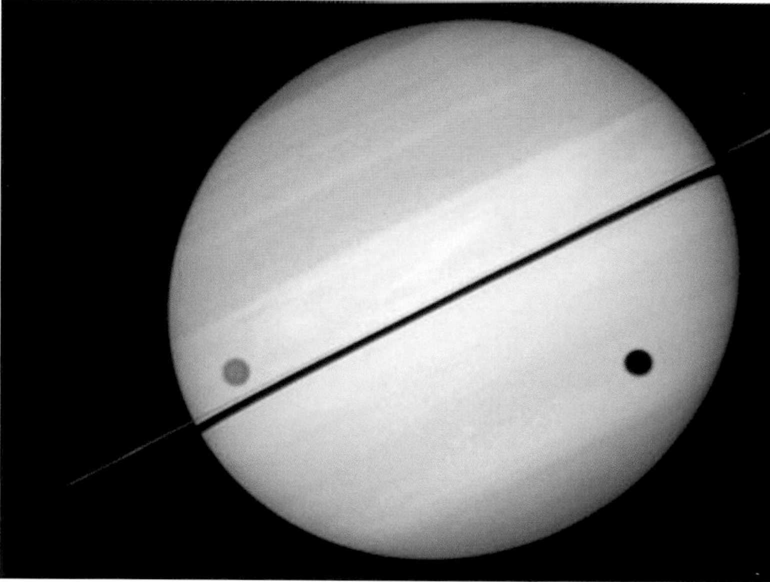

MOONS

Saturn has 57 known moons and a few suspected ones, not counting all the individual particles and chunks within the rings, which technically can all be considered satellites of the beautiful ringed planet.

By far the largest is Titan, which is the second largest moon in the Solar System (only Jupiter's moon Ganymede is larger). Discovered by Christiaan Huygens in 1655, Titan is the only moon in the Solar System to have a substantial atmosphere. Indeed, it is much thicker than Earth's, and is comprised mainly of nitrogen. It also has thick clouds. For a long time astronomers were unable to see below those clouds and find out what was on the surface.

In recent years, telescopes on Earth, using special filters, started to peer through the cloud cover but could make out only the broadest of features. It wasn't until the *Cassini* spacecraft, and its *Huygens* lander, arrived that scientists could start to penetrate that veil. In January 2005, *Huygens* parachuted through the clouds and onto the surface, returning hundreds of images and lots of measurements. The *Cassini* mothercraft has remained in orbit around the Saturnian system, and has instruments that can see through Titan's clouds (including a radar system). Titan seems relatively flat. Its mountains aren't very high, and its impact craters appear to have been filled in, probably by a rain of hydrocarbons.

The presence of lakes on Titan, long suspected, was confirmed by data collected by *Cassini*. It showed large smooth areas—just what you would expect from a body of liquid. The liquid itself is probably methane or ethane (or both), which is fluid at Titan's very cold temperatures.

The *Huygens* probe landed on solid ground that was initially described as being like mud or clay. Mission scientists now prefer to call it a sandy surface, but the sand comprised of icy particles (not silicate, as on Earth).

All the other moons are much smaller than Titan, but the largest of them are still respectable sizes. Moving outwards from the planet we find Mimas (about 125 miles [200 km] in diameter), Enceladus (155 miles [250 km]), Tethys (320 miles [520 km]), Dione (350 miles [560 km]), Rhea (475 miles [760 km]), Hyperion (approx. 95 miles [150 km]), and Iapetus (450 miles [720 km]). In comparison, Titan has a diameter of (3,200 miles [5,150 km]), with its orbit located between Rhea and Hyperion.

Enceladus is particularly interesting. Images sent back by the *Cassini* spacecraft have shown geysers shooting up from its surface in the southern polar region. Previously, only Jupiter's moon Io and Neptune's moon Triton, have been caught in the act of active eruptions. Enceladus has cratered regions and smooth regions, indicating that it has an active geology that is wiping out traces of craters.

Both Mimas and Tethys are notable for having large impact craters that are a substantial fraction of the size of the moons themselves. Both moons are mainly made of ice, as are Dione and Rhea. Iapetus is an enigma. One side is lightly-colored, while the other side is dark. Giovanni Cassini wrote in the seventeenth century that he could see the moon only when it was on one side of Saturn—which, we now realize, must have been when its light-colored side was lit by the Sun.

Iapetus also has a high ridge that runs all the way around its circumference, and several very large impact craters, the largest being around 300 miles (500 km) wide.

All the other moons are quite small, the largest being Phoebe (around 60 miles [100 km] wide), while most are less than 12 miles (20 km) across.

A large bunch of tiny moons were added to the list in 2004 and 2006; many of them are in retrograde orbits—they orbit the planet in the opposite direction to its rotation—and are almost certainly objects captured into orbit by Saturn's gravity.

Only 35 of the 57 moons discovered around Saturn have, so far, been given names, and by tradition, they are always named after the Titans from Greek legends or mythology.

Right Rhea, a serene orb of ice, transits the face of Saturn. Rhea is the second largest of the planet's moons. This view looks toward the unilluminated side of the rings.

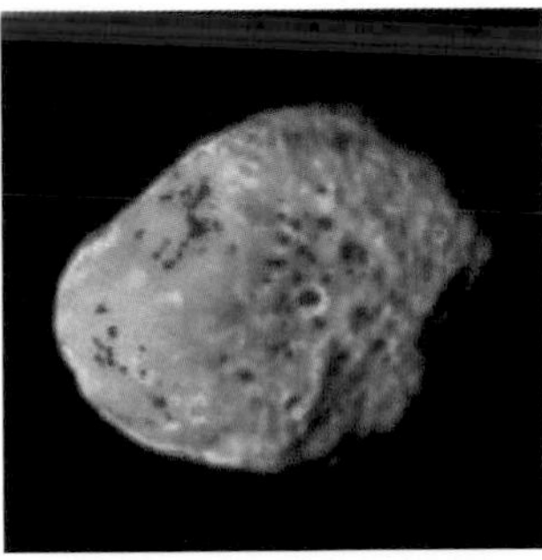

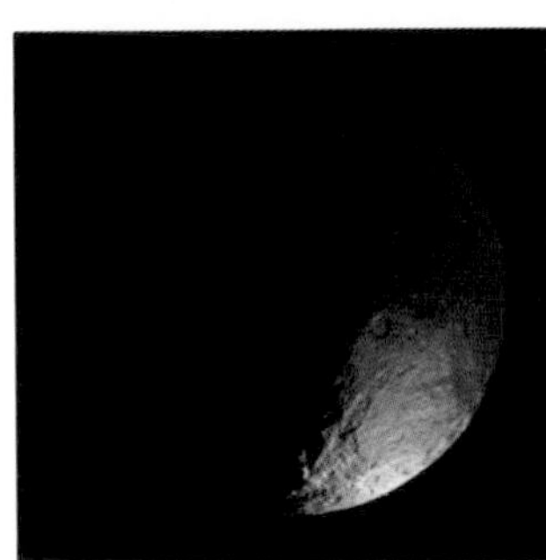

Left Backlit by the Sun, Enceladus, one of Saturn's moons, shows the fountain-like sprays of fine material that tower over the south polar region.

Middle left Saturn's moon Hyperion appears to tumble toward *Cassini* in this image that shows color variations across the surface. The dark areas in the bottoms of craters are seen on all parts of Hyperion.

Bottom left View of the moon, Iapetus, shows the southern part of the dark leading hemisphere—the side that is coated with dark material. The bright South pole is also visible. The dark terrain is known as Cassini Regio.

Right Speeding toward the pale icy moon Dione, *Cassini*'s view shows the tranquil gold and blue hues of Saturn in the distance. The horizontal stripes near the bottom of the image are Saturn's rings, which were edge on.

THE *HUYGENS* PROBE

When the European Space Agency's *Huygens* probe descended through Titan's murky clouds and came to a landing on its icy "sand" surface, it signified only the second time that a spacecraft from Earth had ever landed on a moon—the only other occasions being the many manned and unmanned landings on Earth's Moon.

Released from the *Cassini* mothercraft, *Huygens* was protected by a heatshield when it hit Titan's upper atmosphere. Having slowed, a parachute then deployed and the probe slowly descended. It took two and a half hours to reach the surface. Not knowing what to expect, engineers had built the craft to survive a touchdown on solid land or in an ocean.

On its way down *Huygens* made measurements of the atmosphere and clouds, wind speeds and pressures, and took images of the surface. Some data was lost due to a computing programing error, but sufficient was returned to give scientists a detailed look at this intriguing world. What they saw was a moon with mountain ranges, and what appeared to be rivers and lakes, even oceans. It has now been confirmed that Titan does have lakes of liquid hydrocarbons, as suspected even before the mission was launched.

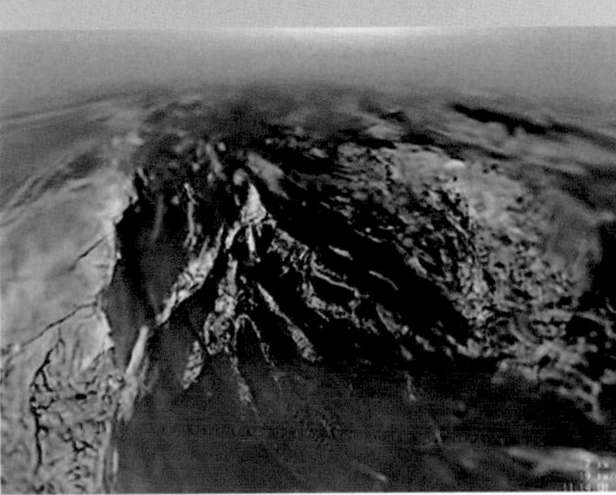

Left Image taken by the ESA's *Huygens* lander during the 147-minute plunge through Titan's thick orange-brown atmosphere to a soft sandy riverbed.

MISSIONS

Four spacecraft have visited Saturn. The first was the NASA deep space probe *Pioneer 11*, which flew past in 1979. It took the first close-up pictures of the planet (closest approach was around 13,000 miles [21,000 km]), discovered the F ring, found two small moons, and studied the planet's magnetic field. With an eye to the future, mission controllers decided to send *Pioneer 11* through the plane of the rings, a potentially dangerous maneuver, but by doing this they could pave the way for the *Voyager* probes that were following it. It survived the passage. The spacecraft then zoomed off into deep space to become one of just a handful of probes that are outward bound from the Solar System.

The *Voyager 1* and *2* probes encountered Saturn in November 1980 and August 1981, respectively. More sophisticated than *Pioneer 11*, they were able to find out much more about the planet. *Voyager 1* sent back highly detailed images of Saturn, its moons, and rings. For the first time, detail could be made out on many of the moons. It made a close approach to Titan, but was unable to see through the clouds. After relaying data, *Voyager 1*, like *Pioneer 11* before it, then headed on to a trajectory that would see it, too, leave the Solar System well behind.

Its twin, *Voyager 2*, also took high-resolution imagery, enabling scientists to study changes that had occurred in Saturn's clouds and its rings since *Voyager 1*'s encounter. Both *Voyagers* were responsible for discovering new moons and new detail in the rings. *Voyager 2* went on to encounter Uranus in 1986, and Neptune in 1989.

Unlike *Pioneer 11* and the *Voyagers*, the *Cassini* spacecraft didn't just pay Saturn a brief visit—it was designed to stay. It went into orbit around Saturn in July 2004—first flying through the gap between the F and G rings. It has completed many long, looping orbits around the planet and its moons.

Below Artist's view of the Saturnian system shows Dione in the forefront, Saturn rising behind, Tethys and Mimas fading in the distance to the right, Enceladus and Rhea off Saturn's rings to the left, and Titan in its distant orbit at the top.

Cassini is a very sophisticated spacecraft, equipped with a large suite of instruments with which to study the Saturnian system. As mentioned above, it also carried the *Huygens* probe that landed on Titan.

Some highlights of *Cassini*'s mission include: Imaging of the spokes in the rings; discovery of four small moons; a close-up look at the moon Phoebe, which appears to be a combination of rock and ice; several flybys of Titan (there will be 45 in all), during which the moon's atmosphere and surface were studied with multiple instruments; and the discovery of geysers shooting water ice from the surface of the moon Enceladus.

The *Cassini* mission still has many years to go, and if, like many other spacecraft before it, it outlasts its original mission duration, doubtless it will go on orbiting Saturn and its moons, sending back valuable information on this fascinating ringed world.

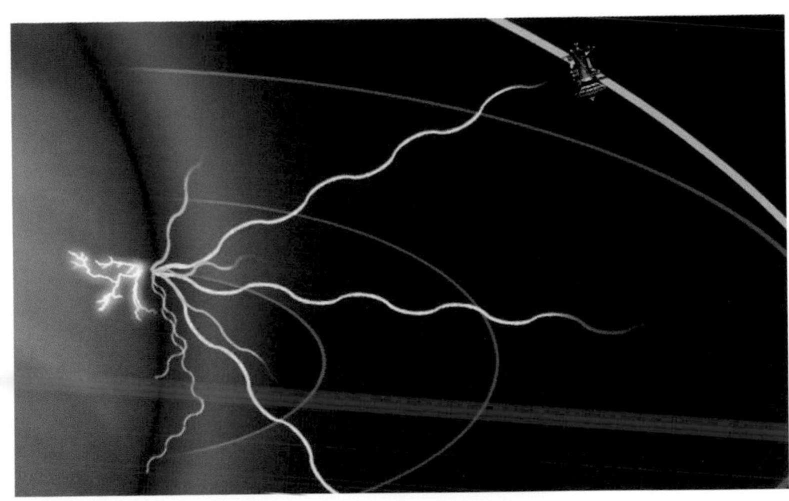

Above Artist's concept shows how *Cassini* can detect radio signals from lightning on Saturn. Lightning strikes emit electromagnetic energy across a broad range of wavelengths, some of which propagate upwards and are detected by the radio and plasma wave science instrument on *Cassini*.

Left Artists impression of *Cassini* during the Saturn Orbit Insertion (SOI) maneuver, just after the main engine has begun firing. The spacecraft is moving out of the plane to the right—firing to reduce its spacecraft velocity with respect to Saturn—and has just crossed the ring plane.

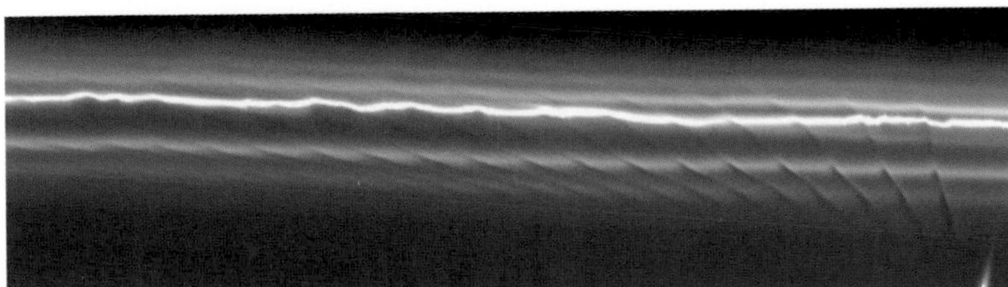

Above This mosaic of 15 *Cassini* images of Saturn's F ring shows how the moon Prometheus creates a ridge in the ring once every 14.7 hours, as it approaches and recedes from the F ring on its eccentric orbit. The channels at the right are the youngest and have near-vertical slopes.

Right Artist's impression of the area surrounding the *Huygens* landing site on Titan, in January 2005. The *Huygens* probe left *Cassini* and reached the upper layer of Titan's atmosphere, landing on the surface after a parachute descent of 2 hours and 28 minutes.

Uranus

FACT FILE

Planetary Order
7th planet from Sun

Average Distance from Sun
19.191AU/1,783,939,400 miles/ 2,870,972,200 km

1 Uranus Day
–17.24 Earth hours (retrograde)

1 Uranus Year
84.02 Earth years

Equatorial radius (Earth = 1)
4.007

Mass (Earth = 1)
14.371

Effective Temperature
-357°F/-216°C

Atmosphere
83 percent hydrogen, 15 percent helium, 1.9 percent methane, trace of ammonia, ethane, acetylene, carbon monoxide, hydrogen sulfide

Moons/satellites
5 major moons—Miranda, Ariel, Umbriel, Titania, Oberon, and 22 others.
2 rings

Notable features
Axis is so tilted (98°) that it is essentially lying on its side.

Above right Ancient Egyptian small bronze statue of Anhur, *Onouris* in Greek, the sky-bearer and warrior god, representing the creative powers of humans.

Although it may appear dull at first sight, the seventh planet from the Sun has hidden depths and peculiarities that make it one of the most intriguing members of the Solar System.

THE MODERN PLANET

Unlike the five planets Mercury, Venus, Mars, Jupiter, and Saturn, all of which are easily bright and prominent enough to be visible to the naked eye—and which therefore have been known since antiquity—Uranus was the first planet to be discovered in the modern era.

The famous English astronomer William Herschel was the first to recognize that the body was not a star, after studying it on March 13, 1781. Uranus had, in fact, been spotted previously by many astronomers from the late 1600s through to the late 1700s, but all took it to be just another star. Being so far away and therefore appearing to be very small, it took Herschel's keen eye to see that it was not a tiny point of light, like a star, but that it had a small and discernable disk.

Noting its round shape, he initially thought that he had found a new comet, and reported it as such on April 26, 1781. But studies by Herschel and the French mathematician and astronomer Pierre-Simon Laplace soon showed that its orbit around the Sun was circular, not elliptical or parabolic—it had to be a new planet, not a comet.

Herschel named it "Georgium Sidus" (the Georgian Planet) in honor of King George III of England. Many other names were proposed however, including "Herschel" in honor of its discoverer, and even Neptune—the name eventually given to the next planet to be discovered.

The name Uranus began to be used in some quarters in the late 1700s, and was in wide use by the early 1800s. It did not become completely official, however, until it was included in England's prestigious *Nautical Almanac* in 1850.

Uranus is the Latinized version of the Greek word *Ouranos*, which was the name given to the god of the sky or heavens. According to many Greek sources, Ouranos was the son of Gaia, the goddess of Earth, and later became her husband. From them both were descended many other Greek deities.

HOT AND COLD

Uranus is the third largest planet (by diameter) in the Solar System. It is larger than Neptune, even though Neptune has more mass. Unlike the inner planets, but like Jupiter, Saturn, and Neptune, Uranus is composed of gas—mostly hydrogen (83 percent) helium (15 percent), with around 2 percent methane plus traces of ammonia, ethane, acetylene, carbon monoxide, and hydrogen sulfide. It's the methane in the atmosphere that gives Uranus its light blue color—the methane absorbs the red wavelengths of sunlight, leaving only the bluer wavelengths to be reflected.

Being located so far from the Sun, the planet's atmosphere is, of course, very cold—the temperature of the cloud tops is a very chilly –360°F (–220°C).

Internally, Uranus has far less heat than Jupiter and Saturn, both of which radiate more heat from their interiors than they receive in the way of sunlight. Uranus's core temperature is around 12,100°F (6,700°C)—compared to Jupiter's 70,250°F (39,000°C) and Saturn's 32,450°F (18,000°C)—which means the convection currents generated in the atmosphere are not as strong as its two larger sister planets. Because of this, Uranus does not have the same sort of strong distinct cloud bands and patterns that Saturn and Jupiter have, although if one looks carefully at the right wavelength—as the *Voyager 2* probe did, and as the Hubble Space Telescope continues to do—some cloud bands can be seen.

UNBALANCED

Uranus is unique in the Solar System in that the tilt of its rotation axis is so large (98°) that it is essentially lying on its side, rolling along on its orbit—most other planets have axial tilts close to a right angle to the ecliptic plane.

Why this should be so is not known, but it could have been that the planet received a massive jolt in a collision with another planetary body in the early years of the Solar System. The effect of this is that Uranus spends much of its 84-year-long orbit with one or the other of its poles facing the Sun, during which periods one polar region at a time gets far more solar heating than the planet's equatorial region.

Despite this, Uranus's equator is hotter than its poles. No explanation for why this should be so has gained wide acceptance in astronomical circles. Also because of Uranus' strange tilt, there is an ongoing argument as to whether the planet's

1690
First known sighting of Uranus by John Flamsteed who thought it was a star and named it 34 Tauri.

1750-1769
Seen at least 12 times by French astronomer Pierre Lemonnier, but considered a star.

1781
Uranus is discovered to be a planet by William Herschel after noticing that the orbit of the body was nearly circular. Herschel originally names it Georgium Sidus (George's Star) after the reigning king of England but pressure soon makes him change it to Georgian Planet.

1781-1787
Pierre Simon-Laplace publishes a series of papers that attempt to show that planetary motions are stable around the Sun. This, along with William Herschel's observations, confirm that Uranus is a planet. The name Uranus is proposed by various publications of the time.

1787
Herschel discovers the first two moons of Uranus, Titania and Oberon.

1821
Bouvard publishes a table attempting to solve the problem of Uranus's orbit.

1845
The mathematicians, British-born John Adams and French-born Jean Leverrier, predict the existence of Neptune. They base their predictions on their studies of the orbital motion of Uranus.

1850-1851
The name Uranus is published in *HM Nautical Almanac* in 1850, giving it official sanction. Two further moons are discovered in 1851 by British astronomer William Lassell, and named Ariel and Umbriel, after characters in Alexander Pope's *Rape of the Lock*.

1948
Uranus' fifth moon is discovered, when the noted American astronomer Gerard Kuiper uses a telescope to distinguish its faint light. It is called Miranda after the character in *The Tempest* by Shakespeare.

URANUS CORE FACTS

Unlike Jupiter and Saturn, the smaller Uranus does not have a huge hydrogen atmosphere. Its core is thought to be rocky and about the size of Earth, surrounded by liquid or frozen water with perhaps some rock mixed in. A relatively thin atmosphere is topped by a bland cloud cover. Although the planet's atmosphere is mostly hydrogen, it also includes some helium and methane. The dull blue appearance we see is caused by the methane molecules that absorb the red wavelengths of sunlight, leaving only the bluer colors to be reflected.

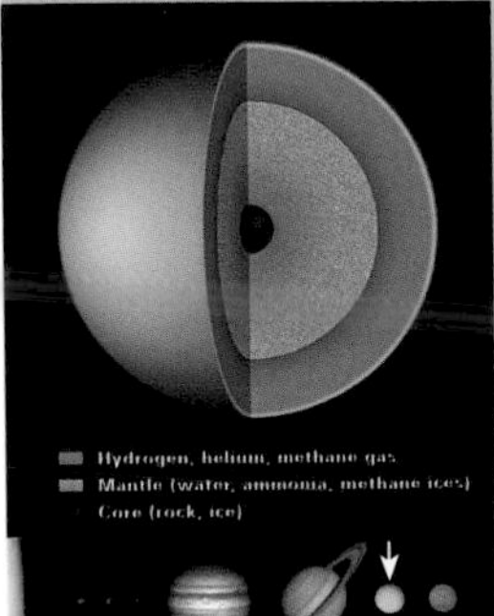

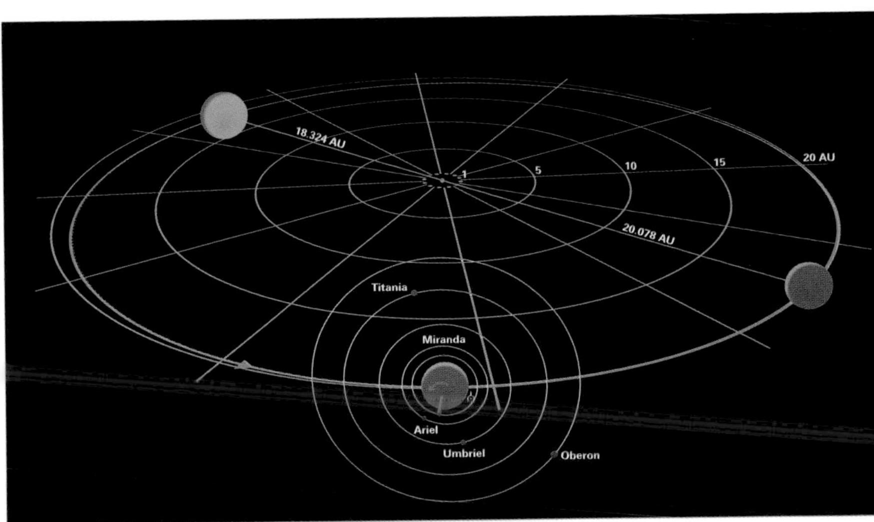

Above Uranus takes just over 84 Earth years to orbit the Sun with an inclination of 6.48° to the ecliptic. Its axial tilt is 97.77° which means it is almost rolling along on its side. Uranus' rotation is prograde—counterclockwise around the Sun when viewed from above. The dash line indicates the degree of tilt of Uranus's orbit from the plane of the Solar System. (not to scale)

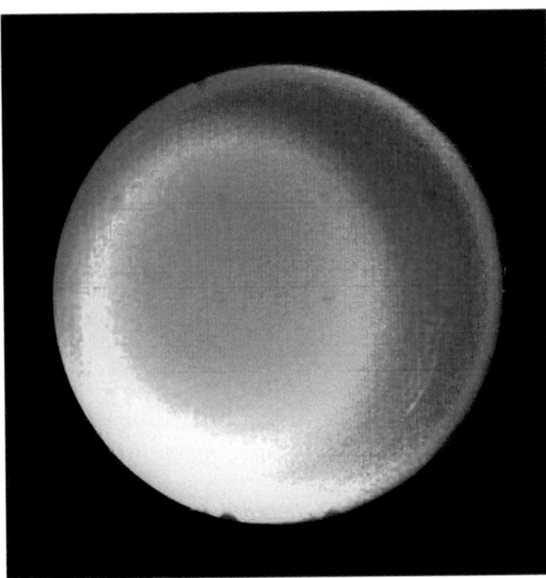

Above Digital imaging and processing enables a view of Uranus's atmosphere. As the absorbed sunlight passes back through a layer of methane gas, it absorbs the red portion of the light, allowing the blue portion to pass through. This image removes the filter caused by the methane.

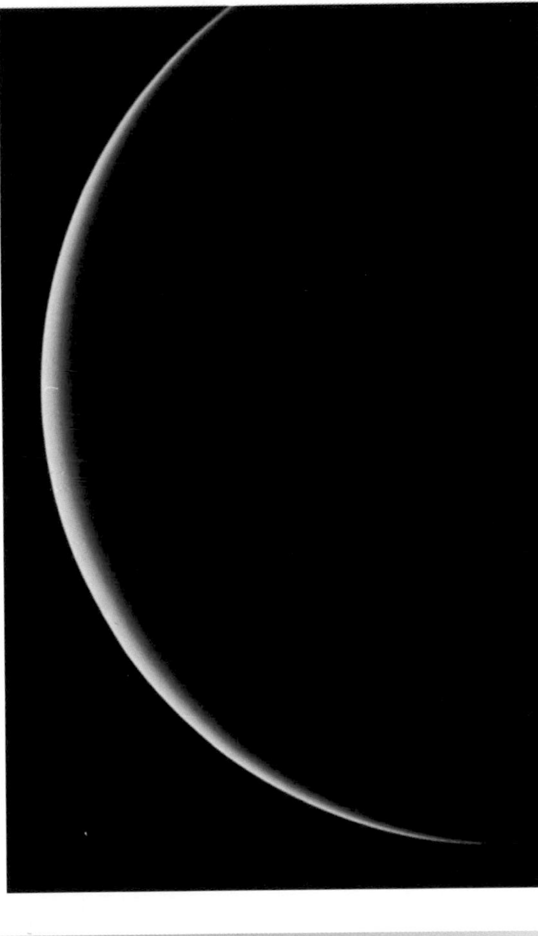

Right Wide-angle 153-degree view of the lit rim of Uranus recorded by *Voyager 2* on its way to Neptune. The pale blue-green color resulting from sunlight reflecting back through the methane in its atmosphere is visible, even at this angle.

1977 Rings are discovered from Earth when Uranus occults—passes in front—a star and it is noticed that there are dips in the brightness of the star before and after it passes behind the body of Uranus.

1977 Twin spacecraft *Voyagers 1* and *2* are launched to explore all the giant planets of our outer Solar System, including Uranus.

1986 *Voyager 2* comes within 50,600 miles (81,800 km) of Uranus's cloud-tops on January 24, 1986. At the time of Voyager's imaging, Uranus's South pole was pointed almost directly at the Sun, according to definitions by the International Astronomical Union.

1986 *Voyager 2* sends back images of 10 new moons, in addition to the five moons already known.

1986 *Voyager 2*'s instruments study the ring system, uncovering the fine detail of the previously known rings and two newly detected rings. It also shows that the banded cloud patterns are extremely bland and faint.

1986 *Voyager 2*'s data shows that the planet's rate of rotation is 17 hours and 14 minutes.

1986 A Uranian magnetic field that is both large and unusual is discovered by *Voyager 2* on its mission.

1990 Hubble Space Telescope is launched to orbit Earth and return images of Uranus and other planets, along with all aspects of our Solar System that it can reach. It shows a much more strongly banded appearance on Uranus than had been previously seen.

2004 Keck Observatory in Hawaii uses advanced optics to capture highly detailed images of Uranus as the planet approaches its southern autumnal equinox.

rotation on its axis is prograde or retrograde; that is, whether it rotates in the same direction as its movement along its orbit (prograde), or the opposite direction (retrograde, like Venus). According to the definition laid down by the International Astronomical Union, Uranus is prograde.

At the time of the *Voyager 2* flyby in 1986, Uranus's South pole was facing the Sun. By the time 2007 came around, the planet was "side-on," meaning that its equatorial regions were then facing the Sun.

The unusual tilt and the planet's long seasons might explain the observed changes in the planet's weather between the time of the *Voyager* flyby and Hubble Space Telescope observations in the late 1990s. Cloud patterns that were revealed only with extensive computer processing in the *Voyager* images (when it was summer in the planet's Southern hemisphere) show much more clearly in the Hubble images—with the planet approaching autumn.

Another mystery is why the center of Uranus's huge magnetic field is not co-located with the geographical center of the planet —it is significantly offset, producing a magnetosphere that looks (if one could see with magnetic eyes) decidedly lopsided. No one seems to know why.

Below Digital illustration of the planet Uranus showing its faint rings. The rings only became visible at the *Voyager* flyby and later by Hubble. The rings appear to wobble, believed to be caused by Uranus's uneven gravitational pull.

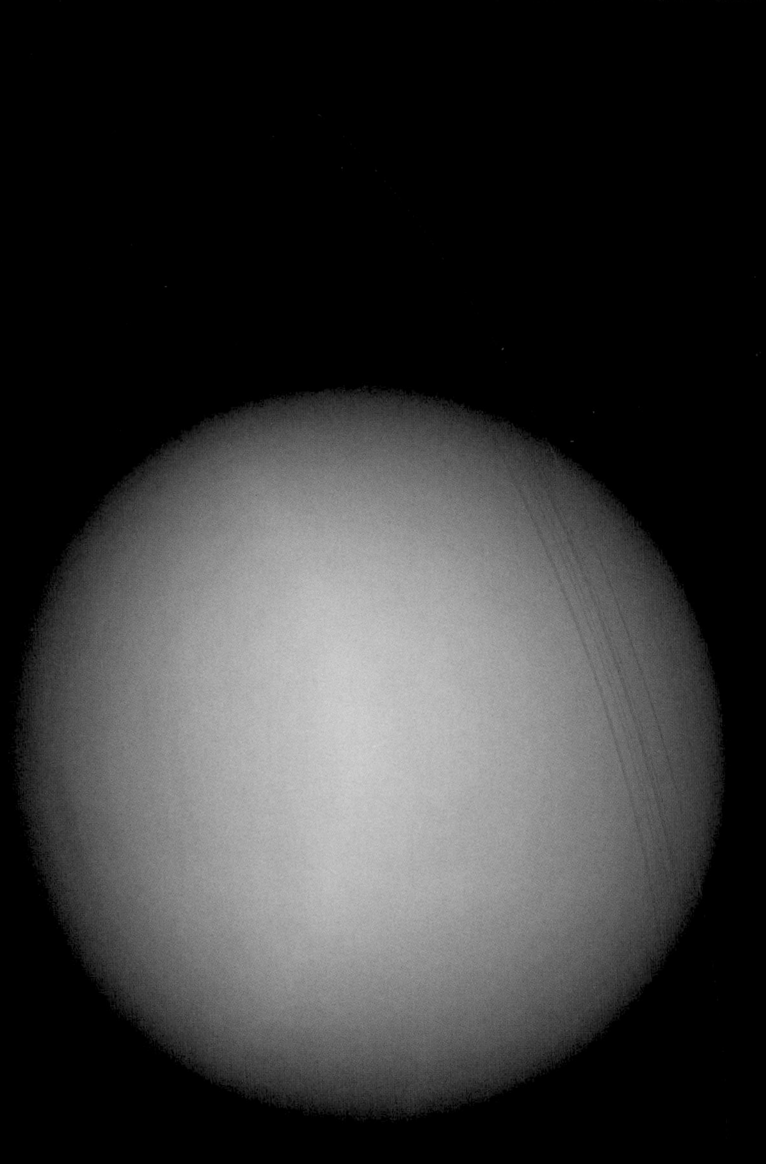

Right Digital Illustration of Uranus and its moons. Twenty-seven moons have so far been observed around Uranus and they are all named after characters in the works of William Shakespeare and Alexander Pope. Uranus's moon system is one of the least massive in the Solar System.

MOONS

We now know of 27 moons circling Uranus, and it's likely there are more to be found. But for a long while, astronomers could be sure of only five moons. The first two moons to be discovered were Titania and Oberon, which were found by William Herschel in 1787, followed by Ariel and Umbriel in 1851 which were both discovered by William Lassell. Next found was Miranda, discovered by Gerard Kuiper in 1948.

That was the full extent of the Uranian satellite system until NASA's *Voyager 2* space probe (the only spacecraft ever to visit Uranus) made its flyby in 1986 during its grand tour of the giant planets. From late 1985 through to closest approach in January 1986, *Voyager 2* found an additional 10 moons—all of them very small. Twelve more were added to the list between 1997 and 2003—one of them found by reviewing old *Voyager* images, and the others found in images taken with the powerful Hubble Space Telescope.

Uranus's five major moons are all thought to be composed mainly of water ice and rock, along with a mixture of organic compounds based on methane. Our only close-up views of these moons were taken by *Voyager 2* that, due to its trajectory, had a good view of their Southern hemispheres but was unable to image their northern sides.

The innermost of the largest moons, Miranda (diameter 300 miles [470 km]), is considered the most geologically active moon in the Solar System—its fractured face is covered in faults and ridges, indicating a "young" surface that has undergone a great deal of upheaval.

Like Miranda, Ariel (diameter 719 miles [1,157 km]), too, has a shattered and bruised surface. *Voyager 2* images also reveal a huge region of plains peppered with craters.

Unlike Miranda and Ariel, Umbriel (diameter 730 miles [1,170 km]) is mostly absent of fracture lines and ridges, but has a seemingly continuous coverage of craters. It is therefore

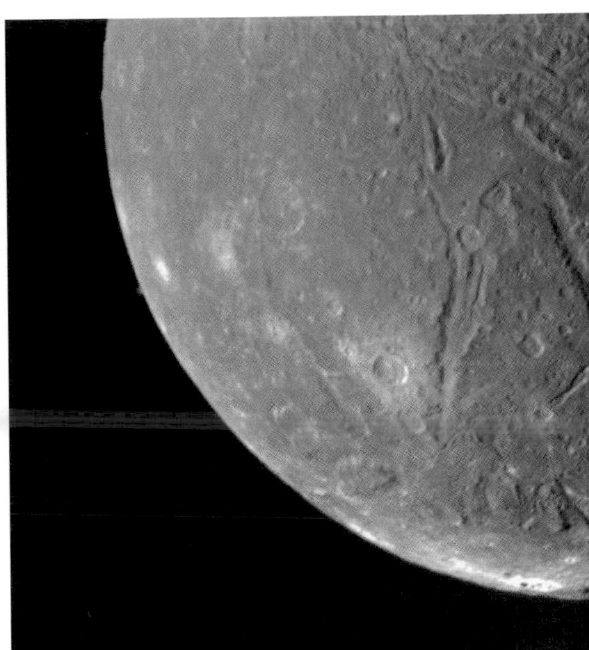

considered to be less geologically active—as any geological activity would have altered the surface by eroding or disturbing those craters. Its surface is also much darker than that of any of its siblings.

Titania (diameter 980 miles [1,577 km]) too has a darker surface, with expanses of craters and a large canyon system that dwarfs Earth's Grand Canyon.

Oberon's (diameter 945 miles [1,522 km]) surface is considered geologically old, with many craters and few signs that might indicate active internal geological processes.

Of the smaller moons, 13 of them orbit the planet closer than the five major moons. They range from the tiny Cupid (diameter around 11 miles [18 km]) to the larger Puck (diameter around 100 miles [160 km]).

The other nine moons are "irregular," meaning that they orbit far from the planet on orbits that are highly inclined to Uranus's equator. All but one of them are in retrograde orbits, circling the planet in the direction opposite to Uranus's rotation.

Irregular moons are thought not to have formed along with the planet, but rather to have been gravitationally captured in the distant past as they were passing by and just got too close.

All of Uranus's moons are named after characters in works of fiction—in this case the works of Alexander Pope and William Shakespeare—unlike most other moons in the Solar System, which are invariably named after the gods and other various beings from Roman and Greek mythology.

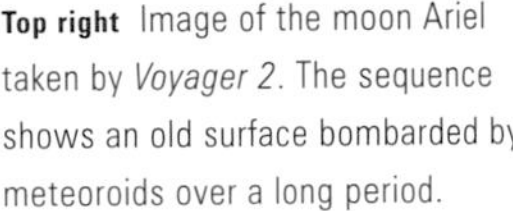

Top right Image of the moon Ariel taken by *Voyager 2*. The sequence shows an old surface bombarded by meteoroids over a long period.

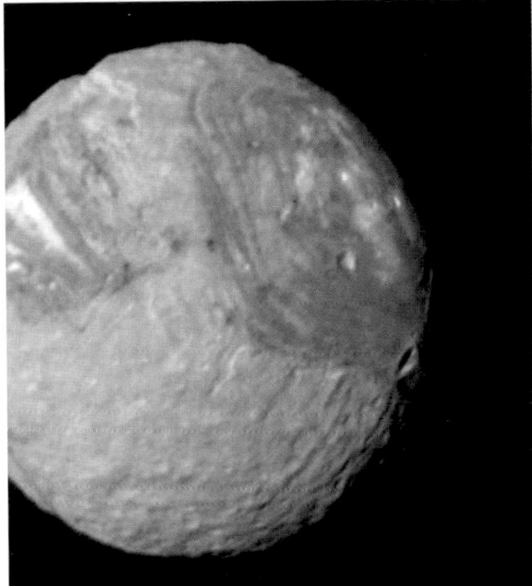

Right *Voyager 2* image of the Uranian moon Miranda, showing many fault valleys and ridges running parallel to the dark- and bright-banded region. Many impact craters also pockmark the surface in this region.

THE WOBBLES

Uranus has two ring systems, the first confirmed in 1986 by *Voyager 2* and the second in 2005 by a team using the Hubble Space Telescope. Tantalizing indications of the presence of the first set of rings were actually gleaned from ground-based observations in 1977, but it wasn't until *Voyager*'s flyby that their presence was confirmed.

A time-lapse movie of Hubble images released in 1999 shows the rings appearing to "wobble" as the planet rotates. Uranus is not a perfect sphere, but rather a slightly flattened globe. The uneven gravitational pull of this shape, together with the gravitational tug of the planet's moons, is thought to be responsible for the wobbly behavior of the rings.

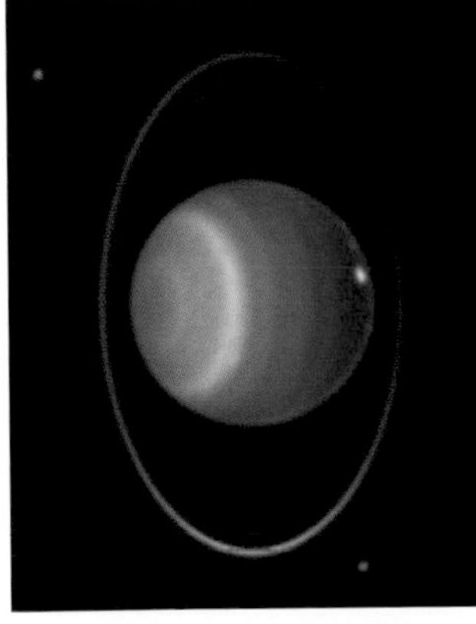

Left Hubble Space Telescope view of Uranus surrounded by its four major rings and by 10 of its 17 known satellites or moons. The Hubble Telescope has also found about 20 clouds.

Neptune

Planetary Order
8th planet from Sun

Average Distance from Sun
30.069 AU/2,795,084,800 miles/
4,498,252,900 km

1 Neptune Day
16.11 Earth hours

1 Neptune Year
164.79 Earth years

Equatorial radius (Earth = 1)
3.883

Mass (Earth = 1)
17.147

Effective Temperature
−353°F/−214°C

Atmosphere
Approx 80 percent hydrogen, 19 percent helium, trace of methane, hydrogen deuteride, ethane

Moons/satellites
13 moons—the largest of which is Triton
2 rings

Notable features
Emits more energy than it receives from the Sun.

Above right Statue of Neptune, the Roman god of the sea. In Greek mythology, Poseidon—Greek for the Latin *Neptunus*—was the god of the sea, as well as of horses, and—as "Earth-Shaker"—of earthquakes.

Right Satellite photograph of Neptune reconstructed from two images taken by *Voyager 2*. In the north, is the Great Dark Spot, accompanied by bright, white clouds. To the south is the bright feature that *Voyager* scientists have nicknamed "Scooter." Still farther south is a feature called "Dark Spot 2," which has a bright core.

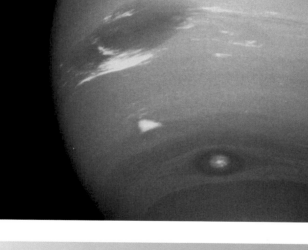

As NASA's *Voyager 2* space probe began to close in on Neptune in 1989, mission scientists were not expecting to find anything particularly startling. They were wrong.

The story of Neptune's discovery actually begins with Uranus, when observations of that planet's motion showed that it was not keeping to its predicted orbit. There was immediate suspicion that the gravity of another, more distant planet, was pulling on Uranus and thereby affecting its orbit—and eventually that planet was found. Neptune became the first planet to be predicted and discovered mathematically.

In the mid-1800s, Englishman John Couch Adams and Frenchman Urbain Le Verrier, both mathematicians and astronomers, independently calculated predictions for where the supposed eighth planet might be located. Initially they had difficulty convincing other astronomers to take the matter seriously, but finally, on September 23, 1846, Johann Gottfried Galle and Heinrich d'Arrest found the planet very near to the spot predicted. An international debate ensued over who should be credited with its discovery, but it was finally decided to jointly honor Adams and Le Verrier. Later, it became clear that the planet had previously been seen by others, including Galileo, but had not been recognized for what it was. The new planet was given the name Neptune, after the Roman god of the sea.

BLUE WORLD

Being so far from the Sun—30 times further than Earth—has made Neptune a difficult planet to study. Ground-based observations revealed little more than the planet's orbit, its size, and the rough composition of its atmosphere. One of its moons, Triton, was found only 17 days after the planet's discovery, but a second would not be discovered until almost 100 years had gone by.

It really wasn't until NASA's *Voyager 2* probe reached Neptune in 1989 that scientists began to unravel some of its mysteries. And much of what was found stunned the waiting world. Far from being a ball of bland inactive gas, the planet was found to have a very dynamic climate system including the fastest winds in the Solar System—up to 1,300 mph (2,100 km/h).

Its blue color comes from small amounts of methane in its hydrogen and helium atmosphere—the methane absorbs the red wavelengths of sunlight, leaving only the blue to be reflected. Presently unexplained is the reason for the darker blue of the large cloud systems; something other than methane absorption is thought to be at work in this area.

As with the other gas giants, what we see is not a solid surface—rather, we see the tops of continuous cloud cover. The temperature of those cloud tops is a very chilly – 346°F (–210°C). Scattered across the upper reaches of Neptune's atmosphere, *Voyager 2* found bands and tufts of

1612 Galileo's astronomical drawings show that he had first observed Neptune on two occasions, both times mistaking it for a fixed star, as the planet was just beginning its yearly retrograde cycle.

1613 In January, Galileo again produces drawings showing he has observed Neptune. However, once again he believes it to be a fixed star.

1821 Alexis Bouvard publishes astronomical tables of the orbit of Uranus. Subsequent observations reveal there are substantial deviations from the tables. This leads Bouvard to hypothesize that perhaps there is some other body or planet that is disturbing Uranus.

1843 John Couch Adams calculates the orbit of an eighth planet. He sends his calculations to Sir George Airy, the Astronomer Royal, for a clarification. Airy begins to draft a reply but never sends it.

1846 John Herschel begins to champion the mathematical approach to a new planet and persuades James Challis to search for it.

1846 After much procrastination, Challis begins his reluctant search in July.

1846 Urbain Le Verrier produces his own calculations that a planet must exist beyond Uranus, causing its unusual behavior.

1846 Using calculations of Le Verrier, Neptune is discovered by German astronomers Johann Gottfried Galle and Heinrich Louis d'Arrest on September 23. Adams and Le Verrier are jointly honored with the discovery. The planet does not get its current name for some time.

1846 Challis later realizes that he had observed the planet twice in August, failing to identify it owing to his casual approach to the work.

NEPTUNE CORE FACTS

Neptune is the smallest of the gas giants but has the greatest proportion of core per planet size—approximately the size of Earth's core. It is believed that the core comprises rock and ice. Above the core is a layer of liquid hydrogen compounds, oxygen, and nitrogen, and surrounding that is a liquid hydrogen layer, including helium and methane. The atmosphere of Neptune consists of mainly hydrogen, methane, and helium. The blue-green color is brighter than Uranus, so scientists believe that there is an unknown compound as well as methane that is intensifying the color.

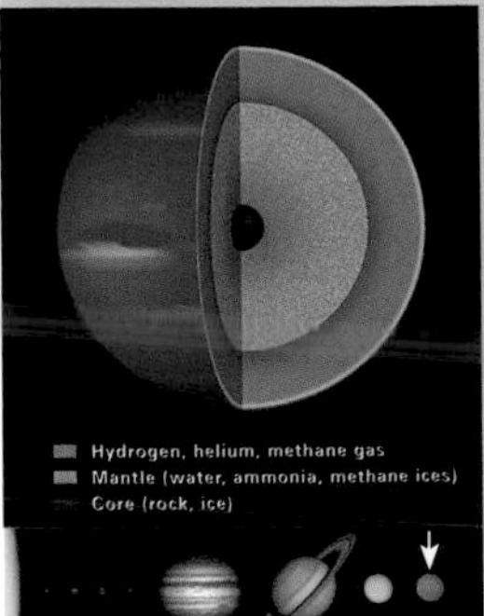

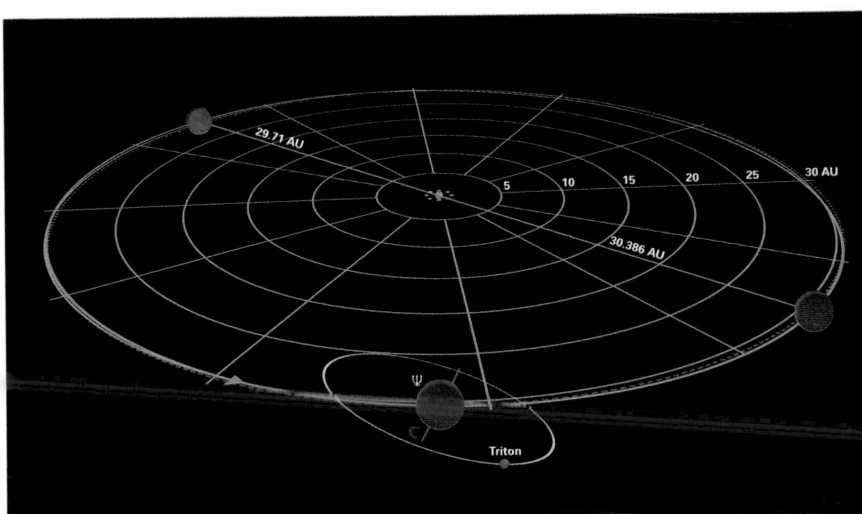

Above Neptune takes 165 Earth years to orbit the Sun with an inclination of 6.43° to the ecliptic. Its axial tilt is 28.32°. Neptune's rotation is prograde—counterclockwise around the Sun when viewed from above. The dash line indicates the degree of tilt of Neptune's orbit from the plane of the Solar System. (not to scale)

white clouds. Also seen was a fast-moving white cloud band nicknamed "Scooter." Most spectacular and unexpected, though, was the presence in the planet's Southern hemisphere of what was dubbed the Great Dark Spot, a huge cyclone similar to Jupiter's Great Red Spot.

Fifteen years later, however, Hubble Space Telescope observations showed that the Great Dark Spot had completely vanished. Taking over the role was a new, smaller, dark spot in the Northern hemisphere. It's uncertain whether the Great Dark Spot has actually dissipated, or whether it is being masked by some temporary affect of the atmosphere.

Neptune is around 4 times wider than Earth, and has more than 17 times Earth's mass. Scientists think it probably has a core of rock and ice—of around one Earth mass.

There is a middle region, or mantle, that is thought to be composed mainly of methane, ammonia, and water. The outer region is made of hydrogen, helium, and methane. In contrast to the frigid cloud tops, the great pressure in the planet's deep interior leads to a temperature of somewhere around 12,650°F (7,000°C).

Just like Jupiter and Saturn, Neptune emits more energy than it receives from the Sun. Neptune is still contracting, following its formation at the beginning of the Solar System around 4.5 billion years ago. It is this contraction that is providing the energy for the planet's wild climate.

An unusual and intriguing possibility is that the high pressures experienced in Neptune's interior could cause the carbon atoms to be split out of the methane molecules; those same pressures could then form diamonds from the carbon. It could be raining diamonds in the depths of Neptune's thick atmosphere.

Left *Voyager 2* image of Neptune's bright cloud streaks. The linear cloud forms stretch approximately along lines of constant latitude, and the sides of the clouds facing the Sun are brighter.

1846
Le Verrier—through the Board of Longitude—suggests Neptune as the name for the new planet. Leverrier Struve comes out in favor of that name on December 29, 1846, to the St Petersburg Academy of Sciences. From then on Neptune became the internationally accepted name.

1846
Only days after Neptune is discovered, amateur astronomer William Lassell thinks he sees a ring around the planet. It turns out to be a distortion caused by his telescope. (The rings we now know of were found by the *Voyager 2* flyby in 1989 and could not have been seen by Lassell.)

1846
Seventeen days after thinking he has discovered a ring, William Lassell discovers Neptune's largest moon, Triton.

1949
Dutch-American astronomer Gerard Kuiper (for whom the Kuiper Belt was named) finds Neptune's third-largest moon, Nereid.

1977
Voyager 2 is launched with a mission to flyby Neptune and other planets on its way to Interstellar space.

Mid 1980s
Evidence for incomplete arcs around Neptune first arise when stellar occultation experiments are found to occasionally show an extra "blink" just before or after the planet occulted the star.

1989
Voyager 2 does a flyby of Neptune on August 24 and 25, skimming the North pole by a mere 3,000 miles (4,800 km).

1989
Voyager 2 finds six new Neptunian moons and three new rings, plus a broad sheet of ring material.

2002–2003
Astronomers using improved ground-based telescopes found five more moons, bringing the known total to 13 moons.

COLD MOONS

Neptune has 13 moons, the largest of which is Triton. Triton is notable for several reasons—its icy surface is very cold and very reflective, and it has a retrograde orbit, that is, it goes around Neptune the "wrong" way.

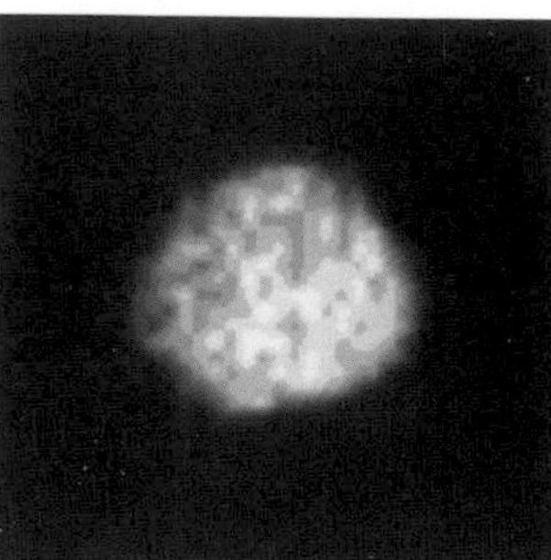

Right First filter images of Neptune's largest moon, Triton, taken by *Voyager 2* in 1989. The different shades reveal some of the surface topography.

Prior to *Voyager 2*'s flyby of Neptune, not much was known about Triton. Using an assumption about how reflective its surface was likely to be, astronomers calculated a diameter for it based on how bright it looked from the distance of Earth.

As *Voyager* got closer, however, it was apparent that Triton was far more reflective than previously believed—this meant the calculation had to be redone, and a smaller diameter figure was arrived at. Triton is, in fact, one of the shiniest bodies in the Solar System. It is also the coldest, at a very chilly – 391°F (– 235°C). *Voyager 2* revealed Triton to be a world with a geologically young surface, with few craters visible. It is thought to comprise around 25 percent water ice, with the remainder being mainly rocky material. It also has a very thin atmosphere of nitrogen.

Most amazingly of all, though, it has active volcanoes, or perhaps geysers, shooting what is thought to be liquid nitrogen high above the surface. Seasonal heating by the Sun is believed to be the energy source driving this very powerful volcanic activity.

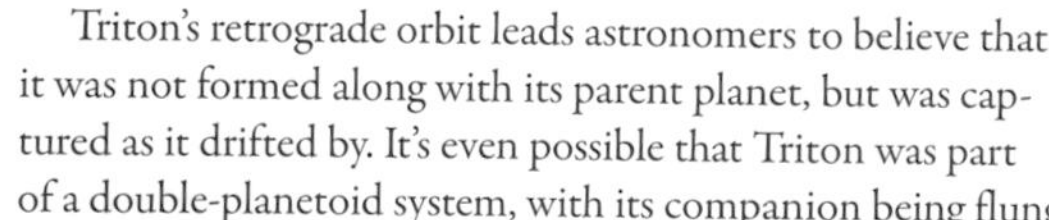

Triton's retrograde orbit leads astronomers to believe that it was not formed along with its parent planet, but was captured as it drifted by. It's even possible that Triton was part of a double-planetoid system, with its companion being flung away at the same time as Triton was captured. Neptune's third-largest moon, Nereid, was the second to be found, in 1949. It wasn't until *Voyager 2*'s encounter in 1989 that six more moons were also discovered. Another five were added by 2003, all found using Earth-based telescopes—three of these, like Triton, have retrograde orbits. Little is yet known about these small moons. A number of them are "irregulars," believed to have been captured rather than formed along with the planet.

RINGS OF DARKNESS

Like the Solar System's other giant planets, Neptune has a system of rings, though they are nowhere near as spectacular as those of Saturn. The first hint of their existence came from ground-based observations in the 1980s, when astronomers watched as a background star winked in and out as Neptune passed in front.

As well as the starlight being blocked when the main bulk of the planet got in the way, the background stars dipped slightly several times even when Neptune wasn't quite in front of them. The interpretation was that the stars could have been blocked by unseen rings. Not only that, the data seemed to indicate that perhaps the rings were broken and incomplete.

It took *Voyager 2*'s arrival at its destination in 1989 for confirmation of this theory to come. *Voyager*'s images revealed a series of rings. One of them, the thin Adams ring, indeed had several ring arcs, which were given the French names Liberté, Egalité, and Fraternité (Liberty, Equality, and Brotherhood).

Recent observations indicate that Neptune's ring system is not stable, and one or more of the rings may break up sometime in the next 100 years or so. The separate arcs in the Adams ring, along with regular ring "ripples," are thought to be kept in check by the gravitational tug of the tiny moon Galatea.

Neptune's rings are composed of a fairly dark material, and contain a much higher concentration of tiny dust particles than do Saturn's rings, for instance.

MAGNETIC FIELD

Neptune's magnetic field is tilted 47° with respect to the planet's axis of rotation, and offset from the planet's center by about half its radius. Neptune doesn't have an iron core like Earth; its magnetic field is probably generated by electrical currents in a water layer somewhere in the middle strata of the planet's atmosphere.

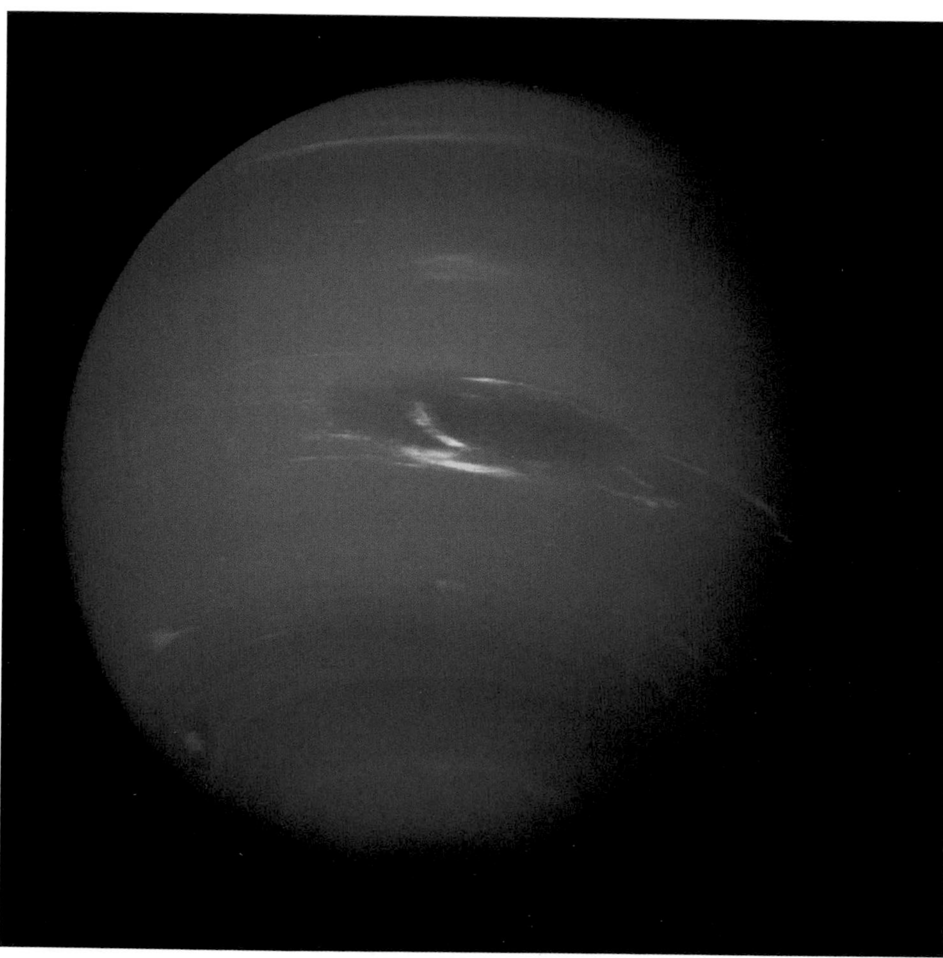

Below This *Voyager 2* image shows the Great Dark Spot and the cloud formations around it. These clouds were seen to persist for as long as *Voyager*'s cameras could resolve them. North of these, a bright cloud band similar to the south polar streak may be seen.

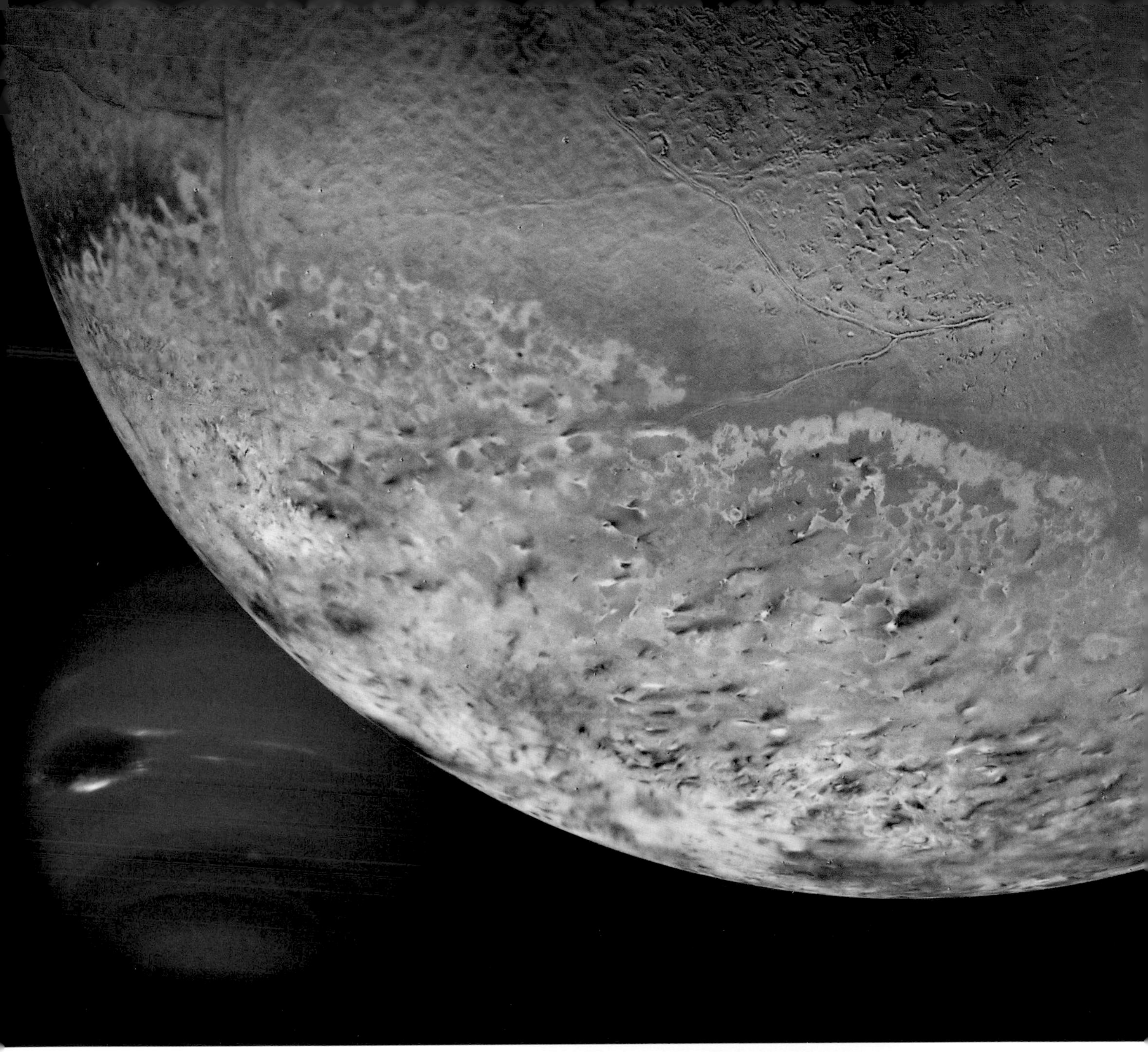

From the surface of the Earth, Neptune is too faint to be seen with the unaided eye. Powerful binoculars or a telescope, plus a chart showing the planet's exact position, are needed to reveal its small bluish-green disk.

Its great distance from the Sun means it has a very slow orbital speed—it takes around 165 Earth years to complete one orbit—and therefore it moves only very slowly through the night sky.

Neptune is so far from the Sun that the sunlight is about 900 times dimmer than it is on Earth.

With Pluto's status now changed to that of "dwarf planet," Neptune has become the most distant "full" planet in our Solar System.

But even without Pluto's recent demotion, Neptune's near-circular orbit—and Pluto's elliptical one—means that for a small part of its orbit, Pluto is actually nearer to the Sun than Neptune, making Neptune on those occasions the farthest planet anyway.

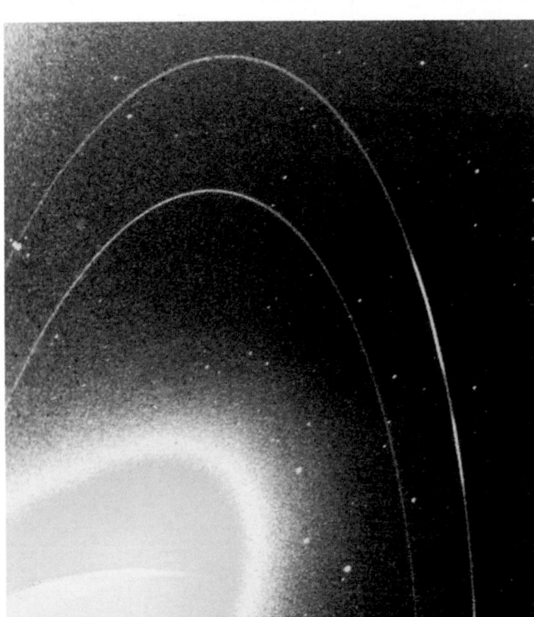

Above Digital composite of Neptune and its moon, Triton. Triton is an icy world colder than Neptune and it is the only large satellite in the Solar System to circle a planet in a retrograde direction—opposite to the rotation of the planet.

Left Wide-angle view taken with a clear filter from spacecraft *Voyager 2* of Neptune's two known rings. Although similar to Saturn's rings, these contain smaller particles of dust.

Dwarf Planets

FACT FILE

Pluto

Discovery
Early 1930, by Clyde Tombaugh

Average Distance from Sun
39.482 AU/ 3,670,050,000 miles/ 5,906,380,000 km

1 Pluto Day
–153.3 Earth hours (retrograde)

1 Pluto Year
247.92 Earth years

Equatorial radius (Earth = 1)
0.180

Mass (Earth = 1)
0.0022

Average Surface Temperature
-387/-369°F (-233/-223°C)

Atmosphere
Nitrogen, methane, carbon dioxide

Moons/satellites
Three moons—Charon, Hydra, Nix.

Notable features
Made of mostly ice with a possible rocky core.

Ceres

Discovery
January 1801, by Giuseppe Piazzi

Average Distance from Sun
2.77 AU/257 million miles/ 414 million km

Equatorial diameter
590 miles (950 km)

Notable features
Largest and most massive known body in the Asteroid Belt.

Eris

Discovery
July 2005, by Mike Brown

Average Distance from Sun
67.7 AU/6.3 billion miles/ 10 billion km

Equatorial diameter
1,500 miles (2,400 km)

Notable features
Most distant object ever seen in orbit around the Sun.

When is a planet not a true planet? It's a question that has exercised the minds of astronomers for years, and in 2006 caused worldwide consternation as astronomers tried to settle the question once and for all.

WHAT IS A PLANET?

It wasn't too many years ago that the question didn't even need to be asked. Everyone knew what a planet was—a large body circling the Sun. And there were nine of them in our Solar System—the four terrestrial worlds—Mercury, Venus, Earth, and Mars, the four "gas giants"—Jupiter, Saturn, Uranus, and Neptune, and finally the oddball—little Pluto, discovered in 1930. As well, the Solar System also had an asteroid belt between Mars and Jupiter (plus some other asteroids scattered on different orbits here and there), plus a vast cloud of comets far, far from the Sun, in a region known as the Oort Cloud. Simple. But Pluto was always considered a problem. It was the ugly duckling of the Solar System; much smaller than the other planets, made of ice instead of rock or gas, and with an orbit that was both highly elliptical, and highly inclined to the orbits of all the other planets. If it formed along with the other planets, how did it come to be in such a different orbit? Or, if it didn't form with the others, where did it come from? Was it a wandering body or had it been captured into a strange orbit around the Sun? And if so, could there be others out there? Pluto had too many unanswered questions.

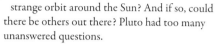

Things began to become a little clearer in recent years with the discovery of more largish icy bodies orbiting the Sun beyond Neptune. One of them was even found to be larger than Pluto itself. So, if Pluto was to be considered a planet, why not some of these others too? Our Solar System might actually have 10, or 15, or 40 planets, instead of nine. Or, if that was too much to take, was it time to change Pluto's status (and its newfound cousins) to something other than a planet?

The crunch came in 2006, when astronomers could no longer put off making a decision. The problem had become glaringly obvious. Pluto was obviously just one of a large number of sizeable icy worlds that form a distinct population of their own in the region out beyond Neptune. A decision would have to be made: Are they all planets, or none of them?

At a major meeting of the International Astronomical Union, the world's peak body for astronomers, a resolution was put which its sponsors hoped would solve the question once and for all. After much discussion, and unexpected worldwide popular interest, a final decision was reached. And, after all the dust was settled, most agreed that it was a reasonable compromise. It went like this. The Solar System has eight "planets"—from Mercury out to Neptune—plus an unknown number of "dwarf planets," which was the new part of the definition into which Pluto and some other bodies would fall.

Above right Bronze statue of Dis Pater, or Pluto, god of the underworld, holding a mallet and goblet, from the second century.

Left The Hubble Space Telescope floats against the background of Earth after a week of repair and upgrade by Space Shuttle Columbia astronauts in 2002. It was Hubble's fourth servicing mission.

Pluto

- **1930** Pluto is discovered by American astronomer Clyde Tombaugh.
- **1978** Charon is discovered by astronomer, James Christy, on June 22, 1978. While examining the highly magnified images of Pluto taken a couple of months before, Christy notices a slight bulge that is confirmed on plates dating back to April 29, 1965.
- **1979–1999** Pluto's highly elliptical orbit brings it closer to the Sun than Neptune, providing rare opportunities to study this distant world and its companion moon, Charon.
- **1985–1989** Pluto makes its most recent close approach to the Sun and during this time Pluto and Charon begin a series of eclipses.
- **1992** Pluto's atmosphere is discovered to be nitrogen and carbon dioxide. Pluto is considered the most prominent member of the Kuiper Belt, the band of icy rocky objects and dwarf planets that orbit the Sun in the outer region of our Solar System beyond the orbit of Pluto.
- **2006** Two additional moons, Hydra and Nix, are discovered from Hubble Space Telescope images. Hubble also records distant images of Pluto and Charon.
- **2006** *New Horizons* spacecraft is launched on January 19, 2006. Its mission is to visit Pluto and Charon. It will arrive in 2015. However, it cannot stop, so its visit will be as a flyby.
- **2006** On August 24, 2006, the International Astronomical Union (IAU) formally downgrades Pluto from an official planet to a dwarf planet.

PLUTO CORE FACTS

Listed as a dwarf planet, Pluto's structure is still relatively unknown. Spectroscopic observation from Earth shows that it has a rocky core surrounded by a mantle of mostly nitrogen ice, with a little methane and carbon dioxide. Due to its frozen conditions, it has very little atmosphere. Scientists do know that small amounts of nitrogen, methane, and carbon dioxide in the thin atmosphere turn from a solid to gas, without the liquid stage in between, as Pluto moves closer to the Sun. On its long orbit away, this thin atmosphere falls back to the surface and refreezes.

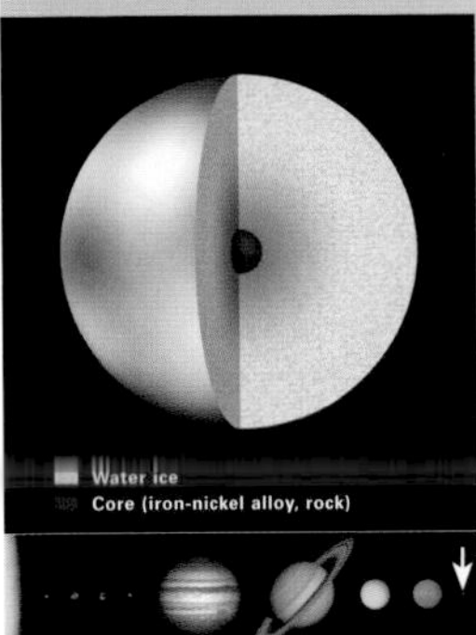

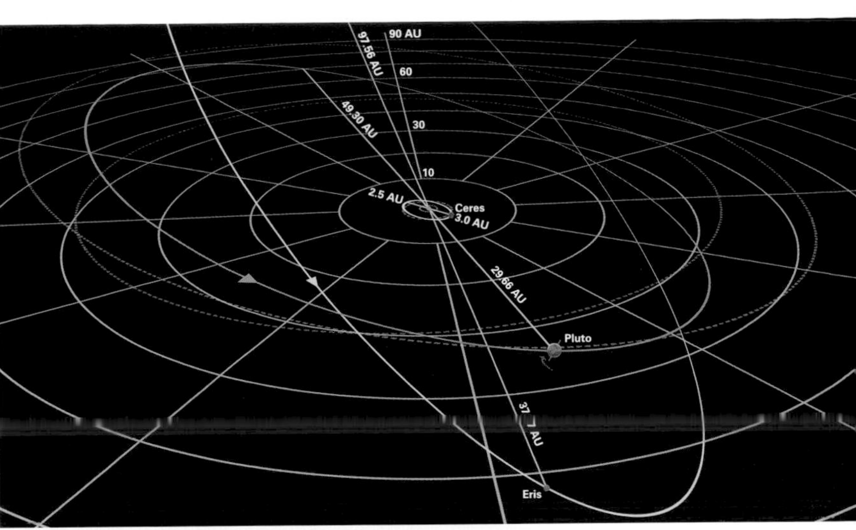

Above Pluto takes almost 249 Earth years to orbit the Sun with an inclination of 11.88° to the ecliptic. Its axial tilt is around 120°. Pluto's rotation is prograde—counterclockwise around the Sun when viewed from above. (not to scale)

Left Computer graphics simulation of a proposed fast flyby of Pluto and its moon Charon in 1991. This mission was proposed but never eventuated as the cost and distance was too great. Since then *New Horizons* has been sent on its way, specifically to investigate Pluto.

Ceres

1801
Ceres discovered on January 1, 1801, by Giuseppe Piazzi, and designated as a planet.

1900s
Ceres is reclassified as an asteroid for over 150 years, as further similar objects were discovered in the area. As the first such body to be discovered, it was given the designation 1 Ceres under the modern system of asteroid numbering.

2005
Hubble images indicate that Ceres may be a planet rather than an asteroid.

2006
After the discovery of another trans-Neptunian object, Eris, a proposition is made by the International Astronomical Union to reinstate Ceres to the status of planet along with Pluto's moon, Charon, and Eris.

2006
Instead, on August 24, 2006, an alternate proposal from the IAU comes into effect labeling Ceres a "dwarf planet." It is not yet clear whether dwarf planet status is, like planet status, a sole defining category, or whether dwarf planets also retain their previous minor body classifications such as "asteroid."

Eris (UB313)

2005
Astronomer Mike Brown, of CalTech, and his team, announces the discovery of yet another Kuiper Belt Object - this one larger than Pluto. This object, provisionally named UB313, or Xena, is officially named Eris by the IAU.

2006
The proposal to accept Eris as a planet, along with Pluto and its moons, leads to the labeling of Eris as a dwarf planet, according to the definition of the IAU. Moon orbiting Eris, discovered in late 2005 by Mike Brown and his team, is officially named Dysnomia by the IAU.

In addition, the Solar System also contains an untold number of "Small Solar System Bodies," such as comets and asteroids.

The definition of a planet became: A body that orbits the Sun on its own (i.e. is not a satellite, or moon of another body), is big enough so that its own gravity has pulled it into a spherical shape, and it has swept away all other contending bodies in its orbital region. Pluto would no longer fit into this definition, as it hasn't cleared its region of similar bodies.

The definition of a dwarf planet became: A body that orbits the Sun, has pulled itself into a round shape, is not a satellite of another body, but which has not cleared the rest of its neighborhood of contending bodies.

Pluto does meet this definition, as does the largest of the asteroids, Ceres, and the recently discovered Pluto-like body, Eris. They were thus reclassified.

Below Digital illustration of Pluto in space. The planet is actually comprised of ice and rock and is the tenth largest body observed directly orbiting the Sun.

Right This view by Hubble Space Telescope confirms the presence of two new moons around the distant planet Pluto. The moons were first discovered by Hubble in May 2005, but the science team probed even deeper into the Pluto system to look for additional satellites and to characterize the orbits of the moons.

THE UGLY DUCKLING

Like Neptune, Pluto's discovery was the direct result of a mathematical prediction—although the prediction actually turned out to be wrong in the end. Astronomers had noticed that Uranus still didn't seem to be keeping to its predicted orbit, even taking Neptune into account. Suspicion grew that another, as-yet-undiscovered body further out from the Sun was giving an extra gravitational tug. If this was right, it should be possible to mathematically predict where it could be found. The race was on again.

At Lowell Observatory in the USA, astronomer Clyde Tombaugh spent many long hours taking, and then examining, photographic plates of the target region of the sky, in the hope of spotting the missing planet. By comparing two plates taken some time apart, any movement of a planet in relation to the fixed background stars would become apparent. On February 18, 1930, he spotted a tiny dot that had moved in the time between taking two plates, made a week apart in January. He had found it!

It wasn't long before it was realized, however, that Pluto was much too small to have had the predicted effect upon Uranus's orbit. That it was found in the place predicted for the extra planet turned out to be just a complete fluke. And it was later determined that it was inaccuracies in the understanding of Neptune's mass that was throwing out the predictions for Uranus' orbit, and there was no need for a further large planet in the dark realms beyond.

Pluto is a world made largely of ice, with possibly a rocky core. The surface is covered mostly with nitrogen ice and a smattering of methane and carbon dioxide. The surface temperature is very cold—around −382°F (−230°C.) Given its small size, frozen conditions, and great distance from the Sun, it is not unexpected that Pluto does not have a substantial atmosphere. But it does have some. Small amounts of nitrogen, methane, and carbon dioxide sublimate (i.e. go straight from a solid to gas without the liquid stage in between), during the periods when the planet's elliptical orbit brings it closer to the Sun. During the long period when it is further away, this thin atmosphere falls back to the surface and refreezes. The *New Horizons* spacecraft hopes to get there before this occurs.

Pluto has long been known to have a large moon, Charon—large, that is, in comparison to Pluto itself. In fact, the ratio of the size of the two bodies has led many astronomers to consider

the duo to be a "binary planet" rather than a normal parent planet with a satellite. Add to the size factor the fact that the barycenter (the center of gravity around which the two planets mutually revolve) is located between them—and not, as in the case of Earth/Moon, actually located within Earth's radius—and the reasoning can be seen to be compelling.

Charon was first spotted in 1978, but it wasn't until 2005 that two more very small Plutonian moons were discovered—Nix and Hydra—using the power of the Hubble Space Telescope. Nix is named for the Greek goddess who was the mother of Charon; Hydra was the many-headed monster that guarded the underworld ruled by Pluto.

Both moons are estimated to be somewhere between 28 and 77 miles (45 and 125 km) in diameter, with Hydra believed to be slightly larger than its sibling.

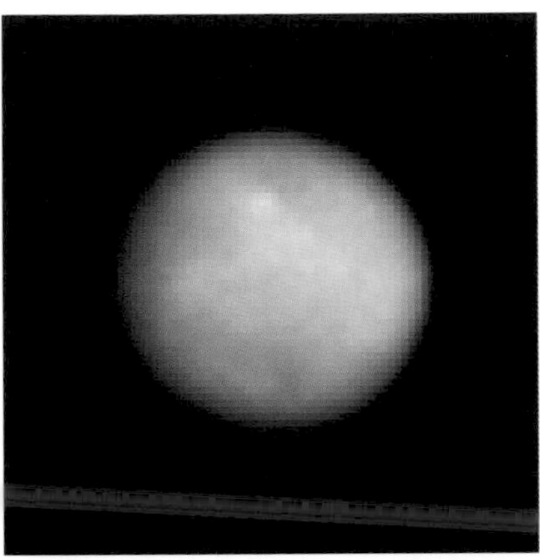

Left Hubble Space Telescope color image of Ceres, the largest object in the asteroid belt. Ceres's round shape suggests that its interior is layered like those of terrestrial planets such as Earth. Ceres is approximately 590 miles (950 km) across and was discovered in 1801.

Below This digital illustration shows the Pluto system from the surface of a possible moon. The other members of the Pluto system are just above the moon's surface. Pluto is the large disk at center right. Charon is the smaller disk to the right of Pluto. Another possible moon is the bright dot on Pluto's left.

Right Artist's impression shows the size of Ceres and Vesta, compared with three other asteroids and Mars which is the large planet at center bottom. Ceres and Vesta reside in the Asteroid Belt, located between Mars and Jupiter.

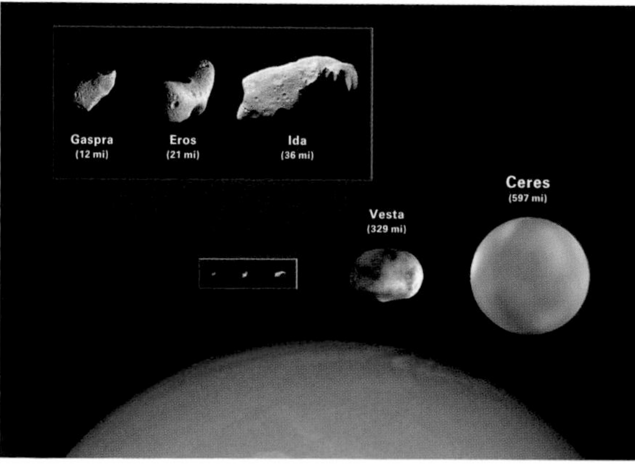

FIRST AMONG EQUALS

Ceres was the first asteroid to be found, in 1801, in the orbital "gap" between Mars and Jupiter. Many astronomers had long wondered about this apparent gap, and speculated that a planet remained to be discovered there. When Ceres was found it was hailed as just that long-lost planet.

Initially its discoverer, Giuseppe Piazzi, thought he had found a comet. But he soon had his suspicions about its true nature, and other observations soon confirmed that its orbit was not comet-like. And so Ceres was hailed by one and all as the "missing" planet. That is, until another body (Pallas) was found in the same gap just over one year later. And then another (Juno) in 1804, and another (Vesta) in 1807.

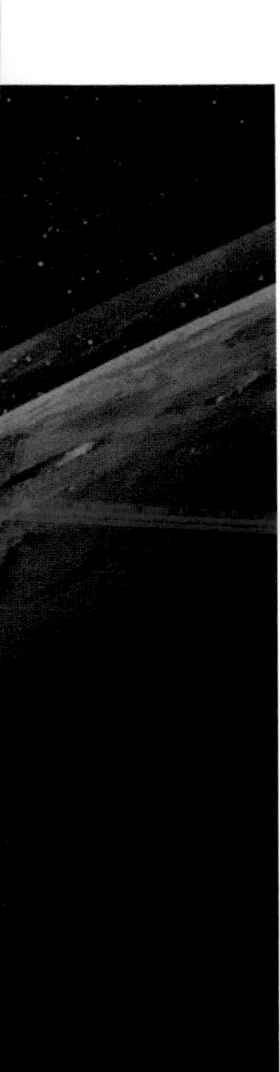

NEW HORIZONS

The main difficulty in studying these dwarf worlds is their great distance from Earth. Even through the very largest and most advanced telescopes, they are little more than tiny disks of light. What's really needed is a good, up-close look, like we've had with all of the main planets.

In the case of two of the dwarfs, scientists are well on their way to getting that up-close look through the medium of planetary space probes. But don't hold your breath—it'll be some years yet before these probes reach their targets.

First up will be Pluto, which is the main target of NASA's *New Horizons* spacecraft. Launched on January 19, 2006, *New Horizons* is on a trajectory that has taken it past Jupiter (for a gravity-assist maneuver) and will see it reach Pluto in July 2015. Unable to stop at the tiny planet, it will conduct detailed observations during a span of a few days—including several hours of closest approach—of both Pluto and its moons. After that, there is the potential to target the craft to one or more Kuiper Belt Objects.

Ceres is the target of another NASA mission, called *Dawn*. This sophisticated spacecraft will first visit the asteroid Vesta before going on to encounter Ceres. Unlike *New Horizons*, which can't stop, *Dawn* will spend time orbiting both Vesta and Ceres, giving us our first detailed look at these representatives of the rocky rubble left over from the Solar System's formation.

Artist's impression of *New Horizons* spacecraft during a planned encounter with Pluto and its moon, Charon. The spacecraft's most prominent design feature is a nearly 8-foot (2.1-m) dish antenna, which will communicate with Earth from as far as 4.7 billion miles (7.5 billion km) away.

It quickly became apparent that there was a population of small worlds orbiting the Sun, at around this general distance. In 1802, the famous English astronomer, Sir William Herschel had coined the term asteroid, meaning "star-like," as seen through even the largest telescope, they look just like stars. And so the region became known as the Asteroid Belt, and Ceres was demoted from planetary status down to lowly asteroid. It is now known that thousands of asteroids reside in this region of space.

Ceres, now newly reclassified as a dwarf planet, is roughly spherical in shape, having a diameter of around 590 miles (950 km). Its mass has allowed it to achieve its round shape, fitting it neatly into the category of dwarfs. In fact, it is the largest of all the bodies in the asteroid belt, comprising around one-third of all the mass of the asteroids combined. Some scientists think Ceres has a "differentiated" interior, meaning that it has separated into layers, with a rocky core surrounding by an icy mantle and crust. Its surface temperature is around −40°F/C. The icy mantle could be as much as 75 miles (120 km) thick and could contain more water than all the oceans on Earth.

Above *New Horizons* spacecraft roars off the launch pad aboard an Atlas V rocket spewing flames and smoke. Liftoff was from Cape Canaveral, Florida, USA, on January 19, 2006. It was the third attempt after bad weather.

THE NEW PLUTO

The Solar System's third dwarf planet—and largest so far—is Eris, which, at 1,491 miles (2,400 km) in diameter is slightly larger than Pluto. Initially hailed as the tenth planet in the Solar System, it was the discovery of Eris that put the cat among the pigeons in terms of the definition of planets, leading to the dwarf planet decision in 2006.

Eris was discovered in January 2005, on images taken back in October 2003. The announcement was made in July 2005. At the time of discovery, Eris was the furthest known Solar System object from the Sun, a whopping 97 times further than Earth, although some other trans-Neptunian objects (TNOs) have orbits that take them further from the Sun. Eris takes around 556 Earth years to complete each highly elliptical, highly inclined orbit.

Like some other trans-Neptunian objects, Eris belongs to a sub-category known as the scattered disk population. These bodies are believed to have their orbits disturbed from close encounters with Neptune, throwing them out of the inner region of the Kuiper Belt, and onto their elliptical orbits. Eris's orbit is inclined at an angle of 44° to the ecliptic plane, which places it well out of the neighborhood of most Solar System bodies, in a location not well studied until recently. This is why it took so long to find it, even though it is larger than Pluto.

Upon discovery, it was given the designation UB313, but it quickly became known as "Xena," a nickname bestowed upon it by its discoverers. This name was completely unofficial, though, and Eris was proposed and approved by the International Astronomical Union committee that assigns names to Solar System bodies. Eris was a Greek goddess of discord and conflict. Eris also has a small moon, named Dysnomia, discovered in September 2005 in images taken with the Keck telescope in Hawaii, USA.

Above Artist's impression of the dwarf planet, formerly known as 2003 UB313, now called Eris, in honor of the Greek goddess of discord and strife.

At present, there are 11 other bodies being seriously considered for classification under the dwarf planet mantle given their size and mass—but dozens more could potentially be added as extra data on them comes to light.

Its quite possible that bodies larger than both Pluto and Eris could be discovered in the coming years, as telescope technologies improve enough, and spacecraft are able to travel far enough, to give astronomers better coverage of the depths of space beyond the realm of the Solar System's major planets.

Other Solar System Bodies

FACT FILE

Kuiper Belt

Average Distance from Sun
About 50 AU/4.6 billion miles/ 7.4 billion km

Notable features
Contains thousands of small icy bodies including Quaoar

Oort Cloud

Average Distance from Sun
Between 5,000 and 10,000 AU/464 billion miles/ 748 billion km

Notable features
Contains billions of icy bodies in solar orbit including Sedna

Far beyond Neptune lives a host of small, mysterious, icy worlds, virtually unmapped and undiscovered, showing conclusively that our Solar System doesn't stop at Pluto.

THE OUTER LIMITS

The Solar System just keeps getting bigger and bigger. No longer do we have just the traditional nine planets, a bunch of asteroids, and an unknown number of comets. Now we have eight planets, three dwarf planets, and sundry thousands of small Solar System bodies.

Many of the latter class live in the dark, cold, distant realm beyond the orbit of Neptune. These Trans Neptunian Objects, or TNOs, have been split into several categories depending on their distance.

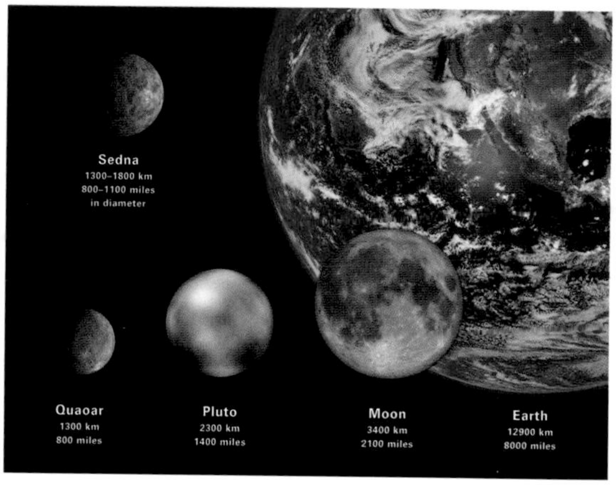

Right Artist's rendition shows Sedna in relation to other bodies in the Solar System, including Earth and its Moon, Pluto, and Quaoar.

First up is the Kuiper Belt, a population of icy bodies that circle the Sun not far beyond Neptune's orbit. Pluto and Eris, both of which also fall into the new dwarf planet category, are considered TNOs because of their location in space. The currently accepted cut-off point for the Kuiper Belt is around 50 astronomical units (AU—the distance between Earth and Sun) from the Sun. Then there is the "scattered disk," a region that extends from the Kuiper Belt much farther out into space. The bodies in this region typically have been "flung" onto highly elliptical and highly inclined orbits through gravitational interactions with the outer planets, mostly Neptune.

Above Artist's concept of Quaoar orbiting our Sun far beyond Pluto. Quaoar is one of the largest objects discovered in our Solar System since Pluto in 1930.

Finally, much further out, is the Oort Cloud, from where most comets are believed to originate. The Oort Cloud is thought to completely enclose the entire Solar System in every direction and contains billions of comets.

Some of the bodies in the Kuiper Belt and scattered disk are potential dwarf planet candidates. One of them is Quaoar, a body believed to be between 620 and 870 miles (1,000 and 1,400 km) in diameter. Discovered in 2002, it has a circular orbit around the Sun, 43 times further from it than Earth. Like many TNOs, Quaoar is believed to be made of a mixture of ice and rock; spectral studies of its surface indicate that it may have undergone some slight heating in the past, with the suggestion also being made that there could be some internal heat due to radioactive decay in the rocks.

Intriguingly, in February 2007, a tiny satellite body was found to be associated with Quaoar. It is probably only around 60 miles (100 km) across.

Another candidate is Sedna, discovered in 2003. It is estimated to be somewhere between 750 and 1,120 miles (1,200 and 1,800 km) in diameter. Sedna was quite an amazing discovery, as it has a highly elliptical orbit that takes it far away from the Sun—up to 975 astronomical units (AU). At the time of discovery, it was thought to be about 90 AU from the Sun; Eris, discovered a little over a year later, was around 97 AU at the time of discovery.

For Quaoar, Sedna, and others to be confirmed as dwarf planets, will require more observations of their characteristics. Already they meet the criteria of being in independent orbits around the Sun, not being satellites of other bodies, and not clearing the rest of their neighborhood of bodies like a major planet has. All that is left is to determine whether they are big enough (in both mass and diameter) to have pulled themselves into a spherical shape. More observations with both ground- and space-based telescopes will no doubt reveal the answers to these many questions.

Other TNOs include Varuna, Orcus, and Ixion. Varuna is what is called a "classical" Kuiper Belt Object, as it has a near-circular orbit that keeps it within the main span of the belt. Varuna, with a diameter estimated to be about 590 miles

1932
Ernst Öpik, an Estonian astronomer, proposes that comets originate in an orbiting cloud situated at the outermost edge of the Solar System.

1950
Dutch astronomer Jan Oort hypothesizes that comets came from a vast shell of icy bodies about 50,000 times farther from the Sun than Earth. This giant swarm of objects, containing billions of icy bodies is named after its discoverer—the Oort Cloud.

1951
Gerard Kuiper proposes the theory of a "belt of comets" in a region beyond Neptune.

1980s
Kuiper's hypothesis of a belt of comets is reinforced when computer simulations of the Solar System's formation predicts that a disk of debris should naturally form around the edge of the Solar System. This belt of debris now takes it name from Kuiper.

1980
Quaoar is photographed in 1980 but is not recognized as a Kuiper Belt Object.

1992
Astronomers detect a reddish speck about 42 AU from the Sun—the first time a Kuiper Belt Object (KBO) has been sighted.

1992
A 150-mile (241-km) wide body is detected, called 1992QB1 at the distance of the suspected Kuiper belt. Several similar-sized objects are discovered, quickly confirming the Kuiper belt exists.

2002
Quaoar, an icy world beyond Pluto, is recognized by Michael E. Brown and Chad Trujillo, as an object dwelling within the Kuiper Belt.

2003
Sedna is discovered by Michael E. Brown, Chad Trujillo, and D. Rabinowitz. It is well out beyond Pluto and within the Kuiper Belt.

2005
FY9, code named "Easterbunny," a very large Kuiper Belt Object, is discovered by the team led by Mike Brown at Caltech. Announcement is made on the same day as the discovery of Eris.

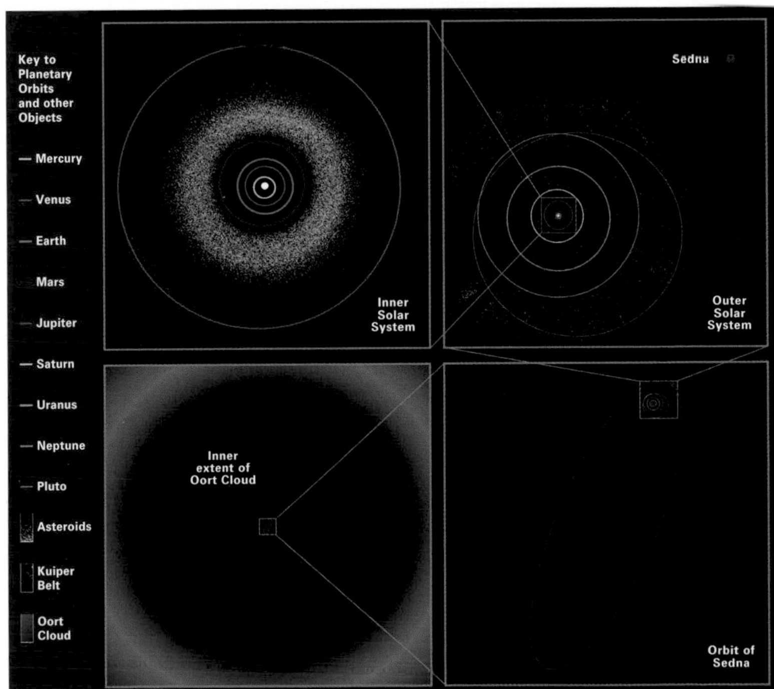

Above Rendering of the farthest object in our Solar System, Sedna, a mysterious planet-like body three times farther from Earth than Pluto.

Right Illustration shows the location of Sedna, which lies in the farthest reaches of our Solar System. Each panel, moving clockwise from the upper left, successively zooms out to place Sedna in context.

(950 km), has potential dwarf planet status. Orcus could be as large as 995 miles (1,600 km) in diameter, again, giving it potential dwarf planet status. Ixion is thought to be no larger than 530 miles (850 km) in diameter. Orcus and Ixion are also examples of a class of bodies called "plutinos" that includes Pluto and its three moons. Plutinos are objects that are locked in a 2:3 orbital resonance with Neptune—that is, for every three orbits of Neptune, they complete two orbits.

Neptune's large moon, Triton, is considered by many astronomers to probably be a Kuiper Belt Object that was captured into orbit. Its compositional resemblance to Pluto and other TNOs, together with its retrograde orbit around Neptune, support this idea. Pluto's large moon Charon, too, is often considered more of a true Kuiper Belt Object, hence the oft-stated position that Pluto and Charon are a binary planet system, rather than a planet and subservient moon.

Asteroids

FACT FILE

Spectral Classification

CARBONACEOUS OR C-TYPE
More than 75 percent of asteroids.

SILICACEOUS OR S-TYPE ASTEROIDS
Approximately 17 percent of known asteroids.

METALLIC OR M-TYPE ASTEROIDS
Mid-region main belt objects that are relatively bright.

Orbital Classification

MAIN BELT
Most minor planets orbit the Sun in the main asteroid belt.

NEAR-EARTH ASTEROIDS (NEAS)
Orbit within 1.3 AU of the Sun.

TROJANS
Located near Jupiter's Lagrange points (60° ahead and behind Jupiter in its orbit).

CENTAURS
Orbit in the outer Solar System.

They might be leftover bits from the Solar System's younger days, but asteroids have much to teach us about interplanetary history. They may even help us build in space.

ROCK STARS

Asteroids are bodies made of rocky and metallic substances, or either, from tens of feet in diameter up to hundreds of miles. Most of them orbit the Sun between Mars and Jupiter in a region known as the Asteroid Belt, although some have orbits that take them further out or close in toward the Sun. Certain asteroids have orbits that cross that of Earth, making them potential Earth-colliders.

At the end of the eighteenth century, many astronomers were hunting for a supposed "missing" planet in the "gap" between Mars and Jupiter. Despite the attention this region of space was given by those involved in the search, the first asteroid was actually discovered by chance in 1801 by Italian astronomer Giuseppe Piazzi.

Named Ceres, it was initially hailed as that missing planet, and given planetary status. But with the discovery of Pallas, Juno, and Vesta, between 1802 and 1807, it became clear that these bodies were in a class of smaller objects that were subsequently classified as "asteroids." The word comes from the Greek, meaning "star-like." For many years the scientifically accepted name had been "minor planet," but with the 2006 decision of the International Astronomical Union on planetary classifications, asteroids now fall within the overall umbrella of "small Solar System bodies."

Ceres, being large enough to have formed into an essentially spherical shape, and with a diameter over 559 miles (900 km), is now classified as a dwarf planet, under the new IAU convention.

More asteroids were found from 1830 onwards and today there are almost 400,000 cataloged; around 14,000 of them have been given names.

Unlike comets, asteroids are not named after their discoverer(s). Rather, an asteroid is initially given a catalog number and, once its orbit has been confirmed by follow-up observations, the discoverer is able to propose a name for the body. This is submitted to a committee of the International Astronomical Union, which accepts or rejects the name. Asteroid names must follow certain rules and standards, such as not being offensive and not being associated in any way with recent political or military leaders.

WHERE DO THEY COME FROM?

For many years, debate raged over whether the bodies in the Asteroid Belt were the fragments of a planet that broke up, or were fragments that didn't get the chance to combine into a full planet. These days the consensus is that the latter option is the correct one. As the Solar System formed from a rotating and contracting

Left Digital impression of an asteroid with the Sun rising behind it. The term asteroid is used to indicate a diverse group of small rocky bodies, probably numbering millions, that drift in the Solar System in orbit around the Sun.

4.6 million years ago Rocky fragments left over from the formation of the Solar System become asteroids.

1801 The first asteroid, 1 Ceres, is discovered by Giuseppe Piazzi, director of the Observatory of Palermo in Sicily. Piazzi names it after Ceres, the Roman goddess of agriculture.

1802 William Herschel first uses the word "asteroid" to describe these celestial bodies. The word comes from the Greek "star-like."

1802 The second and largest asteroid—2 Pallas—is discovered by H. Olbers.

1804 The third asteroid is discovered by K. Harding and named 3 Juno.

1807 Asteroid 4 Vesta is discovered. After eight more years of searching, most astronomers assume that there are no more asteroids to be found, and abandon any further searches.

1891 Max Wolf pioneers the use of astrophotography to detect asteroids. They appear as short streaks on long-exposure plates.

1930s With better technology, a number of new asteroids are discovered. These include 1221 Amor, 1862 Apollo, 2101 Adonis, and 69230 Hermes.

1991 *Galileo* spacecraft is first to observe an asteroid close-up, flying by main-belt asteroid Gaspra on October 29. Shaped rather like a potato, it is riddled with craters and fractures.

cloud of gas, small bodies collided and stuck together to form progressively larger bodies, eventuating in the planets. In some parts of the Solar System, however, the bodies were moving with too much relative velocity to stick together, and tended to smash into one another and fragment further. Asteroids can be thought of as the leftover rubble from the formation of the Solar System. If you added all the belt asteroids together you would end up with as much mass as Earth's Moon. It is thought that there were many more asteroids in the Belt, but that a lot of them have been flung out of their original orbits by the gravitational influence of Jupiter.

DIFFERENT TYPES

Many asteroids are classified into different groups and "families" according to the types of orbits they share—sometimes based on a common origin—and others according to their chemical compositions.

The majority of asteroids are of the C, or carbonaceous type, being heavily endowed with carbon. They tend to be darker with a reddish appearance, and somewhat resemble the type of meteorite known as a carbonaceous chondrite.

A second kind of asteroid is the S or silicate type which make up around 17 percent of the total. Scientists believe

Above Artist's impression of a massive asteroid belt seen by Spitzer Space Telescope in orbit around a star like our Sun. A collision between two asteroids is depicted to the right. Collisions like this replenish dust in the belt, making it detectable to telescopes like Spitzer.

1993
Galileo spacecraft observes Ida, a main-belt asteroid and discovers it has its own moon, Dactyl, in orbit around the asteroid. The tiny body may be a fragment from past collisions.

1995
Hubble Space Telescope maps Vesta, one of the largest asteroids, detailing its basaltic crust. It finds an enormous crater formed a billion years ago.

1996
Asteroid/Comet 95P/Chiron is observed in a chaotic eccentric orbit near Saturn and Uranus.

1997
On the way to its primary mission to Eros, NASA's Near-Earth Asteroid Rendezvous (NEAR) performs a 25-minute flyby of the asteroid 253 Mathilde on June 27. During the encounter, the spacecraft photographs most of the minor planet.

1998
NEAR spacecraft first flies past Eros observing about 60 percent of the asteroid. The minor planet was smaller than expected. NEAR also found that the asteroid has two medium-sized craters, a long surface ridge, and a density similar to the composition of Earth's crust.

1999
Deep Space 1 passes by the near-Earth asteroid 9669 Braille at a range of only 16 miles (26 km). It is the closest asteroid flyby to date.

2000
NEAR mission catches up with asteroid Eros and maintains an orbit for one year.

2000
On March 14, 2000, a month after entering asteroid orbit, NASA renames the NEAR spacecraft NEAR-*Shoemaker* in honor of renowned geologist Eugene Shoemaker. killed in a road accident in 1997, in Australia.

2001
NEAR mission controllers guide the spacecraft to the first-ever landing on an asteroid, Eros. It returns data from the surface for two weeks, including pictures of its rocky surface.

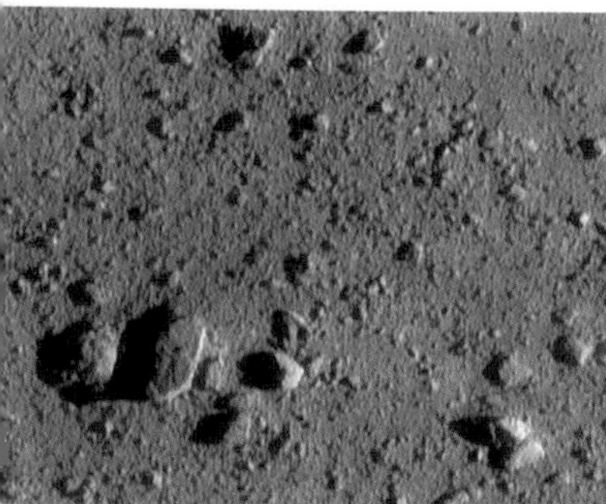

Above NEAR *Shoemaker*'s image of part of the surface of asteroid Eros. The large, oblong rock casting a big shadow measures 24 feet (7.4 m) across.

Above right Location of NEAR *Shoemaker*'s landing site on Eros. The landing site (at the tip of the arrow) is near the boundary of two distinctly different provinces, both of which the spacecraft photographed on descent.

these asteroids have undergone heating and melting. These silicate types of asteroids have a higher measure of reflectivity.

The third category is the M or metallic type, which are believed to be composed largely of an iron-nickel combination. It is thought that these metallic asteroids are actually the central cores of earlier larger bodies that have had their rocky exteriors smashed away during collisions.

Some asteroids have been found to be binary systems, with two bodies gravitationally bound to each other. Some others have been found to have small moonlets.

There are a number of different asteroid families recognized, with more potentially to add to the list. Families are comprised of bodies that follow the same orbital path around the Sun, leading to the conclusion that they are fragments of larger asteroids that have been broken up in the past. Some examples of these families are the Atens and Apollos, which have orbits that cross Earth's orbit, making them potential dangers to our planet. Others types include Trojans, which are bodies that follow either 60° ahead or behind a planet in its orbit—Earth has no known Trojans, Mars is believed to have one, Neptune has five known, while the Trojan king is Jupiter, with almost 2,000 known and probably many more yet to be discovered.

In recent years, the dividing line between asteroids and comets has blurred, with the realization that some non-main belt asteroids might be "burned out" comets. There are even some objects that have been given both asteroid and comet designations, causing further confusion.

An odd kind of asteroid follows what has been dubbed a "horseshoe" orbit, whereby they keep in close proximity to a planet—Earth and Venus have them—while maintaining their orbits around the Sun. For short periods they come close enough to the planet to be considered almost-satellites, but they are not in a true captured orbit such as the Moon.

MISSIONS TO ASTEROIDS

Contrary to the impression given by science fiction movies, with intrepid star captains piloting their craft through huge mazes of tumbling, crashing mountains of rock, the true situation is that the Asteroid Belt is mostly empty space. Notwithstanding the huge number of asteroids discovered so far, with many more expected to be found, the Belt region is so vast that many thousands of miles must lie between each asteroid. A number of spacecraft have made the journey through the belt without any trouble whatsoever.

Some space missions have even given us a close up look at asteroids. First was the *Galileo* spacecraft's encounter with Gaspra in 1991, followed by Ida in 1993. *Galileo* was on its way to Jupiter, and mission planners used the opportunity

Right Artist's impression of the rendezvous of NEAR *Shoemaker* spacecraft with Eros. In 2001, the spacecraft made the first controlled descent to the surface of an asteroid, snapping an incredible series of 69 close-up photos of the rocky surface. After landing, the spacecraft sent back data from the surface of Eros for two weeks.

Left Artist's impression of the *Dawn* spacecraft in the main asteroid belt. Launched in 2007, NASA'S *Dawn* mission will delve into the origins of our Solar System through intense study of Ceres and Vesta, two asteroids that reside in the vast asteroid belt between Mars and Jupiter.

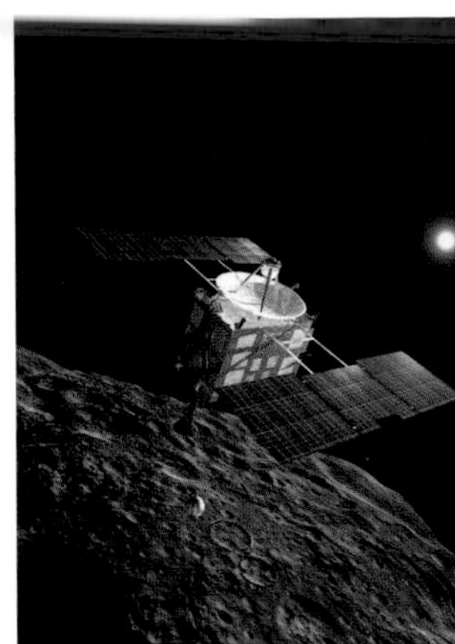

Above Artist's impression of the Japanese *Hayabusa* spacecraft deploying one of the surface target markers that were used to guide the descent onto the surface of the asteroid Itokawa.

of its passage through the Asteroid Belt to visit the two minor planets. (Ida turned out to have a small moonlet, which has been given the name Dactyl.)

Others to be encountered were Mathilde (1997) and Eros (2001), both visited by NEAR *Shoemaker* spacecraft—it made a deliberate hard landing on Eros. The *Deep Space 1* experimental spacecraft visited asteroid Braille in 1999 and the *Stardust* spacecraft encountered Annefrank in 2002.

The Japanese space probe *Hayabusa* arrived at the asteroid Itokawa in late 2005, and attempted to collect a sample of its surface for return to Earth. Various malfunctions of its systems left the craft semi-crippled, and uncertainty remains over whether it was actually able to collect samples. Mission controllers have managed to recover most of the craft's functions, and are hopeful they can return it to Earth.

Other, more ambitious missions are planned, including NASA's *Dawn* mission, which will spend time in orbit around the asteroids Vesta and Ceres in the next decade, and the ESA's *Rosetta* mission, currently on track to meet up with and study two asteroids in 2008 and 2010.

Asteroids might one day turn out to be a source of valuable metals and other substances, for use in building structures in space. The cost of mining them could work out to be considerably less than the cost of lifting materials off Earth's surface in rockets and transporting them in space.

Scientists still have a lot to learn about asteroids. Are they made of solid rock or are they just a jumble of giant boulders loosely stuck together? What threat do they pose? How many are still to be discovered? The expectation is that there are millions still to be found

COLLISION COURSE

Over recent years it has become increasingly clear that asteroids pose a serious risk to life on Earth. Earth has been hit before and, although the likelihood of our planet being hit during our lifetimes is small, it is inescapable that it will be struck again sometime in the future. It might be years, decades, or millennia before it happens, but scientists are taking the threat very seriously indeed.

Surveys are underway to find and characterize those asteroids that might pose a future risk. Astronomers hope to catalog all such large asteroids within the next 10 years or so. The question then becomes, what can be done to reduce the threat? Various schemes have been proposed, including using nuclear bombs, impactors, or "gravity tractors" to deflect such an asteroid when it is still a long way from Earth. If seen early, only a small deflection would be needed.

Comets

FACT FILE

Short-period comets
Orbit the Sun every 200 years or less. Thought to originate in Kuiper Belt.

Long-period comets
Orbit the Sun every 200 years or longer. Thought to originate in Oort Cloud.

Notable features
The solar wind causes the tail to always be away from the Sun so a departing comet will always be chasing its tail.

They grace our skies, could have been responsible for bringing water to our planet and kick-starting life, and may pose a serious threat to our long-term future. No wonder scientists want to know more about comets.

SKY JETS

Comets are small Solar System bodies composed of a mixture of ices, rock, and dust. They've often been called "dirty snowballs." It is thought they formed far out in the distant reaches of the Solar System as the other planets were forming closer in. Because they're made of ice, and have not been significantly heated and changed during their long dark years, it is believed they may preserve, in relatively pristine form, the ingredients from which the Solar System was made. This is one of the main reasons why scientists are very interested in comets—they might actually give us clues about where we came from and what we're made of.

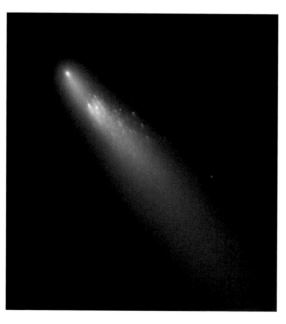

Above right Folio from the *Historia de los Indios*, by Fray Diego Duran, depicts Montezuma, the last king of the Aztecs, sighting a comet in 1519. The comet was believed to foretell the coming of the god Quetzalcoatl, therefore Cortes was assumed to be a god on his arrival soon after.

The solid portion of a comet is known as the *nucleus*, and ranges in size from tens of yards up to tens of miles. At certain times when its orbit brings a comet closer to the Sun, sunlight will cause some of its surface ices to sublimate—that is, convert from a solid straight to a gas without going through a liquid phase in between—forming a huge cloud that spreads out to surround the nucleus. This cloud is called the *coma*. Some of the released gases and particles are "pushed" outward by the pressure of sunlight as well as the solar wind, into *tails*, sometimes extremely spectacularly.

While the nucleus is small, the coma of a big comet can become so large that it ends up being bigger than the Sun! And in extreme cases, a comet's tail can become so long that it could stretch all the way from the Sun to the Earth—93 million miles (150 million km)! The wispy appearance of comet tails is what gave them their name—from the Latin *kome* meaning "hair," and the Greek *kometes*, a star with hair.

Astronomers divide comets into several different groups according the type of orbit they have. The first group is called the long-period comets, and they are believed to originate in the Oort Cloud, a hypothetical swarm of comets that surrounds our Solar System at a huge distance from the Sun (50,000 to 100,000 times further from the Sun than Earth is). Every now and then some of these comets get nudged out of their positions—perhaps by the gravitational influence of another nearby star—and they begin their long, lonely, multi-thousand year treks in toward the Solar System proper, ending up on huge elliptical orbits around the Sun.

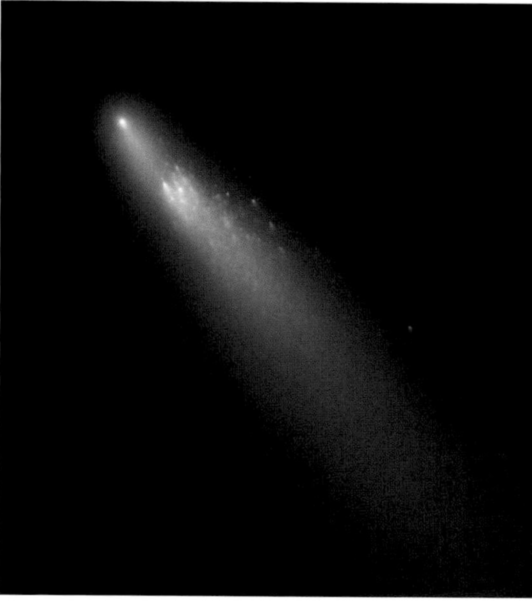

Right Hubble Space Telescope captures extraordinary views of Comet 73P/Schwassmann-Wachmann 3. The fragile comet is rapidly disintegrating as it approaches the Sun.

Opposite Image of Comet C/2001 Q4 (NEAT) was taken at the WIYN telescope at Kitt Peak National Observatory, Arizona, USA, in 2004. Although captured with a wide field of view, only the comet's coma and the inner portion of its tail are visible.

4.5 billion years ago
A solar nebula condenses to form the planets, moons, asteroids, and comets of our Solar System. The planets' gravity propels comets outward to form the distant Oort Cloud, and also funnels most asteroids into a belt between Mars and Jupiter.

613 BCE
A huge comet is recorded in early Chinese records and is probably the earliest record of a comet. It is though to have been Comet Halley.

350 BCE
Aristotle makes the suggestion that comets might exist in the upper atmosphere. This becomes the only view accepted for over a thousand years.

1066
A widely-seen comet, now believed to be Comet Halley, is recorded on the Bayeux Tapestry, a long embroidered cloth that depicts the events leading up to, as well as, the Norman invasion of England, in 1066. It is exhibited in a special museum in Bayeux, France.

1577
Tycho Brahe measures the position of a great comet and compares it with sightings from other places on Earth, proving it is at least four times further away than the Moon.

1609
Johannes Kepler determines that the planets moved about the sun in elliptical orbits, but still believes that comets travel among the planets along straight lines.

1681
Saxon pastor Georg Samuel Doerfel sets forth his evidence that comets are heavenly bodies moving in parabolas about the Sun.

1705
Applying historical astronomy methods, Edmond Halley publishes *Synopsis Astronomia Cometicae*, stating that the comet sightings of 1456, 1531, 1607, and 1682 relate to the same comet, and predicts it will return in 1758. When it did, it became known as Halley's Comet.

1867
Comet Tempel 1 is discovered in 1867 by Ernst Tempel. The comet makes many passages through the inner Solar System orbiting the Sun every 5.5 years.

1950
Explaining the contradiction that comets are destroyed by several passes through the inner Solar System, astronomer Jan van Oort proposes that comets originate thousands of times farther out from the Sun than Earth. The region is now known as the Oort Cloud.

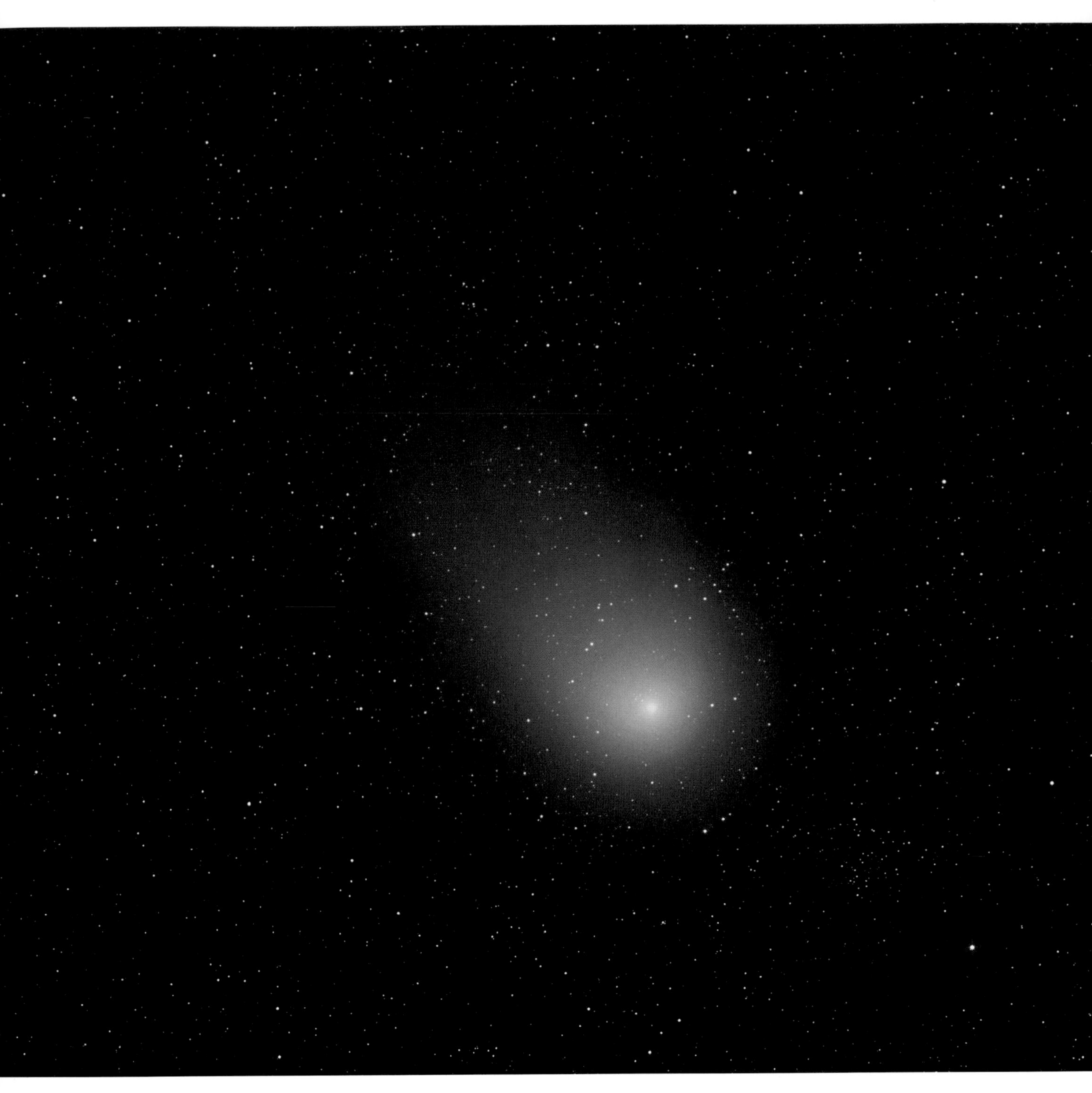

1973
Comet Kohoutek is first sighted on March 7, 1973, by Czech astronomer Luboš Kohoutek.

1975
Comet West is discovered photographically by Richard M. West, of the European Southern Observatory, on August 10, 1975.

1993
Astronomers Carolyn and Eugene Shoemaker, and David Levy, discover the Comet Shoemaker-Levy 9. It was the first comet observed while orbiting a planet other than the Sun.

1994
Comet Shoemaker-Levy 9 hurtles toward Jupiter with the astronomical world watching. The prominent scars from the impacts could be seen on Jupiter for many months after the impact, and observers describe them as more easily visible than the Great Red Spot.

1995
Comet Hale-Bopp is discovered by two independent observers, Alan Hale and Thomas Bopp, both in the US.

1996
The comet, Comet Hyakutake, is discovered on January 30, 1996, by Yuji Hyakutake, an amateur astronomer from southern Japan.

2004
NASA's *Stardust* spacecraft flies by Comet Wild 2, gathering comet dust and taking close-up pictures.

2005
Delta II rocket launches the combined *Deep Impact* spacecraft, which leaves Earth's orbit and is directed toward the comet, Tempel 1, and collects images of the comet before the impact.

2006
NASA's *Stardust* spacecraft returns to Earth on January 15, bearing samples of comet dust. *Stardust* is the first US mission launched to robotically obtain samples in deep space and return them to Earth.

2007
Comet McNaught, the brightest comet in over 40 years, also known as the Great Comet of 2007, is a non-periodic comet discovered on August 7, 2006, by British-Australian astronomer Robert H. McNaught. It is easily visible to the naked eye for observers in the Southern hemisphere.

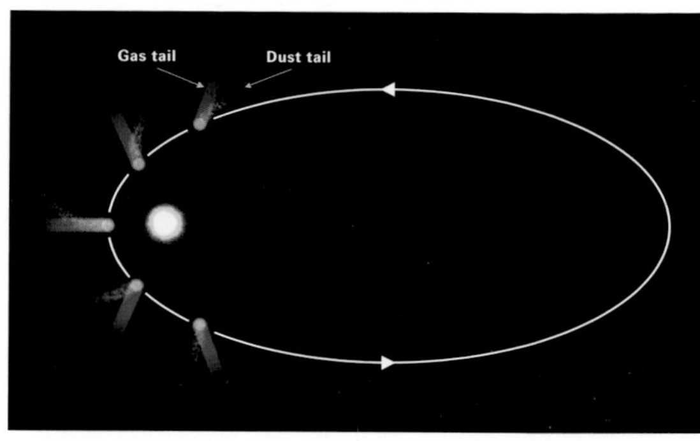

Right Illustration showing the position of the gas and dust tails of comets. Due to the solar wind, the tails always face away from the Sun, no matter what direction the comet takes.

The second group, short-period comets, have orbital periods that are 200 years or less. Many of them are thought to originate in the Kuiper Belt, a zone of icy bodies beyond Neptune that includes Pluto. Comet Halley, with a period of around 76 years, is a famous member of this group.

The third group consists of a handful of comets that orbit the Sun in among the asteroids of the Asteroid Belt—they sometimes show a bit of coma during the period when they are nearest to the Sun.

And a final group is comprised of solitary objects that come in from interstellar space, make one pass through the Solar System and then disappear out the other side, never to return. Not many of this type are seen.

Comets can be very uncooperative when it comes to making predictions about how bright they might become during an *apparition*—viewing season. While it's easy to determine their orbits and trajectories, less easy is predicting whether a newly found comet will become bright and spectacular or remain faint and dull.

American comet discoverer, David Levy, has quipped that comets are just like cats—they both have tails, and they both do whatever they feel like doing at the time.

The problem arises because of the uncertainty over the amount and type of "volatile," (i.e. easily sublimated) ices on their surfaces. A comet with lots of volatiles might form a large coma and sport impressive tails, whereas one with few volatiles will likely be more subdued. The amount and type of material that is sublimated and blasted from the surface of a comet determines the size and density of its coma and tail. Comets that are making their first pass through the inner Solar System often tend to put on a good show, thanks to their pristine covering of volatiles. On the other hand, comets that are on short-period orbits, having gone through many cycles of being heated up by the Sun, are likely to have had much of their surface volatiles eroded away, and put on less of a performance. That said, the opposite sometimes happens in both cases, and no one is really sure why.

COMETS: LIFEBRINGERS?

Where did Earth's water come from? Our planet was forged in the hot inner region of the infant Solar System, where water would have been hard-pressed to survive. So where did it all come from? Quite a few scientists believe Earth's H_2O arrived in the form of comets, bombarding the planet during its formative years. There are tantalizing hints that at least some comets have the same type of water as Earth—based on chemical isotopes. In particular, some studies have suggested that the main belt comets may be the ones that delivered the water. And yet, other studies show the opposite might be the case. Which is right? Another intriguing possibility is that comets seeded Earth with the sort of organic chemicals that eventually led to the emergence of life. It's not as crazy as it might sound at first—once thought to be covered in highly-reflective and relatively pure ice, comets are now known to be extremely dark, covered with a layer of solidified organic sludge. Maybe we owe our very existence to these cosmic wanderers.

Right Satellite enhanced image of Comet Hale-Bopp passing by Earth in 1997. Hubble images determined the comet's diameter to be about 24 miles (40 km).

COMET TAILS

Not all comets develop tails, but when they do they can be spectacular. Tails come in two parts—the gas or ion tail, and the dust tail—accounting for both the gas molecules and the larger clumps of molecules known as dust particles, that are given off by a comet when it is heated by sunlight.

Gas tails form when gas in the coma is pushed out by the pressure of the particles in the solar wind that emanates from the Sun. The gas in these tails is ionized, emitting light of a bluish color. The tails can stretch for vast distances in the case of the largest and most active of comets. Gas tails also tend to be rather straight.

Dust tails form in much the same way, but this time from the pressure of sunlight itself. The dust in the tail does not give off its own light like the gas tail; rather, it reflects the sunlight, which is why dust tails have a yellowish hue. They also tend to be slightly curved.

Because each tail is pushed outward by the solar wind or sunlight, comet tails point away from the Sun—even when a comet is on the outward part of its orbit, away from the Sun, the tails can actually precede it.

HISTORY

Comets have been known to skywatchers since ancient times. There are records of observations going back thousands of years. In those distant times, where mysticism and superstition most often ruled, comets were commonly seen as harbingers of dread or catastrophe. No doubt this was due to their sudden appearance without warning, disrupting the tranquillity and order of the heavenly domain of the gods. Their sometimes long and dramatic tails would have added to the impression of strangeness. There's good evidence to suggest that some comets may have passed very close to Earth during the last few thousand years, making them appear very big and bright—the sight of them in the sky (even the daylight sky) would have been enough to make even the bravest soul wonder if some sort of celestial doom was approaching.

Prior to the invention of the telescope, there was no early warning that a comet was approaching, which is why they seemed to appear "out of the blue." These days, most newly discovered comets never reach naked-eye visibility, and those that do are often spotted years before as nothing more than tiny specks of light. Only the tiniest and most insignificant of comets now go undetected in the inner Solar System.

Although many early stargazers thought comets might be an odd kind of planet, Aristotle thought comets (and, for that matter, meteors) were some sort of strange cloud phenomenon in the upper air. That view largely prevailed until the sixteenth century, when astronomer Tycho Brahe

Left *Adoration of the Magi*, 1304-06, a fresco by Di Bondone Giotta, the Italian artist, shows what is believed to be Comet Halley. It is often thought that the Star of Bethlehem was actually Comet Halley. The fresco is in the Scrovegni Chapel, Padua, Italy.

triangulated measurements he had taken with those of other astronomers, to show that comets must be very far away, many times further than the Moon.

RETURN OF THE COMET

Strange as it may seem today, some scientists of the time believed that comets did not travel in orbits like the planets, but in straight lines through space. But by the early seventeenth century, following work done by Johannes Kepler on planetary motion, some began to wonder if comets followed elliptical or parabolic paths around the Sun. The answer came in 1680, when a bright comet appeared that permitted detailed positional measurements to be made, and calculations showed that it was indeed on a parabolic trajectory. Only seven years later, Isaac Newton showed that comets could obey his law of gravitation and fit the observations made previously.

Less than 20 years later, in 1705, Edmond Halley, the man who would forever become synonymous with comets, used Newton's law to study the orbits of 24 past comets. He noticed similarities in the orbits of three of these comets, and speculated that they might have been the same object seen many years apart as it completed several orbits of the Sun. And he made a bold prediction—this comet would return some time in 1758 or 1759. French mathematicians Joseph Lalande, Alexis Clairaut, and Nicole-Reine Lepaute refined his prediction and narrowed it down to a one-month period in 1759. And sure enough, the comet made its reappearance, just as Halley had said it would. It was given the name Comet Halley, or, under today's comet naming system, 1P/Halley—P for periodic (i.e. it is one of those comets with periods under 200 years), and 1 because it was the first such identified.

Over the next 200 years, many comets of short orbital periods were found and tracked on successive orbits—some with periods of only a few years (e.g. Comet Encke, 3.3 years) and others with periods of decades. Comet Halley's period is around 75–76 years, measured from perihelion to perihelion.

Above Comet Halley on May 19, 1910, photographed at Lowell Observatory, USA. The original photographic plate was digitized and colored. The comet had passed through perihelion one month earlier and was 0.9 AU from the Sun and 0.3 AU from Earth.

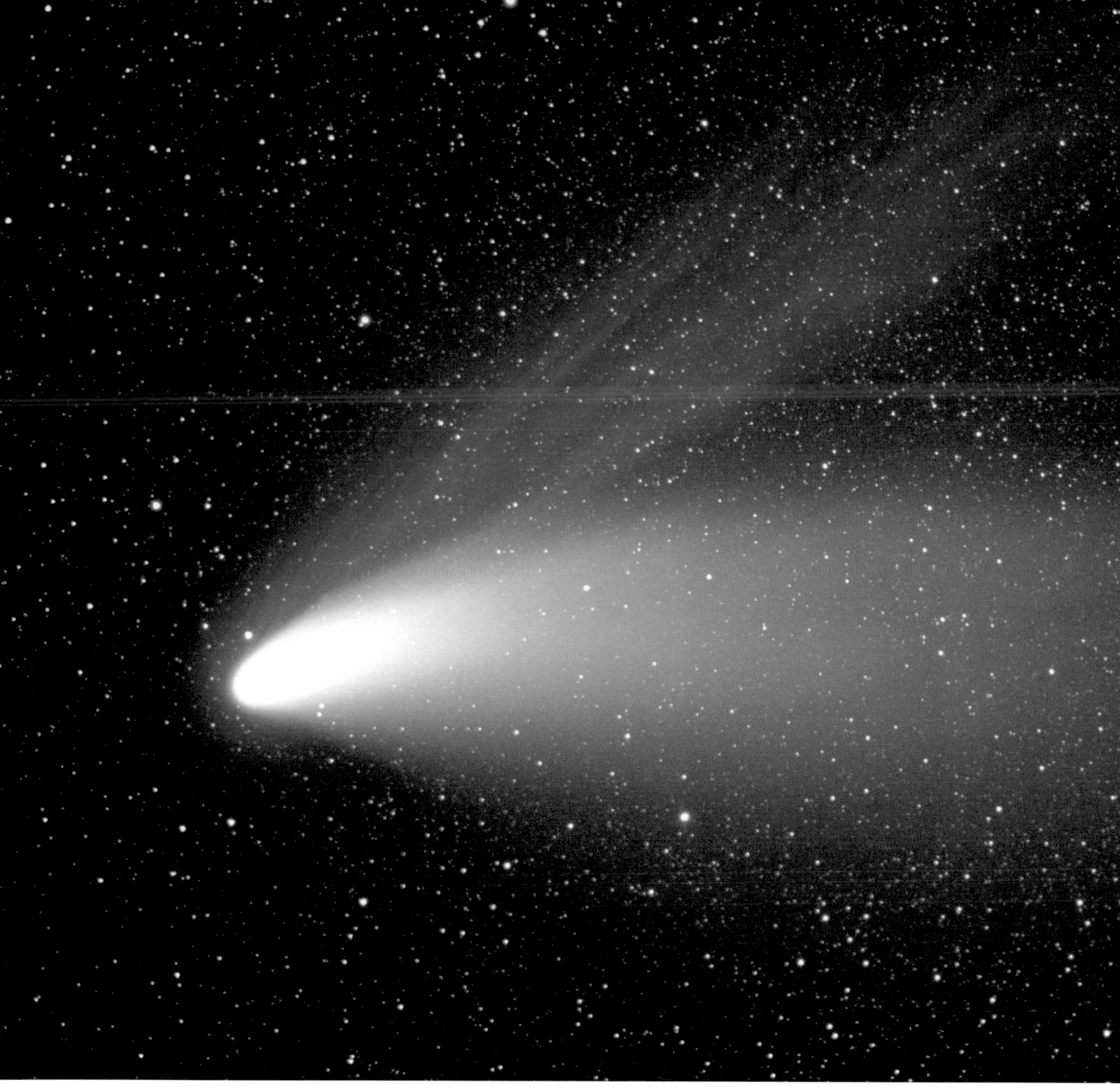

Above Long exposure photograph of Comet Hale-Bopp in 1997. This is the farthest comet ever discovered by amateurs, and appeared 1,000 times brighter than Comet Halley did at the same distance.

Halley has made several notable appearances over the centuries, including one apparition only months before the Battle of Hastings in 1066, during the Norman Conquest. The comet is recorded on the famous Bayeux Tapestry that depicts the battle along its 230 feet (70 m) length.

In April 1910, Comet Halley appeared in Earth's skies and was widely observed. The following appearance, in 1986, disappointed a lot of people, who had been led to believe it would be as good as the 1910 show. The problem was that many of the reports of Comet Halley from 1910 had become confused with those of another 1910 comet, the Great Daylight Comet, which had put on a good show in January of that year. It was so big and bright that it could be seen in daylight, hence the name, and far outperformed poor old Halley. This confusion, combined with unfavorable angles and distances in 1986—as well as more light pollution in cities, which makes it harder to see comets—was what led to a general disappointment in the community.

For scientists, though, 1986 was a golden opportunity to get up close to a comet for the first time. A small flotilla of spacecraft was planned to intercept Halley and take images of its surface. The Europeans contributed the *Giotto* probe, the Soviets had the *Vega 1* and *2* missions, and Japan had two spacecraft called *Suisei* and *Sakigake*. The US had decided not to send a mission, much to the dismay of many scientists. Instruments to study the comet were nevertheless intended to be carried into space aboard the space shuttle, but the destruction of the shuttle *Challenger,* and the subsequent delays put paid to those plans. *Giotto* was the most sophisticated of the spacecraft, and it made a very close flyby, sending

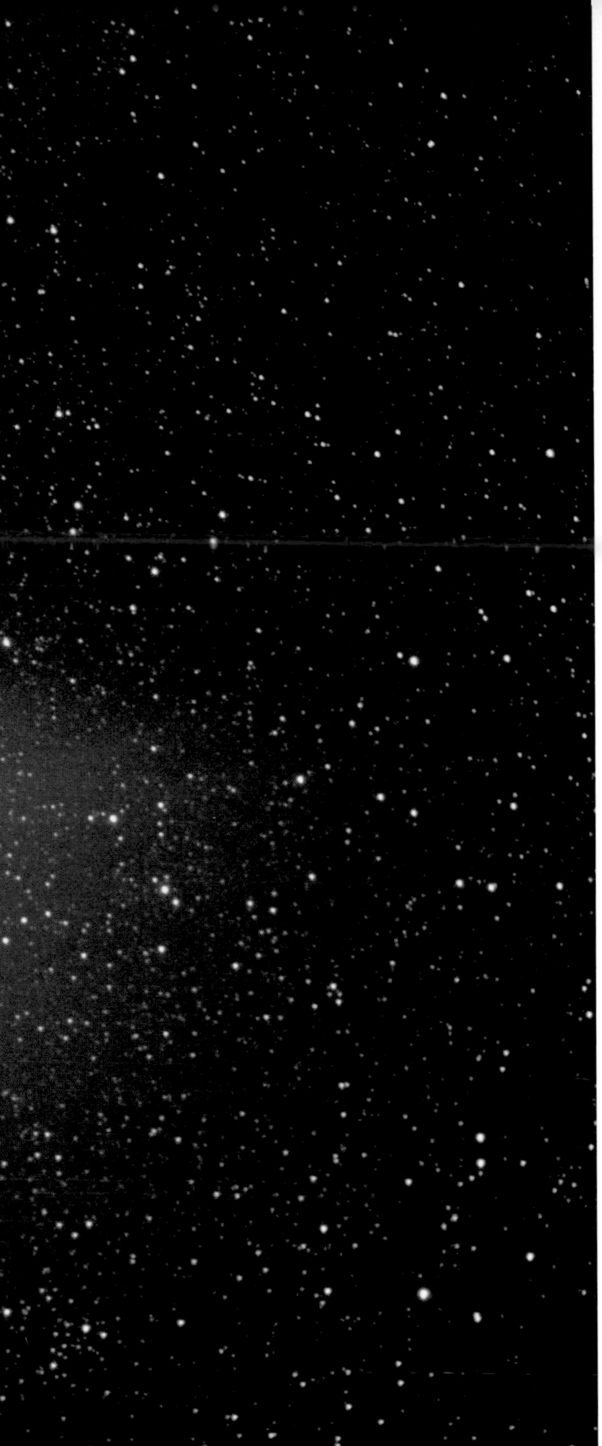

CRASH OF THE CENTURY

On the night of March 24, 1993, astronomers Gene and Carolyn Shoemaker, and David Levy, were astounded to discover what looked like a string of pearls in the night sky. It was a comet that had broken into many pieces, and it was named Comet Shoemaker-Levy 9. The scientific world was even more amazed as it was soon realized that these cometary fragments were on a collision course with Jupiter. Astronomers geared up to catch the never-before-seen spectacle.

The crash happened in July 1994, with each individual collision leaving a dark scar in Jupiter's cloud tops as the fragments exploded upon impact. At a speed of 37 miles (60 km) per second, hitting a cloud is like smashing into a brick wall. Over a period of six days almost 24 pieces hit the planet, ranging in size from hundreds of feet up to about three miles across.

This episode was a perfect example of the role of Jupiter in the Solar System. It acts to alter many comets' trajectories, either by gobbling them up, reducing them to short-period inner-Solar System orbits, or flinging them out of the Solar System altogether.

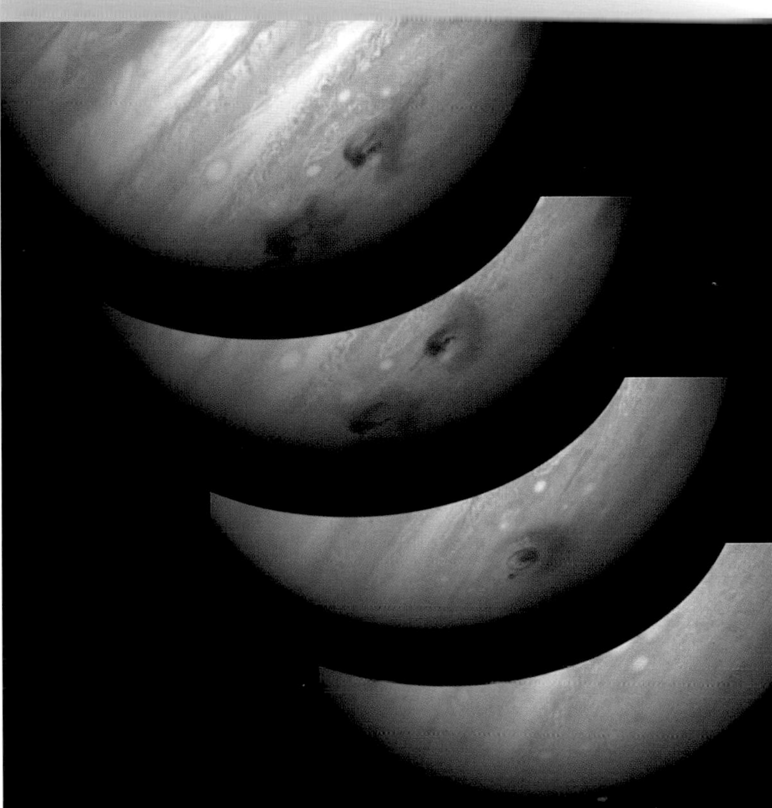

In July 1994, 21 chunks of Comet Shoemaker-Levy 9 slammed into Jupiter. Hubble recorded this spectacular event in a series of time-lapse images, beginning at the bottom. The top image shows three impact sites.

back amazing images of Halley's nucleus, and measurements of the dust density in its coma. The nucleus was revealed to be shaped a bit like a peanut, 9 miles (15 km) long by about 5 miles (8 km) wide. Contrary to most expectations, it did not appear to be a bright, ice-covered object at all. In fact, it was pitch black, about as dark as coal. It appears that the surface is covered with dark dust particles comprised of various organic chemicals, with ice mixed in. The ice is four-fifths water ice, with the rest being mainly frozen carbon monoxide and carbon dioxide, with traces of other substances. The sunlit side of the comet was very active, due to the solar heating. Huge "jets" of material were seen bursting from the surface, like geysers. The unlit side, in contrast, was far less active. With a rotation rate of 52 hours, all parts of the comet go through a slow cycle of heating and cooling.

Left Digital image of *Giotto* spacecraft approaching Comet Halley.The comet is a painting from the Mt Wilson Observatory photograph taken of Halley on May 8, 1910. ESA's first deep space mission, *Giotto,* was designed to help solve the mysteries surrounding Comet Halley by passing as close as possible to the comet's nucleus, which it did on March 13, 1986.

COMET MISSIONS

The same sort of compositions and conditions have now been seen on other comets, following spacecraft rendezvous with several comets during the last decade.

Deep Space 1 was a US spacecraft that made a close flyby of Comet Borrelly in September 2001. It found the comet to be even darker than Comet Halley.

Another US spacecraft, *Stardust*, had an ambitious mission to bring pieces of a comet back to Earth. Launched in February 1999, *Stardust* flew close to Comet Wild 2 at the beginning of 2004, collected samples of the coma as it went by, and delivered them back to Earth in a sealed capsule in January 2006. It also took high-resolution images of the Wild 2's nucleus, showing it to be a potato-shaped object with craters and gouges in its surface.

Another ambitious mission was *Deep Impact,* which intercepted Comet Tempel 1 and launched an impactor into the nucleus. The impactor buried itself under the surface and caused a huge explosion of the ices and other particles there. The *Deep Impact* mothercraft observed the explosion and took readings of the types of ices thrown out from under the surface. The impact was also witnessed by another spacecraft, *Rosetta*, which is itself on track for a comet rendezvous in the year 2014.

Below Artist's impression depicting a view of Comet Wild 2 as seen from NASA's *Stardust* spacecraft during its flyby of the comet on January 2, 2004. During the hazardous traverse, the comet caught samples of comet particles.

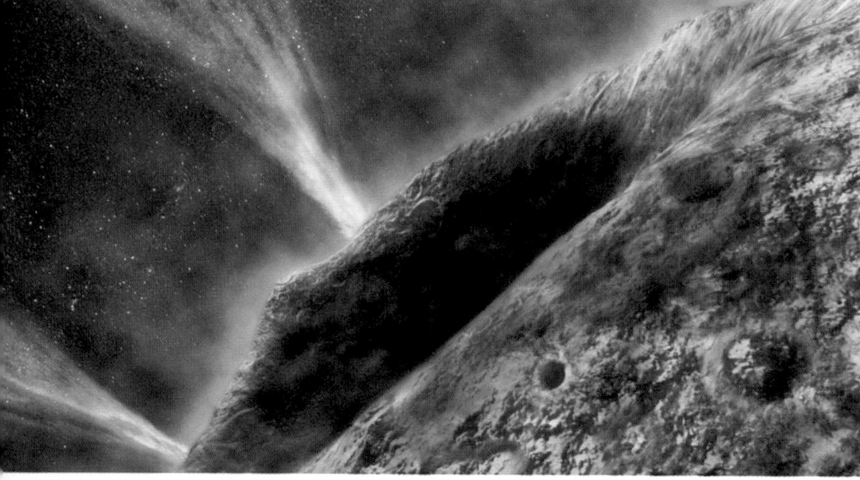

In addition to seeking an understanding of the origin of comets and, through them, the wider Solar System, scientists are keen to study them because of the danger they pose to Earth. Just as asteroids can collide with a planet like Earth, so too, can comets. It's wise, therefore, to learn as much as we can about them, and space missions are the best way of gaining insight into the sizes, shapes, compositions, and so on.

DISCOVERIES

In times past, comet discoveries were largely the domain of amateur astronomers. Several leading amateurs in different countries hold substantial records for the numbers of comets discovered. In the modern age, comets can be found via the Internet! For instance, the Solar and Heliospheric Observatory (SOHO) spacecraft, which monitors the Sun, often captures small comets flaring up and becoming bright as they get very close to the Sun.

These sorts of comets are not normally found in the traditional way. So, by frequent visits to the SOHO web page and keeping an eye on the images received from the spacecraft, amateur astronomers have been able to discover comets without ever having to outside and look at the night sky!

Below Artist's impression of *Deep Impact's* encounter with Comet Tempel 1, on July 4, 2005. This digital image gives us a look at the moment of impact. The large black hole is a large crater on the surface of the comet.

Above Blue-light observation of Comet C/2002 T7 (LINEAR) made by spacecraft *Rosetta*, shows the nucleus from a distance of about 58 million miles (95 million km).

TO CATCH A COMET

Although there have been a number of spacecraft encounters with comets since 1985, with recent missions sending back good pictures of the surfaces of these bodies, so far no spacecraft has successfully landed on one. Hopefully that will change in November 2014 when the European Space Agency's *Rosetta* mission catches up with Comet 67P/Churyumov-Gerasimenko and goes into orbit around it.

Launched on March 2, 2004, the spacecraft's trajectory has already seen it do one flyby of Mars and one of Earth to pick up speed, with two more Earth flybys to go. Along the way to its destination, *Rosetta* will also zip past two asteroids.

Reaching the comet, *Rosetta* will go into a slow orbit and detach a lander, called *Philae*, which will carefully descend to the surface. Using two harpoons, it will secure itself to the comet and begin studies of the icy body's chemistry, and other characteristics as it clings on. The lander's power supply is designed to keep it going for at least one week, although a much longer life is hoped for.

Below Digital impression of *Rosetta* approaching the comet. The spacecraft will be inserted into a parking orbit before deploying its lander. It will obtain data about the comet before being sent on toward the outer Solar System.

Meteors, Meteor Showers, and Meteorites

FACT FILE

What's in a Name?

METEORS
The bright trails coming through Earth's atmosphere—shooting stars

METEOROIDS
The chunks that hurtle through space

METEORITES
The pieces that do not vaporize and reach the surface of Earth

Notable features
Up to 10,000 tons (9,071 tonnes) of meteoritic material falls to Earth each day in microscopic form

Meteors delight the eye with their speedy flash across the night sky. And they delight scientists by bringing traces of other worlds to our planet.

SHOOTING STARS AND FALLING ROCKS

Most people have had the experience of seeing a "shooting star," a brief intense flash of light that zooms across the night sky. They can be a wondrous sight, particularly when they arrive unexpectedly—as most do. But what are they and where do they come from?

The proper name is meteor, which comes from the Greek work *meteoros*, meaning "high up in the atmosphere." Indeed, it used to be thought that they were an atmospheric phenomenon of some kind, such as a fast moving, thin cloud. This is where we get the word "meteorology" from. We now know that meteors are actually the result of a meteoroid entering Earth's atmosphere at high velocity.

Right Labeled meteorite ALH84001. This 4.5 billion-year-old rock was found in Antarctica, in 1984. It is believed to be from Mars, and, scientists believe, may contain fossil evidence that primitive life may have existed on Mars more than 3.6 billion years ago.

A meteoroid is a tiny piece of matter in interplanetary space, anywhere from the size of a grain of sand up to a small boulder. The Solar System is literally littered with meteoroids—many of them are small chips of rock or iron, broken off their larger cousins, the asteroids; others are tiny particles that come from the tails of comets. Each of them is on a long lonely orbit around the Sun.

Sometimes the paths of these meteoroids cross Earth's orbit. And if Earth just happens to be there at that time, the meteoroid collides with our planet's upper atmosphere. Its high velocity causes a brief glow in the air it passes through, and this flash of light is what we call a meteor. That is, the physical object is a meteoroid; the flash of light that betrays its presence is known as a meteor.

Some meteoroids are large enough and strong enough to survive—or partially survive—atmospheric entry, and then fall to the ground, often in pieces. Any such fragments that make it to the surface are called meteorites.

Above This 1951 image shows a museum attendant cleaning the enormous Willamette meteorite found in Oregon, USA, in 1902. It is the largest found in the US and sixth largest in the world.

FIREBALLS AND BOLIDES

Some meteoroids are larger than others, and consequently put on much more of a show when they enter the atmosphere. When a meteor is very bright it is known as a fireball. Fireballs can be so intense that they momentarily light up the observer's surroundings as if they were standing under a full Moon. Even if facing away from the fireball, the flash is so noticeable that it usually prompts the observer to turn around in wonder at the brightness, and where it came from. An ordinary meteor doesn't do that.

Another name sometimes given to very bright meteors is bolide. Although many people equate the terms fireball and bolide, some definitions of bolide narrow it further to insist that it is a bright meteor that explodes or fragments. A bolide can be very spectacular, appearing as if fireworks are underway in the sky. Fireballs and bolides often result in a fall of meteorites to the ground.

METEOR SHOWERS

As mentioned above, some meteors come from comets. As a comet enters that part of its orbit that takes it close enough to the Sun for its ices to sublimate, and tiny particles of comet dust are released. These particles follow the comet, and begin to spread out along its orbit. Eventually, a whole stream of particles can be going around in the comet's orbit, with the densest, "fresher" parts closer to the comet.

When Earth hits one of these streams, it can be like someone turning a "meteor hose" onto the upper atmosphere. Thousands of tiny particles enter the atmosphere, most of them too small to produce any noticeable visual effect. But the few larger ones can put on a display of meteors all seeming to originate from roughly the same spot in the sky (the radiant)—a meteor shower. A particular strong shower with a larger number of bright meteors is called a meteor storm. However, these are quite rare.

Many meteor showers occur every year, each at roughly the same time each occurrence. Depending on where in the sky the radiant is located, some can be visible from both hemispheres, while others are best—or only—seen from one hemisphere or the other.

Meteor showers tend to be given names that reflect the constellation in which the radiant is located. Examples of reliable and very interesting showers are the Leonids (November), Geminids (December), Eta Aquarids (May), and the Perseids (August).

The comets of origin for many showers are known: For instance, the Eta Aquarids come from Comet Halley, while the Leonids come from Comet Tempel-Tuttle.

Note, however, that just because a shower is associated with a particular comet, it does not mean that that comet will be visible at that time of year in any particular year. Showers are usually best in those years after which the comet has passed by Earth.

Countries with the highest number of recorded meteorite landings

Algeria	452
Antarctica	19,884
Argentina	71
Australia	578
Brazil	60
Canada	57
Chile	53
China	92
Czech Republic	21
Egypt	24
England	29
France	77
Germany	69
India	132
Indonesia	18
Italy	55
Japan	64
Kazakhstan	14
Libya	1,302
Mexico	99
Morocco	21
Namibia	18
North West Africa	407
Oman	511
Poland	26
Russia	119
Saudi Arabia	30
South Africa	48
Spain	34
Sweden	19
Switzerland	11
Turkey	14
Ukraine	42
USA	1,346
Western Sahara	22

Left Spectacular time-lapse photograph of Leonid Bolides streaking through the night sky. The Leonids, a prolific meteor shower associated with comet Tempel-Tuttle, get their name from the location of their radiant in the constellation Leo.

1492
Ensisheim meteorite lands in France. It is believed to be the first meteorite whose fall can be dated precisely—November 7, 1492.

mid-1800s
Known and collected since the 1800s, the Canyon Diablo meteorite impacted at Barringer Crater, Arizona, USA, somewhere around 50,000 years ago and was probably known by pre-historic Native Americans.

1894
The Cape York meteorite is discovered by Robert Peary, the famous Arctic explorer, although its existence has been suspected for some years. It is one of the largest meteorites in the world and is believed to have landed around 10,000 years ago.

1920
The Hoba meteorite is discovered near Hoba West Farm, near Grootfontein, Namibia. It is the heaviest meteorite in the world and the largest naturally-occurring mass of iron known to exist on the surface of Earth at this date. Scientists believed that the meteorite landed on Earth approximately 80,000 years ago.

1959
First observation of a meteorite by an automated camera is the Pribram meteorite, which fell in Czechoslovakia (now the Czech Republic) in 1959.

1984
ALH 84001 (Allan Hills 84001) found in Allan Hills, Antarctica, by a team of US meteorite hunters. It is thought to be from Mars.

2002
Sayh al Uhaymir 169 (SaU 169)—a lunar meteorite—is found in the Sayh al Uhaymir region of the Sultanate of Oman in January 2002.

2004
What is believed to be the world's largest meteorite field is found in the region of the Egyptian-Lebanese border. It has more than a hundred traces of crashed meteorites that probably fell to Earth about 50 million years ago.

TYPES OF METEORITES

Meteorites come in many different shapes, sizes, and compositions. Broadly speaking there are three kinds: Stony meteorites (made mostly of a rocky material), iron meteorites (composed mostly of an iron-nickel mixture), and stony-irons (a mixture of the two). Scientists further classify them according to their exact chemical composition. Some meteorites have been matched with spectral analyses of asteroids—strong evidence that they are chips that have been broken away from those asteroids in the past.

Tens of thousands of meteorites have been found over the centuries. Some ancient civilizations used them as sources of metal, or as talismans.

A small number of meteorites are found as a result of people actually witnessing the location of the fall, but most are simply stumbled upon by members of the public, or located in suitable regions of the world by scientific teams.

Meteorites will weather and erode if not found and preserved. For this reason, deserts are ideal locations in which to find them, places such as the Saharan region of Africa, the vast Nullarbor region of Australia, and parts of Antarctica—which, despite all the frozen water, is actually a desert in terms of measured rainfall. While most meteorites are quite small and undistinguished, some are very impressive specimens indeed. The largest known meteorite, the Hoba Meteorite, was found in Namibia and weighs around 66 tons (60 tonnes). Another example is the Willamette Meteorite, found in Oregon, USA, in 1902. It weighs over 16 tons (14 tonnes).

As well as being curiosities, meteorites are very useful scientifically. Apart from the rocks brought back from the Moon by the *Apollo* astronauts and the Soviet space program, meteorites are our only current source of material from asteroids, comets, and even other planets.

Scattered across the face of our planet are large craters that bear testimony to the force with which extremely large bodies can collide with Earth. Perhaps the most famous is Barringer Crater in Arizona, USA, more commonly known as simply Meteor Crater. It is about ¾ mile (1.2 km) wide and around one-tenth of a mile (170 m) deep, and is thought to have been gouged out of Earth's surface around 50,000 years ago by a 160-ft (50 m) wide iron meteorite. The resulting explosion would have been about 150 times more powerful than the atomic bombs used at the end of World War II.

The largest known impact crater is the 185-mile (300 km) wide Vredefort Crater in South Africa, believed to have been made in a collision by a 6-mile (10 km) wide asteroid around two billion years ago. It, and others like it, have been eroded over the millennia and bear no obvious resemblance to a traditional crater; only careful geological study has been able to reveal their presence.

Above Spectacular digital composite of what it might be like to see a meteor shower speeding toward Earth from space.

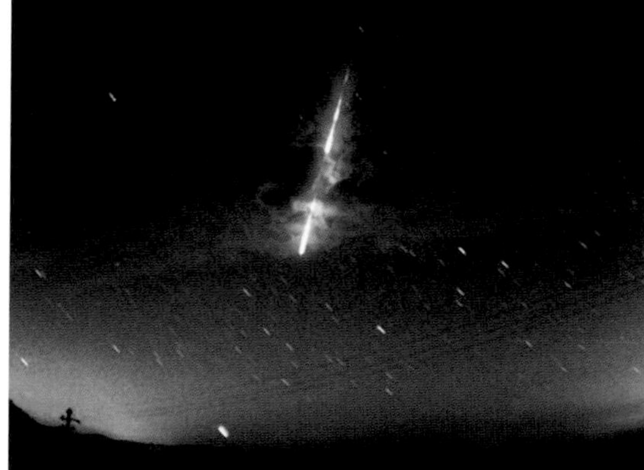

Right Meteor streak running through the Milky Way. Such images are best viewed in the early morning or late night without the interference of lights.

Opposite Aerial view of Barringer Crater, Arizona, USA. Commonly known as Meteor Crater, it is thought to have been caused by a massive meteorite hit around 50,000 years ago.

OBSERVING METEORS

As most meteors are faint, it's best to try and spot them well away from city lights, and on nights when there is no Moon. Artificial lights and moonlight make the sky glow, drowning out all but the brightest meteors (and stars for that matter).

Meteor showers are generally active in the early hours of the morning, which means a late night or early rise for eager meteor observers. You need to find somewhere dark and away from lights. The easiest way to observe is to find somewhere to lie on your back—one of those reclining chairs or pool deck chairs are ideal—and look up. Make sure you allow time for your eyes to adapt to the dark—around 20–30 minutes.

LUNAR AND MARTIAN METEORITES

As well as the thousands of meteorites that come from meteoroids wandering through interplanetary space, occasionally Earth is treated to a very rare and special visitor—a meteorite from the Moon or Mars.

If a very large meteoroid, such as a small asteroid, collides with the Moon or Mars, the impact can be so energetic that pieces of those worlds are flung outward with such velocity that they escape the planet's gravitational grip. After spending many lonely years drifting through space, they can intersect Earth's orbit and fall to the ground just as any other meteorite does. Dozens of examples of lunar and several suspected Martian meteorites have been found.

Meteorites can also be found on other planets! In January 2005, the Mars rover *Opportunity* spotted an unusual rock near the spacecraft's heat shield, close to the rover's landing site. Inspection showed that it was a fairly large iron meteorite. It has been named Heat Shield Rock.

Below Digital composite of Fram Crater in the Meridiani Planum region of Mars.

Right "Heat Shield Rock" observed by *Opportunity* in the Meridiani region of Mars. Examination confirmed it to be an iron meteorite.

Stars

Illuminating the Universe

They are the building blocks of the cosmos, stretching in countless numbers as far as our telescopes can see. They have inspired singers and scientists alike. Stars are our starting point for knowing the universe.

Below Star cluster Pismis 24 in the emission nebula NGC 6357, in the constellation Scorpius. Closer images show the central star to be two stars, halving the solar mass.

If you've ever been fortunate enough to have had the chance to look at the night sky from a really dark location, you'll have been amazed no doubt by the starry splendor overhead. There seem to be countless numbers of stars twinkling away quietly. How many do you think you can see? A thousand? A million? More? The answer often disappoints.

STELLAR FURNACES

Theoretically, over the entire sky there are fewer than 6,000 stars bright enough to be seen with the naked eye, under ideal conditions, away from city lights. But we can see only half of the sky at any one time, of course, and dust and other pollutants, including lights, reduce the number further. So, despite the impression we sometimes get that there are millions of stars above us, we can really only see between 1,000 and 2,500 at any time. Those who live in big light-polluted cities might be lucky if they see a hundred at one time.

But there are many more stars out there in space, of course. It's just that they are so far away that they are too dim to be seen by the unaided eye from the surface of Earth. A telescope is needed to reveal them. Those stars we see in our night sky all seem to be of different brightnesses. It would be logical to assume that the bright ones are simply brighter than the others, but appearances can be deceiving. Yes, some are brighter than others, but some are closer than others. A very bright star a long way away can look a lot fainter than a dim one located much closer.

All those stars are at different stages of their lives. Some are hot and huge, burning their nuclear fuel at a furious rate and probably destined to die young. Others are the tortoises of the heavens, small and cool, but long lived, pacing themselves. These are, no doubt, destined to outlast several generations of the hot flashy upstarts.

The stars are the basic building blocks of the universe. Gathered into giant cities we call galaxies, they are also the celestial bodies around which the planets form—planets like Earth and its many companions in the Solar System.

Our Galaxy, the Milky Way, is alone estimated to contain somewhere between 100 billion and 400 billion stars. And there is at least that number of galaxies in the observable universe. That's a lot of stars. In fact, it has been said that there are more stars in the cosmos than there are grains of sand on all the beaches in the world.

Stars produce the light and heat that illuminate and power the universe. During their lives they produce the wide variety of chemical elements—including, importantly, the heavy elements—that have made life on Earth possible. All the things we take for granted—the metals in our machines, the iron in our blood, the gold in our jewellery—were all forged in stellar furnaces. For a long time, the only star mankind could study in great detail was the one that is closest to us—the Sun. And with the Sun we are largely limited to understanding it as it is now—a snapshot of its

five billion-year-long life, so far. But with advances in theory, observation techniques, and telescope technology over the last century, we have made huge advances in our ability to study and understand all kinds of stars, and examine them at different stages in their lives. From this work, scientists have developed a sophisticated understanding of how stars came to be in the first place, and of their importance to comprehending the progress of cosmic evolution.

In this chapter we'll examine in detail the fascinating lifecycle of the stars—beginning with their birth in dark gas clouds, through their bright and glorious middle years, and then on into old age, and importantly, the process of star death and rebirth. Along the way we'll learn about the many exotic variations in star types, and then examine the strange and violent ways in which some stars end their lives—as black holes, supernovae, white dwarfs, and more. And we'll also consider the age old question of whether some of those stars we see in the night sky have planets where, right now, intelligent creatures like us could be raising their eyes to the heavens and wondering if anyone else is out there.

Above This Hubble Space Telescope image shows N90, one of the star-forming regions in the Small Magellanic Cloud. The high energy radiation blazing out from the hot young stars is eroding the outer portions of the nebula from the inside.

What is a Star?

A star is a rotating sphere of plasma—hot ionized gas. After a newborn star's core is heated by gravitational contraction to 10 million°K, nuclear fusion of hydrogen into helium increases to a significant level. The thermal energy produced eventually builds up enough pressure to counteract the crushing force of gravity, and a normal star, like the Sun, settles into the longest stable period of its existence. Late in a star's life, after the hydrogen fuel in its core is depleted, and if the star has sufficient mass, the fusion of heavier elements occurs.

Above Photograph of Henry Russell, in 1946. Russell, and Ejnar Hertzsprung, independently developed a graph enabling the determination of stellar evolution—the Hertzsprung-Russell diagram.

Mass is a star's most important property. The more massive the star, the greater is the pressure that its upper layers exert on its core—the greater the pressure on the core, the higher its temperature. The core of a star of at least 0.08 solar masses will slowly be heated by gravitational contraction to the threshold for efficient nuclear fusion, 10 million°K. But the rate of fusion will be so slow in such a marginal star that its hydrogen fuel will last for hundreds of billions of years. Such faint stars are called red dwarfs and they are by far the most numerous type observed in the solar neighborhood. They are too faint for us to detect at a great distance, so we can only assume that they are also the most numerous stars elsewhere in our Galaxy and in other galaxies.

Brown dwarfs are less than 0.08 solar masses, so they are not massive enough for their cores to ever reach the critical temperature for stable hydrogen fusion. These failed stars are spheres of gas heated for a while by gravitational contraction. Brown dwarfs are called stars for lack of any other category to place them in, but this may change as astronomers struggle with the definition of an extra-solar planet. They shine mainly in the infrared, and are so faint that we have no idea how many there might be.

THE HERTZSPRUNG-RUSSELL DIAGRAM

A century ago Ejnar Hertzsprung of Denmark, and Henry Norris Russell of the United States, independently developed the Hertzsprung-Russell diagram, one of the most useful tools in astrophysics.

It is a graph that plots stellar luminosity on the vertical axis—in solar units—versus both stellar surface temperature and spectral classification on the horizontal axis. It is possible to plot both on this axis, since spectral type is directly related to surface temperature. Note that temperature increases to the left. This is just the way that it is done and is one of those things cemented by a century of usage, even though it goes against the present-day practice of having the lowest plotted value at a graph's origin.

On the Hertzsprung-Russell diagram the red dwarf stars are of spectral type M, have a low surface temperature and a low luminosity, and thus occupy the lower right corner of the belt of stars labeled "Main Sequence"—those stars whose energy comes from the fusion of hydrogen into helium in their cores.

The core of a one solar-mass protostar will be heated more rapidly by gravitational contraction than a red dwarf's will, and the core temperature will stabilize at 15 million°K. Fusion occurs much more readily at this temperature, and so, even though solar-mass stars have much more hydrogen fuel than red dwarfs, solar-mass stars will run out of nuclear fuel sooner, in approximately 10 billion years. The Sun's position on the main sequence is at the intersection of spectral class G2 and luminosity 1—the vertical axis is labeled in units of solar luminosity.

The most luminous main sequence stars, those of spectral type O, range up to about

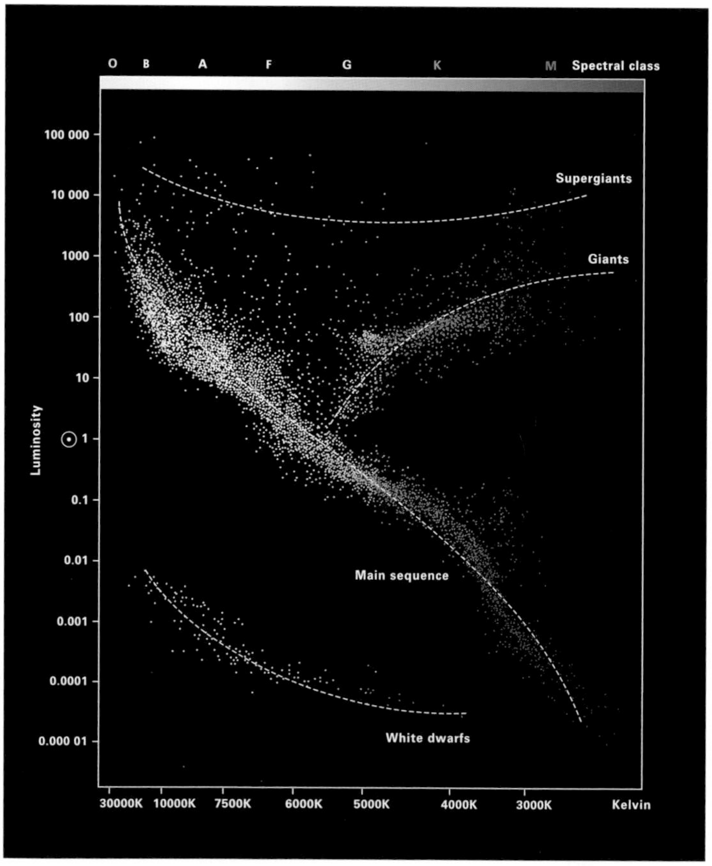

Right The Hertzsprung-Russell diagram (HR diagram) shows the relationship between luminosity, classification, and the effective temperature of stars.

100 solar masses, and are plotted in the upper left corner of the Hertzsprung-Russell diagram. At spectral type O5, magnitude 2.2, Naos, Zeta (ζ) Puppis, has the earliest spectral type of any bright naked-eye O-type star.

Deneb, Alpha (α) Cygni, is 1st magnitude despite being 1,500 light-years away, because it has 60,000 times the solar luminosity. On the Hertzsprung-Russell diagram, Deneb lies at the intersection of that profligate luminosity and spectral type A2, among the rare supergiant stars. Because of its extreme core temperature, fusion proceeds at such a furious pace that its lifespan will be only a few million years. The most massive stars are rare for two reasons. Firstly, relatively few are created. Secondly, they live fast and die young.

The shrinking cores of red giants may eventually get hot enough—100 million°K—to fuse helium into carbon. After they lose the battle with gravity, their crushed and exposed cooling cores are still considered stars, the strange white dwarfs and neutron stars.

Above Hubble Space Telescope view of the core of globular star cluster, NGC 6397. Resembling a treasure chest of glittering jewels, it is one of the nearest clusters, located 8,200 light-years away, in the constellation Ara.

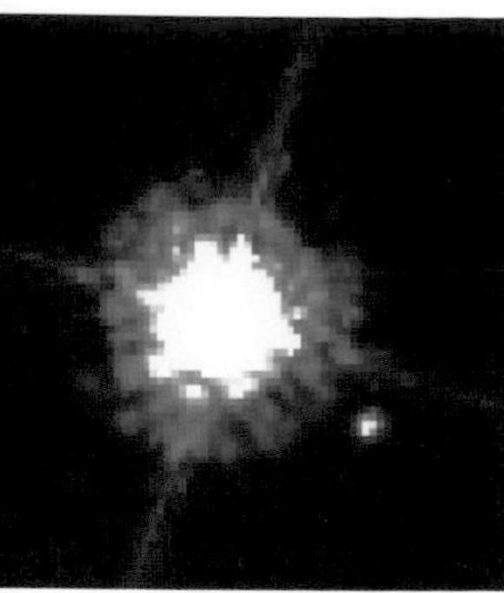

Above Hubble Space Telescope image of what is believed to be a brown dwarf called CHXR 73 B. It is the bright spot at lower right. It orbits a red dwarf star, CHXR 73, which is a third less massive than the Sun, and only two million years old.

NUCLEAR FUSION

The plasma in a star's core consists of uncoupled positively-charged atomic nuclei and negatively-charged electrons, all moving at great speeds. The more massive the star, the greater its core temperature, and thus the greater the speed of its atomic particles. When two positively-charged nuclei meet, the repulsive force of their charges normally deflects them.

However, if they have a head-on collision with enough energy behind it, the two nuclei may get close enough to be bonded by the strong nuclear force.

Within the Sun's core, the success rate for the fusion of two colliding protons—each the nucleus of the common isotope of hydrogen—is one collision in every 10 trillion trillion.

Above Infant stars glowing gloriously in this infrared image of the Serpens star-forming region. The reddish-pink dots are baby stars.

The gamma-ray photons produced during nuclear fusion power a star, although in a solar-mass star it takes about a million years for that energy to reach the star's photosphere and be emitted, mainly at visible, ultraviolet, and infrared wavelengths, rather than as the powerful gamma rays that fusion produced in the core. Every second, our star, the Sun, consumes 661 million tons (600 million tonnes) of hydrogen, producing 656 million tons (596 million tonnes) of helium. The remaining 4.4 million tons (4 million tonnes) are turned into energy in accordance with Einstein's equation $E=mc^2$.

In 1938, German physicist Hans Bethe and his colleagues worked out the first explanation of stellar fusion, the "CNO cycle," in which carbon, nitrogen, and oxygen nuclei act as catalysts for fusion of hydrogen into helium. This is a more efficient process than the proton-proton chain, but it requires the hotter cores of much more massive stars than the Sun. For this Bethe won the 1967 Nobel Prize for Physics.

CLASSIFYING STARS

A star can be classified by the similarities of its spectrum to other stars whose mass and luminosity are known. But first some typical stars had to have their properties measured directly. If we observe a binary star for a significant part of its orbit, then the basic laws of physics allow us to calculate the masses of both stars.

Below Captured by the Galactic Legacy Infrared Mid-Plane Survey Extraordinaire project, this wide image shows a mass of stellar activity in the Milky Way's galactic plane.

Eclipsing binaries are especially valuable because we know that the orbit is edge-on to us and thus the spectroscopically-determined velocities in the orbit are the true speeds, not just a component of the speed. And if we measure a star's magnitude—its apparent brightness in the sky—then we can determine its real luminosity if we can determine its distance.

The distances of the nearest stars can be determined by simple trigonometry by measuring their annual parallax—their apparent displacement against far distant background stars as Earth moves from one side of its orbit to the other over a period of six months.

The known diameter of Earth's orbit is the baseline for this calculation, but the measurement of annual parallax is difficult, since the angles involved are very small, less than one arc-second for even the nearest stars. The precise positional measurements required are best made by the satellites above all of the image distortions caused by our atmosphere.

The most easily measured property on the Hertzsprung-Russell diagram is surface temperature, since it is directly related to the star's color, as long as there is not any obscuring dust in the line of sight to redden the star.

Color can be precisely determined by measuring a star's magnitude through different filters. Spectral type is directly related to the surface temperature, because temperature determines the degree of ionization of the elements at the photosphere of the star.

The coolest stars even show absorption lines in their spectrum caused by molecules. It is important to realize that the surface temperature of a star is not any indication of its core temperature.

Swollen red supergiant stars like Betelgeuse and Antares, whose orange color we can enjoy with the unaided eye, have surface temperatures of about 3,000°K.

The white stars, Sirius and Vega, have surface temperatures of approximately 10,000°K. But the cores of Betelgeuse and Antares, which will eventually be fusing many elements in shells—if they are not already doing so—are immensely hotter than the cores of the ordinary main sequence stars like Sirius and Vega.

FUSION OF HYDROGEN INTO HELIUM

In solar-mass stars, nuclear fusion proceeds by the proton-proton chain. Two protons collide and fuse, emitting a neutrino. One proton gives up its charge and becomes a neutron by emitting a positron (the antimatter equivalent of an electron). The antimatter positron and the first electron (matter) that it encounters annihilate each other, emitting two gamma-ray photons.

The new nucleus, consisting of a proton and a neutron, is deuterium, a rarer isotope of hydrogen. In less than a second, the deuterium nucleus collides and fuses with another proton, forming the rare isotope helium-3 (two protons and one neutron), and a gamma ray is then emitted. If another neutron is added, normal helium-4 is created. There are several ways that this can happen.

Most commonly, two helium-3 nuclei collide, after an average helium-3 life of about a million years, producing one helium-4 nucleus and two freed protons (hydrogen nuclei).

Four hydrogen nuclei have fused to become one helium nucleus, and emitted energy, plus two neutrinos.

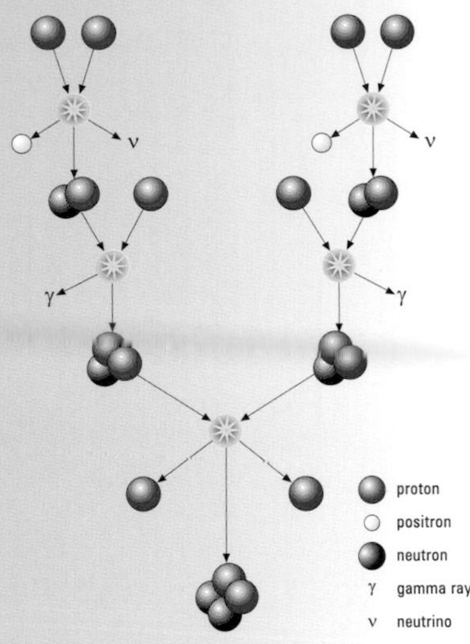

Above Diagram showing the chain reaction of two protons colliding to form a deuterium nucleus, a positron, a gamma-ray, and a neutrino. The process continues down the chain until the last step sets free two protons to start the whole process again.

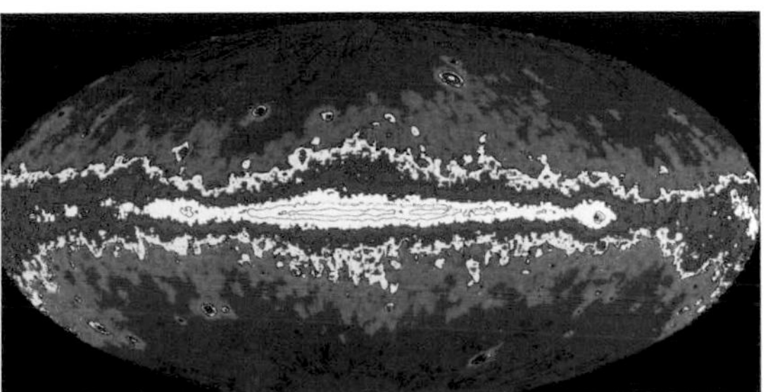

Left Computer image of the entire sky at photon energies above 100 million electron volts. These gamma ray photons are blocked from Earth's surface by the atmosphere. A diffuse gamma-ray glow from the Milky Way Galaxy is seen across the middle.

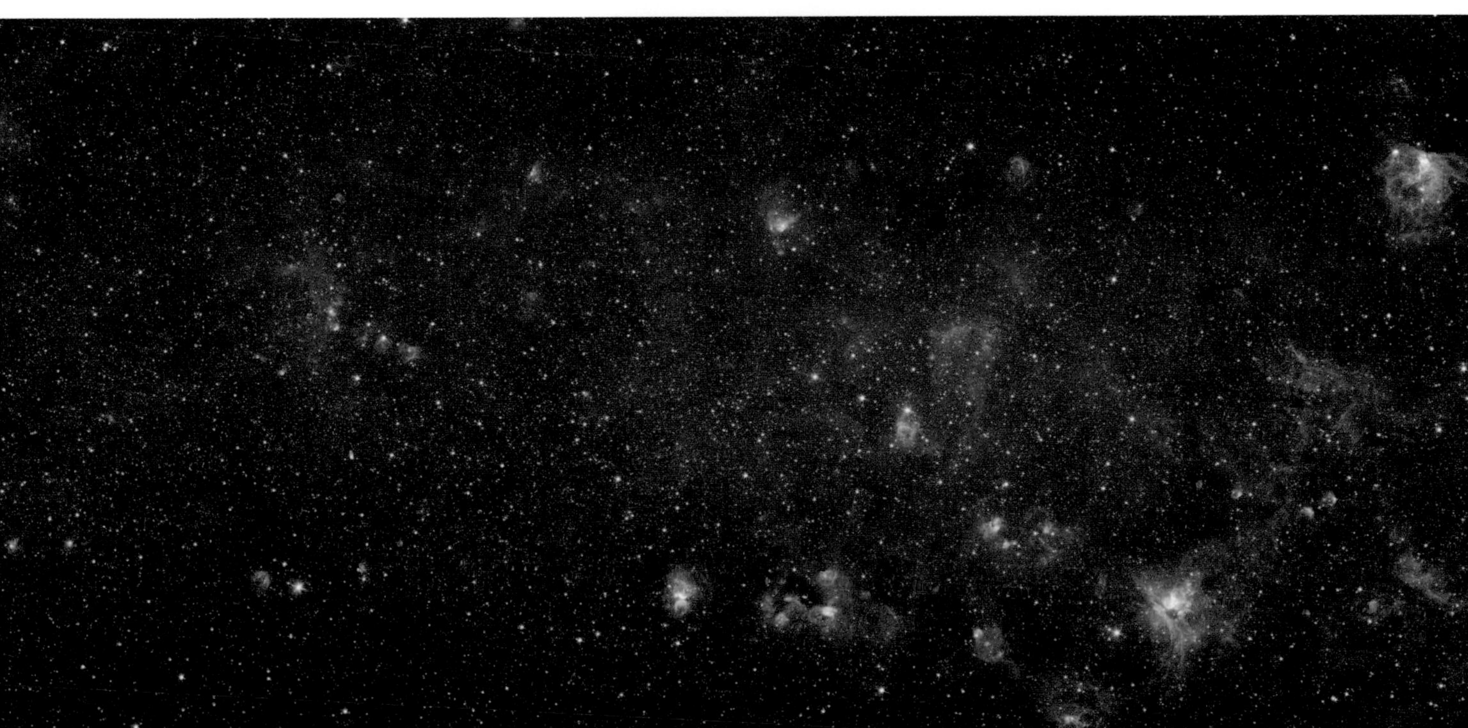

Life and Death of Stars

Big or small, hot or cool, bright or dim, stars come in many different guises, emerging from clouds of dust and gas. And although they're all born in much the same way, their death throes can be very different.

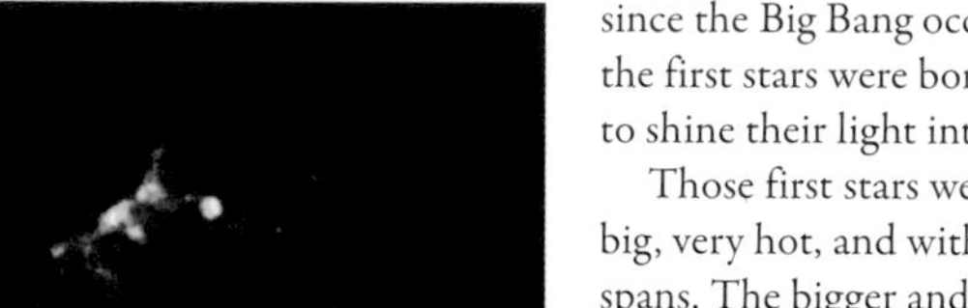

Right Hubble Space Telescope image of a vast cloud of gas being heated by the birth of a new star. Called the Herbig-Haro object #2 (HH-2), the cloud is heated by shock waves from jets of high speed gas being ejected from a newborn star.

Below This Hubble Space Telescope view of the central region of the Carina Nebula reveals a violent maelstrom of star birth. An approximately 1-light-year tall pillar of cold hydrogen towers above the wall of the dark molecular cloud. The 2.5-million-year-old star cluster called Trumpler 14 appears at the right side of the image.

We are all familiar with the stars, twinkling away quietly overhead. The don't seem to change from night to night, year to year, century to century.

Does this mean they've been there since time began? And will they last forever? The answer is no in both cases.

Where do stars come from? How do they form and how long do they last? Stars are huge globes of gas, many billions of times more massive than Earth.

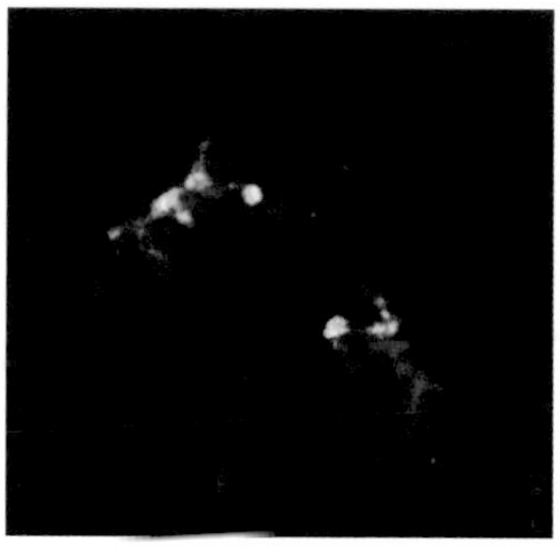

Our Sun is a star, and a fairly ordinary one at that. Some stars are much bigger, and many are much smaller. Some are brighter and some are dimmer. Stars come in a wide variety of sizes, temperatures, and colors.

The story of stars begins not long after the Big Bang, the beginning of the universe. After the intense heat of the Big Bang, the matter created in it began to cool. It also began to collect into huge clouds of gas. Parts of those gas clouds began to get thicker, and in those regions, the gentle gravitational pull of the gas on itself, made the thick regions collapse into huge, dense, hot balls—many times bigger than our Solar System. Eventually, enough gas collapsed into such highly condensed agglomerations that nuclear fusion reactions began to take place. It had taken around 100 million years since the Big Bang occurred, but finally the first stars were born, and they began to shine their light into the cosmos.

Those first stars were probably very big, very hot, and with very short life spans. The bigger and hotter a star is, the longer and faster it burns its fuel, and the shorter it lives.

Those first stars were made only of the lightest elements that had been produced in the Big Bang. But they began the creation of heavier elements, forged in the nuclear reactions in their cores. When this first generation of stars died out in huge supernova explosions, they scattered all these elements into space. This would become the raw material for the next generation of stars.

A STAR IS BORN

Let's use our Sun as an example of the star formation process. Scientists believe that stars such as ours formed from a giant molecular cloud—an enormous region of dust and gas in interstellar space. Galaxies have many of these clouds. Inside, the material in the clouds is not evenly spread; there are parts that are denser than others. Just like the process described above for the first generation of stars, these dense regions

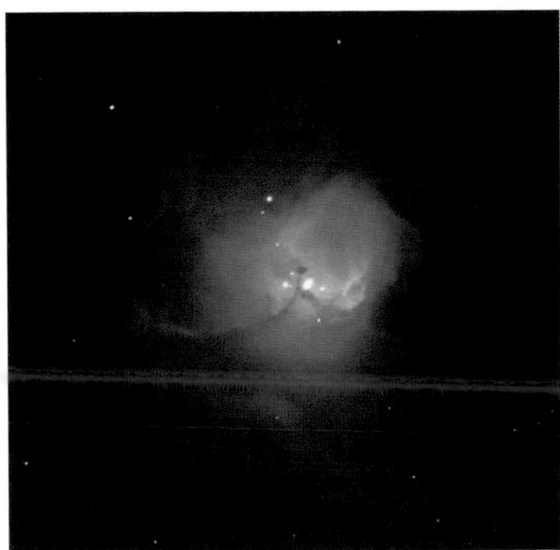

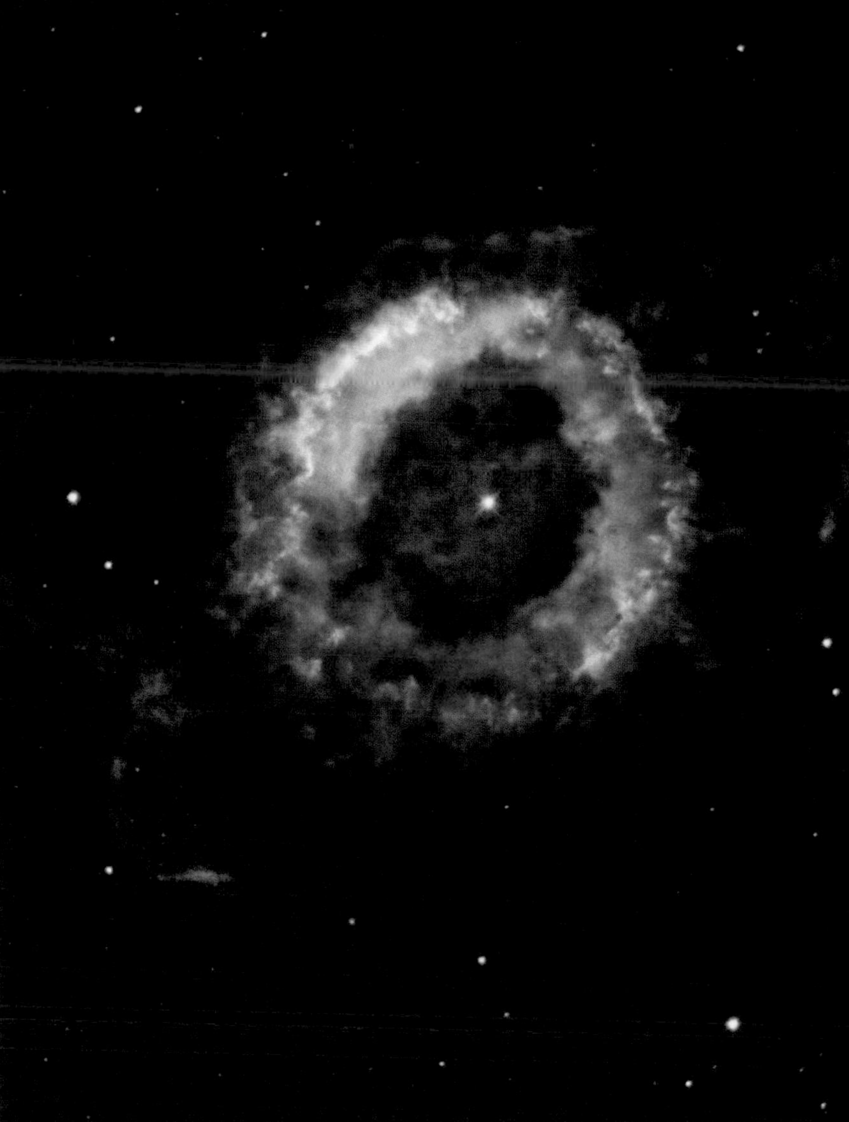

begin to collapse inward under their own gravity, collecting more and more material into themselves. Getting hotter all the time, eventually a huge ball of plasma—an electrically-charged gas—is created. When the internal pressures climb high enough, lighter elements begin fusing into heavier ones, and the star has been born.

It is likely that at least several stars will form out of a single molecular cloud, so it's possible that our Sun has siblings out there somewhere. But so much time has passed since then—around 5 billion years, more than a third of the total age of the universe—that we are unable to identify them.

Our Sun is about mid-way through its lifetime, as calculated by astronomers using their theories of stellar evolution. Because our Sun belongs to a later generation of stars, it contains some of those heavier elements that were formed in the earlier generations. Indeed, the heavy elements contained in the Sun's retinue of planets—including Earth—all came from those early stars. Truly, as it has often been said, we are made of star stuff.

Depending upon how much material was able to gather in the individual star formation regions with the molecular clouds, stars may end up being very massive, very small, or somewhere in between. For a long while scientists were puzzled as to how very large stars could form. As the temperature rose in the cores of the infant stars—while they were still gathering in material—the outward pressure of the energy released should have acted to stop extra material from being pulled in, halting the growth of the star. It is now thought that holes are formed in the material swirling into the growing star, which enable energy to escape along these routes without disrupting the inflow of the material.

Very massive stars burn their nuclear fuel at a furious rate, and can have lifetimes as short as a few million years. The smallest of stars have only feeble nuclear reactions in their cores, and have projected lifetimes of hundreds of billions of years—and the universe is only 13.7 billion years old at present! There are far more of the small type spread throughout the galaxies than there are giant stars.

Astronomers have identified many star-forming regions in our Galaxy—and other galaxies—including the spectacular Orion Nebula. By studying these regions they have been able to get a good handle on the different stages of star birth.

STAR DEATH

There are several ways that stars can end their lives, and it primarily depends on how massive they were to start with.

Using our Sun again as an example, its fate is to become a white dwarf star—a very small dying ember about the size of Earth, but with about 60 percent of the mass of the Sun. It will be about 5 billion years before this process begins to happen, but when it does, Earth and the other inner planets can say goodbye, as we will be engulfed as the Sun first swells to become a red giant star. The Sun will throw off its outer gas layers, and that gas will spread outward eventually

Above Hubble Space Telescope of the planetary nebula NGC 6369, also known as the Little Ghost Nebula. It appears as a small cloud surrounding the faint dying central star. NGC 6369 lies about 2,000 light-years away, in the direction of the constellation Ophiuchus.

Above left Hubble Space Telescope image of young ultra-bright stars nested in an embryonic cloud of glowing gases. The celestial maternity ward, N81, is located 200,000 light-years away, in the Small Magellanic Cloud (SMC).

Right This image of the Sun shows a widely spreading coronal mass ejection (CME) as it blasts more than a billion tons of matter out into space at millions of miles per hour. The Sun is a rather ordinary star halfway through its life at 4.5 billion years. Near the end of its life it will swell to become a red giant star before shrinking to become a fading white dwarf star.

forming into a beautiful planetary nebula. (Planetary nebulae are so called because when first seen through telescopes, they looked a little like planets). The white dwarf left in the middle of the nebula will no longer being able to produce energy through nuclear fusion. All it will have left is its residual heat, which it will lose over billions of years, cooling down and eventually becoming a "black dwarf." This process takes so long that, so far, there has not been enough time since the universe began for any such black dwarfs to form.

If a white dwarf has a companion star in a binary orbit, the dwarf can gravitationally pull material from that star onto itself. Over many years, enough material will have built up on the white dwarf for it to become unstable, and it will explode in a different kind of supernova event. A slightly different process can result in a "nova," where gas from a companion star building up on the white dwarf's surface leads to a surface nuclear explosion, but not strong enough to destroy the star.

Perhaps the most famous white dwarf star known is Sirius B, a companion star to Sirius—a large, bright white star—which is actually the brightest star that can be seen in the night sky. Sirius B would once have been a star about five times the mass of our Sun, before evolving into a white dwarf in the manner described above. Its current size is about the same as the Earth, but with almost as much mass as the Sun packed into that volume. Sirius itself will also one day evolve into a white dwarf star.

CLOUDY SKIES

Stars are formed inside giant molecular clouds, which are huge regions of gas and dust, scattered throughout the galaxy.

Our own Sun was born in one, much like those shown here. Pockets of denser gas form, and then begin to collapse in on themselves. Eventually the pressures become so great that nuclear fusion reactions commence, and a star is born. The clouds themselves are the byproduct of earlier generations of stars which, at the end of their lives, exploded and seeded space with their gas and heavy elements.

Right These three snapshots of cold molecular hydrogen clouds in the Carina Nebula reveal a violent maelstrom of star birth. The glowing edges of some of these objects indicates that they are being photoionized by the hottest stars in the cluster. It is thought that stars may form inside such dusty cocoons.

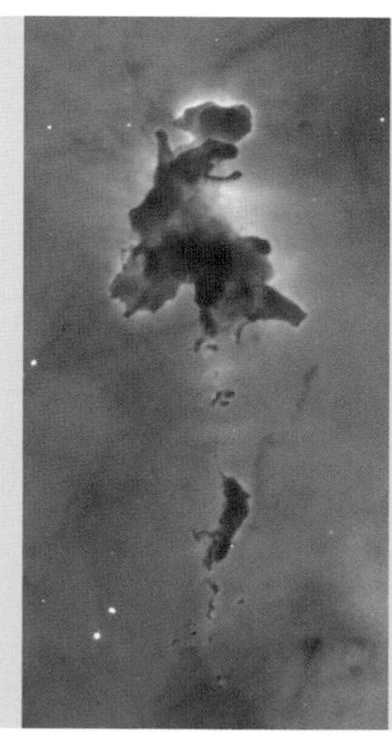

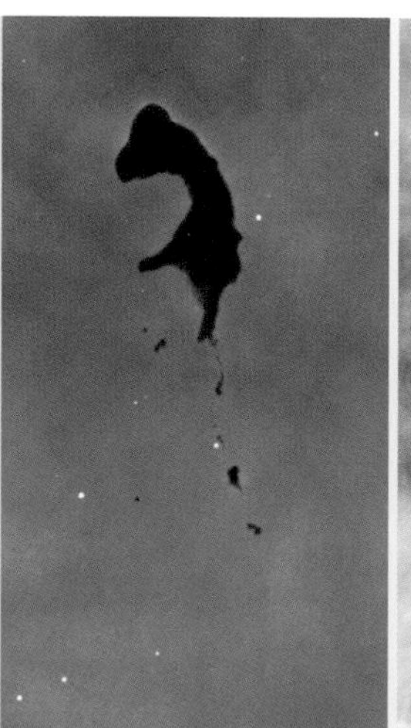

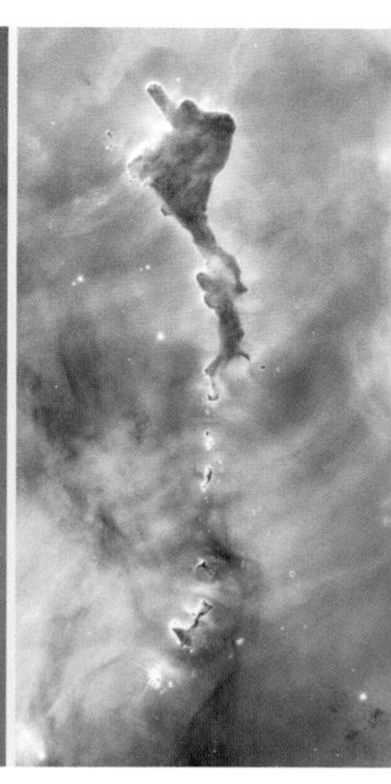

STELLAR NURSERIES

The Great Nebula in Orion is perhaps the best-known example of a nearby star-forming region, being only 1,500 light-years from Earth. Just visible to the unaided eye from a dark site as a tiny smudge of light, a telescope reveals it to be a stunningly beautiful cloud of gas and dust. Deep inside, in regions where the density has increased, new star systems are being born. Astronomers have caught glimpses of this process, in the form of "proplyds"—protoplanetary disks. Studying the Orion Nebula is giving science tremendous insights into the evolution of stars.

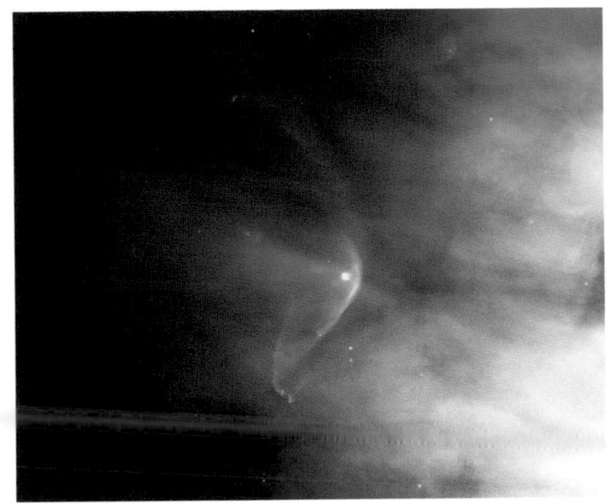

Right This Hubble Space Telescope image reveals various stunning and intricate treasures that reside within the nearby, intense star-forming region known as the Great Nebula in Orion.

The most interesting and violent forms of stellar death happen to stars that are many times the mass of our Sun. Such large stars are destined to end their days in massive explosions, called supernovae.

One outcome of a supernova explosion is a neutron star. When a supernova occurs, the inner part of the star collapses under its own gravity. The stellar material becomes so squashed that, inside many of its atoms, electrons and protons are merged to form neutrons, leaving the dead star composed largely of neutron particles.

Neutron stars are incredibly dense. They can pack more than one and a half times the mass of our Sun into a sphere only 6–12 miles (10–20 km) wide—about the size of a small city. They also have very intense magnetic fields—the strongest in the known universe—many millions of times stronger than Earth's magnetic field.

A well-known example of a neutron star is the one located at the heart of the Crab Nebula, a supernova remnant located about 11 light-years away, in the constellation Taurus. The supernova, that was discovered in the Large Magellanic

Below Hubble Space Telescope image of the tattered remains of a supernova explosion known as Cassiopeia A (Cas A). It is the youngest known remnant from a supernova explosion in the Milky Way.

Below Artist's impression of the so-called Lynx arc, a newly identified distant supercluster that contains a million blue-white stars twice as hot as similar stars in our Milky Way Galaxy. The Lynx arc is 1 million times brighter than the well-known Orion Nebula, a nearby prototypical "starbirth" region. The mega-cluster of stars is 12 billion light-years away and appears as a puzzling red arc behind a distant galaxy cluster, in the northern constellation Lynx.

THE FATE OF OUR SUN

What will be the final fate of our Sun? Is it destined to explode in a huge supernova explosion, perhaps creating a black hole in the process? Alas, no. Our Sun is an average sort of star, and it will have an average sort of ending. Around 5 billion years from now it will swell up into a red giant star, expanding by perhaps 100 times its present diameter. Earth will probably be engulfed in its outer gas layers, but will survive largely intact for a time (albeit with a scorched surface). Eventually the Sun will have flung off its outer layers to form a beautiful planetary nebula, leaving at its core a slowly cooling, dim, small, white dwarf star.

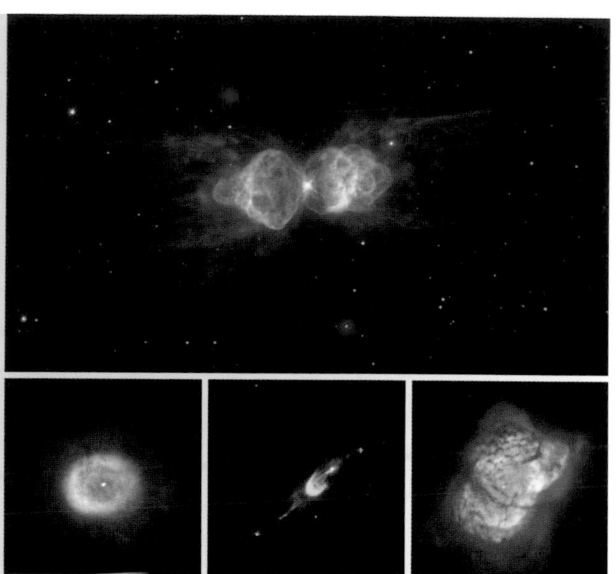

Right This panel of composite images shows part of the unfolding drama of the last stages of the evolution of Sun-like stars. Dynamic elongated clouds envelop bubbles of multimillion degree gas produced by high-velocity winds from dying stars. Planetary nebulae are produced in the late stages of a Sun-like star's life.

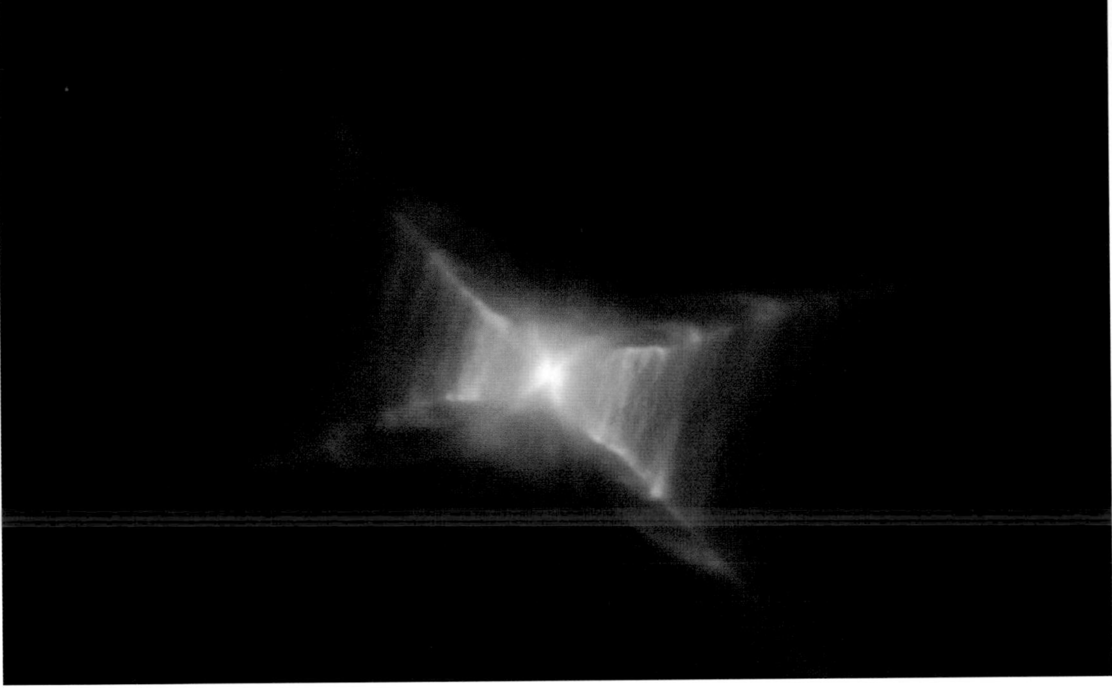

Left This Hubble Space Telescope image shows ladder-like structures within a dying star, cataloged as HD 44179. This is one of the most unusual nebulae known in our Milky Way. It is more commonly called the "Red Rectangle" because of its unique shape and color, as seen with ground-based telescopes.

Cloud galaxy in 1987, is thought to have created a neutron star, although there has not been a direct detection of it yet.

Just as a figure skater spins more rapidly when she pulls in her arms (i.e. gets smaller), a neutron star takes its forebear's rotation and multiplies it many times. Many neutron stars spin on their axes in fractions of a second—quite an incredible thing, when you consider how massive they are.

Some neutron stars are also pulsars. A pulsar occurs when a rapidly spinning neutron star with a huge magnetic field, emits large amounts of radio waves in certain directions. These beams act a bit like a lighthouse, sweeping around and making the pulsar appear to flash on and off—if the observer is in the line of sight. Many hundreds of pulsars have been discovered so far by observers.

The ultimate fate of any star that is more than about eight times the mass of the Sun is a black hole. Just like the formation of a neutron star, the star's core collapses to form a dense object. But in this case, the collapse is so swift and strong that the matter is squashed effectively down to infinity, becoming an object known as a singularity that is the size of a pinhead—a black hole.

The black hole's immense gravity means that it will capture anything that comes too close. The point of no return is called the event horizon. Nothing is fast enough to escape a black hole, not even light, once it has crossed the event horizon.

For many years black holes were simply the playthings of theoretical physicists. Their existence was predicted by Albert Einstein's *General Theory of Relativity*, but few scientists seriously believed that any would ever be found—or even that they existed at all!

But over the years the theory progressed, and observations of unusual objects in deep space called for unusual explanations. Some astronomical processes were seen to involve such high energies that the only thing that could conceivably produce them was something with an enormous gravitational field—and the only candidate scientists had was a black hole.

These days, the existence of black holes is all but confirmed. Dozens of black hole candidates have been found, scattered throughout our Galaxy.

The nearest one to Earth is in a star system called V 4661, about 1,600 light-years away. This system comprises a normal star and an unseen companion. For many years the main star was considered to be a variable star—one whose brightness changes over time according to a recognizable pattern. But a sudden burst of X-rays from it in September 1999 led astronomers to suspect something else was there.

The X-ray burst was a tell-tale sign that the star probably had a black hole companion. Material from the main star was building up in an accretion disk around the black hole; as the material got closer and closer to the black hole, it would give out bursts of X-rays.

THE ODD COUPLE

One of the weirdest star systems so far discovered is known only by its catalog number, SS 433. This binary star system comprises a massive main sequence star, bound in tight orbit with a compact star—either a neutron star or a black hole. Located about 16,000 light-years away, inside our Milky Way Galaxy, SS 433 lies at the center of a supernova remnant called W50. Gas from the main star is being gravitionally pulled onto the compact star; some of that material is accelerated and ejected in two "jets" before it reaches the compact star. The stuff in the jets is moving outward at more than 25 percent of the speed of light.

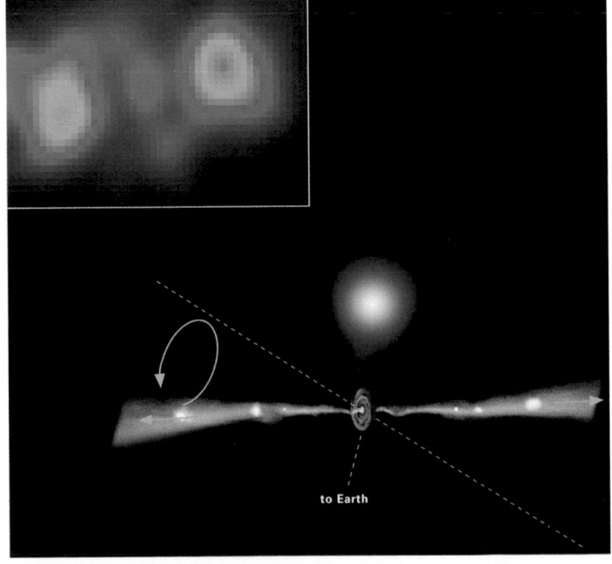

Right Chandra X-ray image of SS 433 (inset). Artist's impression shows two lobes of gas 3 trillion miles apart, emerging from a binary black hole system.

Double Stars

Just over half the stars within our Galaxy are in a binary or multiple star system. This means that about a quarter of the stars in the sky are actually two or more stars orbiting each other. It is only the nearest and most separated pairs that can be seen as distinct stars in a telescope—the visual binaries. The more distant, and/or more closely bound pairs are detected indirectly either by their spectrum—spectroscopic binary, or by the light changes as one star passes in front of the other—eclipsing binary.

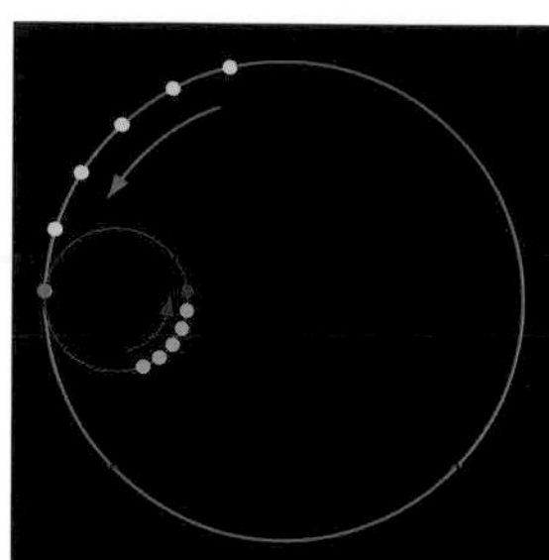

Right Illustration of a binary orbit showing how each star orbits a common center of mass. In close binaries, the stars can transfer matter to each other and change the way the stars look and evolve.

Below Artist's concept shows a debris disk observed around an unusual class of interacting binary stars—in this case a white dwarf–brown dwarf pair.

Nearly any size telescope will reveal that Alpha (α) Centauri (Rigel Kent—the pointer furthest from the Southern Cross) is a pair of yellowish nearly equal brightness stars.

VISUAL BINARIES

Since 1752, repeated measurements of the angular separation and the direction of the line joining these two stars have shown that the fainter star goes round the brighter every 80 years on a path tilted only 11° from edge-on. Both stars are orbiting about their common center of mass. By mapping out the orbits, and knowing the distance to the binary, it is possible to calculate all the properties of the orbits—size, shape, orientation—and then most importantly, the masses of the individual stars. In the case of Alpha (α) Centauri, both stars are similar to the Sun; one slightly more massive and thus hotter and brighter, the other slightly less massive and so fainter and cooler.

SPECTROSCOPIC BINARIES

Most binaries are detected not using an image, but using a spectrograph to measure the spectrum. Each star of the binary produces its own spectrum. As they orbit, the spectral lines shift back and forth in frequency, and so in wavelength, because of the Doppler Effect. As one star approaches us, its spectral lines are shifted to higher frequencies—toward the blue end of the spectrum. The other star in the system would then be moving away from us and so its spectral lines are moved to lower

Right Hubble Space Telescope image of NGC 2346, in the constellation Monoceros. This so-called planetary nebula is remarkable because its central star is known to be actually a very close pair of stars, orbiting each other every 16 days. It is believed that the binary star was originally more widely separated. However, when one component of the binary expanded in size, and became a red-giant star, it literally swallowed its companion star.

frequencies—toward the red end of the spectrum. By measuring the time it takes for the lines to shift back and forth and back again, we know the orbital period. The size of the shift tells us about the orbital velocities, and hence, the total mass of the stars. The Doppler shift gives us no information about the tilt of the orbit. When the orbits are face-on, we would detect no shifts. When the orbits are edge-on, we would measure the actual orbital velocities. Generally we can only put a limit of the total mass.

ECLIPSING BINARIES

If the orbit is edge-on—inclination 90°—each star will alternately pass in front of the other, causing the total light from the system to dim. If the two stars are of the same brightness, then the total brightness drops by half during each possible eclipse. When one star is much fainter and bigger than the other, the effect can be quite dramatic, with a deep and long-lasting primary eclipse, and a shallower less obvious secondary eclipse. The bigger the cool star, the longer the eclipse lasts.

There is a lot of information in the light curve—brightness changes with time—which, when combined with spectra taken throughout the orbits, makes it possible to measure the masses of both stars, and their absolute sizes.

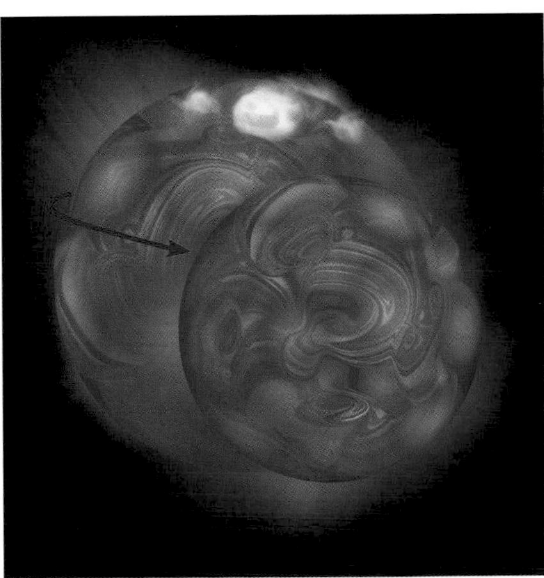

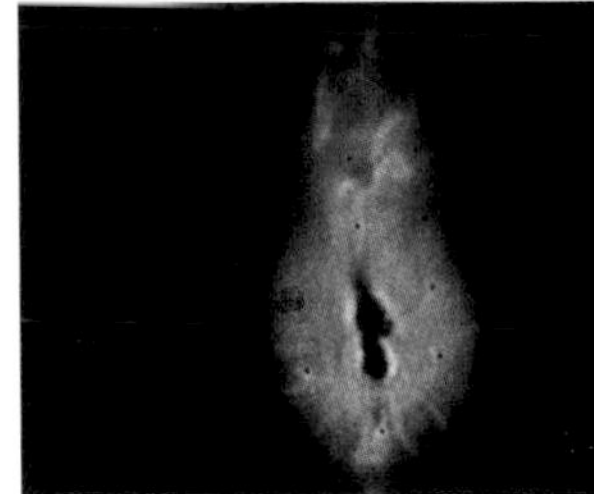

Above Image of the star system R-Aquarii. The two dark knots at the center probably contain the binary star system itself, which consists of a red giant and white dwarf star.

Left Artist's concept of the two closely orbiting stars of 44i Bootis, which pass in front of one another every three hours. The red arrow indicates the direction of orbit.

ACCRETING BINARIES OR INTERACTING BINARIES

If the stars are close enough, mass can be transferred from one star to the other. When mass is transferred between normal stars, the observable effects are slight—emission lines and some X-ray emission. The most spectacular interacting binaries consist of a normal, or giant, star and a compact object—a white dwarf, neutron star, or black hole. As the material leaves the normal star it does not fall directly onto the compact object, but spirals, forming a disk about the compact object. This is because the gas is deflected as the two objects orbit each other.

As the gas falls it releases energy heating up the disk. For white dwarfs, the disk gets hot enough to shine in the IR, optical, and UV parts of the spectrum. For neutron stars and black holes, even more energy is released in the disk, which becomes much hotter, glowing mostly in X-rays. Changes in the amount of matter falling onto the compact object, or changes in the structure of the disk, cause all of these systems to be variable.

Variable Stars

Variable stars are valuable as astrophysical laboratories. Depending on the processes occurring, a variable star's light curve may reveal such properties as its mass, luminosity, distance, or even its diameter. While the data is normally interpreted by professional astronomers, amateurs play a major role in data collection. There is no other area of the science of astronomy in which the work of amateurs is so important.

Opposite This dramatic spiral galaxy was captured by Hubble Space Telescope. Cataloged as NGC 1309, it is part of the Eridanus group of galaxies. Astronomers are searching for variable stars in NGC 1309, particularly pulsating stars known as Cepheids, and eclipsing binary star systems.

Below Image of the supergiant shell LMC-4 in the Large Magellanic Cloud, taken with the Curtis Schmidt telescope as part of the Magellanic Clouds Emission Line Survey.

Changes in the brightness of stars is the most common kind of variability but other types also occur, particularly changes in the spectrum or light that comes from the star.

ERUPTIVE VARIABLES

T Tauri variables have not yet reached the stable main sequence. These very young stars rotate rapidly, generating strong magnetic fields that cause intense bright flares and cool starspots. Their brightness varies irregularly by several magnitudes. Some magnitude drops may be due to their circumstellar disks partially occulting them.

FU Orionis variables, also very young stars, are still located in their birthplace. They may have large amplitude magnitude changes on time scales of years.

UV Ceti stars are the flare stars, red dwarfs of spectral class M. While the mass and luminosity of these dwarf stars is far less than that of the Sun, they have much deeper convection zones. Those that also rotate rapidly twist their magnetic field lines far more than the Sun twists its own magnetic field, and thus they experience more powerful flares than our star does. With the luminosity of these stars being low, and the flares being very powerful, UV Ceti stars can have very brief spikes in brightness, sometimes of many magnitudes. Monitoring such unpredictable variables is an ideal task for amateur astronomers.

R Coronae Borealis variables are cool red giants with significant carbon in their outer layers that has been brought up from the core. Carbon soot occasionally forms in their atmospheres or in their slow-moving stellar winds, and dims them by up to 9 magnitudes for months or years. Easily monitored with binoculars, or even the unaided eye, the prototype of the class, R Coronae Borealis, is a favorite of amateurs. Normally about magnitude 5.8, R CrB fades in a matter of weeks to anywhere between 9th and 15th magnitude at irregular intervals, typically every few years.

CATACLYSMIC VARIABLES

These are close binary stars in which there is an exchange of mass. They include novae and dwarf novae. Novae are binaries in which gas from the outer hydrogen-rich layers of a star, typically a red giant, is streaming into an accretion disk around a white dwarf, and ultimately onto the white dwarf's surface. As hydrogen builds up in a shell on the compact white dwarf, and is compressed by both the new material added on top, and the tremendous gravity of the white dwarf, the critical temperature of 10 million°K is eventually reached at the base of the hydrogen shell, and then the hydrogen ignites and runaway fusion occurs. Novae increase in brightness by 7 to 16 magnitudes in as little as a day or, rarely, as long as several hundred days. Although the main show is usually over in a few weeks, it takes years for the star to completely fade back to its pre-outburst magnitude. Several novae have repeat performances, and it is thought that probably all novae are recurrent, although outbursts might be hundreds to thousands of years apart in some systems.

U Geminorum dwarf novae have sudden increases of 2 to 6 magnitudes, with tens to thousands of days spent sitting at their normal minimum magnitude between outbursts. The similar Z Camelopardalis stars exhibit occasional "standstills"—periods of constant brightness that are about one-third of the way from normal maximum to minimum.

ECLIPSING BINARIES

Eclipsing binaries are oriented in space in such a way that their orbit is edge-on to us, and thus the orbital velocities, as measured by Doppler shift using spectroscopes, are the true velocities, not just components. The eclipse duration and the calculated size of the orbit gives us the diameters of the stars.

Algol, Beta (β) Persei, has an eclipse every 2.867321 days, during which it fades from its normal magnitude of 2.1 down to 3.3 and rebounds again within 10 hours. At maximum eclipse, about 79 percent of the primary star is hidden by its companion. During an evening when an eclipse is predicted, it is enjoyable to watch one unfold by comparing Algol's brightness to that of the neighboring stars every 15 minutes or so, using only the naked eye.

It is even more enjoyable to catch an eclipse without foreknowledge—experienced skywatchers just naturally make a quick scan of the sky whenever they are outside at night. If you do so routinely, you can expect to notice Algol looking rather faint compared to Gamma (γ) Andromedae, about twice a year on average. Checking your almanac will verify your findings, because the magnitude drop is enough to be unmistakable.

A star's Roche lobe is a mathematical surface defining the maximum size to which a star in a binary system can grow, without losing mass. If an evolving swollen star—typically a red giant—overfills its Roche lobe, then the

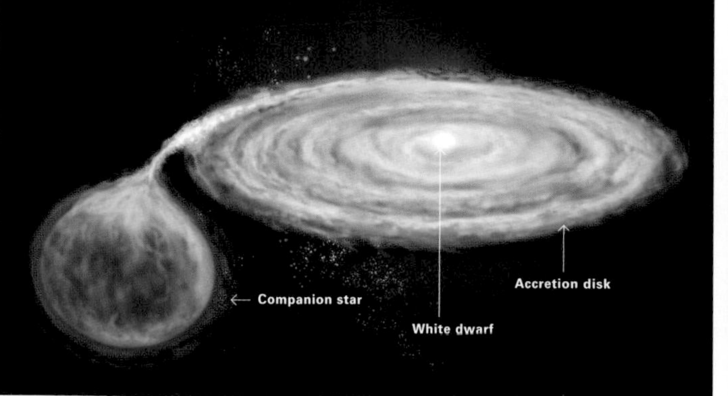
Companion star
White dwarf
Accretion disk

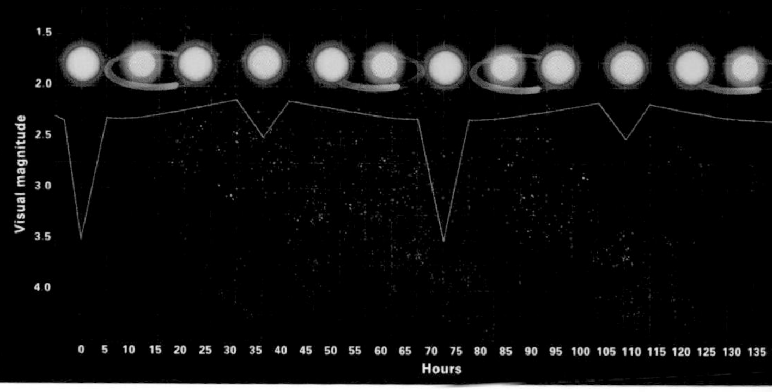
Visual magnitude
1.5
2.0
2.5
3.0
3.5
4.0
0 5 10 15 20 25 30 35 40 45 50 55 60 65 70 75 80 85 90 95 100 105 110 115 120 125 130 135
Hours

companion star's gravity will cause gas to flow from the swollen star onto its more compact companion. While Algol is now classified as a detached system, in which neither star is overflowing its Roche lobe, it is obvious that at some time in the past it was a semi-detached system in which one star overflowed its Roche lobe, and gas accreted onto the other star. The history of mass transfer in Algol is revealed by the current properties of the two stars. Most of the system's luminosity comes from a 3.7-solar-mass main-sequence star that is periodically eclipsed by a cooler 0.8-solar-mass subgiant. We know that if two stars are born at the same time, the more massive star will evolve faster and reach the red giant stage first. The "Algol paradox" asked how the subgiant, the more evolved star, could be so much less massive than the main-sequence star. The answer is obvious to us now: The subgiant—originally the more massive star—evolved to the red giant stage, overfilled its Roche lobe and transferred mass to its, originally, less-massive companion. So much mass was transferred that the red giant shrank to a subgiant.

Meanwhile, the main-sequence star, its mass having been greatly increased by the gas that it accreted from its companion, will evolve faster than it would normally have done. When it, in turn, becomes a red giant, it can be expected to transfer gas back to its original owner!

The most fascinating eclipsing binaries are the contact binaries, like Beta (β) Lyrae. The stars are deformed by gravity, and are within a common atmosphere. The light curves vary continuously as the stars not only eclipse, or partially eclipse each other, but also because of the changing areas of the oval stars that are presented to our view.

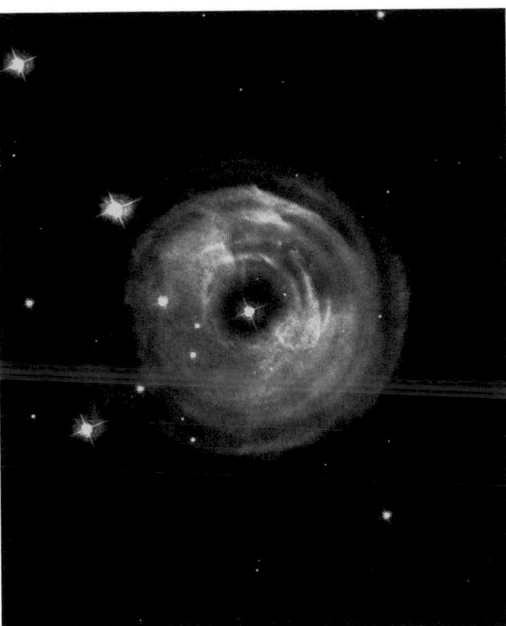

Left Light echo from V838 Monocerotis. The star presumably ejected the illuminated dust shells in previous outbursts. The dust then reflected this light to Earth months later.

Opposite Hubble image of star V838 Monocerotis. The illumination of interstellar dust gave off a pulse of light similar to setting off a flashbulb in a darkened room.

Opposite bottom left Image of a white dwarf as it captures matter from a companion star. As captured matter falls onto the surface of the white dwarf, it accelerates and gains energy.

Opposite bottom right Image of the Algol binary system showing a large percentage of each star is eclipsed during primary and secondary eclipses. The light curve shows that the change in light output corresponds to a change of more than 1 magnitude, visible to the naked eye.

ROTATING VARIABLES

Gamma (γ) Cassiopeiae variables are rapidly-rotating spectral type B stars with strong mass outflow that forms ring or disk structures around the star. These structures can partially occult the star, causing a temporary drop in magnitude.

RS Canum Venaticorum stars have small magnitude fluctuations that are believed to be due to dark or bright spots rapidly rotating across the star's disk.

PULSATING VARIABLES

Long-period variables are red giants, or supergiants, of spectral type M or carbon stars, with periods of 80 to 1,000 days, varying from 2.5 to 11 magnitudes.

The most famous member of the class, Mira, Omicron (ο) Ceti, reaches maximum every 11 months. At minimum, it averages magnitude 9.3. An average maximum is magnitude 3.4, but it can get much brighter. In mid-February 2007, orange Mira peaked at magnitude 2.0, completely dominating its section of the sky. In November 1779, Herschel recorded that Mira surpassed magnitude 2.0 Hamal, "so far as almost to rival Aldebaran." At its maximum, Mira will be well placed in the sky through 2010—a delight for all observers.

Semiregular variables, giants, and supergiants, that vary by 1 or 2 magnitudes over periods of a month to years, show some regularity, but only at times. When red supergiant Betelgeuse, a semiregular variable, occasionally brightens to magnitude 0.2, it will even impress the experienced skywatcher more used to a star that is similar to Aldebaran in brightness.

RR Lyraes are giants, normally of spectral type A, that pulsate with periods of 0.2 to 1.2 days, and an amplitude of between 0.5 to 2 magnitudes. Although many are halo stars, they are commonly called "cluster variables" because of the large number found in globular clusters. These stars are crucial in determining the distances of globulars. All RR Lyraes have nearly the same luminosity, so their apparent magnitude is related to their distance, if the dimming of intervening dust can be properly accounted for.

Supergiant star Delta (δ) Cephei is the prototype of the pulsating Cepheid variable stars. In 1912, Henrietta Leavitt discovered the relationship between a Cepheid's period and its intrinsic luminosity, giving astronomers a powerful tool for determining distances. Larger and more luminous Cepheids take longer to expand and contract. This is called the Period-Luminosity Law. Once a Cepheid's period—typically 1 to 50 days—and apparent brightness have been determined, its known true luminosity allows its distance to be calculated. Because these are among the brightest stars, modern telescopes can resolve Cepheids in galaxies as far away as the Virgo Cluster. Cepheids have been crucial in determining the scale of the universe.

Amateur astronomers contribute variable star observations to bodies including the British Astronomical Association and the American Association of Variable Star Observers.

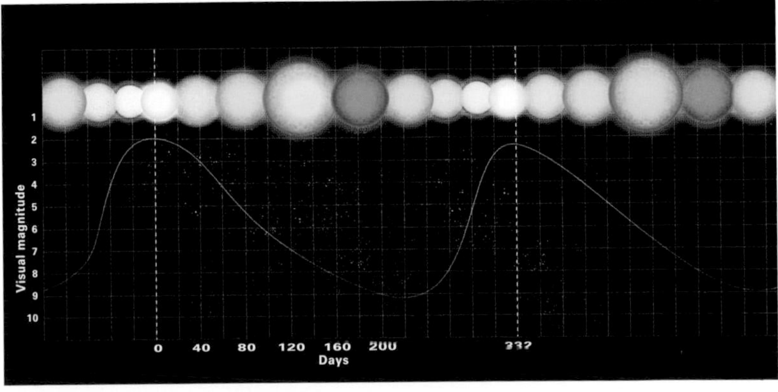

Below Image of the light curve of the Mira variable star during its pulsating stage when the surface layers expand and contract in repeating cycles. The pulsations result in a change in the magnitude.

The Violent Universe

Despite appearances, the universe can be a ferocious and changeable place, full of titanic explosions and star-shattering collisions. The results are supernovae, black holes, and gamma-ray bursts that are continually remodeling the universe in furious splendor.

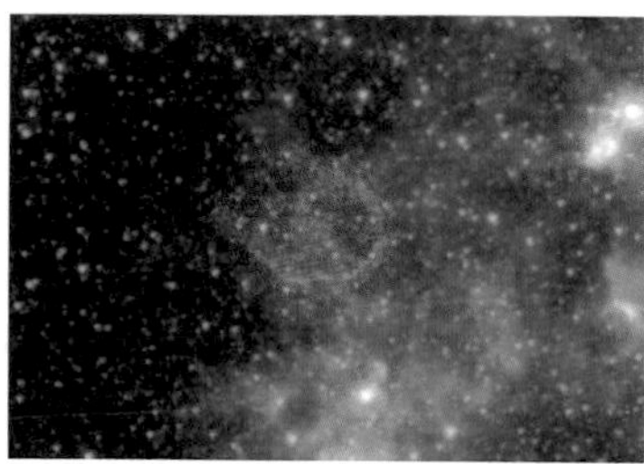

Above This composite image from Spitzer Space Telescope and Chandra X-ray Observatory shows the remnant of an explosion known as N132D. The remnant itself is seen as a wispy pink shell of gas at the center.

Below Illustration of an accretion disk around a black hole. The inset shows a close-up of the central area. Inside the event horizon of the black hole (black), the extreme curvature of space bends the light rays back into the black hole, so no light escapes.

To the unaided eye, the night sky at first seems a very tranquil and unchanging environment. But as astronomers look deeper and further with their giant telescopes, they see that the universe is anything but peaceful and calm. There are events and processes happening out there that simply boggle the mind, with energies and velocities that stretch the limits of theory.

SUPERNOVAE

One of those highly energetic events is a supernova. The word *nova* is Latin for "new." A nova is a star that periodically brightens and fades in the night sky. A supernova is one that brightens dramatically before fading away into perpetual obscurity. The physical processes involved are quite different: A nova doesn't get destroyed, but a supernova does.

Before the advent of modern astronomical observations and theories, it was not really known what a supernova was. It is now realized that, despite the inference implicit in the "new star" tag, supernovae are actually stars that are ending their lives, not starting them.

Supernovae come in two main types: The explosion of a very massive star, or the explosion of a dense compact star.

The first scenario typically occurs when a star that is many times more massive than our Sun reaches the end of its nuclear power cycle, having used up its internal supplies of elements that can be fused into heavier elements. As the nuclear reactions fail, the outward heat pressure that had been sustained by those nuclear reactions also fails, and can no longer compete with the inward pressure of the star's own huge gravity. Gravity wins and the star crushes in on itself, both imploding and exploding in the process.

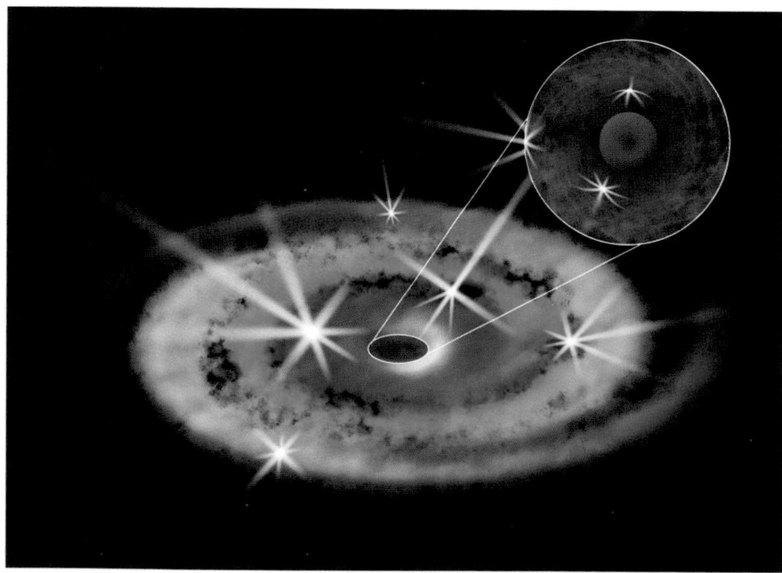

Right This is an artist's enhancement of the expanding remnant of the supernova explostion that created the Crab Nebula, a 6-light-year-wide expanding supernova remnant. Japanese and Chinese astronomers witnessed this violent event in 1054. The Crab Nebula is arguably the single most interesting object, as well as one of the most studied, in all of astronomy. The nebula has a diameter of 11 light-years and is expanding at a rate of about 930 miles per second (1,500 km/s). The Crab Nebula (M1, NGC 1952) is located about 6,300 light-years from Earth, in the constellation Taurus.

The outer shells of the star are thrown off and expand into space to eventually form a beautiful nebula. But the inner regions are squashed down to form a tiny dense object.

Depending on the initial mass of the star, that object might be a neutron star or a black hole.

A neutron star is formed when the electrons in the atoms are squashed into those atoms' nuclei, largely combining with the protons to form extra neutrons. What's left is mostly just a mass of neutrons, packed in incredibly tightly. So tightly, in fact, that a single teaspoon of neutron star material would contain around 110 million tons (100 million tonnes) of matter!

The other main supernova scenario involves a white dwarf star. A white dwarf is the small fading ember of an ordinary-mass star near the end of its life. Our Sun will eventually become a white dwarf. But for one to become a supernova requires the presence of a normal sort of companion star in a close binary orbit (our Sun does not have one of these, so it won't go supernova). The white dwarf continually pulls some of the ordinary star's gas onto itself, slowly building it up until the white dwarf's mass reaches a critical limit, around

1.44 times the mass of our Sun. At this point, the white dwarf literally can't handle the pressure of all that extra mass. It then becomes unstable and suffers a huge nuclear reaction that destroys it.

In each of these two supernova scenarios, the energy given off can, for a short time, make the supernova appear brighter than all the rest of the stars in its host galaxy combined.

Supernovae used to be considered quite rare back in the days when astronomers had only their eyes to rely upon. With the advent of photography, more supernovae were discovered in distant galaxies—but mostly after they had already exploded and faded away. Amateur astronomers have been responsible for discovering many supernovae; some dedicated hunters used to memorize the appearance of faint galaxies, and were able to tell in an instant if a new star had appeared. They would then report the discovery to the wider astronomical community, and professional astronomers would swing their telescopes into action.

These days, there are dedicated automated telescopes searching for supernovae and finding lots of them. The hard slog is now done by computers and automatic systems, although some amateur astronomers keep looking.

INTO THE ABYSS

What would happen if you fell into a black hole? It depends on your point of view. To an outside observer, a person falling into a black hole would seem to slow down when they reached the event horizon, coming to a complete standstill just outside the horizon—frozen in time and slowly fading away. This is due to the effect of strong gravity slowing down time. But if you were the person falling in, it would be very different. To you, your time would seem normal, but everything outside would seem to be going faster and faster. You would see the whole future of the universe flash before your eyes. Oh, and you would also get stretched out by the ever increasing gravity. Scientists call this process "spaghettification."

Right Artist's concept of a black hole at the center of a remote galaxy digesting the remnants of a star. The area around the black hole appears warped because the gravity of the black hole acts like a lens, twisting and distorting light.

Below Artist's concept of a supernova, a stellar explosion that creates an extremely luminous object that is initially made of an ionized form of matter—plasma. The light generated from a supernova can briefly outshine its entire host galaxy before fading over several weeks or months.

DARK STARS

Perhaps the most mysterious object in the astronomical pantheon, and the one that captures the public imagination the most, is the black hole.

Black holes come in two main types: Stellar mass black holes, and supermassive black holes.

Stellar mass black holes form during a supernova explosion, when the core of a massive star collapses in on itself. A star has to be many times the mass of the Sun when it collapses, in order to become a black hole; any smaller and it will become a neutron star instead.

The collapse squashes all of the star's remaining mass into a tiny point called a singularity. This body has such huge concentrated mass that it has an incredibly strong gravitational field. If anything gets too close, it will be inexorably pulled in. The point of no return is called the event horizon. Effectively it is in an invisible spherical boundary surrounding the black hole, inside which there is no hope of escape. Once past the boundary nothing can get out. The more massive the black hole, the bigger the event horizon.

The difficulty with investigating black holes is the fact that nothing, not even light, can escape from one, so there is no way to directly see one. All astronomers can do is study the effects a black hole might have on its surroundings. And this indeed is what has been done. Astronomers have very good evidence for black holes pulling in large amounts of surrounding material, e.g. gas, and even whole stars. This matter spirals in toward the black hole, forming a swirling ring of gas called an accretion disk. The material in this disk moves faster and faster, getting hotter and hotter, and gives off telltale radiation. It is this radiation that astronomers have detected and measured.

The most famous example of this kind of object is the first one that was found, called Cygnus X-1. It is a black hole about 8,000 light-years from Earth, with a companion star in a binary orbit. Matter from the star is being pulled into an accretion disk surrounding the black hole, giving off X-rays in the process—the detection of which is how it was discovered.

Another way that a stellar-sized black hole can be created is in the collision and merger of two neutron stars. Such collisions are thought to be relatively rare.

Astronomers now also have convincing evidence for the existence of supermassive black holes that reside in the cores of many—perhaps all—large galaxies, including our own Milky Way. The one in the center of the Milky Way is thought to have about three million times the mass of our Sun. But that's nothing compared to the black holes at the center of many other larger galaxies. Some of those are thought to have masses billions of times that of the Sun.

Some of these supermassive black holes betray their presence by producing huge "jets" of energy that pour out from the cores of these galaxies. It was only with the confirmation that such supermassive black holes exist that the nature of these jets could be explained.

There is still no solid accepted theory to explain how these enormous black holes could have formed in the first place and become so big, nor at what stage of the galaxies' evolution did it occur. Did the black hole come first, acting as a seed, with the galaxy forming around it? Or did the galaxy form first, with a single black hole growing bigger and bigger by gobbling up other black holes and stars like a runaway monster? No one knows the answer to this yet.

There also is evidence that intermediate size black holes might exist at the cores of globular star clusters.

It has been proposed that countless numbers of mini black holes could have been created during the Big Bang, but that is still conjecture at this stage. There is no observational evidence that such black holes exist.

THE SECOND-BIGGEST BANGS

Gamma-ray bursts (GRBs) are the biggest blasts the universe has seen since the Big Bang. These titanic explosions are seen several times per week by orbiting satellites, and are seen scattered around the universe at huge distances from Earth. Although astronomers have known about GRBs for decades, it is only in recent years that they have come to understand what might be causing them and where they are happening.

The story began in the early 1970s, when declassified US spy satellite data revealed short but intense bursts of gamma-rays coming from random directions in space. What they could be was a complete mystery. Without knowing how far away they were, it wasn't possible to accurately gauge how much energy was involved, and therefore what sort of process might be involved in producing them. Also, many of the

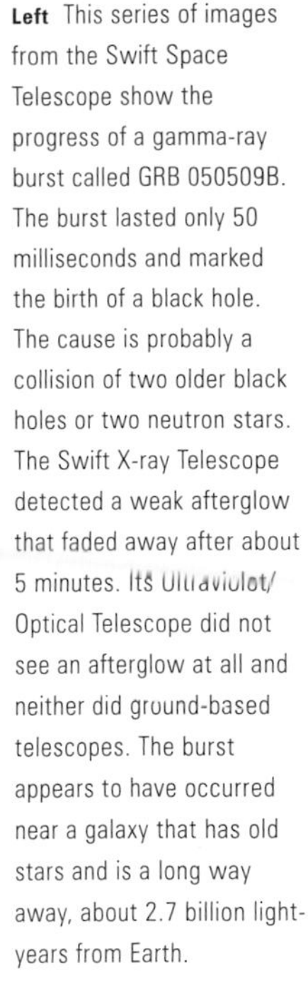

Left This series of images from the Swift Space Telescope show the progress of a gamma-ray burst called GRB 050509B. The burst lasted only 50 milliseconds and marked the birth of a black hole. The cause is probably a collision of two older black holes or two neutron stars. The Swift X-ray Telescope detected a weak afterglow that faded away after about 5 minutes. Its Ultraviolet/Optical Telescope did not see an afterglow at all and neither did ground-based telescopes. The burst appears to have occurred near a galaxy that has old stars and is a long way away, about 2.7 billion light-years from Earth.

SUPERNOVA 1987A

On the evening of February 23, 1987, something was seen in the night sky that had not been witnessed since the telescope had been invented in the seventeenth century. It was an exploding star—a supernova. Although other very faint supernovae had been seen in far distant galaxies, it had been almost 400 years since one bright enough to be seen with the unaided eye had been spotted. It was named supernova 1987A, and it was in the Large Magellanic Cloud, a close-by satellite galaxy of our own Milky Way. This was the first supernova able to be studied with modern instruments, and it gave astronomers a vital insight into the processes of star death.

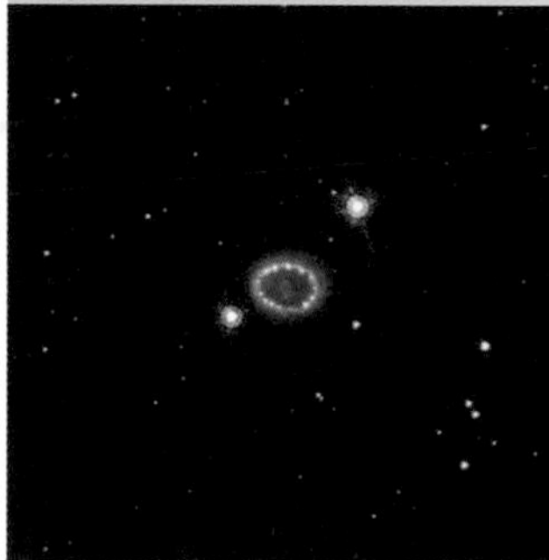

Left Hubble Space Telescope image of Supernova 1987A. The outburst was visible to the naked eye, and being the brightest known supernova in almost 400 years, it gave modern astronomers their first opportunity to see a supernova up close.

Above This is an artist's impression illustrating the explosion of SN 2006gy, a massive star that became what scientists are calling one of the brightest supernovae ever recorded. Supernovae usually occur when massive stars exhaust their fuel and collapse under their own gravity; in this case the star could have possibly been 150 times larger than our own Sun.

bursts were very rapid—a matter of seconds—which didn't permit follow-up observations. Subsequent scientific satellites confirmed that these gamma-ray bursts were real, and not a malfunction of the military satellites.

Most astronomers thought they must have been caused by unknown events within our Galaxy. It was an instrument called BATSE aboard NASA's now-defunct Compton Gamma-Ray Observatory, launched in April 1991, which proved vital in providing clues as to the origin and nature of GRBs.

For a start, BATSE found that GRBs were popping up in random directions around the whole sky; if they had been occurring only in our Galaxy, they should have been confined to only certain regions of the sky. BATSE also confirmed that GRBs come in two basic types: Short duration ones of less than two seconds; and long duration ones, up to a couple of minutes.

The problem with observing GRBs, however, was pinning down their exact direction in the sky. It's hard to do this with gamma-rays. So it wasn't until a later joint Italian/Dutch scientific satellite called BeppoSAX came along that progress resumed. BeppoSAX was equipped with both gamma-ray and X-ray detectors, and used the latter to spot the X-ray afterglow of a GRB. This enabled astronomers to accurately pinpoint its direction in space and turn other telescopes onto it. And they saw that it came from a very faint, very distant galaxy. This settled the case for whether GRBs were a phenomenon within out Milky Way Galaxy or were very far away. The latter was the case.

But the mystery remained. What was causing these huge outbursts of energy?

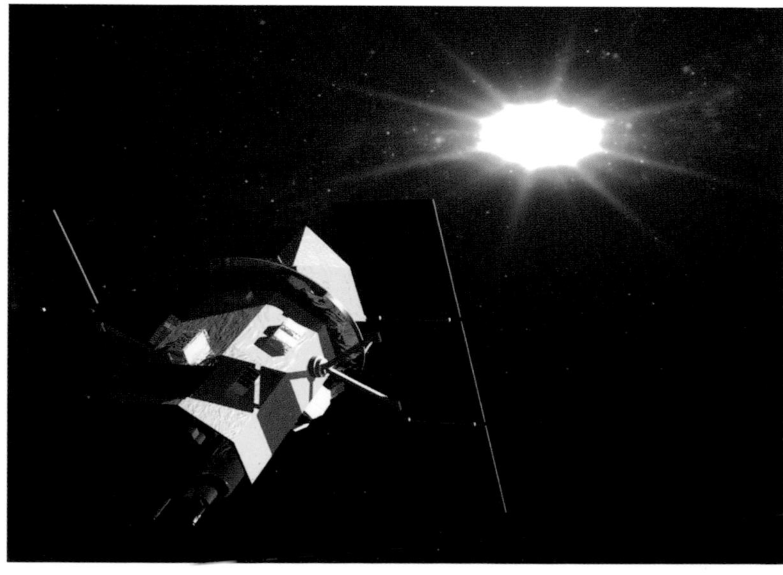

Right This artist's impression shows the Swift satellite observing a gamma-ray burst. Swift is a first-of-its-kind multi-wavelength observatory dedicated to the study of gamma-ray burst (GRB) science. Its three instruments work together to observe GRBs and afterglows in the gamma-ray, X-ray, ultraviolet, and optical wavebands.

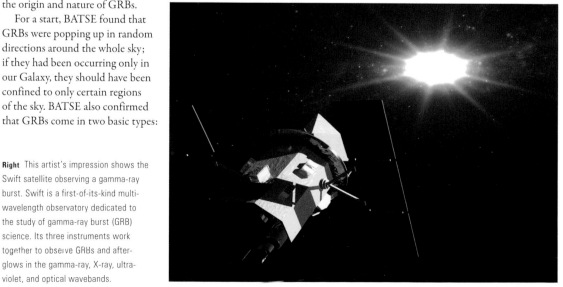

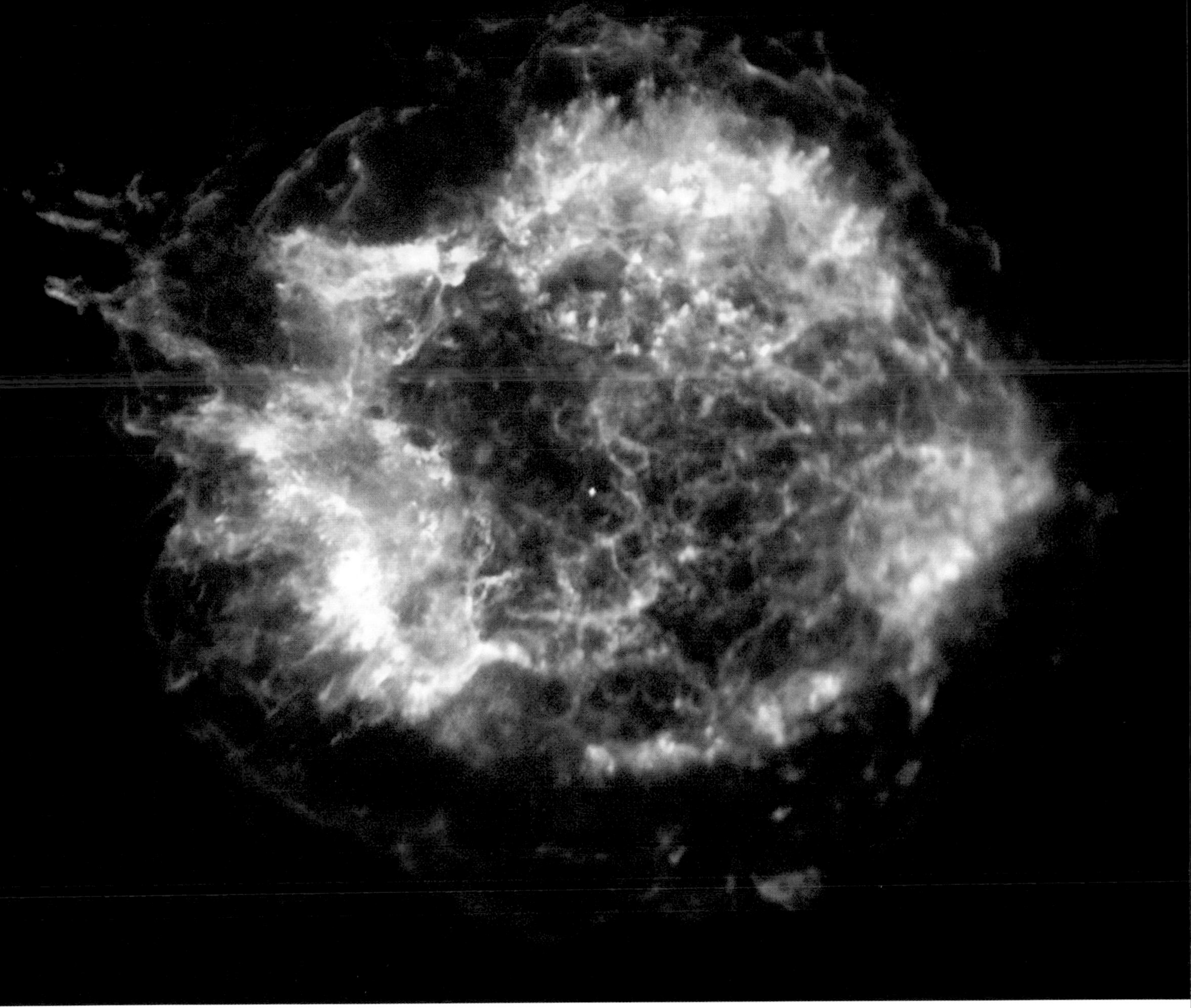

NASA's Swift Satellite mission, launched in November 2004, is a multi-wavelength observatory dedicated to the study of gamma-ray burst (GRB) science. It has the ability to quickly pick up gamma-ray bursts and examine them not only in gamma-rays, but also at X-ray and optical wavelengths. Data collected from Swift has helped clinch the solution to the great gamma-ray burst mystery.

There are two proposed mechanisms for the production of a gamma-ray burst.

One involves an extremely large star that goes supernova, collapsing to form a black hole. In the process, it spews out lots of energy that is concentrated into a narrow beam, which is why they seem to be so overly bright. This process creates the long-duration gamma-ray bursts.

Short duration gamma-ray bursts— generally defined as those lasting less than 2 seconds—are probably also created by supernovae, but smaller ones where the outpouring of energy is spread evenly instead of being confined to a narrow beam, or "jet" as astronomers call it.

COLD WAR PUZZLE

The first detections of gamma-ray bursts came not from scientific observations, but from US spy satellites. The Vela satellite system was designed to pick up the gamma-ray flash from nuclear weapons detonated in space, and it was intended to catch out the Soviets should they try to do a test. No such tests were ever confirmed, but almost from the moment the satellites were launched, they began to pick up gamma-ray flashes from deep space.

The information was kept secret until 1973. From then on, astronomers picked up the trail and the hunt for gamma-ray bursts was on.

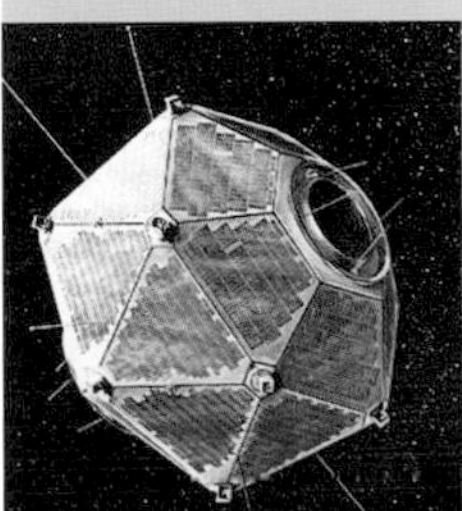

Left The Vela-5B nuclear test detection satellite program was one of the first satellites to report the existence of gamma-ray bursts.

Above This extraordinarily deep Chandra image shows Cassiopeia A, the youngest supernova remnant in the Milky Way. Analysis shows that this supernova remnant acts like a relativistic pinball machine by accelerating electrons to enormous energies. The blue wispy arcs show where the acceleration is taking place in an expanding shock wave generated by the explosion. The red and green regions show material from the destroyed star that has been heated to millions of degrees by the explosion.

Star Clusters

There are two types of star clusters—globular clusters and open clusters. All of the stars in a particular cluster were formed from the same collapsing molecular cloud. While the most massive stars form faster than low-mass stars, all of the stars were formed at approximately the same time—within a few million years—and have the same chemical composition. These two facts make star clusters valuable astrophysical laboratories.

Right The Greek god Perseus with Andromeda, from a fresco, *c.* 50–79 BCE. Perseus was the hero who killed Medusa. According to mythology Perseus rescued and fell in love with Andromeda. Perseus and Andromeda were turned into constellations at the end of their lives.

The observed differences between the members of a cluster must either be because of their different masses, which determine the rate that stars progress through the various stages of a star's life, or because some are single stars, and some are binary or multiple stars.

If you are just starting out as an observer, star clusters will give you the greatest pleasure and begin a marvelous journey into the wonders of our Galaxy.

OPEN CLUSTERS

Open clusters are inhabitants of the Milky Way's disk. They contain anywhere from tens, up to hundreds of stars, and, very rarely, thousands. The unusually rich open clusters, Messier 37 in Auriga, and NGC 2477 in Puppis, each have almost 2,000 stars. Many new open clusters are still being formed in the Milky Way today.

GLOBULAR CLUSTERS

Globulars are great spheres of stars, with a typical membership of hundreds of thousands, all packed into a diameter of about 60 to 150 light-years. Science fiction authors delight in writing about the sky's appearance from a planet located in the core of one of the great globular clusters. The sky would feature thousands of stars rivaling Sirius, and many that are comparable to the brighter planets!

In most of our Galaxy's globular clusters, all the stars that were as massive as the Sun have already become white dwarf cinders, which shows that these clusters are very ancient. Most Milky Way globulars are 12 billion or more years old; some are a few billion years younger. The Magellanic Clouds, nearby satellites of our Galaxy, have a population of younger globular clusters. Globular clusters are also being created in some colliding galaxies, as revealed by dramatic Hubble Space Telescope images. Intriguingly, recent infrared studies of the, perhaps misnamed, Cygnus OB2 Association suggest that it may actually be a very young Milky Way globular cluster.

A few globulars, like Omega (ω) Centauri, are known to contain millions of stars, although there is evidence that some of the more massive globulars were once the nucleus of one of the many dwarf galaxies that the Milky Way has cannibalized as it grew. When the Milky Way ingests a dwarf galaxy, the dwarf's more loosely gravitationally-bound outer stars are stripped off by our massive Galaxy's gravity, joining the Milky Way's halo stars. Like other globular clusters, most of Omega (ω) Centauri's stars are ancient, and have only 2.5 percent of the Sun's proportion of heavy elements, those heavier than helium. But there are two other groups of stars within this globular with enhanced heavy elements. That means that they constitute two much younger stellar populations that formed later, and that then suggests that Omega (ω) Centauri may be the remnant of a dwarf galaxy.

Most, or all, globular clusters lose some of their stars each time that their orbits take them through the Milky Way's core, or too near to a giant molecular cloud as they pass through the disk. There are a few badly stripped globulars that have only a few thousand stars left. One is the very dim cluster NGC 5053, located only 1° southeast of the normal globular M53. Gravitational assist is another mechanism that costs globular clusters some stars over time. When two of a globular's teeming stars pass close to each other, one star may occasionally be accelerated enough to reach the cluster's escape velocity. The other star involved in this encounter sinks toward the cluster's core.

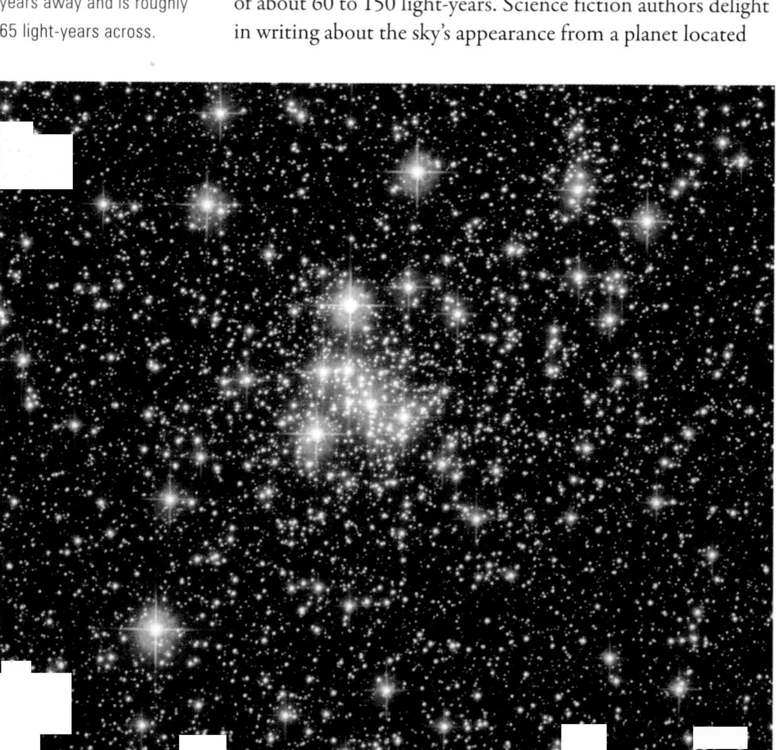

Below Hubble Space Telescope image of open star cluster NGC 290 in the Small Magellanic Cloud. The brilliant open star cluster is located about 200,000 light-years away and is roughly 65 light-years across.

Opposite Hubble uncovers an underlying population of infant star clusters embedded in nebula NGC 346 that are still forming from gravitationally-collapsing gas clouds. The smallest is only half the mass of our Sun.

TWELVE EXCEPTIONAL OPEN CLUSTERS

- NGC 3532 in Carina is a favorite—a huge oblate swarm of fairly equal-magnitude stars. It is one of three naked-eye open clusters that frame the Eta (η) Carinae Nebula.
- Most observers choose the Pleiades as the finest open cluster. A very bright group, it is impressive to the naked eye, and remarkable in binoculars, or a wide-field telescope. If it has a weakness, it is that the stars have a wide range of magnitudes.
- M11 in Scutum is another dense swarm of fairly equal-magnitude stars. This showpiece also gets many deep-sky enthusiasts' votes as the best open cluster to observe.
- Next is Carina's small but bright Gem Cluster, NGC 3293—truly delightful.
- The Hyades, which form the head of Taurus, delight in binoculars, or with the unaided eye.
- The Coma Berenices star cluster is another large group that is best viewed with the unaided eye or binoculars. It is the only deep-sky object that forms its own constellation.
- Magnificent to see is the double cluster in Perseus, NGCs 869 and 884. While neither would make it into the top twelve alone, the pair are striking together. This was one of the "little clouds" known to the ancients, and forms Perseus's upraised scimitar.
- The unaided eye is drawn to M7 in Scorpius, because it is the third-brightest patch in the entire Milky Way. M7 is best observed in wide-field telescopes.
- NGC 2516, another of Carina's many splendors, resembles the Beehive Cluster, but is much richer. Many colored stars highlight this magnificent object.
- The 95 arc-minute-diameter Beehive Cluster, M44 in Cancer, boasts several triple stars. This was another of the "little clouds" that has been known since antiquity.
- M37 is another very rich swarm of fairly equal-magnitude stars. An orange star in its center highlights this finest of the Auriga clusters.
- And last is NGC 457 in Cassiopea, popularly known as the ET Cluster, Airplane Cluster, Owl Cluster, or Kachina Doll Cluster.

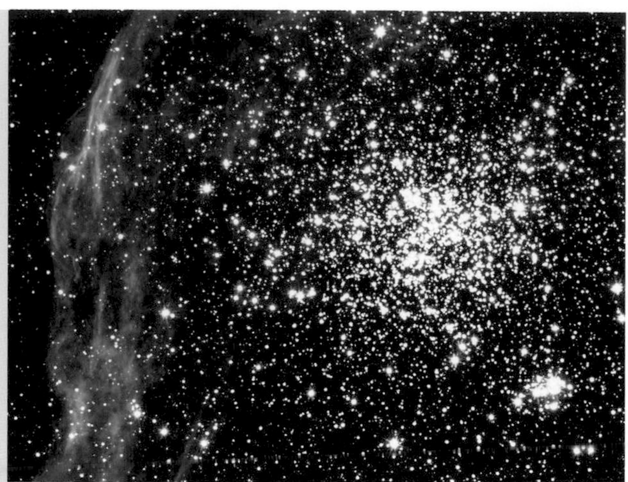

Above Located in the Large Magellanic Cloud, NGC 1850 consists of a main globular cluster in the center and a younger smaller cluster, composed of extremely hot blue stars and fainter red T-Tauri stars.

Below Hubble's sensors have refined the distance to the star cluster Pleiades at about 440 light-years. A circumference about the size of the Moon is overlaid here to give scale to Hubble's very narrow view.

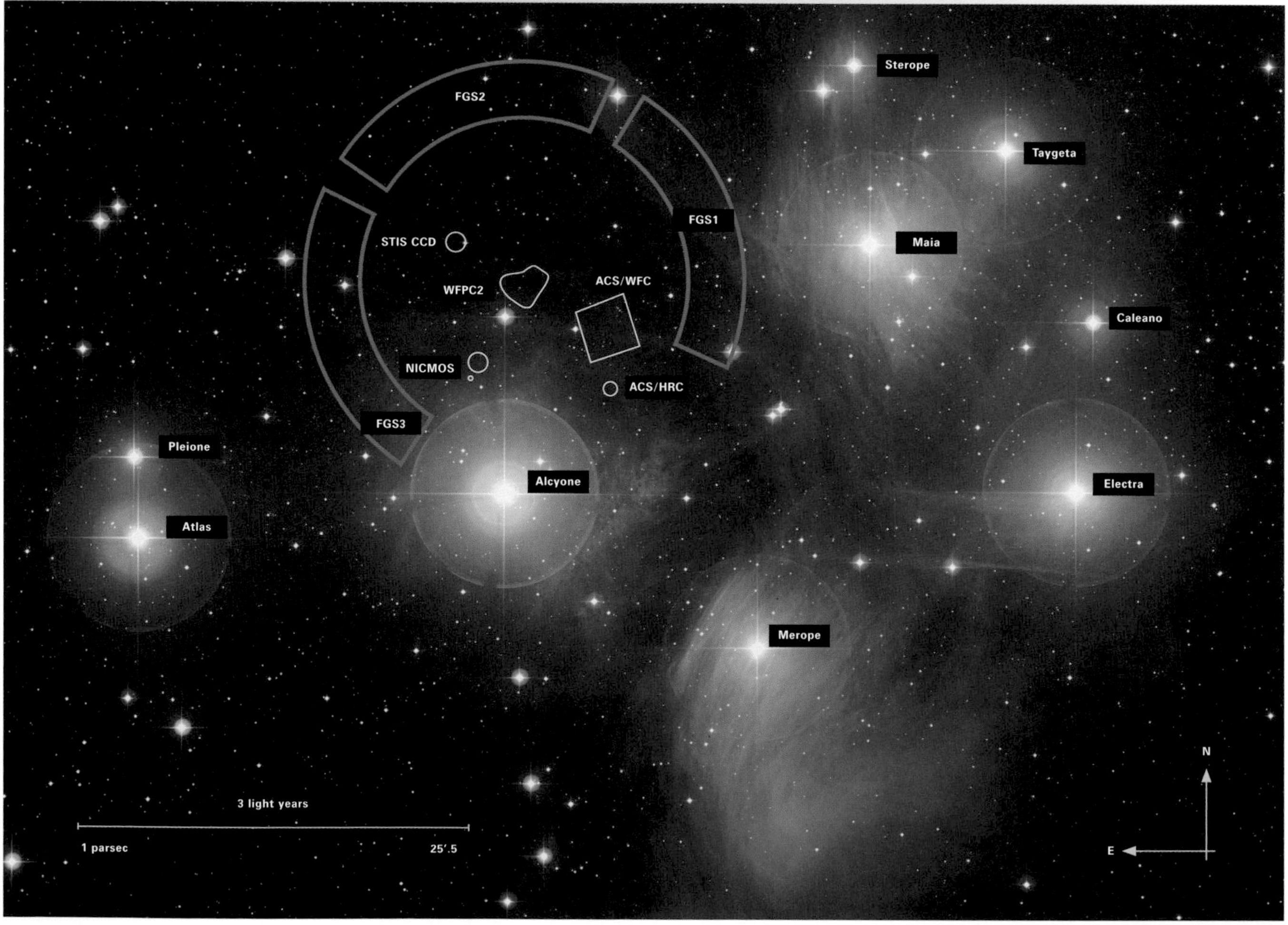

their derived true luminosities, the distance which would dim stars to that degree can be calculated, after allowing for any dimming by interstellar dust.

Hertzsprung-Russell diagrams can also determine a cluster's age by a technique called "main-sequence turnoff." We know that the very luminous stars at the upper left end of the main sequence are short-lived. When a given cluster's stars are plotted on a Hertzsprung-Russell diagram, there will be a spectral type at which giant stars appear above, and to the right, of the main sequence. No hydrogen-burning main-sequence stars will appear to the left of that turnoff point.

The spectral type of the "main-sequence turnoff" point allows the cluster's age to be determined, because astrophysicists know how old a star of that particular spectral type will be when it has just recently evolved into a giant. "Blue stragglers" are found in small numbers in the dense cores of globular clusters and they are a curious exception to the rules above.

In a globular, which all other evidence indicates should be many billions of years too old to have any remaining luminous blue-white stars, a few "blue stragglers" may be found. These are thought to result from recent mergers of binary stars in the swarming core, creating massive stars where none existed before. Such "blue stragglers" will then furiously burn hydrogen, acting like the behavior of the cluster's original allotment of massive stars, all of which expired billions of years ago.

Left Artist's impression shows how the Arches star cluster appears from deep inside the hub of our Milky Way Galaxy. Although hidden from direct view, the massive cluster lies 25,000 light-years away and is the densest known gathering of young stars in our Galaxy.

Above Hubble Space Telescope image of a globular cluster called G1, a large bright ball of light consisting of at least 300,000 old stars. G1, also known as Mayall II, orbits the Andromeda Galaxy (M31), the nearest major spiral galaxy to our Milky Way.

DISTANCE AND AGE

A Hertzsprung-Russell diagram can reveal a cluster's distance. If, instead of luminosity, the magnitudes of a cluster's stars are plotted on a Hertzsprung-Russell diagram's vertical axis, and if the shape of the cluster's main sequence is compared to that of a standard reference model main sequence, then "main-sequence fitting" allows the true luminosity of cluster stars to be determined. By comparing stars' apparent brightnesses to

THE TEN FINEST GLOBULAR CLUSTERS

These clusters are all showpieces, presented in the order of their eyepiece impression.

Unfortunately, for Northern hemisphere observers, most globular clusters reside in the southern half of the sky, including most of the finest globular clusters.

- The second-brightest globular cluster is 47 Tucanae, and the second-largest, but it is the most dramatic in the eyepiece. Its dense core is clearly yellow in bigger scopes, due to the concentration of red giants massed there.
- Omega Centauri is the brightest and biggest, but is rather lacking in individuality, except for the two darker patches within it. It is a huge rich oval of uncountable stars, but some consider it lacks central condensation.
- Pavo's NGC 6752 is the fourth-brightest globular cluster. Arcing star chains converge to a small central pip.
- M13 is only the eighth-brightest, but with its long star chains and propellor-like dark lanes, the Great Hercules Cluster is perhaps the most distinctive in appearance. It is generally considered by observers to be the finest globular in the northern half of the sky.
- M4 in Scorpius is the fifth-brightest and third-largest. Its central bar and loose structure make this the only other candidate for the title of "most distinctive." M4 is currently thought to be the closest globular cluster. Consequently, it is one of the easiest to resolve.
- M22 in Sagittarius is the third-brightest. Even a 2-inch (50 mm) refractor resolves a few stellar pinpoints.
- M15 in Pegasus rises to a remarkable sharp cone, like a classic volcano cone.
- NGC 1851 in Columba is much like M15, but fainter.
- Loose NGC 6397 in Ara is the second-closest, and the easiest to resolve.
- M55 in Sagittarius ranks next, but NGC 2808, M3, M5, and NGC 6541 all have good claims to rank in the top ten.

Nebulae

Nebulae and stars are inextricably intertwined. Stars are born from collapsing molecular clouds. Then, in their old age as red giants, stars enrich their surroundings with carbon that they have created from helium fusion. Heavier elements are produced by supergiant stars when they explode as supernovae, spewing those elements into space, perhaps to form planets, with elements such as iron, silicon, and gold, when another molecular cloud collapses.

Our Galaxy's great molecular clouds are usually detectable only as obscuring masses called dark nebulae or absorption nebulae, made visible by the paucity of stars in comparison to adjacent Milky Way fields, or in silhouette against bright nebulae. While dust grains make their presence known by their absorption and scattering of starlight, they typically make up much less than one percent of the mass of molecular clouds, which are primarily molecular hydrogen and helium, with trace amounts of other gases. The microscopic dust grains are carbon soot, or may be oxides of silicon, titanium, and calcium. These dust grains provide a site for molecules to form upon.

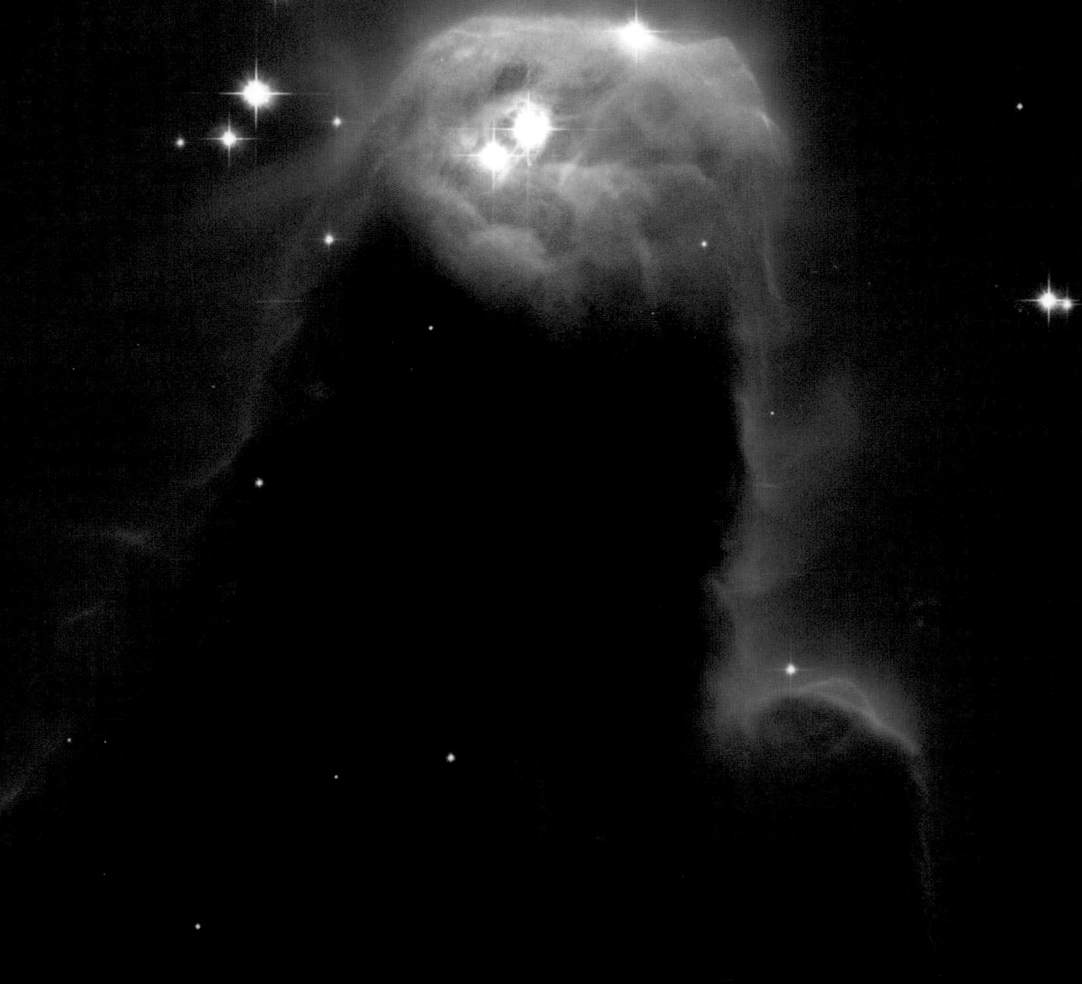

Below The Cone Nebula, a small dark nebula within NGC 2264, is an innocuous pillar of gas and dust—a giant pillar residing in a turbulent star-forming region. The Cone Nebula resides 2,500 light-years away in the constellation Monoceros.

DARK NEBULAE

Many dark nebulae are telescopic objects, but there are some fine ones for the unaided eye. The most obvious is the Coalsack, beside the Southern Cross. The Coalsack forms the head of a much larger object, the dark Emu of the Australian Aborigines. The southern end of the Great Rift, from Alpha (α) Centauri through to Scorpius, forms the bird's body. The Funnel Cloud Nebula, Le Gentil 2, the great dark nebula descending from Cepheus into Cygnus along Right Ascension 21 hours, is the second-best after the Coalsack. The Funnel Cloud Nebula is still a naked-eye object with a 10-day-old gibbous moon visible in the sky. The third-best dark nebula to see is the Pipe Nebula near Theta (θ) Ophiuchi. The Pipe Nebula forms the hindquarters of an even larger complex of naked-eye absorption nebula, the Prancing Horse Dark Nebula. Telescopes and binoculars reveal many smaller dark vacancies, especially in Cygnus and Sagittarius. Two of the finest are oval Barnard 92 and elongated Barnard 93, on the northwestern side of the Small Sagittarius Starcloud, M24. In most telescopes, a single 12th magnitude star is the only point of light within black B92. Another favorite dark nebula complex is called "Barnard's E," formed by Barnard 142 and 143. The dark "E" is easily found in binoculars immediately west of Gamma (γ) Aquilae. In an area of Cygnus that is marbled with dark nebulae, branching Barnard 168 snakes to the Cocoon Nebula, IC 5146.

REFLECTION NEBULAE

Sometimes dust is made visible by reflecting starlight. If stars are embedded in dust, the resulting reflection nebula is usually just an amorphous glow around a star or stars, like M78. One exception that shows lots of detail is the 15 arc-seconds long, bright orange Homunculus Nebula, a double-lobed reflection nebula. The

Above Color mosaic of the Orion Nebula (M42) reveals numerous treasures that reside within the nearby intense star-forming region. The bright star toward the lower left of the image, known as LP Orionis, is surrounded by a prominent reflection nebula. Astronomers believe the star is moving within another veil of material that lies in front of M42.

Right This Hubble Space Telescope photograph captures a small region within Messier 17 (M17). The hotbed of star formation resembles a storm of turbulent gases. M17, also known as the Omega or Swan Nebula, is about 5,500 light-years away in the constellation Sagittarius.

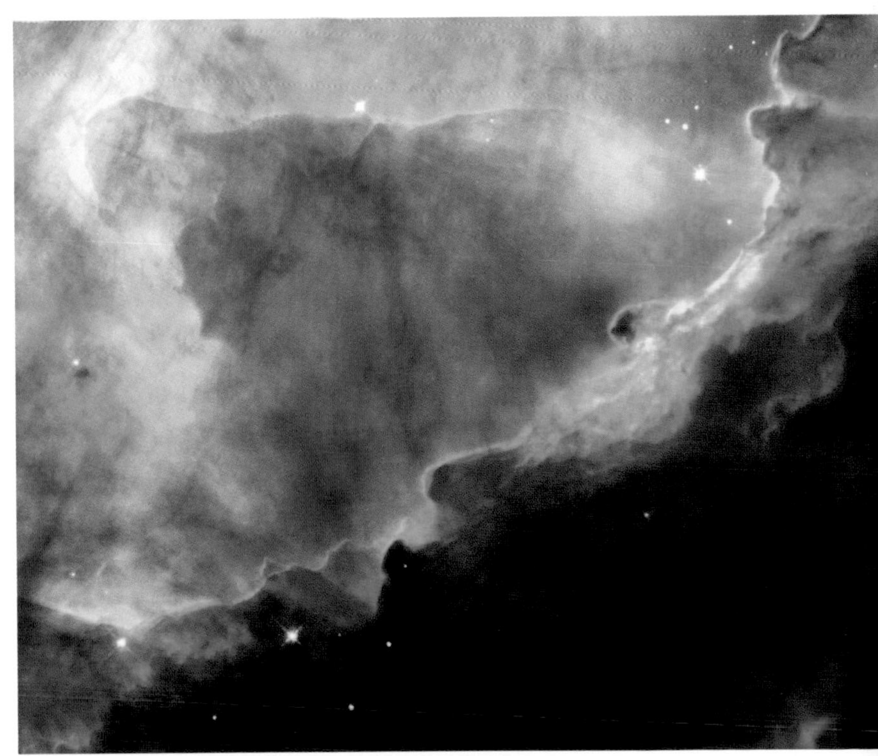

Homunculus surrounds and obscures—to varying degrees at different times—the massive star Eta (η) Carinae, at the heart of NGC 3372, a splendid complex of emission nebulosity and dark nebulae.

A discrete dust cloud physically near to, but not enveloping a star can form a dramatic reflection nebula. An example is IC 2118, the Witch Head Nebula, in easternmost Eridanus. It reflects the light of the supergiant star Rigel, forming a striking 2.5° long subject for astrophotographers, although it is always challenging at the eyepiece of a rich-field telescope. Since dust scatters short wavelengths more efficiently than longer wavelengths, the Witch Head Nebula is bluer than its illuminating star, Rigel. This selective scattering is akin to the reason that Earth's skies are blue.

EMISSION NEBULAE

When star formation has begun in a molecular cloud, ultraviolet radiation from the most massive young spectral type O stars will ionize the surrounding hydrogen gas, exciting it to fluorescence, and destroying nearby volatile dust. We see this as an emission nebula, also called an HII region. The massive young stars' energetic stellar winds and ultraviolet radiation form cavities in the molecular cloud. If a cavity is on our side of the cloud, we see a showpiece object like the Orion Nebula, in which the highly luminous Trapezium stars cause the walls of the cavity to shine. Many of the filaments of nebulosity in images of the Tarantula (NGC 2070), Orion (M42), Swan (M17), and Eta (η) Carinae (NGC 3372) nebulae are also visible at the eyepiece. But these nebulae have too low a surface brightness for an observer to see the bright colors that are on images, with one partial exception, the Orion Nebula. Some observers, especially younger ones, can see greenish-gray in M42's brighter central

Left Like a winged fairy-tale creature on a pedestal, this is actually a billowing tower of cold gas and dust rising from a stellar nursery called the Eagle Nebula, NGC 6611. It is 9.5 light-years high, about twice the distance from our Sun to the next nearest star.

area and, more rarely, a hint of pink in the "Bat Wing." The Tarantula Nebula, in the Large Magellanic Cloud, is one of the largest known. If it were as near to us as the Orion Nebula, and centered there, the Tarantula's spider leg-like filaments would fill the whole constellation of Orion. This huge nebula is visible with the unaided eye, despite being located in another galaxy.

NGC 604, within the galaxy M33, is another huge emission nebula, about 1,000 light years in diameter. A 4-inch (10 cm) telescope reveals NGC 604.

PLANETARY NEBULAE

Planetary nebulae were named by William Herschel for the resemblance that some of them have to the disk of a planet, especially a fainter disk like that of his discovery, Uranus. One planetary, NGC 3242 in Hydra, is even called the "Ghost of Jupiter."

Planetaries are shells of gas that have been ejected from the outer layers of red giant stars at the end of their lives. The very hot exposed core of the star radiates strongly in the ultraviolet, and excites this shell of gas to fluoresce for a relatively brief period —tens of thousands to no more than a million years—before it completely dissipates into space.

Planetaries come in many shapes, but luminous disks or rings are common amongst the brighter ones. Examples are NGC 3918, the Blue Planetary in Centaurus, and NGC 7662, the Blue Snowball in Andromeda. As their names imply, these small planetaries have a high enough surface brightness to show color. Many other small planetaries are blue-green or greenish, partly depending on an observer's personal color sensitivity. IC 418 in Lepus is very unusual—it is called the Pink Planetary because at low power in a large telescope it has a coppery outer rim surrounding the more usual bluish disk.

Larger planetaries have lower surface brightnesses, and are colorless at the eyepiece. Examples are M57, the famous Ring Nebula in Lyra, and NGC 7293, the huge 13 arc-minutes wide Helix Nebula in Aquarius. While many planetaries appear ring-shaped, their actual shape is believed to be more like a very short section of tubing with flared ends. Two-lobed structures are also common. One example is the superb Dumbbell Nebula in Vulpecula, M27, probably the finest of all planetaries. Perhaps the most unusual-looking one is the Spiral Planetary (named for its resemblance to a spiral galaxy), NGC 5189 in Musca.

WOLF-RAYET STAR OUTFLOWS

Massive Wolf-Rayet stars produce so much energy that their strong stellar winds drive off significant amounts of gas from their outer layers. These very luminous stars produce copious ultraviolet radiation, so they sometimes spawn an emission nebula that resembles a planetary nebula. NGC 6888, the Crescent Nebula in Cygnus, and NGC 3199 in Carina look much alike. NGC 2359 in Canis Major was nicknamed Thor's Helmet by astrophotographers because of its resemblance to a Viking helmet with horns. All three of these nebulae respond very well to OIII filters.

SUPERNOVA REMNANTS

Supernova explosions drive much of the star out into space. The original expansion velocity is typically around 3,100–7,500 miles (5,000–12,000 km) per second. The tremendous shockwave compresses the interstellar medium, causing the X-rays from plasma heated to a million degrees Kelvin, radio waves, and sometimes a visible-light emission nebula.

The Veil Nebula in Cygnus, a supernova remnant whose main arcs are NGC 6960 and NGC 6992/5, has a very complex filamentary appearance. A telescope equipped with an OIII filter reveals much of this marvelous detail, making the Veil a showpiece, under the constellation Cygnus.

Although the beautiful Vela Supernova Remnant looks similar to the Veil Nebula on images, Vela's offering is a much more challenging object visually.

Opposite In the most active starburst region in the local universe, lies a cluster of brilliant massive stars known as Hodge 301. Seen in the lower right-hand corner, Hodge 301 lives inside the Tarantula Nebula, which resides within the dwarf galaxy, the Large Magellanic Cloud.

OBSERVING SYNCHROTRON RADIATION

The Crab Nebula, M1 in Taurus, is the remnant from the brilliant supernova observed in 1054 CE. The broad S-shape seen visually is from synchrotron radiation, produced by electrons moving at relativistic speeds in the incredible magnetic field produced by the pulsar spinning at 30 times per second. The pulsar is the core of the star that went supernova. This is the only place in the sky that an amateur can see light produced by this mechanism, since in other supernova remnants synchrotron radiation is mainly confined to radio wavelengths. The reddish shreds of the star that are so dramatic on photographs, and whose rate of expansion can be measured on images taken decades apart, are unfortunately not visible at the eyepiece with normal amateur equipment.

Above Detail of the entire Crab Nebula, a 6-light-year-wide expanding remnant of a star's supernova explosion. Japanese and Chinese astronomers recorded this event nearly 1,000 years ago, as did Native Americans. The blue area on the image is the synchrotron radiation.

Planets of Other Stars

Our Sun is one of around a hundred thousand million stars in the Galaxy. It is a very ordinary star, and has a system of planets—why should other similar stars not have planets of the same type?

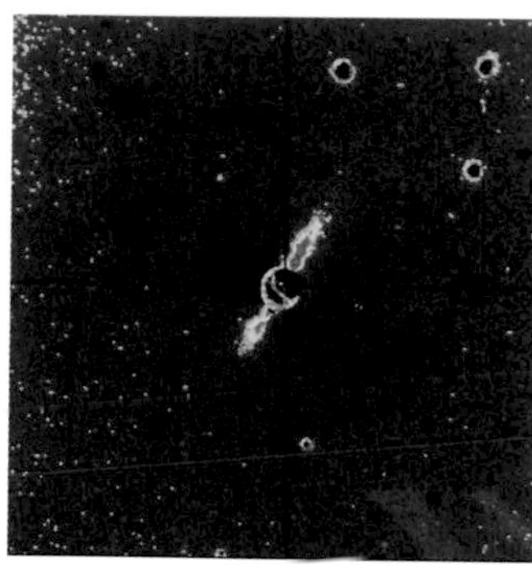

Above Digital photograph of the star Beta (β) Pictoris, 50 light-years from Earth. It appears to have a halo of dust and gas.

This seems logical. If the Solar System has been formed as the result of a near-collision between the Sun and a passing star it might well have been unique, or at least a cosmic rarity, but this theory has long since been abandoned. The Sun is a normal G-type star, classed officially as a yellow dwarf.

THE ROAD TO DISCOVERY

The first developments came in 1983, with the brief but profitable career of IRAS—the Infra-Red Astronomical Satellite. During preliminary testing observations, it was found that the brilliant star Vega (Alpha [α] Lyrae) had what was called "a huge infra-red excess"—in other words, it was associated with a huge cloud of cool, possibly planet-forming material. Other stars were found to have similar surrounds, and before long one huge dust-disk, around the star Beta (β) Pictoris, was actually photographed. In some cases, there were indications that there might be massive orbiting bodies, probably planets. On the other hand there was no definite proof, and there is always the chance of confusing a planet with a brown dwarf star, which is a low-mass star whose central temperature has never become high enough to trigger nuclear reactions such as those which make the Sun shine.

To make a direct observation of an extra-solar planet is obviously very difficult indeed. A planet is much smaller than a normal star, it shines only by reflected light, and it is so close to its parent star that it is drowned in the glare, just as a pocket torch would be overpowered if it lay beside a searchlight. However, various indirect methods can be used. The first of these is astrometric. An orbiting planet will pull upon its parent star, and the star will describe a small circle or ellipse around the center of gravity of the system. This will affect the radial velocity as seen from Earth—that is to say the toward-or-away movement—and spectroscopic observations can make use of the well-known Doppler effect to determine the dimensions of the system and the mass of the perturbing planet.

The first results from this so-called "wobble" technique were decidedly unexpected. In 1992, two astronomers, A. Wolszczan and D. Frail, announced that they had detected a planet moving around a pulsar. Since a pulsar is the result of a colossal outburst involving the collapse of a very massive star, the presence of a planet would indeed be hard to explain. If pulsar planets do exist they must be very strange places. Confirmatory observations have been reported, but it is not easy to see how a planet could have survived the formation of the pulsar itself. Presumably, it would have been captured after the outburst was over.

Right Digital composition of a much younger version of a Solar System similar to our own. Such a planetary system could very well exist in other galaxies.

Then, in 1995, the Swiss astronomers Michel Mayor and Didier Queloz, working at the Haute-Provence Observatory, detected a planet orbiting 51 Pegasi, a solar-type star 54 light-years away; it is slightly more massive and more luminous than the Sun. Its apparent magnitude is 5.5, so that it is nearly visible with the naked eye. The equipment used by Mayor and Queloz was sensitive enough to detect velocity changes down to 40 feet (12 m) per second, but the nature of the planet was surprising; the mass was about half that of Jupiter, the giant of our Solar System, but the distance from the star was a mere 4,350,000 miles (7,000,000 km), about one-eighth the distance between the Sun and Mercury. The orbital period was given as 4.3 days, and the surface temperature had to be around 2,372°F (1,300°C). The planet had to be a gas giant, more like Jupiter than like Earth, and when similar planets were found attending other stars they became commonly known as "hot Jupiters." Others were not so extreme, and more comparable with Uranus or Neptune; a few were less massive still. What is much harder to do is detect planets no more massive than Earth, simply because their effects upon their parent stars would be so small...although astronomers are getting closer. There is no logical reason to doubt that Earth-mass planets exist, and every reason to assume that they do.

The next important method of detection involves the transit of an orbiting planet in front of its star. The result will be a slight temporary drop in the star's brightness as part of its

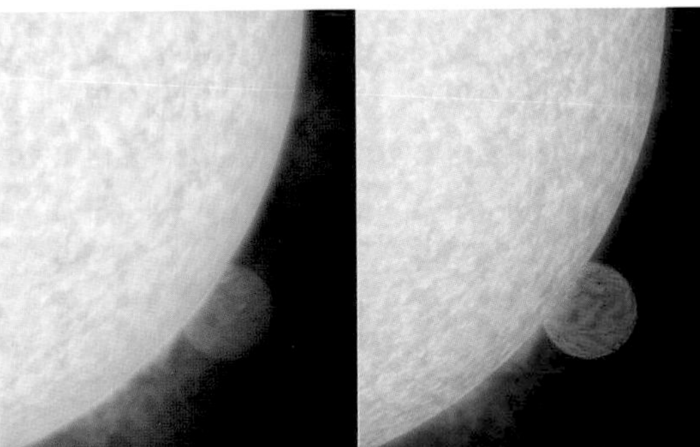

Above This is an artist's conception depicting the pulsar planet system discovered by astronomers Aleksander Wolszczan and D. Frail in 1992.

Left Close-up view of how a fiery hot star and its close-knit planetary companion might look if viewed in visible light (left) and infra-red light (right). Visible light allows the star to outshine its planet; in infra-red, the planet glows.

Opposite Artist's concept of a Jupiter-size planet orbiting the nearby star Epsilon (ε) Eridani. The IAU issued instructions for astronomers to "listen out" to this star for the existence of such a planet.

light is blocked out, but again it is necessary to be wary of confusing a transiting planet with a transiting brown dwarf, and obviously the planet must be large enough to make its presence known. The first success here came in 1999, when observers D. Charbonneaux and M. Brown tracked down a planet moving around the star HD 20645 in Pegasus, 150 light-years away; the magnitude fell by 1.7 percent, which is not very much. Here, too, we have a "hot Jupiter," at a separation of 4,350,000 miles (7,000,000 km) and a period of 3.5 days. The temperature is of the order of 1,382°F (750°C). In 2001, observations with the Hubble Space Telescope made it possible to analyze the atmosphere of the planet; spectra were taken first during transit and then with the planet out of view, so that the spectrum of the star was simply subtracted.

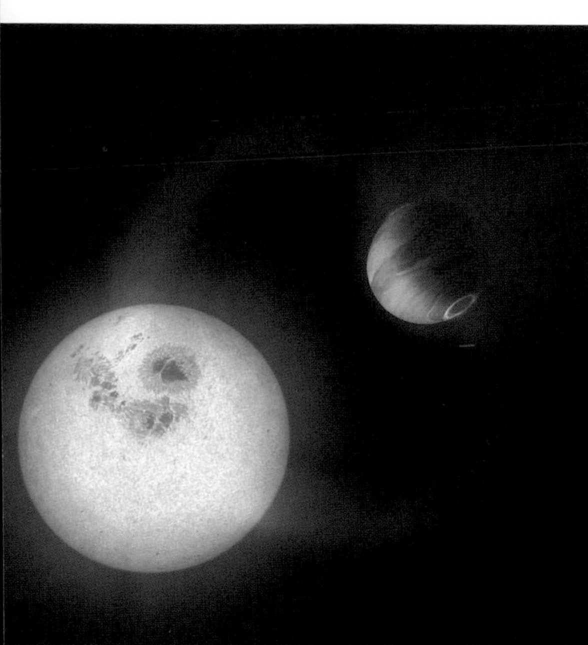

Above Artist's impression of a new type of exoplanet discovered by Hubble Space Telescope. They have been named "hot Jupiters" because they stay close to their stars with short orbital periods. This new class of planet is waiting to be explored.

No signs of water vapor were found, though there were indications of a cloud of silicate dust; however, not too much must be read into this, because investigations are still at a very early stage.

THE SEARCH CONTINUES

By 2007, over two hundred extra-solar planets had been located, mainly by the astrometric technique, together with a few transits. They are by no means all alike—for instance, there are some that are as massive as Jupiter, but less dense than water. There are true systems, such as that of the naked-eye star Upsilon (υ) Andromedae, which is 44 light-years away, and is certainly attended by three planets. The dim red dwarf star Gliese 876, in Aquarius (15.2 light-years away) also has three planets, one of which is no more than eight times as massive as the Earth; the brilliant Fomalhaut, in the Southern Fish, has a dusty disk that seems to be warped by the presence of a planet, and so on—there is plenty of variety.

Searches for Earth-mass planets are continuing. In April 2007, European astronomers announced the detection of a five-Earth-mass planet orbiting Gliese 581, a red dwarf star only 20.4 light-years from Earth. The planet—which would have a diameter probably 1.5 times that of Earth—is at the right orbital distance to be in its star's "Goldilocks Zone," where the temperature is just right for water to exist in liquid form—neither too hot nor too cold. This was the first detection of a close-to-Earth-sized planet in a star's Goldilocks Zone, making it the most "Earth-like" planet found so far. Expect more to follow.

Discoveries such as this inevitably lead on to the question of finding extra-terrestrial life. There have been several programs involving SETI, the Search for Extraterrestrial Intelligence. Radio methods are obviously the most promising, and in 1990, the International Astronomical Union even issued instructions about the procedure to be followed by anyone who picked up a message from deep space. Efforts were also made to "listen out" to two nearby stars that were regarded as promising candidates as planetary centers. Of these two stars, which are both within 12 light-years, one—Tau (τ) Ceti—was a disappointment, but the other—Epsilon (ε) Eridani—undoubtedly does have planets, though as yet we know little about them. If an advanced civilization exists there, radio contact would be possible—though since a there-and-back exchange of signals would take over 20 years, quick-fire repartee would be somewhat difficult.

One day, contact may be possible. At least we have learned a great deal during the last few years, and there is no longer any doubt that planetary systems are abundant in our Galaxy—no doubt in other galaxies too.

Right Digital composition of the planetary system around red dwarf Gliese 581, in the foreground. Gliese 581 is first planet, other than Earth, to be found existing in the "Goldilocks Zone," where the temperature is just right for liquid water to exist.

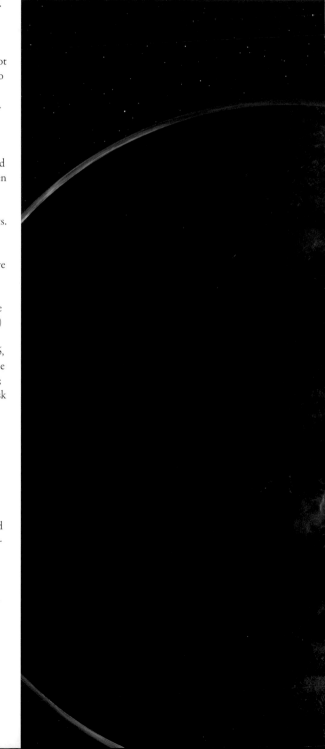

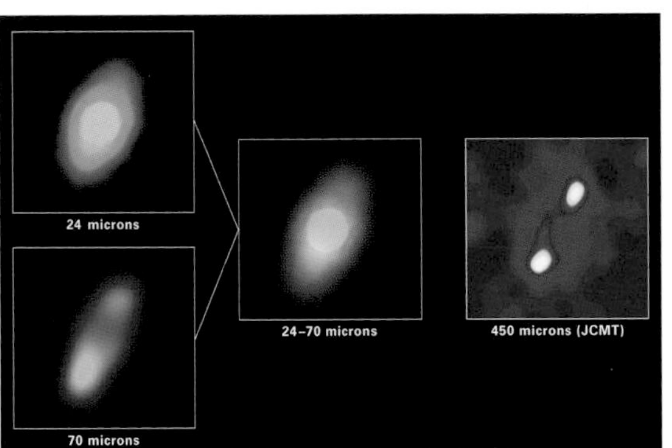

Left First infra-red images of the dust disk surrounding Fomalhaut, the 18th brightest star in the sky. Planets are believed to form from such a flattened disk-like cloud of gas and dust orbiting a star very early in its life.

Astrobiology

Are there other forms of life out there in the cosmos, or are we all alone? How will we ever know? Once considered a frivolous waste of time, these questions are now exercising the minds of leading scientists around the world.

The search for life beyond Earth, by necessity, falls into two broad categories, based on very simple constraints: There are places we can reach with our spacecraft, and there are places we can't. The former category includes the planets and moons of the Solar System, as well as comets and asteroids. The latter includes the environments in interstellar space and around other stars. For these, we are limited to studying them from afar, listening for any signals put out by any putative extraterrestrial civilizations, or analyzing extra-solar planets to see if they have the conditions necessary for life—at least, life as we know it.

IS ANYBODY OUT THERE?

Mankind's attitude to the possibility of life existing beyond our world has changed remarkably throughout the ages. Much of that change has come about from improved knowledge of the conditions that exist beyond Earth.

Below This artist's concept shows the *Cassini* spacecraft flying over Titan's surface with the Huygens probe nearing the surface. Saturn appears dimly in the background through Titan's thick atmosphere of mostly nitrogen and methane. Titan is considered a potential home to microbial life.

At first, the heavens were the home of gods, angels, and other mythical beings. Many religions rejected the belief that beings could exist elsewhere in the universe—after all, god had made Earth alone the place for his creation.

But as the centuries progressed, people were free to form opinions based on evidence and reasoning—even if some of that reasoning turned out to be wrong!

You only have to go back about 100 years, to the turn of the twentieth century, to find widespread beliefs that the planets Mars and Venus were inhabited by alien life forms. Some scientists thought that they could see evidence for changing vegetation patterns on Mars, and huge canal systems transporting water from the polar ice caps to the parched tropics. Venus, being closer to the Sun and covered with clouds, was obviously a jungle world, probably with dinosaurs and other exotic creatures roaming around.

Sadly, all those ideas were shown to be wrong, as improved telescopes and instruments, and early space missions, revealed the true nature of these worlds—cold and barren in the case of Mars, intensely hot and parched in the case of Venus. That didn't stop scientists from trying.

NASA's *Viking* probes landed on Mars in 1976 to directly test the planet's surface for signs of life. Sophisticated biological experiment modules were used by scientists to analyze samples of the surface dust. Despite initial chemical indications that microbial life had been detected, other non-biological explanations were soon found, and the excitement abated. With this result, the notion of life elsewhere in our Solar System went out of vogue.

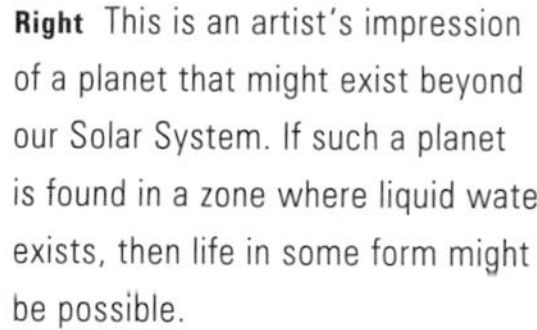

Right This is an artist's impression of a planet that might exist beyond our Solar System. If such a planet is found in a zone where liquid water exists, then life in some form might be possible.

SIGNS OF LIFE

So is it definite that we will never find life on the other planets in our Solar System? Since the time of the Vikings, evidence has been slowly increasing that the conditions necessary for simple life to evolve and perhaps flourish, are to be found in several niches throughout the Solar System. As well, scientists studying microorganisms on Earth had learned just how hardy life can be, surviving in places undreamed of not too long ago—for example, inside nuclear power reactors!

This is not to say that life definitely exists in any of these niches, but at least it raises the chances that it might.

Mars is still the obvious candidate. Its conditions are the most Earth-like of all the planets—it is quite possible that microorganisms could be flourishing just beneath the surface, protected from solar radiation and extreme cold. A significant find in recent years is the presence of small amounts of methane in Mars's atmosphere. Methane is a gas that does not last long; it has to be continually replenished. Volcanoes can do it, but Mars doesn't have any known active volcanism. Another possibility is microbes—some Earth microbes give off methane as part of their biological processes. It's an intriguing thought.

Venus is not considered a likely candidate for life. The temperature and pressure at its surface are far too inclement for any known life to exist.

The other major planets are not considered good candidates for life either, but some of their moons are. For instance, Jupiter's moons, Europa and Ganymede, are thought to have subsurface oceans that could provide a suitable habitat for microbial life.

Left The Very Large Array (VLA) is a radio astronomy observatory located on the Plains of San Augustin, New Mexico, USA. It consists of 27 independent radio antennae, each with a dish diameter of 82 feet (25 m). Astronomers use radio dishes like these to listen for possible extra-terrestrial signals from outer space. None have so far been found.

Above This is an artist's impression is of the *Terrestrial Planet Finder*, a mission due for launch in 2014. It will be used to search for Earth-like planets around other stars that might possibly harbor life.

SETI AT HOME

The search for extraterrestrial life is no longer confined to scientific institutions and remote observatories. Ordinary people can now get involved, through a groundbreaking program called *SETI@home*. Over the internet you can download a small program that runs on your computer when it is not being used for anything else. The program downloads small packets of data from radio telescopes that SETI searches, and sifts the data for signals that might be coming from outer space. Millions of people now participate in this program. It's safe, fun, and may just lead to the first evidence for life beyond Earth! More information at *http://setiathome.berkeley.edu/*

Above The Arecibo radio telescope in Puerto Rico is currently the largest single-dish telescope in the world. It is used to listen for signals from areas of the Galaxy that might contain intelligent extraterrestrial life.

Saturn's moon Titan, although freezing cold, and covered in very un-Earth-like chemicals—ethane and methane—could also be a potential hiding place for unusual microbes.

Comets are not seriously considered candidates for life as such, but there are serious suggestions that comets—which are covered with complex organic chemicals—might seed planets with some of the ingredients necessary for life to evolve, along with copious supplies of water, of course.

Studying the bodies in our Solar System using robotic probes is just one way of going about trying to find extraterrestrial life. But finding life is one thing; finding intelligent life is another. There is no serious suggestion that intelligent life exists in our Solar System elsewhere than Earth. For this search, scientists need to look further afield and use a totally different technique.

RADIO SEARCHES

The stars, and any planets they might have, are so far away that we cannot hope to devise any means of getting there any time soon. But scientists are able to study them from afar, and try to learn something about their nature, and whether intelligent life exists on them.

The distinction between simple life and intelligent life is important. The universe could be teeming with simple life, but we'd have a hard time trying to find it. Simple life does not betray its presence over astronomical distances. But intelligent life is likely engaged in activities that would make it easy to spot—radio signals, perhaps major engineering works on planetary scales, and so on. Powerful radio signals from Earth—television, radar, etc.— have been beaming out

THE DRAKE EQUATION

How many planets might have life? SETI—Search for Extraterrestrial Intelligence—pioneer, Frank Drake, came up with an equation to estimate the number. It involves some guesswork, but as the years go by the numbers become a little firmer. The equation is:

$$N = R_* \times f_p \times n_e \times f_l \times f_i \times f_c \times L$$

R_* is the rate at which stars form in our Galaxy; f_p is the fraction of stars that have planets; for each star that has planets, n_e is the number of those planets that can support life; f_l is the fraction of those planets where life actually gets started; f_i is the fraction of those where the life develops what we would call intelligence; f_c is the further fraction of those that develop the technology to communicate across interstellar distances (eg. radio transmitters); and L is a measure of how long those civilizations last before dying out or destroying themselves. Depending on the numbers you plug into the Drake Equation, the universe is either very empty of life or teeming with it. As the years progress, the latter option is looking better and better.

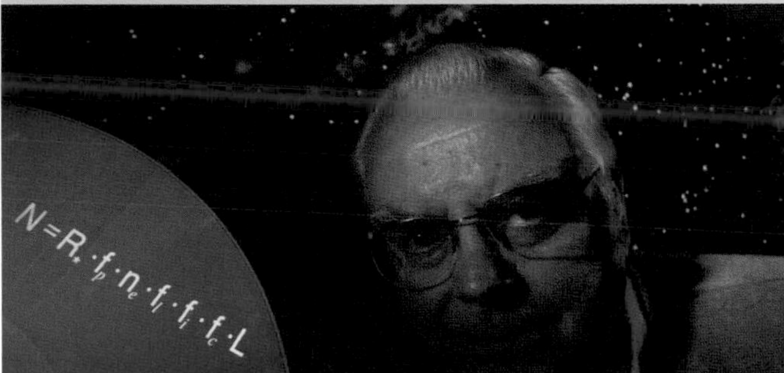

Left Frank Drake, the American astronomer and astrophysicist, is famous for founding SETI, the method of using home computers to process information from possible radio signals, and for creating the Drake equation in the 1960s. The equation allows scientists to quantify the uncertainty of factors which might determine the number of possible extraterrestrial civilizations.

into space for around 70 years. This means that any being within 70 light-years of us, equipped with an appropriate radio receiver, will already know that we exist.

Can we do the same? Can we pick up signals from an extraterrestrial civilization?

In 1967, scientists thought they had done it. British astronomers Jocelyn Bell and Antony Hewish were using a radio telescope to study the heavens, when they picked up a strong, regular, pulsating signal—too regular to be produced by natural processes. Initially they thought some sort of local man-made interference must have caused it, but they were soon able to eliminate that possibility. Not knowing what else it might be, they concluded that it might, just might, be a signal from an extra-terrestrial race. They called the signal LGM-1—Little Green Men 1.

But the excitement was short-lived. It turned out there was a natural process that could cause such a regular signal. LGM-1 was, in fact, the first pulsar ever to be discovered. Pulsars are rotating neutron stars that emit incredibly regular pulses of radio waves. These pulsing radio waves are more accurate than most of the clocks on Earth.

Since that time, many deliberate searches have been conducted using radio telescopes to try and pick up any signals that might be put out by extraterrestrials. No confirmed candidates have yet been found. This could mean that there are none to be found, but more likely it's simply because those searches have only scratched the surface. There are so many directions in which to look and listen, and so many different radio frequencies from which to choose, that it will take many decades yet before scientists can say that they have covered the whole sky. But the efforts continue.

PLANET FINDERS

Maybe the best hope of detecting life will come from new space telescopes, such as NASA's proposed *Terrestrial Planet Finder*. This satellite, and others, will try to spot planets in orbit around other stars and analyze the light reflecting from their atmospheres. This could reveal whether certain gases are present, such as ozone or methane, which could be clues to whether those planets sustain even simple forms of life.

Below Artist's concept of complex organic molecules comprised of carbon and hydrogen, and considered among the building blocks of life. The Spitzer Space Telescope has detected these molecules in galaxies that existed back when our universe was only one-quarter of its current age of about 14 billion years.

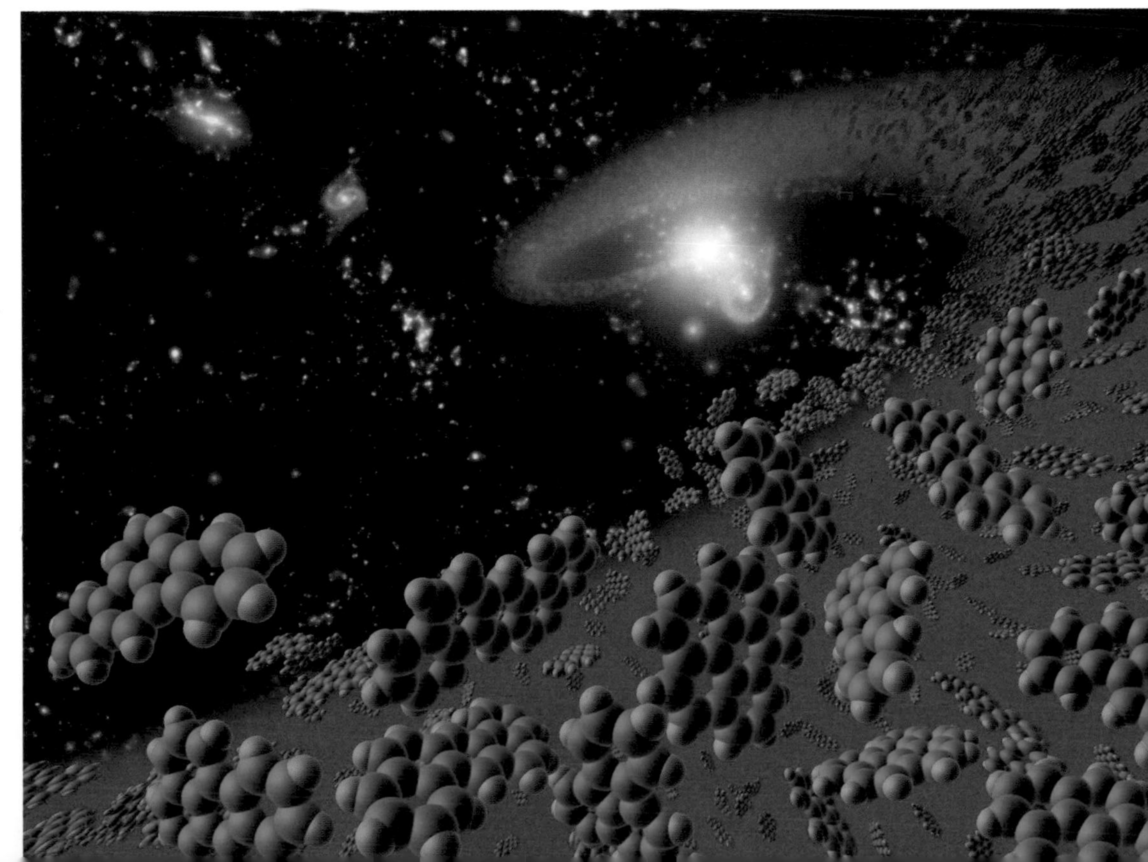

Galaxies

Cities of the Cosmos

As eighteenth-century astronomers surveyed the night sky with the very first telescopes, they came upon patches of light that resembled little clouds. At that time, these objects were all called *nebulae*, Latin for clouds, and all were thought to be within the Milky Way, believed to be synonymous with the entire universe. But they were wrong.

Below Hubble Space Telescope view of the disk galaxy, NGC 5866, in the northern constellation of Draco. Catching the galaxy tilted nearly edge-on, the image highlights a reddish bulge around a bright nucleus, a blue disk of stars parallel to the dust lane, and a transparent halo.

In 1845, amateur astronomer William Parsons, the third Earl of Rosse, built a giant telescope with a 72-inch (183 cm) diameter mirror, called the Leviathan of Parsonstown, in Ireland. It enabled astronomers to see that some nebulae had spiral shapes. The "spiral nebulae" were considered especially interesting, since they were thought to be perhaps other solar systems forming. The introduction of the spectroscope to astronomy in the 1860s showed that some nebulae, such as the Great Orion Nebula, were composed of gas, and such objects are still called nebulae today. But the spectroscope showed that the spiral nebulae had similar spectra to those of many stars. This led some astronomers to the bold idea that "the spiral nebulae" were distant island universes, each comparable to the Milky Way.

In the 1920s, Edwin Hubble, after whom the telescope is named, took very long exposures of the "Great Nebula in Andromeda" using the new 100-inch (254 cm) reflector on Mt Wilson in California, USA. His images resolved it into stars, and by the 1960s the word "nebula" ceased to be used to describe what we now call galaxies. The "Great Nebula in Andromeda" then became the "Andromeda Galaxy." Hubble also devised the first classification system for galaxies, based on their appearance. The system in use today had its origins with Hubble's, but it was later modified and made more complex by Gérard de Vaucouleurs.

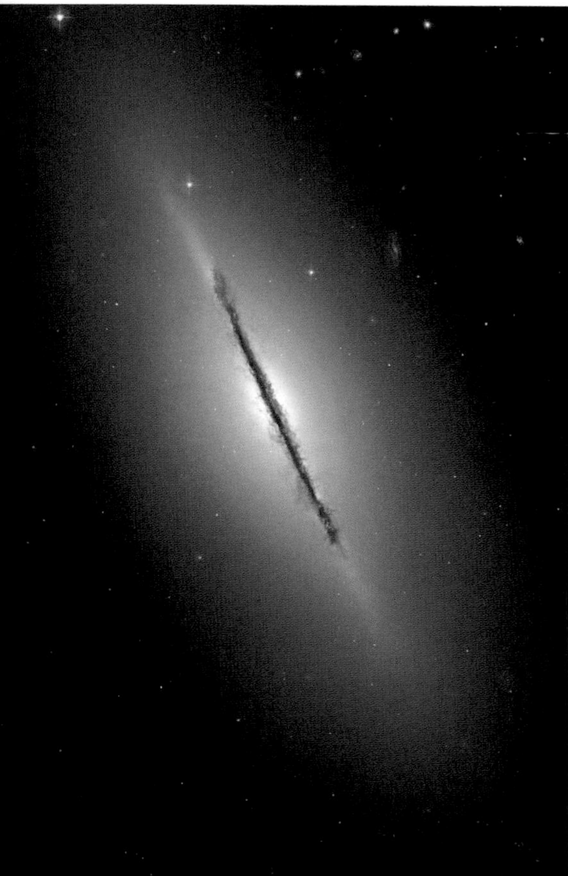

The classification system described below is the de Vaucouleurs' system. It is important to remember that every one of the billions of galaxies is different from every other one.

A classification system should aid in the discussion of galaxies by grouping them into similar types, but a convenient system will only work well to describe most galaxies. Some active galaxies like Centaurus A, NGC 5128, resist being slotted into models.

ELLIPTICAL GALAXIES

Elliptical galaxies look rather bland, but they include both the most massive galaxies, the giants found at the heart of large galaxy clusters, and the least massive, the barely detectable dwarf spheroidals. Elliptical galaxies are mostly composed of old stars, and have little cool gas and dust available to create new stars.

Elliptical galaxies are classified by their degree of flattening. An E0 galaxy appears nearly round, while an E7 is extremely elongated. A galaxy's classification depends partly on its orientation to us, since even a cigar-shaped galaxy seen end-on would appear round. Most elliptical galaxies have a star-like nucleus, and their surface brightness falls off steadily to the outer fringes. When two large spiral galaxies merge, as the Andromeda Galaxy and the Milky Way are predicted to do in three billion years or so, the result is believed to be a giant elliptical galaxy.

Examples of giant elliptical galaxies are M87 and M49 in the Virgo Galaxy Cluster. NGC 147 in Cassiopeia, a dE5 satellite of the Andromeda Galaxy, is an example of a dwarf elliptical. Dwarf spheroidals are even smaller.

SPIRAL GALAXIES

Spiral galaxies have the most attractive form. Normal spirals have an elliptical bulge of old stars at their core, surrounded by star-forming spiral arms. Some, like the Whirlpool Galaxy, M51, have only two major arms; these are called "grand design spirals." Others, like the Sunflower Galaxy, M63, have many thin arms, sometimes with branches.

Normal spirals are indicated by "SA," followed by a lower case "a," "b," "c," or "d," indicating how tightly wound the arms are, and the size of the central bulge relative to the disk. Intermediate cases are indicated by a combination, such as "cd." Thus, M51 is an SAbc. Barred spirals, indicated by "SB," are an even more beautiful variation. Two splendors are NGC 1365 (SBc) in Fornax and NGC 1300 (SBbc) in Eridanus. Both have quite prominent bars and two major arms, one extending from each end.

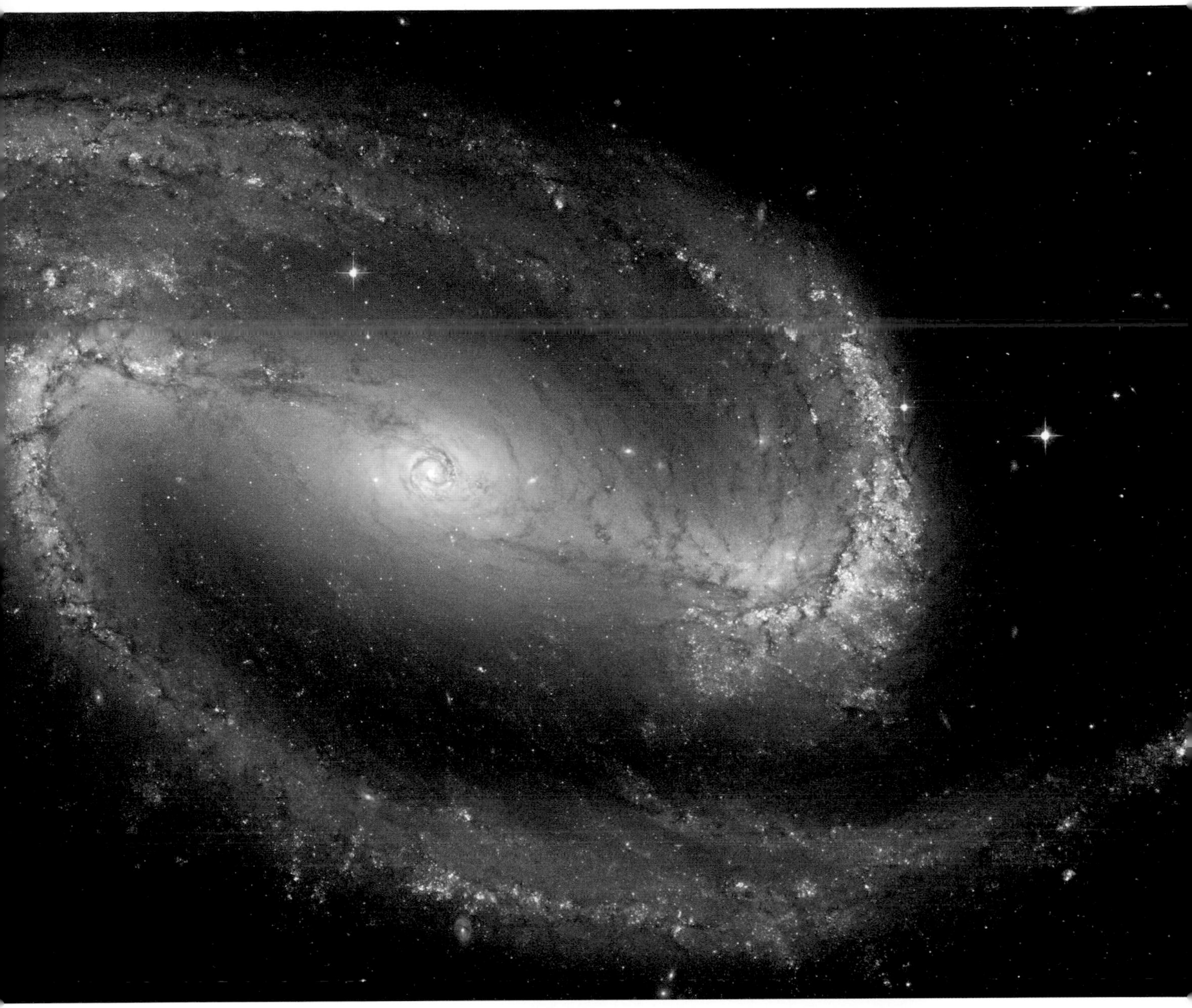

LENTICULAR GALAXIES

Lenticular galaxies, the prototype being Sextans's NGC 3115, the Spindle Galaxy, are an intermediate type between ellipticals and spirals. They are indicated by "SA0," or "SB0" if they display a bar. Lenticulars somewhat resemble spirals, having a central bulge and a disk, but they lack spiral arms. They are common in galaxy clusters.

IRREGULAR GALAXIES

Irregular galaxies, labeled "IA," or "IB" if barred, do not fall into any other classification. Most are relatively small, the prototype being the Small Magellanic Cloud. Many have active star-forming regions.

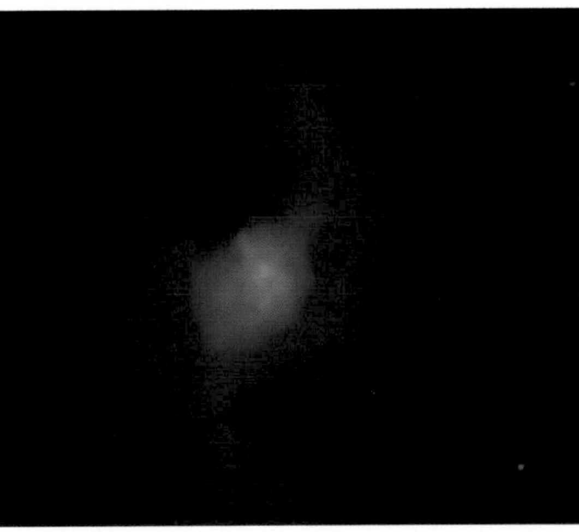

Above Hubble Space Telescope image of the barred spiral galaxy NGC 1300. This galaxy is considered to be prototypical of barred spiral galaxies. It is 69 million light-years away in the constellation Eridanus.

Left X-ray image of galaxy cluster MS0735, located about 2.6 billion light-years away in the constellation Camelopardalis.

Milky Way

Our Galaxy is called the Milky Way, because its appearance on a dark night reminded the ancient Greeks of spilled milk. The word "galaxy" has its root in the Greek word for milk, *galactos*, because astronomers knew of only one galaxy, our own, until the 1920s. The words "Milky Way," "galaxy," and "universe" were used interchangeably back then. Even today, the "Galaxy" usually refers to the Milky Way.

Right This painting titled *Birth of the Milky Way,* depicts Hercules and Juno. It was painted by Paul Rubens in 1636–37 and now resides in the Museo del Prado, Madrid, Spain.

While our home Galaxy—once thought to be the entirety of the universe—is now known to be only one of many billions of galaxies, we can still be proud of the Milky Way. Not only is our galaxy a spiral, the most beautiful kind, but it is also one of the two dominant members of the Local Group of Galaxies.

Within a diameter of about 100,000 light-years, the Galaxy has over 100 billion stars. The stars in the small central bar, in the huge spherical halo, and in the globular clusters, are mostly ancient. Most star formation now occurs within clouds of gas and dust in the spiral arms, which form a thin disk, only about 3,000 light-years thick.

The surrounding disk of atomic hydrogen gas has a diameter of about 165,000 light-years, while the mysterious dark matter, detectable only by its gravitational influence, extends out to at least the two major satellite galaxies—the Magellanic Clouds—and probably well beyond them.

A BEAUTIFUL EDGE-ON SPIRAL

Learning the nature of the Milky Way, and our place in it, has been a slow process of looking into our telescopes and studying the other spiral galaxies laid out before us. Those that are face on to us display a brighter central region that usually contains a prominent, almost starlike, nucleus.

With enough aperture, the larger, but fainter disk of some nearby galaxies can be resolved visually into mottled spiral arms. Long-exposure photographs make the arms much more obvious, and reveal supergiant stars and little knots of light. Spectroscopes show these knots of light to be star clusters and nebulae, just like those in the Milky Way. The sky offers a continuum of spiral galaxies that are oriented at angles increasingly farther away from face-on, all the way to some that are almost edge-on.

The edge-on spirals, like NGC 4565 in Coma Berenices, looked very familiar to astronomers who had studied the naked-eye appearance of the Milky Way from the isolated dark mountaintops where observatories are located. NGC 4565's very thin disk has dark dust clouds along its length, and there is a round and brighter central bulge. The equatorial dust lane of NGC 4565 is most prominent where it crosses this central bulge in silhouette.

In order to view these wonders of the Milky Way, you must do so from a true wilderness site, one completely free of light pollution or moonlight, when the constellation Sagittarius is at its highest. The effect is more impressive from the Southern hemisphere, where Sagittarius passes through the zenith. You will see the mottled Milky Way stretch across the sky from horizon to horizon, with a band of dark dust clouds, called the Great Rift, marching down its center, all the way from Deneb to Alpha (α) Centauri. The Milky Way widens significantly in Sagittarius, southern Ophiuchus, and Scorpius. Thus, from naked-eye observation alone, you can deduce that the Milky Way is an edge-on spiral galaxy.

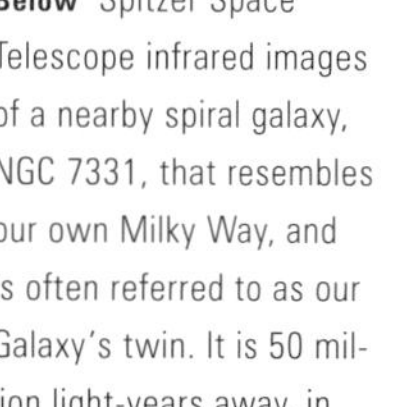

Below Spitzer Space Telescope infrared images of a nearby spiral galaxy, NGC 7331, that resembles our own Milky Way, and is often referred to as our Galaxy's twin. It is 50 million light-years away, in the constellation Pegasus.

Opposite Hubble Space Telescope image of the Sagittarius dwarf irregular galaxy. The brightest stars in the picture, easily distinguished by the spikes radiating from their images, are closer stars lying within our own Milky Way Galaxy.

Above Based on scans made by the second Palomar Sky Survey, this image shows NGC 2237, the Rosette Nebula, in the constellation Monoceros, on the outer end of the Local Arm of the Milky Way.

Right This is a composite radio-light image and rendition of our Galaxy, as seen in visible light. It shows high-velocity clouds of gas high above the plane of the Milky Way, which rain gas into the Galaxy, seeding it with the heavy elements necessary for star birth.

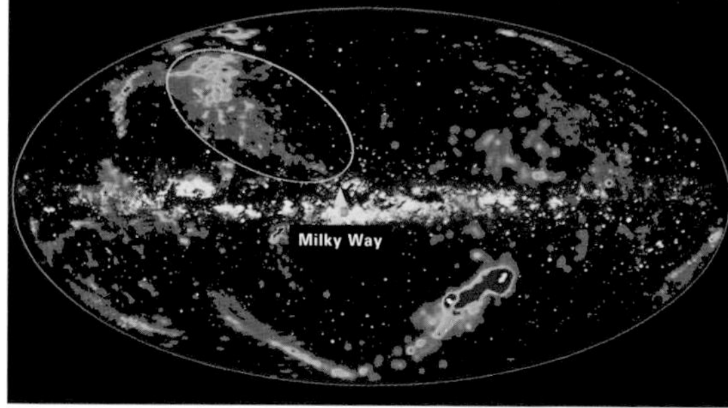

MAPPING THE GALAXY

The same dust clouds that provide these visual clues also hide our Galaxy's core. While an 8-inch (200 mm) telescope easily reveals the nuclei of hundreds of galaxies, the largest optical telescopes on the planet cannot see the Milky Way's nucleus at visual wavelengths.

Radio telescopes made the first breakthroughs in mapping the Galaxy. The distribution of atomic hydrogen was mapped by its strong signal at a wavelength of 21 cm. Next, the clouds of molecular hydrogen were mapped, using the knowledge that molecular hydrogen is always accompanied by trace amounts of carbon monoxide. Although molecular hydrogen cannot be directly detected at the temperatures of molecular clouds, typically about 10°K/–441°F/–263°C, carbon monoxide radiates at millimeter wavelengths. Radio telescopes eventually charted the Milky Way, revealing a barred spiral with many arms that, perhaps, resembles NGC 6744 in Pavo.

At the Sun's distance from the Galaxy's center—28,000 light-years—each orbit around the core at a velocity of about 135 miles (220 km) per second takes 230 million years. The Sun has no connection with any spiral arm—it just passes through them while orbiting the Galaxy.

THE SPIRAL ARMS

The Galaxy's spiral arms have been named. The Sun is on the inner edge of the Local Arm, frequently called the Orion Arm. When we view this arm, we are looking away from the galactic center. Three bright star clusters, the Beehive, the Pleiades, and M7 all lie near to us. Looking toward the outer end of the Local Arm, we see the supergiant star Rigel, the three stars of Orion's Belt, the Orion Nebula, and Monoceros's Rosette Nebula. Looking inward along the arm, we see the supergiant star Deneb, the Veil Nebula supernova remnant, and the North America Nebula. Supergiant stars are always associated with a spiral arm, because their short profligate lives are over before their orbit takes them far from their birthplace.

The next arm outward is the Perseus Arm, including the famous Double Cluster. Beyond, lies another spiral arm, called the Outer Arm.

Looking toward the galactic center, we see the Sagittarius Arm. As we look inward along this arm, we see the active star-forming regions M8, the Trifid Nebula, the Swan Nebula, and the Eagle Nebula. Looking outward toward the tip, we view the Eta Carinae Nebula and its many attendant open clusters.

The molecular clouds on the inner edge of the Sagittarius Arm hide most of the Galaxy's central regions. But we do get glimpses, between dust clouds, of the next two inner arms, the Scutum–Crux Arm and the Norma Arm.

The dense stars in M24, the Small Sagittarius Starcloud, are a small section of the Norma Arm, seen through two intervening spiral arms.

Current research indicates that the Norma Arm contains the most massive molecular clouds, and exhibits the greatest rate of massive star formation within the disk.

Although we have named six spiral arms, the Milky Way is currently thought to have only four major spiral arms. Firstly, the Local Arm is not considered to be one of the major arms—it is suspected to be just a spur, attached to either the Perseus or the Sagittarius arms. Secondly, many galaxies display nearly-encircling spiral arms, and that is suspected to be the case with the innermost arm, the Norma Arm, and the Outer Arm—they are thought to be the root and the tip of the same arm.

THE SPHERICAL HALO

The Milky Way's star-forming disk is embedded within its spherical halo. The halo includes the globular clusters, as well as a thin population of stars. However, the halo is far from empty. Its mass, consisting primarily of dark matter, is ten times that of the disk.

Most halo stars are red dwarfs, much older and fainter than the Sun. An exception is Arcturus, the fourth-brightest star in the sky, a spectral K-type giant with a high space velocity relative to the Sun. This is because Arcturus is not a disk star like the Sun, but is instead a halo star that happens to be passing through the disk at this time, as it follows its highly-inclined orbit.

While all disk stars rotate in the same direction around the core, halo stars can orbit in any direction. Some halo stars, including Arcturus, have a chemical composition that differs slightly from most Milky Way stars. There is good evidence that Arcturus was born elsewhere, in one of the many smaller galaxies that have been captured and assimilated into the growing Milky Way over time, just as the Sagittarius Dwarf Spheroidal Galaxy is presently being incorporated—the exceptionally luminous globular cluster M54 is believed to have been its nucleus.

Above Hubble Space Telescope's remarkable view of a perfectly edge-on galaxy, NGC 4013, similar to the Milky Way, lying about 55 million light-years away in the direction of the constellation Ursa Major. The image reveals exquisite detail of huge clouds of dust and gas extending along, as well as far above, the galaxy's main disk.

THE CORE

The Milky Way is no longer thought to have an elliptical galactic bulge, like the Andromeda Galaxy does. Recent composite images of our Galaxy, made in the infrared at wavelengths of 1 to 4 microns, show the galactic center to be peanut-shaped, which implies that there is a small bar in the Milky Way's core. The bar, mainly composed of the old stars that dominate the centers of spiral galaxies, slowly rotates within an annulus of ionized and molecular gas called the 3–5 Kiloparsec Ring.

Infrared studies have revealed a dense cluster of millions of stars at the Galaxy's nucleus. The measured orbital velocities of the stars closest to the center can only be explained if a million-solar-mass black hole lurks at the heart of the Milky Way. Consistent with how well our Galaxy has hidden its other secrets from us, black holes were proven to lie at the hearts of many other galaxies before the evidence for one at the heart of our own was found.

Above Two-micron survey centered on the core of the Milky Way Galaxy, toward the constellation Sagittarius. The reddish stars seemingly hovering in the middle of the disk trace the densest dust clouds in our Galaxy.

Local Group of Galaxies

The Local Group of Galaxies stretches across about ten million light-years, with all its galaxies gravitationally-bound together, forming a luminous neighborhood within our vast universe.

Our Milky Way Galaxy and the Andromeda Galaxy, M31, are the dominant members of the Local Group of Galaxies. Both are large spirals. The Andromeda Galaxy is the queen of the Local Group. It has about 120 percent of the Milky Way's mass, so their sizes are of the same order of magnitude. The Triangulum Galaxy, M33, is a much smaller spiral.

The Large and Small Magellanic Clouds, both satellites of the Milky Way, are the fourth and fifth largest galaxies in the group. Twenty-one other galaxies are members, but all range from small to inconsequential, and the majority of them are in orbit around one of the two dominant spirals.

Despite the well-known expansion of the universe, the fact that the Local Group is gravitationally-bound together means that the galaxies within the group are not moving away from each other. However, the distance between our group and every other clump of galaxies is increasing.

Not surprisingly, there have long been questions about whether various galaxies on the outskirts are gravitationally-bound bonafide members of the Local Group, and some sources will list more galaxies than the, perhaps conservative, number of 26 given here. Those nearby galaxies that most sources no longer consider to be members are certainly gravitationally tugged by the Local Group, but do not seem to be in orbit around its center of mass.

Below This image, by ground-based telescopes, shows the neighboring spiral galaxy M33. The galaxy resides 2.2 million light-years from Earth, in the constellation Triangulum.

THE ANDROMEDA GALAXY

The Andromeda Galaxy is classified as an SAb galaxy, where "SA" means it is a normal spiral, and "b" means that its central bulge is average-sized. M31's only unusual characteristic is that it sports two nuclei, compelling evidence that it has ingested a significant galaxy "recently." (As we have come to know, the astronomical version of "recently" might mean many millions of years ago.)

All large galaxies are thought to have grown to their present size by assimilating their smaller neighbors. Resistance is futile.

Color images of M31 show a golden elliptical bulge of elderly stars, surrounded by thick spiral arms, with stellar associations of young blue supergiant stars dotting the arms of the galaxy.

The galaxy is quite inclined to our line of sight, only 12.5° from edge-on, and this limits the detail that can be detected visually in a telescope. However, an 8-inch (20 cm) telescope easily shows a nucleus, the bright elliptical central bulge, a stellar association that is prominent enough to have its own NGC number—NGC 206—as well as two long dust lanes on the side of the galaxy nearest to us.

Seeing two dust lanes means, by definition, that the brighter band between the dust lanes is one of M31's thick spiral arms. With an appropriate chart, an 8-inch (20 cm) telescope can track down M31's two brightest globular clusters—they look like slightly fuzzy stars at high power. One of them, known as G1, is the most massive globular cluster anywhere within the Local Group. A 16-inch (40 cm) telescope reveals the 18 brightest members of M31's retinue of 300 to 400 globular clusters, plus several open clusters. The necessary charts for hunting M31's clusters are readily available on-line and at any university library.

ANDROMEDA SATELLITES

The Andromeda Galaxy has four major satellite galaxies. The two brightest, M32 and M110, are found in the same field of view as M31, and are so bright that they can be observed with 7×50 binoculars. However, the only detail shown by telescopes is a nucleus for each, and their degree

Above This artist's concept shows a view across a mysterious disk of young blue stars encircling a supermassive black hole at the core of the neighboring Andromeda Galaxy (M31).

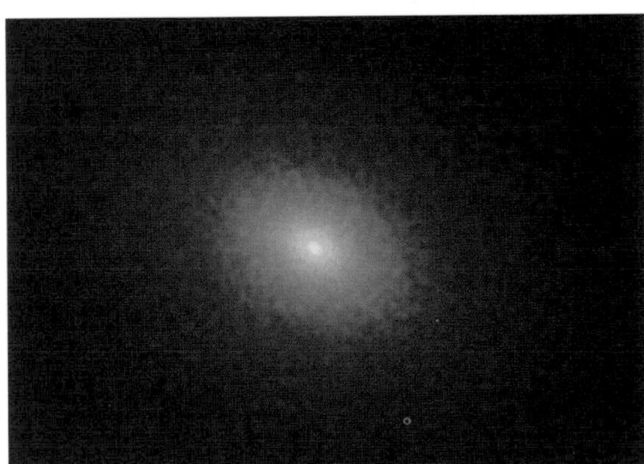

Above Hubble Space Telescope image of the central core of the elliptical galaxy M32. Theoretical models suggest that the structure of M32 is consistent with a central three million solar-mass black hole.

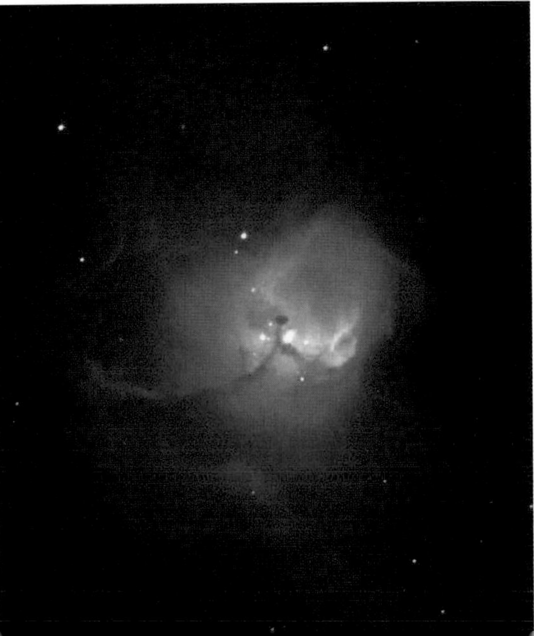

Left Hubble Space Telescope image of young ultra-bright stars nested in their embryonic cloud of glowing gases. Called N81, it is located 200,000 light-years away, in the Small Magellanic Cloud.

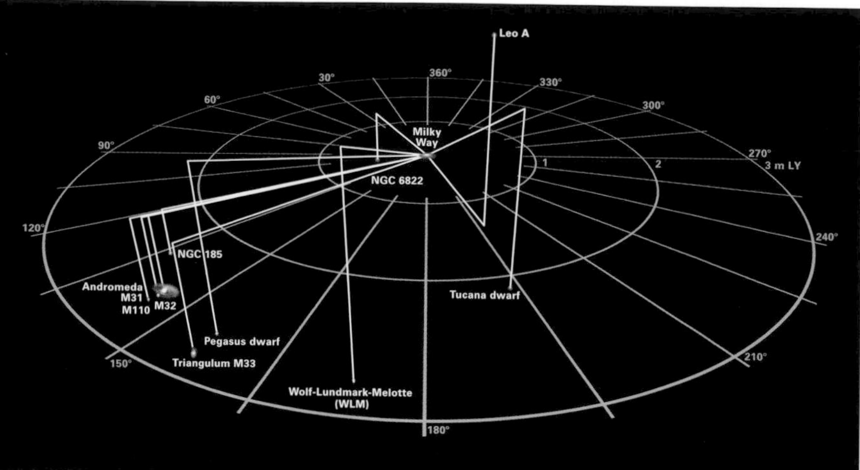

Above Schematic drawing of the Local Group of Galaxies that includes our Milky Way. The galaxies cover a 10 million light-year diameter and have a binary (dumbbell) shape.

of elongation. M32 is a type E2 galaxy, meaning that it is elliptical, but not too far from round. It has a very high surface brightness. Surprisingly, little M32 holds a larger black hole at its heart than our own Milky Way Galaxy.

M110 is much larger than M32, but has a correspondingly lower surface brightness, making it elusive under light-polluted skies. M110 is classified as a "S0/E5 pec." The "E5" means that it is a very elongated elliptical, but the "S0" means that it has some characteristics of a lenticular galaxy as well. The addition "pec" means peculiar, and that is because photographs of M110 show two narrow dust lanes, features that elliptical galaxies should not possess.

An 8-inch (20 cm) telescope can find two smaller satellite galaxies, NGC 185 and the elongated, very faint NGC 147. Both are dwarf ellipticals. While they are located in Cassiopeia, 7°N of M31, they are its satellites. NGC 185 also has the addendum "pec" to its type, because it has a dust patch too. Recently, three more companions to the Andromeda Galaxy have been discovered; all three are very small dwarf ellipticals, and are cataloged as And I, And II, and And III.

The Triangulum Galaxy, M33, the third-largest spiral galaxy in the Local Group, is about 2,200,000 light-years from us, but only about 700,000 light-years from M31. The Triangulum Galaxy has only about one-fifth as many stars as the Milky Way does. It is classified as an "SAcd". "SA" means that it is a normal spiral; "cd" means that it has a very small core. With a telescope, it is one of the easiest galaxies in which to discern spiral arms, and careful observation reveals many small knots that are stellar associations and nebulae.

OUR NEIGHBORS

At least 11 satellite galaxies orbit the massive Milky Way. Most are minor dwarf spheroidal galaxies, named after the constellation that they are located in. They have names like the Ursa Minor Dwarf, Draco Dwarf, and Sculptor Dwarf. Most were discovered photographically, when an excess of stars was noted scattered across a large area of sky.

As an example, the Sculptor Dwarf is about 300,000 light-years away, is only 8,000 light-years in diameter, has only about two million solar masses, and lacks any central concentration. Only the most skilled amateur astronomers have ever observed its very faint glow.

The easiest of these minor galaxy systems to observe is probably Leo I, detectable in a 10-inch (25 cm) telescope with clean optics. On a transparent night, it will appear as a smear of light just ⅓°N of Regulus.

Right Hubble Space Telescope image of the supernova remnant (SNR), known as "E0102." It is located almost 50 light-years away from of the edge of the massive star-forming region, N 76, also known as Henize 1956, in the Small Magellanic Cloud.

Left Digital impression of a satellite view of the Small Magellanic Cloud (SMC), a dwarf galaxy in orbit around the Milky Way Galaxy. The SMC contains about five billion stars.

The Fornax System has five globular clusters that are much more obvious than their host galaxy; four of these globulars are visible with 8-inch (20 cm) telescopes.

In contrast to the dwarf spheroidal galaxies, the Milky Way has two much larger satellites, the Large Magellanic Cloud (LMC) and the Small Magellanic Cloud (SMC). They are prominent naked-eye showpieces located in the far southern sky. The LMC, only 150,000 light-years distant, has about 15 billion stars, and the SMC has one-third as many. Both are wonderfully detailed in amateur telescopes, offering a scattering of nebulae and globular clusters, and myriad open star clusters, some of which are easily resolved into extragalactic stars. The LMC's glorious Tarantula Nebula is one of the largest known. The Small Magellanic Cloud is classified as an irregular galaxy. The Large Magellanic Cloud is classified as a barred spiral that is missing most of the usual attributes of such a galaxy, except for its very prominent bar. While hints of faint structures, somewhat resembling spiral arms, have been photographed, they are more likely tidal tails drawn out by gravitational interactions with the nearby SMC and Milky Way.

A few independent loners complete the Local Group. These include both a scattering of dwarf elliptical galaxies, and four somewhat larger irregular galaxies—NGC 6822, IC 10, IC 1613, and LGS 3.

The most interesting irregular is NGC 6822, known as Barnard's Galaxy for its discoverer, the very keen-eyed American comet-hunter E. E. Barnard. At a dark observing site, a 4-inch (10 cm) telescope shows the faint elongated galaxy. Large Dobsonians will reveal two nebulae within Barnard's Galaxy, at its northern end.

STAR PARTY

At a distance of about 2,400,000 light-years, the Andromeda Galaxy (M31) is the most distant object that can be seen easily with the unaided eye, and it was one of the "little clouds" cataloged by ancient astronomers.

Amateur astronomers enjoy showing the splendors of the night sky to the public, and this is a showpiece object. On a public star night held at a site well away from city lights, most people can discern the fuzzy patch that is M31 with their unaided eyes, without difficulty, if they are carefully led on the naked-eye star hop.

Nothing seen through a telescope will impress observers as much as seeing another major spiral galaxy without needing any optical aid. They are actually seeing back in time, viewing "fossil light" that left the Andromeda Galaxy at about the same time that the first humans appeared in Kenya. Such a star party will leave a life-long memory!

Above Digital enhancement of a star field surrounding the Andromeda Galaxy (M31), 2.5 million light-years from Earth. Visible with the naked eye on a clear night away from city lights, M31 is the nearest major spiral galaxy to our own Milky Way.

Distant Galaxies

The galaxies are not rushing into new areas of the universe; it is the space between galaxy clusters that is expanding. And it seems that the expansion rate is accelerating.

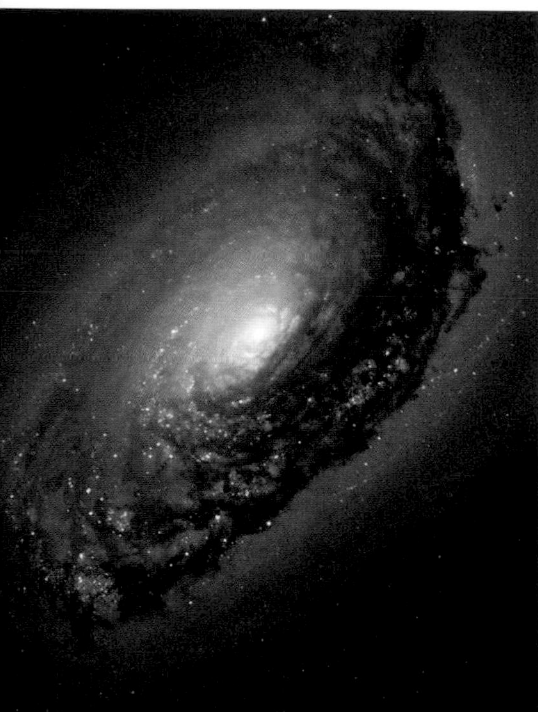

Above This Hubble Space Telescope image of M64 shows a spectacular dark band of absorbing dust in front of the galaxy's bright nucleus, giving rise to its nicknames of the "Black Eye" or "Evil Eye" galaxy. M64 is located 17 million light-years from Earth, in the northern constellation Coma Berenices.

In 1913, V. M. Slipher of Lowell Observatory in Arizona, USA, took spectra of many "spiral nebulae"—the word "galaxies" was not yet in general use—and found that recognizable spectral lines were, in most cases, further toward the red end of the spectrum than they were on the laboratory spectra.

This type of a Doppler shift is called redshift, which means that the spectral lines are shifted toward a longer wavelength.

If lines that are normally found in the ultraviolet part of the spectrum are shifted into the blue, or if lines that are normally found in the infrared part of the spectrum are shifted into radio wavelengths, those shifts to a longer wavelength are also referred to as redshift, just for convenience.

It was immediately realized that "spiral nebulae" with redshifts were moving away from us, but the full import of these observations was not understood, because most astronomers still thought these objects were within our own Milky Way Galaxy

After Edwin Hubble resolved the "Great Nebula in Andromeda" into supergiant stars with the new 100-inch (2.5 m) reflector on Mount Wilson in California, USA, in 1924, some of the supergiants proved to be Cepheid variable stars. The Cepheids study allowed him to prove that Andromeda was far beyond the Milky Way, and the existence of galaxies beyond our own became firmly established.

Hubble then applied his genius to studying galactic redshifts. While the easily-studied Andromeda Galaxy had a Doppler shift that indicated it was approaching us, most other galaxies' Doppler shifts were redshifts, indicating recession. To photograph the spectra of the more distant galaxies with the technology of the day, required that Hubble take exposures over several consecutive nights.

By 1929, Hubble had laboriously established the redshifts of a large sample of galaxies. He announced that the more distant the galaxy, the greater is its measured redshift, and therefore the faster it is moving away from us.

He further declared that the universe was expanding, one of the most startling concepts ever announced by an observational astronomer. His contribution to astronomy has been immense and has been rewarded by having the Hubble Space Telescope—arguably the most successful scientific satellite ever—named for him.

To understand the concept of expansion of the universe, imagine galaxy clusters being represented as dots painted on a balloon. As the balloon is inflated, every dot moves away from every other dot, so the expansion is not centered upon any particular point on the surface of the balloon. If dots A, B, and C, are originally each located 4 inches (10 cm) apart along a line, then if the distance between A and B increases by about ½ inch (1 cm), it follows that the distance between B and C will also increase by ½ inch (1 cm). The distance between A and C will thus increase by about 1 inch (2 cm) in the same period of time. So from the perspective of dot A, the more distant dot C is moving away at twice the speed that nearby dot B is.

GALAXIES BEYOND COUNTING

The famous Hubble Deep Field image, a 10-day exposure taken with the Hubble Space Telescope, and a more recent southern deep field, indicate that the number of galaxies in the universe is comparable to the hundred billion or so stars in our own galaxy.

THE NEAREST GALAXY GROUPS

The nearest clump of galaxies to the Local Group is the Sculptor Group. Its closest significant galaxy, the amorphous face-on spiral NGC 300 of type SAd, lies 8 million light-years away. The brightest and largest members, each half a degree long, are the showpiece spiral NGC 253, of a mixed type SABc, and edge-on NGC 55, of type SBm—similar to the Large Magellanic Cloud. NGC 55 is a diffuse splinter in a small scope. Larger apertures add several knots. Very elongated NGC 253, the Silver Coin Galaxy, is a starburst galaxy that shows mottling and dust lanes in the eyepiece.

The small amorphous face-on spiral NGC 7793 is a type SAd that appears comet-like in a telescope. NGC 7793 and NGC 253 lie on the far side of the Sculptor Group, almost twice as far from us as NGC 300. The group extends beyond

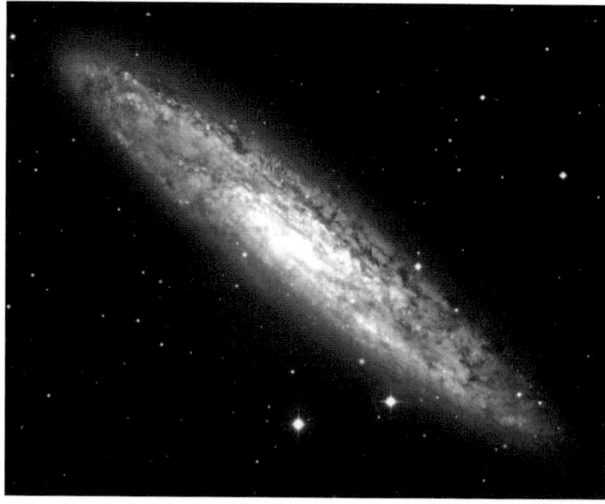

Right Ground-based image of the magnificent spiral galaxy NGC 253. The galaxy is located about 14 million light-years from Earth, in the constellation Sculptor.

the constellation Sculptor, northward into Cetus, where an SABd spiral is a member—the very elongated and mottled NGC 247. There are, of course, fainter galaxies in the Sculptor Group, as well.

The Messier 81 Galaxy Group in Ursa Major and Camelopardalis is about 12 million light-years distant. The oval spiral galaxy M81, a type SAab, and the disturbed irregular galaxy M82, are an interacting pair. Only 37 arc-minutes apart, they are obvious in binoculars. Near to M81, an 8-inch (20 cm) telescope reveals two smaller members of the group, spiral NGC 2976 (a type SAc pec, where "pec" means "peculiar"), and irregular NGC 3077 (yet another peculiar galaxy).

The remaining spiral galaxy in the M81 Group, NGC 2403 in Camelopardalis, is far enough away from all of the gravitational interactions around M81 to appear quite normal. A type SABcd, it is visible in binoculars.

Below Hubble image of the merging pair of Antennae galaxies. As they smash together, billions of stars are born, mostly in groups and clusters of stars. The orange blobs are the cores of the original galaxies.

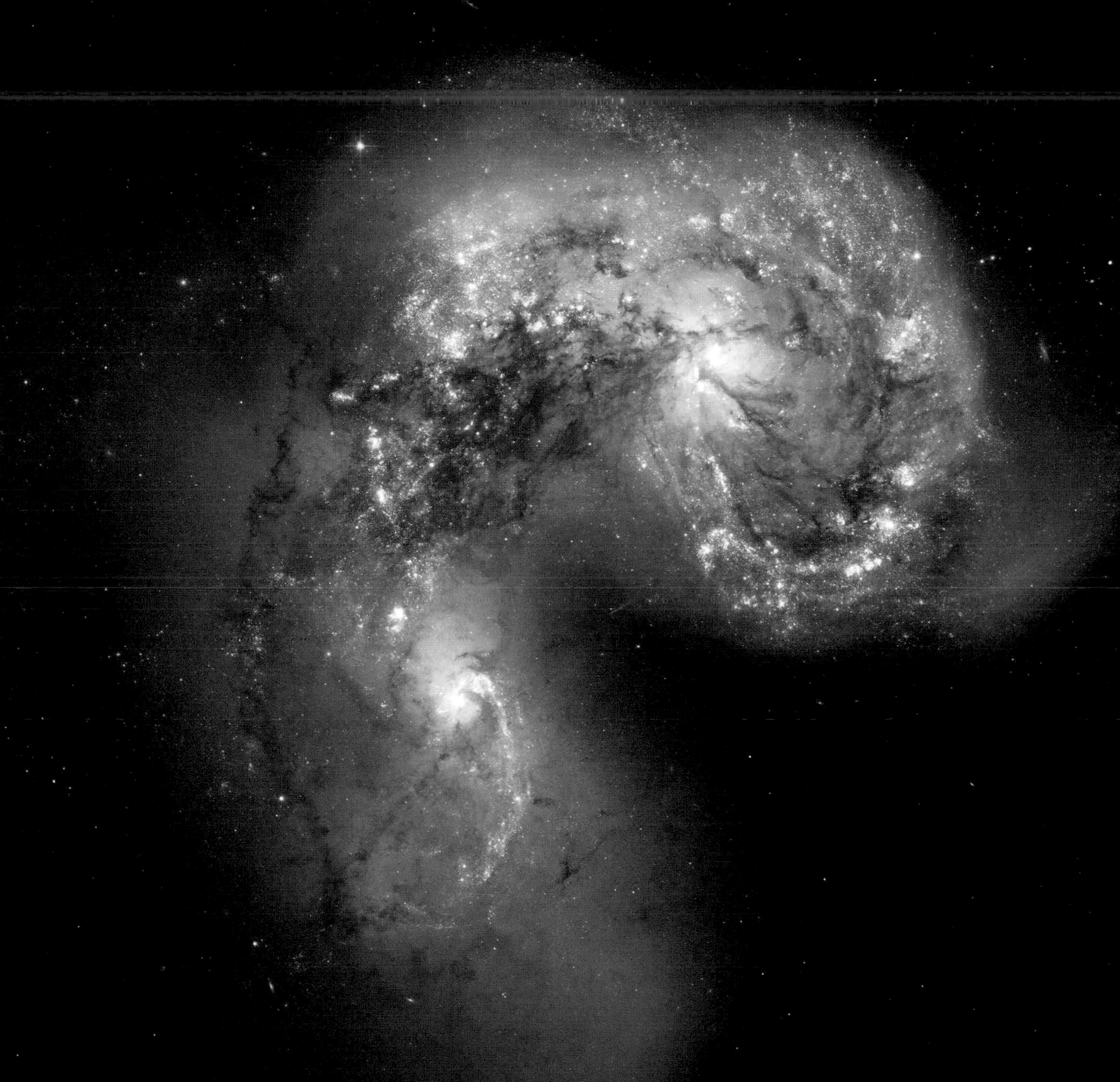

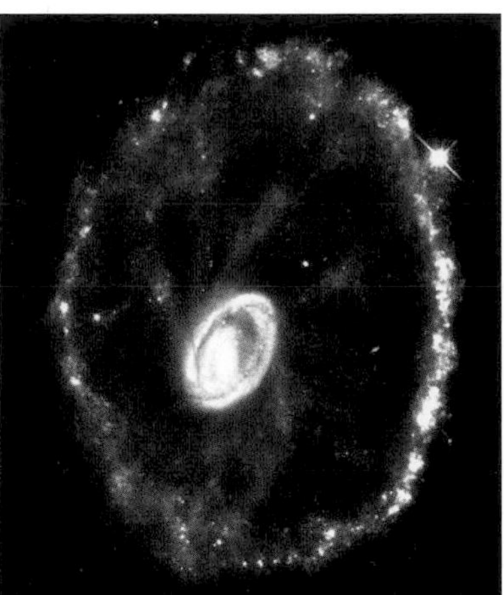

Right Aptly named the Cartwheel Galaxy, this Hubble image shows the galaxy's nucleus has wisps of material connecting the nucleus to the outer ring of young stars. The unusual configuration was created by a nearly head-on collision with a smaller galaxy about 200 million years ago. It is located 500 million light-years away, in the constellation Sculptor.

STARBURST GALAXIES

M82 has undergone a burst of star formation due to a close encounter with its massive neighbor, M81. An 8-inch (20 cm) reflector shows a diagonal dust lane crossing the long axis of the edge-on irregular galaxy M82; a 25-inch (60 cm) reflector reveals a second dark band, plus five condensations. Images of M82 show hot pink hydrogen gas being violently blown out of both sides of the central area of the disk to a distance of 10,000 light-years—the images rather resemble fountains of erupting lava! (M82 used to be referred to as "the exploding galaxy" before we understood what the images showed.) Tidal effects from a close passage by M81 caused so many molecular clouds to collapse over a short period of time that M82 temporarily became a starburst galaxy. The birth of unusual numbers of supergiant stars soon resulted in a great many supernova explosions in the same area. Each supernova's shockwave creates a bubble of hot gas and when many supergiant stars explode over a short period, a superbubble is created. Since M82 is not a particularly massive galaxy, the superbubbles were able to blow out through the galaxy's disk and far beyond. However, M82 probably is massive enough to recapture much of the hot gas eventually.

Smaller starburst galaxies may not have enough gravity to recapture their gas. In such a case, the starburst followed by the inevitable surge in supernovae, will probably drive all of their interstellar gas off into space.

Ironically, a starburst could thus turn a small galaxy into a sterile one, where star formation had ceased completely.

Both of the nearest galaxy groups have starburst galaxies. However, NGC 253 in the Sculptor Group is a more massive

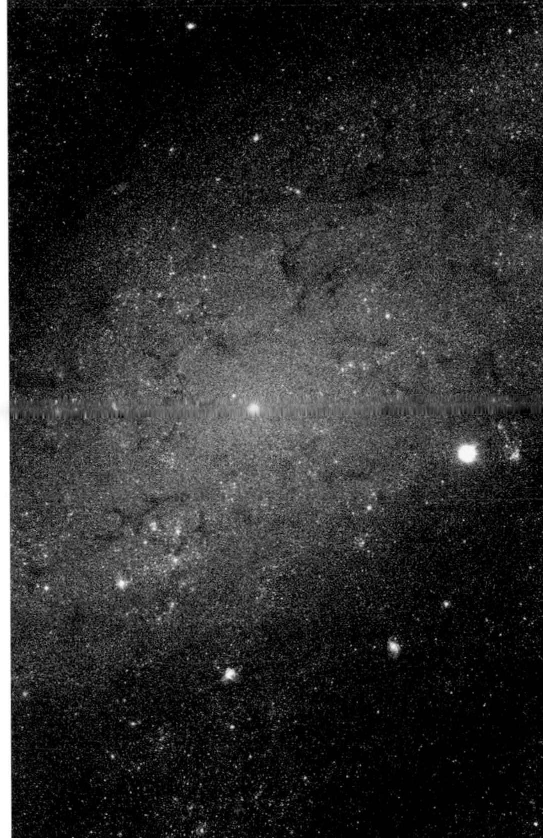

THE NAKED-EYE DISTANCE RECORD

While the Andromeda Galaxy, at a distance of about 2,400,000 light-years from Earth, is the most distant object that can be seen easily with the unaided eye—an observation that even completely inexperienced observers can make if the galaxy is pointed out to them—some experienced amateur astronomers have seen five times farther than that! More than a dozen of the most talented and keen-eyed observers in North America have spotted magnitude 6.9 Messier 81 with the unaided eye, from remote wilderness observing sites, after lengthy dark adaption, gazing across 12 million light-years without optical aid.

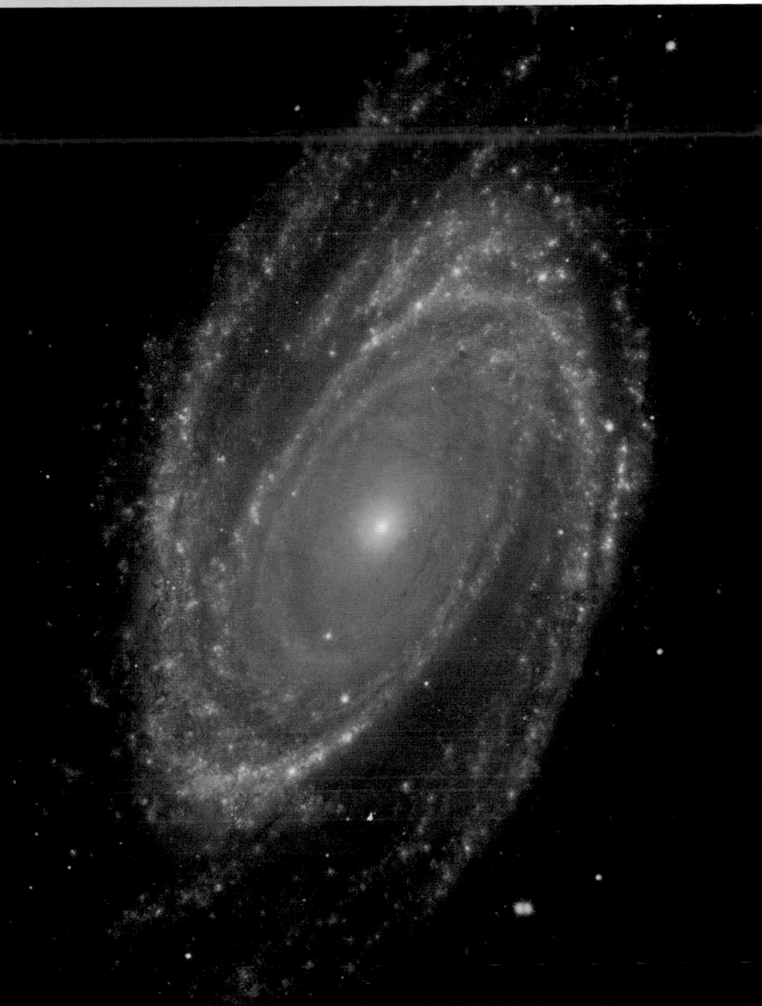

Above Hubble Space Telescope image of NGC 300. The luminous blue specks are young and massive stars called blue supergiants, and are among the brightest stars seen in spiral galaxies.

Left Hubble Space Telescope, Chandra X-Ray Observatory, and Spitzer Space Telescope combined image of active galaxy M82. Hubble's observation of hydrogen emission appears in orange. Hubble's observation appears in yellow-green. X-ray data recorded by Chandra appears here in blue; infrared light recorded by Spitzer appears in red.

Above Hubble Space Telescope image of M81 high in the northern sky in the constellation Ursa Major. At an apparent magnitude of 6.9, it is just at the limit of naked-eye visibility and is one of the brightest galaxies that can be seen from Earth. The galaxy's angular size is about the same as that of the full Moon.

galaxy than M82, and nothing quite as dramatic is visible in NGC 253. The starburst has produced a great deal of dust and elevated some of it above the galactic plane, accounting for NGC 253's rather strange appearance—it probably has the dustiest face of any nearby large spiral galaxy. Infrared images make it obvious that a great number of stars are forming behind that dust, and heating it. X-ray telescopes show extremely hot gas on both sides of NGC 253's disk. Scientists presume this originates from supernovae.

The most dramatic starbursts occur when spiral galaxies are involved in a collision or a merger. The Hubble Space Telescope (HST) captured an unforgettable image of Sculptor's Cartwheel Galaxy, cataloged as ESO 350-G40. One of the two smaller galaxies in the image is believed to have plowed through what had been a normal spiral galaxy. The head-on collision resulted in a bright starburst ring (the wheel), connected by faint "spokes" to the galaxy's core—the hub of the wheel. The Cartwheel is 500 million light-years distant, and its spectacular details appear only on images, although a 16-inch (40 cm) telescope in the Southern hemisphere shows the core and tantalizing hints of the ring.

The HST image of the Antennae Galaxy, NGC 4038/9 in Corvus, shows spectacular starburst activity in two merging spirals. A 6-inch (15 cm) telescope shows only an amorphous blob, but in a 25-inch (60 cm) telescope the pair look like a broken heart, and eight of the super star clusters are visible. Wide-field images show two long tidal tails of stars that have been drawn out of the galaxies. (The tails are outside of the small area included in the HST image.) Computer simulations predict that the two spirals will merge into a giant elliptical galaxy in a process taking about a billion years.

Galaxy Clusters

Almost all galaxies are found within clusters, which, typically, are bound by hot X-ray emitting gas and large amounts of "dark matter."

Above Illustration from *Organum Uranicum,* an explanation of the theory of the universe, by Sebastian Munster (1489–1559).

Galaxy clusters range from small groups of a few galaxies like the Local Group, the Sculptor Group, and the M81 Group, on up to large irregular clusters like the Virgo Cluster, and even larger spherical clusters like the Coma Cluster.

VIRGO CLUSTER

The heart of the Virgo Cluster, the nearest large cluster, lies about 65 million light-years away. Determining this distance, and thus the scale of the universe, was one of the primary science goals of the Hubble Space Telescope (HST), but the first images from orbit showed that its mirror had not been ground correctly. Corrective optics were not installed on the HST until the Space Shuttle completed the first Hubble servicing mission.

This delay gave astronomers, using the excellent mirror of the Canada–France–Hawaii Telescope on Mauna Kea, Hawaii, USA, the opportunity to image Cepheid variable stars in Virgo Cluster galaxies before the Hubble Space Telescope optics were corrected.

There are at least 2,000 galaxies in the Virgo Cluster—frequently called the "Virgo Clutter" by frustrated beginners trying to pick out the 16 Messier galaxies there from all of the rest. Stars are relatively thin in this part of the sky, and observers frequently give up star-hopping—they find that galaxy-hopping is more effective!

However, at this distance, few galaxies show much detail in the eyepiece. Most display a nucleus, and the distribution of light across the galaxy usually allows the observer to guess whether it is a spiral or an elliptical. Most of the time, the choice will be confirmed by the galaxy's cataloged type. Large elliptical galaxies are common in the Virgo Cluster, and observers will soon become adept at visually classifying an elliptical, as its degree of flattening warrants.

Below Ultraviolet image of a small area of the Virgo Cluster of galaxies. The Virgo Cluster spans more than 12° in the sky, and is so massive that it is noticeably pulling our Galaxy toward it.

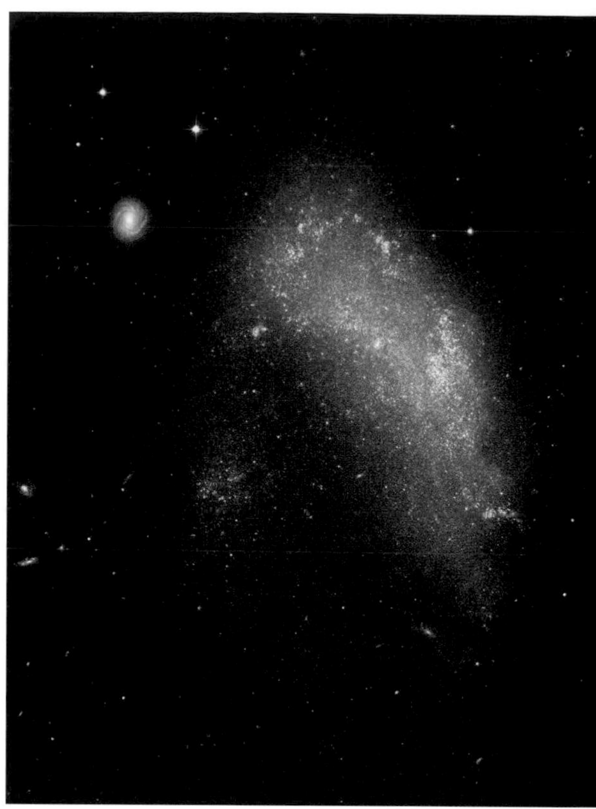

Above NGC 1427A, which is located some 62 million light-years away from Earth in the direction of the constellation Fornax. Under the gravitational grasp of the Fornax Cluster, the small bluish galaxy is plunging headlong into the group at 400 miles (600 km) per second.

The only galaxies within the Virgo Cluster that show significant detail, even with a 16-inch (40 cm) telescope on an excellent night, are two face-on spirals, type SAc M99, whose spiral arms are discernable, and type SABbc M100, which shows knots identifiable as emission nebulae when compared to an appropriate image.

A very challenging observation is the monster elliptical M87's jet from the central black hole, as it is barely discernable at high power. Other than the seldom-observed jet, M87 looks rather bland in the eyepiece. But, as is often the case in astronomy, the huge E0.5 elliptical galaxy's interest lies in understanding what you are looking at, and M87 is all about superlatives. By capturing and assimilating a great many smaller galaxies, the monster has grown to three trillion solar masses. More than ten thousand globular clusters orbit it. Observations of the gas orbiting the nucleus leads to the conclusion that the black hole at its heart is of two to three billion solar masses.

Nearby Markarian's Chain, running from M84 and M86 through to NGCs 4458 and 4461, is the most interesting part of the Virgo Galaxy Cluster. Here an 8-inch (20 cm) telescope can view 10 galaxies simultaneously in one low-power field of view: Two large ellipticals (M84 and M86), three spirals, three lenticulars, and two small ellipticals.

As an irregular galaxy cluster, the Virgo Cluster has multiple clumps of galaxies, each clustered around a large elliptical. There are clumps of galaxies around each of M60, M49, and the nearby NGC 4365.

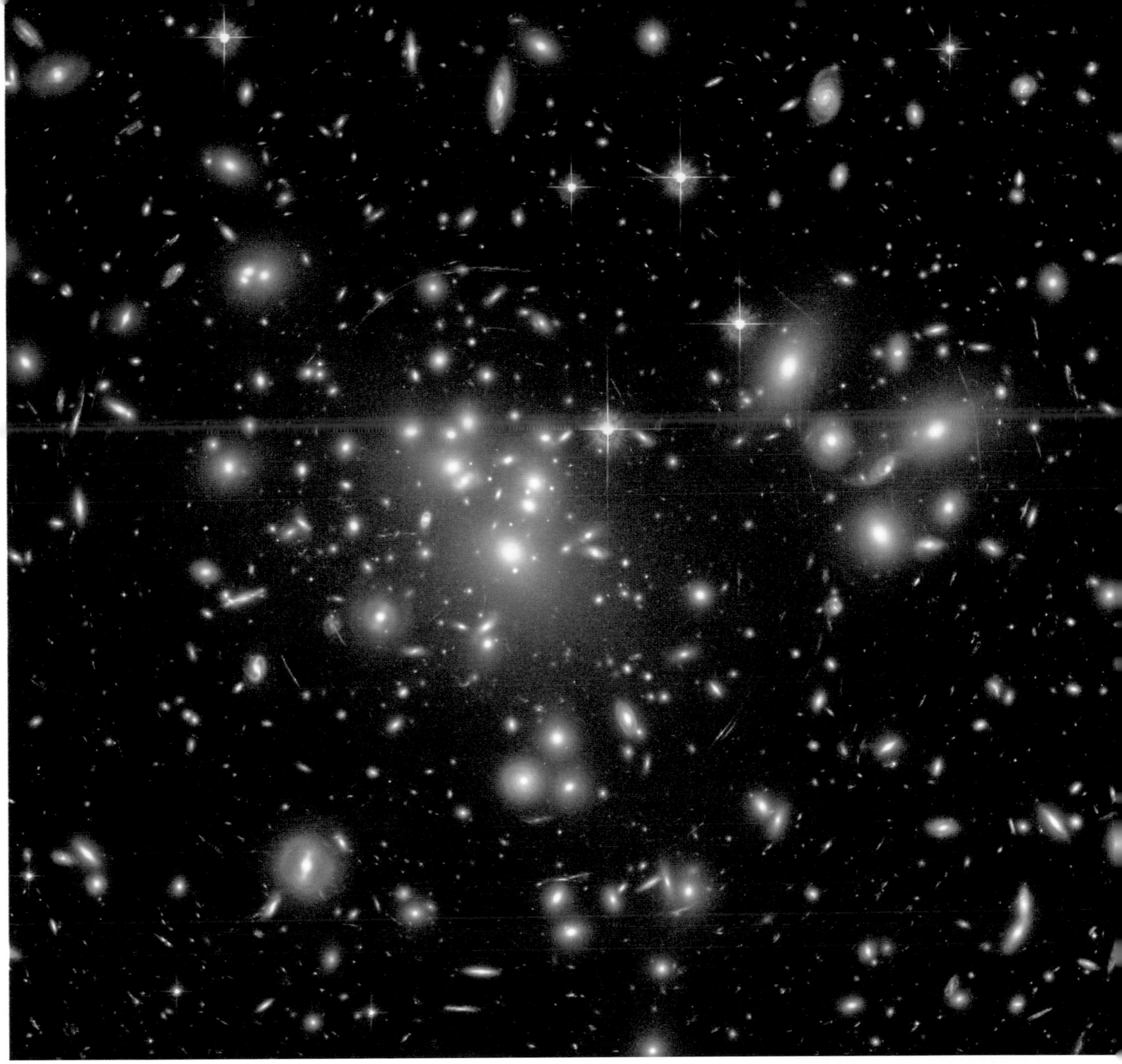

Above Hubble Space Telescope image of the center of one of the most massive galaxy clusters known, Abell 1689. The gravity of the cluster's trillions of stars, plus dark matter, acts as a 2-million-light-year-wide "lens" in space. The faintest objects are probably over 13 billion light-years away.

FORNAX CLUSTER

The Fornax Cluster also offers rich fields of galaxies. The brightest member, NGC 1316 or Fornax A, is a peculiar galaxy. It is magnitude 8.8, and several other members are magnitude 10, so this cluster is readily accessible to most observers.

CORONA BOREALIS GALAXY CLUSTER

Abell 2065, the very challenging Corona Borealis Galaxy Cluster, lies 1.5 billion light-years away. On a superb night, a 16-inch (40 cm) telescope reveals eight of the cluster's 400-plus swarming galaxies. These are the most distant normal galaxies visible in the eyepieces of amateur telescopes. This huge galaxy cluster is so distant that it covers only half a degree of sky.

COMA CLUSTER

The larger Coma Cluster, Abell 1656, is about 400 million light-years away, in the constellation Coma Berenices. A spherical cluster, it boasts at least 10,000 members.

Two giant elliptical galaxies, of the cD (central dominant) type lie at Coma's core.

The pair, NGC 4874 and NGC 4889, are two of the only five galaxies that a 6-inch (15 cm) telescope shows to an observer. However, with large apertures the view is marvelous and many of the small galaxies orbiting the two giants become visible. A 12-inch (30 cm) telescope reveals about 30 cluster members.

It is extremely difficult for a spiral galaxy to survive in the mid-regions of a giant cluster like this as its disk will most likely be stripped of its interstellar gas, and it will become a lenticular.

Above Coma Cluster, Abell 1656, contains two giant ellipticals, NGC 4874 (right) and NGC 4889 (left). It is about 400 million light-years from Earth.

Active Galactic Nuclei

Einstein's *General Theory of Relativity* predicts that if mass is squeezed into a small enough volume of space, the normal dimensions of space and time become so warped that nothing—not even light—can escape. It's there, but you can't see it! Hence a "black hole."

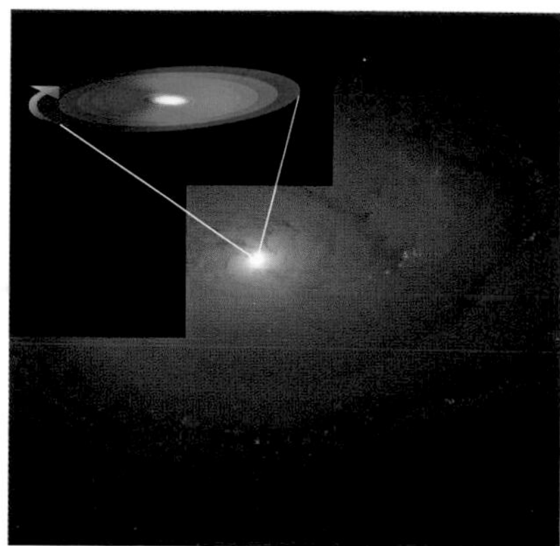

Above Hubble Space Telescope image of the planetary nebula Henize 3-1475, and its bizarre jet. It has been nicknamed the "Garden-sprinkler" Nebula.

There is good evidence that black holes with a few times the mass of our Sun—a few "solar masses"—exist as the final evolutionary phase of very massive stars. There is also good evidence that supermassive black holes—with millions or more solar masses—exist in the centers of galaxies.

Our Galaxy apparently harbors a black hole of some four million solar masses. We cannot see it directly, but the Hubble Space Telescope has been able to trace the orbits of stars circling it, and from their speeds we can estimate its mass. Similarly, measuring general stellar speeds in the nuclei of other galaxies has let us indirectly detect many other black holes, some ranging up to around a billion solar masses.

In general, the larger the bulge of stars in the center of a galaxy, the more massive the black hole it contains. But whether the bulge or the black hole came first is just like the chicken or the egg, and a matter of much debate in the research world. Also, not all black holes in the centers of galaxies exist as peacefully as ours. In a few percent of giant galaxies, interactions or disturbances are allowing material to be sucked toward the black hole, like bath water swirling toward the plug hole. While a black hole itself does not emit any light or radiation, it creates ferocious gravitational forces in its immediate vicinity—enough to power a cosmic fireworks show. Hence, an "active galactic nucleus."

Material will no more travel in a straight line into a black hole than water will travel in a straight line into a plug hole. Instead, like the bath water, the material picks up rapid circular motion. Mutual interactions force the material into a disk. Within the disk, as the material spirals ever closer to its fate, its motion becomes ever more rapid and energetic. The very final moments are effectively beyond our perception. Even a black hole of a billion solar masses is contained with an event horizon no larger than the orbit of Uranus—way too small for our telescopes to resolve details. So the active nucleus itself is far too small for us to discern, and what we see of active nuclei seems at first to be quite an assortment.

There are two different kinds of active galactic nuclei. The first kind squirts out material, but the active nuclei responsible are relatively inconspicuous—these are the radio galaxies. The other kind has luminous nuclei, ranging from those barely discernable, to the most continuously luminous objects in the universe—these are the quasars, Seyfert galaxies, BL Lac objects, and Liners.

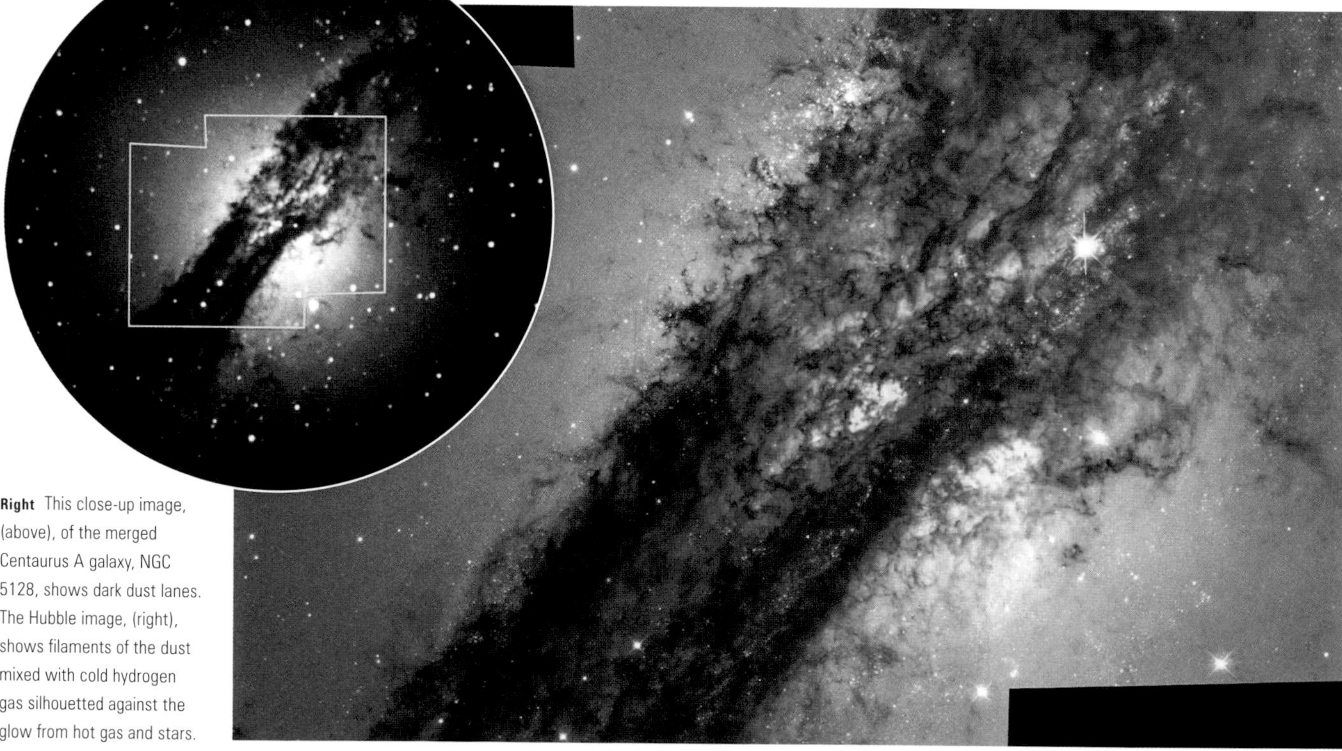

Right This close-up image, (above), of the merged Centaurus A galaxy, NGC 5128, shows dark dust lanes. The Hubble image, (right), shows filaments of the dust mixed with cold hydrogen gas silhouetted against the glow from hot gas and stars.

JET PROPULSION

Not everything that falls in, enters the black hole itself. Active nuclei also eject material, almost always in two oppositely directed jets. The material in these jets—whatever it is—has not come out of the black hole, but must be some of the infalling material turned around and greatly accelerated. We do not properly understand how this happens, but we have some obvious insight. We are aware that as the material gets closer and hotter, it becomes ionized. Ionized material—electrically charged—racing around in orbit makes an enormous electrical generator. Probably it is the high voltages produced that divert and eject some of the infalling material, but quite how that ejected material is formed into the two opposing jets is still not clear. One idea is that the accretion disk is thick enough to almost totally wrap around and enclose the black hole, leaving only two small orifices top and bottom. The orifices then form the ejected material into the two jets.

Below Artist's impression of the energy source within galaxy PKS 0521-36. The extraordinary high pressure and temperature generated near the hole would cause some of the infalling gas to be ejected along the direction of the black hole's spinning axis to create the optical jet.

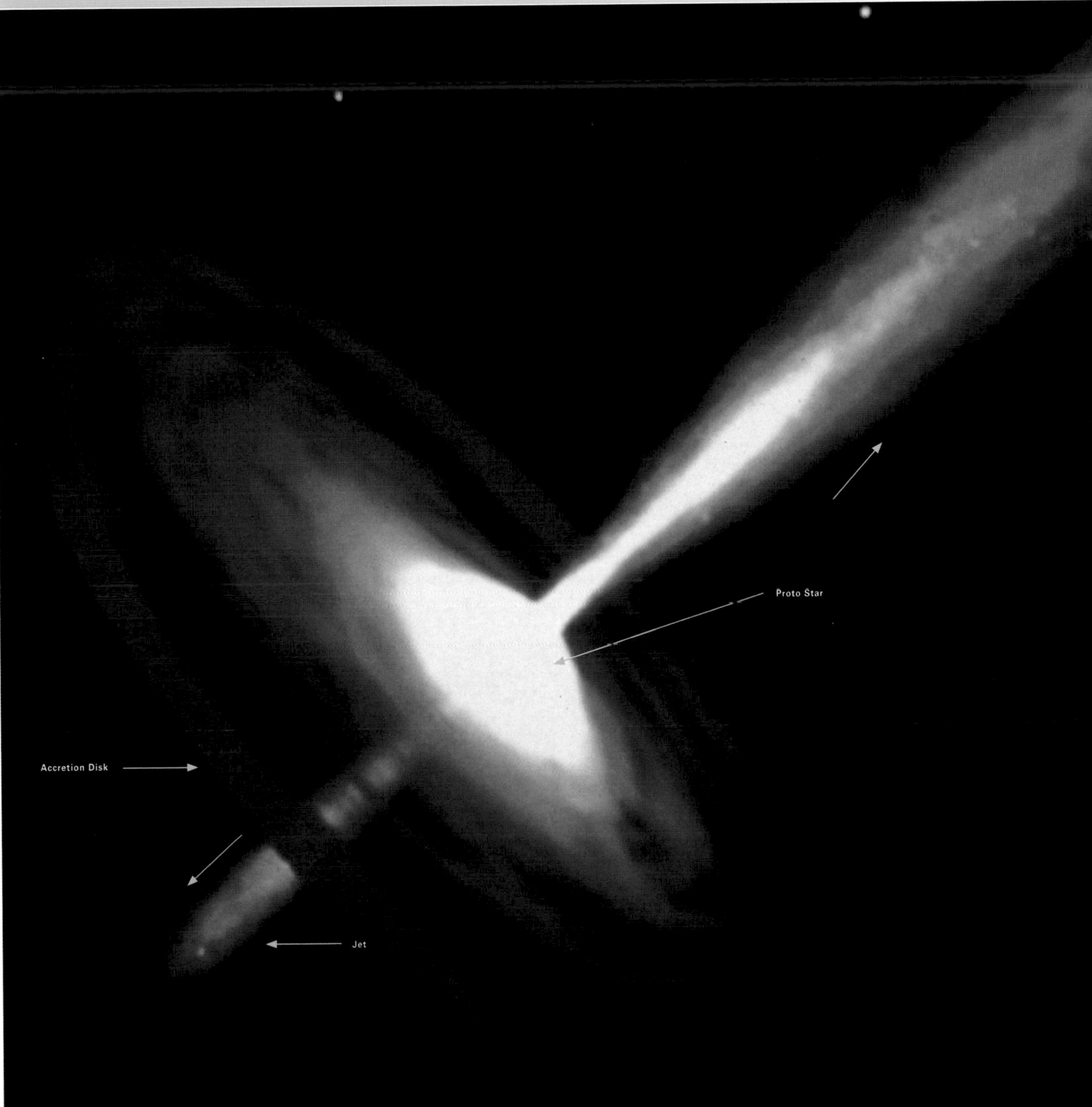

Radio Galaxies

Most giant elliptical galaxies—the heavyweights of the cosmos—look deceivingly benign. In visible light they are seen as gigantic spheroidal conglomerations of old stars, clearly dominant in mass, but seemingly mild in character. This is not so.

Opposite Composite image of MS0735.6+7421, a galaxy cluster located about 2.6 billion light-years away. The red glow is radio emission ejected from the central galaxy of the cluster.

Radio observations may show a very different picture. Astonishingly, two oppositely directed jets have been shot from the tiny central nucleus, with enough power to carry their contents far clear of the parent galaxy, before losing their stability and finally splattering into gigantic radio lobes.

DOUBLE RADIO LOBES

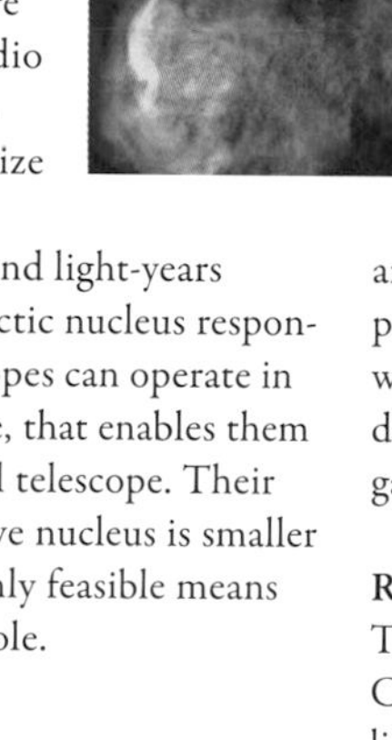

Right False-color radio image of galaxy M87 shows the intensity of the radio energy being emitted by the single jet of subatomic particles. M87 is located 50 million light-years away, in the constellation Virgo.

Double radio lobes—more or less symmetric about the parent galaxy—are therefore the signatures of radio galaxies. Their radio structure, sometimes spreading over millions of light-years, dwarfs the physical size of the parent galaxy. Yet the size of the parent galaxy—ten- or a hundred-thousand light-years across—dwarfs the size of the active galactic nucleus responsible. We know this because radio telescopes can operate in a very long baseline interferometry mode, that enables them to resolve far finer detail than any optical telescope. Their observations have shown us that the active nucleus is smaller than a couple of light-years, where the only feasible means of obtaining so much energy is a black hole.

EJECTORS

We have no direct way of telling what the material is in the jets, but the character of radio emission from jets and lobes—a so-called "power law"—matches that of "synchrotron emission," whereby high-speed electrons whirl in magnetic fields. We do know that the energy emitted from the lobes is usually enormous, and it ought to be that the electrons responsible soon lose their speed and power to radiate. Clearly, the jets continually replenish the energy in the lobes, but quite how they do so is not properly understood. The speed of material in the jets may also vary. Some jets may show curvature, apparently due to motion of the parent galaxy within the tenuous intergalactic medium, and lobes that are brightest in their centers. Generally known as Fanaroff-Riley Type I, these are low velocity ejectors. The other type—Fanaroff-Riley Type II—are fast ejectors, with very straight jets, and lobes brightest around the edges. There are also cases of one-sided jets, most probably caused by very high speed (relativistic) ejection, where a jet directed toward us is greatly enhanced, but a jet directed away is greatly diminished in brightness. A few radio galaxies exhibit complex radio structure.

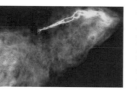

Below Composite image of galaxy Centaurus A. Radio emission is shown in green, while very hot gas (X-ray emission) is shown in deep blue; the rest is optical emission.

RADIO SOURCES

The nearest radio galaxy is NGC 5128 in the New General Catalog or "Centaurus A," to give it its radio name. In visible light, it is seen as a giant elliptical galaxy, but with dust lanes and star-forming regions wrapped around its waist, it is almost certainly the remains of a smaller spiral galaxy that has been shredded by its strong gravity. Radio observations show a pair of inner radio lobes and an enormously distended pair of outer radio lobes as well. If "radio" eyesight was possible, it would appear as a conspicuous double object in the southern sky. It is obvious that the activity in Centaurus A is somehow related to its "cannibalism" of a neighboring galaxy.

The same is certainly true of other cases. Though it is far more distant, the parent galaxy of Cygnus A—one of the brightest radio sources in the sky—has a similar appearance. NGC 1316, another well-known giant elliptical in the southern skies, exhibits loops and telltale signs of "recently digested" smaller galaxies. and is a double lobe radio source.

Some distant quasars also show the classic double lobe radio structure, though it is not easy to see what sort of galaxy is their host. In some cases, very long baseline interferometry has also been able to monitor the progress of blobs of radio emission as they are ejected.

Astonishingly, the apparent speed of motion of these blobs is many times the speed of light, seeming to violate Einstein's fundamental law. However, the situation can be explained if the blobs travel close to the speed of light, but come almost straight toward us.

Such blobs would also have their radiation enhanced, whereas blobs traveling in any other direction might not be visible. But does this mean that blobs may be emitted in all directions, and not only in the two opposed jets? The situation is by no means clear, and neither is our understanding.

Quasars

Quasars are active galaxies with highly luminous nuclei. Their phenomenal luminosity somehow originates in the immediate vicinity of the supermassive black holes that they harbor. They stand out as the beacons of the cosmos.

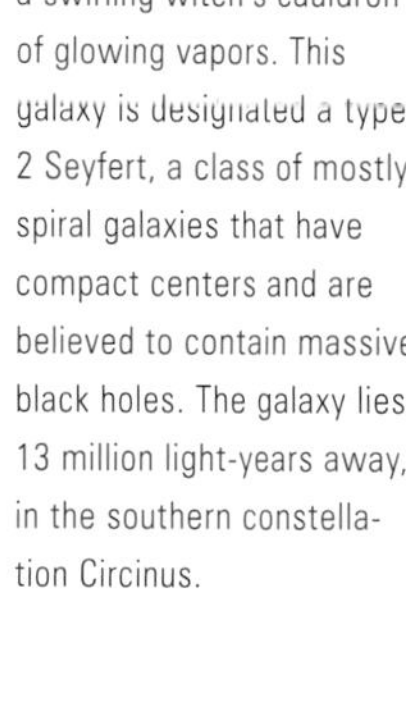

Right This Hubble Space Telescope image shows the the black hole core of the Circinus Galaxy resembling a swirling witch's cauldron of glowing vapors. This galaxy is designated a type 2 Seyfert, a class of mostly spiral galaxies that have compact centers and are believed to contain massive black holes. The galaxy lies 13 million light-years away, in the southern constellation Circinus.

The energy output is not only visible light; it probably originates as high energy X-rays, and cascades down the electromagnetic spectrum, making the nuclei apparent in X-ray, ultraviolet, visible, infrared, and occasionally radio wavelengths. While they often exhibit jets, like radio galaxies, clearly most of the energy released has been diverted into luminosity.

OUTPUT

The radiation output often shows variations on time scales as short as a week. To show such coherent variation, the region responsible cannot be larger than light-weeks in size, further indication that something as compact as a black hole must be involved. In physical size, the active nucleus must be tiny compared to the galaxy that hosts it.

Yet, astonishingly, the light from the nucleus may outshine all the stars in the galaxy many times over, so much so that the host galaxy is often lost in its glare. Suffice to say that when quasars were first discovered in the 1960s, they were first mistaken for foreground stars within our own Galaxy.

The word "quasar" is an acronym for "quasi-stellar radio source," named because the first quasars were identified by their radio emission. Nowadays, we know that the majority, ironically, are radio quiet.

QUASAR TYPES

Not all quasars are "super luminous." Seyfert galaxies—first identified by Carl Seyfert in 1943—are recognizable as normal galaxies, predominantly spirals, but possessing highly luminous nuclei. The so-called Seyfert 1 nuclei are identical in all respects to quasars, save for their more modest luminosity.

Today we recognize that the same phenomenon extends down to low luminosity—very mild quasar nuclei are sometimes barely detectable in nearby spiral galaxies. There are also related galaxies with low luminosity nuclei known as Seyfert 2 galaxies, and low-excitation Liners.

Our best insight into the physical conditions in the immediate vicinity of the nucleus is by spectroscopy, displaying the amount of emitted light against wavelength—effectively color. Active nuclei stand out because of their conspicuous emission lines of hydrogen, oxygen, and nitrogen. Broad hydrogen emission lines come from the densest environment, and tell us that the gas clouds responsible are moving at thousands of miles a second, vastly more than they would do anywhere else in the galaxy, in a region within light-weeks of the nucleus itself. Such high velocities indicate the presence of an enormous, yet condensed mass that could only be a supermassive black hole. Narrower emission lines indicate much lower velocities from a more extensive region, even hundreds of light-years in size.

Quasars and Seyfert 1 galaxies show both broad and narrow lines. Seyfert 2 galaxies and Liners have only narrow lines, while BL Lac objects have no emission lines at all, but show variations on time scales as short as hours. (The curious name "BL Lacertae" was given to the prototype object, when it, too, was mistaken as a foreground star.)

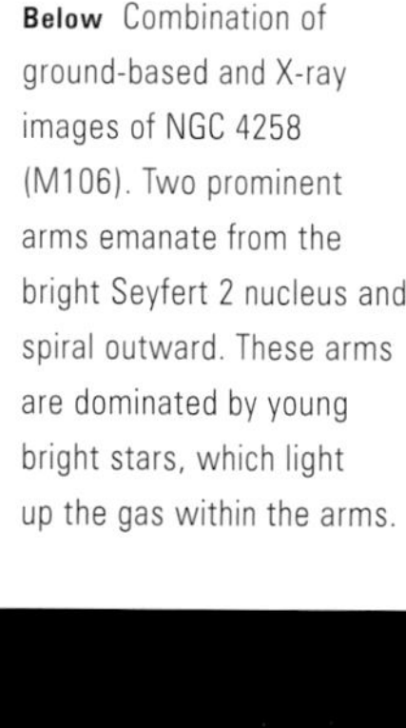

Below Combination of ground-based and X-ray images of NGC 4258 (M106). Two prominent arms emanate from the bright Seyfert 2 nucleus and spiral outward. These arms are dominated by young bright stars, which light up the gas within the arms.

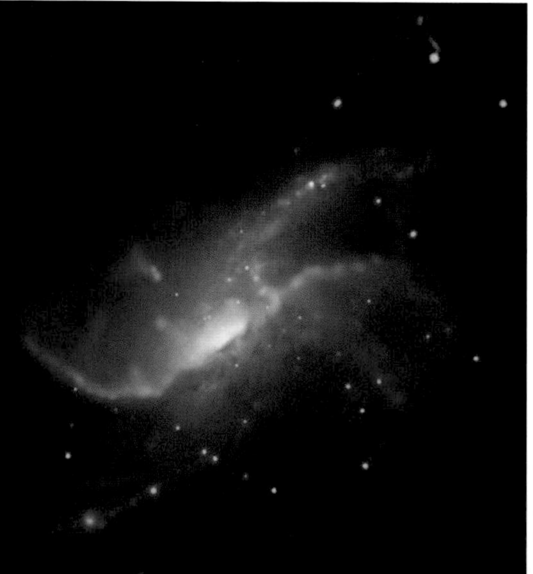

UNIFICATION

Recent years have seen a model emerge that unifies these different types. The key element of this model is that the accretion disk close to the nucleus forms an obscuring torus—same shape as a ring doughnut. If you happen to view this torus side-on, the nucleus and its immediate surroundings will be hidden.

To see a quasar or Seyfert 1 galaxy, you need to see the torus flat-on, so the nuclear regions are not obscured.

As the line of sight moves from flat-on toward edge-on, the increasing obscuration makes the object looks like a Seyfert 2 galaxy and then a Liner.

If the active nucleus is emitting jets, and a jet is seen end-on, the outcome is a BL Lac object.

There are not as many quasars as there used to be. When we look deep into space—and therefore back in time to when the universe was about a third of its present age—we see a much higher population of bright quasars. But the galaxies then were much closer to one another and interactions far more common. As with the giant elliptical radio galaxies, it is interactions that seem to feed material into the supermassive black holes, so turning on quasar activity. Possibly then, all giant spiral galaxies have the potential to have periods of quasar activity. Although our own Galaxy has only a modest supermassive black hole, perhaps its nucleus once shone brightly, though from Earth it would have largely been obscured by the dust clouds of the Milky Way.

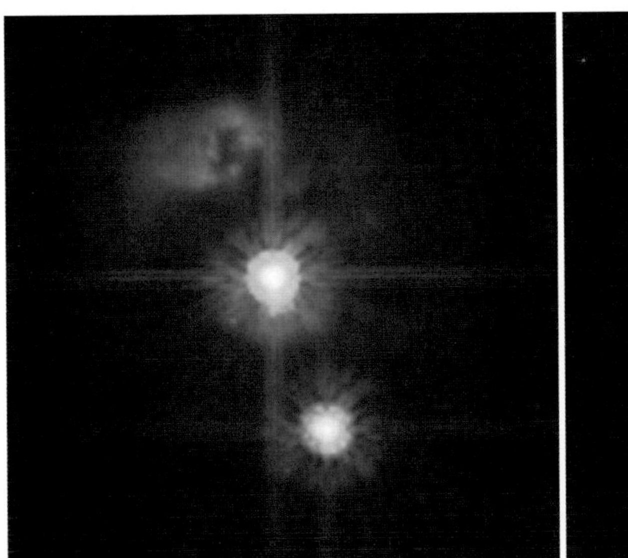

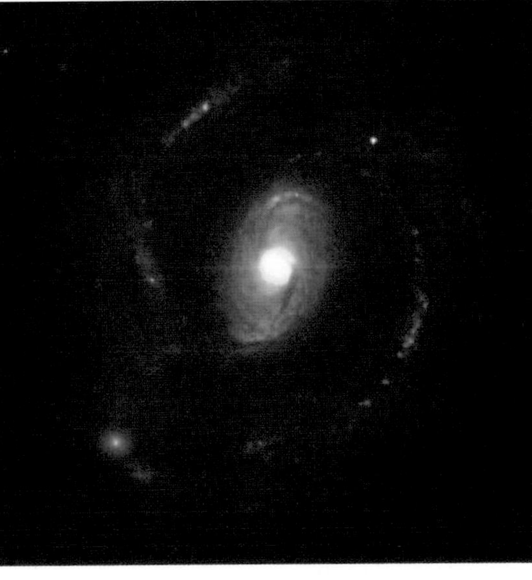

Above Artist's impression of a quasar located in a primeval galaxy (proto-galaxy) a few hundred million years after the Big Bang. Observations suggest the first stars formed before the supermassive black holes that power the quasar engines in galaxy centers.

Left Hubble images of two quasars. Left, HE0450-2958 (about 5 billion light-years away) does not have a massive host galaxy. Right, HE1239-2426 (about 1.5 billion light-years away) has a normal host galaxy.

Large Scale Structures

On a very large scale—hundreds of millions of light-years—the universe displays a remarkable texture. Like holes in a bath sponge, bubbly cosmic voids pervade the distribution of galaxies.

Above This stunning fresco, *The Chariot of Mercury and Virgo,* marks the direction toward the center of our Local Supercluster. It forms part of the fresco in the cupola of the Old Library at the University of Salamanca in Spain.

Opposite Called the Millennium Run simulation, this diagram shows how galaxies and large-scale structures have evolved following the Big Bang, 13.7 billion years ago. The image shows a projected density field for a 15 Mpc/h thick slice of the redshift z=0 output. The overlaid panels zoom in by factors of four in each case, enlarging the regions.

Almost all the galaxies gather close to one another around the perimeters of the voids, only a few are scattered inside. As in a bath sponge, the congregations of galaxies are connected in some way, as they form an interconnected labyrinth. The same interconnection appears to be true with the cosmic voids. However, the labyrinths are not regular, with some regions having a high density of galaxies with only small percolating voids, while other regions have a lower density and much larger voids. High-density regions are often elongated into filaments, or even irregular "walls," and are usually termed superclusters.

Within such structures, clusters of galaxies may condense. The recognition of these large-scale structures, late in the twentieth century, was completely unexpected.

Surprisingly, the universe didn't seem old enough for such gigantic entities to form. Only if the universe contained much more mass would it have been possible. The existence of the structures is therefore one of the indications that there may be considerable amounts of dark matter present. But scientists wondered what seeded the sponge-like pattern. Here, the likely answer is bizarre.

If the Inflationary Theory is correct, then the smallest fluctuations imaginable at a quantum level, would have been explosively enlarged to cosmic scales.

COMPUTER SIMULATIONS

Human life spans are far too fleeting for us to witness the evolution of the cosmic texture. But we have had remarkable success with so-called "n-body" simulations in super computers. The illustration opposite is from the Millennium Run simulation, constructed by a team of researchers and scientists known as the Virgo Consortium.

Initial ingredients—Gaussian fluctuations, cold dark matter, and the Lambda factor—have led to the development of a network of filaments, termed the "Cosmic Web." The texture is very similar to that seen in the distribution of galaxies, and gives us some confidence that we may at least partially understand what physical processes have shaped the cosmos. However, some details still remain unclear. Simulated giant galaxies seem to have far too many satellites, the simulated voids are not as empty as the real ones seem to be, and the real data reveals some condensations of galaxies far denser than the simulations can do.

SCHEMATIC DISTRIBUTION

Where then do we fit in? The nearest high-density region to us is the Virgo Cluster, though it is not particularly rich as clusters go. Nevertheless, it is the dominant structure in an irregular somewhat flattened large-scale structure known as the Local Supercluster or Virgo Supercluster. Our Galaxy lies on the fringe of that supercluster, in a protrusion that contains our Local Group of galaxies and neighboring groups. We are some 50 million light-years from the Virgo Cluster. Close to us is also the Local Void, stretching into a cosmic void almost 150 million light-years across. Surprisingly, Sir John Herschel first described the Local Supercluster in the mid-nineteenth century. In the mid-twentieth century, it was rediscovered by Gerard de Vaucouleurs. Jaan Einasto and Mihkel Joeveer were the first to ask whether the universe had a "cellular" texture in the later 1970s. It was only in the 1980s that the broader research community accepted these findings.

PLOTTING THE SKY

The recently completed 6dF Galaxy Survey (6dF for six-degree field) made use of the Anglo-Australian Schmidt Telescope at Siding Spring, near Coonabarabran, Australia. The telescope could observe almost 140 galaxies simultaneously, and the survey has covered the entire southern sky. Plots show greater detail of nearby large-scale structures and voids than has previously been possible.

This particular plot—a cross section through a volume of space—reveals the general texture of bubbly voids surrounded by galaxies. More particularly it shows high-density regions permeated by smaller voids, with intervening low-density regions having large, more anemic structures, and larger voids.

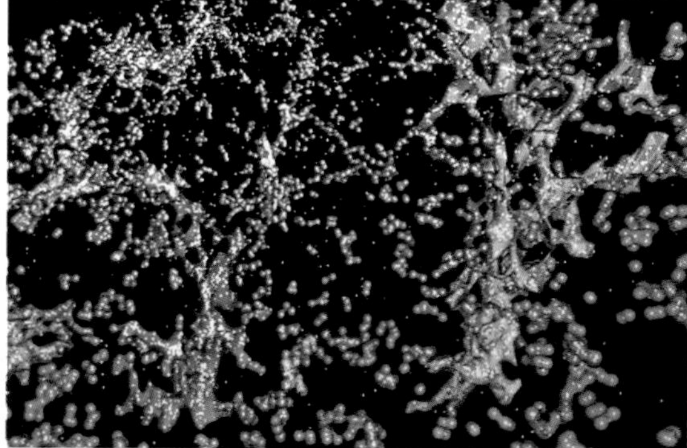

Above Visual of data of the 6dF (six-degree field) survey showing the clumpy voids between the galaxies. The survey collected more than 120,000 redshifts over the southern sky over a five-year period.

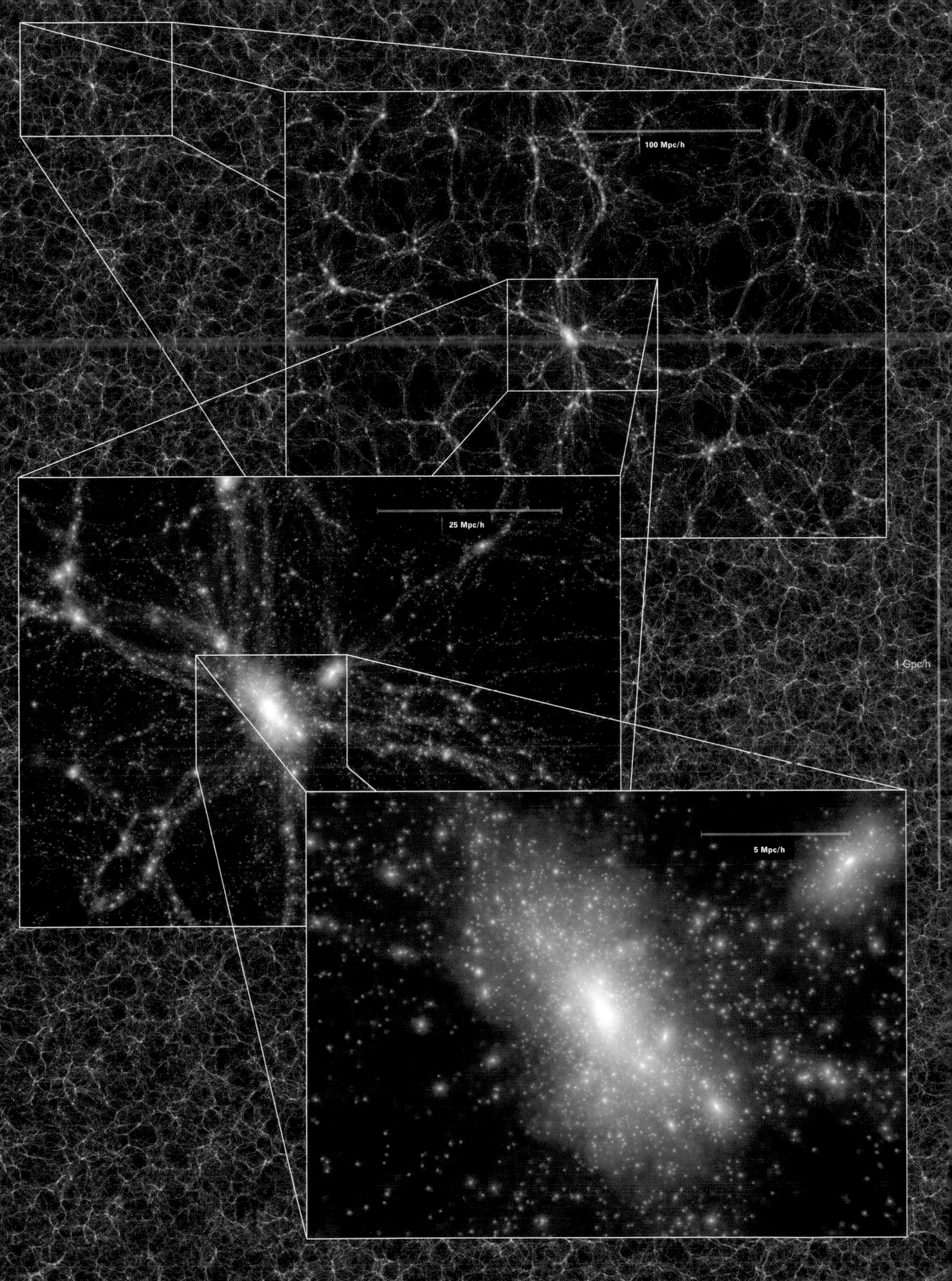

100 Mpc/h
25 Mpc/h
5 Mpc/h
1 Gpc/h

Modern Cosmology

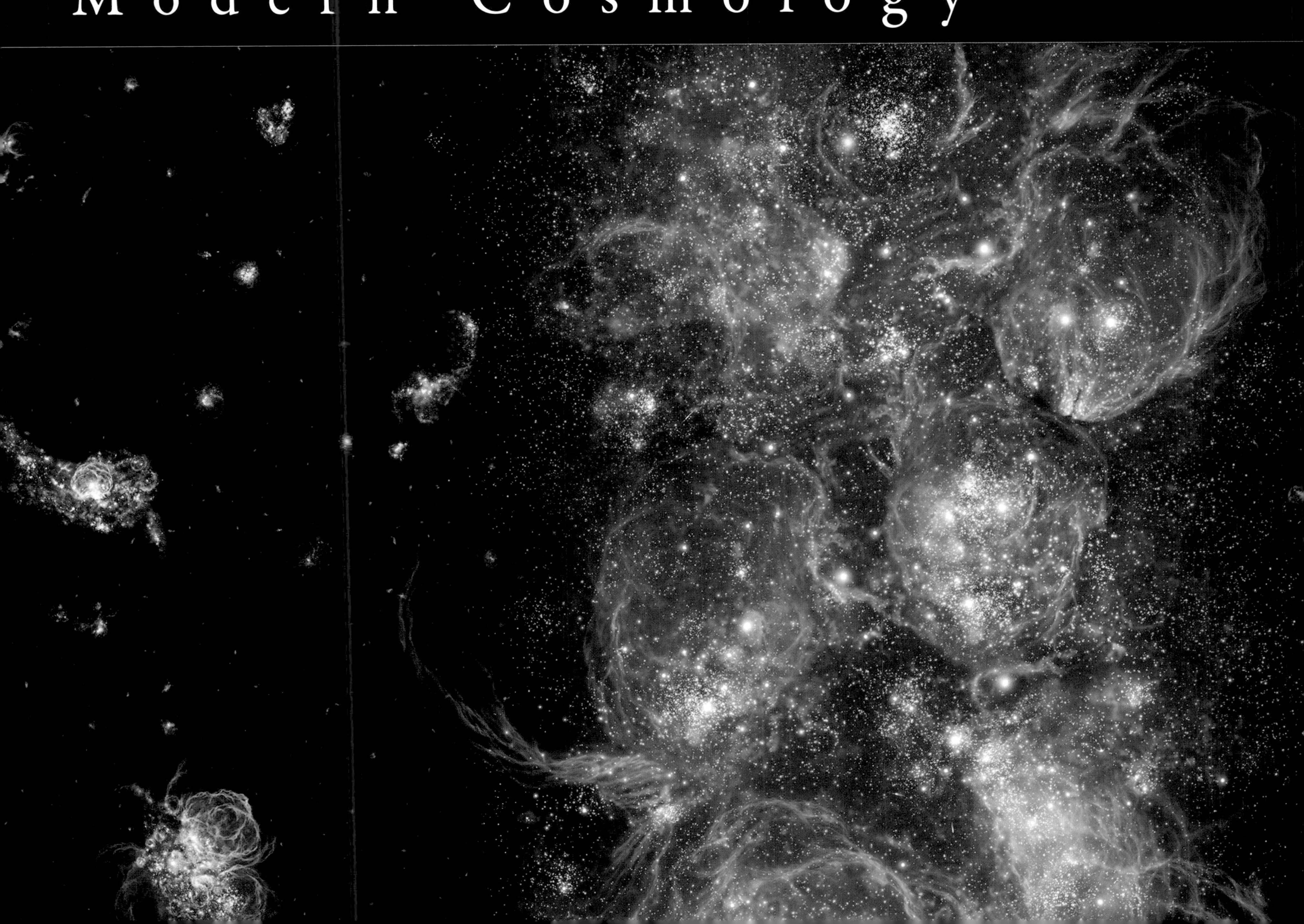

The State of the Universe

Cosmology is the study of the origin, structure, and evolution of the universe. It is a science that is always under construction as new discoveries cause revisions to old theories. Cosmology tries to answer the big-picture questions: how and when was the universe born, what is its overall shape, what is it made of, and how will it ultimately end? Trying to solve these mysteries can make your head hurt!

Right Controversial British astronomer Sir Frederick Hoyle (1915–2001) did much to popularize the study of astronomy with his ventures into science fiction, the best known examples of which are *The Black Cloud* (1957) and *A for Andromeda* (1962).

By 1925, astronomers had realized that the universe consisted of hundreds—maybe thousands—of galaxies, many of which were rushing away from us at enormous speeds; that is, the universe seemed to be expanding in all directions.

COMPETING THEORIES

The first scientist to take these findings and extrapolate backward was Abbé Georges Eduard Lemaître (1894–1966). In 1927, he theorized that the universe had a definite beginning, a time in which all its matter and energy were concentrated into a single point. When that point exploded like a burst of fireworks, it marked the beginning of time and space and caused the universe to expand.

Twenty years later, an alternative idea arose. Called the Steady State theory, it assumed that the cosmos was the same at all times, in all places, and in all directions. The universe had no beginning and would have no end, and the expansion that had been observed was simply an artefact of the creation of new matter that pushed old matter away. British astronomer Fred Hoyle, a vociferous proponent of the Steady State model, coined the term "Big Bang" as a sarcastic comment on the opposing theory.

Opposite The Hubble Space Telescope spent 11 days focused on one particular part of the sky to obtain this image, known as the Hubble Ultra Deep Field. These galaxies are so far away that we see them as they appeared when the universe was less than a billion years old.

In the same year that the Steady State theory was published, physicist George Gamow proposed that some of the chemical elements observed today were created within the first few minutes of the birth of the universe. Further, he argued that the early universe was very hot, and cooled as it expanded. If the Big Bang theory was correct, Gamow predicted that today's cosmos should be filled with just a little of the heat left over from its birth. The Steady State theory made no such prediction.

Astronomers called this predicted remnant heat the cosmic microwave background, or CMB radiation. When it was accidentally discovered in 1965, CMB radiation hammered the final nail into the coffin of the Steady State theory.

ALL IS NOT WHAT IT SEEMS

Even though it became the accepted hypothesis, the Big Bang theory left many questions unanswered. Exactly what happened during the first milliseconds of its birth that led to the universe we see today? What is the universe made of, and what is its ultimate fate? Cosmologists have since come up with some answers, but many mysteries remain.

In the early 1980s, Alan Guth proposed that in the minuscule fraction of a second immediately after the birth of the universe, the size of the cosmos expanded by a factor of at least 10^{30}. In other words, space itself inflated dramatically. His theory helps to explain why the universe is so large, even though it is only 13.7 billion years old. At the same time, astronomers were concluding that most of the mass of the universe is invisible, consisting of so-called dark matter. What this is composed of nobody knows, but if there is enough dark matter, the expansion of the universe will stop and reverse direction, leading to a "Big Crunch" in the far-distant future.

To further muddy the cosmological waters, in1998 it was discovered that the rate of expansion of the universe is actually speeding up. It seems the universe is full of an unknown force called dark energy that is pushing everything apart. According to cosmologists, the stuff we can actually see—from atoms and molecules to planets, stars, and galaxies—comprises a mere four percent of the universe. No wonder it is hard to understand our cosmos: We can't see most of it!

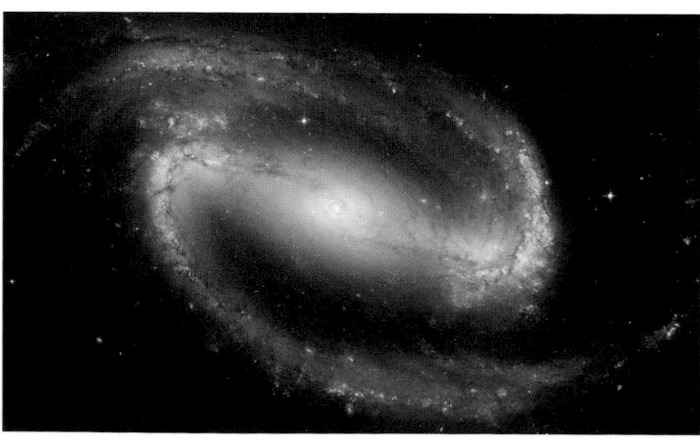

THE BIRTH OF COSMOLOGY

It's hard to imagine, but two hundred years ago astronomers believed that the universe consisted of our Solar System and the stars, which were an indeterminate distance away. Less than a hundred years ago the Milky Way was thought to be the entire universe, though there were questions about the nature of the many dim fuzzy patches of light scattered throughout the sky.

In 1923, Edwin Hubble, using the 100-inch (250 cm) Mt Wilson telescope, discovered that the fuzzy Andromeda "nebula" was actually a huge city of stars about a million light-years away—a distance more than three times the known diameter of the Milky Way. Astronomers quickly found that other fuzzy patches were also distant galaxies (such as NGC 1300 seen here). But where did they come from? When were they born? And what was their fate? In attempting to answer these fundamental questions, astronomers "invented" the modern science of cosmology.

The Origin of the Universe

Much about the science of cosmology is counter-intuitive, and completely alien to our everyday ways of thinking. Astronomers have now realized that the Earth and our Solar System constitute a minute fragment of an immense long-lived universe, the nature of which is turning out to be almost completely different from all that we find familiar.

In a 2007 *New York Times* article, theorist Lawrence M. Krauss put it another way: "We're just a bit of pollution. If you got rid of us, and all the stars and all the galaxies and all the planets and all the aliens and everybody, then the universe would be largely the same. We're completely irrelevant."

THE EARLY UNIVERSE

The Big Bang is the theory that explains how the universe came to be, but it does not describe the precise moment of cosmic birth. Cosmologists freely admit that they do not know what caused our universe to come into existence, and any discussion of this topic is speculative. So let's assume that the universe just is—a tiny speck, perhaps a millionth of a foot across—and go on to the next fraction of a millisecond of its life.

If the universe didn't "explode" into existence, what happened? Think of the birth of the cosmos as an unfolding of matter, energy, and time that took place everywhere simultaneously. One of the problems that faced cosmologists late in the twentieth century was explaining how the universe expanded rapidly enough to end up as big as we see it today. In the 1980s, Alan Guth and others proposed that a period of extremely rapid expansion—inflation—occurred in the first 10^{-34} seconds of its life. During this inflationary burst, the universe doubled in size at least 85 or 90 times, creating a super-hot, super-dense mixture of matter and energy.

Below In an artist's impression, the COBE satellite orbits the earth. In 1992, this satellite made the first whole-sky map, above right, of the Cosmic Microwave Background radiation that was first found accidentally by Arno Penzias and Robert Wilson in 1965.

The driving force behind this rapid expansion is believed to be vacuum energy. Although it defies common sense, physicists know that empty space has energy. In a universe as small as ours at its birth, a minuscule amount of vacuum energy would be powerful enough to inflate the cosmos many times over. But as the universe grew, the influence of vacuum energy waned, inflation weakened, and the rapid expansion of the universe slowed.

As the universe expanded, it cooled and became less dense. One second after its birth, the cosmic temperature was roughly 10 billion degrees. Two seconds later, the universe was a scorching plasma of photons (radiation) and subatomic particles smashing together to form protons and neutrons.

Astronomers call the next several hundred thousand years of cosmic history the Radiation Era. During this time the universe was dominated by the remnant energy from its birth. The cosmos was a searing sea of energetic electrons and protons, too hot to allow matter to form. Little is known about this high-energy era because the plasma that filled the universe prevented radiation (in the form of light) from escaping, leaving the newly born cosmos completely opaque to our prying eyes.

THE COSMIC MICROWAVE BACKGROUND

Some 380,000 years after the Big Bang, the universe had cooled to less than 4,000 degrees and hydrogen and helium nuclei begin capturing electrons to form electrically neutral atoms. This stable form of gas is not nearly as good as plasma at blocking radiation, and so, as the amount of hydrogen and helium increased, the cosmos slowly became transparent, which meant that astronomers could finally "see" what was going on.

As the universe expanded, the radiation from its early days was stretched (red shifted) and cooled. Today it can be observed only as a cold (2.7° above absolute zero or −459°F, −263°C), universe-filling, faint glow of microwaves: Cosmic Microwave Background radiation. The presence of CMB radiation was predicted by the Big Bang theory, but its discovery was accidental. Arno Penzias and Robert Wilson, at the Bell Telephone Laboratories in Murray Hill, New Jersey, found it in 1965. They were building a specialized antenna, and an annoying hiss that they could not eliminate turned out to be CMB radiation. (They shared the 1978 Nobel Prize in Physics for their discovery.)

In the early 1990s, the COsmic Background Explorer satellite (COBE) measured the distribution and temperature of the microwave radiation, and detected faint irregularities in both. So sensitive were the satellite's instruments that they recorded temperature differences of a mere one hundred thousandth of a degree between the "hot" and "cold" regions.

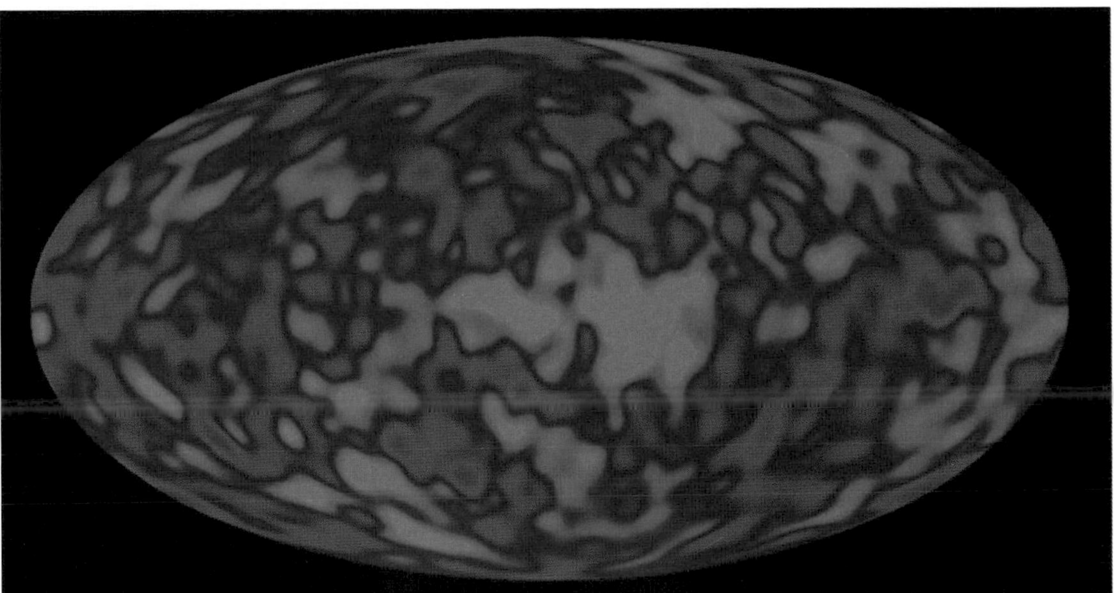

Left This map of the Cosmic Microwave Background radiation comprises two years of data from the COBE satellite. The blue and red spots represent regions of greater or lesser density in the early universe. These variations led to the formation of the first stars and galaxies in the young cosmos.

(For their work on Cosmic Microwave Background radiation, the COBE team leaders, John Mather and George Smoot, won the 2006 Nobel Prize in Physics.)

Smoot and others argued that these differences were fluctuations in the density of matter in the early universe: Fluctuations that eventually led to the formation of galaxies and galaxy clusters. But in the homogeneous 400,000-year-old universe, what caused those variations? Even better data was needed, so in 2001 a follow-up mission was launched: WMAP, or the Wilkinson Microwave Anisotropy Probe.

ENTER DARK MATTER

Even as Alan Guth was proposing his inflation theory to explain how the universe got to be so big so quickly, other astronomers were grappling with another vexing question: Why can't we see all of the universe?

Back in 1933, Swiss astrophysicist Fritz Zwicky pointed out something odd about the mass of the Coma cluster of galaxies. Zwicky estimated the cluster's total mass in two ways: By the motions of galaxies near its edge, and by the number of galaxies in, and the total brightness of, the cluster. He discovered that it possessed about 400 times more mass than expected. In other words, the gravity of all the visible galaxies was too small to keep the cluster together, and yet there it was—a stable cohesive group. Zwicky concluded that there must be a significant amount of "invisible" matter with enough mass (and therefore gravity) to prevent the Coma cluster from flying apart.

For forty years nobody paid much attention to Zwicky's findings. Then, in 1975, astronomer Vera Rubin announced that stars living outside the core of many spiral galaxies orbit at roughly the same speed. Rubin calculated that for this to happen, upward of 50 percent of the mass of these galaxies must be contained in a dark halo enveloping each galaxy. In fact, by the turn of the century astronomers had concluded that 85 percent of the mass of the universe consists of invisible dark matter, with the remaining 15 percent being visible baryonic matter—the protons, electrons, and neutrons that comprise all the "stuff" we see in the universe. But what is this dark matter made of, and where did it come from?

Some of it could be baryonic matter that is literally too dim to see: Multitudes of brown dwarf stars, dead stars, supermassive black holes, and even massive gas clouds that astronomers simply haven't detected yet. These objects have been nicknamed MACHOs (MAssive Compact Halo Objects). The non-baryonic possibilities are known as WIMPs: Weakly Interacting Massive Particles. (Who says physicists have no sense of humor?) Potential WIMP candidates include exotic subatomic particles that are as yet undiscovered, and neutrinos with significant mass (their mass is presently thought to be very small). If they exist, high-mass neutrinos might account for most of the dark matter in the universe, because enormous numbers of neutrinos were created during the Big Bang.

DARK MATTER AND COSMIC BIRTH

While some small percentage of this invisible material could be MACHOs, it is now generally accepted that dark matter was present very early in the life of the universe. In fact, for something that is invisible, it probably played a key role in the birth of the first galaxies.

The Big Bang distributed matter (once it began to form) evenly in all directions. The problem is, the theory doesn't explain how this matter might have started clumping together in a young homogeneous universe that was rapidly expanding in all directions. There certainly wasn't sufficient baryonic material to exert enough gravitational influence to

Below The Coma cluster is teeming with galaxies. Yet it has 400 times more mass than can be accounted for by all the galaxies seen here. The invisible mass is made up of dark matter, which comprises about 85 percent of the mass of the universe.

BIG BANG MISCONCEPTIONS

The Big Bang did not occur as an explosion somewhere in space. In fact, the theory doesn't even explain how the universe actually began, only how it expanded and cooled immediately after its birth. It is difficult to talk about what came before the Big Bang, or what caused it, because if something existed previously, astronomers don't know of any way for data from the pre-universe era to reach us. Since the universe encompasses all of space and time, the question "What's outside the universe" has no meaning. The observable universe is finite in size, but beyond what astronomers can see is, simply, more universe. That we can see galaxies moving away from us doesn't mean we are at the center of the universe. Any observer, anywhere in the cosmos, will see the same thing: It is simply an artefact of the expansion of space.

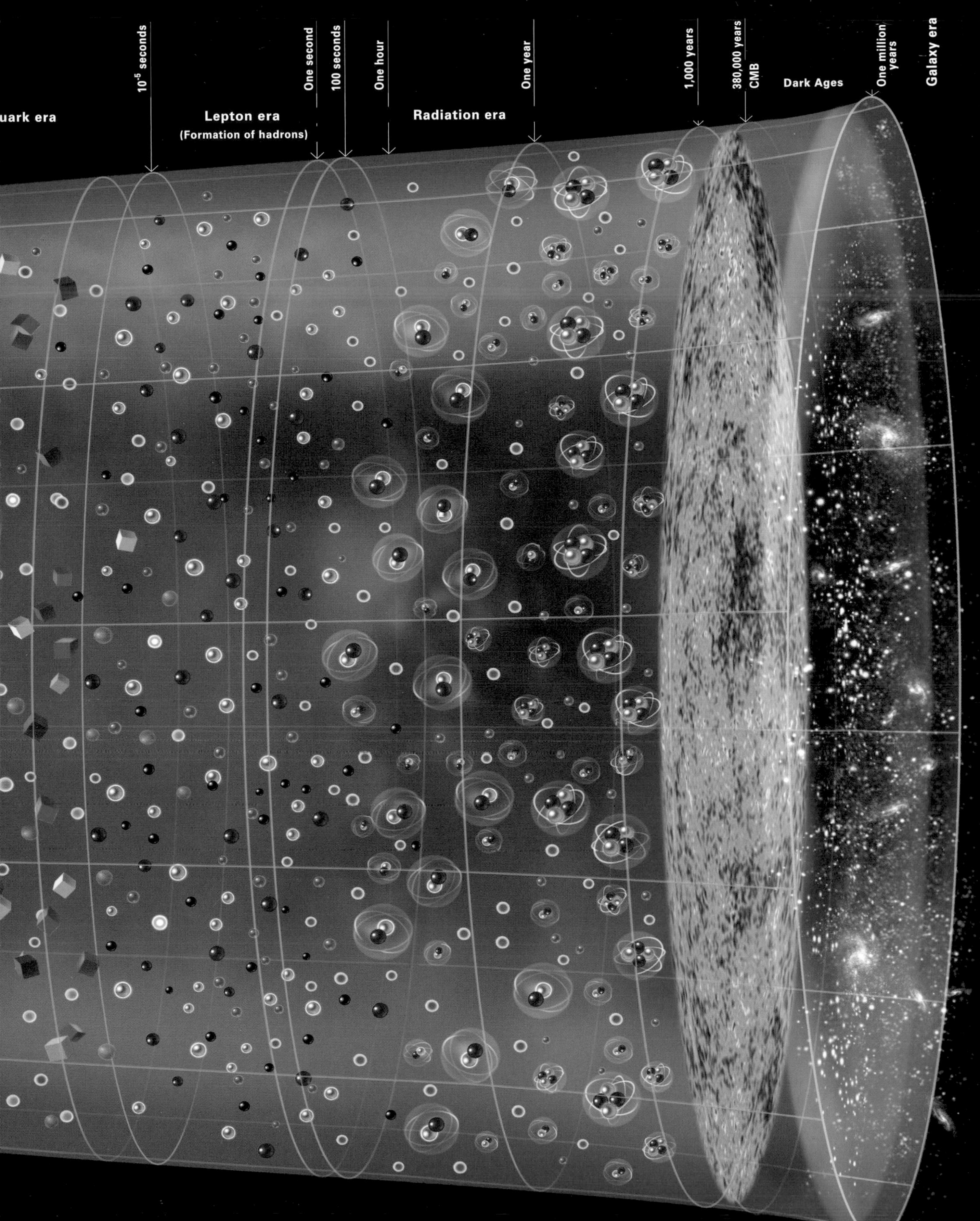
uark era
10^{-5} seconds
Lepton era
(Formation of hadrons)
One second
100 seconds
One hour
Radiation era
One year
1,000 years
380,000 years
CMB
Dark Ages
One million years
Galaxy era

Opposite The first stars began shining less than 400 million years after the birth of the universe, filling the young cosmos with their light.

start the clumping process. Yet something supplied enough gravity to create some minor cosmic lumps, which eventually turned into galaxies. That something was probably dark matter—and lots of it.

Cosmologists had theorized that buried in the Cosmic Microwave Background radiation might be signs of this initial clumping. The COBE results saw it on a large angular scale, but astronomers hoped that WMAP's better angular resolution would detect much finer features in the CMB, allowing them to fill in the details.

WMAP: PULLING IT ALL TOGETHER

WMAP was beaten to the punch by a variety of Earth-based observations made with instruments on mountaintops and in the Antarctic, so instead of groundbreaking discoveries, WMAP results were mostly confirmations of the findings of others. But the satellite did have two advantages: It could survey the entire sky, and its instruments were extremely sensitive. Consequently, its measurements were eagerly anticipated, even if some of its results were anticlimactic.

Among its many findings, WMAP gave the most precise estimate ever of the age of the universe (13.7 ± 0.2 billion years); confirmed that a mere 16 percent of the matter in the cosmos is normal baryonic material; and determined that dark matter, which makes up the remaining 84 percent of the mass of the universe, is cold. This means that neutrinos, which move at high speeds and prefer warm temperatures, are not the main constituent of dark matter.

WMAP also found that space everywhere is flat—confirming a prediction made by inflation theory. In addition, astronomers can use data obtained by WMAP to begin to distinguish between different versions of inflation. Already a number of exotic variations have been ruled out, but much work remains to be done.

One prediction of inflation theory is that there were minuscule fluctuations—at the subatomic (quantum) level—in the plasma that was present immediately after the birth of the universe. When inflation occurred, those tiny ripples were magnified enormously. These fluctuations ultimately became slightly denser regions of the new cosmos, which eventually developed into the galaxies and the galactic clusters that we see today. The data that has been obtained by WMAP matches these predictions very well.

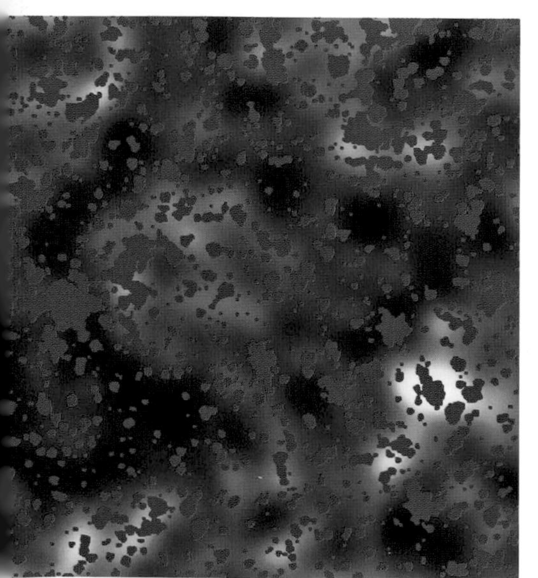

Below This Spitzer Space Telescope infrared image shows background light from a period when the universe was less than one billion years old. (Nearby stars and galaxies were masked out.) Scientists don't yet know if this light is coming from the first stars or first galaxies.

THE GROWTH OF THE UNIVERSE

Some time between 400,000 and 400 million years after its birth, the cosmos "turned on," and the first stars bathed the young universe in their light. But astronomers are not sure how this happened. Cosmologists have dubbed this period the Dark Ages because, to date, astronomers have not been able to directly observe what was going on. In fact, this period is so mysterious that no one is even sure which structures appeared first, stars or galaxies.

Most people think of a galaxy as an assemblage of stars, but astronomers view it as a large mass, of which stars are merely the obvious members. So, during those early times, did hydrogen and helium atoms slowly come together to form the original suns? Or did enormous accumulations of these atoms, brought together under the influence of dark matter, subdivide into proto-galaxies that, in turn, began to collapse and produce the first stars?

Recent studies of several very distant quasars, whose light left them when the universe was only about 900 million years old, revealed clear signs of the presence of iron—an element that can be made only in the hearts of exploding stars. But before they explode, stars have to be born and live out their lives, which means that the iron in the quasar's stars probably came from a previous generation of suns that had already come and gone. That earlier generation may perhaps be those elusive first stars, born just a few hundred million years after the Big Bang.

This idea is in line with WMAP's findings. By measuring polarization patterns in the microwave background, WMAP found that the first stars probably appeared about 400 million years after the birth of the cosmos.

By the time the universe was one billion years old, it was ablaze with galaxies. The Hubble Space Telescope has found hundreds of what appear to be normal-looking dwarf galaxies, though they were producing stars at a furious rate, about ten times faster than is happening today. These early galaxies were small—roughly 10,000 times less massive than our Milky Way. At about this time, massive galaxies began to appear—including, it seems, our own Milky Way. (Some of the oldest stars in our galaxy are more than 12.7 billion years old.) The merger of two or more dwarf galaxies probably created some of these big star cities. But there is also evidence that those first small galaxies literally changed the chemistry and composition of the young universe, which in turn affected the birth of the next generation of stars and galaxies.

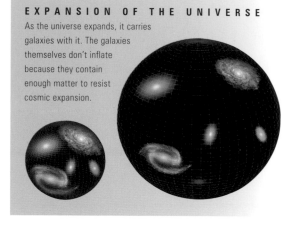

EXPANSION OF THE UNIVERSE

As the universe expands, it carries galaxies with it. The galaxies themselves don't inflate because they contain enough matter to resist cosmic expansion.

THE VIEW TODAY

By the time of its second billionth birthday, the cosmos was becoming quite set in its ways. Many of the normal galaxies that astronomers study today were being assembled during this time. The universe continued to expand, though the expansion rate was slowing. But some four billion years after its birth, the cosmos began to undergo another change, a change that would seriously affect its future. Dark energy began to make its presence felt.

The Future of the Universe

We all want to know what the future holds, but we're not good at predicting it. Even trying to forecast tomorrow's weather is a challenge, so pity cosmologists who try to peer billions of years into the future to foresee the ultimate fate of the universe. Until recently there were several clearly defined possibilities. Now everything has been thrown into disarray thanks to something we can't even see—dark energy.

In the mid-1990s, two teams of astronomers were measuring the brightness of very distant Type 1a supernovae. These stellar explosions are among the brightest events in the universe, which makes them an ideal tool for determining the distances to their remote galactic hosts.

DISTANT BEACONS

The light from a Type 1a supernova follows a predictable path, always peaking at the same level of brightness. This lets astronomers compare the supernova's real brightness (its intrinsic luminosity) against how bright it actually appears (its apparent luminosity) in the image of the explosion. From this, the distance to the supernova can be calculated. As a side benefit, cosmologists can use Type 1a supernovae to estimate the expansion rate of the universe at different times in the past.

Astonishingly, the astronomers discovered that very remote Type 1a supernovae were significantly fainter than they should have been, based on their distances estimated using other techniques. This suggests that they are farther away than originally thought, which implies that the cosmos has expanded more than the standard models of the universe's expansion say it should have. And that means that the rate of expansion of the universe must be accelerating.

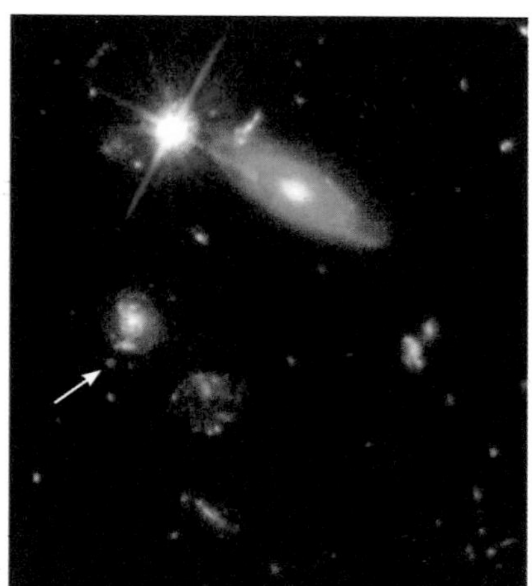

Above The galaxy and supernova (arrowed) shown here are estimated to be 8 billion light-years away. These remote starblasts help astronomers probe the time when the cosmic rate of expansion switched from deceleration to acceleration due to the repulsive force of dark energy.

Opposite Any star that passes too close to a black hole, as in this artist's concept, is torn to shreds and swallowed. Black holes, although they gradually evaporate over vast periods of time, may eventually be among the last bits of matter left in the dark and ever-expanding universe.

ENTER DARK ENERGY

So what could be causing the universe to expand at an accelerating rate? Astronomers call it dark energy because, whatever it is, it's a repulsive energy or force and is not made of matter. WMAP data revealed that 74 percent of the mass–energy composition of the universe is dark energy—22 percent is dark matter, and 4 percent is baryonic matter.

Recent studies have shown that matter and gravity dominated the early cosmos, causing its expansion to slow. But about nine billion years ago dark energy began to make its presence felt, and five to six billion years ago its repulsive force overcame the force of gravity. At this point the expansion of the universe began to accelerate.

It is possible that dark energy is simply a property of space that is spread uniformly throughout the universe and is unchanging over time. Because it is part of the fabric of space–time, cosmologists also call it vacuum energy. Another possibility is that dark energy is associated with an energy field whose density can vary across time and space. In this theory it is called quintessence, and if this idea is correct there are profound implications for the future of the universe.

THE FUTURE LOOKS DIM

As the universe expands more dark energy is generated, but it is not yet clear what this means. According to quintessence, if the density of dark energy increases over time, the universe will expand ever more rapidly and everything—galaxies, stars, and even atoms—will eventually be torn apart in a destructive frenzy nicknamed the "Big Rip." If the dark-energy density decreases, gravity might ultimately win the battle and everything be drawn back together into a "Big Crunch."

But WMAP data suggests that the density of dark energy in any particular volume of space does not appear to be changing with time, which implies that it is an inherent property of space–time. This suggests that cosmic expansion will continue to gradually accelerate, the universe will expand forever, and our cosmos will go quietly into the night. Even as the universe inflates, its constituents will gradually age. Stars will die, and star formation will decline as interstellar gas is consumed and star-forming nebulae disappear. Galaxies will dim as their stellar lights fade.

As billions of years pass, the distance between clusters of galaxies will increase, as will the speed at which they move away from each other. Eventually they will be so remote that their light will no longer be visible to each other. Ten trillion years from now, any inhabitants of a planet in the Milky Way will look out into a cosmic ocean of near blackness, empty of all but the local cluster of feebly shining galaxies. Trillions and trillions of years hence, our universe will consist of evaporating black holes, hulks of dead stars, decaying particles, and dark energy...or perhaps not.

At the beginning of this chapter, cosmology was described as a science that is always "under construction." These last few pages have summarized our current understanding of the universe. Stay tuned for updates.

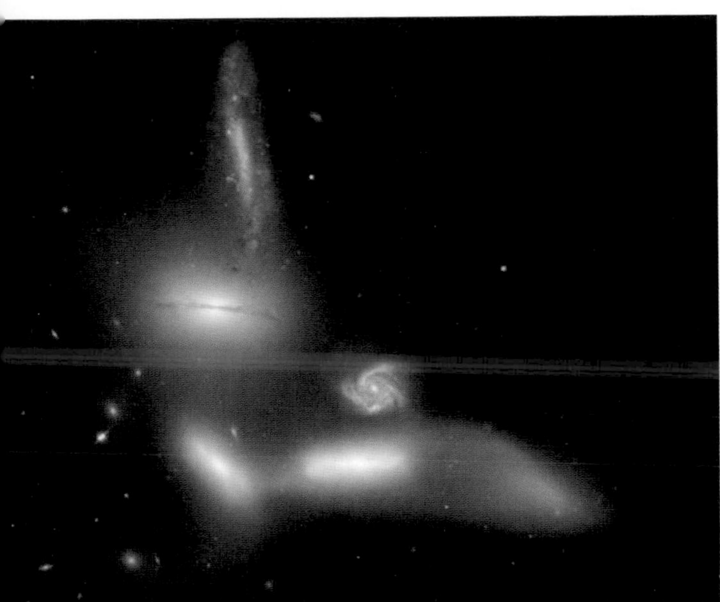

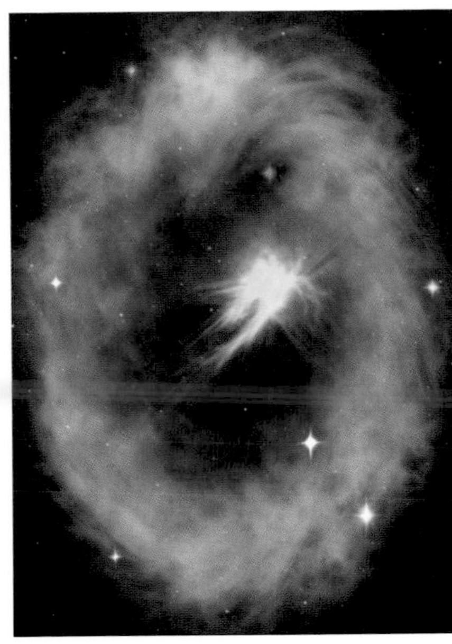

Far left While close-knit galaxies will manage to cling together in the future, dark energy will cause the clusters themselves to accelerate away from each other, and the Milky Way, in the expanding cosmos.

Left Type 1a supernovae are an excellent tool for measuring astronomical distances; however, astronomers were surprised to find that distant supernovas were fainter than predicted, leading to the theory of dark energy.

AQUARIUS
PISCES
ARIES
Triangulum
TAURUS
Caput Medusæ
Andromeda
Pegasus
Equuleus
CAPRICORNUS
Delphinus
Caßiopea
Cepheus
Cygnus
Perseus
Antinous
Auriga
Erichtonius
Aquila
SAGITTARIUS
Lira
Polus Eclipticæ Boreus
Circulus Arcticus
Stella Polaris
Colurus Solstitiorum
Polus Arct.
Vrsa minor
Engonasi
Hercules
GEMINI
Vrsa maior
Calisto
Castor
Pollux
Dubhe
Draco
Corona Bor.
Serpentarius
Ophiuchus
Coma Berenices
Colurus Æquinoctiorum
Bootes
Serpens
CANCER
Arcturus
SCORPIUS
LEO
Tropicus
Circulus Æquinoctialis
VIRGO
LIBRA
Regulus
HYPOTH PTOLEM
HYPOTH COPERNIC
Sol

History of Astronomy

Looking to the Heavens

Astronomy is considered by some to be the oldest of all sciences. Since the first emergence of *Homo sapiens,* approximately 200,000 years ago, thousands of cultures around the world have noticed the repetitive changing cycles of celestial objects and terrestrial seasons, and sought to make sense of them.

Opposite The massive pillars of Stonehenge in England were deliberately set out to mark alignments of the Sun and Moon. This sacred site was used for thousands of years by people who were very closely connected to astronomical cycles.

Below In 1660, the Dutch-German mathematician and cosmographer Andrea Cellarius published his *Harmonia Macrocosmica,* an atlas of how contemporary astronomers viewed the heavens. This color plate from the book illustrates Ptolemy's system of the universe.

It is not so difficult to understand why the heavens created, and still create, so much interest and enquiry. The phenomena which shape the natural world, and which dictate the fate of humanity, originate in the sky. Lightning provided the precious gift of fire; floods had the power to take it away. The seasons, which affected the availability of food, were accompanied by predictable and repetitive movements of the Sun, Moon, stars, and planets. Understanding these things greatly assisted humanity's survival.

This chapter deals with astronomical beliefs, research, and discovery, from pre-history through to the twenty-first century. It notes the differences and similarities in the conclusions reached by different cultures, traces the gradual separation of astronomy from astrology, and records the terrible fates that befell some early astronomers who had the audacity to question the accepted wisdom. We discover why so many ancient cultures independently created calendars and zodiacs that were based on a 12-month cycle.

Further, we map out and explain the important milestones and breakthroughs in astronomy, and the people who made them, from Plato to the under-recognized Aristarchus, to Copernicus, Galileo, and beyond. The significance of various discoveries and their effects on our view of the world are explored.

We also trace the development of observatories, from ancient times up to the present day. And we record the latest advances in astronomy and astrophysics, which shed increasing light on issues such as the history of time, the nature of light, the future of the universe, and the sub-atomic nature of matter.

THE DANGERS OF ASTRONOMY

Astronomy has often been a battleground between the forces of reason and superstition. During the European Renaissance—which took place from about the fourteenth to the seventeenth centuries—the science of astronomy was a particularly dangerous profession. As it became apparent that the Earth was not the center of the universe, the reaction of most religious and political rulers was to "shoot the messengers" whose proofs rendered their geocentric, God-based worldview untenable, for the basis of their power was exposed as a sham, built on a series of illusions. Astronomers (and other pioneers of science), under threat of loss of livelihood, excommunication from the Church, torture, or death, learned to disguise their observations in cryptic form, or to withhold them until after their deaths. The "heretical" (and correct) revelations of the seventeenth-century astronomer Galileo Galilei were not recognized by the Catholic Church until 1993.

THE BIG QUESTIONS

Astronomy has enabled us to explain many events that have occurred since 10^{-43} of a second after the Big Bang, which—about 13.5 billion years ago—set in motion the universe as we know it. Yet much remains unknown.

How did the Greek scholar Democritus, some 2,000 years before science proved him right, know that the universe is made of atoms, and that the Milky Way galaxy is not just a glow, but is made of billions of stars too small to be seen separately? Why did astronomical discovery stall for over 1,000 years after Ptolemy's (largely incorrect) observations and theses of the second century CE?

Where is most of the matter in the universe hiding? (We know it's there, but we can't find it.) What happened before the Big Bang, if anything? Are there other universes? What lies beyond our universe, if anything? We know that the universe is expanding at an ever-increasing rate—how could this possibly be? Can we find a way to escape before our Sun explodes, vaporizing Earth about 50 million years from now? Some of these important questions are answered here; others will perhaps remain forever unanswered, but there are obviously excellent practical reasons for humanity to persevere with astronomical research, quite apart from our propensity to seek knowledge for its own sake.

HOW DID THEY DO IT?

We can only surmise what people knew or believed before the advent of writing some 5,000 years ago; however, numerous artefacts dating from that era reveal that the science of astronomy was already advanced in many cultures. For example, Stonehenge and the Egyptian pyramids, both constructed about 4,500 years ago, reveal a pre-existing, highly detailed, and precise astronomical knowledge. We know that their construction required devices capable of measuring earthly and celestial distances to an amazing degree of accuracy, and complex machines to enable the transportation and positioning of rocks of enormous weight. Sadly, no evidence remains as to how these awe-inspiring feats were achieved.

Ancient Astronomy

All ancient peoples studied the sky, and gave names to the pictures they envisaged in groups of stars. Separated by vast distances and time, many cultures developed similar beliefs: The Babylonians, Greeks, Chinese, Egyptians, and others assigned god-like powers to the stars, planets, and constellations; the Babylonians and Mongols both saw the Milky Way as a seam sewing together the two halves of heaven, while the Vikings, Sumerians, and some Native North Americans believed it was a bridge between the living and dead.

Below In Greek and Roman mythology, various names are attributed to a Sun god, including Helios, Phoebus, and Apollo. Here we see Apollo represented in the German publication, *War Habits of the Romans*.

Many early civilizations created calendars to help make decisions about the timing of various activities: Religious festivals, planting and harvesting, even wars, were guided by both the regular, and irregular, movements of celestial bodies. Understanding astronomy could be a matter of life and death.

In high latitudes, the lengthening days of spring forecast the return of the growing season with its promise of continuing life. Near the equator, the changing stars indicated the coming of the monsoon rains and a shift in wind direction, crucial to agriculture and sea trade.

All the ancients revered the Sun as the primary god. The Babylonian sun-god Shamash provided human society with laws. The Greeks saw Helios (Apollo), god of the Sun, music, poetry, prophecy, and medicine, daily drive his flaming chariot across the sky. Egypt's Ra protected the human realm: King Akhenaton (*c.* 1500 BCE) claimed that Ra was the only god, making him the first monotheist in recorded history.

ASTRONOMICAL OBSERVATIONS

In ancient times, the Egyptians knew that the first appearance of Sirius—the brightest star in the sky—alongside the morning sun heralded the imminent annual flooding of the River Nile and the salvation of their parched fields.

Where the ancient Greeks observed the constellation Orion ("The Hunter"), the ancient Japanese saw Betelgeuse and Rigel, the two brightest stars, as samurai warriors preparing to fight; they were separated by the three stars between, which the Greeks knew (as we do still) as "Orion's Belt."

The Egyptian pyramids, and Stonehenge on England's Salisbury Plain (both dating from around 2500 BCE), display through the extreme accuracy of their alignments a comprehensive knowledge of celestial movements over a long period of time, proving that the study of astronomy was already very advanced. The pillars of Stonehenge (known as sarsen station stones) are arranged to note important events such as solstices and equinoxes, and the site was used by people from near and far for various rites of hope and thanksgiving.

The Great Pyramid at Giza in Egypt reveals surveying procedures of a very high order. The base, each side of which measures 745 ft (227 m), is level to within less than an inch (2.5 cm), while the difference between the pyramid's longest and shortest sides is less than 2 in (5 cm). Its alignment with the cardinal points of the compass is almost perfect, the error being less than 0.02 percent. Unfortunately, there is nothing left to tell us how its creators were able to make such accurate calculations, nor what machines they used.

CALENDARS

By the first century CE, the Mayans of southeast Mexico were using the "Long Count Calendar," which was an accurate comprehensive method of recording the monthly and annual cycles that formed the basis of their political, agricultural, and religious life. Ceremonial buildings—often pyramids—were precisely aligned with compass directions and also, frequently,

CHINESE ASTRONOMY

Chinese astronomers, studying the stars as long ago as 2100 BCE, envisaged 283 constellations. Rather than mythological heroes and beasts, these constellations revealed scenes from social and regal life. The astronomers also categorized "guest stars," which we now know as novae, supernovae, and comets. By the third century BCE, they were aware that Earth was in motion, and that this motion was responsible for the seasons. They saw the relationship between Earth and the heavens as similar to that between the two parts of an egg, the yolk (Earth) floating on a sea while the white (the heavens) revolved around the North Pole. By the first century BCE, they were using the armillary, a sphere with lines showing various celestial trajectories, and had built a number of planetarium observatories.

with special positions of the Sun, Moon, and various constellations. At the equinoxes, for example, the Sun might radiate through small openings in the walls, lighting up sacred areas in a building's interior.

The Babylonians, in what is now Iraq, were the first to divide the year into 360 days, the sky into 360 degrees of arc, the degree into 60 minutes, and the day into 24 hours, with each hour also divided into 60 parts. Around 700 BCE they produced the extraordinary baked clay artefacts known as the Mul-Apin Tablets, which summarize thousands of years of observations of the motions of stars and planets, and list constellations including Scorpius and Leo.

In ancient Assyria, Rome, South America, China, and India, a year also consisted of 360 days, divided into 12 months of 30 days. The Egyptian and Mayan calendars featured an undivided 360-day year.

From a human viewpoint, the Sun, Moon, and planets appear to move in a large circle, passing through the same constellations over the course of a year. The Babylonians invented the 12 signs of the zodiac based on these constellations, the Sun passing through one constellation per month as Earth revolved around it. Their system was subsequently adopted in Egypt, Greece, India, and China. (The Aztecs later created a similar system, probably independently.)

ANCIENT GREECE

Babylonian knowledge reached Greece around the sixth century BCE. But where the Babylonians used their observations of the stars to divine omens (astrology), the Greeks were more concerned with understanding the physical principles that underpinned the workings of the universe (astronomy). It is hard to overstate the importance of this difference in emphasis, which marked the beginning of the separation of superstition and science in the quest for knowledge.

The philosopher-astronomer Thales gained great fame for correctly predicting the solar eclipse of May 25, 585 BCE.

Democritus (460–370 BCE), who correctly predicted the atomic structure of matter, also surmised that the Milky Way

Above The Aztec solar calendar was carved into this great stone, which is now held in a museum in Mexico. The pre-Columbian Aztecs based their calendar on those used throughout ancient Mesoamerica.

Left The fifteenth-century book, *Illuminated Treatise on Astronomy,* describes and illustrates the constellations. On this page we can see, in the bottom right-hand corner, the Greek Sun god Helios driving his fiery chariot across the sky.

Right Here we see the great Greek astronomer Hipparchus at the observatory in Alexandria, Egypt. Hipparchus drew on the knowledge of the Egyptians and the Babylonians, but it is not known if he actually visited Alexandria.

galaxy is not a fog of light, as it appears, but comprises stars so faint and numerous that they appear to merge together. How could he envisage such things, millennia before they were proved? We can only hazard a guess.

The father of Western philosophical thought, Plato (*c.* 428–348 BCE), also turned his mind to astronomy, contending that Earth was round, and that its motion caused day and night. Plato's empirical, geometry-based methods were a giant step toward astronomy being dealt with in a scientific manner.

Heraclidus Ponticus (390–310 BCE) was the first to deduce that Mercury and Venus are satellites of the Sun, while Heraklides of Pontus (388–315 BCE) proposed that Earth spins on its axis.

Aristotle (384–322 BCE) was an unsurpassed genius in the field of astronomy (as well as in philosophy, psychology, history, physics, and logic). His observations led him to believe that Earth is spherical, as are the stars and planets—for if Earth were not spherical, lunar eclipses would not show a curved outline, nor would the stars change position when one traveled north or south. However, Aristotle also surmised that, because falling objects are attracted to it, Earth must be at the center of the universe.

Below This detail of a fresco by the Italian artist Raphael depicts two prominent Greek thinkers. Aristotle and Plato both contributed greatly to the world's understanding of the science of astronomy.

The astronomer and mathematician Eudoxus of Cnidus (fourth century BCE) outlined a system of 27 crystal concentric spheres, rotating at different speeds on different axes. Celestial bodies embedded in these spheres rotated around the spherical Earth. Later Greek astronomers revised this system, but Eudoxus's idea of perfect circular movements and a geocentric universe was generally accepted in Western astronomical thought until the seventeenth century. (Eudoxus traveled widely; he even studied astronomy with the priests at Heliopolis in Egypt.)

The esteemed scholar Aristarchus of Samos (*c.* 310–230 BCE) proposed that the Sun was much larger than previously believed. Observing that Mercury and Venus never strayed far from the Sun, and that Mars, Jupiter, and Saturn sometimes appeared to travel backward across the sky, Aristarchus explained these phenomena by concluding that Earth and the planets must all revolve around the Sun. He reasoned that the seasons were caused by a tilt in Earth's axis. Aristarchus is also famed for his treatise *On the Sizes and Distances of the Sun and Moon*, which provides an astounding geometric argument, based on observation, to calculate that the Sun was approximately 20 times as distant from Earth as the Moon, and 20 times the Moon's size. (While the true

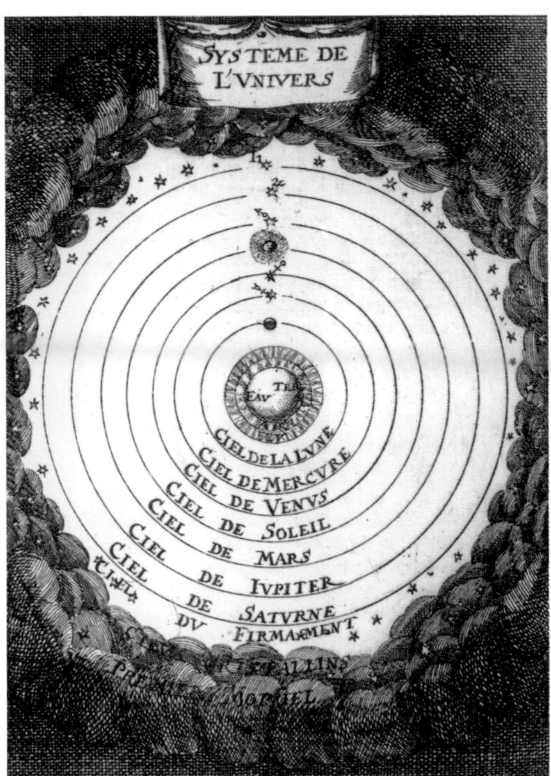

multiple is about 400 in each case, Aristarchus's reasoning was correct, the error being caused by his lack of accurate instruments.) His ideas, all since proved correct, were too radical to be generally accepted for some 2,000 years.

In 129 BCE, Hipparchus, ancient Greece's most influential astronomical observer of all, created the first accurate star catalog of 850 stars visible to the naked eye, and classified them into six different categories of brightness, a system of magnitude that is still in use. Sadly, most published works by Hipparchus have been lost, including an astronomical calendar, an array of writings on the subjects of optics, astrology, arithmetics, and geology, as well as his treatise called *On Objects Carried Down by their Weight*.

PTOLEMY

Ancient astronomy reached its zenith with the publication of Claudius Ptolemy's extraordinary work, *Almagest* (meaning "the greatest" in Arabic) in approximately 150 CE. A Greek-Egyptian citizen of Rome, Ptolemy made his astronomical observations from the great cultural center of Alexandria in the years 127–151 CE. His 14-part text discussed mathematical and astronomical concepts, summarizing and refining contemporary knowledge, and applied geometric principles to show that the heavens moved in a logical and predictable pattern. It contained an expanded version of Hipparchus's star chart, and listed the 48 constellations that are the basis of today's constellation system.

Although Ptolemy could not accept that Earth revolves around the Sun, he nonetheless accurately explained the uneven motion of the seven known planets by refining Eudoxus's theory that they were embedded in revolving concentric crystal spheres. Ptolemy pictured the heavens as a revolving sphere with a stationary rotating Earth at the center. His reasoning was erroneous, but his observations were so accurate that his geocentric view of the universe was accepted, with occasional minor refinements by Hindu and Arab astronomers, for more than 1,000 years.

THE ANTIKYTHERA MECHANISM

By 100 BCE the Greeks had invented the astonishing Antikythera Mechanism, a bronze apparatus about the size of a brick. Recovered in 1901 from a ship sunk off southern Greece around 65 BCE, the mechanism can calculate the past and future positions of the Sun and the Moon over long periods with great accuracy. It utilizes 22 gears in four layers and includes a differential gear (which up until this discovery was believed to have been invented in 1575).

Above This page from a fifteenth-century illuminated manuscript depicts the astronomer Ptolemy looking at an astrolabe, a device used to show the positions of the Sun and stars at a given time and place.

Above left A. M. Mallet's 1683 work, *Description de L'Univers,* has this representation of Ptolemy's geocentric system of the universe.

Millennia of Discovery

After the decline of the Greek and Roman civilizations, astronomy in Europe degenerated into astrology, divination, and alchemy. The study of astronomy was not reawakened until the ninth century in Baghdad, with an Arabic translation of Ptolemy's *Almagest*. Over the next 300 years, *Almagest* was also translated into Spanish and Latin, and astronomy found its way back to Europe via Arab-occupied Spain.

Opposite This seventeenth-century Ottoman miniature shows astronomers observing the Moon and the stars. At this time, the astrolabe quadrant was used by many astronomers and cartographers across the extensive Ottoman Empire.

The astrolabe—a forerunner of the sextant—is a device that first appeared in the Arab world around the year 1000. It consisted of a flat disk with a movable sighting-rod that was aimed at the Sun or stars to provide precise angles, which could then be translated into highly accurate navigational and time-keeping calculations. Around 1070, Spanish astronomer Arzachel invented an improved model.

By 1088, the Chinese government had established an entire department devoted to astronomy. The headquarters included an observatory, a four-story astronomical clock tower, and a vast bureaucracy to carry out research based on the continuous stream of new information. It also featured the first compass—an iron needle suspended on a silk thread. Scholars discovered the magnetic poles, produced a highly accurate luni-solar calendar, and created star and planet charts of such high quality that they are still useful today.

In 1277, Chinese scientist Guo Shoujing invented a new mathematical system that facilitated a quantum leap in precision of measurement and breadth of discovery. His records survive to this day at the Purple Mountain Observatory near Nanjing.

In the early 1400s Ulugh Beg, scientist and ruler of much of East Asia, constructed a three-story, 164-foot (50 m) diameter, underground observatory in Samarkand. It featured state of the art instruments such as a huge astrolabe embedded in the mountainside on a perfect north–south aspect, and stone arcs of great precision embedded in the earth. Large teams of visiting scholars produced the most accurate and detailed star charts and calendars to date, as well as producing an abundance of raw data that would soon prove useful to scholars in Europe—which was about to experience an unprecedented explosion in scientific research, invention, and knowledge.

Above Nicolaus Copernicus declared in his famous work *De revolutionibus orbium coelestium*: "At the middle of all things lies the sun."

THE RENAISSANCE PIONEERS

At the time of the rebirth of serious astronomical studies in Europe, the Church wielded enormous power, and the idea that the Earth is not the center of the universe was regarded as heresy; the geocentric work *Almagest*, by the second-century Greek-Egyptian scholar Ptolemy, was still the standard text.

Nicolaus Copernicus (1473–1543)

The first European astronomer of note in this period was a Polish clergyman, Mikolaj Kopernik, Anglicized as Nicolaus Copernicus. A student of law, mathematics, and medicine at Kracow, Bologna, Rome, Padua, and Ferrara, he developed a passionate interest in astronomy. He made no observations of his own, but in 1512 he published *Commentariolus*, an account of his heliocentric view of the world based on his fastidious mathematical analysis of Ptolemy's findings.

In 1530, Copernicus completed his landmark work *De revolutionibus orbium coelestium* ("The Revolution of the Celestial Spheres"), which proved beyond reasonable doubt that the Earth revolves around the Sun, and that humans are therefore not supreme in the scheme of things. Fearing Church backlash, Copernicus restricted circulation of his tome to a few select scholars, forbidding its publication until he was on his deathbed in 1543.

Copernicus believed that Ptolemy's observations and mathematics were correct but rejected his principles. His system was taught in some universities in the 1500s, but did not filter through to the academic world until around 1600. Some, including John Donne and William Shakespeare, feared Copernicus's theory, believing that it would destroy the natural order and lead to anarchy.

Although Copernicus was an important player in the development of the heliocentric cosmology, his work might have remained in relative obscurity without the accurate observations of Tycho Brahe, the meticulous mathematics of Johannes Kepler, and the all-round genius of Isaac Newton. These pioneers took Copernicus's theory as a starting point, and garnered from it the basic laws governing celestial mechanics.

Tycho Brahe (1546–1601)

Danish astronomer Tycho Brahe was a brilliant mathematician who analyzed Ptolemy's observations to determine whether the Ptolemaic or the Copernican system was correct.

THE BOWL WE CALL THE SKY

In the tenth century CE, the astronomer and mathematician al-Khujandi constructed a huge observatory in Persia (modern-day Iran). By observing the Sun, al-Khujandi calculated the tilt of the Earth's axis relative to the Sun. Around the same time, the scholar and poet Omar Khayyám compiled a calendar that was more accurate than the Julian calendar, and almost as accurate as the Gregorian one. Amazingly, he calculated a year to be 365.24219858156 days long, which was accurate to the sixth decimal place.

The Indian astronomer Bhaskara (1114–1185) proposed a heliocentric model in his astronomical treatise, *Siddhanta-Shiromani*. He also mentioned the law of gravity, stated that the planets do not orbit the Sun at a uniform speed, and accurately calculated many astronomical constants, such as solar and lunar eclipses. Latin translations of his work were available from the thirteenth century, so he may well have influenced the work of Copernicus.

وَيُقْذَفُونَ مِنْ كُلِّ جَانِبٍ دُحُورًا اى يَرْمُونَ من كل جا

In 1576 his sponsor, Frederick II of Denmark, built him a castle called Uraniborg—the first real astronomical observatory in Europe—on the island of Ven, between Denmark and Sweden. Brahe invented large accurate instruments and made nightly observations of the stars and planets that yielded measurements far more accurate than previous observations.

Numerous scholars visited Uraniborg, and Tycho trained a generation of young astronomers in the art of observation. Previous observations had been accurate to no more than 15 arc minutes, but Brahe's were accurate to 2 arc minutes, and his best observations were accurate to about half an arc minute. He was aided by Swiss Jost Bürgi's 1577 invention of a clock that measured minutes, rather than rough fractions of an hour, leading to a significant leap in accuracy and enabling Brahe to discover some serious inaccuracies in existing astronomical tables.

Brahe compiled an extensive table of planetary positions and a star catalog, and made the most accurate naked eye astronomical measurements of his day. He developed his own "geoheliocentric theory" of planetary motions, positing that the Sun orbited the Earth and the other planets orbited the Sun—a step backward from Copernicus. Brahe could not accept heliocentrism because he could not abandon Aristotelian physics—the only physics then in existence—in which heavy bodies fall to their natural place, the Earth, which is the center of the universe. If the Earth were not the center of the universe, Aristotelian physics was fatally undermined.

Brahe profoundly changed observational practice. Whereas earlier astronomers had merely observed the positions of the planets, the Sun, and the Moon at certain important points of their orbits (for example, zenith, opposition), Brahe meticulously observed these bodies throughout their orbits and identified a number of hitherto unnoticed irregularities. Without these uniquely accurate observations, Brahe's protégé, the gifted young Johannes Kepler, could not have discovered that planets move in elliptical orbits. Brahe hired Kepler to help with analysis of his extensive data, a task that Kepler continued after Brahe's death in 1601.

Brahe was also the first astronomer to discover and allow for atmospheric refraction; the phenomenon that causes light to bend as it passes through air. He is reported to have written his own epitaph: "He lived like a sage and died like a fool."

Below Danish astronomer Tycho Brahe is depicted here in 1587. From his remarkably well-equipped observatory, Uraniborg (meaning "Castle of the Heavens"), Brahe and his team carried out their most influential work.

Johannes Kepler (1571–1630)

Kepler was a brilliant theoretician who succeeded Brahe as the Imperial Mathematician to the Holy Roman Emperor Rudolf II. He was wrestling with his pet theory of planetary motion, which he based on circular orbits and perfect geometric forms. But his ideas were contradicted by the observations of Copernicus and Brahe, and were eventually destroyed by Galileo's 1610 discovery of four moons orbiting Jupiter. Undeterred, Kepler turned his attention to the irregular movement of Mars and discovered that it, and all the known planets, had elliptical orbits.

In 1609, Kepler published *Astronomia nova*, which contained two revelations that marked the starting point of

Below right Johannes Kepler's work *Astronomia nova* marked the beginnings of modern astronomy. Born into a poor family, Kepler received a scholarship to study at the University of Tübingen, and became a mathematics teacher.

Above This detail from an 1841 fresco depicts Galileo presenting his invention, the telescope, to the Doge and Senate of Venice. The remarkable discoveries Galileo made using this and later telescopes proved too controversial for the Church.

modern astronomy. Kepler's First Law states that planets move in elliptical paths with the Sun at one focal point of the ellipse. His Second Law states that a planet revolves around the Sun such that the line that connects the planet and the Sun sweeps out equal areas in equal times. Thus a planet's velocity increases as it moves closer to the Sun. Kepler speculated on a reason for this law—and almost beat Newton to discovering gravity. In 1619, while studying the relationship between the planets' velocity and their distance from the Sun, he came up almost by accident with his Third Law, which holds that the ratio of the square of a planet's orbital period to the cube of its average distance from the Sun is the same for all planets.

These mathematical certainties provided a springboard for the information explosion that was about to follow. Kepler's mathematical genius and his belief that astronomy must be grounded in mathematics and physics mark him as one of history's greatest applied mathematicians.

Galileo (1564–1642)

Galileo Galilei was arguably the greatest astronomer who ever lived. He discovered and published numerous proofs that the Sun is the center of the Solar System, and pioneered experimentation as a process blending scientific observation with quantitative mathematization and theoretical conceptualization. He discovered, among other things, that in free-fall the distance fallen increases as the square of the time elapsed, and that the velocity acquired is directly proportional to the time.

In 1609, hearing rumors of a newly invented Dutch spyglass, Galileo invented a much more powerful telescope and aimed it at the heavens. He discovered that the Moon was not perfectly smooth, as held by the Church, but was covered in mountains and valleys and pockmarked with craters. He found that the Milky Way is composed of millions of stars, and observed four of Jupiter's moons. He proved that the planets are much closer to Earth than the stars, because the telescope enlarged them to disks, whereas the stars remained tiny. In 1610, he published these heretical discoveries in *The Sidereal Messenger*; this marked the beginning of his difficulties with the Church, which maintained that the scriptures taught that the Earth does not move.

Galileo built ever-more powerful telescopes, up to a maximum magnification of 30×, and became more circumspect, to the point of withholding news of many of his discoveries. But in 1632 he published *Dialogue on the Two Chief World Systems*, in which a simpleton whose ideas are clearly not Galileo's articulates the Pope's beliefs. As a consequence, he was hauled before the Inquisition in Rome the following year and forced to live under house arrest for the rest of his life. Galileo speculated on the force that made planets orbit the Sun, and—like Kepler—almost discovered gravity.

GALILEO'S GIFT

To this day, the unit of acceleration due to gravity (9.8 meters per second) is known as "the Galileo." In 1989 the spacecraft *Galileo* was launched from a space shuttle on a journey to Venus and Jupiter. The Catholic Church did not absolve Galileo for his "blasphemies" until 1993—more than three and a half centuries after the publication of the *Dialogue*.

Isaac Newton (1642–1727)

Sir Isaac Newton was a brilliant astronomer, mathematician, optician, and inventor, who was fascinated by the nature of matter, cosmic order, light, colors, and sensations. A loner, he was considered eccentric for inventing such strange devices as a mouse-powered windmill. Speculating on gravity in the light of Galileo's discoveries, he was supposedly inspired by the sight of a falling apple to deduce that the force that attracts objects to the Earth is the same force that keeps the planets in orbit around the Sun.

Below Isaac Newton's 1672 reflecting telescope had a plane mirror inside, set at a 45 degree angle, and a concave mirror at the end.

His groundbreaking *Philosophiae naturalis principia mathematica* ("Mathematical Principles of Natural Philosophy," 1687) merges celestial and terrestrial mathematics into a single scheme based on the inverse-square law of gravitational attraction. It also contains his three revolutionary laws of motion, based on his new concept of mass and the principle of inertia.

Newton discovered that an object's gravitational attraction depends on its mass, and that the strength of the attraction diminishes with the square of the distance from the object. This laid the groundwork for a whole new field of science: Rational mechanics. He refined Kepler's discoveries about planetary motion, and deduced that the tendency toward the center in elliptical orbits varies as the inverse square of the distance from the Sun. He further noted that planets in orbit must cause perturbations in the orbits of other planets, and explained why our Moon causes ocean tides on Earth. And he invented the reflective telescope, which overcame the problem of light being broken up into its component parts by the refractive telescope pioneered by Galileo.

THE HERSCHELS

Kepler, Galileo, and Newton had discovered the mechanical laws that govern the universe. Further revelations would depend on the advent of new and improved technology.

Below Frederick William Herschel constructed his Great Forty-Foot reflecting telescope in his garden in England. In 1789, using this telescope, Herschel discovered Saturn's sixth and seventh moons, Enceladus and Mimas.

Frederick William Herschel (1738–1822)

Herschel was born in Germany, and moved to England in 1757. There he began to build telescopes, and on March 13, 1781, he observed what he thought was a nebula, or perhaps a comet. It turned out to be Uranus, and instant fame followed: George III rewarded him with an annual pension sufficient for him to devote all his time to astronomy. In 1787, he discovered two moons around Uranus, and, in 1789, Saturn's sixth and seventh moons. He also resolved indistinct nebulae into clusters of stars, and discovered hundreds of double stars.

Herschel was the first astronomer to realize that starlight takes a long time to reach Earth, and that star-watchers are gazing into the past. In 1800, he discovered infrared radiation while measuring the heat from the Sun at different wavelengths. His increasingly powerful telescopes reached further into space than ever before, and he began to outline the composition of the Milky Way.

Caroline Lucretia Herschel (1750–1848)

Caroline, Herschel's sister, moved to England in 1772 and joined her brother in his observations. She polished telescope mirrors and used her brilliant mathematical ability to make calculations from their observations. She discovered eight comets, and, after William's death, compiled a catalog of his observations of nebulae and star clusters. She was the first notable female astronomer.

John Frederick William Herschel (1792–1871)

In 1816, John Frederick, Herschel's only child, abandoned a brilliant university career to help his ailing father with his observations. He made award-winning observations of double stars and, after Frederick's death, used his father's telescope to update his catalog of clusters and nebulae, and produced a catalog of 5,075 double stars, of which he discovered 3,347. He was Secretary of The Royal Society from 1824 to 1827, then President of The Astronomical Society. In 1833 he published his brilliant *Treatise on Astronomy* for the lay reader.

From 1834 to 1838, Herschel used his father's telescope to observe the southern skies from the Cape of Good Hope in South Africa. Returning to England, he analyzed and published the results of his researches and served another three-year term as president of the (now Royal) Astronomical Society. He subsequently refused numerous offers of prestigious academic and political posts, preferring to work quietly on his own researches. From 1849 to 1873, he expanded his *Treatise on Astronomy* through 12 editions and made a failed attempt to convert British currency to the decimal system. On his death, he was hailed as the century's greatest scientist, and was interred near Isaac Newton in Westminster Abbey.

THE TWENTIETH CENTURY

In 1925, the American astronomer Edwin Hubble proved that the Milky Way is not the entire universe when he discovered that Cepheid variables in spiral nebulae proved that they were far beyond the boundaries of our galaxy.

Edwin Powell Hubble (1889–1953)

Hubble's discovery meant that the universe was at least ten times larger than had previously been supposed. He used the new 100-inch (254 cm) telescope at Mount Wilson Observatory, just east of Los Angeles, to identify the "red shift," or Doppler Effect, which causes objects moving away from us to appear red, while those approaching us appear blue. A similar effect occurs when the siren on a passing emergency vehicle

Opposite Edwin Hubble is seen here at the Mount Wilson Observatory telescope in 1937. Hubble joined the observatory in 1919, and used the telescope to study spiral nebulae.

seems to lower in pitch as it passes. Color and sound both travel in waves, which are compressed as they approach, thus decreasing the wavelength. (Blue has a shorter wavelength than red, as do higher-pitched musical notes than lower.) In 1929, Hubble observed 46 galaxies whose distance from Earth had been confirmed by his observations at Mount Wilson, and discovered that the red shift was indeed greater at greater distances.

Hubble's most significant discovery, it proved that the universe is not only expanding but accelerating. This relationship could be applied to entire galaxies whose stars were too faint to be viewed individually. The revelation increased the size of the observable universe by yet another factor of ten. The Hubble Space Telescope is, aptly, named for him.

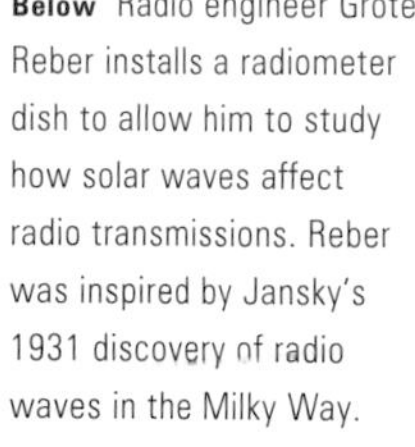

Below Radio engineer Grote Reber installs a radiometer dish to allow him to study how solar waves affect radio transmissions. Reber was inspired by Jansky's 1931 discovery of radio waves in the Milky Way.

Henry Norris Russell (1877–1957)

Russell was for many years regarded as the leading theoretical astronomer in America. He pioneered the use of atomic physics in the analysis of the stars, laying the fundamentals of astrophysics. He analyzed the physical conditions and chemical structure of stellar atmospheres and calculated the relative abundance of the elements. His assertion (plagiarized from Cecilia Payne-Gaposchkin) of the overwhelming abundance of hydrogen in the universe became accepted as one of the basic facts of cosmology.

Harlow Shapley (1885–1972)

Shapley was an American astronomer who studied under Henry Norris Russell at Princeton University. He was among the first to realize that the Milky Way galaxy is much larger than previously believed. He argued (correctly) against the theory that the Sun is at the center of our galaxy, and believed (incorrectly) that globular clusters and spiral nebulae are within the Milky Way. He correctly claimed that the Milky Way galaxy is centered near the constellation Sagittarius.

From 1921 to 1952 Shapley was director of the Harvard College Observatory, where he studied the Magellanic clouds and made catalogs of galaxies. He wrote many books, was an important popularizer of astronomy, and helped to form UNESCO.

Karl Guthe Jansky (1905–1950)

American physicist and radio engineer Karl Jansky is best known for his 1931 discovery of radio waves emanating from the Milky Way. A founding father of radio-astronomy, Jansky built an antenna designed to receive radio waves at a frequency of 20.5 MHz (wavelength about 14.6 m). It consisted of a circular series of masts, and was around 100 feet (30 m) across and 20 feet (6 m) tall.

After accounting for known radio waves emanating from the Sun and thunderstorms, Jansky concluded that the remaining radiation must be coming from the Milky Way, and was strongest in the direction of the center of the galaxy, toward Sagittarius. He wanted to follow up on this discovery, but could not obtain funding, and so radio-astronomy lay dormant for some years.

Two men who learned of Jansky's discovery had a great influence on the development of radio-astronomy. Grote Reber (1911–2002) was a radio engineer who built a radio telescope in his Illinois, USA, back yard in 1937 and undertook the first systematic survey of stellar radio waves. John Kraus (1910–2004) founded a radio observatory at Ohio State University after World War II. His textbook on radio-astronomy is still regarded as a standard text by radio astronomers.

Ralph Asher Alpher (1921–)

Asher is an American cosmologist who is best known for his correct prediction in 1948 of residual cosmic microwave background radiation from the Big Bang.

Stephen Hawking (1942–)

Stephen Hawking is best known for his iconic best-seller *A Brief History of Time: From*

Above Vera Rubin continues to challenge established theories on astronomy. Her extensive work has shown that we still know very little about our universe.

Above left The British mathematician Stephen Hawking suffers from motor neurone disease, and since 1905 has used a computer and speech synthesizer to communicate his brilliant ideas to the world.

Left Caroline Herschel paved the way for later female astronomers. In 1787, King George III gave her a pension of 50 pounds per year to continue her brother William's work. For the first time, a woman had been officially recognized for a scientific position.

Big Bang to Black Holes, which has sold millions of copies in more than 20 languages since its publication in 1988. In the 1960s, he developed a mathematical proof that space and time *must* begin in a Big Bang and end in black holes.

Einstein's general relativity theory had predicted black holes, as well as super-dense neutron stars and points of infinite density, which were also discovered in the 1960s and named pulsars and quasars. Hawking's mathematical calculations, along with astronomical observation, confirmed—and explained—Einstein's improbable scenarios. Hawking subsequently provided mathematical explanations for pulsars and quasars, and combined relativity theory with quantum physics to shed further light on our knowledge of black holes.

A writer of genius, Hawking is regarded by most as the greatest mathematician since Einstein, or possibly of all time. Asked in October 2005 on British TV to explain his statement that the question, "What came before the Big Bang?" is meaningless, he replied that it was like asking, "What lies north of the North Pole?"

WOMEN ASTRONOMERS

In the eighteenth century, Caroline Herschel had demonstrated that women had a contribution to make to the study of astronomy, and the twentieth century produced a number of distinguished women astronomers.

Henrietta Swan Leavitt (1868–1921)

In 1908, American astronomer Henrietta Leavitt observed that Cepheids—giant stars visible over vast distances—varied in "absolute magnitude," their real brightness, and in their "blink rates," or periods between maximum and minimum brightness. She noted that the brightest Cepheids were blinking slowly, while the dimmer ones were fast, and that this was a consistent graduated phenomenon. This meant that a Cepheid's distance can be calculated because, knowing its absolute magnitude or real brightness from its blink rate, we can deduce that any reduced luminescence is due to increased distance.

Establishment science still thought that the Milky Way galaxy was the entire universe; Leavitt had discovered the way to look beyond the Milky Way.

Cecilia Payne-Gaposchkin (1900–1979)

British astronomer Payne-Gaposchkin gained the first PhD in astronomy ever awarded by Harvard University. She had studied at Cambridge (UK), but females were not permitted a degree so she went to America. A brilliant physicist, she was also denied a degree in physics (because she was female) and turned instead to astronomy. Her thesis correctly deduced that hydrogen is the major component of the Sun, the stars, and in fact the entire universe, but her supervisor, Dr Henry Norris Russell, insisted that her analysis of the stellar spectrum was incorrect. Establishment opinion held that all stars consisted mainly of iron, as Earth does.

Russell refused to allow Payne-Gaposchkin to graduate until she wrote a postscript to her thesis, stating that her results were "clearly untrue." Even more disgracefully, Russell and his male colleagues conspired to hinder her career when it became clear that she had been right all along. The Sun was a mystery at that time; however, Payne-Gaposchkin revealed a fundamental truth about universal processes, laying the groundwork for research that looked into nuclear reactions.

Vera Rubin (1928–)

In 1970, Vera Rubin began studying the movements of Andromeda, the nearest galaxy to the Milky Way. The galactic vortex should have been moving fast at the center and slower toward the edges, but it wasn't. Rubin concluded that the galaxy extended much further out and largely consisted of "dark matter" that did not radiate light. This was a revolutionary concept, and was at first derided by the establishment. She has since been proven correct. She also concluded that our local galaxies were being drawn toward the constellation Pegasus; this was also treated with scorn by the establishment, but other astronomers confirmed it more than 20 years later.

Rubin was the first to realize that invisible materials and forces exist, and are probably the major influence in the universe. In 1996, she became only the second woman to be awarded the Gold Medal of the Royal Astronomical Society; Caroline Herschel received the award in 1828.

Observatories of the World

Isaac Newton wrote a famous passage in *Opticks* (1730), where he reasons that telescopes "cannot be so formed as to take away that confusion of rays which arises from the tremors of the atmosphere." Adaptive optics—unknown to Newton—allows a modern telescope to greatly reduce the blurring or tremors caused by the atmosphere.

Right English scientist Isaac Newton (1642–1727) is portrayed here investigating the properties of light. His findings on the color spectrum led to his invention of the reflecting telescope.

Below Developments in telescope technology through the ages include Johannes Hevelius's Large Astronomical Telescope. This 1670 engraving from his *Machina Coelestis* serves as a visual record, as this device—with many others—was lost following a fire at his observatory.

To Newton "...[t]he only remedy is a most serene and quiet air, such as may be found on the tops of the highest mountains." Following this idea, the Astronomer Royal of Scotland, Charles Piazzi Smyth, visited the Canary Islands in 1856.

Using a 7-inch (18 cm) telescope, Smyth spent 65 days on El Teide—the highest mountain in the Spanish-owned islands—making astronomical observations at heights of 8,850 feet (2,700 m) and 10,500 feet (3,200 m). He found he could see stars as faint as 14th magnitude—that's four magnitudes fainter than the faintest star of 10th magnitude visible at sea level in Scotland. He also found the image formed by his telescope to be significantly sharper.

The majority of the world's largest, most powerful telescopes are concentrated in a handful of high-altitude sites. These locations have clear dry skies, free from cloud and dust; a stable transparent atmosphere above to give the sharpest images possible; and are so dark that the light from the Milky Way glow casts shadows across the ground on a moonless night. Such sites are located in the Andean mountains in Chile, the Canary Islands, Hawaii, and the Rocky Mountains in western USA.

Like modern astronomers, the fifteenth-century astronomer Ulugh Beg carried out research to understand the universe. Not having a large telescope forced Ulugh Beg to minimize observational error by using the largest instruments possible. Using only a sundial with a 165-foot (50 m) high gnomon, he measured the length of the sidereal year as 365.2570370 days, with an error of less than one minute. He also measured the position of about 1,000 bright stars with a typical error of 16 arc minutes. Even today, one of the most common activities of an observatory is measuring the position and brightness of objects.

IMAGING INTERFEROMETERS

One of the most exciting developments in astronomy is the use of imaging interferometers at optical and near infrared wavelengths. Two dedicated observatories exist—the CHARA (Center for High Angular Resolution Astronomy) Array, a Y-shaped arrangement of six 40-inch (1 m) telescopes on Mount Wilson, California, USA, and the Magdalena Ridge Observatory Interferometer, a Y-shape arrangement of ten 55-inch (1.4 m) telescopes in New Mexico, USA, which is currently under construction. For both instruments the maximum spacing between two telescopes is about 1,115 feet (340 m), giving them the resolution of a telescope with a 1,115-foot (340 m) aperture. This allows details as small as one six-thousandth of an arc second to be seen at a wavelength of 1 micrometer (μm)—equal to the angular size of a car seen at 870 million miles (1,400 million km) away.

To date, similar interferometers take many hours to build up an image of the disk of a bright nearby star to measure its size and brightness. Both the CHARA array and the Magdalena Interferometer could achieve the same result in a matter of minutes and can image faint objects with complex detailed structure.

In 2006, the CHARA array managed to solve the long-standing mystery of why the star Spica emits 50 percent more energy than theory predicts. Measurements of the star's diameter and brightness show it was dimmer around the edge and brighter in the middle. The cause was found to be that Spica rotates very fast while one of its poles points toward Earth. Calculations show Spica is rotating so rapidly—about once every 12 hours—that it is being squashed into an oval shape. This causes Spica's poles to have a temperature of 17,850°F (9,900°C) while the equator is 13,700°F (7,600°C), making the star appear bright when viewed from Earth.

One of the ultimate possibilities for an interferometer is to directly image an earth-sized planet orbiting around a nearby star, revealing an image of a planet with a resolution similar to that which the human eye sees when looking at the Moon in the night sky.

NORTHERN HEMISPHERE OBSERVATORIES

The best located observatory is found on the summit of a 13,800-foot (4,200 m) high volcano named Mauna Kea, in Hawaii. Known as the Mauna Kea Observatory, it is home to four of the world's ten largest telescopes, the twin 400-inch (10 m) Keck telescopes, the 325-inch (8.3 m) Subaru telescope, and the 320-inch (8.1 m) Gemini North telescope. The second best site is the Roque de los Muchachos Observatory in the Canary Islands on the island of La Palma. The site is located at an altitude of 7,875 feet (2,400 m) and is home to the 410-inch (10.4 m) Gran Telescopio Canarias and the Swedish Solar Telescope.

The Swedish Solar Telescope is a 40-inch (1 m) refracting telescope designed to record the sharpest images ever taken of the Sun. In theory, as the aperture of a telescope increases so does its ability to see smaller details. In practice, a telescope's image is blurred and distorted by the chaotic mixing of warm and cold air in the atmosphere. The resulting image has, at best, a resolution achievable with a 20-inch (50 cm) aperture. For a 40-inch (1 m) telescope to achieve its theoretic resolution

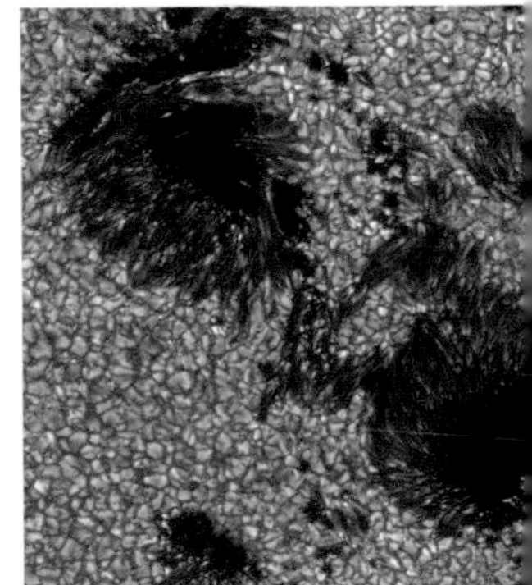

Top Located at Mauna Kea's summit, the Hawaiian observatory, named for its location, is home to a number of telescopes in a range of types—optical, infrared, radio, and submillimeter

Above The Swedish Solar Telescope—located on the Canary Islands—provides astronomers with some of the clearest images of the Sun to date.

LIST OF LARGEST TELESCOPES

Telescope	Size	Opened	Altitude	Observatory/Country
Gran Telescopio Canarias	410 inch (10.4 m)	2006	2.4 km	Roque de los Muchachos Observatory, Canary Islands
Southern African Large Telescope	400 inch (10.0 m)	2005	1.8 km	South African Astronomical Observatory, South Africa
Keck 1 and 2	400 inch (10.0 m)	1993/96	4.2 km	Mauna Kea Observatory, Hawaii
Hobby-Eberly Telescope	368 inch (9.2 m)	1997	2.0 km	McDonald Observatory, Texas
Large Binocular Telescope 1 and 2	330 inch (8.4 m)	2007	3.2 km	Mount Graham International Observatory, Arizona
Subaru	325 inch (8.3 m)	1999	4.2 km	Mauna Kea Observatory, Hawaii
Very Large Telescope 1, 2, 3, and 4	322 inch (8.2 m)	1997–99	2.6 km	Paranal Observatory, Chile
Gemini North	320 inch (8.1 m)	1999	4.2 km	Mauna Kea Observatory, Hawaii
Gemini South	320 inch (8.1 m)	2001	2.7 km	Cerro Pachón, Chile
Magellan 1 and 2	256 inch (6.5 m)	2000–02	2.4 km	Las Campanas Observatory, Chile

Above The Southern African Large Telescope, located at the South African Astronomical Observatory, is the largest in the Southern hemisphere. Its camera is now producing high-quality images of the night sky from its southerly perspective.

requires the use of adaptive optics. Light entering the telescope is directed onto a flexible mirror that changes its shape up to 1,000 times a second to remove atmospheric blurring. To minimize heating by sunlight, all the air inside the telescope is also removed. In the very best images of the Sun, details as small as 0.1 arc seconds or 40 miles (70 km) are visible on the surface. That's equivalent to reading the smallest line in an eye test chart from a distance of 2 miles (3 km). These images have helped to solve a long-standing puzzle as to why the Sun's atmosphere gets hotter rather than colder as one moves upward, away from its visible surface. In simple terms, regions of strong magnetic fields were found near the surface, which regulate the release of sound waves from the Sun's interior that propagate up into the atmosphere, causing the observed heating.

The world's first 400-inch (10 m) class telescope was completed in 1993 as part of the W. M. Keck Observatory on Mauna Kea; an identical telescope was completed in 1996. Each telescope weighs 300 tons (275 tonnes), stands 8 stories tall, and has a 400-inch (10 m) primary mirror using an innovative arrangement of 36 hexagonal segments that are precisely aligned to work together as a single solid mirror. Each telescope uses an altitude-azimuth mount to precisely position the telescope and track objects across the sky. As the telescopes move, each of the mirror segments is continuously adjusted to maintain the ideal shape. To achieve the sharpest images, Keck uses an adaptive optics system. A 6-inch (15 cm) flexible mirror changes its shape up to 670 times per second to cancel out atmospheric distortions, making images 20 times sharper, with details as small as 0.03 arc seconds visible. In 2005, the adaptive optics system was used to look at the dwarf planet Eris. The sharper images revealed a small bump next to Eris—a moon. The new moon, about 100 miles (150 km) wide and named Dysnomia, allowed the mass of Eris to be calculated for the first time.

The 368-inch (9.2 m) Hobby-Eberly Telescope at the McDonald Observatory in Texas was built for around 20 percent of the cost of a single Keck telescope. This was achieved using a simple telescope mount that fixes the altitude of the telescope at 55 degrees but allows the entire telescope to rotate around the vertical. The end result is an inexpensive telescope able to see 70 percent of the night sky.

SOUTHERN HEMISPHERE OBSERVATORIES

In the Southern hemisphere is a site considered to be the best in the world—even better than Mauna Kea. Known as Dome C, it is located in the highlands of Antarctica and tests show the site outperforms every other observing site on Earth. The next best sites are the high mountain peaks in Chile, home to the 320-inch (8.1 m) Gemini South telescope, the

two 256-inch (6.5 m) Magellan telescopes, and the four telescopes of the Very Large Telescope (VLT). The European Southern Observatory's Very Large Telescope is located in the Atacama desert of northern Chile. The VLT is composed of four 322-inch (8.2 m) telescopes that can be used separately, as an interferometer, or in combination. When combined together, the array of four telescopes would have the light-gathering power of a 630-inch (16 m) telescope. Four smaller 70-inch (1.8 m) telescopes are used in combination with the four larger telescopes to form an interferometer, allowing high resolution imaging of bright sources. The suite of instruments used by the VLT covers the spectrum from near ultraviolet to mid-infrared. At infrared wavelength, the adaptive optics system used by each of the VLT telescopes can produce images three times sharper than the Hubble Space Telescope.

The largest telescope in the Southern hemisphere is located in South Africa, on the edge of the Kalahari desert. Named the Southern African Large Telescope (SALT), it is based on the design of the Hobby-Eberly Telescope and uses a 400-inch (10 m) spherical mirror made from 91 hexagonal segments.

The telescope mount is fixed at an altitude of 53 degrees and can rotate about the vertical axes. Located 43 feet (13 m) above the mirror is a tracking platform that moves across the mirror surface as Earth rotates to track an object for up to two hours. While tracking, the telescope mount and mirror remain fixed.

Above This image shows one of the four domes of the Very Large Telescope (VLT) opening up as the sun sets. The VLT is located atop Cerro Paranal in Chile, ideally located—both geographically and climatically—for observing space.

Into the Future

The future holds endless possibilities, although forecasting the future is fraught with peril. Still, there are fascinating lines of astronomical research that should bear fruit in the new millennium. Almost everything described in this book was discovered in the twentieth century, when cosmological knowledge grew immensely. Just as we wonder at how little astronomers knew a hundred years ago, will our descendants wonder at our inadequate understanding of the universe?

A hundred years ago, the idea of life beyond Earth belonged in the realm of science fiction. Today, astrobiology—the study of life in the universe—is a respected science, even though we have yet to find evidence of a single organism, living or dead, on a world other than our home planet.

THE SEARCH FOR LIFE

Mars has been the focus of the search for extra-terrestrial life for decades. In January 2003, two robotic space probes, *Spirit* and *Opportunity*, landed on opposite sides of the Red Planet to look for signs of water. *Opportunity*'s visit to the craters Eagle and Endurance showed evidence of an ancient desert-drainage basin where the water had evaporated to form salt-rich sands. In the Columbia Hills, *Spirit* discovered strong indications that early Mars (at least in this region) was molded by impacts, volcanism, and subsurface water. At the same time, cameras on-board spacecraft orbiting Mars spotted signs that, within the past few years, dirt-laden water may have flowed down Martian gullies. In theory, where there is water there is, or was, life.

In the next two decades, the search for life on Mars will be completed. NASA plans to send a Mars Sample Return mission to collect samples of Martian soil and return them to Earth. A follow-up mission, the Astrobiology Field Lab, will roll across the Martian plains searching for life.

Meanwhile, scientists are looking for water (and life) in some unusual places. The ice-covered surface of Europa, the third-largest moon of Jupiter, may be harboring a vast reservoir of liquid water. On Saturn, the ice-moon Enceladus has geysers of ice spewing from pockets of water in liquid form.

INTO THE TWENTIETH CENTURY

At the beginning of the twentieth century, the Solar System was thought to consist of eight planets, 21 moons, 452 asteroids, and assorted comets; the planet Pluto had not been detected, nor had Kuiper Belt objects. The Earth was considered to be only 50 million years old, and the distances to only 70 of the nearest stars were known. Red giants, white dwarfs, pulsars, supernovae, and black holes were unheard of. The Milky Way galaxy was the entire universe. The universe was believed to be static, and there was no accepted theory of how it came to be.

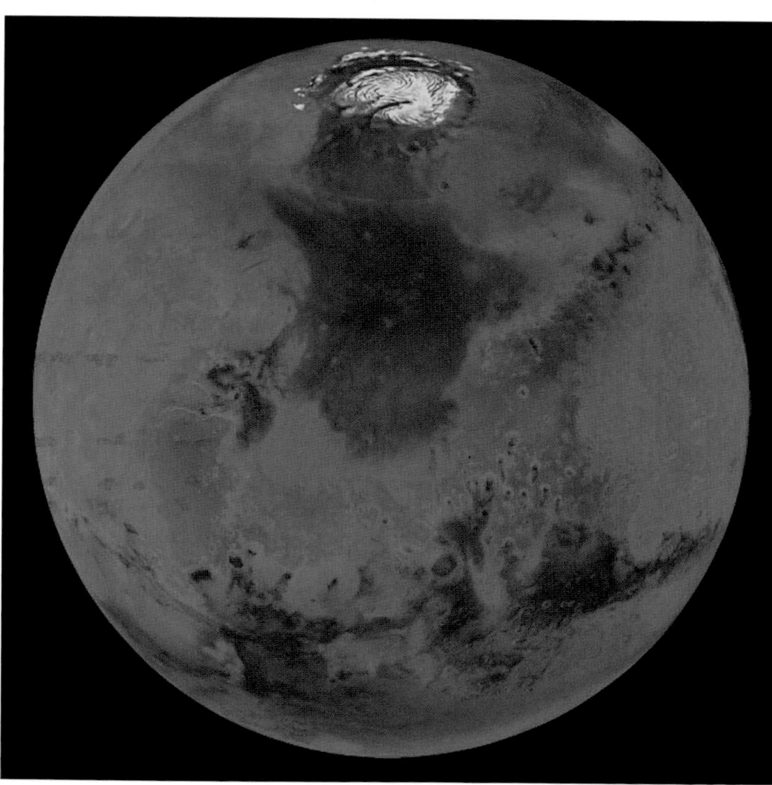

Below Mars may be the first place astronomers discover evidence of life existing on a planet other than Earth. The light area in this Mars Global Surveyor image suggests that water has flowed on Mars.

Below right A Martian sample is sent back to Earth. This is an artist's impression of the Mars Sample Return mission that NASA plans to launch perhaps as early as 2011.

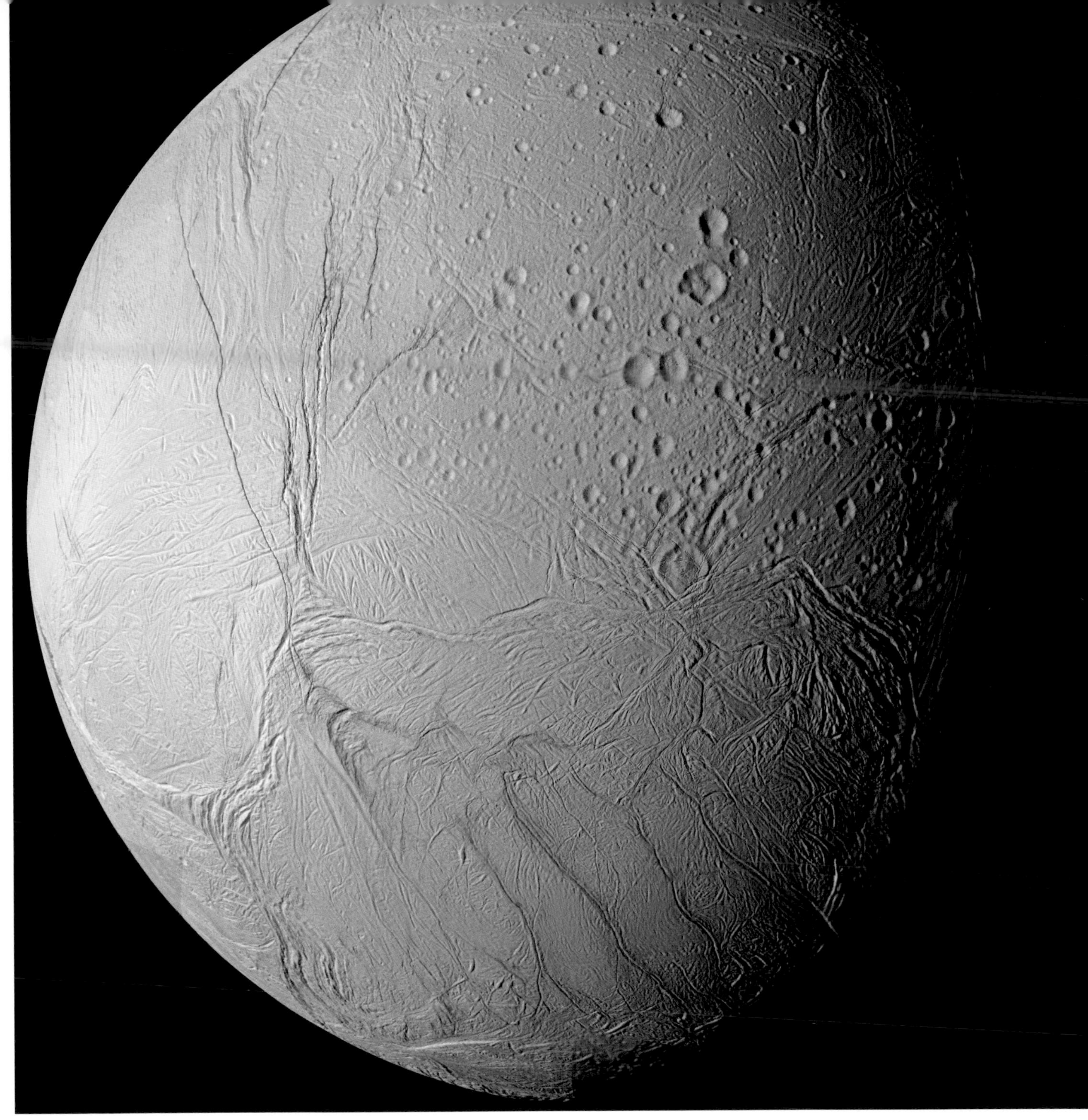

And Saturn's largest moon, Titan, which is 3,200 miles (5,150 km) in diameter, intrigues astronomers, who see it as a primitive Earth, now frozen. All places are prime candidates for the existence of life—past or present—in our Solar System.

LOOKING FOR EARTH II

In 1995, the first planet orbiting a star other than our Sun was discovered. Twelve years later, the number of these known extrasolar worlds, or exoplanets, had risen to more than 200 gas-giant bodies larger than Jupiter. That total is now set to grow enormously.

Most of the discoveries to date have used an indirect method of observation—the Doppler velocity-shift technique. Astronomers watch over many years to see if a star wobbles back and forth as it moves through space, responding to the gravitational tug of unseen planets orbiting it. This painstaking observation takes many years to complete, and is very difficult to interpret.

Recently, however, astronomers have begun seeking exoplanets that transit (that is, pass in front of) their star. When this happens, the star's light is slightly dimmed for a short time until the transit is complete. Once a transiting planet is found, it is possible to isolate its light, study its spectrum, and determine the chemistry of its atmosphere. During the next decade, ground-based planet-transit surveys are expected to discover many dozens of exoplanets. However, the technique is capable of discovering only the larger planets, most of which will be bigger than Jupiter.

Above This high-resolution view of Saturn's ice-moon Enceladus was taken from the *Cassini* spacecraft in 2005. The picture reveals craters and cracks in the moon's surface; scientists believe the pull of Saturn's gravity cause these cracks to open and close each day.

Top This artist's representation depicts NASA's Kepler mission, which will search the Milky Way Galaxy for Earth-size and smaller planets.

Above The infrared Spitzer Space Telescope (illustrated here) was the first mission in NASA's Origins program, which seeks to explain where we come from, and determine if we are alone.

Innovative space-based research techniques will extend the research. In 2007, the French launched COROT, a satellite designed to study stars and search for extrasolar planets. From 2008, NASA's Kepler mission will begin the hunt for planets as small as Earth using a photometer, a specialized telescope that is 3 feet (1 m) in diameter. Astronomers expect both spacecraft to find hundreds, perhaps even thousands, of new worlds. Later still, Darwin (European Space Agency) and the Terrestrial Planet Finder (NASA) will be able to directly scrutinize Earth-size exoplanets.

Before long, astronomers will have pictures of Earth-like worlds orbiting distant stars. Direct imaging will not show signs of an alien civilization, but other means can be employed to detect the presence of life-indicating molecules (such as oxygen) in a planet's atmosphere. When this happens, discussion of life beyond Earth will cease to be theoretical.

EYES ON THE STARS

In 1609, when Galileo used new technology (the telescope) to look at the heavens, he observed sights undreamt of by his peers. Since then, every advance in astronomical technology has brought new revelations about our universe. For astronomers, bigger is definitely better when it comes to telescopes.

In 2002, on a mountaintop in northern Chile, the European Southern Observatory's Very Large Telescope project—Paranal Observatory—became fully operational. For the first time, its four 26-foot (8 m) telescopes were linked to provide the light-collecting power equivalent to a 52-foot (16 m) single telescope. In Arizona, at the Large Binocular Telescope (LBT), computers control two 27½-foot (8.4 m) mirrors to create the equivalent of a single instrument with a resolving power equal to that of a 75½-foot (23 m) telescope.

However, these giants will be dwarfed by some ambitious upcoming projects. The Americans and Canadians are working on a Thirty-Meter Telescope (TMT). The Europeans are considering two giants—the 138-foot (42 m) European Extremely Large Telescope (E–ELT) and the 164-foot (50 m) Euro50 telescope. And the European Southern Observatory is pondering a 328-foot (100 m) behemoth, appropriately named the OverWhelmingly Large telescope (OWL), with 40 times the resolution of Hubble Space Telescope.

Not all the planned telescopes are optical. In Chile's Atacama Desert, the first two 39-foot (12 m) antennas for ALMA (Atacama Large Millimeter Array) are now in place and functioning. The final array of 50 moveable radio dishes should be completed by 2012. Construction on the Square Kilometer Array (SKA) of radio dishes should start about 2011, with completion by 2020. Both radioscopes will peer deep into the early universe, looking for clues about its birth.

Despite the size of these proposed Earth-bound giants, telescopes in space still have their place. Even though the Spitzer Space Telescope, which was launched in 2004, has a mirror only 2½ feet (0.8 m) across, its views of the infrared universe far surpass anything obtainable from the ground.

The Galaxy Evolution Explorer (GALEX), also launched in 2003, has piled up an impressive list of discoveries as it explores the ultraviolet cosmos. The Swift GRB Explorer has been helping astronomers to probe mysterious gamma-ray bursts from distant galaxies.

SOLVING THE MYSTERIES OF THE UNIVERSE

In the coming decades astronomers hope to use these new eyes to solve many of the vexing questions that arose from twentieth-century discoveries, including the big-picture questions that strike a chord with everyone: How did the universe begin? How will the universe end? Is there more than one universe? Why is 96 percent of the universe invisible to us? What is dark matter made of? And what exactly is dark energy? Why does the universe exist?

In the first years of the twenty-first century, the Wilkinson Microwave Anisotropy Probe (WMAP) began mapping cosmic microwave background radiation—the energy emitted by the universe from a time when it was less than 400,000 years old. WMAP's preliminary findings suggest we live in a flat universe, 13.7 billion years old, that contains lots of cold dark matter. It also discovered that the first stars were born when the universe was a mere 100 to 400 million years old. This probe may also help astronomers to distinguish between different versions of inflation, which is the theory that tries to explain exactly what happened during the very first moments in the life of the universe.

STILL MORE TO DISCOVER

Imagine a pair of 2½-mile (4 km) long, 13-foot (4 m) wide tubes set at 90 degree angles to each other, with beams of laser light constantly racing down each arm. This is LIGO, the Laser Interferometer Gravitational-wave Observatory. Its target is to observe gravitational waves that are invisible and still only theoretical. Einstein's 1916 theory of General Relativity predicted them—and almost a hundred years later, scientists are finally preparing to observe them.

Gravitational waves are ripples in the fabric of space. Any celestial object can produce them, but only very large-mass objects like exploding stars, colliding neutron stars, and massive black holes produce ripples strong enough for LIGO to detect. As they travel, these ripples literally warp space by shrinking it ever so slightly in one direction and stretching it in another. The hope is that LIGO's right-angle array will detect this minute warping and give scientists more insights into the massive objects that create it.

In the next decade, a three-spacecraft mission called LISA, which stands for Laser Interferometer Space Antenna, will be launched to detect gravitational waves from space. There will be a 3-million-mile separation between each component, creating a "ripple detector" far more sensitive than that of LIGO. LISA's goal is to detect gravitational waves from the birth of the universe itself.

All of this new technology is designed to answer the questions astronomers are asking today. But if history is any guide, the answers will inevitably lead to more questions.

Above Taken by the Spitzer Space Telescope in May 2007, this image shows baby stars "hatching" in Orion's head. It is believed this "birth" resulted from the explosion of a massive star 3 million years ago.

Below This time-exposure depiction of LIGO, the Laser Interferometer Gravitational-wave Observatory, shows laser beams running down each tube. LIGO researchers include 500 scientists at universities in the USA and eight other countries.

Space Exploration

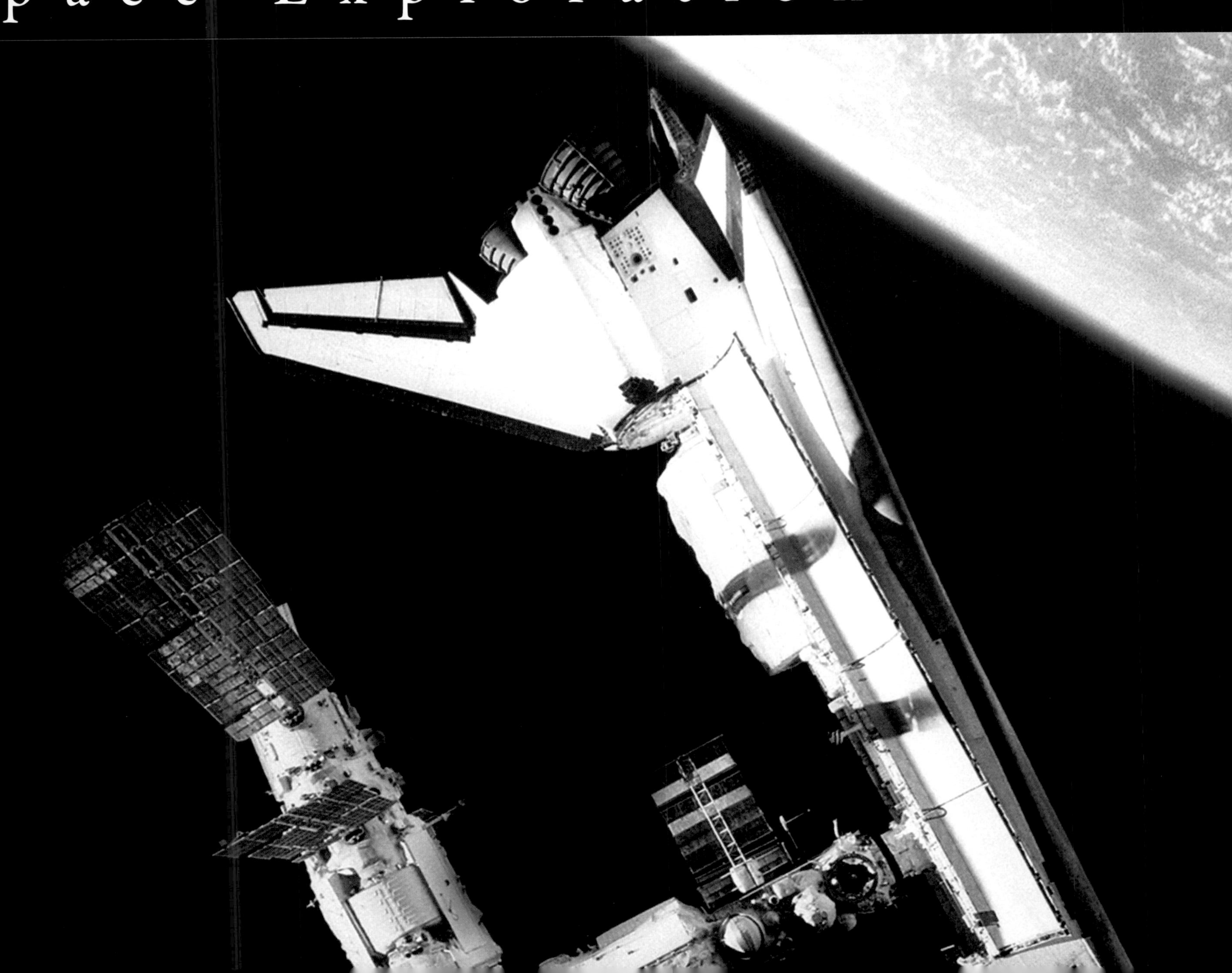

Exploring Space

The dream of space exploration is as old as the human imagination. Ancient myths found in cultures around the world tell tales of heroes attempting to conquer the skies. But to make the dream of space flight a reality, a technology had to be developed that could actually propel space explorers beyond Earth and into space—that technology was rocketry.

Right English author Robert Anderson discusses the production of rockets and propellants in his 1696 treatise *The Making of Rockets*. It was one of several texts on rocketry published in Europe in the seventeenth century.

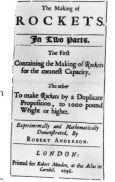

The Making of
ROCKETS.
In Two Parts.
The First
Containing the Making of Rockets for the meanest Capacity.
The other
To make Rockets by a Duplicate Proposition, to 1000 pound Weight or higher.
Experimentally and Mathematically Demonstrated, By
ROBERT ANDERSON.
LONDON:
Printed for Robert Morden, at the Atlas in Cornhil. 1696.

Why is rocketry the ideal technology to achieve space flight? The common forms of propulsion for flight—propellers and jet engines—must be surrounded by an atmosphere in order to work. Rockets, however, operate on the reaction principle described in Newton's Third Law of Motion ("for every action there is an equal and opposite reaction"), and can function equally well in an atmosphere or in the vacuum of space. All rockets, whether fireworks or space launchers, burn fuel, which produces exhaust gases. Forced out through a narrow opening in the rocket's base, these gases produce an outward thrust, and it is the opposite reaction to this thrust that pushes the rocket forward.

HOW IT ALL BEGAN

Although Hero of Alexandria described the principles of reactive motion around 2,000 years ago, the first practical use of reaction propulsion was the development of firework rockets in China around the sixth century CE. At first used for festival celebrations, rockets were quickly recognized as useful weapons of war: A treatise written in 1045, *Wu-ching tsung-yao (Complete Compendium of Military Classics)* describes "fire arrows" that may have used firework rockets. The first confirmed use of war rockets is in 1232 CE, when China used them against the Mongols, who were besieging the city of Kai-feng-fu.

Above A Chinese soldier launching a rocket-propelled spear, one of the many early "rocket weapons" developed in China from the eleventh century. The long stick provided stability in flight.

Knowledge of rocketry spread quickly throughout Asia, with rocket weapons found in Japan, Java, Korea, and India by the end of the thirteenth century. The Mongols are believed to have carried the knowledge of rocketry westward into Europe: They used rockets in the Battle of Legnica in Silesia (Poland) in 1241, and against Baghdad in 1258. Arabic and European works from the mid-thirteenth century describe rocket-like weapons and "black powder," and the earliest use of the Italian word *rocchetta* (rocket) dates from 1379.

War rockets were popular with European armies throughout the Middle Ages and the Renaissance, but by the eighteenth century they had mostly been replaced by firearms and cannon.

However, war rockets remained popular in Asia, and advanced Indian rockets with a range of almost 1¼ miles (2 km) were used in battle against the British by Hyder Ali (1781) and Tipu Sultan (1792–1799).

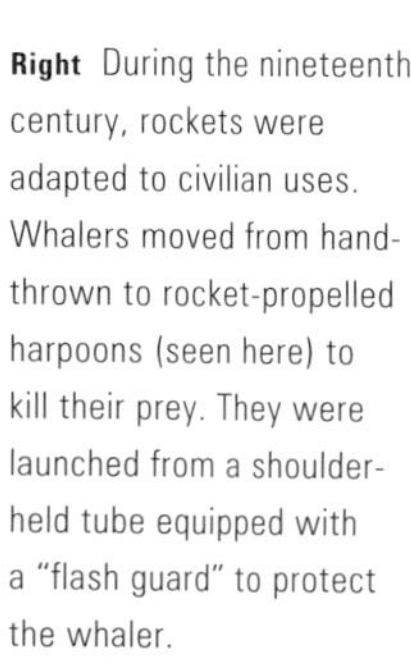

Right During the nineteenth century, rockets were adapted to civilian uses. Whalers moved from hand-thrown to rocket-propelled harpoons (seen here) to kill their prey. They were launched from a shoulder-held tube equipped with a "flash guard" to protect the whaler.

ROCKETRY TAKES OFF

The success of the Indian rockets inspired Colonel William Congreve to develop a new war rocket for the British army. His rockets used black powder, had an iron casing and a 16-foot (4.9 m) guide stick for stability, and had an average range of about 1¾ miles (2.8 km). Introduced in 1804, the rockets were more accurate and versatile than their predecessors, able to be used in land and sea battles. They were easier to transport and deploy than earlier models, were available in a range of sizes, and could be used as incendiaries, artillery, and signals. Deployed around the world by British forces, Congreve-style rockets were quickly adopted by European and US armies, who worked to improve them. This resulted in advances such as a stickless rocket, invented by American William Hale in the 1840s.

The success of military rockets sparked an interest in rocketry for civilian use, which saw the introduction of rocket-powered whaling harpoons, rocket-powered maritime rescue apparatus, and rain-making rockets. By the late 1800s, improvements in artillery and the development of radio had made rockets largely obsolete as weapons and signals, but some visionaries were beginning to realize their potential for space travel; indeed, as early as 1649 the French writer Cyrano de Bergerac, in his *Voyage dans la lune (Journey to the Moon)*, had imagined the use of rocket propulsion for space travel.

FLIGHTS OF FANCY

Fictional space travel tales date from as early as the second century CE, when the Greek satirist Lucian of Samosata (120–180 CE) wrote two space travel stories, *Vera Historia (True History)* and *Icaro–Menippus*. The astronomical discoveries revealed by telescopes in the seventeenth century inspired a new crop of space travel tales, which were used to present new scientific knowledge and explore philosophical ideas while avoiding censure from the Church. Without the technology to achieve space flight, and without any true notion of the nature of space, many fanciful means of space travel were envisaged for these fictional explorers, such as "astral travel," migrating birds, giant springs, and evaporating dew.

Left Icarus is killed when he flies too close to the sun and his wax wings melt. This classic Greek tale illustrates our long-held fascination for flight.

Below One of the first campaigns using Congreve rockets was the British attack on Copenhagen in 1807. The bright trails made by the rockets are depicted in this painting of the battle by William Sadler II.

The Pioneers of Spaceflight

At the beginning of the twentieth century, the technological and scientific knowledge required to make the dream of spaceflight a reality was developing. By the end of the century, the simple black-powder rocket had been transformed into massive space launch vehicles, capable of taking satellites into orbit, people to the Moon, or probes to the planets. Scientists and engineers from around the world have played a part in this transformation.

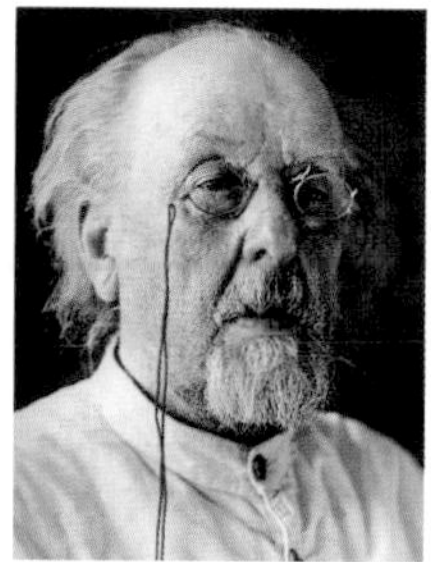

Above The great "space dreamer" Konstantin E. Tsiolkovski once said: "Earth is the cradle of humanity, but one cannot remain in the cradle forever."

The scientific romances of such writers as Jules Verne and H. G. Wells inspired people to dream about traveling into space: Two of those dreamers were Konstantin E. Tsiolkovski (1857–1934) and Robert H. Goddard (1882–1945), who sought to turn the vision into reality. Ironically, these pioneers came from the two countries that would later be bitter rivals in the Space Race.

THE FOUNDERS OF SPACEFLIGHT

Konstantin Tsiolkovski, a Russian schoolteacher, was the first person to fully realize that rockets were the ideal vehicles to make space travel a reality. Tsiolkovski's theoretical work provided the scientific foundation for astronautics by establishing all the basic mathematical laws of spaceflight and demonstrating that only a liquid-fuel rocket would have the thrust to put a rocket into Earth's orbit or to journey to another planet. In 1903, he published the first major work on astronautics, *The Exploration of Space with Reactive Devices*.

A man of vision and imagination, Tsiolkovski was an important popularizer of space travel in the USSR. He had wide-ranging astronautical interests and envisaged the first designs for multistage rockets and space stations, for life-support systems and spacesuits, and for the use of satellites and solar energy.

Robert Goddard was an American physicist and rocket experimenter. After becoming interested in rockets at an early age, he chose physics as a career, and in 1919 published *A Method of Reaching Extreme Altitudes*, which described the first practical sounding rocket for the investigation of the upper atmosphere. In 1926, Goddard designed and flew the world's first successful liquid-fuel rocket, powered by liquid oxygen and petrol. Interested not only in all aspects of rocketry but also other fields such jet propulsion, electronics, and solar heating, Goddard was a reclusive but prolific inventor and experimenter, who continued to build and launch rockets until his death in 1945.

The results of Goddard's research and experimentation have been recognized as fundamental to the development of modern rocketry. NASA's Space Flight Center in Maryland, USA, is named after him.

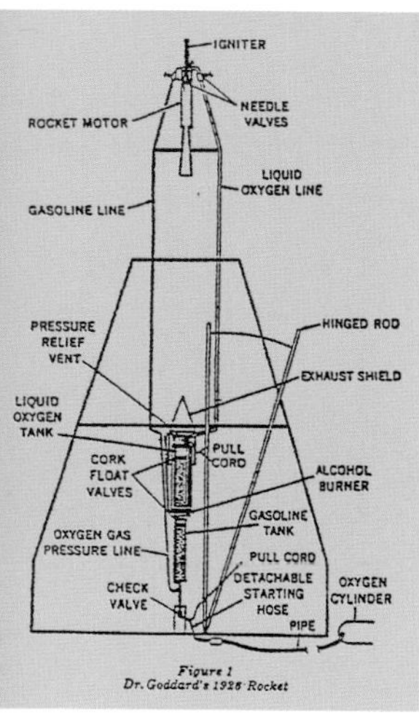

Above and right American Robert Goddard poses next to the world's first liquid-fuel rocket, which was launched on March 16, 1926. Flying for just two and a half seconds, the rocket rose to 41 feet (12.5 m). Goddard placed the motor at the top of the 10-foot (3 m) tall rocket, believing it would improve flight stability (see plan shown above).

Right An illustration from Jules Verne's *A Trip From the Earth to the Moon* (1865), which inspired many space pioneers. Verne read widely on the subjects of astronomy, geology, and engineering.

THE SPACEFLIGHT MOVEMENT

Inspired by the work of Tsiolkovski and Goddard, an international "Spaceflight Movement" began to develop in the 1920s and 1930s. Enthusiasts in many countries formed space travel and rocket societies with the clear purpose of undertaking theoretical and practical research in rocketry and spaceflight. Some of these groups made very significant contributions to the development of rocketry.

A leading figure in this movement was Hermann Oberth (1894–1989), a Romanian-born German rocket experimenter and spaceflight publicist. Independently of Tsiolkovski and Goddard, he developed spaceflight concepts and publicized them in his visionary 1923 book, *Die Rakete zu den Planetenraümen* (*The Rocket into Planetary Space*). By presenting spaceflight as an engineering problem to be solved, Oberth's writings and his promotional work encouraged the development of rocket experimentation groups in Europe, particularly in Germany. In 1927, he founded the *Verein für Raumschiffahrt* (Society for Space Ship Travel), or VfR, a very important pre-war spaceflight society. This group included many young engineers who would become leaders in the development of rocket technology during and after World War II. The most prominent of these was Wernher von Braun, who would eventually lead the early US space program.

Above Members of the GIRD rocketry research group fuelling the Soviet GIRD-09 rocket with liquid oxygen. A hybrid rocket, using solidified gasoline, GIRD-09 reached an altitude of 1,312 feet (400 m) during a 13-second flight in August 1933.

As in Germany, interest in space travel was high in the USSR, with the first laboratory for rocket research established in 1921. The world's first exposition of "interplanetary machines and mechanisms" was held in Moscow in 1927, and the first astronautical encyclopedia was published between 1929 and 1935. In 1931, the Group for the Study of Reactive Devices (GIRD in Russian) was formed. Its space-enthusiast engineers worked with government support in conjunction with government laboratories. The GIRD group designed and launched the USSR's first liquid-fuel rocket, GIRD-X, in 1933, and laid the foundations for a strong home-grown development of rocket technology after World War II. Among the group's members was Sergei Korolev, the "Chief Designer" of the Soviet space program.

Rocket societies and experimenters also existed in many other countries. Some, like the British Interplanetary Society, could carry out only theoretical research because of local ordnance regulations; others achieved varying degrees of success in building and flying rockets.

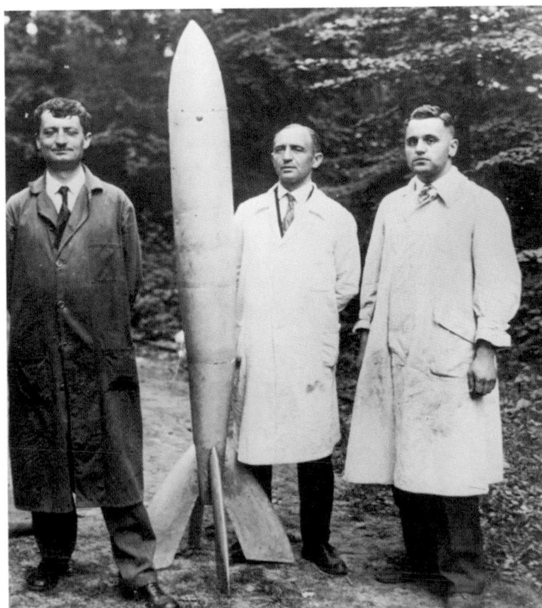

Left Prominent spaceflight researcher and promoter Hermann Oberth with two colleagues from the Verein für Raumschiffahrt. They are posing with a wooden model of a proposed streamlined rocket design. Many members of the VfR would later participate in the V-2 rocket program.

The V-2 and the Cold War

The ancestry of all modern missiles and space-launch vehicles can be traced back to the V-2 rocket, one of the "vengeance weapons" developed by Germany during World War II. The V-2, also known as the A-4, was the world's first long-range missile, its technology representing a major leap forward in rocketry.

Right On May 15, 1959, a crowd gathered at Coney Island, New York City, to see the first American ICBM (Intercontinental Ballistic Missile) be lifted into its launching position.

Opposite The German V-2 long-range missile is seen here in 1944 before its launch at Cuxhaven. From this site the rocket could be fired over the North Sea.

Until the 1930s, there was little military interest in the rocketry experiments inspired by the Spaceflight Movement. However, as Nazi Germany began to re-arm under Hitler, the German Army became interested in rockets for military use as a way to circumvent the Treaty of Versailles, which was imposed at the end of World War I and prohibited Germany from possessing heavy artillery. In 1932, Wernher von Braun was employed by the army to work on rocket projects. He was soon joined by other VfR members, who would form the nucleus of the army's rocket research team.

THE VENGEANCE WEAPON AND ITS IMPACT

In 1937, the German Army established a rocket research center on the island of Peenemünde in the Baltic Sea. Here, a cadre of scientists and engineers, with von Braun as technical director, worked on developing larger, more advanced rockets with greater range and payload capacity for use as weapons. Their work resulted in the Aggregat-4 (A-4), which made its first test flight on October 3, 1942.

Above German rocket engineer and inventor, Wernher von Braun (left) and his brother Magnus (right) surrendered to the US Army in May 1945.

Traveling at supersonic speed, with an impressive range of 200 miles (320 km) and an altitude potential of 60 miles (96 km), the A-4 represented a tremendous step forward in rocket design and technology: Its basic design concepts for motor, fuel, guidance, and steering systems remain at the heart of even today's most advanced rockets.

First fired operationally on September 8, 1944, Nazi propaganda dubbed the A-4 "Vengeance Weapon 2" (V-2). More than 3,000 of the rockets were fired operationally during the war at targets including London, Antwerp, and Paris. Its warhead packed with 2,250 lbs (1,000 kg) of amatol high explosive, the V-2 claimed more than 5,000 lives—some 2,700 in Britain alone. Nevertheless, the V-2 was not particularly successful as a military weapon, since it could not be guided with precise accuracy, its complex technology was unreliable, and it was extremely expensive compared with conventional weapons. Its main value was psychological: It generated fear and trepidation because it could not be detected or intercepted before it struck.

COLD WAR MISSILE MEN

There were two leading figures in the Cold War missile race, one Soviet and one in the USA. Sergei Korolev (1906–1966), seen below, was a member of GIRD who escaped imprisonment by Stalin to become the anonymous Chief Designer of the Soviet space program. With his identity concealed from the public, Korolev oversaw the development of the USSR's long-range missile program after World War II and was the driving force behind its space program until his death.

Wernher von Braun (1912–1977), who had been instrumental in the development of the V-2, was taken to the USA after World War II to work on missile programs; he later developed the Saturn V rocket. A lifelong supporter and popularizer of space travel, he spearheaded the early US space program, but his reputation is tarnished by his association with the Nazi regime and the horrors of the forced labor used to build the V-2s.

THE COLD WAR ACCELERATES ROCKETRY

Despite its shortcomings as a weapon, the V-2 demonstrated that it was possible to develop rockets that were capable of long-distance travel while carrying a strategically useful payload. After World War II, many countries scrambled to acquire German technical expertise: Wernher von Braun and more than a hundred of his team were spirited into the USA and employed in US Army missile programs. The USSR acquired its own cadre of former Peenemünde personnel, who worked with Soviet rocket engineers such as Sergei Korolev before being returned to Germany in the 1950s.

With the onset of the Cold War between the USA and the USSR, there was a rush to develop long-range missiles capable of delivering nuclear warheads over intercontinental distances. Intermediate Range and Intercontinental Ballistic Missiles (IRBMs and ICBMs) became part of strategic planning, and massive resources were poured into improving missile technology. Achievement in missile technology came to be perceived as a status symbol, proclaiming the power and influence of nations that possessed it. Though this rapid development was driven by the military advantages of the guided missile rather than by the desire for spaceflight, by the mid-1950s the largest missiles had achieved the capability to act as space-launch vehicles, able to lift scientifically useful payloads.

Anticipating the Space Age

Until the end of World War II, little was known about Earth's upper atmosphere, since the maximum altitude that a research balloon could reach was around 25 miles (40 km). Because long-range missiles would traverse this region in flight, their developers became just as interested as scientists and meteorologists in the nature of the upper atmosphere and the physical relationship between Earth and space. Consequently, "sounding rockets"—small sub-orbital rockets that could take scientific measurements within the upper atmosphere—were developed to explore the fringes of space.

Above The first purpose-built sounding rocket, the US *WAC Corporal* was launched in October 1945. Although not in use for long, it led to the development of the very successful American *Aerobee* sounding rocket.

Right Developed by the US Naval Research Laboratory, the first *Viking* sounding rocket was launched on May 3, 1949. *Viking* rockets pioneered the use of gimballed motors for steering. The *Viking* rocket was later developed into the *Vanguard* satellite launcher.

TO SPACE WITH SOUNDING ROCKETS

Robert Goddard had proposed the concept of a sounding rocket as early as 1919, and had launched a prototype in 1929. GIRD experimenters in the USSR also flew some small rockets carrying scientific instruments in the 1930s. However, the first true sounding rocket, specifically designed for exploring the upper atmosphere, was the US *WAC Corporal* (derived from the US Army's Corporal missile). It was first launched in September 1945. At the same time, the US Navy began work on its *Viking* sounding rocket, which would later be adapted to become the *Vanguard* launch vehicle. This was an unusual step because most early space launchers were derived from ICBMs, which had the greater thrust necessary to launch satellite payloads into orbit.

Comparatively cheap and quick to build and launch, and able to fly a wide variety of payloads—from radiation detectors to telescopes and even small animals—sounding rockets became a popular scientific tool in the 1950s, with many countries developing their own. Thousands have been launched over the past 60 years, and they continue to make significant contributions to scientific research.

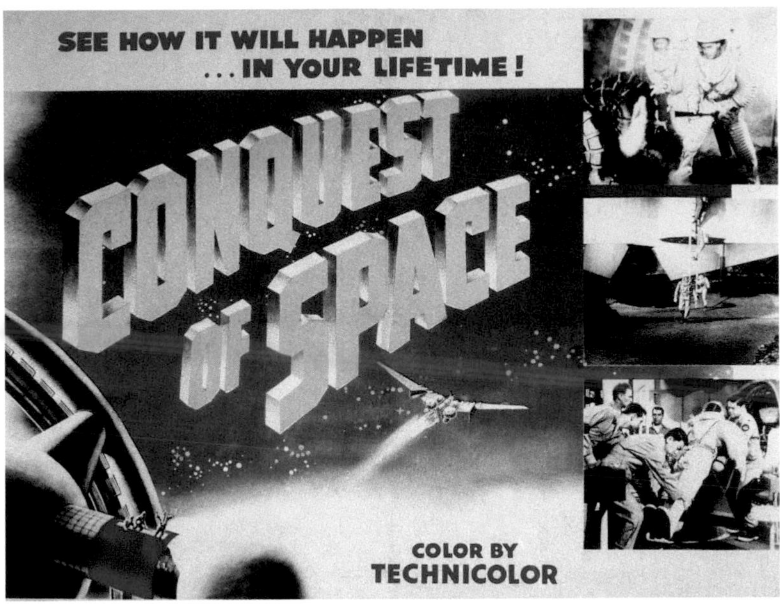

Above The 1955 spaceflight film *Conquest of Space* was derived from a classic book by space promoter Willie Ley and space artist Chesley Bonestell. It tells the story of the first American mission to Mars.

THE INTERNATIONAL GEOPHYSICAL YEAR

Data from sounding rockets contributed invaluable information to our knowledge of the space environment; in fact, upper atmospheric research provided a trigger for the Space Race. In 1952, the world scientific community proposed the International Geophysical Year (IGY) for the 18 months from July 1957 to December 1958, which corresponded with a period of maximum solar activity. The IGY would be an international year of scientific research that would investigate Earth's relationship with the space environment.

Sounding rocket programs were proposed for the IGY, and by 1954 there was also recognition that satellites could be developed by 1957 to make a further contribution to the research program. The USA declared in 1955 that it would launch scientific satellites during the IGY as part of its research program, but a similar Soviet announcement was largely overlooked because western nations did not believe that the USSR had the technology to achieve this goal. With the USA experiencing difficulties in launching its proposed *Vanguard* satellite, it was therefore an immense shock and blow to the prestige of the west when the Soviet Union lofted the world's first artificial satellite, *Sputnik 1*, on October 4, 1957—just 15 years after the launch of the first A-4/V-2. *Sputnik 1* was a great propaganda victory for the USSR, even though it was technologically more primitive than Vanguard. The Space Age and Space Race had begun.

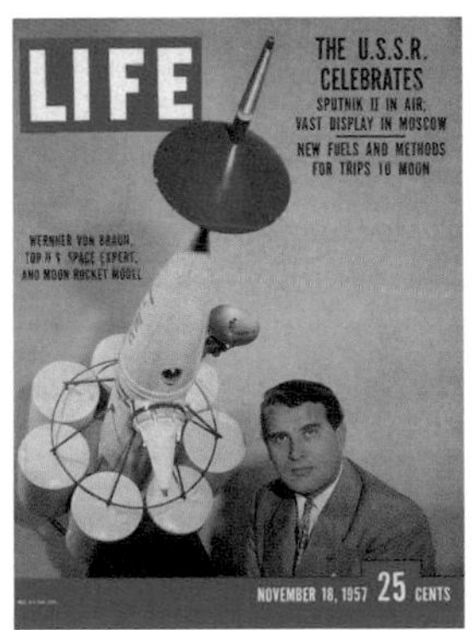

Above This 1957 issue of LIFE magazine, published shortly after the launch of the USSR's *Sputnik 1*, features Wernher von Braun with a model of one of his 1950s rocket designs.

A DECADE OF SPACE PROMOTION

As rocket and missile technology developed during the Cold War, space promoters began to encourage the belief that space travel would soon be a reality. In the USA, Wernher von Braun and Willy Ley produced a stream of books and articles on future space projects: Among the most influential were the lavishly illustrated articles that appeared in *Colliers* magazine between 1952 and 1954, based on von Braun's ideas for a future space program. Drawing inspiration from these articles, Walt Disney produced a set of space documentaries that were seen around the world. In the UK, the British Interplanetary Society actively promoted future spaceflight, and there were similar space publicists in other countries.

Attracted by the excitement and adventure of the conquest of space, the entertainment industry produced speculative spaceflight films and radio plays derived from the scenarios of the space promoters and the realms of science fiction. The films relied heavily on special effects and modelwork to create their space technology and environments, with varying degrees of accuracy. Some classics of this genre include *Destination Moon* (1950), *Conquest of Space* (1955), and *The Road to the Stars* (1957), a visionary Soviet space promotion film. Their use of real and fictional space technology and the theme of space exploration and adventure helped to create a strong public interest in spaceflight and the general belief that space travel was just around the corner.

Left Scientists at the 1957 IGY conference inspect a model of *Sputnik 1*, the world's first artificial satellite. The International Geophysical Year provided the impetus for the development of the first satellites.

Sputnik, Explorer, and Vanguard

The launch of *Sputnik 1*, the world's first artificial satellite, ushered in the Space Age, in which the age-old dream of traveling into space finally became reality. For the first time, humanity had the know-how and the technology to move beyond its home world to explore the cosmos, by proxy with satellites and space probes, and even in person.

However, in the Cold War environment of the time, political propaganda equated technological advancement with ideological superiority, and space exploration was immediately seen by the USA and the USSR as another arena in which to compete for political supremacy through technological superiority. Both nations sought propaganda victories through the achievement of status-conferring space "firsts": The Space Race had begun. This contest drove the development of spaceflight at an astounding rate; only 12 years would elapse from the launch of the first satellite to the first human landing on the Moon.

Above This full-size model of the 23-inch (58 cm) diameter *Sputnik I* was proudly displayed in the Soviet Union's pavilion at the 1958 World's Fair (Expo 58) in Brussels. It proclaimed Soviet space pre-eminence to the world.

A BEEP HEARD AROUND THE WORLD

The Space Age began on October 4, 1957, when the USSR stunned the world by launching the first Earth-orbiting artificial satellite, *Sputnik 1*. Weighing a little more than 186 lbs (83 kg), *Sputnik* ("traveling companion") was a polished metal sphere with four long antennae. It contained a radio transmitter and atmospheric measuring instruments, and orbited Earth every 90 minutes. Seen as a bright, fast-moving "star," *Sputnik 1* was visible in many countries, and huge crowds waited to see it cross overhead. *Sputnik* transmitted its famous "beep beep" signal for 21 days, and remained in orbit for 96 days before disintegrating on re-entering Earth's atmosphere.

Sputnik 1 was massive compared with the tiny *Vanguard* satellite proposed by the USA, and a shockwave shook the west at its launch: It was assumed that if the USSR could put such a large satellite into orbit, it must have the technology to attack America using ballistic missiles. The USA's immediate response was to scramble to get a satellite into orbit as quickly as possible.

Right Leonid Sedov (left), seen here at a conference with rocket pioneer Hermann Oberth (right), was a leading Soviet physicist. At the 1955 International Astronautical Congress in Copenhagen, he announced the USSR's plans to launch a satellite during the IGY.

EXPLORER 1

Launcher development difficulties with the Vanguard satellite program had allowed the USSR to take the honors for launching the first satellite. This was a particularly bitter pill for the USA, as an alternative proposal by Wernher von Braun's Army rocket team, which would have been ready to fly before *Sputnik*, had previously been turned down. To recover national prestige, President Eisenhower authorized von Braun's team to undertake a crash program to develop a new satellite, *Explorer*, as an alternative to *Vanguard*. Von Braun and his associates, which included scientists from the Jet Propulsion Laboratory in California, constructed the *Explorer* satellite and developed its *Juno* launch vehicle, which was adapted from the US Army's Redstone missile (itself a direct descendant of the World War II V-2 guided missile designed by von Braun) in just 85 days. Even before the American scientists had completed their work, the USSR scored another coup when *Sputnik 2* took the first living animal, Laika the dog, into space in November 1957.

Explorer 1 was successfully launched from Cape Canaveral, Florida, on January 31, 1958. It carried 18 lbs (8 kg) of instruments designed to gather data on cosmic rays, meteorites, and orbital temperatures. Among other results, these instruments revealed the existence of the Van Allen radiation belts around Earth.

VANGUARD 1

Beset by launcher development difficulties, the Vanguard program failed to achieve its goal of launching the world's first satellite. However, when it finally reached orbit on March 17, 1958, the tiny *Vanguard* satellite made significant scientific discoveries. Weighing barely 3⅓ lbs (1.5 kg)—50 times less than *Sputnik 1*—*Vanguard 1* was packed with early miniaturized electronics and used the first spacecraft solar cells, which enabled it to continue transmitting until 1964. *Vanguard* tracking data enabled the true shape of the Earth to be determined: It is slightly pear-shaped, with the Southern hemisphere larger than the Northern hemisphere.

Below William H. Pickering, director of the Jet Propulsion Laboratory (left), James A. Van Allen, designer of the instruments on *Explorer 1* (center), and rocket team leader Wernher von Braun (right) proudly display their creation.

Above *Explorer 1*'s Juno 1 launch vehicle was a Jupiter-C missile, with a technological lineage going back to the V-2 via the US Redstone missile from which the Jupiter missile series was derived.

Right The *Vanguard 2* satellite was launched in February, 1959. The tiny satellite was designed to measure cloud-cover distribution, making it a prototype weather satellite.

THE ORIGINS OF NASA

Both Vanguard and Explorer were "civilian" US armed services projects, conducted openly in contrast to the secrecy that had surrounded the development of *Sputnik 1*. Although the USA had plans for secret military space programs, President Eisenhower wanted to make a propaganda point by operating the US national space program without apparent military interest or control.

On October 1, 1958, the National Aeronautics and Space Administration (NASA) commenced operations as a civilian government agency to oversee the US space program. It was created from the restructuring of the former National Advisory Committee for Aeronautics, taking over its personnel and facilities together with research centers and laboratories, such as the Jet Propulsion Laboratory, that had been associated with the armed services.

Animals in Space

As the post-war development of powerful new rockets edged us ever closer to a long-anticipated era of human space travel, animals would play a critical role. Before any person could venture into space, scientists had to determine whether the human body could withstand the crushing forces of acceleration, exposure to deadly cosmic radiation, and physiological effects associated with weightless spaceflight—all of which could cause life-endangering structural and functional changes.

Right This Albanian postage stamp celebrates Laika the dog's pioneering voyage in the Soviet spacecraft *Sputnik 2*. Laika was one of many Russian stray dogs to take part in the Space Race.

Through studying a limited number of V-2 rockets and parts seized toward the end of World War II, and by recruiting captured German designers and technicians involved in war-related booster research, the United States and the Soviet Union each began developing its own ballistic technology and arsenal. Military and civilian scientists would soon request space in the nose cones of these rockets in order to conduct vital upper-atmosphere biological experiments using simpler life forms such as insects, plants, and seeds.

RIDING THE ROCKETS

In 1948, a series of V-2 flights was carried out at the White Sands Proving Ground in New Mexico, USA, to explore the possibilities of ejecting capsules from the missiles and recovering them by means of a parachute. The US Air Force decided to take advantage of these flights by sending monkeys aloft as "simulated pilots," so a rudimentary pressurized capsule was constructed and several monkeys were hastily trained for the flights. The first of these rocket tests took place on June 11, 1948, lofting an anesthetized rhesus monkey named Albert into the skies, but the animal did not survive. Other flights were launched from White Sands over the next three years, carrying a numeric series of monkeys named Albert; however, like the first attempt they ended in parachute failure and the deaths of the animals.

In 1951, Aerobee sounding rockets replaced the V-2, but parachute failures and the loss of the animals continued until May 21, 1952. On that date, two Philippine macaque monkeys named Patricia and Michael, together with two white mice, survived a rocket flight to a height of 39 miles (62 km), and all of the animals were successfully recovered after the flight.

SOVIET DOGS IN SPACE

Meanwhile, the Institute of Aviation Medicine, based in Moscow, had also realized a need for biological rocket research. Unlike their American counterparts, the Russians decided against using monkeys, believing them to be skittish by nature and vulnerable to disease. Earlier experiments had demonstrated that dogs were preferred test subjects, calmer and better able to endure monotonous training. This was especially true of stray mongrels already inured to the rigors and harsh climates of the streets. With this in mind, teams were sent out to gather up small suitable candidates around Moscow in the spring of 1951.

On August 15 that year, two dogs named Dezik and Tsygan were launched on a ballistic flight in the ejectable nose cone of a Soviet R-1 rocket. Both dogs were recovered unharmed after reaching an altitude of just over 62 miles (99 km), making

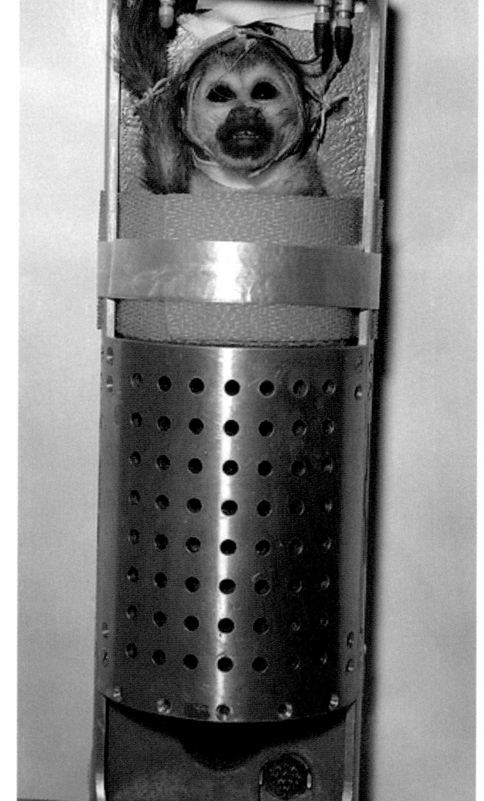

Left NASA's May 28, 1959, Jupiter Intermediate Range Ballistic Missile launched a nose cone carrying a squirrel monkey named Miss Baker (seen here in her biopack couch) and a rhesus monkey named Able into sub-orbit. Both monkeys survived re-entry; however, Able died a few days later.

Right Laika was placed in a purpose-built contraption for her 1957 journey into space. She died hours after take-off, from overheating and stress. At the time, officials falsely claimed that she was poisoned so she would die quickly rather than starve to death.

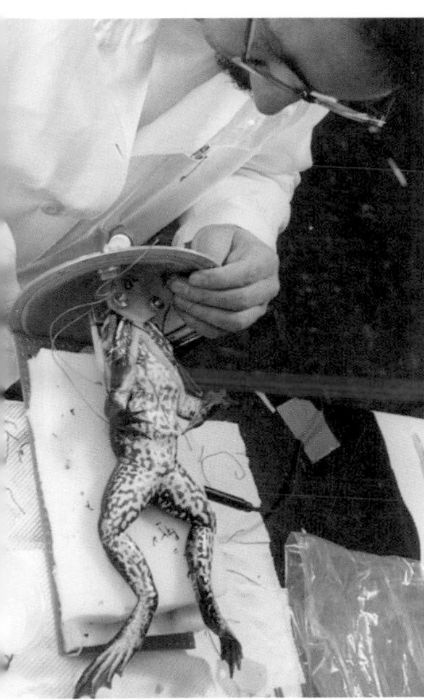

Above A NASA scientist implants micro-electrodes into a bullfrog in preparation for the 1970 Otolith Experiment. Two bullfrogs were placed in a weightless environment to test how gravity affects balance.

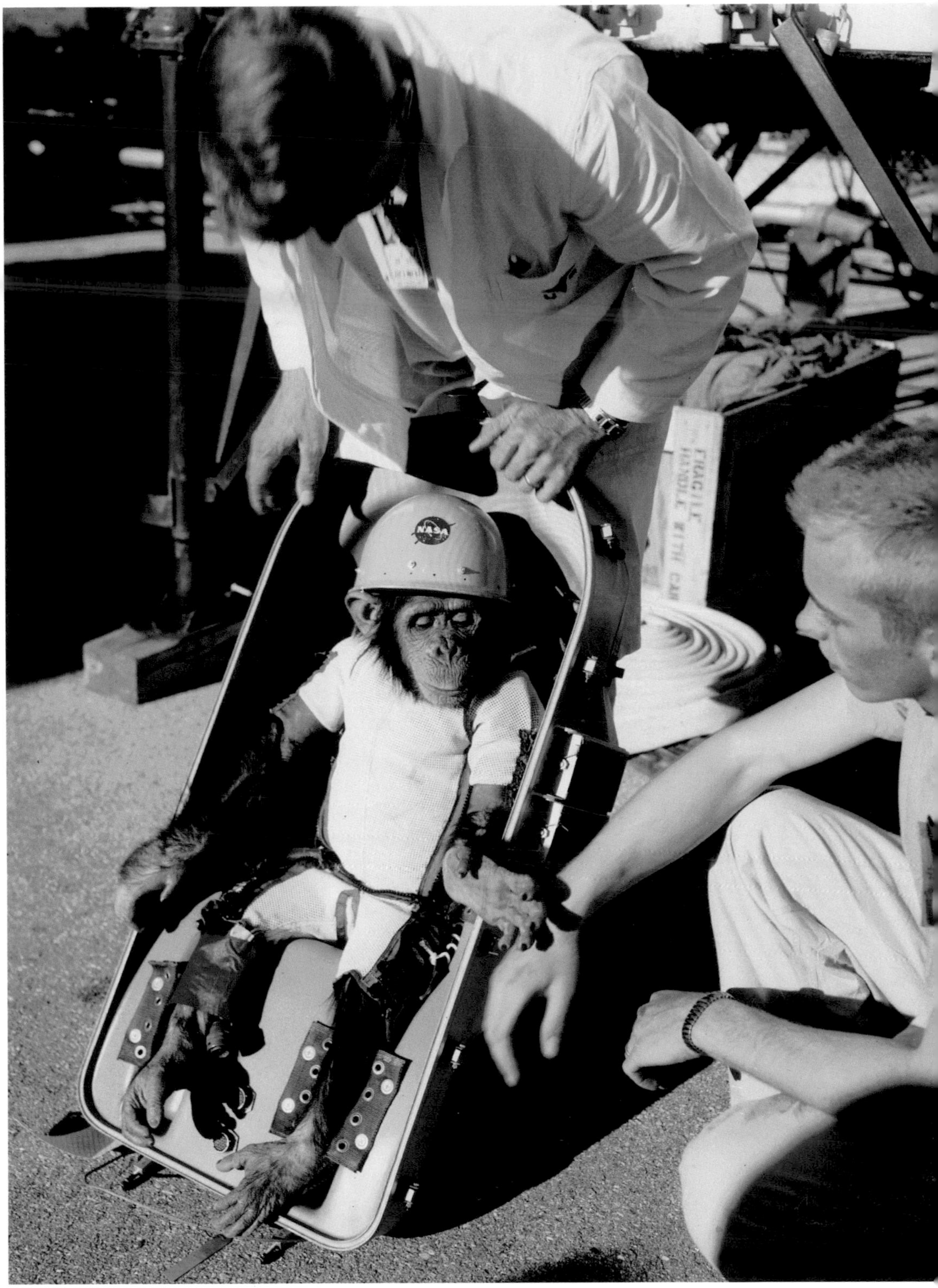

them the first animals known to have survived a flight into space. Other dogs would subsequently fly sub-orbital space shots, several of them more than once.

In October 1957, the USSR stunned the world by launching the world's first satellite, *Sputnik 1*. Four weeks later, on November 3, there was a further shock with the launch of *Sputnik 2*. This time there was a passenger on board—a small dog named Laika. Sadly for Laika she would die within hours of being launched, but the little dog had made history as the first creature ever to orbit Earth.

In the lead-up to the first human spaceflight, several Soviet dogs flew precursory orbital missions, while on January 31, 1961, a chimpanzee named Ham was launched on a sub-orbital mission ahead of America's first manned spaceflight by Alan Shepard. On November 29 of the following year, a chimpanzee named Enos flew into orbit aboard another Mercury spacecraft, this time paving the way for a successful three-orbit flight by astronaut John Glenn.

France was also experimenting around that time with sub-orbital animal spaceflights. White rats and a cat named Felicette would be sent aloft, while other flights would carry monkeys named Martine and Pierrette; both monkeys were safely recovered post-flight.

Later Apollo, Skylab, Mir, and shuttle missions would involve carrying vast numbers of biological subjects into space, including pocket mice, tortoises, frogs, fish, spiders, and a variety of non-human primates.

As plans evolve to send humans on protracted journeys to asteroids and Mars, it is expected that animals may once again act as pathfinders to help assess any potential risks to these pioneering space explorers.

Above Ten months before Alan Shepard's sub-orbital flight, a chimpanzee named Ham flew a test mission aboard a Mercury capsule. Ham recovered well from the flight, during which he was weightless for 6.6 minutes.

The Space Race Begins—Mercury and Vostok

Having put the first satellites into orbit, manned spaceflight was the next goal. With animal astronauts paving the way to ensure that living things could survive and function in the microgravity of space, both the USA and the USSR began to plan manned spaceflight programs that would put the first space travelers into orbit.

Right First person in space, Yuri Gagarin, appeared on the cover of the April 21, 1961, edition of *TIME* magazine. The young Soviet cosmonaut captured the imaginations of people both East and West.

VOSTOK

The Vostok (Russian for "east") program, was the USSR's first-ever manned space program, inaugurated in 1959 in response to NASA's announcement of Project Mercury. In early 1960, twenty military pilots were selected to become the nucleus of the Soviet cosmonaut corps. While the Mercury astronauts were accorded public adulation before their flights, the Soviet cosmonauts trained in strict secrecy and were unknown to the public until they made their spaceflights. Cosmonaut training took place under Soviet space program "Chief Designer," Sergei Korolev, who was also responsible for the development of the Vostok spacecraft (derived from the design of the Zenit spy satellite) and its identically named launch vehicle—a derivative of Korolev's R-7 ICBM. Five test flights, under the name Korabl-Sputnik, were made between May 1960 and March 1961, carrying dogs and test dummies fitted with instruments.

Above In 1963, the Soviets claimed another "first" in the Space Race—the first woman to fly in space. Valentina Tereshkova, a former textile worker, was selected for her parachuting experience. Vostok cosmonauts parachuted from their spacecraft before it landed.

Then, on April 12, 1961, 27-year-old Yuri Gagarin became the first person in space, riding the *Vostok 1* spacecraft into history for a single-orbit flight of 108 minutes. His lift-off shout of "*Poyekhali!*" ("Let's go!") heralded another huge propaganda coup for the USSR. There were six flights in the Vostok program, which tested basic spacecraft systems such as heat shields and life support, and demonstrated that humans could survive launch and re-entry into Earth's atmosphere, and could withstand weightlessness for several days. Successively, the Vostok program achieved a day-length flight (*Vostok 2*); a "joint" flight, with *Vostok 3* and *Vostok 4* launched only a day apart into similar orbits; and a five-day mission (*Vostok 5*) that was longer than the total duration of the US Mercury flights. The USSR sought to prove its claims to equality in its citizens by flying the first woman in space, Valentina Tereshkova, on the final Vostok mission (*Vostok 6*) in June 1963.

PROJECT MERCURY

Initiated in 1958, Project Mercury, which was named for the Roman messenger of the gods, was the first manned US space program. By orbiting a manned spacecraft around Earth, investigating human ability to function in microgravity, and recovering both astronaut and spacecraft safely, it aimed to discover whether humans would be able to survive in space.

Seven Mercury astronauts were announced in April 1959, but only six would eventually fly in the program. ("Deke" Slayton was grounded for many years with a heart condition.) The Mercury spacecraft, designed by the prolific engineer Max Faget and NASA's Space Task Group, was much smaller than Vostok but more versatile, with more advanced instrumentation and electronics than its Soviet counterpart. Several precursor flights, some carrying primates (including the cosmonaut chimp, Ham), tested the spacecraft and launch systems. However, considerable difficulties were experienced with the Atlas booster required for orbital missions. Consequently, the first two US spaceflights were sub-orbital, using a modified Redstone missile as the launch vehicle.

The US approach, both more public and more cautious than that of the USSR, enabled *Vostok 1* to be launched before the first US astronaut, Alan Shepard, who made the first US spaceflight (MR-3)—a 15-minute sub-orbital lob in *Freedom 7*—just 23 days later, on May 5, 1961. A second sub-orbital mission (MR-4) was marred by the loss of the spacecraft, *Liberty Bell 7*, after splashdown. The first US orbital flight finally occurred on February 20, 1962, when John Glenn orbited Earth for just over five hours in *Friendship 7* (MA-6). Glenn's mission was the first to utilize the ICBM-derived Atlas launch vehicle. The next two Mercury flights in 1962, MA-7 and MA-8, extended the duration of each mission and the complexity of the experiments undertaken. The final Mercury flight, MA-9, took place in May 1963, when Gordon Cooper, in *Faith 7*, spent a full day in space.

"WE CHOOSE TO GO TO THE MOON"

To counter the apparent Soviet lead in the Space Race, especially after Gagarin's history-making flight, US President John F. Kennedy wanted a spectacular space feat that would outweigh the USSR's space achievements. Several options were considered, and on May 25, 1961, barely three weeks after Shepard's flight, Kennedy inaugurated the Apollo lunar program in a speech to the US Congress in which he declared that the United States would "land a man on the Moon and return him safely to the Earth" before the end of the decade. The race to the Moon was on.

Right Mercury spacecraft *Freedom 7* lifts off from Cape Canaveral, Florida, at 9:34 A.M. EST on May 5, 1961. On board is Alan Shepard, the first American to fly into space.

Below A full-scale replica of *Vostok 1* supported by a giant gantry in Moscow's All-Russia Exhibition Center. The Soviet Union announced the historic flight to the world while Yuri Gagarin was still in orbit.

Left The Project Mercury astronauts were: Front row, left to right, Walter M. Schirra, Jr., Donald K. "Deke" Slayton, John H. Glenn, Jr., and Scott Carpenter; back row, Alan B. Shepard, Jr., Virgil I. "Gus" Grissom, and L. Gordon Cooper.

The Space Race Heats Up—Gemini and Voskhod

In the wake of Mercury, NASA moved quickly to develop the equipment and techniques to meet President Kennedy's goal of a lunar landing by the end of the 1960s. The USSR, although slow to join the Moon race, pursued the propaganda game by focusing on achieving further space firsts.

Above Edward White II, the first American to make a spacewalk, floats in the vacuum of space, on June 3, 1965, during the *Gemini IV* mission. White can be seen here holding a small nitrogen-powered maneuvering gun intended to assist him in moving around.

GEMINI

To accomplish a Moon landing by the close of the decade, NASA needed to rapidly expand its spaceflight experience, practicing the techniques required for a lunar mission and ensuring that astronauts could physically and psychologically survive space missions the length of the proposed Apollo Moon flight. When the second manned US space program, Gemini, was announced in January 1962, its objectives were to subject both astronauts and equipment to spaceflights of up to two weeks; to rehearse the rendezvous and docking skills for orbiting spacecraft to dock with each other and maneuver while docked (this was essential for the lunar rendezvous planned for Apollo); and to practice extra-vehicular activity (EVA).

Named for the constellation the Twins, the two-person Gemini was a larger and heavier version of Mercury, with engineering improvements that simplified maintenance and made it more maneuverable. It was launched by a Titan II rocket, a modified ICBM. Ten Gemini missions occurred between March 1965 and November 1966, extending NASA's expertise in space operations.

VOSKHOD

Sergei Korolev had scheduled longer Vostok projects, but when the Apollo program was announced, Soviet Premier Khrushchev demanded more space triumphs. This led to the development of the Voskhod ("Sunrise") program, which was largely a propaganda exercise aimed at defeating the USA in the Space Race. The USSR took dangerous risks during this period. To beat Gemini to the first multicrew spaceflight, *Voskhod 1* was launched on October 12, 1964. It was a modified version of Vostok, stripped down for three cosmonauts, in conditions so cramped that they could not wear spacesuits.

Khrushchev was deposed during *Voskhod 1*, but his thirst for space victories survived with *Voskhod 2*, launched on March 18, 1965, to beat *Gemini IV* to the world's first spacewalk. It too was a modified Vostok, with a crew of two and an inflatable airlock. Alexei Leonov became the first person to walk in space, but he nearly died because his spacesuit ballooned and stiffened, making it difficult for him to re-enter the airlock.

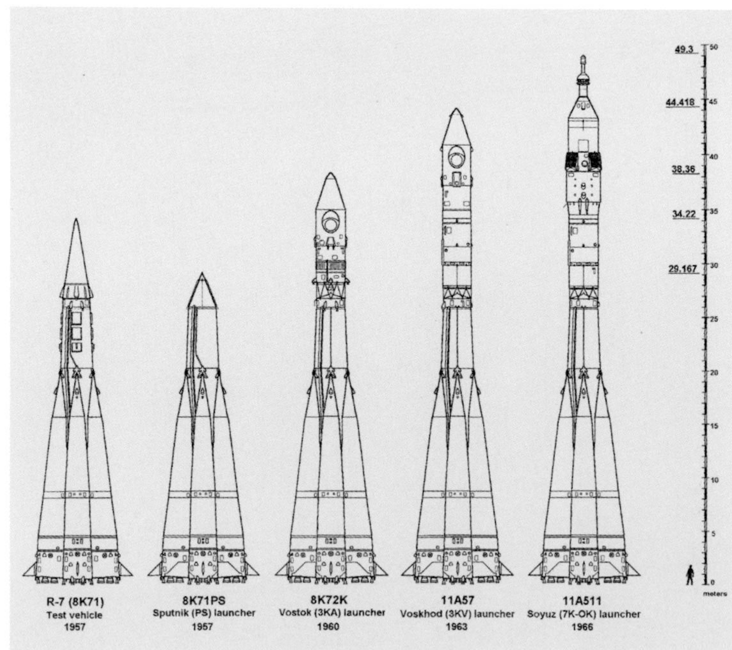

Right Diagram illustrating the family of Soviet launch vehicles derived from Korolev's original R-7 ICBM.

Above On March 18, 1965, Soviet cosmonaut Alexei Leonov became the first person to make an excursion outside a spacecraft. His 10-minute spacewalk almost ended in disaster. This still is from the Soviet documentary film *Man Walks in Space*.

GEMINI PROGRAM MANNED MISSIONS

Mission	Crew	Launch date	Duration	Highlights
Gemini III	Virgil Grissom John Young	March 23, 1965	4 hours 52 minutes 31 seconds	First manned Gemini flight Completed three orbits
Gemini IV	James McDivitt Edward White II	June 3–7, 1965	4 days 1 hour 56 minutes 12 seconds	First EVA by an American (White): 22 minutes
Gemini V	Gordon Cooper Charles Conrad	August 21–29, 1965	7 days 22 hours 55 minutes 14 seconds	First use of fuel cells for electrical power
Gemini VII	Frank Borman James Lovell	December 4–18, 1965	13 days 18 hours 3 minutes 1 second	Spaceflight duration record Operated as rendezvous target for *Gemini VI*
Gemini VI	Walter Schirra Thomas Stafford	December 15–16, 1965	1 day 1 hour 51 minutes 24 seconds	First spacecraft rendezvous (with *Gemini VII*)
Gemini VIII	Neil Armstrong David Scott	March 16, 1966	10 hours 41 minutes 26 seconds	First docking with another spacecraft (unmanned Agena stage) Malfunction led to first emergency landing of a manned US mission
Gemini IX	Thomas Stafford Eugene Cernan	June 3–6, 1966	3 days 2 hours	Three different types of rendezvous 2-hour EVA Docking attempt abandoned due to target vehicle problems
Gemini X	John Young Michael Collins	July 18–21, 1966	2 days 22 hours 46 minutes 39 seconds	First use of Agena target vehicle propulsion systems Rendezvoused with *Gemini VIII* target vehicle 49-minute EVA standing in hatch 39-minute EVA retrieving experiment from Agena stage
Gemini XI	Charles Conrad Richard Gordon	September 12–15, 1966	2 days 23 hours 17 minutes 8 seconds	Record Gemini altitude: 743$^1/_3$ miles (1,189.3 km), reached using Agena propulsion system after rendezvous and docking 33-minute EVA and 2-hour stand-up EVA
Gemini XII	James Lovell Edwin Aldrin	November 11–15, 1966	3 days 22 hours 34 minutes 31 seconds	Final Gemini flight Rendezvous docking with Agena target (kept station with it during EVA) EVA record: 5 hours 30 minutes

Below In June 1966, the *Gemini IX* mission experienced problems with the Augmented Target Docking Adapter (ATDA) and was unable to dock. *Gemini IX*'s crew members, Stafford and Cernan, nicknamed the ATDA the "angry alligator."

Robotic Lunar Explorers

As our nearest celestial neighbor, and the focus of space travel fantasies since ancient times, the Moon was a natural target for the earliest attempts to send a space probe beyond the orbit of Earth.

Right After just four months as president, John F. Kennedy declared that "...I believe this nation should commit itself to achieving the goal, before this decade is out, of landing a man on the Moon and returning him safely to the Earth."

Far right The Lunokhod automated lunar rover was controlled by a five-man team on Earth. Lunokhod's equipment included four television cameras, a soil mechanics tester, a solar X-ray experiment, and a magnetometer.

Following President Kennedy's proclamation in 1961 that the USA would land a man on the Moon by 1970, lunar exploration by robotic probes that would photograph and land on the Moon to establish potential landing sites for manned missions became a priority for both the USA and the USSR. Although the USA suffered a succession of failures in its early attempts to explore the Moon using automated craft, the USSR launched three successful lunar probes in 1959; these were important early propaganda victories in the Space Race.

LUNA

The Soviet probe *Luna 1* made the first lunar flyby in January 1959, and *Luna 2* was the first to impact the Moon's surface. Even more spectacularly, *Luna 3*, in October 1959, transmitted back to Earth the first images from the ever-hidden far side of the Moon, revealing that it was somewhat different from the Earth-facing side.

The Luna program experienced some failures, but in January 1966 *Luna 9* made history by becoming the first spacecraft to touch down safely on the Moon and transmit images from its surface. The landing revealed that the lunar surface dust was not so deep as to prevent a manned landing, as had been feared. Later Luna missions—some in lunar orbit and equipped with cameras and scientific instruments, others designed for soft landing—had only mixed success.

In 1970, *Luna 16* not only landed on the Moon but returned a sample of lunar soil to Earth in a small remote-controlled capsule. This was part of a new direction in Soviet lunar exploration, taken to bolster the face-saving claim after the *Apollo 11* landing that the USSR had never been racing the USA to the Moon but had always intended a program of safe robotic exploration. Two further Luna missions also successfully returned samples of lunar soil to Earth before the program was terminated in 1976. In addition, the USSR developed and launched the first remote-controlled roving vehicles, known as Lunokhod ("moonwalker"). Originally intended for advance exploration in anticipation of the Soviet manned lunar program (had it gone ahead), they were remotely controlled by operators back on Earth. *Lunokhod 1* was launched in 1970 and *Lunokhod 2* in 1973.

Below The first photograph of Earth seen from the Moon, transmitted back home by *Lunar Orbiter 1*. This significant image was taken on August 23, 1966, at 16:35 GMT, just before the American space probe passed behind the Moon.

Below Spacecraft in the Ranger program launched in the mid-sixties transmitted more than 17,000 pictures back to Earth before they impacted on the lunar surface. These images provided invaluable information on the nature of the Moon's surface.

RANGER

Although the USA initially faced many disappointing failures in its attempts to explore the Moon with space probes, it planned three series of automated lunar missions, in parallel with its three manned programs, in order to acquire detailed information about the lunar surface and high-resolution imagery to establish potential landing sites for the Apollo missions. The earliest series of probes, Ranger, was designed to crash-land into the lunar surface while taking images all the way to impact. The first Ranger missions were unsuccessful, but in 1964 *Ranger 7* returned an awe-inspiring series of detailed surface images. *Ranger 8* and *Ranger 9*, in 1965, were equally successful, with images from *Ranger 9* broadcast live on television as it impacted the lunar surface.

SURVEYOR

Launched between 1966 and 1968, the American Surveyor program was designed to demonstrate the feasibility of actually landing a spacecraft on the Moon. Before *Luna 9* and Surveyor, no-one knew how deep the lunar dust was, nor whether the lunar crust had properties that might prevent a manned landing. The Surveyors carried a suite of instruments to evaluate the suitability of the sites for the manned Apollo landings. Several Surveyor landers carried small scoops designed to test the mechanical properties of the lunar soil. Some also had alpha scattering instruments that helped to determine the chemical composition of the soil. The Surveyor program, with five successful landings out of seven attempts, paved the way for the Apollo missions to proceed.

LUNAR ORBITER

Complementing the Surveyor landing program were the Lunar Orbiter probes, which were launched throughout 1966 and 1967 with the express objective of mapping the surface of the Moon before the proposed Apollo landings. The five missions in the series were all successful, and 99 percent of the Moon was photographed with a resolution of 60 m or better. The first three missions were dedicated to imaging 20 potential lunar landing sites—chosen on the evidence of Earth-based observations—while the later missions had broader scientific objectives. The Lunar Orbiters featured an ingenious imaging system that permitted on-board processing of high-resolution film images, which were then scanned and transmitted back to Earth.

Above *Lunar Orbiter 2* returned this image of the Moon's surface. Because they were taken at low to moderate sun angles, the photographs from the Lunar Orbiter probes provided scientists with a wealth of information about the Moon's physical form.

The Rush to Disaster—*Apollo 1* and *Soyuz 1*

The race to the Moon drove both the US and the Soviet space programs at a frantic pace. As a result, flawed design decisions and political pressures sowed the seeds for disaster, which struck both programs in 1967 and brought the lunar landing rush to a tragic halt.

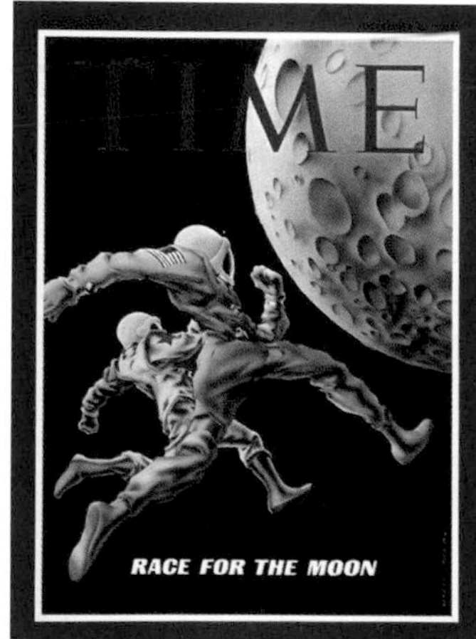

Above The Space Race featured regularly as a cover story in issues of *TIME* magazine in the 1960s. The heat of the Moon race led to the disasters in both the US and USSR space programs in 1967.

Both programs went into hiatus for reassessment and redesign. The US Apollo program emerged stronger and went on to land the first men on the Moon with *Apollo 11* in 1969. However, the Soviet program had to struggle to regain its momentum.

APOLLO 1

On January 27, 1967, during a pre-flight test for what would have been the first manned mission of the Apollo program, astronauts Virgil "Gus" Grissom (MR-4, *Gemini III*), Edward White (*Gemini IV*, the first US spacewalker), and Roger Chaffee were aboard their spacecraft on the launch pad. They were conducting a launch simulation to ensure that their Command Module was fully operational. Various problems delayed the test for many hours. As the simulation was about to resume, the crew suddenly radioed ground control that there was a fire in the craft. Within 17 seconds, all transmissions from the crew ended.

The astronauts attempted to open the spacecraft hatch to escape, but the procedure was an awkward one and the fire spread quickly, generating thick smoke and fumes that overcame them before they could release the hatch. Spacecraft technicians on the pad ran toward the Command Module, but it ruptured before they could reach it, releasing flames and clouds of black smoke. The intense heat and dense smoke drove the rescuers back again and again. At last they managed to open the hatch, but sadly the crew were already dead, asphyxiated by the toxic smoke inside the craft.

The investigation that followed showed that the fire had probably started around one of the wire bundles near Grissom's seat on the left side of the cabin. Investigators concluded that it was likely caused by a spark resulting from damaged insulation around the wires. In the pure-oxygen, high-pressure atmosphere of the Command Module, which was constructed of many flammable materials that gave off toxic gases, the resulting fire exploded into a conflagration. Coupled with the inward-opening hatch, which was difficult to operate against the pressure inside the cabin, these conditions gave the crew of *Apollo 1* little chance of escape.

Above right Scorched exterior of the *Apollo 1* spacecraft after the fire. Pressure build-up inside the spacecraft during the fire was so great that it ruptured the hull. Outrushing smoke and flames prevented rescuers from approaching the spacecraft for critical minutes.

Right This grim photograph shows the burnt interior of the Command Module in which the three Americans were killed in 1967. They had been training for an Earth orbiting mission scheduled to be launched in February.

As a result of the fire, the design of the Command Module was extensively revised to include less flammable materials, a lower-pressure atmosphere using a less dangerous mixture of oxygen and nitrogen during launch, and a new hatch that could be opened more easily and rapidly.

SOYUZ 1

On April 24, 1967, only months after the loss of the *Apollo 1* crew, the USSR experienced its own space disaster with the crash-landing of *Soyuz 1*. Soyuz ("Union") was a new three-person craft intended for the lunar program as the equivalent of Apollo. Despite various failures on three previous test flights, and although its designers and cosmonauts knew that the craft was not yet fully ready for flight, political pressure from the highest levels forced them to proceed with a most ambitious "spacewalk spectacular" that aimed to achieve new Soviet firsts in the Space Race.

It was intended that Vladimir Komarov, who had previously flown on *Voskhod 1*, would launch alone in *Soyuz 1* and then rendezvous and dock with *Soyuz 2*, launched the following day. Two cosmonauts from *Soyuz 2* would then spacewalk to *Soyuz 1* and return to Earth with Komarov. Thus, in one joint mission, the USSR's space program would accomplish and surpass many of the Gemini achievements while the USA was still recovering from the loss of *Apollo 1*.

However, trouble struck *Soyuz 1* soon after it was launched when a solar panel failed to deploy. Attempts at maneuvering the spacecraft also failed, forcing the decision to cancel the mission and return *Soyuz 1* to Earth. Re-entry was successfully accomplished and a small drogue parachute was released. However, due to the failure of a pressure sensor, the main parachute, needed to slow the descending spacecraft to a safe touchdown speed, could not be activated, and when Komarov released his reserve chute it became entangled with the drogue chute. Without any available means to slow it down, the descent module crashed into a field, obliterating the spacecraft and killing Komarov instantly. He was posthumously awarded the title of Hero of the Soviet Union.

Like Apollo, the Soviet lunar program was delayed by about 18 months while the design of Soyuz was improved. Despite the later cancellation of the Soviet Moon program, the Soyuz spacecraft would go on to a long and successful career in the Soviet space station program.

Above US astronauts Virgil Grissom (right), Edward White (center), and Roger Chaffee (left) are seen here during an earlier simulation for the disastrous *Apollo 1* mission.

Left Vladimir Komarov (left) waves to a Russian crowd in Moscow, with his fellow cosmonauts after the successful conclusion of the *Voskhod 1* mission. After the *Soyuz 1* crash, US President L. B. Johnson declared that "The death of Vladimir Komarov was a tragedy in which all nations share."

Destination Moon—*Apollo 7–10*

The loss of *Apollo 1* caused a hiatus of some 20 months while the Apollo program was reassessed and the Command Module redesigned. America's lunar landing project emerged from this period with a renewed vigor and confidence that would enable it to achieve the first Moon landing just nine months after the first successful manned Apollo flight.

Above *Apollo 7* blasts off from Kennedy Space Center in 1968. Almost 6 minutes into the mission, commander Walter M. Schirra reported that "She is riding like a dream."

Right This mission profile plan for the manned *Apollo 8* mission shows its path to the Moon and return. The mission achieved all of its objectives and scored a propaganda victory by claiming a significant space first.

With understandable caution after the *Apollo 1* disaster, *Apollo 7* was the first manned Apollo mission—designed as a shakedown flight to test the Apollo Service Module and redesigned Command Module.

APOLLO 7—TESTING IN EARTH ORBIT

Under the command of Walter Schirra, the only astronaut to fly all three US spacecraft—Mercury, Gemini, and Apollo—the crew also included Walter Cunningham and Donn Eisele, both of whom were making their first spaceflights.

Launched at 11:02:45 A.M. EST, on October 11, 1968, *Apollo 7* spent 11 days in Earth orbit, during which all its hardware and systems were tested; none showed any significant problems. The Service Module's main propulsion engine was successfully tested and the maneuver of extracting the Lunar Module from its launch cradle was also practiced. The crew caught severe colds and became irritable and non-cooperative with Mission Control, but the mission nevertheless demonstrated that the Apollo spacecraft was "Go!"

APOLLO 8—AROUND THE MOON

Even before *Apollo 7*, NASA was aware that the delivery of the Lunar Module would be considerably delayed and that it would not be ready for testing in Earth orbit following the *Apollo 7* flight. This, coupled with intelligence information suggesting (correctly) that the USSR was planning to launch a circumlunar mission in order to upstage Apollo by flying cosmonauts around the Moon, led NASA to revise its flight program and take the risky decision to make its second Apollo mission also its first flight to the Moon. This was a bold move because *Apollo 8* would also be the first manned flight of the massive Saturn V launch vehicle (*Apollo 7* had used the smaller Saturn IB), which had had only two previous test flights.

Launched on December 21, 1968, *Apollo 8*, carrying Frank Borman, James Lovell, and William Anders, made the first crewed flight to the Moon, arriving in lunar orbit three days later. The crew of *Apollo 8* made 10 orbits of the Moon, and on Christmas Eve (US time) made a moving telecast from lunar orbit, reading from the book of Genesis. The breathtaking images taken during this six-day mission, showing the Earth rising over the Moon and our world floating in the blackness of space, are some of the most important space images ever taken; they would help to encourage environmental awareness and give rise to the concept of Spaceship Earth.

APOLLO 9—TESTING THE LUNAR MODULE

Although the Lunar Module—the landing vehicle that would carry two astronauts to the surface of the Moon—was not ready to fly the *Apollo 8* mission, it was tested in Earth orbit in March 1969, during the *Apollo 9* mission. This 10-day flight was the first to test the complete Apollo spacecraft that would journey to the Moon, putting all three Apollo vehicles (Saturn V, Command Module, Lunar Module) through their paces. The crew, James McDivitt, Russell Schweickart, and David Scott, practiced the undocking and docking maneuvers that would occur in lunar orbit, and also fired the Lunar Module's descent engine, and later its ascent engine in a simulated liftoff. *Apollo 9* would also be the first Apollo mission in which the Command and Lunar modules had individual names—*Gumdrop* and *Spider*, respectively—since the craft needed separate call signs while they were maneuvering undocked from each other.

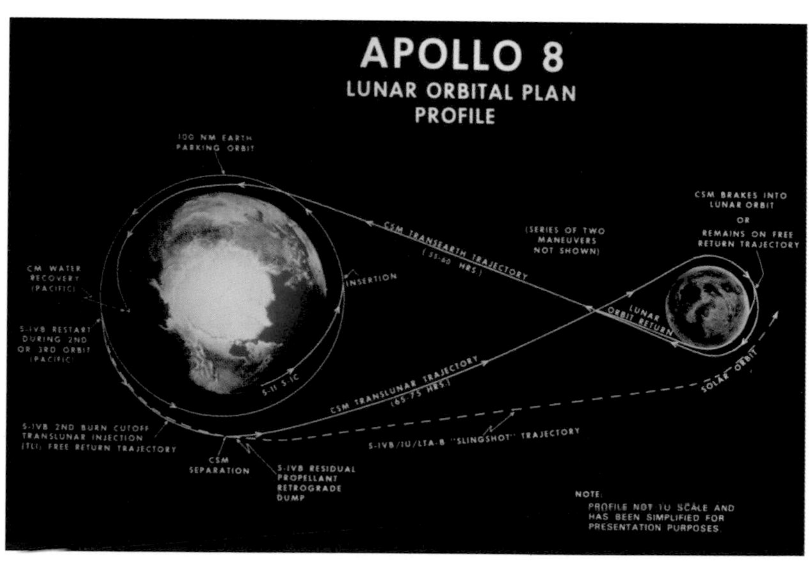

APOLLO 10—DRESS REHEARSAL

With the Lunar Module successfully tested in Earth orbit by *Apollo 9*, the eight-day *Apollo 10* flight became the full dress rehearsal for the planned *Apollo 11* landing mission. If all went well with *Apollo 10*, the historic first Moon landing would be cleared to proceed. The experienced crew of Thomas Stafford, John Young, and Eugene Cernan, all of whom had flown during the Gemini program, chose light-hearted names for their Command Module (*Charlie Brown*) and Lunar Module (*Snoopy*), which would attract some criticism as being "too flippant."

Launched on May 18, 1969, *Apollo 10* closely followed the *Apollo 11* flight plan, orbiting the Moon 31 times, and twice took its Lunar Module down to within 9 1/10 miles (14.5 km) of the lunar surface. The mission was not a mere technical success; it also provided spectacular coverage of spacecraft operations and views of the Earth and Moon with 19 color television transmissions. The stage was now set for the first landing on the Moon.

Above The *Apollo 8* mission returned more than 800 photographs, including the first pictures of Earth from deep space. This awe-inspiring photograph of the Earth rising over the Moon is one of the iconic images of the Space Age.

Left The *Apollo 9* mission's Lunar Module, named *Spider*, is put through its paces in Earth orbit. The lunar contact probes at the end of its landing legs can be clearly seen

Apollo's Triumph

Kennedy's goal of a US Moon landing by the end of the 1960s was achieved when *Apollo 11* made the first landing on the Moon and a man stepped onto another world. Accomplishing the first lunar landing ahead of the USSR won the propaganda battle of the Space Race for the USA. It was a historic event—one of humanity's greatest technological achievements and a defining moment of the twentieth century.

Above *Apollo 11* mission official crew portrait. Neil A. Armstrong (left), Lt. Col. Michael Collins (center), and Col. Edwin E. "Buzz" Aldrin (right) were all veterans of Gemini program flights.

Apollo 11, the fifth manned Apollo mission, was launched on July 16, 1969. Over a million people crowded the vicinity of Kennedy Space Center in Florida to watch the launch live, and more than 600 times that number witnessed it on television at home and abroad, as *Apollo 11*—carrying commander Neil Armstrong, Command Module pilot Michael Collins, and Lunar Module pilot Edwin "Buzz" Aldrin—thundered into space.

ONE GIANT LEAP

Apollo 11, with its Command Module *Columbia* and Lunar Module *Eagle*, passed behind the Moon on July 19, firing its Service Module propulsion engine in order to enter lunar orbit. After several orbits, the Command and Lunar modules separated, and Michael Collins was left alone in orbit in *Columbia* as Armstrong and Aldrin descended to the lunar surface in *Eagle*. The landing site was to be the Sea of Tranquility, which had been selected using images from Ranger and Surveyor space probes. As the landing began, however, it was discovered that the Lunar Module was further along its descent trajectory than planned and would have to land some distance west of the intended site.

During the descent to the lunar surface, several unusual "program alarms" occurred. These were triggered by the navigation and guidance computer reporting that it was overloaded (because an unnecessary radar had been left on). In fact, there was nothing wrong with the spacecraft, and the landing attempt continued.

It soon became apparent that *Eagle* was descending toward a field of sizable rocks scattered around a large crater. It was a dangerous place to land, so Armstrong took manual control of the Lunar Module and, with Aldrin's assistance, guided the spacecraft to a landing at 20:17 UTC on July 20, 1969, with only seconds' worth of fuel left. Armstrong's first words after landing were: "Houston, Tranquility Base here. The *Eagle* has landed." In Mission Control in Houston, where everyone was anxiously aware of the fuel situation, the reaction was jubilant.

Above right Because Armstrong was the main photographer on the lunar surface, there are few pictures of him on the Moon. Aldrin took this rare photograph of Armstrong packing rock and soil samples into *Eagle*'s Modular Equipment Stowage Assembly (MESA) during the first-ever EVA on the Moon's surface.

Left *Columbia*, the *Apollo 11* Command Module, is seen here in lunar orbit. As the Lunar Module descended to the Moon, Command Module Pilot Michael Collins cautioned Armstrong and Aldrin: "You cats take it easy on the lunar surface."

ONE SMALL STEP

With the astronauts too keyed up to rest as planned, the first lunar EVA was moved forward. Six and a half hours after the landing, Armstrong, as commander of the mission, prepared to put the first human footprint on the Moon. At 2:56 UTC on July 21 (still July 20 in the USA), Armstrong left the cramped Lunar Module and climbed down the nine-rung ladder fixed to one of its legs. As he did so, he deployed the Modular Equipment Stowage Assembly (MESA) that was folded against *Eagle*'s side, and activated the TV camera. The signal was picked up at both the Goldstone tracking station in the USA and the Honeysuckle Creek Tracking Station in Australia, where the image quality was better: The pictures from Honeysuckle were broadcast to an audience of more than 600 million people worldwide.

After describing the surface dust as "fine ... like a powder," Armstrong stepped off *Eagle*'s footpad and became the first person to set foot on another world. As his boot touched the surface of the Moon, Neil Armstrong uttered his now famous declaration: "That's one small step for (a) man: One giant leap for mankind." (About 8 minutes after Armstrong stepped onto the lunar surface, the Parkes Radio Telescope, also in Australia, received a still higher quality signal, and its feed was used for the rest of the telecast from the Moon. The single-dish telescope has a diameter of 210 feet [64 m].)

WHAT DID HE REALLY SAY?

Those millions of people who were listening to Armstrong's words in 1969 heard him say, "That's one small step for man: One giant leap for mankind"—an apparently tautologous remark. Armstrong intended to say "a man" and has always believed that he did actually say it, although it has been generally assumed that he mistakenly forgot the "a." In 2006, a digital audio analysis of Armstrong's words seemed to indicate that he had actually said "a man," but that the indefinite article was inaudible, lost to static noise and the technical limitations of contemporary communications. Armstrong prefers the saying to be written as you see it in the main text, with the "a" in parentheses.

Above Buzz Aldrin photographed his own bootprint in the lunar soil, as part of an experiment to study the nature of lunar dust and the effects of pressure on its surface.

MOONWALKERS

Immediately after stepping onto the lunar surface, Armstrong inspected the Lunar Module, took some photographic panoramas and collected a contingency sample of lunar soil (in case *Eagle* had to make an emergency departure).

When Aldrin joined him 18 minutes later, the two astronauts set up the television camera on the surface in order to broadcast their activities; they also tried out some different methods of moving around, including bouncy "kangaroo hops." They planted a US flag and took a congratulatory phone call from US President Richard Nixon.

Armstrong and Aldrin deployed the Early Apollo Scientific Experiment Package (EASEP), which included a laser ranging retroreflector and a passive seismometer. Armstrong moved about 400 feet (120 m) away to take photographs at the rim of East Crater, while Aldrin took two geological core samples. Both astronauts also collected rock samples. Because many of the surface activities took longer than expected, Armstrong and Aldrin were unable to complete their documented lunar sample collection.

After two and a half hours on the lunar surface the astronauts returned to *Eagle*. They loaded two sample boxes containing 48 lbs (21.8 kg) of lunar surface material into their craft, but lightened the ascent stage for return to lunar orbit by leaving behind their life support backpacks, lunar overshoes, and other equipment.

BACK TO EARTH

After lifting off from the Moon and rendezvousing in orbit with *Columbia*, precautions were taken to prevent any germs or contamination from the Moon from being brought back to Earth. Before transferring to the Command Module, Armstrong and Aldrin vacuumed their clothes and equipment thoroughly to remove any traces of lunar dust. *Eagle* was jettisoned and left in lunar orbit, to crash back later onto the Moon's surface.

On July 24, 1969, *Apollo 11* returned to Earth, splashing down in the South Pacific. Although they were welcomed as heroes, the crew of *Apollo 11* were immediately placed into quarantine for three weeks in case the Moon might contain any unknown pathogens that could affect humans. On August 13, Armstrong, Aldrin, and Collins were finally released from isolation to begin a long series of celebrations marking their historic achievement—and the defeat of the USSR in the Space Race.

MORE THAN A REPEAT PERFORMANCE

Just four months after the *Apollo 11* mission, in November 1969, *Apollo 12* flew to the Moon. This 10-day, manned mission began sensationally when lightning struck the craft's Saturn V launch rocket, raising the possibility that the flight would have to be aborted. However, as all the spacecraft systems checked out as unaffected, the mission, under the command of Charles "Pete" Conrad, with Command Module pilot Richard Gordon and Lunar Module pilot Alan Bean, was able to proceed. The all-Navy crew of *Apollo 12* named their Command Module *Yankee Clipper* and their Lunar Module *Intrepid*.

The lessons learned from *Apollo 11* enabled Conrad and Bean to put *Intrepid* down on the lunar surface with such pinpoint accuracy that the craft landed just 600 feet (183 m) from the *Surveyor 3* lunar probe that had arrived on the Moon some 31 months earlier. The *Apollo 12* astronauts made two excursions onto the lunar surface, totaling seven and a half hours. Unfortunately, television coverage of their surface activities was lost when Al Bean accidentally burned out the television camera; nevertheless, during these EVAs Conrad and Bean collected more than 75 lbs (34 kg) of Moon rocks and soil samples, and deployed the first ALSEP (Apollo Lunar Surface Experiments Package), which was a set of six scientific experiments, powered by a small nuclear-powered battery, that was more advanced than *Apollo 11*'s experiment package.

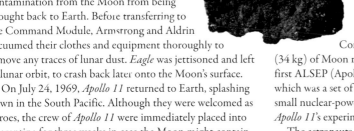

The astronauts also walked over to the Surveyor spacecraft and retrieved its television camera and some other components to be taken back to Earth so that the effects on their materials of long-term exposure on the lunar surface could be studied. In addition, after Conrad and Bean returned to lunar orbit, *Intrepid* was deliberately crashed back onto the Moon's surface so that the seismic shock waves created by its impact would register on the ALSEP seismometer and provide some indication of the structure of the Moon.

After spending a day in lunar orbit photographing the surface of the Moon, *Apollo 12* returned to Earth. During the splashdown, near American Samoa, a 16 mm camera broke free from storage and struck Bean on the forehead, injuring him slightly. This and the lightning strike at launch were the only two untoward events of the mission. Like the crew of *Apollo 11*, the crew of *Apollo 12* was quarantined for three weeks as a precaution, in case they were carrying lunar germs.

Above *Surveyor 3* had been sitting in a lunar crater in the Ocean of Storms for some time when *Apollo 12*'s Lunar Module *Intrepid* (seen in the background) landed nearby in November 1969.

Left This rock sample was part of the 48 lbs (21.8 kg) of lunar geological material brought back to Earth by *Apollo 11*. The astronauts found two main types of rock at their landing site—basalts and breccias.

Opposite Neil Armstrong and *Eagle* are reflected in Aldrin's visor in this classic *Apollo 11* image—one of the most recognizable photographs of the twentieth century. The Moon's lack of atmosphere means that the astronauts' footprints visible in this picture will survive for millions of years until eroded by micrometeorite impacts.

Left *Apollo 11* splashed down some 812 nautical miles southwest of Hawaii. The astronauts and their recovery team wore biological isolation outfits to prevent any contamination from lunar germs.

Exploring the Moon

Apollo 13, the third lunar landing mission, became the Apollo program's first serious in-flight space emergency. Launched on April 11,1970, with its crew of James Lovell, Fred Haise, and Jack Swigert, *Apollo 13* was intended to be the first of three lunar missions devoted to geological research.

Above The badly damaged *Apollo 13* Service Module, photographed from the linked Lunar Module/Command Module after it had been jettisoned. One panel on the Service Module was completely blown away by the explosion.

Opposite A sunlit trail leads away from the *Apollo 14* Lunar Module *Antares*. The trail was made by the two-wheeled cart known as a Modularized Equipment Transporter (MET), which the astronauts used on their two lunar EVAs.

After a crew change only a week before launch, when it was discovered that the astronauts had been exposed to German measles and that Command Module pilot Ken Mattingly might not be immune, the liftoff also had its drama, with an early engine cut-off on the second stage. Then, on April 13, two days after the launch, the spacecraft was crippled by an explosion in the Service Module, caused, it was later discovered, by a short-circuit across a wire with damaged insulation inside a liquid oxygen tank.

The explosion resulted in a rapid loss of power and oxygen to the spacecraft; plans for the lunar landing were quickly abandoned and the systems in the Command Module, *Odyssey*, were turned off so as to preserve its ability to re-enter the atmosphere upon return to Earth. The Lunar Module, *Aquarius*, with its independent power, water, and oxygen supplies, became the crew's "lifeboat" while *Apollo 13* made a looping trajectory around the Moon and returned to Earth. Under horribly trying conditions, with little power, drinking water, or cabin heating, the crew managed to conduct the necessary maneuvers and engine burns critical to their safe return to Earth. On the ground, the flight controllers and their support teams displayed considerable ingenuity under extreme pressure, working out jury-rigged solutions to sustain the spacecraft and ensure the crew's safe return.

On April 17, *Apollo 13*, having jettisoned *Aquarius* shortly before re-entry, splashed down safely; it was a "successful failure" that demonstrated the resilience of the Apollo spacecraft and the program as a whole.

INTO THE LUNAR HIGHLANDS

The *Apollo 13* accident investigation delayed the US space program for almost 12 months. However, eventually *Apollo 14* was launched in February 1971, its target the Fra Mauro Highlands, which had been *Apollo 13*'s goal. The crew, consisting of Alan Shepard (the first American in space and the only Mercury astronaut to fly to the Moon), Stuart Roosa, and Edgar Mitchell, was the least experienced of any Apollo lunar mission. They named their Command Module *Kitty Hawk* and their Lunar Module *Antares*.

Despite a string of technical problems, including initial difficulties in docking with the Lunar Module, the mission was accomplished without major incident. Shepard and Mitchell carried out two lunar EVAs, totaling almost nine and a half hours, and collected 98 lbs (44 kg) of lunar samples. They deployed another ALSEP package, and also trialed the use of a small cart to help carry equipment. Shepard, an avid golfer, smuggled a makeshift six-iron club and two golf balls onto the Moon and made several one-handed swings there. The *Apollo 14* crew was the last to have to endure the quarantine period when they returned to Earth.

ROVING THE LUNAR SURFACE

Apollo 15, which was launched in July 1971, was the first of three "J Series" missions. These were designed for longer stays on the Moon, with a greater focus on science thanks to various equipment upgrades. Unlike the crews of previous Apollo flights, the astronauts on *Apollo 15* and later missions would be able to extend their exploration of the lunar surface over much greater distances using the Lunar Roving Vehicle (LRV), an ingenious electrically powered "Moon car" that could carry the crew and their equipment much further than they could traverse on foot.

Commanded by David Scott, with Alfred Warden as Command Module pilot and James Irwin as Lunar Module pilot, *Apollo 15* had an all-Air Force crew. They chose the name *Falcon* for their Lunar Module, after the Air Force mascot, and their Command Module was called *Endeavour*. Landing in the geologically interesting area of Hadley Rille, Scott and Irwin made three EVAs on the lunar surface and collected 170 lbs (77 kg) of lunar samples. They traveled a total of 17⅖ miles (28 km) in the LRV during their excursions.

In addition to the research carried out by the two astronauts on the lunar surface, the *Apollo 15* Service Module carried a bay full of scientific instruments that were trained on the Moon from orbit, and a small satellite that was released into lunar orbit before *Apollo 15* departed. This tiny probe would spend a year studying the "mascons" (strange concentrations of mass under the lunar surface that could perturb spacecraft orbits) and other lunar phenomena.

Right Mt Hadley provides the backdrop as *Apollo 15* astronaut James Irwin works on the Lunar Roving Vehicle during the first EVA of the mission. The "Moon car" was carried to the lunar surface folded and attached to the exterior of the Lunar Module.

The Soviet Lunar Program

Despite its early lead in the Space Race, the USSR was slow to respond to the challenge of President Kennedy's lunar goal for the US space program. It was not until 1964 that a comprehensive manned lunar program was finally approved; however, the project was fraught with delays and developmental difficulties and was ultimately unable to mount a serious challenge to the Apollo missions.

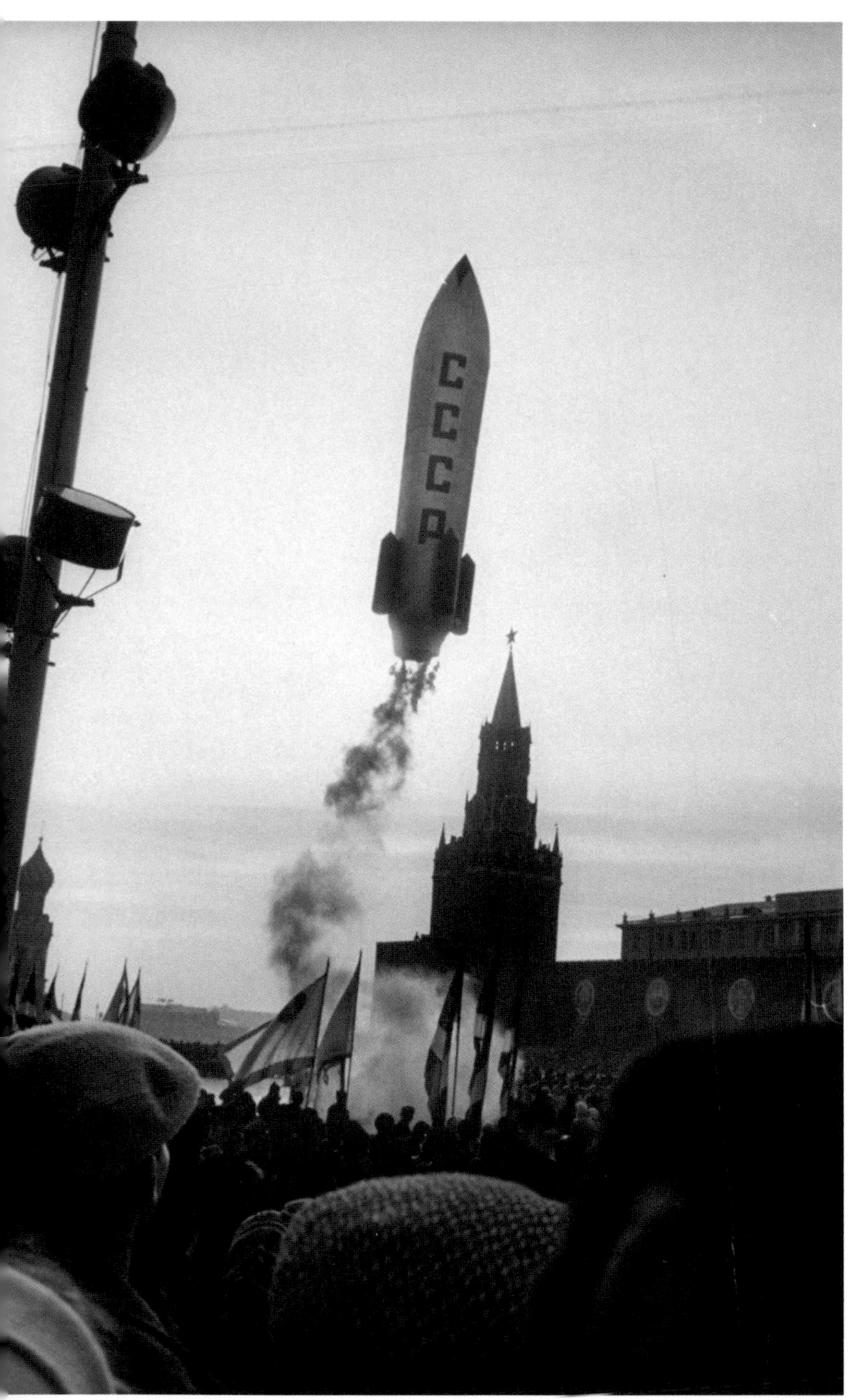

The USSR had no equivalent of NASA: Its space activities were divided among different design bureaus, attached to different ministries, that were responsible for both designing and constructing spacecraft. There was considerable rivalry between these institutions, and Sergei Korolev had various conflicts with other bureau leaders, particularly Vladimir Chelomei, who used his political connections to obtain support for his projects. As a result, the Soviet lunar program was bedeviled from the start, with rival groups lobbying for, and working on, different projects at the same time.

THE BATTLE OF THE DESIGNERS

During 1961, Korolev and Chelomei were both designing launchers to carry cosmonauts to the Moon: N-1 (Korolev), and Proton (Chelomei). In 1962, Chelomei's design bureau was charged with developing a spacecraft, designated LK-1, to be used with Proton for a circumlunar flight. In August 1964, it was officially decided to put a single cosmonaut on the Moon in 1967–1968 (before Apollo), as well as conducting a circumlunar mission with two cosmonauts by October 1967, the fiftieth anniversary of the Bolshevik Revolution.

THE LUNAR PROGRAM TAKES SHAPE

After Nikita Khrushchev was deposed in October 1964, the lunar landing program was again reconsidered, and by 1965 had taken shape around two different projects: A manned lunar landing program led by Korolev's design bureau, using his Soyuz spacecraft and N-1 rocket, with the goal of a first lunar landing in 1968, and a separate manned program of circumlunar flights under the leadership of Chelomei, still targeting the 1967 anniversary. The robotic Luna program that was already underway would be extended to include a remote-controlled lunar rover (later to become Lunakhod) that would be used to reconnoitre the planned landing sites one month before the manned mission arrived. The Lunakhod would be delivered by a Ye-8 spacecraft.

In 1965, the Soviet space program was placed under the control of the newly created Ministry of General Machine Building. The lunar landing program adopted the same lunar rendezvous technique as Apollo. Its spacecraft, known as the L3 Complex, consisted of a Soyuz-based lunar orbiting spacecraft (LOK), analogous to the Apollo Command Module, as well as a one-man LK "lunar ship," which would carry the Moon-landing cosmonaut to and from the lunar surface.

Left A mock rocket soars above a parade during the fiftieth anniversary of the 1917 Bolshevik Revolution, in 1967, as an expression of pride in Soviet space achievements. There were unsuccessful plans for cosmonauts to orbit the Moon as part of the anniversary celebrations.

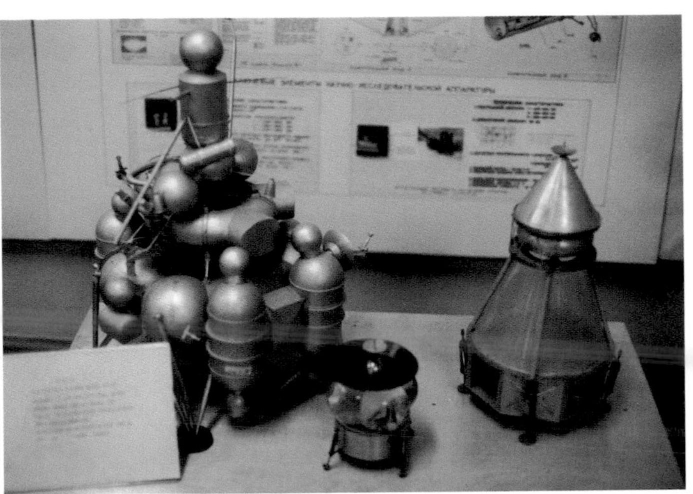

Left The single crew LK spacecraft was the Soviet equivalent of NASA's Lunar Module. It featured a viewing porthole facing upward to be used during docking, and a side porthole for observing the lunar surface.

Far left Model of the Luna Ye-8 spacecraft (left) which was used for lunar sample return missions and also for landing the Lunokhod rovers on the Moon. The Ye-8 was a product of the Lavochkin design bureau.

In the circumlunar program Chelomei's LK-1 spacecraft, was replaced by an adapted version of the Soyuz, designated L-1, which would fly atop Proton.

DISASTER AND DELAY

In 1966, just as the lunar effort was finally gathering momentum, Korolev died suddenly. This was a severe blow to the program, already beset with development difficulties with the N-1 launcher, the Soyuz spacecraft, and the L-1 circumlunar craft. The *Soyuz 1* disaster, in April 1967, delayed both the Soyuz and L-1 programs considerably, and the original goal of launching a crewed circumlunar flight in October 1967 had to be abandoned. However, since achieving a manned circumlunar flight before the USA would still have considerable propaganda value, the L-1 program strove to match NASA's plans to send *Apollo 8* around the Moon in December 1968. Despite a massive effort, however, the Russsians were unable to beat *Apollo 8* to lunar orbit, and the L-1 program was subsequently wound down.

Hoping that the first Apollo landing might be delayed, the Russian lunar landing program pressed on, despite the failure of the first N-1 rocket launch in February 1969. At best, it was hoped that a first Soviet lunar landing would occur in 1970–1971. As a fallback propaganda position, an automated return probe was prepared that would recover a sample of lunar soil and return it to Earth before the first American landing. Space station development was also pushed forward as an alternative spaceflight goal, and the USSR began to deny that it had plans to go to the Moon.

When pre-*Apollo 11* Soviet sample return missions failed, along with a July 1969 N-1 test, the USSR recognized that it had lost the Moon Race, though it still hoped to mount an advanced manned lunar-landing project, L3M, which would include extended stays on the Moon. However, funds for the program were significantly reduced and, with continuing failures of the N-1 rocket, the Soviet lunar landing program slowly tapered off and was finally cancelled in 1974, having failed to win the Space Race for the USSR.

Above The Proton launcher was developed by Chelomei from an ICBM design to become the launch vehicle for the planned Soviet circumlunar flights. After the lunar program was abandoned, it went on to become a workhorse heavy launcher for the Soviet and Russian space programs.

Right N-1 rocket on the pad before the disastrous July 1969 test. The rocket crashed back to the pad just after launch and destroyed it in a massive explosion. The second N-1 in the background was a ground test mock-up used for rehearsing parallel launch operations.

The End of Apollo

Apollo 16 was launched in April 1972 to explore the highlands of the Moon's Descartes region. The commander was John Young, the first person to travel twice to the Moon (he had been in the *Apollo 10* crew), Command Module pilot was Ken Mattingly, and Charles Duke was Lunar Module pilot. The landing almost had to be scrubbed when the main propulsion system of the Command Module, *Casper*, malfunctioned, causing concerns about adjusting its lunar orbit. The landing went ahead, but the mission was shortened by one day.

Above *Apollo 16* LRV and Lunar Module *Orion*. John Young and Charles Duke made three excursions using the Lunar Roving Vehicle, setting a lunar speed record of 11 miles per hour (18 km/h). During this mission, an ultraviolet camera/spectrograph was used for the first time on the Moon.

Young and Duke spent three days exploring the lunar surface, traveling 16⅘ miles (26.9 km) with the LRV during three EVAs. Among the 211 lbs (95.7 kg) of samples they collected was the largest single rock returned by Apollo astronauts, a chunk weighing 25 lbs (11.3 kg). Young and Duke also tested the LRV, reaching a top speed of 11 mph (18 km/h)— the lunar speed record to this day. Like *Apollo 15*, the *Apollo 16* Command Module carried a scientific package to study the Moon from orbit and launched another small satellite. While returning from the Moon to Earth, the crew carried out an EVA to bring in film from exterior cameras and conduct an experiment on microbial survival.

THE LAST MAN ON THE MOON

By 1972, the Apollo program had been greatly curtailed from the plans of the 1960s. Though only the sixth successful lunar flight, *Apollo 17* was the last crewed mission to the Moon in the twentieth century. Launched in December 1972 in a spectacular night liftoff, it was the only mission to include a scientist in its crew—geologist Harrison "Jack" Schmitt, Lunar Module pilot. Eugene Cernan (one of the *Apollo 10* crew) was commander, and Ron Evans piloted the Command Module, *America*.

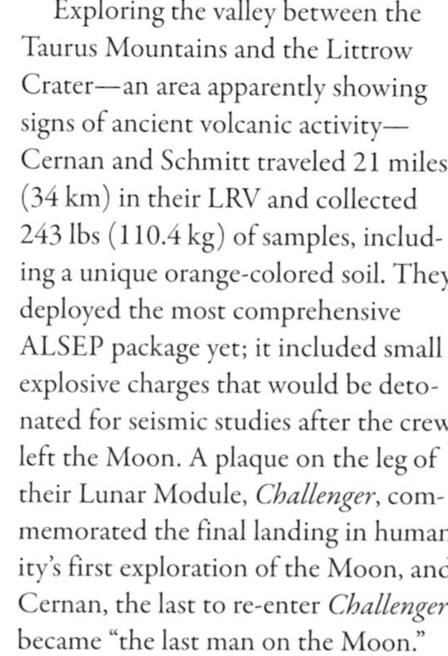

Exploring the valley between the Taurus Mountains and the Littrow Crater—an area apparently showing signs of ancient volcanic activity—Cernan and Schmitt traveled 21 miles (34 km) in their LRV and collected 243 lbs (110.4 kg) of samples, including a unique orange-colored soil. They deployed the most comprehensive ALSEP package yet; it included small explosive charges that would be detonated for seismic studies after the crew left the Moon. A plaque on the leg of their Lunar Module, *Challenger*, commemorated the final landing in humanity's first exploration of the Moon, and Cernan, the last to re-enter *Challenger*, became "the last man on the Moon."

Apollo 17 broke several records—lasting 12½ days, it was the longest manned lunar landing flight; it also boasted the longest total lunar surface EVAs, the largest lunar sample return, and the longest time in lunar orbit.

DÉTENTE IN ORBIT

In the 1970s, when US–Soviet political relations began to thaw, détente replaced the Cold War and the last Apollo hardware was used in a joint Soviet–US mission, known as the Apollo–Soyuz Test Project (ASTP). This historic mission saw, for the first time, astronauts and cosmonauts docking a US and a Soviet spacecraft together in space to conduct symbolic exchanges and carry out joint experiments. The Apollo crew consisted of commander Thomas Stafford, Command Module pilot "Deke" Slayton, and Docking Module pilot Vance Brand. *Soyuz 19*, the Soviet craft, was commanded by Alexei Leonov, the world's first spacewalker, with Valery Kubasov as flight engineer.

Both spacecraft were launched on July 15, 1975, docking together on July 17. Stafford and Leonov met at the entrance to the linking docking module to exchange the first international handshake in space. The two craft remained together for 44 hours; the crews exchanged ceremonial gifts and visited each other's ships to eat and talk. The project also included docking and re-docking maneuvers and joint experiments. When the meeting in the skies ended, *Soyuz 19* remained in space for another five days and Apollo for nine days. A political and technical success, ASTP was the last crewed US space mission until the launch of the first space shuttle in 1981.

Right Geologist Harrison Schmitt was the first scientist to be sent to the Moon on an Apollo mission. He is seen here in front of a large boulder, at a location called Station 6, during the third *Apollo 17* EVA.

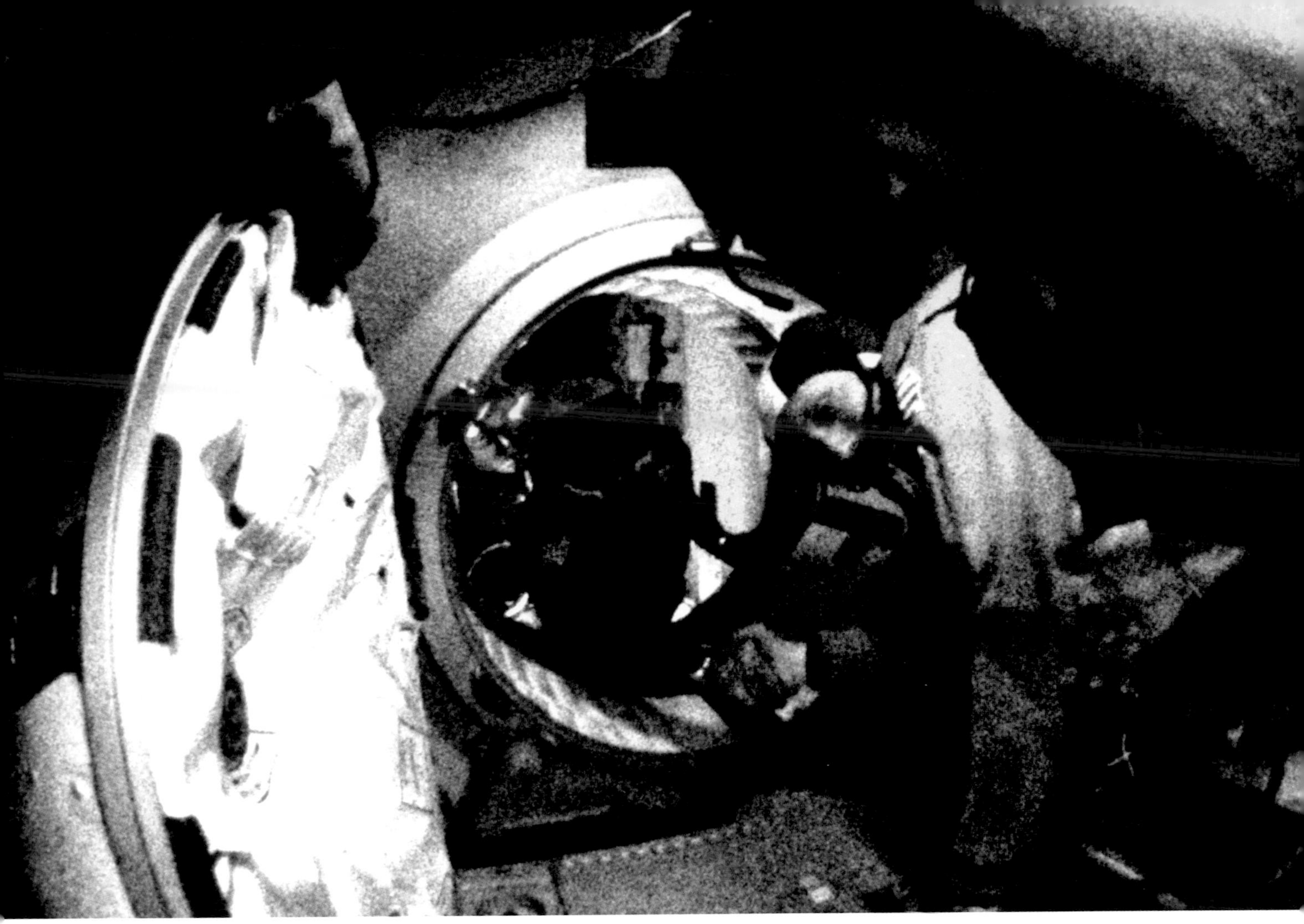

INTO SPACE AT LAST!

Command Module pilot Donald "Deke" Slayton (1924–1993) was one of the original "Mercury Seven" astronauts. Initially selected as pilot of Mercury flight MA-7, in 1962, he was grounded because of a heart problem.

Unable to fly in space, Slayton stayed with NASA as the head of the Astronaut Office, which was responsible for selecting the crews for each mission. Twenty-three years later, restored to flight status, he became the oldest person to fly into space up to that time—he was 51 years of age.

Donald "Deke" Slayton (right) and Thomas Stafford try out some Russian space food during the Apollo–Soyuz Test Project. Stafford holds a tube of borscht, jokingly labeled "vodka," on board the Soyuz spacecraft.

Above Thomas Stafford (right) and Alexei Leonov shake hands during the US–Soviet space mission, ASTP. Stafford reaches out from the specially designed docking module; Leonov is still inside *Soyuz 19*. Leonov greeted Stafford in English, Stafford spoke in Russian.

Below This artist's impression shows *Soyuz 19* and the Apollo spacecraft docking at 2:17 P.M. US Central Time on July 17, 1975, as part of the Apollo–Soyuz Test Project. ASTP signalled the end of the Space Race as the Cold War thawed with the détente of the 1970s.

Space Stations

While the concept of giant wheel-like space stations was undoubtedly popularized through the epic Stanley Kubrick film *2001: A Space Odyssey*, the reality is that our first Earth-orbiting space stations were ungainly looking but highly functional space laboratories. Literally packed floor to ceiling with a crowded labyrinth of research modules, apparatus, and experiments, these orbiting outposts would greatly advance our knowledge of science, medicine, weather patterns, and the Universe. That research continues to this day.

SALYUT—THE FIRST SPACE STATION

In 1869, American author Edward Everett Hale's fictional short story *The Brick Moon* contained what is recognized as the first published mention of an Earth-orbiting outpost. Hale wrote of a sphere 200 feet (60 m) in diameter built of heat-shielding bricks, flung into orbit 4,000 miles (6,400 km) above the planet by gigantic rotating flywheels.

The following century Romanian-born physicist Hermann Oberth, one of the founding fathers of rocketry and astronautics, coined the term "space station" in discussing orbiting space platforms. After World War II, German rocket designer Wernher von Braun set down a credible design for a human-tended space station. His creation, published in the popular *Collier's* magazine, consisted of circular modules linked in a wheel-like design, which would rotate to simulate gravity. Fittingly, within the lifetimes of these European visionaries, massive space stations sent aloft by two nations would begin productive lives orbiting Earth.

Above In a 1952 edition of *Collier's* magazine, visionary designer Wernher von Braun described the rotating wheel as the most practical design for an orbiting space station.

With the grudging realization that they had lost any chance of beating America to the first manned lunar landing, the Soviet Union instead diverted its attention to the long-term occupancy of space. Plans for their first space station had been conceived in 1964, within a top-secret military space program known as Almaz (Diamond). Three years later, construction would be approved.

Originally, there were three main hardware components—a 20-ton (18.14 tonne) space station, tended by crews of three, and equipped with an Agat optical camera for reconnaissance imaging of military installations on Earth; a reusable transport logistics spacecraft used to ferry military cosmonauts and cargo to and from the station; and Proton rockets for launching all of the components into space. Additionally, the station would be equipped with a single small capsule capable of being ejected for a rapid return to the ground, which might contain surveillance film and other small payloads.

There would be three elements to each Almaz station—habitation quarters for crew, a large-diameter work section containing surveillance equipment and other instruments, and a transfer section that would house a docking port, a capsule ejection system, and an EVA (spacewalk) hatch.

Right Visitors to Russia's Rocket and Space Corporation Energia museum can see the *Salyut 4* space station both inside and out.

Soviet engineers had actually completed two flight-rated Almaz station hulls by 1970, but then the Soviet leadership decided to concentrate on developing a more open, civilian space station project, using elements from the Almaz design. One of the Almaz hulls was transformed into a space station that would support genuine scientific studies, although work would continue on the military Almaz program. In order to confuse Western observers, both programs would be simultaneously developed under the name of Salyut.

THE FIRST TRUE SPACE STATION

At 01:34 A.M. (GMT) on April 19, 1971, a week after the tenth anniversary of Yuri Gagarin's flight, the Soviet Union launched the world's first space station, *Salyut 1* (officially DOS-1), into orbit atop a Proton rocket. While the Salyut (Salute) series would usher in an entirely new era of space exploration and discovery for both spacefaring nations, *Salyut 1* was plagued with failures and would come to be regarded adversely by the superstitious cosmonauts.

Four days after the launch of *Salyut 1*, a three-man *Soyuz 10* crew was launched for an intended docking and occupancy of the station. To their frustration, they were unable to enter the space station following a problem with the docking mechanism. After more than five fruitless hours, *Soyuz 10* had to undock and return home.

Two months later, another three-man crew flew *Soyuz 11* to a manual docking with *Salyut 1*, and this time the crew was able to enter and occupy the station. Like the previous crew, they were not wearing cumbersome space

Right A view of the Almaz space station displayed in its original configuration. It was secretly developed as part of a military reconnaissance program in tandem with the civilian Salyut space station.

suits in order to save precious space aboard the cramped *Soyuz*. Cosmonauts Georgi Dobrovolsky, Vladislav Volkov, and Viktor Patsayev would live and work aboard the station for three weeks, successfully completing a full program of scientific, astronomical, and biomedical studies. On June 29, 1971, the crew strapped themselves into their *Soyuz* craft for the return to Earth, and separated from the *Salyut* station. An automated re-entry and parachute landing took place, but when recovery crews opened the spacecraft they found to their horror that all three cosmonauts had perished within seconds, when the air escaped from their cabin due to a jammed valve. All subsequent crews were once again required to wear lifesaving pressure suits. *Salyut 1* was abandoned, and would burn up on re-entry in October 1971.

Salyut 2, a military version of the space station *Almaz*, was launched in April 1973, but within two weeks it had broken up in Earth orbit. The following month, DOS-3—a true *Salyut*—was launched. However, problems on its very first orbit caused the newly launched station to waste all of its attitude control fuel. It too was abandoned, although the Soviet leaders tried to save face by describing it as a routine science satellite, *Kosmos 557*.

Salyuts 3, *4*, and *5* would support a total of five crews. Of these three, only *Salyut 4* was a genuine civilian station—*Salyuts 3* and *5* were military outposts, carrying powerful telescopic cameras and film pods that could be ejected back to Earth.

A much-modified civilian *Salyut 6* was launched in September 1977, and would host five long-duration crews over the next four years. The longest occupancy lasted six months, but ten additional short-duration visits were made by other cosmonauts, including guest crew members from Warsaw Pact countries as part of the Interkosmos program. A stripped-down, disposable, and unmanned version of the *Soyuz* spacecraft called *Progress* would also begin regular services to the station, ferrying up supplies, fuel, mail, water, and other consumables.

Due to delays in the Mir program, the back-up station for *Salyut 6* was launched into orbit in April 1982. *Salyut 7* would eventually operate for four years, playing host to 10 crews. *Salyut 7*'s final two-person crew had already visited the new orbiting *Mir* station, and on June 24, 1986, after 50 days aboard *Salyut 7*, they boarded their *Soyuz T-15* craft, detached, and a day later docked again with *Mir*.

The abandoned *Salyut 7* station would eventually be destroyed as it re-entered Earth's atmosphere over South America on February 7, 1991.

Above During its four-year orbital lifetime, the *Salyut 7* space station provided a quantum leap over previous Salyut capabilities, and would be home to numerous Soviet and international crews.

SKYLAB—A LABORATORY IN SPACE

In its nine months of operation, *Skylab* provided many complex, program-threatening engineering problems. Conversely, America's first space station would also gain renown as an orbiting laboratory of immeasurable scientific value. During *Skylab*'s 3,896 revolutions of the planet, its three visiting crews confirmed that the resources of space can offer new and unique approaches to science and technology, and the way in which we look at Earth and our universe.

THE BEGINNING OF *SKYLAB*

In 1963, the US Air Force made plans for a small orbiting station known as the Manned Orbiting Laboratory (MOL), derived from NASA's Gemini program but principally used for covert reconnaissance purposes. Seventeen military candidates were selected and began training to occupy the station, two at a time, but the station was cancelled in 1969 before there were any operational flights.

NASA created the Apollo Applications Program (AAP) in 1965, tasked with developing long-term uses for Apollo program hardware to maintain an American presence in space beyond the manned lunar landing missions. It would eventually be decided to convert a Saturn IVB second stage, once intended for use on an Apollo Earth orbital mission, to an orbital workshop configuration. Various launch options for placing the massive cylindrical station into orbit were considered, but then the cancellation of the final three manned Apollo lunar flights freed up a Saturn V booster. The fully-fitted AAP station (later renamed *Skylab*) could now be placed into orbit in its entirety.

The unmanned station, designated *Skylab 1* or SL-1, was launched from the Kennedy Space Center on May 14, 1973, atop a two-stage version of the Saturn V launch vehicle. It was inserted into a circular orbit 269 miles (430 km) above Earth. Once a stable orbit had been achieved, ground controllers waited for confirmation that the workshop's solar arrays had deployed, but it was a signal they never received. Later analyses of the launch data showed that a protective micrometeoroid shield had come loose soon after launch, and was torn apart in the supersonic air stream. Debris impact also caused the loss of a solar array needed to provide electrical power to the station. Meanwhile, the second solar panel was jammed closed by an aluminum strap and had failed to unfurl. NASA and contractor personnel now had to salvage the mission in the face of massive problems. Apart from a severe reduction in power, the loss of the meteoroid shield not only meant a loss of thermal protection for the crew members, but solar heating would soon raise the crippled station's internal temperature to an unsustainable 126°F (52°C).

The first manned mission to *Skylab* (SL-2) had been scheduled for launch the next day, but was rescheduled to allow repair equipment and techniques to be developed and tested. Following 11 days of intense activity and contingency training, the SL-2 crew was launched aboard an Apollo spacecraft on May 25, a Saturn IB rocket carrying them to an orbital link-up with the overheated space station.

Once docking had been achieved and preparations made for the first repair attempt, mission commander Charles "Pete" Conrad undocked the Apollo spacecraft, and, watchfully, flew over to the unfurled array. The craft's hatch was opened, and while scientist-astronaut Joe Kerwin clung onto his ankles to prevent him floating away, pilot Paul Weitz stood up through the open hatch and tried to manually drag the solar array panel free using a hooked pole. When this failed, Weitz tried to cut the panel straps using a modified branch lopper. Seventy-five minutes later, Conrad called off the attempt and redocked with the station, before attempting some alternative methods.

Once inside *Skylab* the following day, the crew was able to remedy the problem of the missing meteoroid shield through the novel method of pushing a specially-created parasol device through a small aperture in *Skylab* that had been designed to expose scientific samples to the space environment. Once the metallic sheet was through, the parasol apparatus was unfurled, and brought down to blanket and protect much of the exterior. The deployment operation was a success, as results soon showed.

The parasol device worked, and by the fourth day, temperatures within the station had been reduced to a comfortable working level. Three days later, Conrad and Kerwin carried out a risky four-hour EVA, during which they managed to free the remaining solar array using wire cutters. It then deployed fully, the station powered up sufficiently, and the *Skylab* project had been saved through a combination of ingenuity, planning, and sheer courage.

Above The launch of the *Skylab* space station (SL-1) from the Kennedy Space Center, USA, in 1973. Three crews would occupy the station on long-duration science missions.

Top *Skylab 2* commander Charles "Pete" Conrad (left) and pilot Paul Weitz enjoy a light-hearted moment with their spacesuits aboard the orbiting laboratory.

Opposite With the Earth as a backdrop, *Skylab* shows the effects of damage caused during its launch. A solar panel is missing, and the first crew has erected a gold-colored shield to protect and cool the station.

Right This pencil sketch of Comet Kohoutek by *Skylab 4* astronaut Edward Gibson was part of a coordinated observation effort that comprised ground-based observatories, unmanned rockets, balloons, and satellites as well as aircraft and *Skylab*.

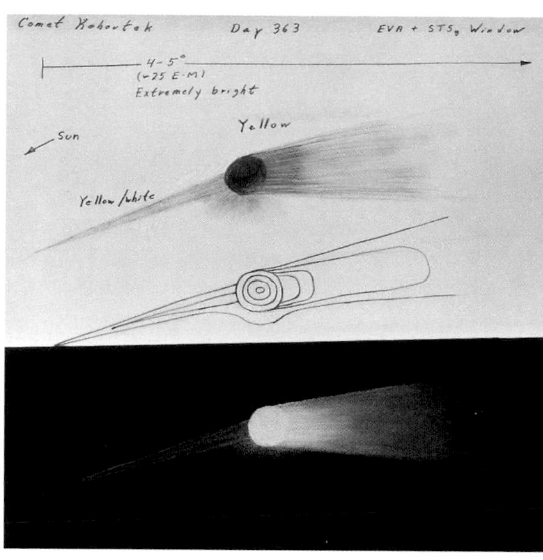

WORK IN SPACE CONTINUES

On June 22, 1973, having successfully completed their 28-day mission, the crew strapped themselves into their Apollo spacecraft ready to return to Earth. Repairing the station meant they only managed to complete around 80 percent of their assigned scientific tasks, but they had saved a multi-million dollar spacecraft, and an entire research program. They had brought the station back to life and made it safe for the next two crews to occupy, while proving that humans can live and work in space for nearly a month with only minimal adverse physical effects.

The second *Skylab* crew (SL-3) was launched to the unoccupied space station on July 28, and docked with *Skylab* the same day. Mission commander Alan Bean, together with pilot Jack Lousma and scientist–astronaut Owen Garriott, would occupy *Skylab* for a record-breaking 59 days. This crew performed, and even exceeded their assigned tasks, with such efficiency that they also made up for any experiment shortfalls from the first crew's dramatic tenancy.

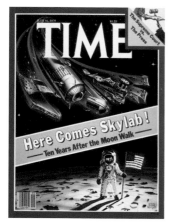

Above With the imminent re-entry of *Skylab* in 1979, there was intense worldwide speculation as to where remnants of the space station might shower down on the Earth.

After their departure, *Skylab* would remain unoccupied for 52 days before the SL-4 crew of commander Gerry Carr, pilot Bill Pogue, and scientist–astronaut Ed Gibson was launched on November 16, to continue the work on what would be the longest and most productive of all three *Skylab* crews. Among vital experiments they would conduct over the next 84 days was tracing the development of spots on the sun that can cause radio interference and have other influences on life back on Earth. In addition to completing numerous solar and Earth observations, two crew members also performed a lengthy spacewalk, during which they photographed the comet Kohoutek, then looping around the Sun, before retreating into the far reaches of the Solar System. This crew would also be called on to make a number of minor repairs to station equipment, but when faced with an impossibly demanding experiment schedule, they had to discuss the excessive workload with ground controllers, and a more realistic schedule was put in place. It was an important lesson for future space station planners.

On February 8, 1974, the final *Skylab* crew departed the station, leaving it parked in a drifting Earth orbit. They had traveled 34.3 million miles (55.2 million km), with the three crews chalking up an overall total of around 71.5 million miles (115 million km) in *Skylab*'s 171 days of human occupation. In all, the nine *Skylab* crew members performed 10 EVAs, totaling more than 42 hours, to carry out external repair and maintenance work, retrieve experiment packages and film cartridges from telescope cameras, and photograph celestial objects. They had conducted around 2,000 hours of vital experiments covering a vast and varied array of medical, scientific, and astronomical disciplines, in the meantime proving that human beings could safely adapt to living and working for protracted periods in the microgravity of space.

With the development of the space shuttle under way, it was hoped that an early shuttle flight would dock with *Skylab* and elevate it into a higher orbit, allowing for the retrieval of several exterior-mounted experiments, and a possible rehabitation by future crews. But the shuttle schedule slipped, causing these and other possible rescue plans to be abandoned as orbital decay slowly dragged *Skylab* closer to a fiery re-entry.

On July 11, 1979, *Skylab* slipped out of orbit and was incinerated over an area covering the Indian Ocean and some sparsely occupied areas of Western Australia, where scattered debris would later be recovered.

SKYLAB EXPEDITIONS

Expedition	Commander	Pilot	Scientist–Astronaut	Launch	Landing	Duration (days)
Skylab 1 (SL-1)	Unmanned	—	—	May 14, 1973	—	—
Skylab 2 (SL-2)	"Pete" Conrad	Paul Weitz	Joe Kerwin	May 25, 1973	Jun 22, 1973	28.03
Skylab 3 (SL-3)	Alan Bean	Jack Lousma	Owen Garriott	Jul 28, 1973	Sep 25, 1973	59.46
Skylab 4 (SL-4)	Gerry Carr	Bill Pogue	Ed Gibson	Nov 16, 1973	Feb 8, 1974	84.50

Right *Skylab 3* astronaut and scientist Owen Garriott conducts a spacewalk outside the *Skylab* space station to deploy a twin-pole solar shield. Dr Garriott's spacewalk lasted a record seven hours.

Above Inside *Mir*'s science module, the crews of STS-71, *Mir-18* and *Mir-19* gathered for the traditional group photograph.

Above right Once described as resembling a large dragonfly, the space station *Mir* is framed against the blueness of Earth.

A STATION CALLED *MIR*

On February 20, 1986, the core module of a mighty Soviet space station called *Mir* (meaning peace or world) was launched into orbit to begin an epic journey. Over the next 15 years it would not only be occupied by numerous Soviet and Russian expedition crews, but would play host to astronauts from several countries and space agencies, including the United States, Europe, and Japan. America's space shuttles would rendezvous and dock with *Mir*, and seven NASA astronauts would carry out long-duration, internationally co-operative missions working alongside their Russian counterparts. *Mir* truly became a significant touchstone for Russian national pride before it was de-orbited in 2001, coming to a spectacular if ignominious end over the South Pacific, as it tore a final blazing path through Earth's atmosphere.

FLAGSHIP IN SPACE

Mir's core module would provide visiting crews with living quarters, life support systems, power sources, and areas for research. Additionally, it was equipped with docking ports for manned *Soyuz-TM* spacecraft, and unmanned Progress-M supply ships. The *Soyuz-TM* craft would ferry crews and cargo to and from *Mir*, while the Progress-M craft carried water, consumables, mail, data, and equipment to the station, and was used to rid the crews of space-consuming waste material. The non-recoverable Progress ships would eventually burn up during re-entry.

In the 10 years following the launch of the core module, *Mir* would incrementally expand with the addition of six further modules. Two of the most important were *Kvant-1*, an astrophysics laboratory allowing research into the physics of active galaxies, quasars, and neutron stars, and *Kvant-2*, a scientific and airlock module, in which experiments were carried out in such fields as biology and life science, Earth observation, and the effects of exposing electronic and construction materials to raw space.

The specialized modules and other components added to the station during its operational lifetime raised its combined mass by more than 100 tons (90.72 tonnes), and its volume to around 14,125 cubic feet (400 m^3).

Mir was home to a record stay by cosmonaut Valery Polyakov, who occupied the station from January 8, 1994, to March 22, 1995, a total of 438 days. While the massive station and its crews held more records than most other space ventures combined, *Mir* was also plagued by an increasing number of near-catastrophic accidents—a life-threatening fire, major power outages, onboard computer crashes that caused the ageing station to gyrate out of control, life-support breakdowns, and basic plumbing problems. One of the worst incidents occurred in June 1997, when a remotely-controlled Progress supply ship veered off course and crashed several times into the *Spektr* science module, puncturing it and

Right Plagued by technical problems and a life-threatening fire, *Mir* was severely damaged when a remote-controlled supply ship collided with the space station, causing the damage evident in this photograph.

causing a rapid depressurization of the station. It would cause the *Spektr* module to be permanently sealed off and abandoned. Four months earlier, a serious fire had filled the station with thick choking smoke, threatening to burn a hole through the station's hull before it could be extinguished.

Operating well beyond its life expectancy, *Mir* would be plagued by glitches. NASA began urging Russia to discontinue efforts to keep the station in orbit, believing that much of the money used by the Russian Space Agency in maintaining *Mir* came from a dwindling budget already earmarked for their participation in the International Space Station.

On August 27, 1999, two cosmonauts and a French researcher closed the hatch on *Mir* and were ferried back to Earth, abandoning the station to its fate. The Russian Space Agency subsequently turned control of the station over to MirCorp, which hoped to reinvigorate *Mir* as a commercial space laboratory. They even had plans to turn the station into an orbiting hotel for space tourists prepared to pay $20 million (£10 million/15 million Euro) for the privilege of a few days in orbit. MirCorp did help fund the launch of two cosmonauts to the station in April 2000, but this would be the last manned visit. The prohibitive cost associated with keeping *Mir* in orbit eventually ended any further plans, and the Russian government finally agreed that *Mir* should be de-orbited, with any surviving segments splashing down in a remote area of the Pacific Ocean.

Below An historic moment is captured by cosmonauts aboard a Russian *Soyuz-TM* spacecraft as shuttle *Atlantis* docks with the *Mir* station for the first time on mission STS-71, June 1995.

MIR MODULES AND MAIN PURPOSE

Module	Main Purpose	Launch Vehicle	Launch Date
***Mir* core module**	Living quarters	Proton 8K82K rockets	February 20, 1986
Kvant-1	Astronomy	Proton 8K82K rocket	March 31, 1987
Kvant-2	Life support systems	Proton 8K82K rocket	November 26, 1989
Kristall	Technology, science, and astrophysics laboratory	Proton 8K82K rocket	May 31, 1990
Spektr	Experiment laboratory	Proton 8K82K rocket	May 20, 1995
Docking module	Shuttle docking port	*Atlantis*	November 12, 1995
Priroda	Remote sensing module	Proton 8K82K rocket	April 23, 1996

MIR SPECIFICATIONS

Weight	20 tons (18 tonnes) (core module), 130 tons (117 tonnes) when completed
Length	107 feet (33 m)
Width	90 feet (27 m)
Orbital altitude	145 miles (233.5 km)
Orbital velocity	17,500 mph (28,163 km/h)

Right This photograph of *Mir* space station was taken by astronauts aboard shuttle *Atlantis* as it closed in for a docking with *Mir*. *Atlantis* delivered water, supplies, and equipment, including two new solar arrays to upgrade *Mir*.

END OF A DREAM

When *Mir* was finally de-orbited on March 23, 2001, the station had exceeded its life expectancy by a decade. It had provided a temporary home for scores of cosmonauts, astronauts, and visitors from other space agencies, who had performed a total of 78 spacewalks. Overall, more than 16,500 science experiments were conducted on *Mir*, including those designed to prepare for future technologies and procedures such as space tether and laser communications systems, and the construction of future space stations. In its 15-year operational life, *Mir* had completed 89,067 orbits, covering 2,260,840,632 miles (3,638,470,307 km), roughly three times the distance from the Earth to Saturn.

The de-orbit began with the launch of an unmanned Progress M1-5 cargo ship on January 24, carrying additional fuel for the braking procedure several weeks later. At the designated time, three firings of the Progress propulsion system slowed *Mir*, at the same time altering its orbit from a circular to an elliptical circumnavigation of Earth. A final firing then decreased *Mir*'s speed by 56.8 feet (17.3 m) per second, sufficient to lower it into the upper fringes of the atmosphere.

FREEDOM, *ALPHA*, AND THE INTERNATIONAL SPACE STATION (ISS)

It is a project so vast and complex that no single nation could tackle it alone. Sixteen nations have been involved in the creation of the International Space Station (ISS), a massive craft that became our newest star in the night sky; growing increasingly brighter as additional modules, solar arrays, and other components were delivered and attached over several years and several missions. With assembly planned for completion in 2010, the ISS is a long-term laboratory in space, established in a realm where gravity, temperature, and pressure can be manipulated to assist in performing long-duration research in the medical, materials, and life sciences areas, while allowing vital studies and experiments that could never be achieved in ground based-laboratories. It will also serve as the first step to future space exploration.

A WORLD-CLASS ORBITING LABORATORY

On January 25, 1984, during his annual State of the Union address to a joint session of Congress, President Ronald Reagan announced that he was directing NASA to develop a permanently manned space station within the decade. He then added: "We want our friends to help us meet these challenges and share in their benefits. NASA will invite other countries to participate so we can strengthen peace, build prosperity, and expand freedom for all who share our goals."

By the middle of 1985, Canada, Japan, and Europe had accepted in principle the president's invitation, thus initiating the most extensive experiment in international cooperation ever undertaken. NASA's administrator at that time, James Beggs, would refer to the creation of the space station as "the next logical step" in space exploration, and future shuttle manifests would include spacewalks to test and demonstrate space station construction techniques. The proposed American space station would be named *Freedom*.

Between 1984 and 1993, however, *Freedom* suffered seven major cost-cutting redesigns, forced by initial underestimates of the project, a lack of clear direction, and a US Congress no longer willing to appropriate additional funding for the project. This not only meant lost capacity, fewer facilities, lesser crew numbers and capabilities for the station, but caused considerable unease among the international partners. By June 1993, unsustainable cost over-runs resulted in a bill to terminate the entire program being introduced in the House of Representatives. It failed by a single vote.

That October, members of the Russian Space Agency held urgent talks with NASA officials, including then administrator Dan Goldin, reaching an agreement to plan and create an international space station, while inviting active participation from other countries. This agreement effectively ended any plans for space station *Freedom*. International partners from four continents who would become involved in the ISS were Belgium, Brazil, Canada, Denmark, France, Germany, Italy, Japan, the Netherlands, Norway, Russia, Spain, Sweden, Switzerland, the United Kingdom, and the United States. NASA unofficially gave the new scaled-back international station a provisional designation of *Alpha*, but the lackluster name has never really caught on.

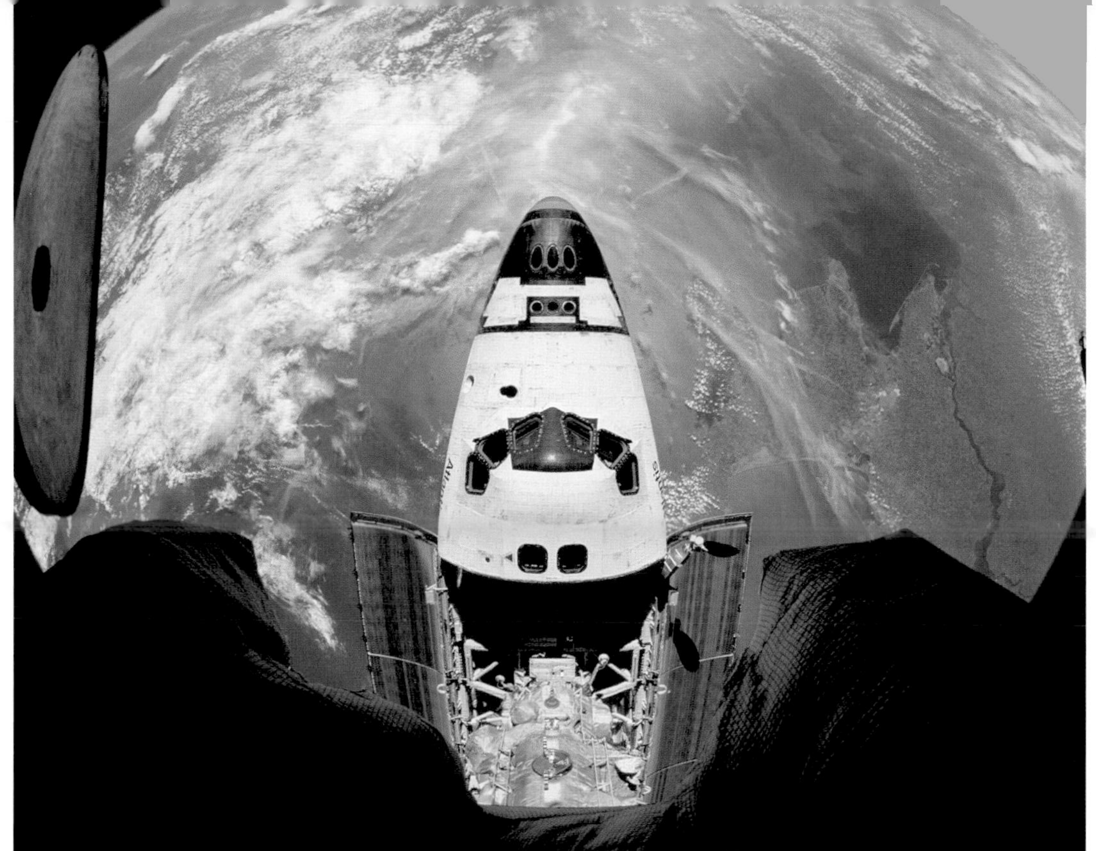

Left A fish-eye lens allows a unique view of the space shuttle *Atlantis* docking with the *Mir* space station. The space shuttle fleet supplied and serviced *Mir* during 11 flights until the station was deliberately de-orbited on March 23, 2001, breaking apart during atmospheric re-entry over the South Pacific Ocean.

Below Below, an artist's concept of the cancelled *Freedom* space station. Lack of direction and funding led to its demise in favor of the successful International Space Station.

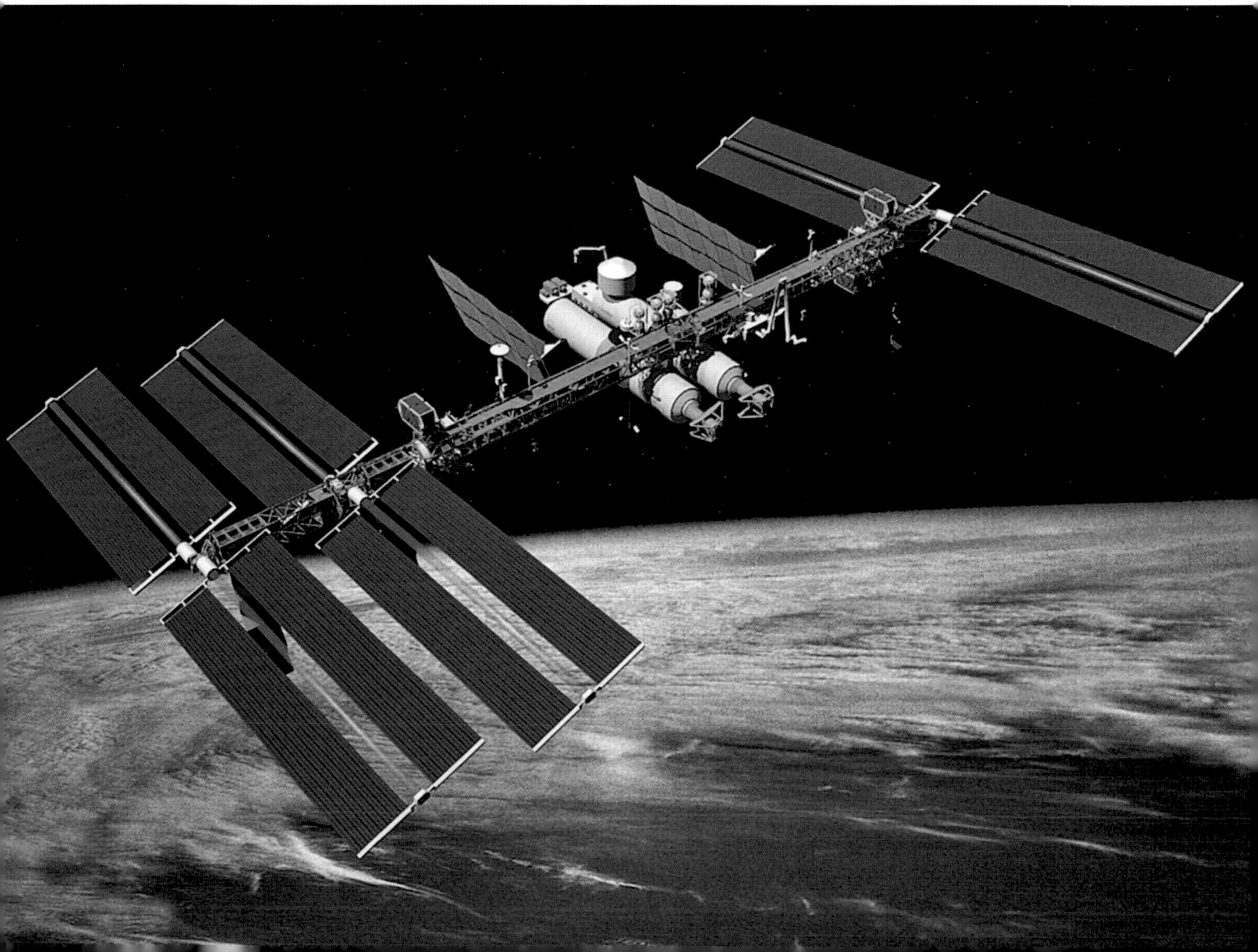

Right During construction work on the International Space Station, mission specialist Christer Fuglesang takes a spacewalk on day six of *Discovery*'s December 2006 mission.

THE ISS AND THE FUTURE

The next planned generation of American manned spacecraft scheduled to visit the ISS will be those in the Orion series, formerly known as the Crew Exploration Vehicle (CEV). Similar in shape to an Apollo Command Module, the much roomier and reusable Orion spacecraft is being designed for launch atop an Ares rocket (also undergoing development), with the first manned mission planned for no later than 2014. The powerful propulsion system on Orion spacecraft can also be used to regularly boost the ISS into a higher orbit, while a future variant of the Orion/Ares combination will one day carry astronauts on to the moon.

In the interim period, the reliable Russian Soyuz spacecraft will continue ferrying crews and supplies to the ISS.

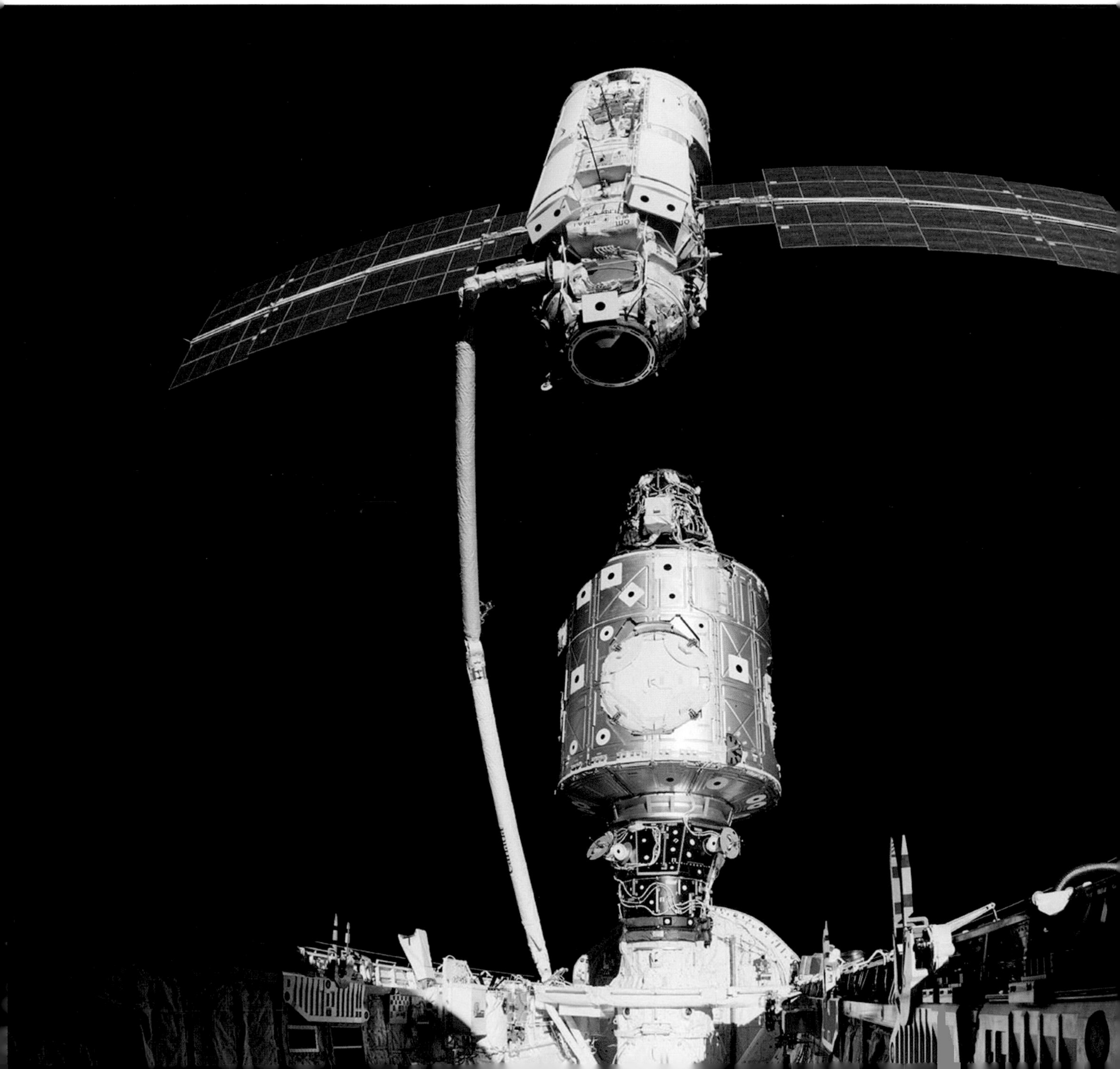

BUILDING A HOME IN SPACE

Construction of the ISS began with the launch of the American-owned but Russian-built *Zarya* control module from the Baikonur cosmodrome in Kazakhstan on November 20, 1998. Two weeks later, shuttle *Endeavour* lifted off from the Kennedy Space Center carrying the *Unity* connecting module, which spacewalking astronauts would connect to *Zarya*.

Shuttle *Discovery* flew to the ISS in June 1999, carrying tools and cranes to aid in the assembly process. In May 2000, this would be followed by a visit from shuttle *Endeavour*, whose crew carried out crucial maintenance work in preparation for the arrival of the *Zvezda* Service Module—the third major component of the station. Russia had constructed the *Zvezda* module, which would provide living quarters for the crews soon to inhabit the ISS, and it was launched to a successful link-up with the ISS on June 25, 2000. Three months later, shuttle *Atlantis* brought further supplies and a crew that would prepare *Zvezda* for occupancy by the first long-duration crew, or Expedition 1.

Another docking flight would carry further components, and then, on November 2, 2000, the Expedition 1 crew—one American astronaut and two Russian cosmonauts—arrived aboard a Soyuz spacecraft, ready to begin a full-time occupancy of the station by successive expedition crews.

With the loss of shuttle *Columbia* and its crew, construction of the ISS was abruptly halted in February 2003, although *Soyuz* flights to the station would continue in order to complete regular crew changeovers. Shuttle flights to *Alpha* would resume in July 2005, when the "return-to-flight" mission of STS-114 delivered hardware and supplies, but it would be another year before crews would recommence active construction of the station.

NASA is committed to completing the ISS in 2010, at which time the space shuttle fleet is scheduled to be compulsorily retired. The completed ISS—around four times the size of Russia's *Mir* station and five times larger than *Skylab*—will measure 356 feet (108.5 m) across by 192 feet (58.5 m) long, and 100 feet (30.5 m) high. It will maintain an internal pressurized volume of 34,700 cubic feet (983 m^3). The 310-foot (95 m) long integrated Truss Structure and solar arrays, with an active area of 32,528 square feet (3,022 m^2), will generate more than 80 kilowatts of electrical power for the station.

A fully assembled ISS could house rotating international crews of six or seven in an area roughly equivalent to the interior of a 747 jet liner. More than 100 station elements comprising structures, equipment, and supplies will have been delivered on 45 space shuttle missions, Soyuz and Progress spacecraft, and atop Russian Zenit and Proton rockets, while multiple assembly and maintenance spacewalks will have been carried out by successive crew members.

With five laboratory modules sponsored by participating international partners, the ISS will have sufficient room and electrical power for occupying crews to conduct a vast amount of scientific research and experimentation, and to support the Vision for Space Exploration announced by US President George W. Bush in 2004.

Above A Russian Progress cargo ship, carrying everything from spare parts for an exercise treadmill to oxygen for the astronauts aboard the ISS, lifts off from Baikonur, Kazakhstan.

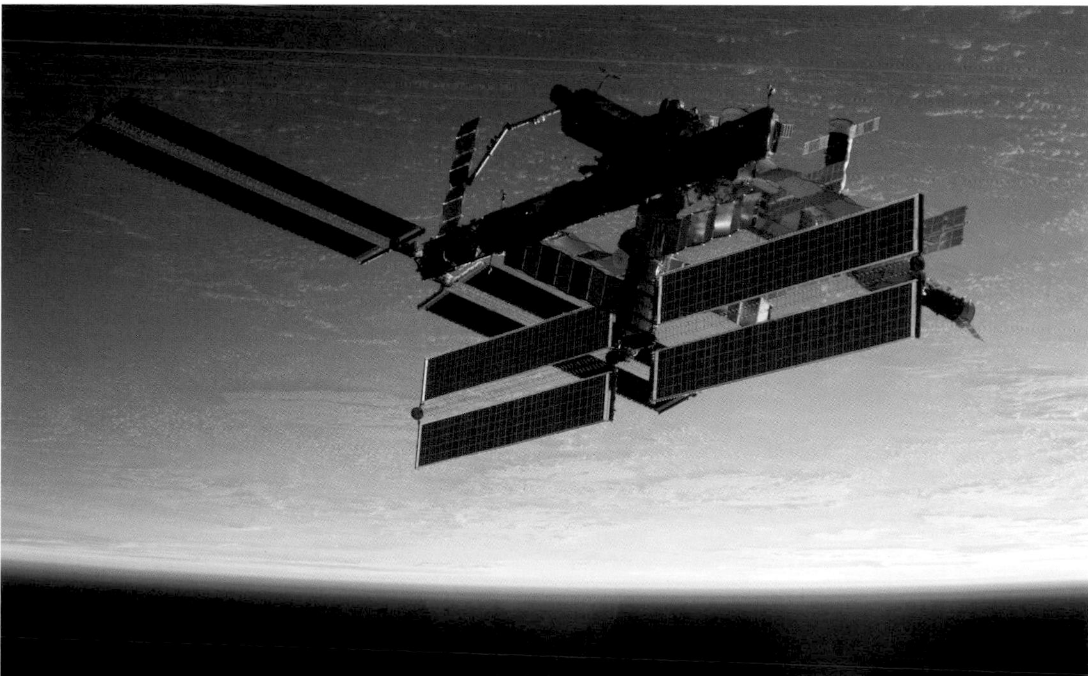

Left At the beginning of construction work on the International Space Station, this photograph shows the joining of the US-built *Unity* node to the Russian-built *Zarya* module.

Above The International Space Station was photographed by a crew member aboard *Discovery* on the STS-114 mission, following the shuttle's undocking on August 6, 2005.

Space Shuttles

For NASA, the notion of a reusable winged space shuttle was born in the midst of the Apollo moon-landing euphoria. The space agency was eager to press ahead and create a new generation of spacecraft, but was faced with crippling cuts in funding. As NASA's Congressionally allocated budget tightened and public enchantment with lunar exploration rapidly dwindled, the space agency took the unpalatable step of scrapping the final three manned Apollo missions in order to apply these savings to future space ventures.

Right *Columbia* (seen here on its second flight) was NASA's first space shuttle. Its maiden voyage took place almost a decade after President Nixon approved the development of a new kind of spacecraft.

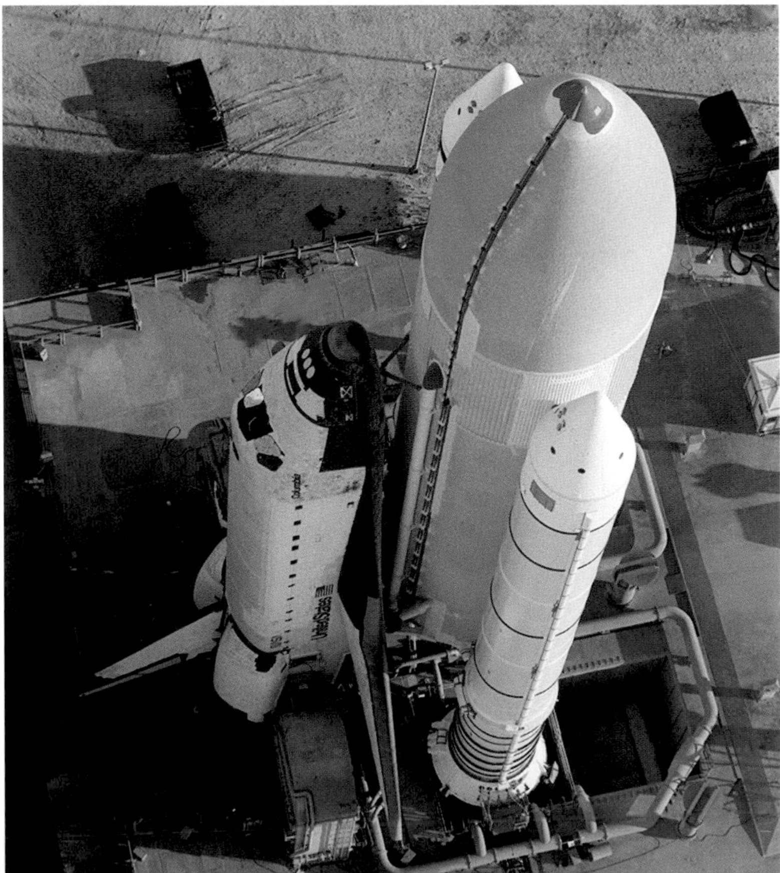

Below In readiness for the STS-90 mission, *Columbia* is rolled out to Launch Pad 39B at the Kennedy Space Center on March 23, 1998. The space shuttle is set to perform yet more scientific investigations in space.

Despite numerous development and budgetary problems, and a flawed promise of reliable, inexpensive, and frequent access to orbit, the space shuttle would usher in a new era of space travel. Tragically, however, it would also be involved in catastrophic accidents that resulted in the loss of two orbiters as well as the lives of 14 astronauts.

A WINGED SPACECRAFT

Against a need to look beyond the three-man single-use spacecraft that would service the Apollo, Skylab, and ASTP manned missions, NASA's management also faced massive problems in developing an entirely new and reusable spacecraft, or orbiter, with severely limited funding as well as a greatly diminished workforce.

Congress, while churning billions of taxpayer dollars into an unpopular war in Vietnam, found it difficult to support a space program whose purpose remained obscure. Meanwhile a pragmatic President Richard Nixon wanted to relegate the entire space program to a lesser role in what his administration called "realigning national priorities."

Initially, NASA considered using a piloted rocket that would carry the orbiter and its crew into space before disengaging and returning to a runway landing. When this option was dismissed as unworkable, it was decided that the orbiter would be the only piloted component of the launch assembly. Instead a brace of solid-fuel rocket boosters, or SRBs, would provide a combined 3.3 million lbs (14,679 million newtons) of thrust to help lift the shuttle from the launch pad. With all their fuel expended, these boosters would be jettisoned and parachute down to an ocean splashdown, where they would be recovered and refurbished for use on later launches.

The plan was that the orbiter would be mounted on a large external fuel tank positioned between the SRBs. This tank would supply liquid oxygen and liquid hydrogen propellant to the shuttle's three main engines, aiding in the massive thrust needed to lift the shuttle assembly off the ground. The fuel tank would also be jettisoned once all the propellant had been consumed, or a little over eight minutes into the launch. However, unlike the rest of the launch assembly the tank was expendable and would not be refurbished.

On January 5, 1972, after preliminary plans had been submitted by NASA, President Nixon announced that the development of America's next generation of spacecraft would proceed. In his speech he spoke of transforming the space frontier into familiar territory and making transportation into space a routine venture. He further predicted that the space shuttle, or Space Transportation System (STS), would "take the astronomical cost out of astronautics."

In July that year, NASA made North American Rockwell the prime contractor for the orbiter and integrator for the entire launch system. Plans then called for a total of five orbiters to be developed and manufactured over six years. Unpowered glide tests would be carried out in 1976, with the first manned orbital test flight slated for 1978.

Right NASA saw the future of the American space program in a reusable winged space shuttle. This satellite image is just one of countless pictures that document the grand adventure of space travel.

AN ORBITER TAKES SHAPE

The stubby delta-winged shuttle orbiter was destined to become the most complex vehicle ever designed and constructed, incorporating 49 rocket engines, 23 communications, radar and data-link antennae, five computers, separate sets of controls for maneuvering in two elements—space and the atmosphere—and fuel cells to produce electricity.

Each orbiter would be more than 122 feet (37 m) long, 78 feet (24 m) wide, and 57 feet (17 m) high. One crucial feature was a 60-foot (18 m) long payload bay with outward-opening doors, a huge area capable of transporting up to 65,000 lbs (29,500 kg) of cargo, experiments, or several satellites to and from space. Mounted along one side of the cargo bay would be a remote-controlled articulated arm for the deployment and retrieval of satellites. Spacewalking astronauts would gain access to the unpressurized cargo bay by means of a central air lock.

Another innovation was in the area of thermal protection. Previously, ablative heat shields had protected American spacecraft during the fiery heat of ballistic re-entry, but with the size and configuration of the shuttle this was no longer an option. Instead, more than 32,000 silica thermal insulation tiles would be produced, each no more than 6 inches (15 cm) square and purpose-made to match the contours of the hull and the wings. Each tile would be meticulously glued to the shuttle's vulnerable exterior. Key areas requiring extra insulation, such as the nose and underbelly of the orbiter, would be protected by black thermal tiles that could withstand temperatures of around 2,170°F (1,188°C), while reinforced carbon composite tiles would shield the leading edges of the orbiter's stubby wings. Less critical areas would be covered with thousands of heat-resistant white tiles.

While earlier American spacecraft had splashed down in the ocean at the end of each mission, the reusable orbiters were specifically designed to land on lengthy runways. Unlike commercial jetliners, however, the orbiter did not carry a fuel supply, and would return as an unpowered glider. This meant the crew would be committed to a first-time landing.

Right During a test landing in 1977, on the Edwards Air Force Base concrete runway, a problem with *Enterprise*'s flight control system was discovered. The potentially dangerous error was corrected using other NASA aircraft prior to the first orbital flight of a space shuttle.

At the end of its orbital mission the shuttle's descent would begin like that of a normal spacecraft, with the firing of the main engines slowing the vehicle's momentum and pitching it slowly out of orbit into a long shallow descent through the Earth's atmosphere at around 13,000 mph (21,000 km/h). Although the option of a fully automatic landing was available, at 40,000 feet (12,000 m) the shuttle commander would choose to assume control of the craft and manually guide it to a landing. The final approach angle would exceed 20 degrees, more than ten times steeper than a commercial jetliner, leading some astronauts to playfully compare the flight characteristics of the shuttle to that of a brick. The shuttle would touch down at a speed of around 200 mph (322 km/h).

NASA's optimism ran high, and it was announced that the shuttle was destined to become the workhorse of America's space program over the next two decades, carrying astronauts and payloads into orbit on approximately 40 missions per year.

Left *Enterprise* sits piggyback style atop NASA 905, NASA's first Shuttle Carrier Aircraft (SCA), during one of the Approach and Landing Tests carried out on the orbiter prototype. Scientists studied the aerodynamics of the "mated" space vehicles.

Below The orbiter separates from its SCA in a free-flying mission. This second phase of tests was designed to investigate *Enterprise*'s glide and landing characteristics.

A SPACECRAFT NAMED *ENTERPRISE*

On September 17, 1976, the first orbiter vehicle (officially OV-101) rolled out of Rockwell's assembly building in California, USA, ready to begin a series of glide flight drop tests (known as Approach and Landing Tests or ALTs) the following year. Originally, this orbiter was to be called *Constitution*, but following a spirited letter-writing campaign by fans of the TV series *Star Trek*, it was decided to change the spacecraft's name to *Enterprise*, after the telegenic star ship.

The glide tests would be conducted at Edwards Air Force Base in California's Mojave Desert; the 75-ton (68 tonne) orbiter would be mounted on top of a specially modified 747 in order to check the spacecraft's aerodynamic characteristics.

The first in a series of captive test flights took place on February 18, 1977, with the unmanned shuttle's main engines shrouded in a drag-reducing fairing. In the second phase of these tests, the free flight of the craft, two pilots would be on board *Enterprise*. As the 747 reached 22,800 feet (6,900 m) above the California desert, explosive bolts separated the piggybacking orbiter from the carrier aircraft, which then banked away in a slight dive. The two pilot-astronauts, Fred Haise and Gordon Fullerton, then flew *Enterprise* to an unpowered touchdown five and a half minutes later at Edwards Air Force Base. Altogether, four pilot-astronauts would be involved in these five flights.

The last of the manned free-flying test flights took place that October, minus the tail fairing. With a first launch then expected in June 1979, veteran astronaut John Young and rookie pilot Bob Crippen were named as the crew to take Orbiter 102 (*Columbia*) on the maiden orbital flight of a space shuttle.

Originally *Enterprise* was to have been completely overhauled for orbital operations; however, NASA would later scrap these plans. There would be just four operational orbiters in the original shuttle fleet—*Columbia*, *Challenger*, *Atlantis*, and *Discovery*.

THE SHUTTLE FLEET

In total, there have been six American space shuttles:
Enterprise: Test vehicle; no orbital flights
Columbia: First flight April 12, 1981; lost February 1, 2003
Challenger: First flight November 11, 1982; lost January 28, 1986
Discovery: First flight August 30, 1984
Atlantis: First flight August 8, 1985
Endeavour: First flight May 2, 1992

Below The space shuttle *Columbia* sits on its launch pad at the Kennedy Space Center. After every mission, each thermal insulation tile on the shuttle's exterior is re-waterproofed: A solution is injected into a tiny hole in the center of each tile.

A NEW BREED OF ASTRONAUT

The popular concept of NASA's all-American male astronauts had been shaped by the early Mercury and Gemini flights, and fully molded by the time of the Apollo moon landings. But with the advent of the space shuttle, and the advantages it offered for conducting science in orbit, a new generation of shuttle astronauts was selected by NASA in August 1978. Pilot-astronauts would be assigned to fly the innovative spacecraft, but there would also be a new category of space explorers known as Mission Specialists. For the first time, women would not be precluded from becoming NASA astronauts in this category, resulting in six of the 35-strong group joining as its first female trainees, or ASCANs (astronaut candidates).

COLUMBIA—FIRST SPACE SHUTTLE

The maiden voyage of space shuttle *Columbia* began from the Kennedy Space Center just after 7:00 A.M. EST on April 12, 1981. Considerable risk was involved because never before had a crew-rated American spacecraft carried crew members on its first flight into space. Coincidentally, on that same date 20 years earlier, cosmonaut Yuri Gagarin became the world's first space traveler.

Incredible progress was made in those first two decades of human spaceflight, but *Columbia* was far more than a winged spacecraft of astonishing complexity. That first shuttle flight into space, despite being two years behind schedule, provided a much-needed affirmation of America's technological prowess at a time when the nation's citizens were questioning that very capability. The success of *Columbia*'s 37-orbit mission finally gave Americans something to cheer about. As a

NUMBERING THE SHUTTLE FLIGHTS

At first, each shuttle flight was given a simple numeric designation. This would continue until STS-9, when NASA introduced a new method of identifying shuttle missions.

Under this system, STS-9 had a second designation of STS 41-A. The first number (4) indicated the NASA fiscal year in which the mission was scheduled to fly, with each fiscal year beginning in October. The next number (1) indicated the launch site, in this case, the Kennedy Space Center (KSC). Prior to the tragic loss of *Challenger* in 1986, a second launch facility was being prepared at Vandenberg Air Force Base in California, which would have been launch site 2. Construction of this site was later abandoned due to persistent technical problems. The letter (A) indicated the first scheduled flight for a particular fiscal year. Even if missions were delayed until the following year, or flew out of sequence, they would retain their flight designation.

This confusing system remained in place until the final *Challenger* mission, STS 51-L (fiscal year 1985; launch at KSC; twelfth planned flight). With the resumption of shuttle flights in September 1988, the earlier system was reintroduced, and the twenty-sixth mission was designated STS-26.

Shuttle missions may still operate out of numeric sequence, but this system will remain in place until the planned retirement of the shuttle fleet in 2010.

relieved President Ronald Reagan told Young and Crippen after their flight, "Through you, we feel as giants again."

The principal mission objectives on this test flight had been to demonstrate the safe launch into space and return of *Columbia* and its crew, and to verify the combined performance of the entire shuttle "stack;" the orbiter, SRBs, and external fuel tank. Most of *Columbia*'s orbital systems were checked, while a double opening and closing of the wing-like payload bay doors confirmed the integrity of their operation in weightlessness. The crew also noticed that several of *Columbia*'s protective heat tiles were missing, dislodged during the launch. After consultation with the ground, it was decided that they were from non-critical areas, and their absence should not affect a safe re-entry. This proved to be the case.

Above Earth provides the backdrop for this stunning image of *Columbia* in space. The shuttle was named after the sloop *Columbia*, which undertook the first American circumnavigation of the world.

Right This photograph shows *Discovery* in a dramatic liftoff. The shuttle's maiden mission carried measuring devices and sensors to record any stresses on the spacecraft, from launch to landing.

Columbia would continue to make history, flying into orbit again seven months later carrying its first commercial payload. By the end of the northern summer of 1982, four manned orbital test flights had been successfully completed and the shuttle was declared fully operational. The inclusion of Mission Specialists on subsequent crews was then approved. That November, STS-5 became the first commercial shuttle mission, which was highlighted by the deployment of two communications satellites by *Columbia*'s four-man crew.

HIGHS AND LOWS

In April 1983, a second shuttle named *Challenger* was launched into orbit. Two months later, on *Challenger*'s second flight (STS-7), Sally Ride became America's first woman in space.

NASA was soon having problems maintaining its shuttle schedule, with technical and weather problems causing numerous delays. On several missions there had been ominous warnings of problems with the shuttle and the launch

USA
NASA
Discovery

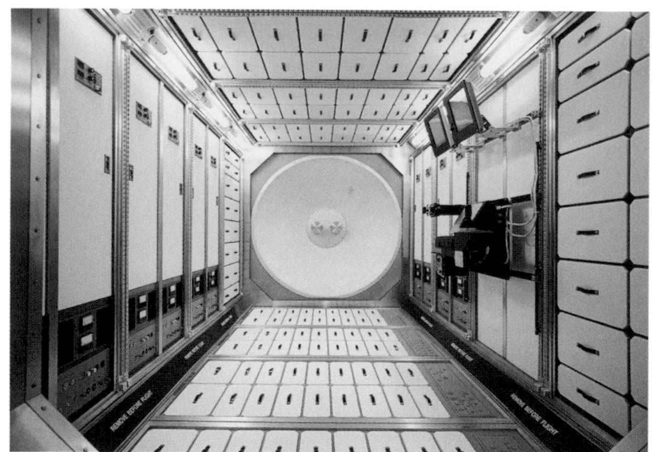

Right The payload bay is located near the center of the space shuttle. Thermal seals on the 60-foot (18 m) long payload bay doors give a relatively airtight compartment when the doors are closed and latched.

Far right Mission Specialist Bruce McCandless II uses a power tool while conducting an experiment in space. His feet are secured in mobile foot restraints.

system, and some flights returned with indications of serious malfunctions. Despite the delays, there were many mission highlights—satellites were retrieved, repaired by spacewalking astronauts, and redeployed into orbit. On mission STS 41-B, Bruce McCandless performed an untethered spacewalk, effectively becoming the first ever human satellite.

A new category of astronaut called Payload Specialist was also introduced on shuttle missions. Trained representatives of companies, the military, or other nations, they would accompany payloads into space to assist in their deployment, or conduct specific experiments in microgravity. Those who flew included industrial scientists, an oceanographer, politicians, and even a Saudi prince. By 1986, in an attempt to garner some positive publicity, NASA was pressing ahead with plans to send private citizens into space, including a teacher and a journalist. These plans came to an abrupt end with the loss of the space shuttle *Challenger* and its crew in January 1986.

Following the *Challenger* tragedy, NASA was forced to make considerable changes in its accountability and the way in which it operated. Problems with the O-rings that had led to the disaster were resolved, and shuttle flights resumed on September 29, 1988, with the launch of *Discovery* on mission STS-26. Congress also allocated funds to build a replacement orbiter, which was named *Endeavour*, in commemoration of Captain James Cook's eighteenth-century ship of exploration in the Pacific region.

Over the next few years shuttle flights into orbit became almost regular occurrences. Some notable highlights were: The deployment of the *Magellan* Venus probe from *Atlantis* in 1989; the launch of the Jupiter-bound *Galileo* spacecraft, also from *Atlantis,* in 1989; the deployment of the Hubble Space Telescope (HST) from *Discovery* in 1990; the maiden spaceflight of the replacement shuttle *Endeavour* in 1992; the dramatic repair mission to the Hubble Space Telescope in 1993; and the historic moment when Sergei Krikalev became the first Russian cosmonaut to participate in a US spaceflight in 1994.

The *Mir* space station was the culmination of the Russian space program's efforts to maintain long-duration human presence in orbit, and the first major Russian–US partnership following the fall of the Soviet Union. In 1995, the Shuttle–Mir cooperative program became operational when *Discovery* rendezvoused with the massive Russian space station, however, no docking was planned for this mission. This flight was also significant in that Eileen Collins became the first woman shuttle pilot. In June of that year, *Atlantis* docked with *Mir*, delivering supplies and beginning a program of planned crew transfers. Subsequent shuttle missions would continue these Russian–American crew transfers as a prelude to the next major collaborative venture—the International Space Station.

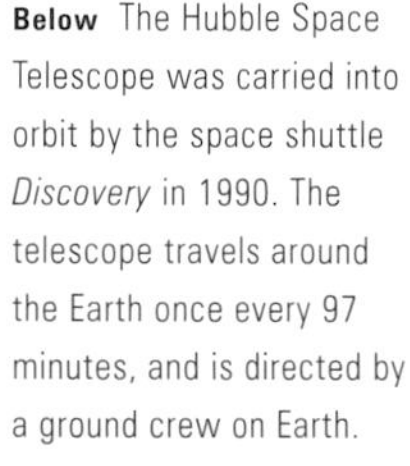

Below The Hubble Space Telescope was carried into orbit by the space shuttle *Discovery* in 1990. The telescope travels around the Earth once every 97 minutes, and is directed by a ground crew on Earth.

Opposite Bruce McCandless floats in space during the STS 41-B mission. He uses the joysticks on his Manned Manuevring Unit (MMU) for steering, or to activate nitrogen jet thrusters to propel himself in any direction.

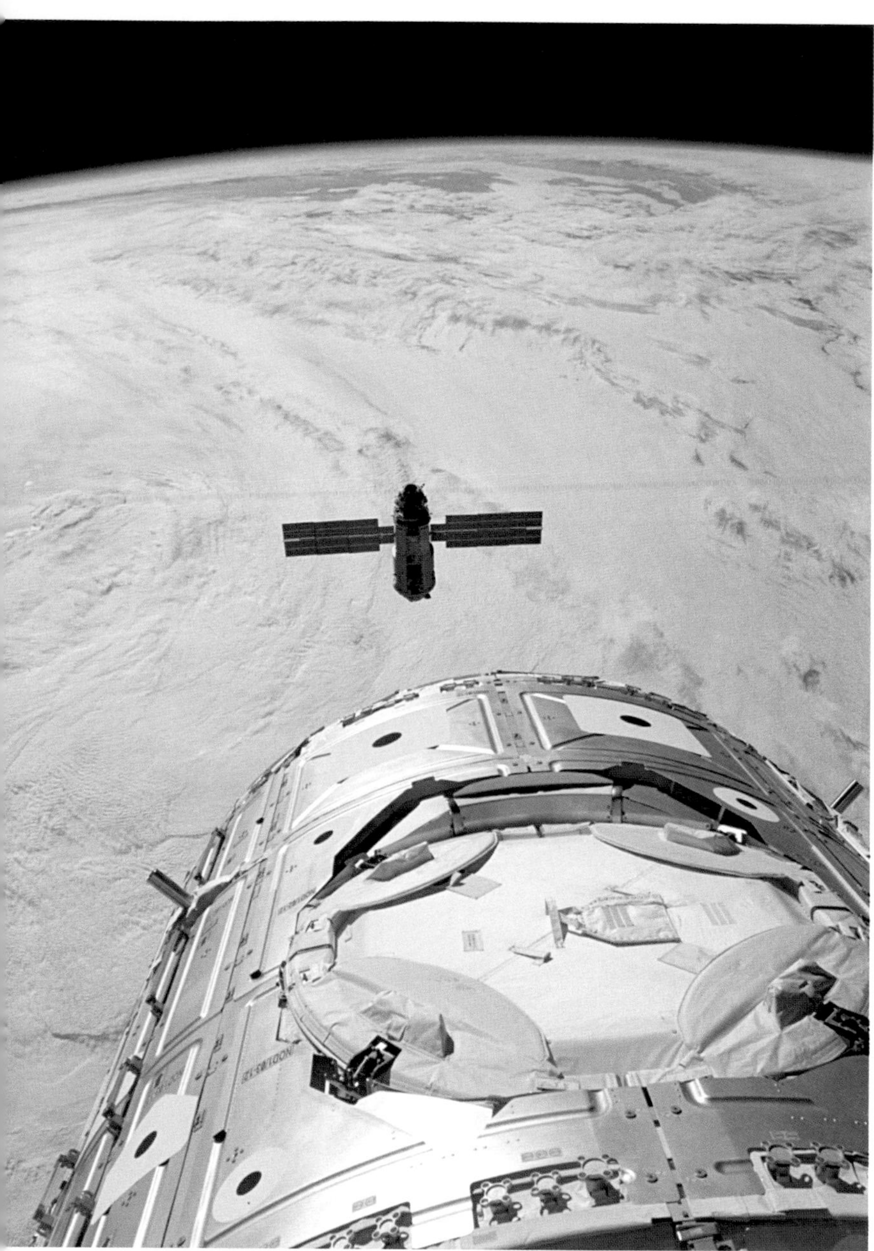

Above The Russian-built, US-funded *Zarya* module nears the US spacecraft *Endeavour* with Node 1 *(Unity)* in the foreground. Inside *Endeavour*, the STS-88 mission crew are preparing the Remote Manipulator System to capture *Zarya*.

THE INTERNATIONAL SPACE STATION (ISS)

One high-profile shuttle flight took place in October 1998, when former Mercury astronaut John Glenn was launched as part of a seven-person crew aboard shuttle *Discovery*. Thirty-six years had elapsed between his two flights into orbit. Glenn's flight was followed just five weeks later by the first flight of a shuttle to the International Space Station. Two of *Endeavour's* crew would conduct three intricate spacewalks, connecting the US-built *Unity* module to the Russian-built *Zarya* module, which had been placed in Earth orbit the previous month.

What is hoped to be the permanent human occupation of space began in March 2001, with the eighth shuttle mission to the ISS. On that flight, *Discovery* docked and picked up the Expedition One crew, launched to the station aboard a Russian Soyuz spacecraft five months earlier, and delivered the Expedition Two crew to the station.

Following the loss of *Columbia* in February 2003, the next shuttle flight (STS-114) would launch on July 26, 2005, under the command of Eileen Collins, on her final space mission. Despite strict safety measures, some foam insulation broke off the external fuel tank during liftoff, but an in-flight inspection revealed no significant damage, and the flight was successfully completed. The delivery of provisions and further construction flights to the ISS resumed on July 4 of the following year.

The shuttle program is scheduled to end in September 2010, with the ISS assembly completed, and a final servicing flight to the Hubble Space Telescope. Beyond this a new US space program is planned, in which *Orion* spacecraft will be launched aboard Ares I and Ares V launch vehicles, carrying future astronauts to the ISS, the moon, and one day beyond.

CHALLENGER'S FINAL VOYAGE

Apart from the tragic 1967 launch pad fire that claimed the lives of three Apollo astronauts, the loss of space shuttle *Challenger* and its crew of five men and two women provided the most serious setback to America's manned spaceflight program since it had begun 25 years earlier. In one terrible moment on the morning of January 28, 1986, the dream of space travel for everyone disappeared in a violent explosion 46,000 feet (14,000 m) above the Atlantic Ocean. As well as being a particularly diverse crew, including the first Jewish woman, first Hawaiian, first Japanese-descent, and second African–American astronaut, most public attention had been focused on a 37-year-old American schoolteacher, ostensibly demonstrating how far space exploration had advanced with the space shuttle. Plans were even underway for launching the next civilian shuttle passenger—a journalist.

In a sense, *Challenger's* fate was sealed by a sequence of events that conspired to stretch the launch date well beyond the original target of July 1985. The shuttle program was inexorably slipping well behind schedule, and lengthy postponements were becoming more frequent than launches. For an increasingly desperate NASA, the flight of a breezy, articulate, New Hampshire teacher named Christa McAuliffe offered a rare chance for some positive public reaction.

The late January weather at Cape Canaveral, Florida, was brutally cold on launch day; at 7:00 A.M., it was measured at 23°F (–5°C). Thick stalactites of ice were reported by prelaunch inspection teams to have formed beneath Launch Complex 39B, and deep concerns were being expressed about lifting off in such icy conditions. But that evening, President Ronald Reagan was due to give his televised State of the Union address, in which he would talk about the long-anticipated flight of a civilian teacher on board *Challenger*. Whether this factor pushed the space agency too hard in making a decision to launch that fateful morning remains unclear, but NASA allowed the countdown to continue.

Seventy-three seconds into its tenth mission, roaring into a deep azure sky, *Challenger* was suddenly engulfed in a huge fireball. The failure of a solid-fuel rocket booster O-ring to seat properly at a crucial joint (later attributed to the extreme cold at the launch pad), had allowed hot combustion gases to spew like a blowtorch from the joint onto the mounting structure of the external fuel tank. Eventually it collapsed, igniting the propellants in a massive explosion.

Great shock and worldwide mourning for the seven crew members were soon transformed into a compulsion to discover the cause of the accident. Apart from physical faults that had resulted in the catastrophe, deep flaws were found in the way NASA operated. Complacency and unsafe practices

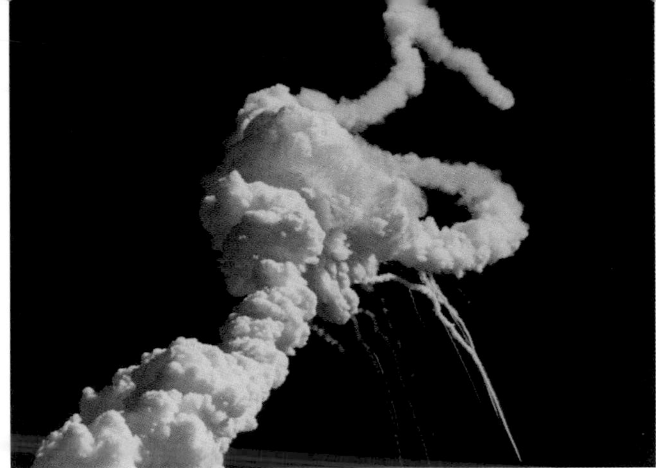

Left *Challenger*'s lost crew were (clockwise from top left): Ellison Onizuka, Christa McAuliffe, Gregory Jarvis, Judith Resnik, Ronald McNair, Dick Scobee, and Michael J. Smith.

Far left Just over a minute after launch, disaster struck the space shuttle *Challenger*. Here we see the exhaust plumes from its main engine and solid-fuel rocket booster wrapped around a ball of hot combustion gases.

were rife, and schedule-pressed managers were routinely overruling the advice of engineers. In its findings, a specially convened Presidential commission would heavily criticize NASA's management structure and safety program, and a number of changes would be implemented.

COLUMBIA'S FINAL VOYAGE

By February 1, 2003, while nearing the end of its twenty-eighth space mission (STS-107), the space shuttle *Columbia* had spent a combined total of 300 days in space, completed 4,808 orbits, and, including its final crew of seven, had transported 160 humans into orbit.

That morning, *Columbia* was bound for a landing at the Kennedy Space Center, heading east on its landing approach, and sweeping supersonically high above San Francisco. Suddenly, anomalies began to appear on ground monitors, suggesting a problem with the temperature of the hydraulic system in *Columbia*'s left wing. It was also reported that the same wing was encountering unexpected drag, or wind resistance. Six minutes later the spacecraft's commander, Rick Husband, was contacted about an alarming rise in tire temperatures within the wing. Husband began to reply, but his transmission was abruptly terminated as *Columbia* went out of control and broke up in the thickening atmosphere.

The primary cause of the tragedy, which also claimed the lives of seven astronauts, was traced to the shedding during launch of a 28-oz (800 g) chunk of foam insulation from *Columbia*'s external fuel tank. The foam struck the left wing 82 seconds after liftoff, and is thought to have left a hole or crack of indeterminate size in one of the most critical areas of the shuttle's protective surfaces. During re-entry, superheated gases seeped through this breach, melting the wing frame, which eventually caused the wing to collapse and shear off. *Columbia* was torn apart just 16 minutes from landing.

Once again, an accident review board was set up. As with the *Challenger* disaster, it would find that NASA's organizational culture had almost as much to do with the loss of *Columbia* as the foam insulation impact. It cited poor management communication, practices, and information barriers, suggesting that the agency had actually done little to address shuttle crew safety after the loss of *Challenger*. NASA in turn blamed a spiralling decline in the White House budget, which resulted in a drastic cut in its workforce as well as an excessive use of outside contractors.

Prior to its tragic loss, *Columbia* had become something of a workhorse of the shuttle fleet, hauling experiments into space on extended science missions. During a US$145 million (£75 million/110 million Euro) overhaul completed in 2001, the pioneering orbiter had finally received most of the modifications it would need to rendezvous and link up with the orbiting ISS, but that goal would never be achieved. Scheduled to make its first flight to the station late in 2003, *Columbia*'s crew for that mission had included teacher-astronaut Barbara Morgan, who 17 years earlier had served as the backup for Christa McAuliffe on her ill-fated *Challenger* mission.

Below This video still shows the piece of foam insulation (ringed in red) that fell from *Columbia*'s external fuel tank, hitting its left wing. The damage led to the shuttle's disintegration on re-entry.

Left The astronauts who died on *Columbia*'s final flight were: (back row) Ilan Ramon, Michael Anderson, and David Brown; (middle) Kalpana Chawla and Laurel Clark; (front row) William McCool and Rick Husband.

Far left Members of the *Columbia* Reconstruction Team place pieces of debris on an outline of the shuttle. NASA has loaned some pieces to The Aerospace Corp. for research purposes.

Europe in Space

Perhaps the most remarkable aspect of Europe's space activities is the coexistence of both national and multinational projects. As well as the European Space Agency (ESA), which embraces 17 European nations, most European countries have a national space research institution.

Right The *SMART-1* lunar probe was designed to provide information about what the Moon is made of, as well as searching for water ice at the south lunar pole.

Uniquely, space activities in Europe take place on three quite separate levels: Independent national programs, cooperative bilateral programs, and European collaborative ventures.

At all of these levels, European space research operates in four main fields: Access to space, space science, space applications, and manned space flight. Of these, access to space and space science are two areas that are of prime importance to European space-related interests.

Above An Ariane 5 rocket launches into space from Kourou, French Guyana, carrying a military communications satellite for the USA and Spain and a Dutch research satellite.

ACCESS TO SPACE

In 1973, after successful satellite launches by France and Britain, the Continental nations decided that Europe must have guaranteed access to space, and embarked on the development of a joint launch system. After six years of extensive tests, the first Ariane rocket lifted off from French Guyana for its maiden flight on December 24, 1979. Since then, Europe has given birth to a family of launchers.

The first Ariane generation was designed primarily to put two telecommunications satellites into orbit simultaneously, thus reducing costs. More powerful versions of Ariane developed from the original design, and in the late 1980s Ariane 4 became the workhorse of the family: Between June 1988 and February 2003, no fewer than 113 of these satellites were launched, capturing 50 percent of the market for commercial satellites. In 1986, Ariane 5 entered the market. It was designed to carry two large telecommunications satellites into orbit, to launch scientific and government payloads into medium or low orbit, and to send probes to outer space.

Ariane launchers are marketed by Arianespace, a private company whose shares are held by the industrial interests that participate in the program. This company—the first commercial space carrier—commands about half the world market for commercial telecommunications satellites. In the future, it may be that the payload capacity of Ariane 5 will rise by the end of the decade. And, after 2010, new launch systems could be developed to satisfy market requirements.

EUROPEAN SPACE SCIENCE ACHIEVEMENTS

From humble beginnings with small sounding-rockets in the very early 1950s, the nations of Europe have made many breakthroughs in the space science arena.

In the mid-1970s, Germany and the USA collaborated to launch *Helios I* and *Helios II*—space probes for measuring the Sun. Traveling at 157,995 miles per hour (252,792 km/h), these probes set a speed record for space flight.

Launched by Ariane in 1985, the space probe *Giotto* was designed to explore Comet Halley. The probe passed the comet a year later, at a distance of only 375 miles (600 km). It took spectacular pictures of the comet's nucleus, which proved to be darker than soot, and revealed that the comet was a peanut-shaped body with three gas jets spurting on its sunlit side. The data gathered showed that the comet took shape about 4.5 billion years ago, and has remained virtually unchanged ever since.

The *X-ray Multi-Mirror Newton* is an orbiting X-ray observatory, named in honor of Isaac Newton. It was launched by an Ariane 5 rocket on December 10, 1999, on a very eccentric orbit around Earth, with an apogee of 71,000 miles (114,000 km) and a perigee of a mere 4,000 miles (7,000 km).

INTO THE TWENTY-FIRST CENTURY

SMART-1 was part of ESA's program of Small Missions for Advanced Research in Technology. Its main objectives were to test a solar-powered plasma engine and to gather information about the origin and chemical composition of the Moon. It also mapped the lunar surface in three dimensions by taking images from several different angles. Launched in September 2003, *SMART-1* was Europe's first Moon mission.

ESA's first planetary mission, *Mars Express*, was so named because its interplanetary voyage was relatively short—it was launched when Earth and Mars were closer than ever before in history—and also because of the speed with which it was designed and built. It was launched on June 2, 2003, and entered the orbit of Mars on December 25, 2003. Its main scientific objectives were to take high-resolution photographs of the planet's surface and to perform mineralogical mapping. In November 2005, scientists reported that the data that *Mars Express* had gathered suggested the presence of subterranean water or ice on the planet.

The probe *Rosetta* was launched in 2004 to study the comet Churyumov–Gerasimenko. Scientists hope that the outcome of this mission will be to discover more about what our Solar System looked like before the planets formed. A "lander" will probe the nucleus of the comet and attempt to establish its chemical composition.

Above This is an artist's impression of the comet-bound probe *Rosetta*, sitting on the surface of a comet as it hurtles through our Solar System. *Rosetta* cost approximately one billion euros to develop and launch.

Top *Mars Express*, pictured in its orbit around Mars in this artist's impression, has been transmitting high-resolution images and mapping data about the red planet's surface since it entered orbit in late 2003.

Left The *SMART-1* lunar probe, launched from French Guyana, was the first space probe ESA has ever sent to the Moon and also ESA's first small mission for advanced research in technology, designed to demonstrate key capabilities for future missions.

Emerging Spacefaring Nations

Many of the most memorable achievements of space exploration have been undertaken by two nations—the USA and Russia—along with major contributions by European nations. However, other countries are now beginning to stake their claim on the world stage.

A number of countries around the world are currently taking up the challenge of space exploration and establishing independent space programs, including rocketing their own citizens into orbit.

Above A Chinese Long March 3-A carrier rocket leaves the launchpad.

THE CHINESE SPACE PROGRAM

When the Space Age began in the late 1950s, the USSR and China had a close political relationship. It was therefore not surprising that Chinese involvement in spaceflight began with assistance from its Communist ally. At the time, this technical cooperation was given for military reasons. China wanted its rocketry program to grow so that it could launch missiles.

Plans had to change in 1960, when relations between the two nations cooled dramatically and China pursued its rocketry research alone. The lack of allies and the concentration on missiles instead of space rockets meant it was not until April 1970 that China launched its first satellite, named *Dōng Fāng Hóng 1*, meaning "Red East 1." With a short working life of only 28 days, the satellite made atmospheric readings, but was primarily launched to study how well the satellite itself could function. It also allowed China to join the exclusive "space club" of nations that had independently launched their own satellites. While many nations had already launched satellites atop American or Soviet rockets, China was only the fifth nation to do so independently, coming two months after Japan had achieved the same objective.

China's rocketry successes were due in great part to Tsien Hsue-shen, a co-founder of California's Jet Propulsion Laboratory, who was forcibly repatriated to China during the "red scare" era of the 1950s, despite having been of immense help to the USA during World War II. Once back in the country of his birth, Tsien directed China's very successful rocket and missile programs for the next three decades.

Following its initial success, the Dōng Fāng Hóng series of satellite launches continued, with diverse objectives including weather observation, science studies, and remote sensing. By 1985, Chinese rocketry was sufficiently advanced to allow the nation to offer its rockets as a launch service for other countries, such as Pakistan and Sweden. While the Chinese also worked with a number of American satellite companies, these relationships were complicated by the US government's fear that advanced American technology would fall into Chinese hands. A US ban on specific technological exports has severely limited China's ability to launch non-Chinese satellites.

Right Colonel Fèi Jùnlóng (left) and Colonel Niè Haishèng manned the *Shenzhou 6* mission on its five-day flight.

Such limitations have not stopped the Chinese from pursuing an even more impressive program. Not content to launch only unmanned spacecraft, in 2003 they became the third nation to independently launch a person into space, and have since followed up that feat with a second manned spaceflight in 2005. Their approach has been very cautious—the *Shenzhou* spacecraft was first test-flown in 1999, with long gaps between missions. The Chinese have also been careful to build on already-proven spacecraft and spacesuit technology, particularly by incorporating design lessons learned from the Russians. This slow and careful approach may be due to the high financial cost of human spaceflight, but perhaps also because the well-paced propaganda value of the flights exceeds the scientific returns. By placing humans into orbit, China has gained a new prominence on the world stage, including a greater respect for its military missile capabilities.

Chinese space officials have stated that the country's plans for the future include a space station, and sending humans to the Moon. With such a slow, carefully paced program, it is difficult to know whether these objectives will ever be achieved. However, the propaganda value of such statements has already been immense.

Above A Chinese technician runs toward the unmanned spaceship, *Shenzhou 4*, which returned safely to Earth six days after its launch into space, opening the way for independent manned spaceflight.

Right A Long March II F rocket carrying the *Shenzhou 6* space craft sits on its launch pad at the Jiuquan Satellite Launch Center in Gansu Province.

CHINESE MANNED SPACEFLIGHT

On October 15, 2003, a Long March rocket carried Lieutenant-Colonel Yang Liwei into orbit from a launch site in China's Gobi Desert. While Chinese-born astronauts had previously flown in orbit as citizens of other countries, this marked the first time that China had sent a citizen into space. Even more impressively, the *Shenzhou 5* spacecraft he flew was a Chinese creation. The mission, which followed a series of unmanned spacecraft test flights, lasted over 21 hours and attracted enormous worldwide media attention.

China followed up this first flight with a second, on October 12, 2005. This time, the *Shenzhou 6* spacecraft carried two crew members, Colonel Fèi Jùnlóng and Colonel Niè Haishèng, on a flight that lasted over four days.

THE AMBITIOUS DARING OF THE JAPANESE

The Japanese space program began with modest rocketry experiments a decade after World War II, but in February 1970 the nation launched their first satellite, *Ohsumi*. This satellite was very basic, as the rocket that launched it was the focus of the test. More sophisticated satellites soon followed, with emphasis on X-ray astronomy and regular launches over the next few decades. Japan has also been an active partner with the American manned space program, selecting and training Japanese astronauts to fly on space shuttle missions, and building an impressively large Japanese Experiment Module, named *Kibō*, to be launched by the shuttle and attached to the growing International Space Station.

Right Japanese astronaut and space shuttle Mission Specialist Mamoru Mohri prepares for his second space shuttle voyage in 2000. In his two missions, he completed more than 300 orbits of Earth.

The first Japanese astronaut, Mamoru Mohri, launched aboard the shuttle *Endeavour* in 1992, but prior to this a Japanese cosmonaut became the first citizen of his nation to fly into space. Television newsman Toyohiro Akiyama flew as part of the Soyuz TM-11 mission to the *Mir* space station in December 1990, his seat paid for by his television network employers. Akiyama spent a week aboard the station, sending regular special reports back from space to the Japanese viewing public.

Japan's independent unmanned efforts in space exploration have grown more ambitious over the decades. Although the Americans chose not to fund a spacecraft to fly by Comet Halley in 1986, Japan sent two probes—*Sakigake* and *Suisei*.

Left The first pressurized section of the Japanese Experimental Module (JEM)—named *Kibo* ("Hope")—waits for launch at the Kennedy Space Center Space Station Processing Facility, along with the Italian-built US module Node 2.

With this expanded ambition, however, came a number of costly and embarrassing satellite and rocket failures, leading many observers to wonder if the Japanese space program was stretching itself too thinly with too little funding. For example, the *Hiten* spacecraft was launched to the Moon in 1990 and released a smaller lunar-orbiting spacecraft called *Hagoromo*, but overexpenditure on fuel and transmitter failures meant that the mission was not as successful as was hoped.

In 1998, Japan launched its first interplanetary probe, *Nozomi*, which was designed to study the atmosphere of Mars. Sadly, there were problems with changing the probe's trajectory, low fuel levels, and electrical circuit difficulties. The satellite was no longer able to reach Mars orbit, and the plans for it had to be abandoned.

The *Hayabusa* space probe, which was launched in 2003, was an extremely ambitious mission to collect a surface sample from an asteroid and return it to Earth. While the probe did manage to reach the asteroid in 2005, an attached mini-lander failed, and it is unclear whether the main probe was able to obtain a sample. The results will not be known until the probe completes its long journey back home.

It is hoped that these difficulties are mere growing pains for a space program that has undertaken a major reorganization in recent years. While it may never be able to compete with the larger budgets of other spacefaring nations, Japan has shown that it has not lacked ambition and daring in its efforts to expand scientific knowledge of the solar system, and further lunar missons are planned.

INDIA'S GROWING SELF-CONFIDENCE

There is an element of controversy about India's strong showing in global space programs. Many economists argue that developing nations such as India should concentrate more on lifting all of their citizens out of poverty before embarking on ambitious space programs. Others argue that these nations need to engage in the global marketplace, including in space efforts, to raise their overall living conditions. Against the backdrop of this debate, India has pursued an impressive space program. Their first satellite, *Aryabhata*, was launched in 1975 by the Soviet Union. Soon India was capable of launching spacecraft independently, and in July 1980 became the seventh nation to do so, launching the *Rohini 1B* satellite. Since then, a large number of Indian satellites have followed, aiding the country's development in the areas of communications, weather monitoring, and disaster warnings. A probe to the Moon is now planned, and Indian politicians have declared this as a symbol of the nation's growing scientific self-confidence.

In the field of manned spaceflight, there are tentative plans for an independent program, and India has also collaborated with the Russians and Americans. An Indian citizen, Rakesh Sharma, flew as part of a Soviet mission to the *Salyut 7* space station in 1984, spending a week there performing scientific experiments. An Indian space scientist was also scheduled to fly aboard the space shuttle in 1986, before the *Challenger* disaster caused the mission to be cancelled. Although she was an American citizen when she flew, Indian-born NASA astronaut Kalpana Chawla was also given much media attention in India, and her death in the *Columbia* accident was treated as a national tragedy. A series of Indian satellites has been named for her.

Below An unidentified person offers a floral tribute before a poster of Indian-born astronaut Kalpana Chawla, who was one of seven astronauts killed when the US space shuttle *Columbia* disintegrated in 2003.

Left The Indian Polar Satellite Launch Vehicle C7 rocket, launched in January, 2007, carried four payloads, including India's indigenous remote sensing satellite named *CARTOSAT-2*.

Robotic Explorers

Space exploration is an expensive and hazardous activity, which has made it difficult so far to extend our exploration in person beyond the relatively close Moon. However, our desire to learn more about our cosmic backyard has encouraged us to explore the Solar System by proxy through the development of automated probes, which act as our eyes and ears in traveling to distant worlds.

Below The launch of NASA's *Lunar Prospector* spacecraft from Cape Canaveral. At mission end it was deliberately crashed onto the lunar surface seeking evidence of frozen water.

These robotic explorers have revolutionized our knowledge of our nearest neighbors in space, revealing a Solar System that is profoundly different from what we thought we knew about it just 50 years ago.

CHECKING OUT THE NEIGHBORS

Almost as soon as the Space Age began, space probes were launched to the Moon and the nearer planets, yielding important information that could not be discerned from Earth-based observations. The early Soviet and American lunar probes provided vast quantities of new data about the Moon, even before the first astronauts set foot there.

Interest in lunar exploration revived during the 1990s, when research suggested that the Moon could well be a source of valuable new resources. The US probes *Clementine* (1994) and *Lunar Prospector* (1998) explored the Moon with new multispectral imaging technology in order to identify potential mineral resources and to search for evidence of frozen water. Since 2004, with the renewed US emphasis on manned lunar exploration, the Moon has

once again become a target for automated spacecraft from a number of spacefaring nations, as well as from the USA.

LIFTING THE VEIL OF VENUS

Venus was once thought to be a sister planet of Earth, but its thick cloud layer prevented direct telescopic observation of its surface: It became an early target of planetary exploration in an attempt to penetrate its veil of mystery. America's *Mariner 2* (1962) was the first successful Venus probe. Its data disproved theories about a verdant or oceanic Venus, and information from successive Mariner craft and Soviet Venera probes revealed that Venus is a hellish world of high temperatures and pressure, created by a runaway greenhouse effect.

Some highlights of exploration of Venus include the Soviet spacecraft *Venera 9* (1975) and *Venera 13* (1981), which landed on the surface, sending back images and taking soil samples, and the US spacecraft *Pioneer-Venus* (1978–1992) and *Magellan* (1989–1994), which radar-mapped 98 percent of the planet's cloud-obscured surface.

THE RED PLANET

Mars was once thought to be home to intelligent life, populated in science fiction by advanced civilizations. Even after telescopic observations proved this unlikely, it was still believed that there might be vegetation and simple life forms on the planet. But the first successful Mars probe, the US *Mariner 4* (1965), revealed a very different planet, with a cratered surface and an atmosphere that was mostly carbon dioxide. Later Mariner and Soviet "Mars" spacecraft confirmed Mars to be a much harsher environment than was expected, but the first successful landers, the US *Viking* craft (1976), still took samples of Martian soil and analyzed them in onboard laboratories for signs of life, although the results were inconclusive.

After the announcement in 1996 of possible evidence for ancient microbial life on Mars, the Red Planet again became a focus for exploration, beginning with the first Mars rover, *Sojourner*, during the 1997 Mars *Pathfinder* mission. Since then, every launch window has seen an American spacecraft sent on its way to Mars, though not always successfully. The *Spirit* and *Opportunity* rovers have been exploring the planet's surface since 2004, revolutionizing our detailed knowledge of Martian geology, while orbiters, including ESA's *Mars Express*, are finding more telling evidence for the existence of water—and perhaps life—on Mars today.

THE SUN AND MERCURY

Our other neighbors in the inner Solar System have not escaped scrutiny by robotic explorers. Spacecraft have studied the Sun to learn more about how it affects our world. Since it was launched in 1990, the *Ulysses* spacecraft, a joint NASA–ESA project, has orbited the Sun's poles to provide a completely new view of our closest star. More recently, the *Genesis* spacecraft (2001–2005) has, amazingly, trapped and returned samples of the solar wind in special collectors for study.

Until quite recently Mercury, the closest planet to the Sun, was largely a mystery, the details of its surface almost impossible to distinguish from Earth-based telescopes. But in 1974–75, *Mariner 10*, the first two-planet space probe—it also explored Venus—gave us a close-up view of this harsh and Moon-like world. In the coming decade, new spacecraft will return to Mercury to explore the planet in greater detail.

Left The *Magellan* spacecraft is nestled in the shuttle *Atlantis*'s payload bay prior to launch in 1989. *Magellan* used radar to map 98 percent of the surface of Venus.

Below left Maat Mons, a volcano standing a massive 5 miles (8 km) above the surface of Venus, as seen by *Magellan*. It is the highest volcano on Venus.

Below The *Pathfinder* spacecraft delivered the *Sojourner* robot rover to the surface of Mars in 1997. The small rover operated for 90 Martian days, covering about 2,691 square feet (250 m^2).

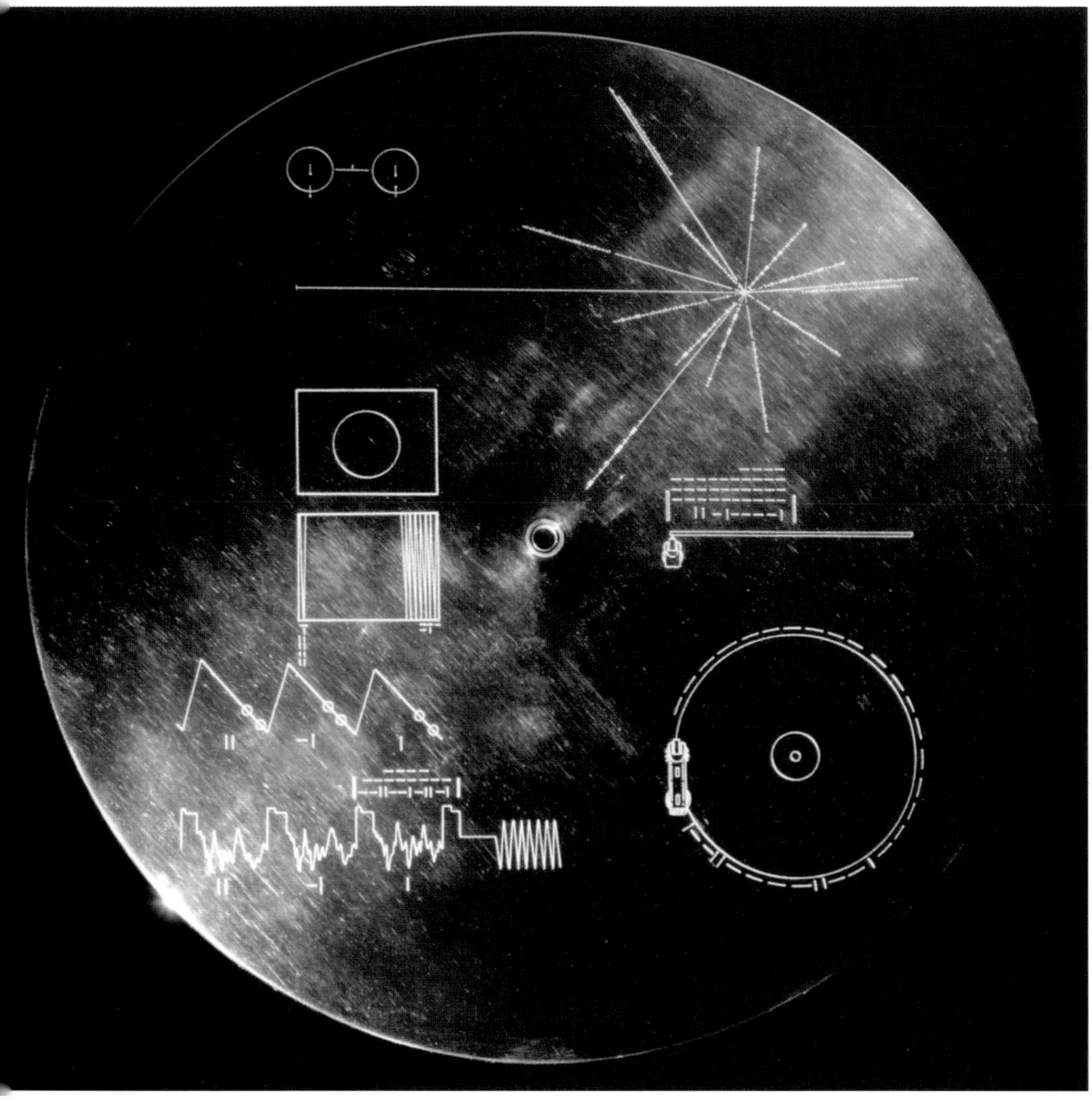

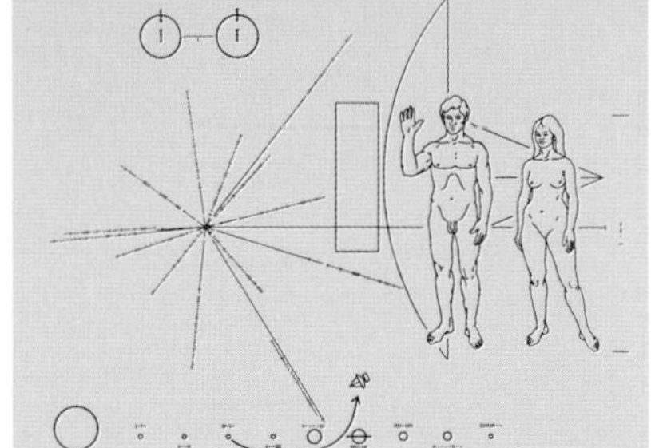

Above Plaques aboard the *Pioneer 10* and *11* craft were designed to show what planet they were launched from and by what kind of beings. The radiating lines represent the positions of 14 pulsars, arranged to locate the Sun as the home star of our civilization.

OUTWARD BOUND

The exploration of the outer Solar System needed spacecraft more advanced than those initially used to explore Mars and Venus: They would have to navigate the hazardous Asteroid Belt and function autonomously for the long periods required to reach the outer planets. It was not until 1972 that the US *Pioneer 10* spacecraft was launched toward Jupiter, followed by *Pioneer 11* in 1973. *Pioneer 11* visited both Jupiter and Saturn, allowing us to see Saturn's glorious rings up close for the first time. Both spacecraft carried a small plaque informing any extraterrestrial beings that might find them after they traveled beyond the Solar System about their origin and the people who had sent them.

By the end of the twentieth century, robotic probes had explored every planet in the outer Solar System except Pluto. And in 2006 the *New Horizons* spacecraft was launched to finally explore Pluto and its moons, and possibly other trans-Neptunian objects in the Kuiper Belt.

Left and below Mounted on the outside of the spacecraft, the "Sounds of Earth" records carried on *Voyagers 1* and *2* will provide information about the Earth and its people to any civilization that finds them. The images on the record cover include instructions for playing the disc.

A GRAND VOYAGE

After the Pioneers, NASA planned a "grand tour" of the outer Solar System to explore not only Jupiter and Saturn in greater detail but—if the spacecraft survived long enough—to continue on to Uranus and Neptune. The twin spacecraft *Voyagers 1* and *2* were launched in 1977 and reached Jupiter in 1979. Their images from Jupiter provided unprecedented detail of the turbulent Jovian atmosphere and the first close-up views of Jupiter's family of moons, revealing the larger ones as surprising worlds in their own right. Among the unexpected discoveries at Jupiter were the existence of a ring system and the discovery of active volcanoes on Io, one of the planet's moons.

At Saturn, *Voyagers 1* and *2* had different scientific goals. Arriving in 1980, *Voyager 1* flew by Saturn, its research focused on studying the moon Titan, at that time the only moon in the Solar System that was known to have an atmosphere. The trajectory needed to approach Titan meant that *Voyager 1* was then thrown out of the plane of the ecliptic. However, *Voyager 2* arrived at Saturn in 1981, and was able to conduct research on its rings and moons and then journey on to Uranus, arriving there in 1986, and thence to Neptune, which it reached in 1989.

Between them, *Voyagers 1* and *2* revolutionized our knowledge of the giant gas planets, their ring systems and moons. They shed new light on how their atmospheres operated; revealed amazing and beautiful complexities in their ring systems; found new moons; and provided detailed images of others. Both Voyagers were also fitted with a message for any future finders: a special "phonograph record" of sounds and images from Earth, protected by a message-bearing cover similar to the Pioneer plaque.

So spectacular and tantalizing were the scientific discoveries at Jupiter and Neptune that other spacecraft have since returned to explore them in even more detail. *Galileo*, launched by NASA in 1989, arrived at Jupiter in 1995 and explored the Jovian system until it was retired in 2003. *Cassini-Huygens*, a NASA–ESA joint mission launched in 1997, has been exploring the Saturn system since 2004. Its *Huygens* probe successfully landed on the moon Titan.

Left Artist's conception of a Voyager spacecraft traveling in interplanetary space. Having completed their original mission to the outer planets, *Voyagers 1* and *2* are still transmitting data as they reach the outer limits of the Solar System.

EXPLORING ASTEROIDS AND COMETS

Robotic spacecraft have also visited the smaller members of our Solar System. On its way to Jupiter, *Galileo* provided the first close-up images of asteroids, and spacecraft en route to other destinations have also photographed asteroids that they have encountered. But the first dedicated asteroid probe was *NEAR Shoemaker*, which photographed Mathilde in 1997 before entering into orbit around Eros, finally landing on its surface in 2001. Japan's *Hayabusa* probe, launched in 2003, attempted to take a sample of asteroid Itokawa in 2005, but news of its success or otherwise will have to wait until *Hayabusa* returns to Earth in 2010. Future missions to the asteroid belt are being planned, such as the *Dawn* mission to the dwarf planet Ceres.

Cometary exploration began with the armada of spacecraft launched to explore Comet Halley on its visit to the inner Solar System in 1986. The ESA's *Giotto*, the USSR's *Vega 1* and *2*, and Japan's *Suisei* and *Sakigake* all investigated the comet. Their data endorsed the generally accepted account of a comet as a "dirty snowball," and later cometary probes have confirmed the description. Launched in 1999, the *Stardust* spacecraft succeeded in collecting samples from the coma of Comet Wild 2 in 2004 and brought them back to Earth in 2006, and the European probe *Rosetta*, which is currently on its way to Comet Churyumov–Gerasimenko, will put a small lander on that comet's surface in 2014.

Below The space shuttle *Atlantis* released the *Galileo* spacecraft on its voyage to Jupiter, shown in this artist's rendering. Named for the man who first identified Jupiter's four largest moons, the craft spent eight years exploring the Jovian system.

Space Telescopes

Light from even the closest star to our Solar System has to travel for over four years to reach our eyes. In doing so, it passes through the vacuum of deep space. However, having traveled billions of miles, in the very last fraction of a second it journeys through our atmosphere. Clear points of light then lose their clarity, and twinkle due to the distorting effect of the air.

Right NASA's three Great Observatories—Hubble, Spitzer, and Chandra—joined forces to probe an expanding supernova remnant, first seen 400 years ago by the astronomer Kepler, for whom it is named.

Opposite Astronaut F. Story Musgrave, on the end of the space shuttle's Remote Manipulator System arm, services the Hubble Space Telescope, while Jeffrey A. Hoffman (in the payload bay) assists.

As Earth's population grows, light pollution from expanding cities has also been making ground-based astronomy more difficult. Astronomers had long dreamed of putting a telescope above the planet's atmosphere, to better see the stars and other distant objects.

While new discoveries in adaptive optics are making the twinkling effect easier to counter in ground-based images, Earth's atmosphere does not only distort visible light. Ultraviolet, infrared, and X-ray wavelengths are also extremely difficult to investigate from the ground. Valuable work can be done by sending scientific payloads on quick rides above the atmosphere on sounding rockets, and other atmospheric effects can be lessened by flying equipment in high-altitude aircraft, or placing ground-based telescopes atop high mountains. For the least distortion, however, the most effective observatories are currently in Earth orbit.

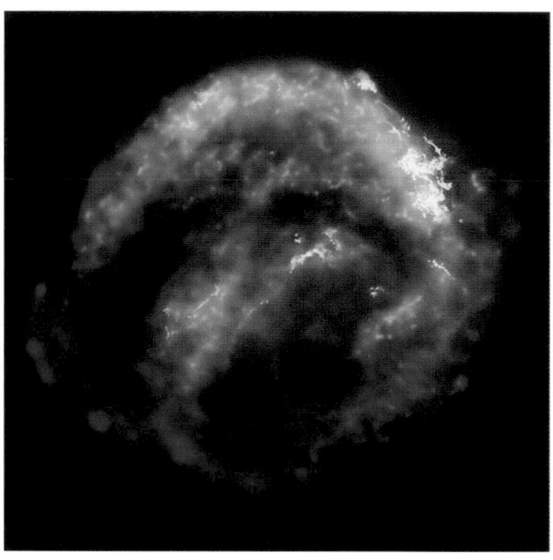

AN EYE IN SPACE

Instruments to make astronomical measurements have been launched aboard satellites almost since the beginning of the Space Age, and have grown increasingly sophisticated over time. France, the United Kingdom, Canada, the former Soviet Union, Japan, and the combined countries of the European Space Agency are just some of the nations to have sent astronomy-dedicated satellites into orbit. The COROT space telescope is one such mission—an international project led by the French to observe stars for signs of exosolar planets. The SOHO spacecraft is a joint European–American mission to study the sun. Space telescopes do not have to be large to be effective; the Canadian MOST spacecraft, which studies variation in starlight, is only 2 feet (60 cm) tall and 1 foot (30 cm) wide.

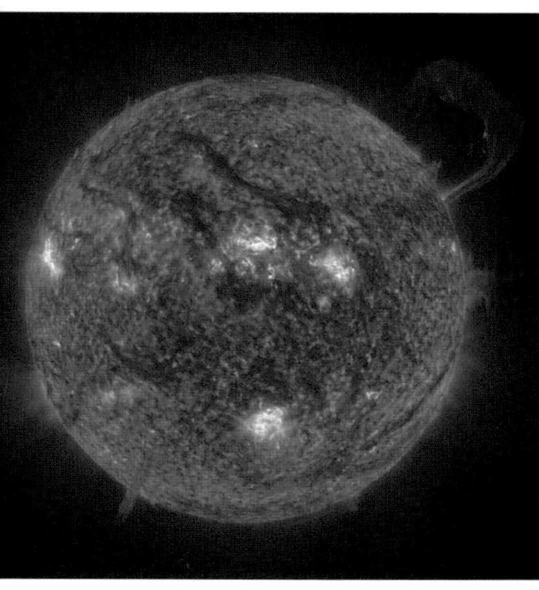

Above Launched in December, 1995, the Solar and Heliospheric Observatory (SOHO) is designed to study the sun, its extensive outer atmosphere, and the origin of the solar wind.

NASA, in collaboration with international partners, has launched a number of space-based astronomical observatories with varying scientific roles. The Spitzer Space Telescope, launched in 1993, studies the infrared range of the spectrum. Another three large observatories were launched and released by the space shuttle. The Compton Gamma-Ray Observatory, launched in 1991, studied the universe's gamma-ray emissions, while the Chandra X-Ray Observatory was sent into orbit in 1999 to advance the field of X-ray astronomy. The Hubble Space Telescope left Earth in 1990 to study a range of wavelengths including light visible to human eyes.

THE HUBBLE SPACE TELESCOPE

Undoubtedly the most famous of the orbiting observatories, the Hubble Space Telescope has made a large number of extremely important astronomical discoveries. They include a more accurate measurement of the expansion of the universe, observation of a comet impact with planet Jupiter, evidence of planets around other stars, and images of remote galaxies never previously observed. The telescope has also been an inadvertent proving ground for astronaut repair missions. Shortly after launch, it was discovered there was a tiny flaw in the shape of the primary mirror. A mission to install correcting optical equipment, and subsequent servicing missions, have been among NASA's most critical and successful spacewalking tasks. Following the second shuttle disaster in 2003, it was decided that it was too risky to continue to service Hubble. The resulting outcry from astronomers and the public meant that this decision was eventually reversed.

Astronomical observations have also been made from manned spacecraft. The Soviet Union conducted studies from their *Salyut* space stations, and the *Mir* space station had a dedicated astronomy module, *Kvant-1*, specializing in X-ray astronomy. NASA's first space shuttle mission dedicated exclusively to astronomy was *Astro-1*, launched in 1990 and carrying a number of ultraviolet and X-ray telescopes. A follow-up mission with similar equipment, *Astro-2*, was flown in 1995. The Apollo Telescope Mount, a large section of the

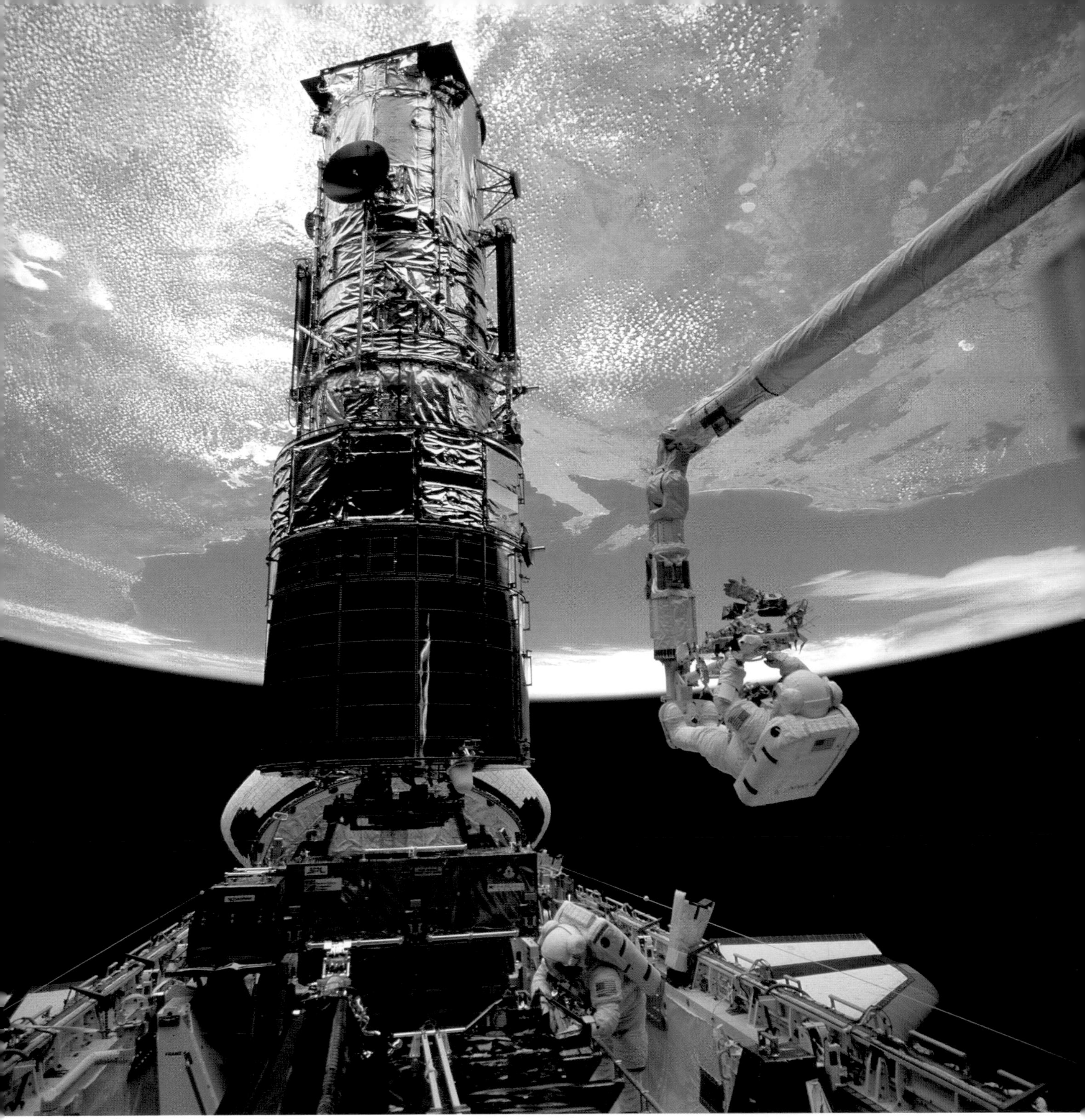

Skylab space station, also made a number of important discoveries about the Sun during its 1973–74 lifetime. Many other NASA missions have incorporated astronomy experiments; these include the first telescope to be used from the surface of another body in space. During the *Apollo 16* mission in 1972, a telescope was set up on the surface of the Moon, and images were collected of nebulae, star clouds, and Earth's atmosphere in the far-ultraviolet range.

The next generation of space telescopes is being planned. The James Webb Space Telescope is currently under development, with a larger mirror and greater observing power than Hubble, although it will only study the infrared range.

Left An artist's impression of the next generation of space telescopes: when its launch goes ahead, the James Webb Space Telescope will observe the cosmos in the infrared range of the spectrum.

Satellite Applications

It is hard to imagine a time when images of news events from the other side of the world would take days to reach us. Yet, before the age of satellites, this was the world we lived in.

In many ways, the world has grown closer because of satellite applications, which allow us to send information and ideas across the world as if we were all sitting in the same room.

While it is possible to transmit information around the globe using undersea cables, in many ways it is much easier to send and receive the information using a satellite orbiting the planet. This also allows the sender and receiver to be mobile, such as on a ship or aircraft, and they can also be on the remotest parts of our planet's surface and still stay in contact.

Some early satellites had a reflective surface, allowing signals to be bounced off their coatings, but more sophisticated satellites receive and retransmit messages electronically. Television signals, telephone calls, radio communications, and even internet services utilize their capabilities.

Satellites also allow humans to have an overview of activities on the Earth below. This capability is extremely important for weather reporting and forecasting. Rather than attempting to combine a large number of scattered reports from the ground, entire weather systems can be seen from orbit, and their direction and changes recorded. Large storms forming in remote parts of the ocean, that may otherwise have been unknown, can be constantly monitored. In this way, natural disaster alerts can be issued with much earlier warning, and loss of life averted. Hurricanes, forest fires, dust storms, and volcanic activity can be observed for short-term effects. In addition, longer-term changes such as ocean current direction, ice and snow cover, pollution, ozone depletion, and global warming can be accurately monitored over time.

Above Radio telescopes are used to receive data from satellites in orbit around other planets, such as Mars, enabling space agencies to produce images of alien landscapes such as the one opposite.

Right Hurricane Alberto (August, 2000), off the coast of Bermuda, was photographed and monitored by NASA's Sea-viewing Wide Field-of-view Satellite (SeaWiFS).

Accurate maps can be made by using imagery from space, and long-term observations can also show human and natural changes to Earth, such as the growth of cities, changes in agricultural land use, increase in the amount of city lights, and changes in natural vegetation. While some of this information could be slowly and carefully collected on the ground, a satellite vantage point in Earth orbit allows for a consistent place from which a wider picture can be constantly taken and updated. In a similar way, satellites have been sent into orbit around other planets in our solar system, such as Mars, Saturn, and Jupiter, to study cloud and surface features, and how they change over time.

RECONNAISSANCE SATELLITES

Not all observations of Earth from space are done for peaceful purposes. Earth orbit is an ideal place from which to spy on other countries. Unlike flying aircraft over foreign nations, considered a violation of their airspace if done without permission, Earth orbit is considered a more neutral area where satellites can cross over many nations every hour. With sophisticated cameras and other sensing devices, activities such as troop moments and the building of new military facilities can easily be observed from space, unless extraordinary efforts are made on the ground to hide them. At one time, photographs were taken on board satellites and then ejected in canisters to be returned to Earth, but now this information is generally returned electronically. In addition to visual analysis, some military satellites can eavesdrop on the communications of other nations, and also allow operatives in those countries to safely send out messages. There are even satellites that are specifically designed to destroy other military satellites.

While some question the use of space for military purposes, satellites are often needed to ensure that international agreements are being kept. Missile and nuclear weapons tests can be more easily monitored from space, as can other evidence of military buildup and hostile intentions. Some military satellite technology has also transferred into civilian usage. The most notable example is the Global Positioning System, or GPS. By measuring minute time differences between a number of satellites, a person holding an object as small as a cell phone can know their precise location on the planet. Still maintained by the military, this system can be used by civilians for navigation of vehicles, mining operations, surveying, mapping, and emergency service assistance. GPS can not only give a user their geographical position, but also provide their speed and direction if they are in a moving vehicle. Hikers and climbers can use GPS for safety and navigation in remote areas. Systems are even being developed to assist people who are blind with navigation assistance, creating a spoken-word map to direct them.

Left Military satellites used for reconnaissance can cross many nations every hour, observing and recording potential threats without violating any country's airspace.

Below This image of Mars's Grand Canyon (the Valles Marineris) shows the surface of Mars in high resolution. The detailed information gathered by the satellite enabled the construction of a three-dimensional rendering, shown at the bottom.

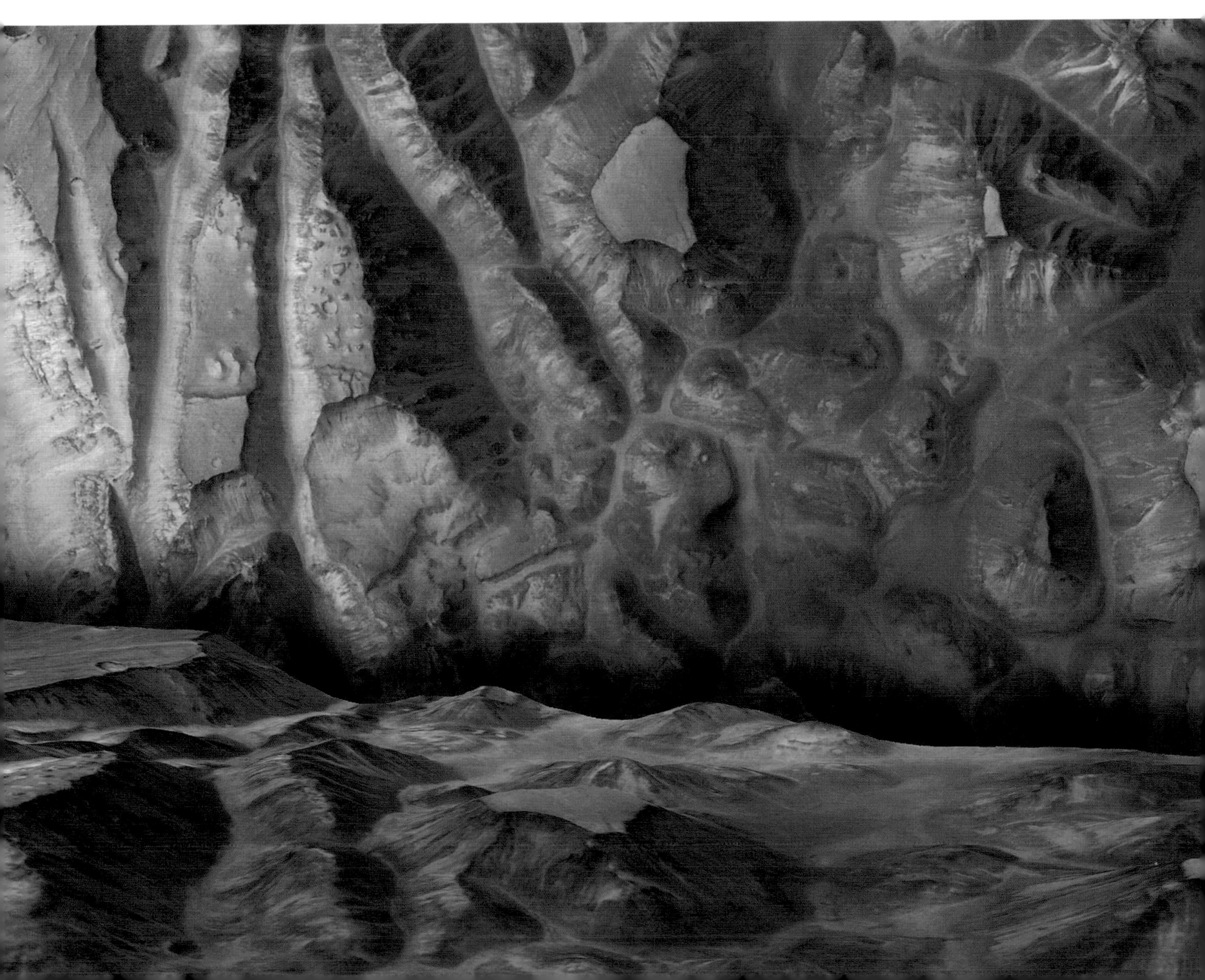

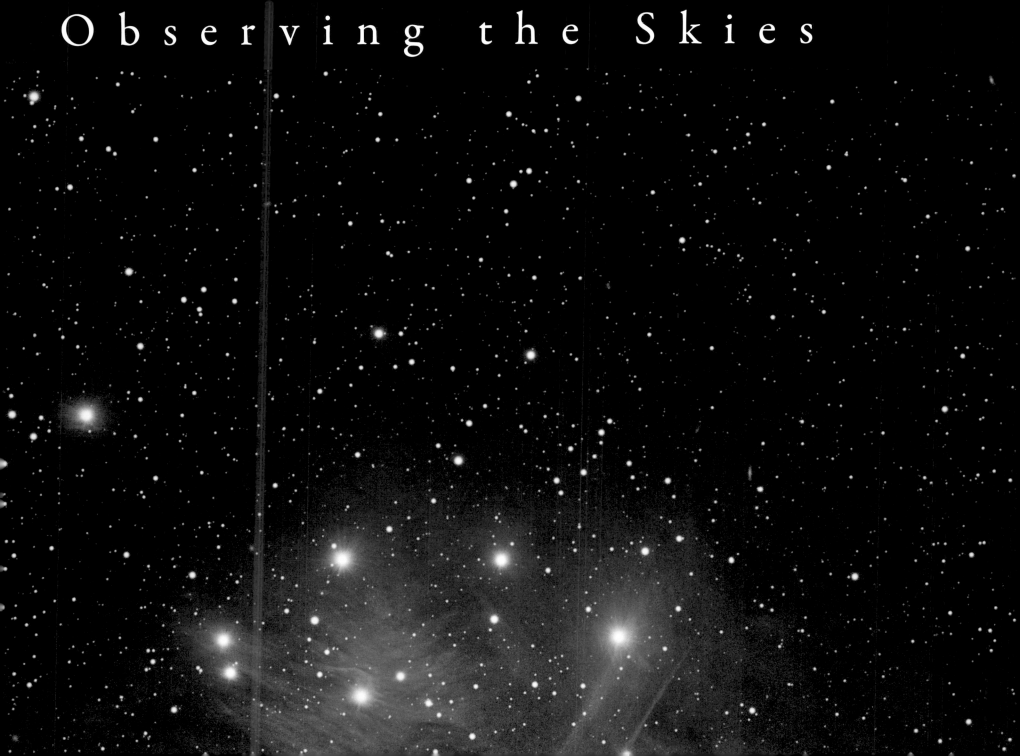

Observing the Skies

Stargazing

German-born British amateur astronomer Frederick William Herschel (1738–1822) taught music by day and observed the sky on clear nights from his home at Bath in England. Using reflecting telescopes that he built himself, he made organized sweeps across the sky while his sister Caroline recorded his descriptions of the objects that came into his field of view.

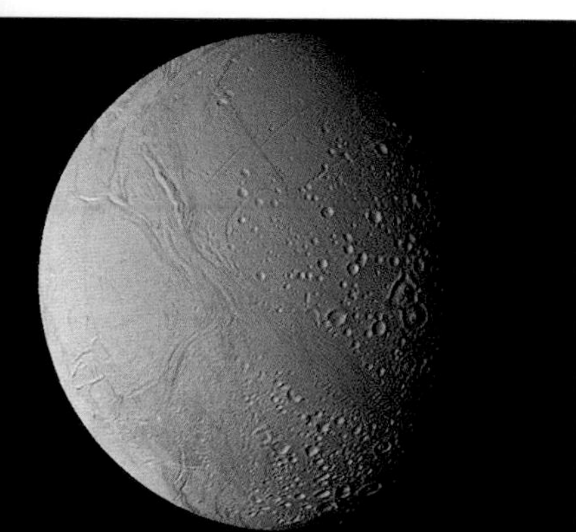

Above This *Voyager 2* mosaic of Enceladus, a moon of Saturn, was taken from 74,000 miles (119,000 km) away. This moon is about 300 miles (500 km) in diameter and has the brightest and whitest surface of any of Saturn's moons.

When Herschel began his observations, fewer than 200 deep-sky objects were known, but he discovered over 2,000 more. During a sweep in 1781, he discovered Uranus, thereby doubling the size of the known Solar System. In 1787, he discovered Uranus's two largest moons and, two years later, added Saturn's moons Mimas and Enceladus. He also discovered many double stars.

As they follow in his footsteps, modern amateur astronomers feel a special kinship with Herschel, even as they envy his opportunity to see what had never been seen before.

In the entire sky there are about 145 deep-sky objects that are true splendors. Most of these wonders are described in this chapter. The thousands of other deep-sky objects that amateurs observe are just fuzzy glows in the eyepiece, with perhaps a few details such as a brighter core. With such objects, it is understanding the true significance of what we are seeing that makes the observation worthwhile. This is the one advantage that modern amateurs have over Herschel: Knowing the nature of the quarry that we hunt down and see.

GALAXIES

Galaxies may be spirals, lenticulars, ellipticals, or irregulars. If a spiral is face-on to us, its surface brightness may be low, perhaps making it difficult to see at all. Many neophytes have trouble finding M33, for example, because there is not enough contrast in brightness between this large galaxy and their sky, especially if they are viewing from a light-polluted suburb. Yet perseverance pays off with M33. Since it is face-on, its subtle details are laid out before us. As observers gain experience in detecting low-contrast details, learn the importance of viewing from sites far from light pollution and air pollution, and gain access to larger-aperture telescopes, they will eventually be able to see not only this galaxy's spiral arms, but even emission nebulae, stellar associations, open clusters, and also a globular cluster within it.

Lenticulars and ellipticals rarely show anything beyond a brighter core and perhaps a star-like nucleus. However, seeing a giant elliptical galaxy at the center of a galaxy cluster with numerous small galaxies in orbit around it, eventually to be cannibalized by the growing giant, makes a lasting impression. One irregular galaxy, the Small Magellanic Cloud, displays a wealth of bright detail, but more distant irregulars show only a few knots.

OPEN CLUSTERS

Brighter open-star clusters are easily resolved into dozens of gems arranged in interesting patterns. Frequently the diamonds will be offset by a ruby or a topaz. Double and triple stars are common within open clusters.

GLOBULAR CLUSTERS

Globular clusters vary in their degree of concentration. Many have star chains in their halos, some have mysterious dark lanes in their cores, and a few rise to a brilliant central pip.

When visiting a star party, there are few things more dramatic to an amateur observer than watching the ever-increasing concentration of stars as a showpiece globular drifts into the field, at high power, in a big Dobsonian.

Right M33, the Triangulum Galaxy, is often referred to as the Pinwheel Galaxy because of its shape. Able to be seen with the naked eye under good conditions, M33 is the most distant object that can be viewed without aid.

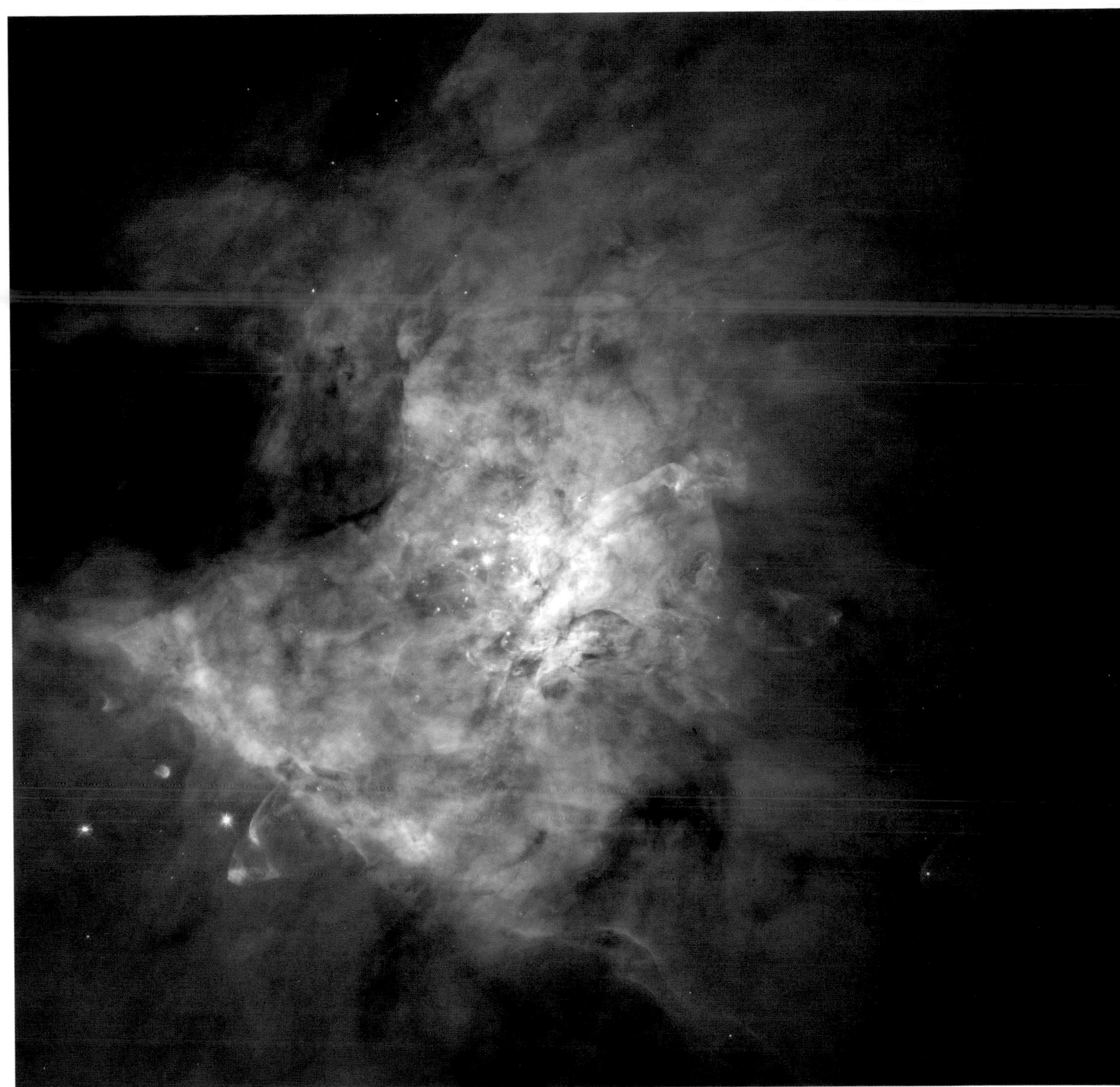

Above Hubble Space Telescope image of the Orion Nebula, located 1,500 light-years away, along our spiral arm of the Milky Way. It lies in the middle of the sword region of the constellation Orion—the Hunter—which dominates the early winter evening sky at northern latitudes.

DOUBLE STARS

Most double stars are white, but the celebrated ones often have a striking color contrast. Sometimes an orange primary will make the secondary look like its complementary color, green, but knowing that the color is mainly in our brain does not make the pair any less attractive. Some binaries are resolved by binoculars; others are so close that they are used as tests of optical quality on the steadiest nights.

NEBULAE

The brightest emission nebulae give very dramatic views. Many of the swirls of nebulosity seen on photographs of the Tarantula, Orion, and Veil nebulae can be seen at the eyepiece. But emission nebulae are too faint to show the bright colors that appear on images, with one exception: Some observers, especially younger people, can see greenish and, more rarely, pinkish tints in the Orion Nebula.

However, many small planetary nebulae have very high surface brightnesses and exhibit obvious color, usually blue-green or blue. Through a large telescope, the Pink Planetary in Lepus sports a coppery outer rim. At the center of the Eta (η) Carinae Nebula, a gray complex of emission nebulosity and dark nebulae, lies a tiny bright orange reflection nebula, the Homunculus Nebula.

Equipment

The simplest ways of observing the stars, and frequently the most pleasurable, are with the naked eye or binoculars. Casual sky-watching can be completely spontaneous, unlike telescopic observing, which usually requires a block of time set aside and frequently requires advance planning.

Right Wide-field view of the Milky Way between Sagittarius and Scorpio. An image such as this is possible by mounting the camera at the focal point of a large telescope. Accurate tracking of the subject to compensate for the rotation of Earth during the exposure is also necessary.

Once the constellations have been learned, they become old friends reappearing at the same season every year. On an evening stroll your eyes automatically turn skyward, glancing briefly at each familiar constellation. "What's that star doing there near Regulus?" you may ask yourself when you see Leo rising in the east after some months behind the Sun, but then you remember that Saturn or Mars has moved into Leo since you last saw it. Over the following months you will note the planet's slow progress through the constellation.

NAKED-EYE AND BINOCULAR OBSERVATION

Only the naked eye is needed to find bright constellations such as Leo. But the binoculars will probably be uncapped as soon as you look for nearby Sextans or Crater. Binoculars are a natural extension of naked-eye observing, because both are generally used for the same kinds of objects. Your unaided eyes give the widest possible field of view; the next-widest field, 7°, is provided by ordinary astronomy binoculars such as 7x50s. It is seldom a question of whether your eyes or your binoculars should be used—usually you will use both in turn, whether you are enjoying a conjunction of Venus with the crescent Moon and earthshine, a naked-eye comet, or the Milky Way's splendid star clouds and dark nebulae. Meteor showers, auroras, and the zodiacal light need the naked-eye's larger field of view.

Binoculars larger than 10x50s require tripods. Binoculars are available with 80 mm, 100 mm, and even 150 mm lenses. While such tripod-mounted behemoths give marvelous views of the Milky Way and some other astronomical objects, they lack the 7° fields and instantaneous portability provided by ordinary handheld binoculars.

Image-stabilized binoculars, available from 10x30s to 15x50s, are for those who appreciate and can afford the finest optics. Few people can hold normal 15x binoculars steady, but 15x works fine with image-stabilized binoculars, at the cost of a smaller field of view. They give superior resolution while retaining the convenience of handheld binoculars.

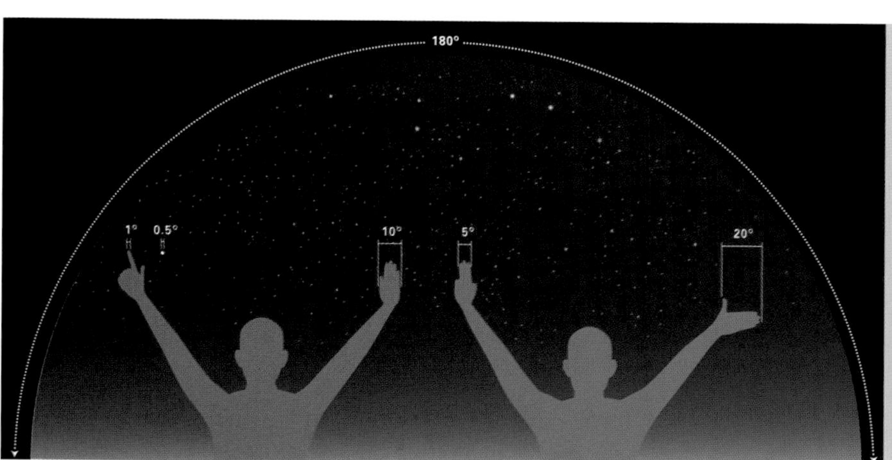

ANGLES IN THE SKY

It is necessary to know how to measure degrees in observing, in order to find out how far away stars, planets, and other objects are from each other.

The difference in the size of people's fists is generally proportional to the length of their arm. So, at arm's length, where the moon equals 0.5°, then the tip of the index finger measures about 1° and the three middle fingers about 5°. The width of your hand held at arm's length equals about 10°. To measure other angles between objects, use the index finger and small finger stretched out to find 15°; stretch your hand sideways and extend your thumb to find 20°; and to find 25°, measure the distance between the small finger and thumb stretched apart.

TELESCOPES

Telescopes come in two basic types: Refractors collect light with an objective lens at the forward end of a tube; reflectors collect light with a mirror. Telescopes are mounted on either an altazimuth mount—one that rotates about two different axes, one for altitude and one for azimuth—or an equatorial mount which has an axis parallel to Earth's axis, so that only motion around that axis is needed to follow the stars as Earth rotates beneath them.

Once, only equatorial mounts could automatically track objects, whether for astrophotography or visual observing, but computers can now accurately follow objects with altazimuth mounts. Computers can also automatically slew telescopes to targets, an option called "Go-To."

Simple lenses do not focus all colors at the same distance. This leads to a smearing of the image, called "chromatic aberration," and loss of contrast. The solution is to make refractor objectives out of two lenses made from types of glass having different refractive indices. Standard refractors are called achromatic refractors, meaning "color-corrected." They

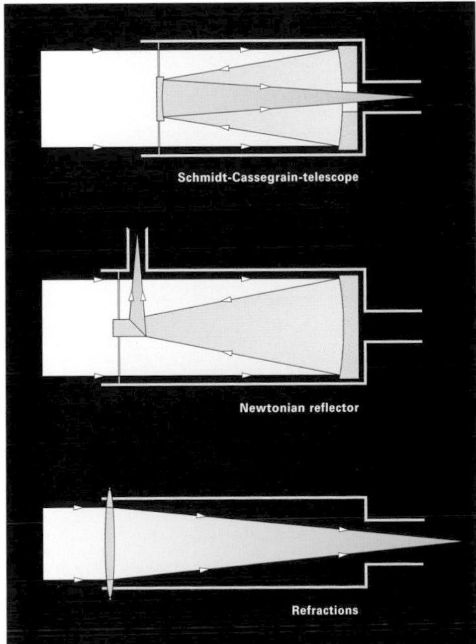

Left Illustrations showing the different ways that light is passed through three different telescopes. Top, the light passes across a primary concave mirror to correct for spherical aberration. Middle, the light rays hit a curved mirror which focuses them all to the same point. Bottom shows a refractor lens where light rays are directed through glass which often bends the rays resulting in false color.

Above This illustration shows a refractor telescope that resolves high contrast, fine resolution images ideal for planetary viewing. However, since the light passes through glass, which bends the wavelengths, it produces false-color images on bright objects.

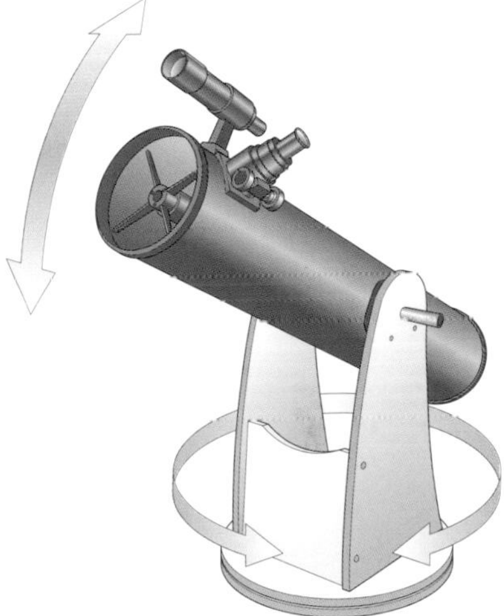

Above This illustration shows a reflector telescope, or Newtonian, which reflects light off the surface of a curved mirror. As the light rays never pass through glass, these telescopes resolve true-color images.

Right Nebula filters have revolutionized observing, turning many nebulae that used to be considered challenge objects into richly detailed showpieces. An OIII filter should be your priority (see the Veil Nebula in Cygnus). A hydrogen-beta filter is best for a limited number of objects, notably the Horsehead, California, and Cocoon nebulae. A UHC or similar filter passes both the OIII and H-beta wavelengths, and performs almost as well on the Horsehead Nebula.

Above A Newtonian 10-inch (25 cm) telescope with a f/5.1 lens. It can be used to take images of stars, constellations, and planets.

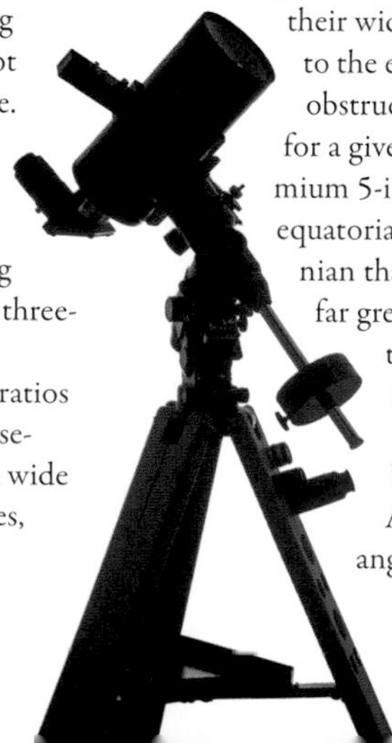

Right Reflector telescopes such as this, produce true-color resolutions but if the reflector is large the spherical mirrors can distort the focus resulting in images that lack contrast.

need a long focal ratio to perform well—typically f/12 to f/15. This results in a long heavy tube requiring a sturdy mount, so achromatic refractors are not very portable beyond a 4-inch (10 cm) aperture. Although a color fringe remains—normally violet—achromats perform well on planets, double stars, and clusters.

Apochromatic refractors go further, utilizing very expensive types of glass to make a two- or three-element objective that is free of false color.

Apochromatic objectives work well at focal ratios as short as f/5.2, allowing a short tube and consequently a lightweight portable telescope with a wide low-power field of view, typically several degrees, while offering superb high-power performance on planets as well. Astrophotographers treasure "apos" for their wide fields with star images that are round nearly to the edge. Since refractors have no secondary mirror obstruction in the light path, they give the finest images for a given aperture. The choice is really between a premium 5-inch (12 cm) apochromatic refractor on a Go-To equatorial mount or a premium 25-inch (63 cm) Dobsonian that collects almost 25 times as much light and has far greater resolution—on those very rare nights when the atmosphere is steady enough to permit ½ inch (1.2 cm) resolution, or better.

Newtonian reflectors focus light with a parabolic mirror located at the lower end of the tube. A flat secondary mirror near the upper end, angled at 45°, diverts the converging cone of light

to a focuser mounted outside the tube, where the image is magnified with an eyepiece. The mirrors occasionally have to be optically aligned, a procedure called collimation. Usually only a few tweaks are required to achieve optimum performance.

Newtonians are by far the least expensive telescope design, yet their performance is comparable with that of far more expensive telescopes. Some are on heavy equatorial mounts, usually in an observatory, but most are now used on simple and robust altazimuth mounts called Dobsonians. A keenly priced 8- or 10-inch (20 cm or 25 cm) Dobsonian that performs quite creditably on deep-sky objects costs about as much as a premium eyepiece.

Cassegrain telescopes have a convex secondary mirror that redirects the light cone back down the tube and through a hole in the primary mirror. Classical Cassegrains tend to be equatorially mounted large-aperture observatory telescopes.

Catadioptric telescopes are compound designs based on Cassegrains, but with a corrector lens at the front and a secondary mirror mounted on the lens.

The greatest virtue of catadioptrics is their folded light path. This allows a short tube, making them quite portable in aperture sizes up to 8 inch (20 cm). Their portability makes them popular with city-dwellers.

Catadioptrics have relatively small fields of view, however, and the front corrector is notorious for dewing up, requiring a heater in most climates.

Astrophotographers like the high-quality mounts that are offered with factory-made instruments, and the round star images, free of the diffraction spikes caused by the secondary support vanes of Newtonians and classical Cassegrains.

Which telescope to buy? All have their strong points, and all are compromises, so most long-time hobbyists own several telescopes of different apertures.

Right Before purchasing your own telescope, it's best to attend a star-party at a local astronomy club and try out the different types, and the assorted eyepieces available. You should take into account not only what you can see but the portability and reliability of the different types of telescopes available. Computerized telescopes scan the sky and locate an object for you.

DOBSONIANS

John Dobson, an impoverished San Francisco monk, revolutionized amateur astronomy in the early 1980s when he built 18 to 24-inch (45 to 60 cm) telescopes out of scrap. He ground his own mirrors, used sonotubes (cylindrical forms for pouring cement) for tubes, and made mounts out of plywood with bearings made of kitchen countertop material riding on teflon. Since then, thousands of amateurs have followed his lead and, while most now purchase their mirrors, mirror cells, secondary support vanes, and focusers, the rest of a functional large-aperture Dobsonian with silky-smooth motions can be built in only one or two weekends. However, many amateurs spend a winter on their telescope-making projects, producing furniture-quality Dobsonians that they proudly enter into competitions at star-parties. In large-aperture Dobs the solid tube is frequently replaced with aluminum truss tubes, allowing the scope to break down into several sections for easy transportation.

Attend a star-party or join an astronomy club and go observing with their members—either will allow you to try many different types and brands of telescope. And take the advice of knowledgeable amateur astronomers.

In the meantime, most clubs have telescopes for rent and many have a club observatory, typically with a 14- to 24-inch (35 to 60 cm) telescope that any member is able to use after some training.

ACCESSORIES

New telescopes commonly include two Plossl eyepieces, which are a perfectly serviceable design. Over time you will undoubtedly expand your collection. Try different eyepieces at a star-party to ensure that they work with your telescope's particular optical configuration.

Premium low-power eyepieces that offer exceptionally wide fields of view, sharp to the edge, are so heavy that they cause balance problems with many telescopes, as well as costing as much as an 8-inch (20 cm) Dobsonian.

Logging your observations from the beginning is strongly recommended as it enhances your experience.

To preserve your dark adaption, buy the dimmest red LED flashlight available—one that barely illuminates a star chart.

Above Sky maps are available in printed form, online, or as software. Used with the right telescopic eyepieces, it's possible to easily locate most objects in the sky.

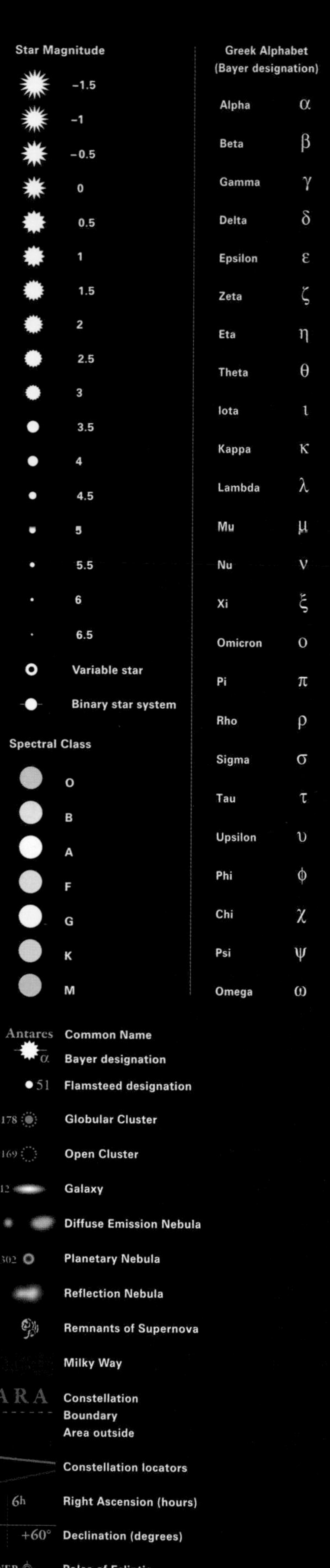

Using the Constellation Charts

Browsing the night skies for stars and constellations can bring much enjoyment to the amateur and professional alike. Using the maps that follow in this section will aid in navigation around the night sky.

The detailed constellation maps on the following pages show the stars and objects of the 88 constellations. The maps can be used to pick out the constellation in the night sky, and the small globes at bottom right will give a quick general guide to the approximate location of the constellation in the Northern and Southern hemispheres. In these locator maps, New York, USA, and Sydney, Australia, are used as examples for the Northern hemisphere and Southern hemisphere respectively. In Sydney, a typical constellation will rise in the east and move counterclockwise across the sky before setting in the west, while in New York, the constellation moves clockwise across the sky. A number of constellations do not set and so remain in the sky at all times. These circumpolar constellations move about the celestial pole, tracing out a circle on the chart that is always above the horizon and so never sets. Such constellations are visible in either New York or Sydney, but not in both. Each constellation rotates around a point (celestial pole) and has a radius equal to the distance of the constellation from the nearest celestial pole.

These detailed maps show the brighter stars of the constellation, from magnitude 6.5 upward, and all other significant features within the constellation. Oriented North–West–South–East, the featured constellation stands out around its lighter colored neighbors. A "stick-figure" configuration shows the typical outline of the constellation, which relates—often loosely—to the figure that represents the constellation—some constellations bear a much closer resemblance to their representative than others. At left is an extensive legend detailing all the elements used on the maps, as well as a handy guide to the Greek letters used in the Bayer designation of stars.

Accompanying the maps and fact files is descriptive text that will enable observers to determine the features to look for in the constellation, while the images give an indication of some of the constellation's outstanding features.

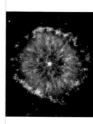

314 Observing the Skies

Aquila

Above This is a Hubble Space Telescope image of the strikingly unusual planetary nebula, NGC 6751. Glowing like a giant eye in the constellation Aquila, the nebula is a cloud of gas ejected several thousand years ago from the hot star visible in its center.

Right This image, taken by the Cerro Tololo Inter-American Observatory in Chile, shows a wide angle view toward the center of our own Milky Way Galaxy. The teapot shape of the stars in the constellation Sagittarius can be seen but the central region of the Galaxy is obscured from sight by intervening gas and dust clouds, and by the myriad stars between Earth and the galactic core.

Fact files

These fact files give a brief profile of each constellation and provide handy information about the constellation—including a pronunciation guide, the figure representing the constellation, the genitive, the abbreviation, the right ascension and declination, a general visibility range, a brief list of notable objects, and a list of named stars.

Mythology and History
This section provides information on the mythology related to the ancient constellations. For the modern constellations, their historical background is discussed.

Constellation Stars
In this section, the major stars are discussed, and attention is drawn to any special qualities or unusual points in the stellar composition of the constellation.

Objects of Interest
Notable objects in the constellation are given attention in this section, accompanied by descriptions and interesting facts.

Symbols
The symbols offer a handy guide to the observation requirements (naked-eye, binoculars, or telescope) of the interesting objects that can be seen in the featured constellation.

Constellation configuration
Constellation "stick-figures" are shown, though these often bear little resemblance to the constellation figure that they are intended to depict.

eagle. Zeta (ζ) and Epsilon (ε) form one wingtip, with Theta (θ) forming the other; Lambda (λ) is at the tail. Star atlases from the sixteenth to the eighteenth century showed considerable variance in how the figure was depicted.

Altair is the 13th brightest star in the night sky and is the lucida of Aquila. Its name originates from the Arabic for "flying eagle." Altair is a nearby A-type star, only 16.8 light-years distant and ten times brighter than our Sun.

OBJECTS OF INTEREST

The Great Rift, a long patch of obscuring dust and gas in the plane of the Milky Way running from Sagittarius to Cygnus, is at its most prominent in Aquila. Under a dark sky it is visible cleaving the Milky Way into two separate streams. The open cluster **NGC 6709** is a bright and beautiful triangular-shaped assemblage of medium-faint stars, of which about 50 are visible in an 8-inch (20 cm) telescope. In a 6-inch (15 cm) telescope, **NGC 6755** is a smaller, cloud-like cluster with many faint stars.

Aquila has several excellent planetary nebulae, of which the finest, **NGC 6751**, is a small, round, grayish, disk-shaped patch of fog. **NGC 6772** is a little larger—but fainter—with a translucent grayish smoke-ring and a darker center. **NGC 6803** has a tiny bright blue disk, and **NGC 6804** is a small, oval, grayish disk of haze with nearby stars.

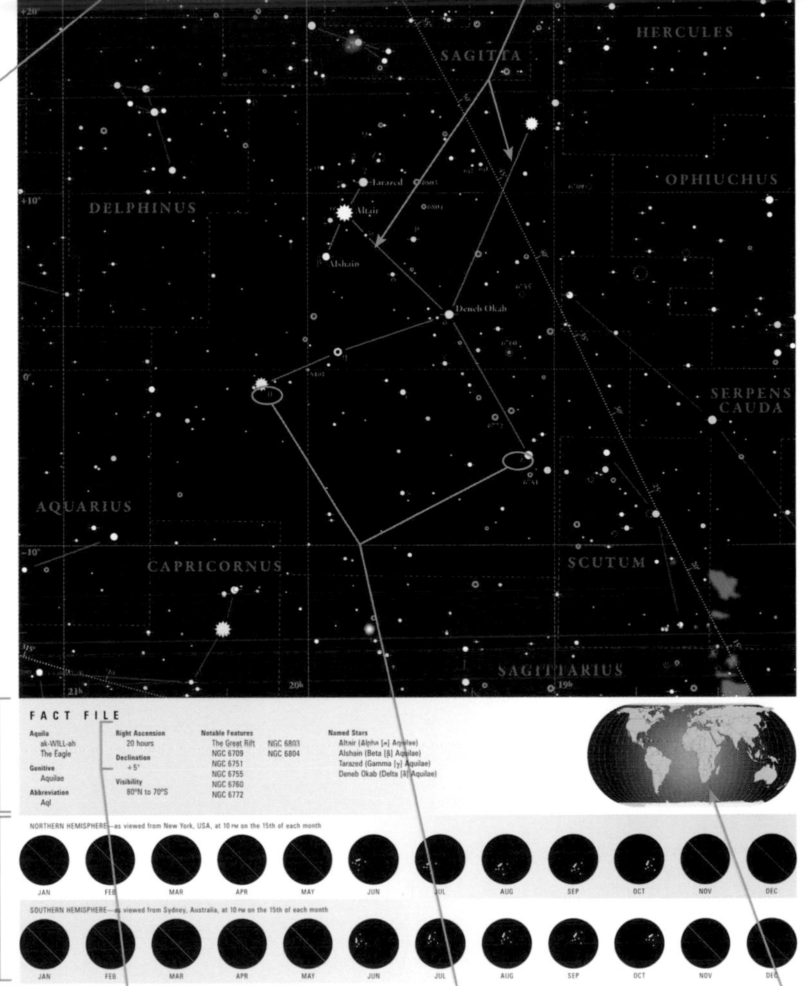

Constellation locator charts
The small globes indicate the general area that the constellation can be found in the night sky each month. The cardinal points are oriented (clockwise from top) North–West–South–East in the diagrams, rather than the North–East–South–West orientation used on atlas world maps. Guides are included for both hemispheres.

Right Ascension and Declination
Working in a similar manner to the lines of longitude and latitude on a world map, the stars and other objects of interest in the constellation are plotted in the sky along the lines of Right Ascension (RA) and Declination (Dec).

Bayer designation
Johannes Bayer instituted an identification system for the stars of each constellation in 1603. Using the letters of the Greek alphabet, and usually applied to the constellation stars in corresponding order of brightness, the Greek letter is coupled with the genitive, for example Alpha (α) Canis Majoris. On the feature maps, the Greek letters alone are used.

World map
A general guide to the visibility of the constellation is shown on the world map. The whole of the main pattern of the constellation can be seen from at least 5 degrees above the horizon in those parts of the world indicated by the non-shaded area on the map.

Andromeda

Andromeda is one of the best known constellations, more so because of the nearby galaxy it contains than the prominence of its stars.

PRINCESS IN CHAINS

Andromeda is just one player in a mythological story with many of its characters represented in the sky.

Andromeda was the daughter of Cepheus and Cassiopeia, rulers of ancient Ethiopia. Both can be found just to her north in the sky, with Cassiopeia the more obvious as a W-shaped pattern of bright stars. Cassiopeia angered the sea god Nereus by boasting that Andromeda was more beautiful than his daughters, the fifty sea nymphs known as the Nereids. Nereus's response was to dispatch Cetus to ravage Ethiopia. Cetus, variously identified as a sea monster or a whale, can be found a little to the south of Andromeda in the sky.

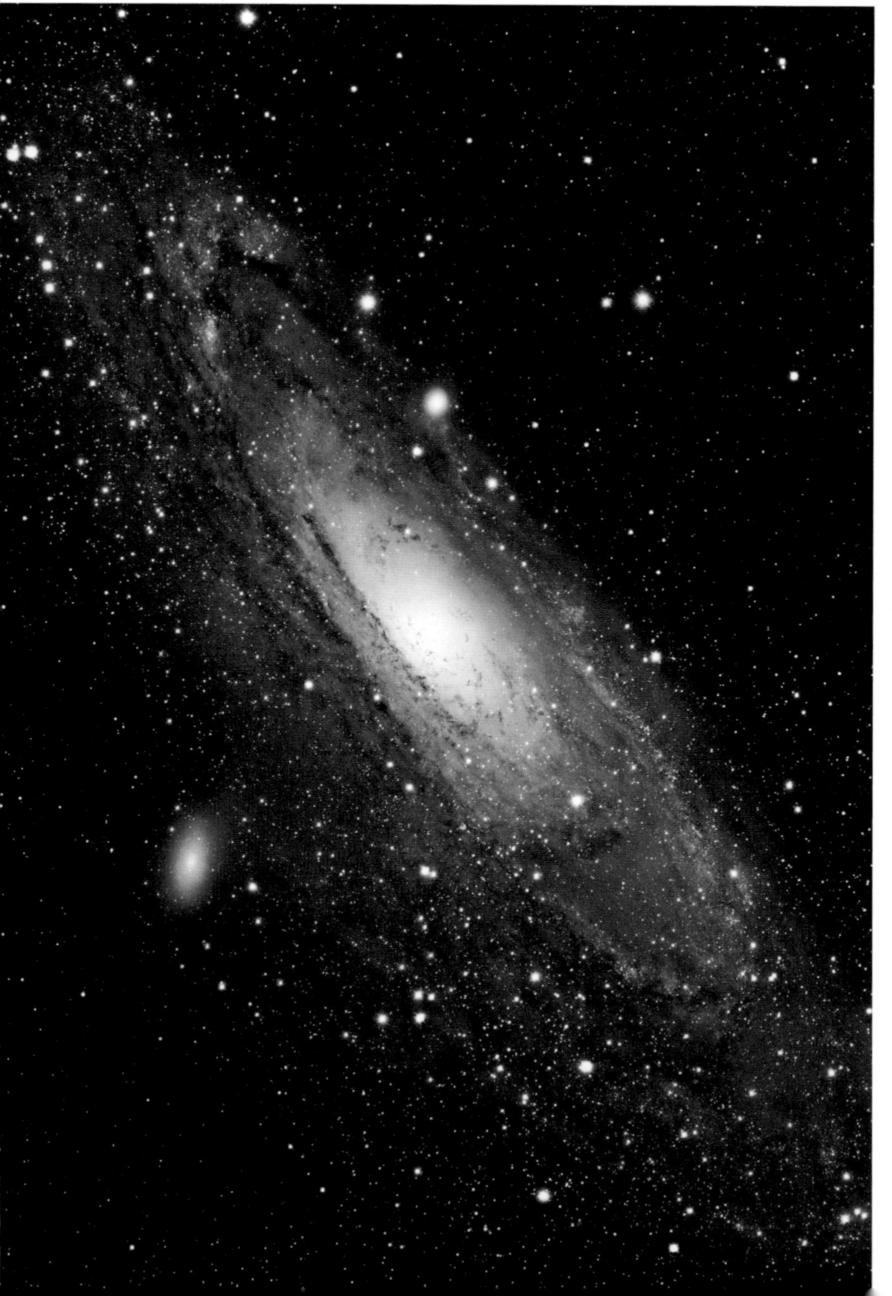

Seeking a way to appease the angry god, the royal couple sought advice from an oracle. The only way to save their country was to sacrifice their daughter to Cetus. Andromeda was chained to a rock near the sea, but was saved by the hero Perseus who arrived on the winged horse Pegasus. Perseus saved Andromeda by turning Cetus to stone using the hideous head of Medusa. The stars of Perseus lie just to the east of Andromeda. Pegasus, prominently marked by the stars of the Great Square of Pegasus, is found to her southwest.

THE STARS OF ANDROMEDA

The major stars of the Andromeda constellation form a long V-shape, with the southernmost leg containing the brighter stars. The vertex of the V is Alpha (α) Andromedae, which, although in Andromeda, acts as a corner star of the Great Square of Pegasus.

OBJECTS OF INTEREST

M31 (NGC 224), the **Andromeda Galaxy**, is the closest major galaxy to the Milky Way, lying 2.5 million light-years away. Along with the Milky Way, it dominates the Local Group of galaxies. M31 has the distinction of being the most distant object visible to the unaided eye. However, a dark sky is needed in order to see it well, since being so close, its light is spread over a patch of sky more than five times the size of the full moon. As a result, the view of the galaxy is often more impressive through a large pair of binoculars rather than a telescope.

M32 (NGC 221) and **M110 (NGC 205)** are small elliptical satellite galaxies to M31. Both are visible in a 4-inch (10 cm) telescope, with M32 being smaller and more compact, and therefore easier to see.

Another galaxy found within the borders of Andromeda is **NGC 891**. This is a spiral galaxy seen edge-on, but requiring the use of at least a 6-inch (15 cm) telescope in order to track it down.

Among the stars of this constellation is **Gamma (γ) Andromedae**, often regarded as one of the best double stars in the sky for a small telescope. There is an attractive contrast in color between the magnitude 2.3 golden-yellow primary and the magnitude 5.1 greenish-blue companion lying 10 arc-seconds away.

NGC 7662 is a planetary nebula that is visible through a 6-inch (15 cm) telescope as a glowing blue spot of light 30 arc-seconds across. The **Blue Snowball Nebula**—as it is sometimes known—looks more like an eye in images from the Hubble Space Telescope, peering at us across 5,600 light-years.

Left With its readily visible formation, records of the Andromeda Galaxy date back over the centuries. This nearest neighbor to our own galaxy is a member of the Local Group of galaxies and a major feature within the constellation of Andromeda.

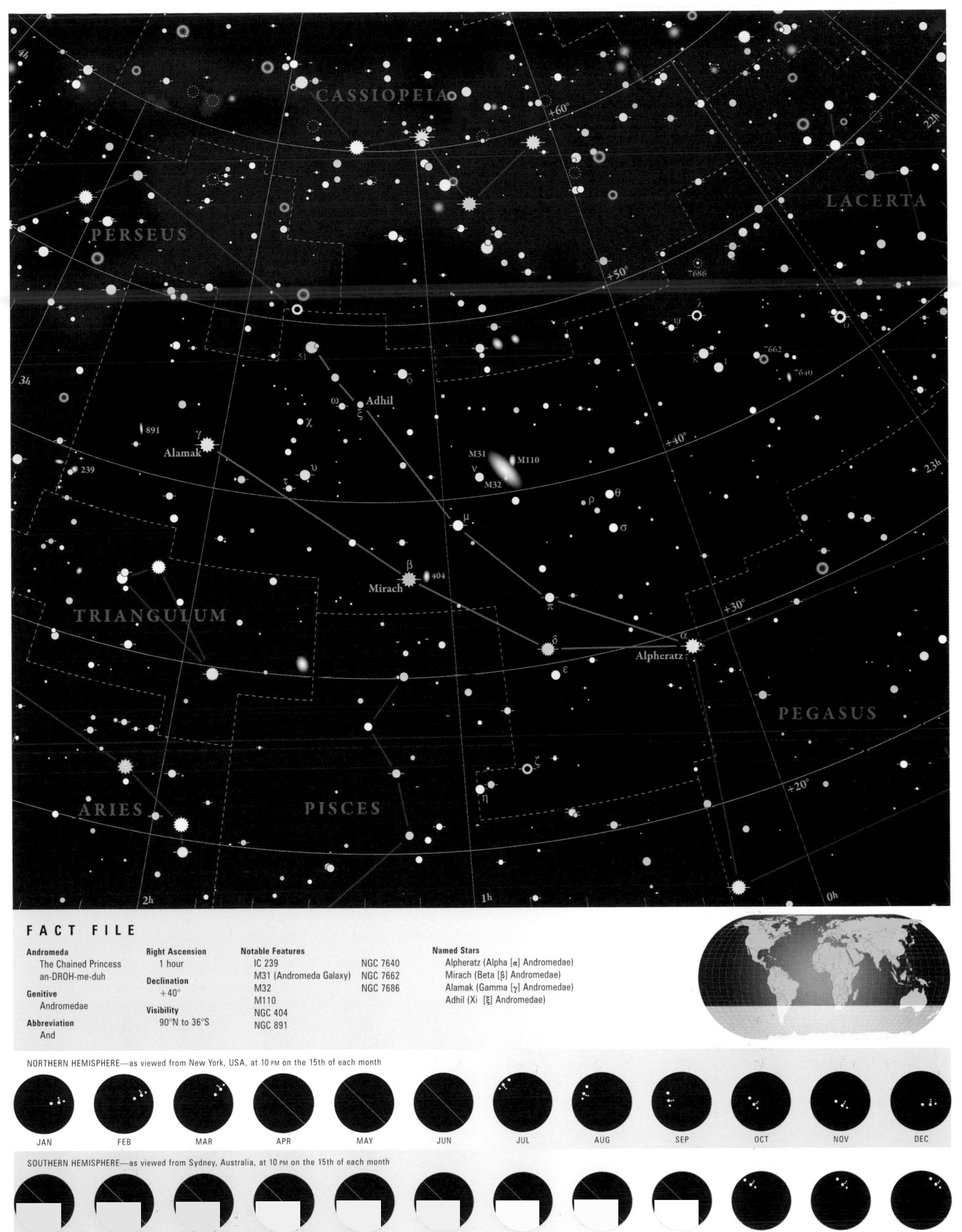

FACT FILE

Andromeda
The Chained Princess
an-DROH-me-duh

Genitive
Andromedae

Abbreviation
And

Right Ascension
1 hour

Declination
+40°

Visibility
90°N to 36°S

Notable Features
IC 239
M31 (Andromeda Galaxy)
M32
M110
NGC 404
NGC 891
NGC 7640
NGC 7662
NGC 7686

Named Stars
Alpheratz (Alpha [α] Andromedae)
Mirach (Beta [β] Andromedae)
Alamak (Gamma [γ] Andromedae)
Adhil (Xi [ξ] Andromedae)

NORTHERN HEMISPHERE—as viewed from New York, USA, at 10 PM on the 15th of each month

JAN FEB MAR APR MAY JUN JUL AUG SEP OCT NOV DEC

SOUTHERN HEMISPHERE—as viewed from Sydney, Australia, at 10 PM on the 15th of each month

JAN FEB MAR APR MAY JUN JUL AUG SEP OCT NOV DEC

Antlia

The dim stars of the small constellation called Antlia, located just off the southern Milky Way, play host to a multitude of galaxies.

PUMP IT UP

The stars wedged between Vela and Hydra that constitute the modern constellation called Antlia were not recognized by ancient cultures as representing a group of their own.

Abbé Nicolas Louis de Lacaille formed Antlia into a constellation while working at an observatory on the Cape of Good Hope. He named it "la Machine Pneumatique" (later Latinized to "Antlia Pneumatica" in the second edition of the map published in 1763) to commemorate the development of the air pump by Robert Boyle and his assistant Robert Hooke at Oxford University. (Interestingly, Robert Hooke is credited with the discovery of Jupiter's Great Red Spot in 1650.) Since the time of Lacaille, the stars and form of Antlia have undergone no significant change.

THE STARS OF ANTLIA

Antlia is located fairly close to the plane of the Milky Way, and its background is rich in faint stars. The principal stars, Alpha (α), Theta (θ), Epsilon (ε), and Iota (ι), all 4th and 5th magnitude points, make up a large trapezoid that bears little resemblance to an air pump. Behind, a throng of barely visible points provides an excellent setting for Antlia's galaxies.

Alpha (α) Antliae, at magnitude 4.28, is the lucida of the constellation. It is a K-type orange giant star, 365 light-years distant. With about 2.5 times the mass of our Sun, it is over five hundred times more luminous. As an ageing star near the end of its life, it will soon become a red giant, and after a period of pulsating as a Mira-type variable, it will die as a white dwarf.

OBJECTS OF INTEREST

Although it is located close to the Milky Way, no star clusters or nebulae adorn Antlia. The fascinating planetary nebula **NGC 3132** lies exactly on the boundary with Vela and is treated as part of that constellation.

Antlia is awash with galaxies. Many are faint but a few are bright and interesting. The best is the beautiful two-armed spiral **NGC 2997**. Visible in a 4-inch (10 cm) telescope, it is a quite large in an 8-inch (20 cm) as a low-surface brightness haze with a round, small, slightly brighter core. Under a dark rural sky, the slightly brighter spiral arm pattern within the halo can just be discerned.

NGC 3347, which includes **NGC 3354** and **NGC 3358**, is an excellent group for an 8-inch (20 cm) telescope. The two larger, brighter hazes are –47 and –58, with a tiny –54 squeezed between. A similar-looking group surrounds the bright elliptical galaxy **NGC 3268**. **Zeta (ζ^1) Antliae** is a fine double for virtually any telescope, with a close pair of almost matching yellow stars shining out in a star-sprinkled field.

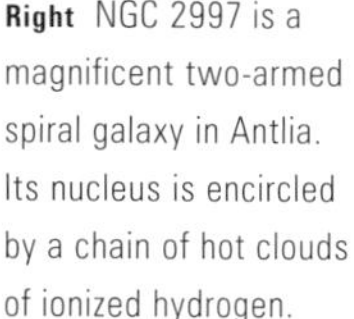

Right NGC 2997 is a magnificent two-armed spiral galaxy in Antlia. Its nucleus is encircled by a chain of hot clouds of ionized hydrogen.

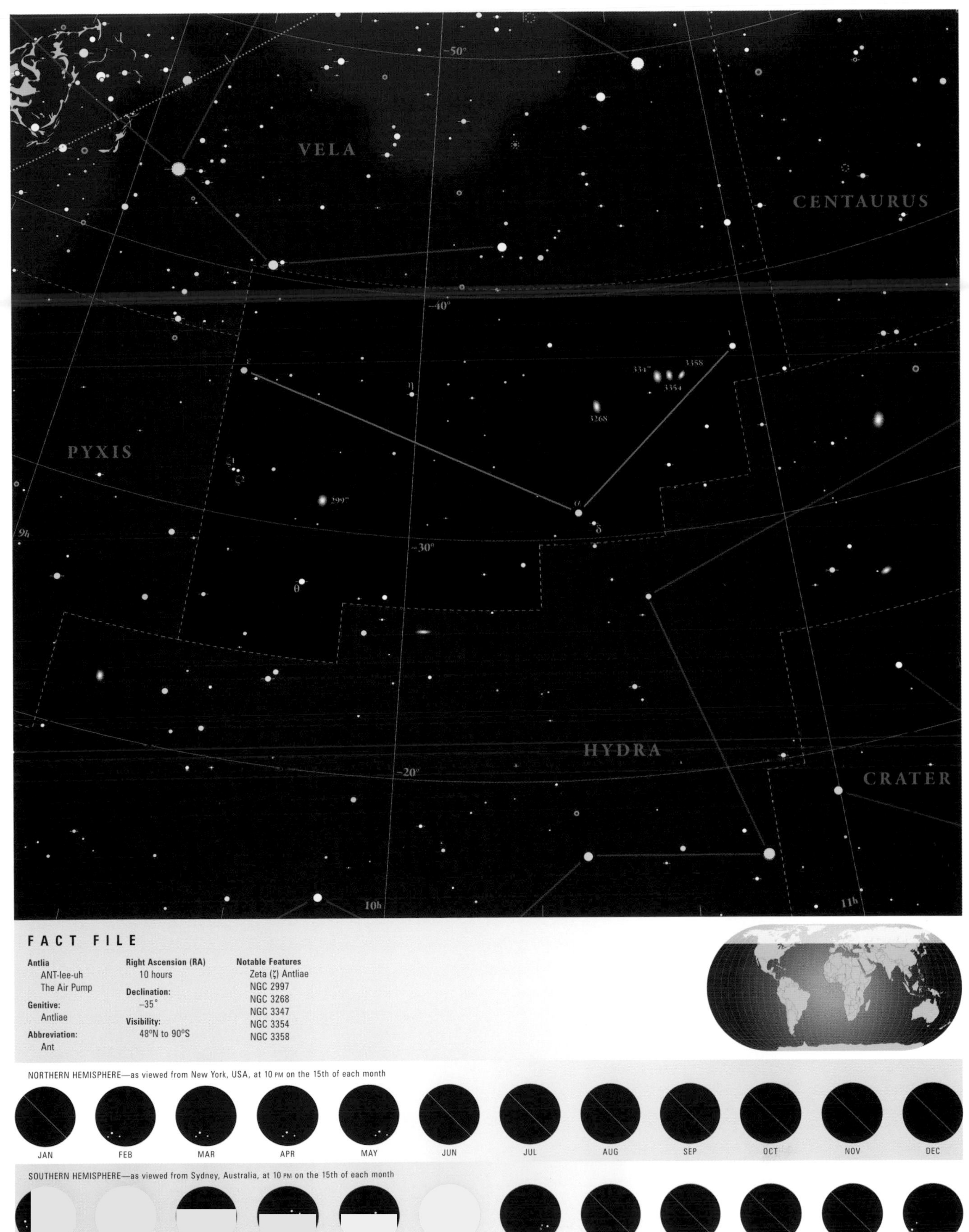

FACT FILE

Antlia
ANT-lee-uh
The Air Pump

Genitive:
Antliae

Abbreviation:
Ant

Right Ascension (RA)
10 hours

Declination:
–35°

Visibility:
48°N to 90°S

Notable Features
Zeta (ζ) Antliae
NGC 2997
NGC 3268
NGC 3347
NGC 3354
NGC 3358

NORTHERN HEMISPHERE—as viewed from New York, USA, at 10 PM on the 15th of each month

JAN FEB MAR APR MAY JUN JUL AUG SEP OCT NOV DEC

SOUTHERN HEMISPHERE—as viewed from Sydney, Australia, at 10 PM on the 15th of each month

JAN FEB MAR APR MAY JUN JUL AUG SEP OCT NOV DEC

Apus

The faint obscure stars of Apus commemorate one of the world's most colorful and attractive birds—the beautiful Bird of Paradise.

LIGHT AND LEGLESS

The small and faint asterism of stars now recognized as Apus was too far south to be known by early European observers. The figure represents a bird of paradise, sometimes offered as a gift (after its apparently unsightly legs were removed) by natives in the East Indies to European explorers. The name Apus is derived from the Greek word for "footless."

The Dutch explorers Pieter Dirkszoon Keyser and Frederick de Houtman introduced the group initially at the end of the sixteenth century. The constellation was first depicted on a star-map by Petrus Plancius in 1598 and was called "Paradysvogel Apis Indica." Soon afterward, Johannes Bayer similarly portrayed it as "Apus Indica" in his star atlas *Uranometria*, published in 1603. Since Bayer's time there has been no significant change to the form or stars of Apus.

THE STARS OF APUS

Apus lies between the distinctive outline of Triangulum Australe and the South Celestial Pole. All of the principal stars of Apus are of 4th or 5th magnitude. A small right-angled triangle—made up by Beta (β), Gamma (γ), and Delta (δ) Apodis—is the most visible part of the grouping, while Alpha (α) is somewhat removed, and lies near the border of Chamaeleon. All four of these stars are either red or orange giant stars that exhibit yellowish or orange tints when observed with binoculars.

OBJECTS OF INTEREST

Lying well away from the Milky Way, Apus has no open clusters, nebulae, or planetary nebulae to offer to the amateur observer. There are many galaxies in Apus, but all are faint and tiny. The constellation also has two globular clusters, but neither is particularly noteworthy.

NGC 6101 is the better of the two clusters, and in an 8-inch (20 cm) telescope appears as a hazy patch with a couple of faint stars sewn in. **IC 4499** is slightly smaller and fainter, but holds the distinction of being the closest globular cluster to the South Celestial Pole. It is a challenging object to see in an 8-inch (20 cm) telescope under a rural sky. Both clusters show little compression toward the center.

Delta (δ) Apodis is a wide optical double star consisting of slightly uneven orange-hued stars that are easily resolved in binoculars. **Theta (θ) Apodis**, 3 degrees northwest of Alpha, is a pulsating red giant that varies irregularly in brightness between magnitude 6 and magnitude 9. Its fluctuations can be followed simply in binoculars.

Right NGC 6101 is one of two globular clusters in the constellation Apus, appearing as a hazy patch in an 8-inch (20 cm) telescope.

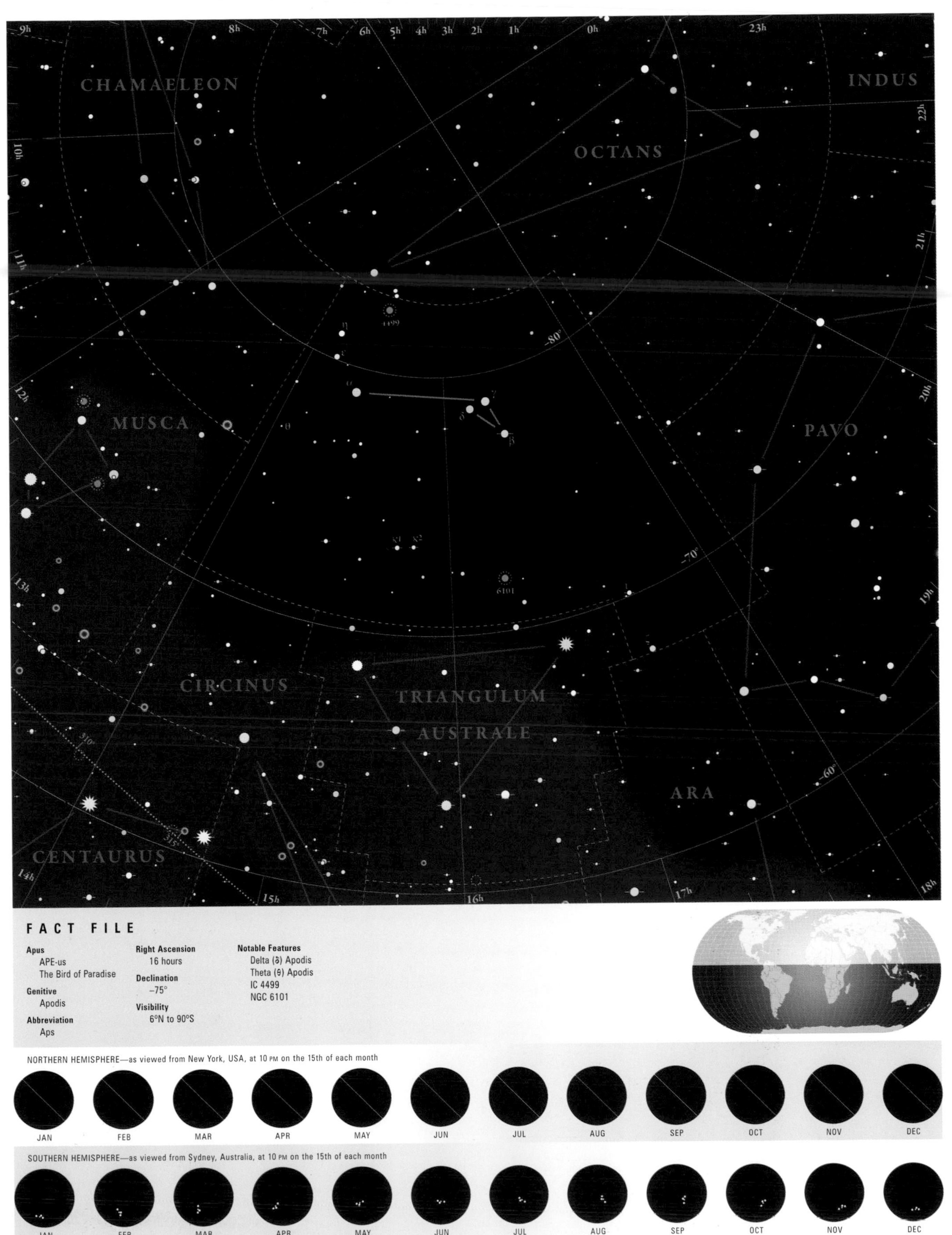

FACT FILE

Apus
APE-us
The Bird of Paradise

Genitive
Apodis

Abbreviation
Aps

Right Ascension
16 hours

Declination
–75°

Visibility
6°N to 90°S

Notable Features
Delta (δ) Apodis
Theta (θ) Apodis
IC 4499
NGC 6101

Aquarius

One of the oldest known constellations, Aquarius depicts the cup-bearer of Zeus. It is sometimes drawn with the mythological river Eridanus cascading from the water-bearer's jug.

LET IT FLOW

The word "Aquarius" is the Latin form of the constellation name meaning "water." The ancient Egyptians identified the faint stars of Aquarius with their gods of the Nile, believing that the constellation gave rise to the life-giving annual flood. To the Greeks, these stars represented Ganymede, son of King Tros, who gave Troy its name. Ganymede was reputedly the most beautiful mortal alive and Zeus became infatuated with him. As Ganymede watched over his father's sheep, Zeus swooped down in the form of an eagle and carried him off to Mount Olympus, where he served water, nectar, and ambrosia to the gods. The adjoining constellation, Aquila, commemorates the eagle.

THE STARS OF AQUARIUS

Generally, the stars of Aquarius are fairly faint—only two are brighter than 3rd magnitude. The most easily recognized part is the asterism that makes the "water jar," composed of 5th and 6th magnitude stars in the southeast of the constellation. Although these stars are faint, they stand out reasonably well in a rural sky. Star atlases of the sixteenth to nineteenth centuries depict Aquarius pouring water from this jar into the mouth of Piscis Austrinus.

Beta (β) Aquarii, or Sadalsuud (Arabic for "the luckiest one of all"), is the brightest star in Aquarius, edging out Alpha by just a fraction. It is a yellow G-type supergiant star, six times heavier and 50 times larger than our Sun. Shining at us from over 600 light-years away, it is also 2,200 times more luminous than the Sun. Alpha (α) Aquarii is similarly named Sadalmelik, meaning "the luckiest star of the kingdom." The origin and significance of both names is lost in the mists of time.

OBJECTS OF INTEREST

Despite being far from the Milky Way, Aquarius has a fair measure of interesting objects for the amateur sky-watcher to view. The largest easily visible planetary nebula in the sky is **NGC 7293**, or the **Helix Nebula**. Over a half-moon in diameter, it can be difficult to see from suburbia, but under a rural sky it is visible, even with binoculars, as a large, round, faint patch of haze. A 6-inch (15 cm) telescope shows a huge grayish smoke-ring of gas with a slightly darker center. **NGC 7009** is the **Saturn Nebula**—a tiny but bright oval planetary nebula with strong aqua coloration and two small opposing rays projecting from its sides, making it look like Saturn with the rings edge-on.

M2 is much the better of two globular clusters in Aquarius, and in an 8-inch (20 cm) telescope resolves into a myriad of faint sparks with a sharp concentration to the center. By comparison, **M72** is a poor, much smaller globular that is difficult to resolve. Of the multitude of galaxies, **NGC 7184** is the easiest to see and an 8-inch (20 cm) telescope shows a fairly bright, small, spindle-shaped halo with a brighter core.

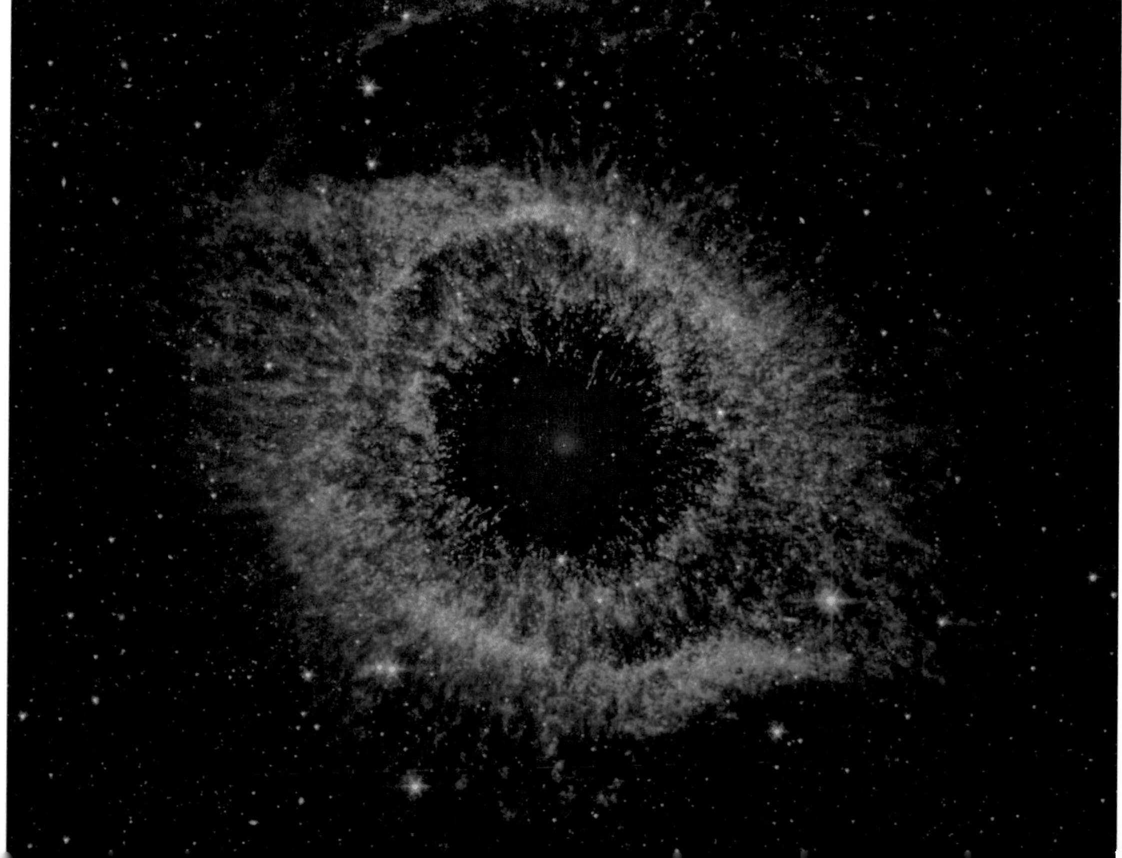

Right The Helix Nebula (NGC 7293), sometimes referred to as the "Eye of God," is about 700 light-years away from Earth and one of our closest planetary nebulae. A favorite among amateur astronomers for its scintillating colors, the dust cloud enveloping the central white dwarf measures 2 light-years across. This infrared image was taken from NASA's Spitzer Space Telescope in 2007.

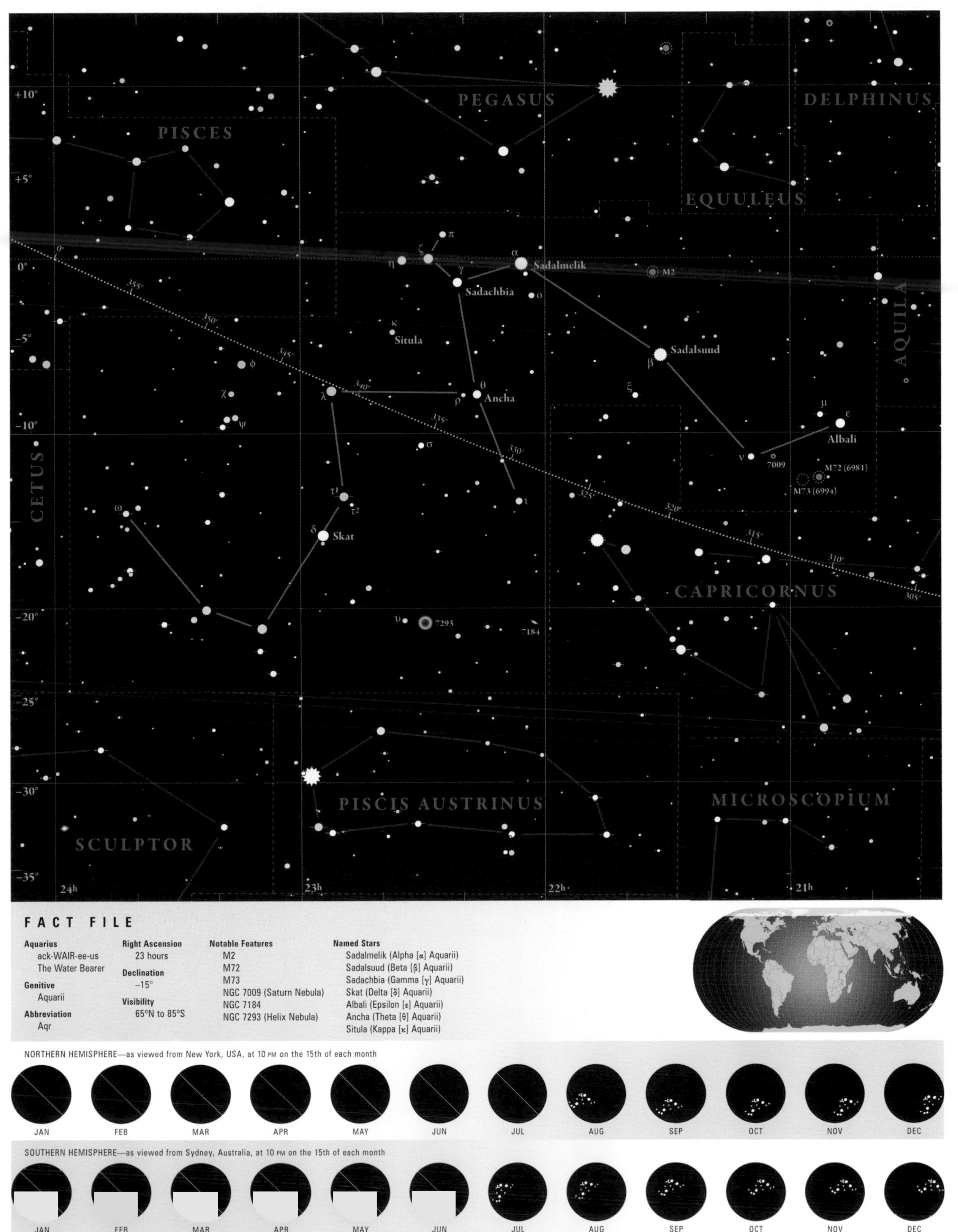

FACT FILE

Aquarius
ack-WAIR-ee-us
The Water Bearer

Genitive
Aquarii

Abbreviation
Aqr

Right Ascension
23 hours

Declination
–15°

Visibility
65°N to 85°S

Notable Features
M2
M72
M73
NGC 7009 (Saturn Nebula)
NGC 7184
NGC 7293 (Helix Nebula)

Named Stars
Sadalmelik (Alpha [α] Aquarii)
Sadalsuud (Beta [β] Aquarii)
Sadachbia (Gamma [γ] Aquarii)
Skat (Delta [δ] Aquarii)
Albali (Epsilon [ε] Aquarii)
Ancha (Theta [θ] Aquarii)
Situla (Kappa [κ] Aquarii)

NORTHERN HEMISPHERE—as viewed from New York, USA, at 10 PM on the 15th of each month

JAN FEB MAR APR MAY JUN JUL AUG SEP OCT NOV DEC

SOUTHERN HEMISPHERE—as viewed from Sydney, Australia, at 10 PM on the 15th of each month

JAN FEB MAR APR MAY JUN JUL AUG SEP OCT NOV DEC

Aquila

Taking its name from the royal eagle of Zeus, the constellation of Aquila swoops through the clouds of the Milky Way.

ON EAGLES' WINGS

The best known mythology relating to Aquila is Greek, but it is likely that the eagle is a considerably older constellation. Aquila represents either Zeus himself, in the form of an eagle, or possibly his own royal eagle plunging to Earth to abduct Ganymede, a young prince of Troy. Zeus became infatuated with Ganymede and bought him to Olympus to be his lover and cup bearer to the gods. In time, Ganymede was placed among the stars as the nearby constellation of Aquarius.

Some of the stars that are now part of modern Aquila were once used to form the now obsolete constellation of Antinous—The Youth. This group dates back to the second century BCE when the Roman emperor Hadrian—perhaps inspired by the tale of Zeus and Ganymede—created it in remembrance of a favorite youth in his court. According to legend, the misguided young man sacrificed himself with the aim of prolonging the life of his emperor. In some sixteenth- to nineteenth-century star atlases Aquila is depicted as carrying Antinous in his talons, as though representing Ganymede.

THE STARS OF AQUILA

Many of the stars comprising Aquila are quite faint and some imagination is required to visualize the figure. The trio of stars—Alpha (α) flanked by Beta (β) and Gamma (γ)—is an easily recognized asterism forming the neck and head of the eagle. Zeta (ζ) and Epsilon (ε) form one wingtip, with Theta (θ) forming the other; Lambda (λ) is at the tail. Star atlases from the sixteenth to the eighteenth century showed considerable variance in how the figure was depicted.

Altair is the 13th brightest star in the night sky and is the lucida of Aquila. Its name originates from the Arabic for "flying eagle." Altair is a nearby A-type star, only 16.8 light-years distant and ten times brighter than our Sun.

OBJECTS OF INTEREST

The Great Rift, a long patch of obscuring dust and gas in the plane of the Milky Way running from Sagittarius to Cygnus, is at its most prominent in Aquila. Under a dark sky it is visible cleaving the Milky Way into two separate streams. The open cluster **NGC 6709** is a bright and beautiful triangular-shaped assemblage of medium-faint stars, of which about 50 are visible in an 8-inch (20 cm) telescope. In a 6-inch (15 cm) telescope, **NGC 6755** is a smaller, cloud-like cluster with many faint stars.

Aquila has several excellent planetary nebulae, of which the finest, **NGC 6751**, is a small, round, grayish, disk-shaped patch of fog. **NGC 6772** is a little larger—but fainter—with a translucent grayish smoke-ring and a darker center. **NGC 6803** has a tiny bright blue disk, and **NGC 6804** is a small, oval, grayish disk of haze with nearby stars.

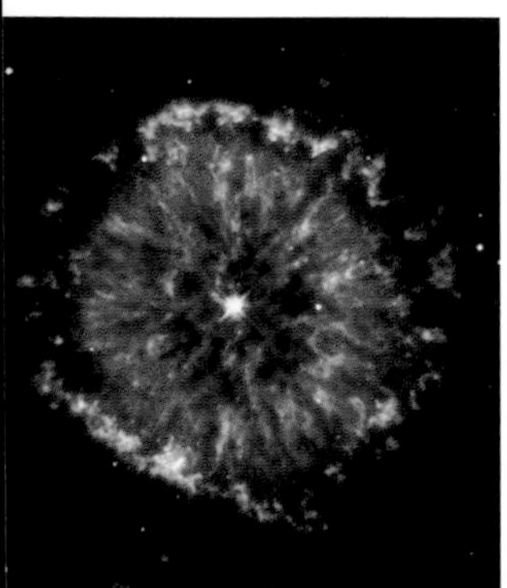

Above This is a Hubble Space Telescope image of the strikingly unusual planetary nebula, NGC 6751. Glowing like a giant eye in the constellation Aquila, the nebula is a cloud of gas ejected several thousand years ago from the hot star visible in its center.

Right This image, taken by the Cerro Tololo Inter-American Observatory in Chile, shows a wide angle view toward the center of our own Milky Way Galaxy. The teapot shape of the stars in the constellation Sagittarius can be seen but the central region of the Galaxy is obscured from sight by intervening gas and dust clouds, and by the myriad stars between Earth and the galactic core.

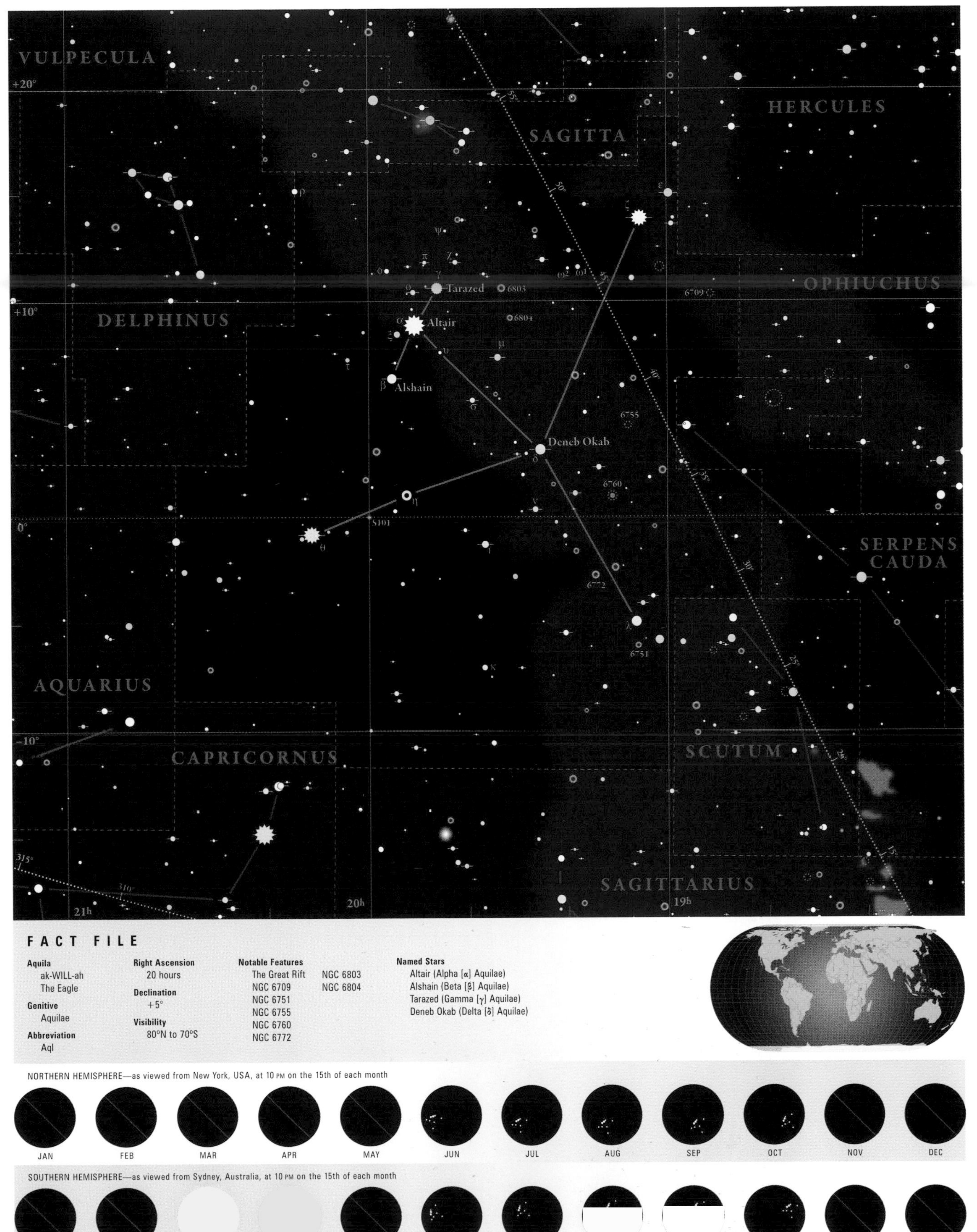

FACT FILE

Aquila
ak-WILL-ah
The Eagle

Genitive
Aquilae

Abbreviation
Aql

Right Ascension
20 hours

Declination
+5°

Visibility
80°N to 70°S

Notable Features
The Great Rift
NGC 6709
NGC 6751
NGC 6755
NGC 6760
NGC 6772
NGC 6803
NGC 6804

Named Stars
Altair (Alpha [α] Aquilae)
Alshain (Beta [β] Aquilae)
Tarazed (Gamma [γ] Aquilae)
Deneb Okab (Delta [δ] Aquilae)

NORTHERN HEMISPHERE—as viewed from New York, USA, at 10 PM on the 15th of each month

JAN FEB MAR APR MAY JUN JUL AUG SEP OCT NOV DEC

SOUTHERN HEMISPHERE—as viewed from Sydney, Australia, at 10 PM on the 15th of each month

JAN FEB MAR APR MAY JUN JUL AUG SEP OCT NOV DEC

Ara

Many early cultures recorded this constellation in their skies. The Altar is the most southerly constellation that was well known to ancient astronomers.

HEAVENLY OFFERINGS

The Hebrews interpreted the stars of Ara as either the altar that Noah built after surviving the flood, or the one that God ordered Abraham to build upon which to sacrifice his only son, Isaac. The Greeks saw Ara as the altar where the gods swore their allegiance to overthrow the Titans. To the Romans, this constellation was an altar where incense was burned for the dead.

In some of the early "modern" sky-atlases, Centaurus is depicted carrying the carcass of Lupus the wolf on a sword or spear in the direction of Ara, apparently for sacrificial burning. There is, however, no mythological history to support that picture. If the portrayal were accurate, Lacaille ruined it when he used a blank space between the three constellations to create Norma, the level and square, which became an odd obstacle for Chiron (Centaurus) to traverse on his way!

THE STARS OF ARA

Directly south of the Scorpion's tail, the main stars of Ara are easy to make out. One oddity is the lack of any common names for its stars, especially given that it is an ancient figure. Two opposing curved lines of stars mirroring each other's shape mark the sides of the altar—Alpha (α), Beta (β), Gamma (γ), and Delta (δ) on the west, with Epsilon (ε), Zeta (ζ), and Eta (η) on the east. The bright Milky Way flowing through the adjacent constellations of Norma and Scorpius could be interpreted as celestial smoke from its burning offerings.

Alpha (α) Arae, shining from a distance of 240 light-years, is a hot B-class star that is seven times greater than the mass of the sun and more than two thousand times as luminous.

OBJECTS OF INTEREST

The northern end of Ara is quite close to the plane of the Milky Way, and there are many open clusters and planetary nebulae there. The best of these is the cluster **NGC 6193**, a nebulous white double star at the center of the group that has a collection of stars in loops and sprays to its east with some faint nebulosity. The cluster is easy to see in a 4-inch (10 cm) telescope, while the nebula needs an 8-inch (20 cm) telescope to be readily visible.

There are three globular clusters in Ara, but **NGC 6397**, just visible to the naked eye, is possibly the closest globular to our sun at 7,500 light-years distant. Looking almost like an open cluster in a 6-inch (15 cm) telescope it is an uncompressed, large, sprawling assemblage nearly half a moon diameter across.

The southern reaches of Ara have many galaxies, and near Eta (η) Arae, **NGC 6215** and **NGC 6221** are bright galaxies paired in a star-studded field.

Right Resembling a treasure chest studded with glittering jewels, NGC 6397 is one of our closest globular clusters at 7,500 light-years distant, just behind M4 in Scorpius. It contains approximately 400,000 stars.

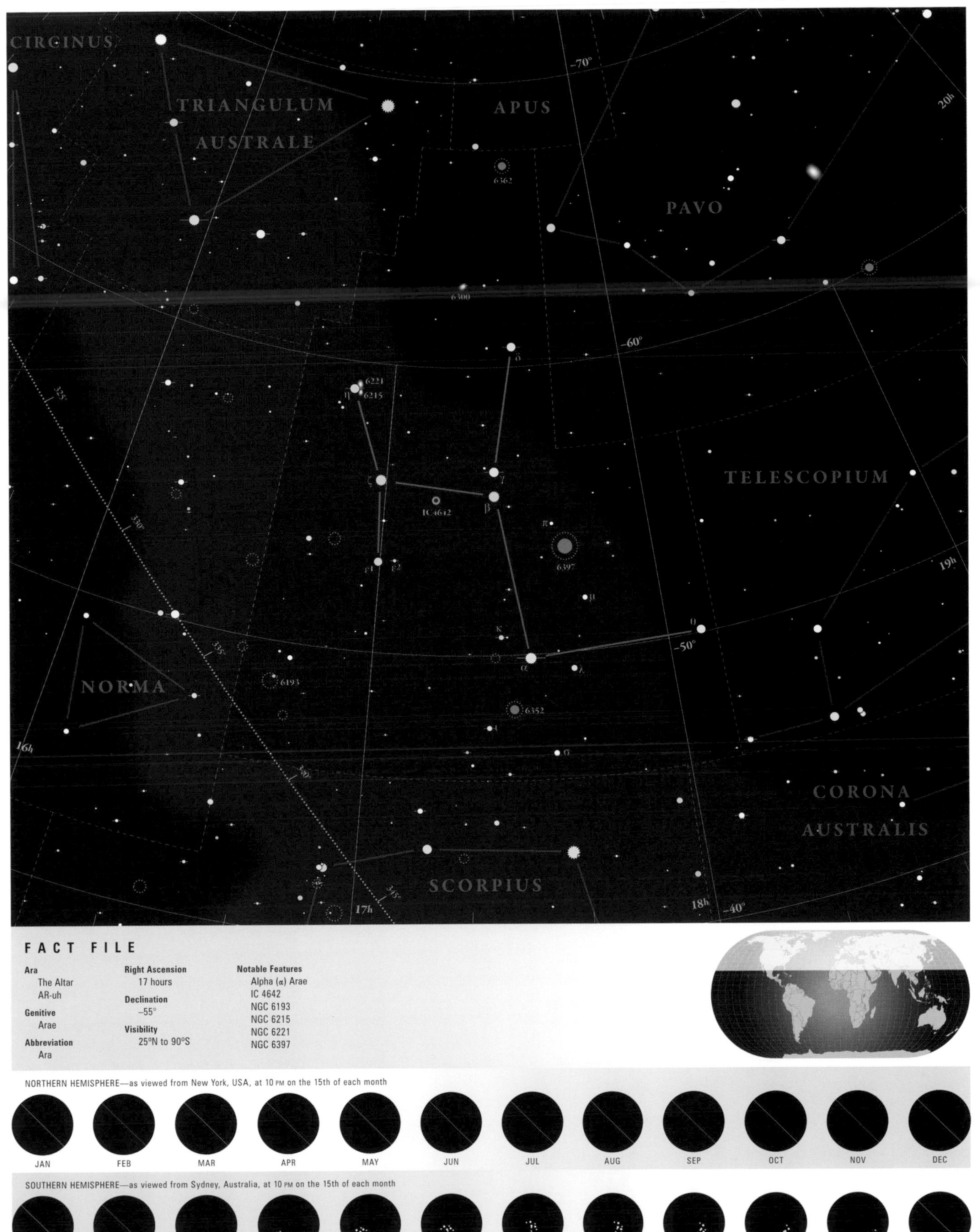

FACT FILE

Ara
The Altar
AR-uh

Genitive
Arae

Abbreviation
Ara

Right Ascension
17 hours

Declination
–55°

Visibility
25°N to 90°S

Notable Features
Alpha (α) Arae
IC 4642
NGC 6193
NGC 6215
NGC 6221
NGC 6397

NORTHERN HEMISPHERE—as viewed from New York, USA, at 10 PM on the 15th of each month

JAN FEB MAR APR MAY JUN JUL AUG SEP OCT NOV DEC

SOUTHERN HEMISPHERE—as viewed from Sydney, Australia, at 10 PM on the 15th of each month

JAN FEB MAR APR MAY JUN JUL AUG SEP OCT NOV DEC

Aries

The stars of Aries represented renewal for ancient cultures because this is where the vernal equinox occurred. Later, the stars came to represent the fabled ram with the Golden Fleece.

A FLEECE OF GOLD

The interpretation of Aries as a ram most likely originated with the Sumerians, and subsequent civilizations incorporated it into their own mythologies. The Greeks assimilated the ram into the epic tale of Jason and the Argonauts.

Hermes (Mercury), the messenger of the gods, saw that King Thebes's children, the twins Phrixus and Helle, were being mistreated by their step-mother, so he sent a mystical winged ram to rescue them. They escaped by clinging to the ram as it flew to the kingdom of Colchis. There, King Aeetes received them gladly and gave his daughter into the hand of Phrixus. After the ram was sacrificed in a consecrated grove of trees, its fleece turned to gold and it was hung on a tree to be guarded by a dragon that never slept. Jason's adventures in recovering the fleece from King Aeëtes are outlined in the constellation description for Carina.

THE STARS OF ARIES

A few thousand years ago, the Sun reached its vernal equinox point (for the northern hemisphere) in Aries. Due to precession of the ecliptic, that point is now in Pisces, and in 600 years it will move into Aquarius. The most easily recognized part of Aries is the small flat triangle formed by 2nd magnitude Alpha (α), 3rd magnitude Beta (β), and 4th magnitude Gamma (γ) Arietis, which precedes the V-shaped Hyades asterism in Taurus by about 2.5 hours. During the seventeenth century, the German astronomer Jakob Bartsch took several stars from Aries and formed a new constellation known as Vespa, the Wasp. It progressed through several incarnations as a fly or a bee until the twentieth century, when astronomers returned its stars to Aries.

Alpha (α) Arietis, or Hamal, from the Arabic meaning "the sheep" is the lucida of Aries. At 66 light-years distant, it is a nearby K-type giant star about twice as massive as the Sun and 90 times brighter.

OBJECTS OF INTEREST

The showpiece of Aries is **Gamma (γ) Arietis**, otherwise known as Mesarthim, a name derived from the Arabic for "first star in Aries." It is an excellent binary with nearly perfectly matched B-type and A-type stars that, curiously, look pale yellow to the eye. A 4-inch (10 cm) telescope shows them beautifully.

Only one low-power field from Mesarthim is **NGC 772** —the best of Aries's galaxies. With an 8-inch (20 cm) telescope, this lovely spiral galaxy has one unusually strong spiral arm, a faint oval halo, and a small, much brighter core. **NGC 691** is barely visible in a 6-inch (15 cm) telescope, but in an 8-inch (20 cm) it displays a small, faint, round halo with a weakly brighter spot in the center.

Below The giant unbarred spiral galaxy NGC 772 resides in Aries. During 2003, two supernovae were discovered here within three weeks.

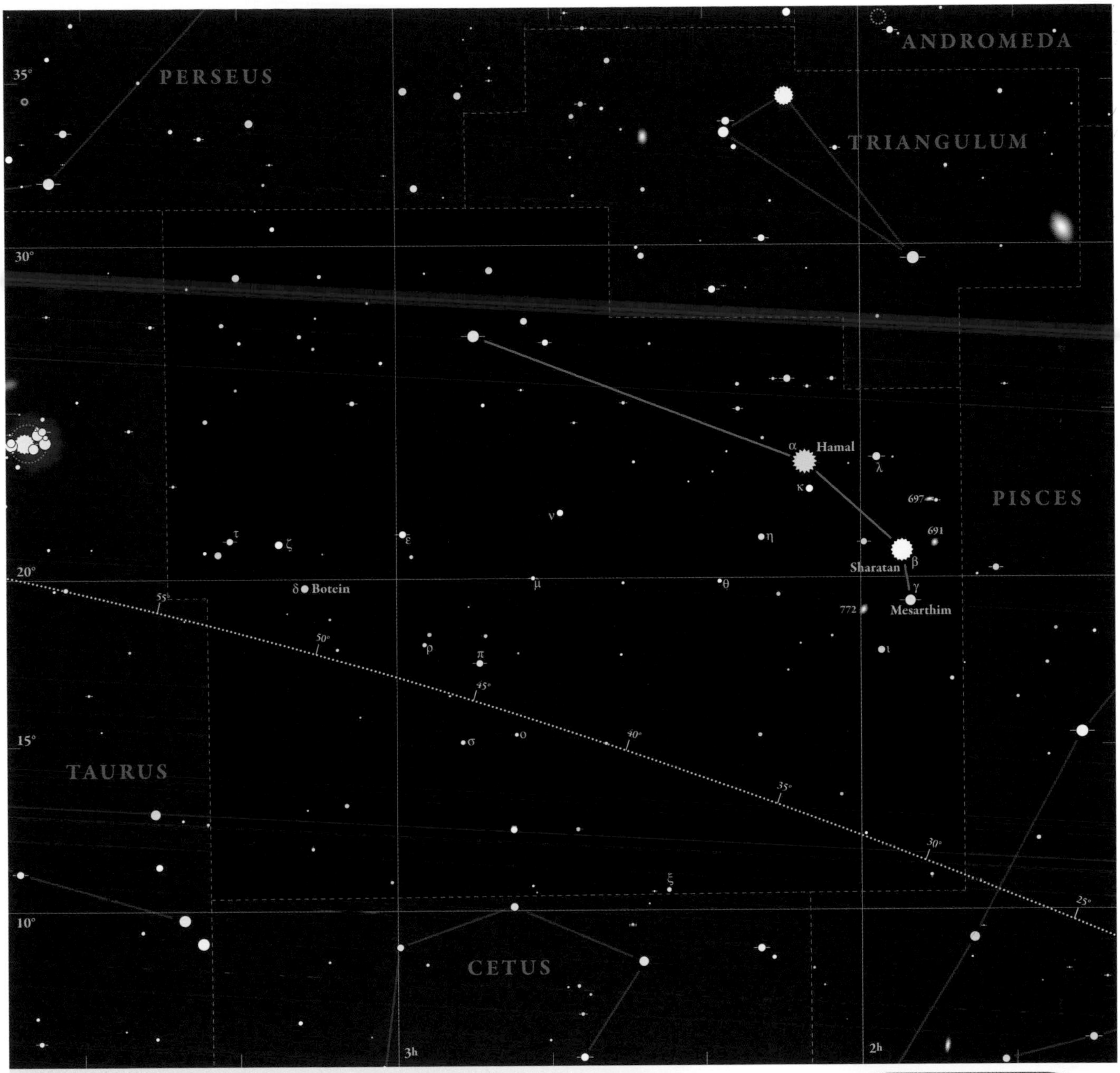

ANDROMEDA
PERSEUS
TRIANGULUM
35°
30°
20°
15°
10°
α Hamal
λ
κ
697
PISCES
ν
691
τ
ζ
ε
η
Sharatan β
γ
Mesarthim
772
δ Botein
μ
θ
ρ
π
ι
σ
ο
TAURUS
ξ
CETUS
3h
2h

FACT FILE

Aries
AIR-eez
The Ram

Genitive
Arietis

Abbreviation
Ari

Right Ascension
3 hours

Declination
+20°

Visibility
90°N to 62°S

Notable Features
Alpha (α) Arietis
Beta (β) Arietis
Gamma (γ) Arietis
NGC 691
NGC 772

Named Stars
Hamal (Alpha [α] Arietis)
Sharatan (Beta [β] Arietis)
Mesarthim (Gamma [γ] Arietis)
Botein (Delta [δ] Arietis)

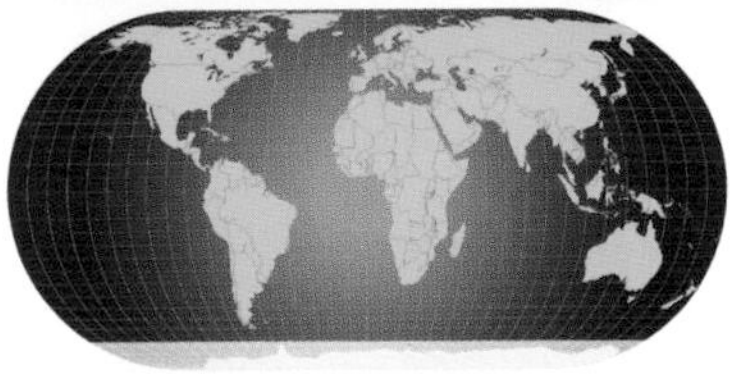

NORTHERN HEMISPHERE—as viewed from New York, USA, at 10 PM on the 15th of each month

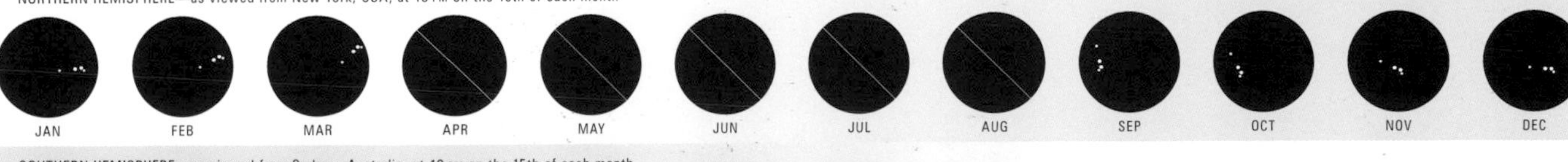

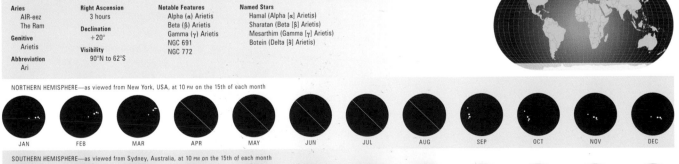

SOUTHERN HEMISPHERE—as viewed from Sydney, Australia, at 10 PM on the 15th of each month

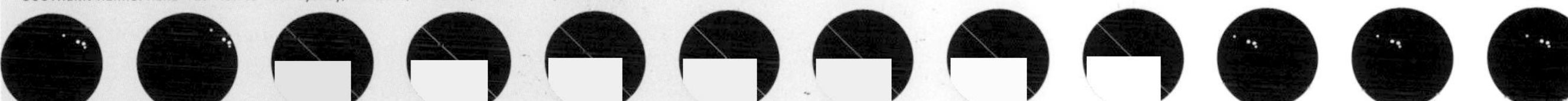

Auriga

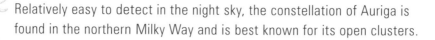

Relatively easy to detect in the night sky, the constellation of Auriga is found in the northern Milky Way and is best known for its open clusters.

RIDING THE CHARIOT

Auriga is known as a charioteer, but there is no significant mythology associated with the figure. Indeed, the mythology is rather muddled because the charioteer is depicted holding a she-goat (represented by the bright star Capella) and three kids (represented by the thin triangle of stars just to the southwest of Capella).

THE STARS OF AURIGA

The brightest stars in the constellation form a pentagon, although the southernmost star of the figure has been shared with Taurus since antiquity and professional astronomers have officially assigned it to Taurus. Alpha (α) Aurigae, the golden-yellow magnitude 0.08 Capella, is the sixth brightest star in the sky.

Epsilon (ε) Aurigae, one of the three "kids," is a supergiant and one of the most luminous stars known. With an estimated absolute magnitude of –8, it shines brightly despite being perhaps 2,000 light-years distant. Epsilon is also a famous eclipsing binary star, fading from magnitude 2.92 to 3.83 once every 27 years. Lasting a year, these eclipses are caused by a far larger but less massive semi-transparent companion.

OBJECTS OF INTEREST

Of Auriga's many open clusters, the three best lie in an arc across the central part of the pentagon. These clusters are a challenge for the unaided eye, but are obvious in binoculars, and are showpieces for small telescopes. **M37 (NGC 2099)**, one of the sky's dozen best open clusters, offers a dense cloud of similar-brightness stars, highlighted by a central orange gem that lies at the northern tip of a S-shaped star-chain. Of M37's more than 1,800 suns, about 150 are revealed by moderate apertures.

M36 (NGC 1960) is a fairly sparse galactic cluster whose brighter members form an oblique cross. A double star lies at its heart.

M38 (NGC 1912) looks like the Greek letter π (pi) or a miniature replica of the constellation Perseus. A much smaller cluster, the rich and compressed **NGC 1907**, lies in the same field of view. Such a tiny group would usually be much more distant, but both clusters are approximately 4,000 light-years away.

IC 410 is an emission nebula whose gas is excited to fluorescence by the stars of the associated open cluster, **NGC 1893**. At a dark site, an 8-inch (20 cm) telescope equipped with an OIII filter reveals the horseshoe-shaped nebulosity, bejeweled with stars. IC 410 looks like a thicker version of the better known Crescent Nebula in Cygnus.

A fine-matched double star for small refractors is **41 Aurigae**, featuring white and lilac components at 8 arc-seconds apart. Then turn the telescope to the wider lemon and purple pair, **14 Aurigae**. Larger apertures add a third member of magnitude 11.

Left The spectacular lilac hue of the Flaming Star Nebula, known also as IC 405, is a mixture of blue and red light emitted by the star AE Aurigae reflected to Earth by the surrounding dust. About 1,500 light-years distant, it is visible in a small telescope.

LYNX
CAMELOPARDALIS
PERSEUS
Capella
Menkalinan
Almaaz
Hoedus II
Hoedus I
Hassaleh
Elnath
TAURUS
GEMINI
ORION

FACT FILE

Auriga
or-RYE-gah
The Charioteer

Genitive
Aurigae

Abbreviation
Aur

Right Ascension
6 hours

Declination
+40°

Visibility
90°N to 31°S

Notable Features
Alpha (α) Aurigae
Epsilon (ε) Aurigae
M36
M37
M38
NGC1893
IC 410

Named Stars
Capella (Alpha [α] Aurigae)
Menkalinan (Beta [β] Aurigae)
Almaaz (Epsilon [ε] Aurigae)
Hoedus I (Zeta [ζ] Aurigae)
Hoedus II (Eta [η] Aurigae)
Hassaleh (Iota [ι] Aurigae)

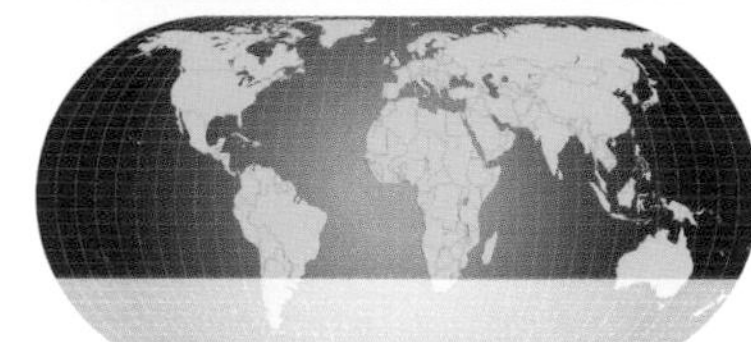

NORTHERN HEMISPHERE—as viewed from New York, USA, at 10 PM on the 15th of each month

JAN
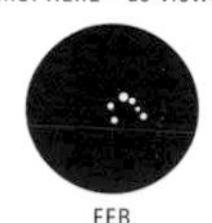
FEB
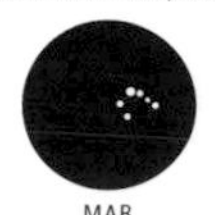
MAR
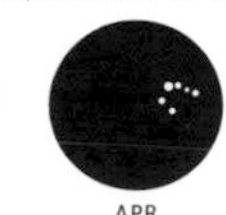
APR

MAY

JUN

JUL
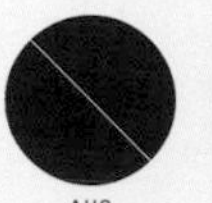
AUG
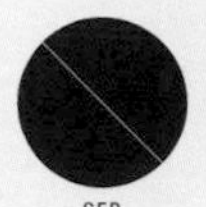
SEP

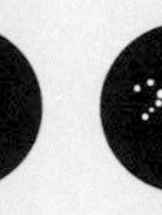
OCT
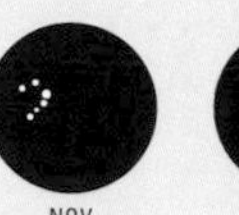
NOV

DEC

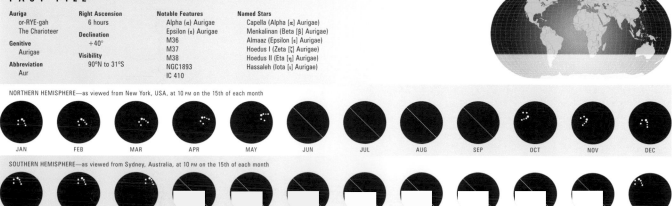

SOUTHERN HEMISPHERE—as viewed from Sydney, Australia, at 10 PM on the 15th of each month

JAN

FEB

MAR

APR

MAY

JUN

JUL

AUG

SEP
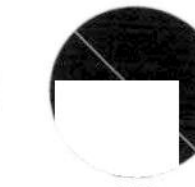
OCT

NOV

DEC

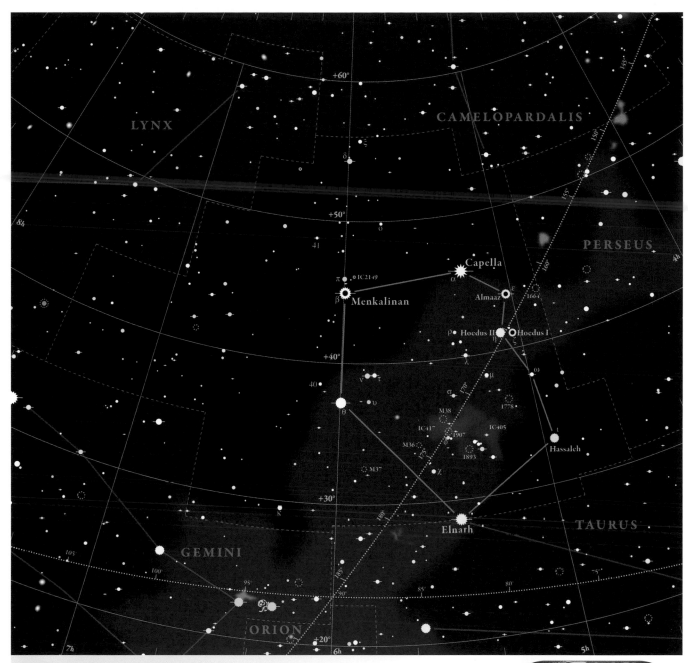

Boötes

The large constellation of Boötes, the Herdsman, is best known for its attractive double stars. In countries where the Big Dipper is known as the Plow, Boötes is the Plowman.

HUNT AROUND THE POLE

In Greek mythology, Boötes, the son of Zeus and Callisto, continually drives Ursa Major (the Great Bear) around the North Pole, aided by two hunting dogs that are represented by the constellation Canes Venatici.

THE STARS OF BOÖTES

At magnitude –0.05, champagne-colored Arcturus is the fourth brightest star in the sky, and the brightest in the northern half of the sky. It is just passing through our Sun's neighborhood in our galaxy's thin disk, since Arcturus is actually a member of the galaxy's halo. This implies that this giant star was originally a member of one of the many small galaxies that the Milky Way has consumed over the life of the universe.

Following the arc of the Big Dipper asterism's handle will lead to Arcturus—remember the phrase "arc to Arcturus." The arc passes along the western side of Boötes, which is shaped like an icecream cone.

OBJECTS OF INTEREST

Although there is a score of galaxies here bright enough for moderate apertures, none can compare with the showpieces found in the four constellations along the western border of Boötes. The brightest, at magnitude 10.1, is **NGC 5248**, an elongated spiral galaxy with a brighter core. Larger telescopes show some mottled structure.

The magnitude 9 globular cluster **NGC 5466** can easily be found with a 4-inch (10 cm) telescope, and an 8-inch (20 cm) will resolve about a dozen sparks scattered across a faint background glow. It is a loose concentration Class XII globular, resembling an open cluster. A 16-inch (40 cm) telescope brings out more stars, arranged in three arcs.

Mu (μ) Boötes is a wide, yellowish white and deep yellow double star of magnitudes 4.3 and 6.5, separated by 108 arc-seconds. Its secondary resolves into a 2.2 arc-seconds double of magnitudes 7.0 and 7.6, featuring yellow and orange components.

All of the following binaries can be enjoyed with a good-quality 2½-inch (6 cm) refractor. **Pi (π) Boötis** is a yellowish pair, of magnitudes 4.9 and 5.9 at 5.5 arc-seconds apart. **Epsilon (ε) Boötis**, or Izar (from the Arabic for "veil"), is a glorious gold and blue binary of magnitudes 2.3 and 4.5 at a separation of 2.9 arc-minutes. Some observers see the companion as greenish in small refractors. The magnitude difference prevents splitting this famous pair with any telescope unless the atmosphere is steady, but the reward will be worth the effort of trying on several nights.

Xi (χ) Boötis is a yellow and reddish-orange pair, of magnitudes 4.7 and 6.9 at 6.6 arc-seconds. Their separation varies from 1.8 to 7.3 arc-seconds during their 152-year orbit. **Struve 1910** is a classic headlight double: Two magnitude 7.5 yellow suns at 4.3 arc-seconds.

Above The stars of the constellation Boötes are shown against a background of approximately 300,000 fainter galaxies and stars, revealed by the National Optical Astronomy Observatory's Deep Wide-Field Survey.

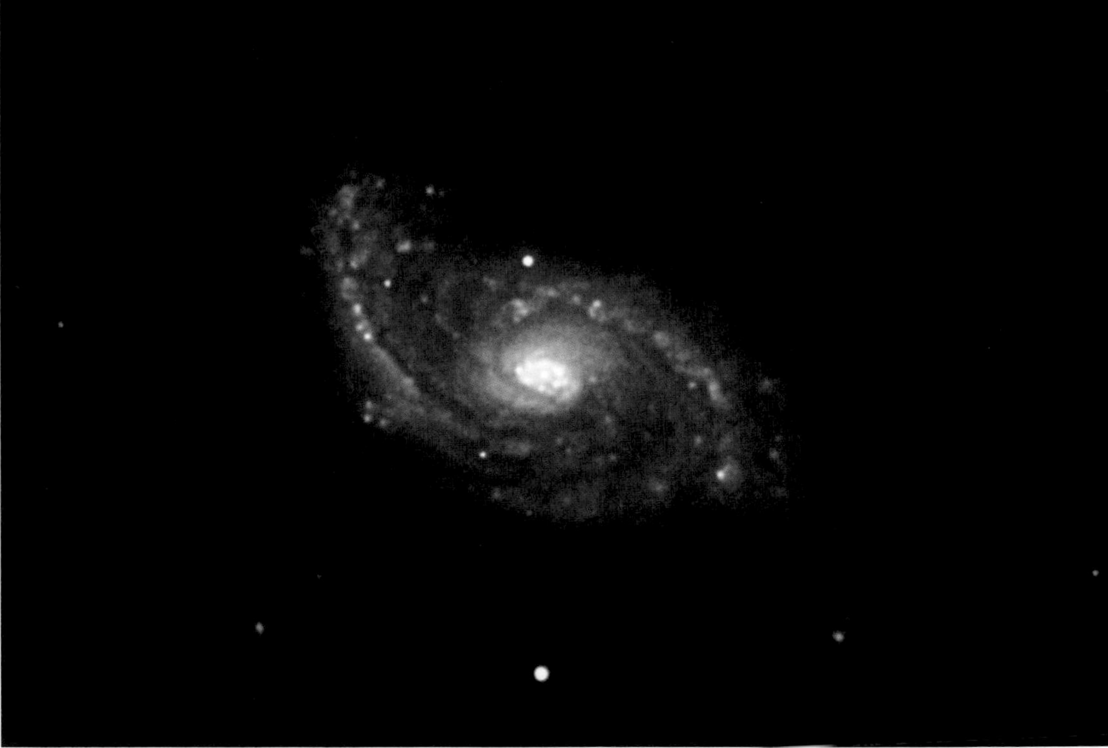

Right The elongated spiral galaxy NGC 5248 is the brightest galaxy in the Boötes region, and a great object for a telescope with a moderate aperture.

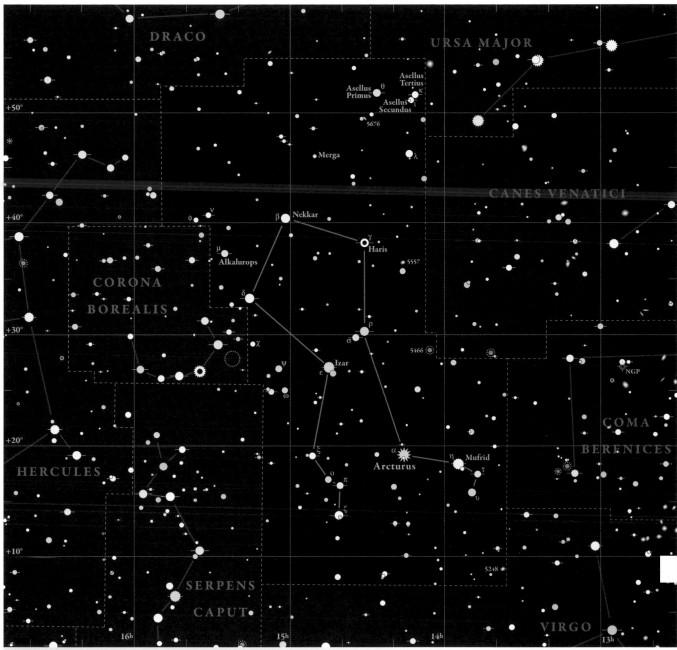

DRACO
URSA MAJOR
CANES VENATICI
CORONA BOREALIS
HERCULES
SERPENS CAPUT
COMA BERENICES
VIRGO
Asellus Primus θ
Asellus Tertius κ
Asellus Secundus ι
5676
Merga
λ
β Nekkar
γ Haris
5557
μ Alkalurops
δ
ρ
σ
χ
5466
ψ
ε Izar
ω
NGP
ξ
α Arcturus
η Mufrid
τ
υ
ο
π
ζ
5248
+50°
+40°
+30°
+20°
+10°
16h
15h
14h
13h

FACT FILE

Boötes
The Herdsman
bo-OH-teez

Genitive
Bootis

Abbreviation
Boo

Right Ascension
15 hours

Declination
+30°

Visibility
90°N to 45°S

Notable Features
NGC 5248
NGC 5466
NGC 5557
NGC 5676

Named Stars
Arcturus (Alpha [α] Bootis)
Nekkar (Beta [β] Bootis)
Haris (Gamma [γ] Bootis)
Izar (Epsilon [ε] Bootis)
Mufrid (Eta [η] Bootis)
Asellus Primus (Theta [θ] Bootis)
Asellus Secundus (Iota [ι] Bootis)
Asellus Tertius (Kappa [κ] Bootis)
Alkalurops (Mu [μ] Bootis)
Merga (38 Bootis)

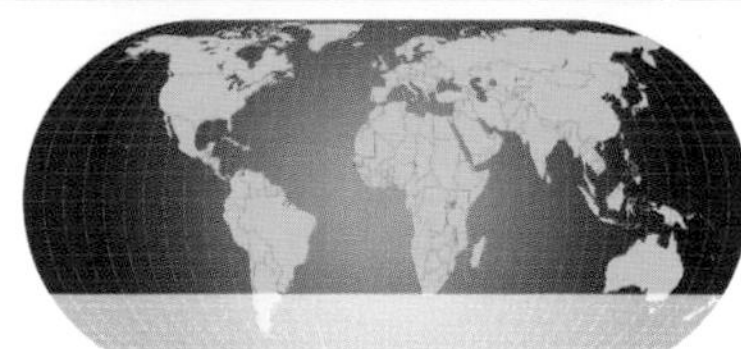

NORTHERN HEMISPHERE—as viewed from New York, USA, at 10 PM on the 15th of each month

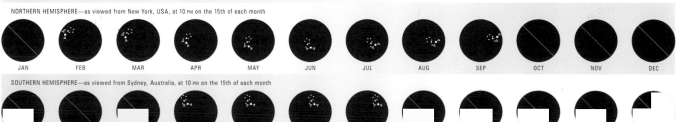

JAN

FEB

MAR
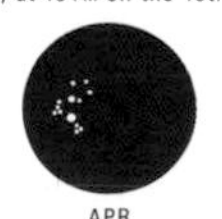
APR

MAY

JUN

JUL

AUG

SEP

OCT
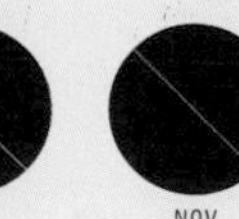
NOV
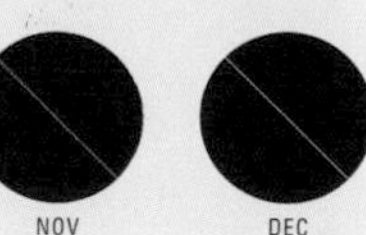
DEC

SOUTHERN HEMISPHERE—as viewed from Sydney, Australia, at 10 PM on the 15th of each month

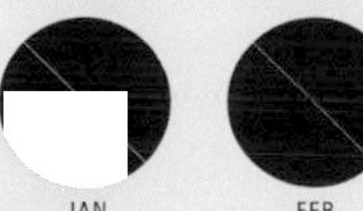
JAN

FEB

MAR

APR

MAY

JUN

JUL

AUG
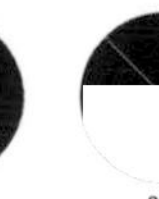
SEP

OCT

NOV

DEC

Caelum

For the amateur sky-watcher, the stars of the sculptor's chisel are among the most difficult to discern.

ETCHED IN THE HEAVENS

This small desolate patch of sky, jammed between Columba and Eridanus, had to wait until Abbé Nicolas Louis de Lacaille devised a figure to fill it during his visit to the Cape of Good Hope Observatory between 1751 and 1752. Lacaille initially named it "les Burins," and drew it as a pair of engraving tools tied together with ribbon. Although its stars were visible to ancient astronomers in Europe and the Middle East, there are no legends or myths associated with it.

Several years later, Lacaille Latinized the name to Caelum Scalptorium, now shortened simply to Caelum; it represents a burin—a type of chisel used to engrave metal. Since that time, its stars and form have undergone no significant change.

THE STARS OF CAELUM

The dim scattered stars of Caelum offer little support for their depiction as a chisel. Only one 4th magnitude star and a few of 5th magnitude adorn the group.

Alpha (α) Caeli is at the dim end of 4th magnitude, while Gamma (γ) and Beta (β) are just a little brighter than 5th magnitude. Alpha is a relatively nearby star, just 66 light-years away, and not too dissimilar to the Sun. It is an F-type star about five times brighter and one-and-a-half times as massive as our luminary. It has a tiny M-class red dwarf companion in a very wide orbit. This diminutive mote, just visible in a large telescope, has about 30 percent of our Sun's mass but only one-hundredth of its brilliance.

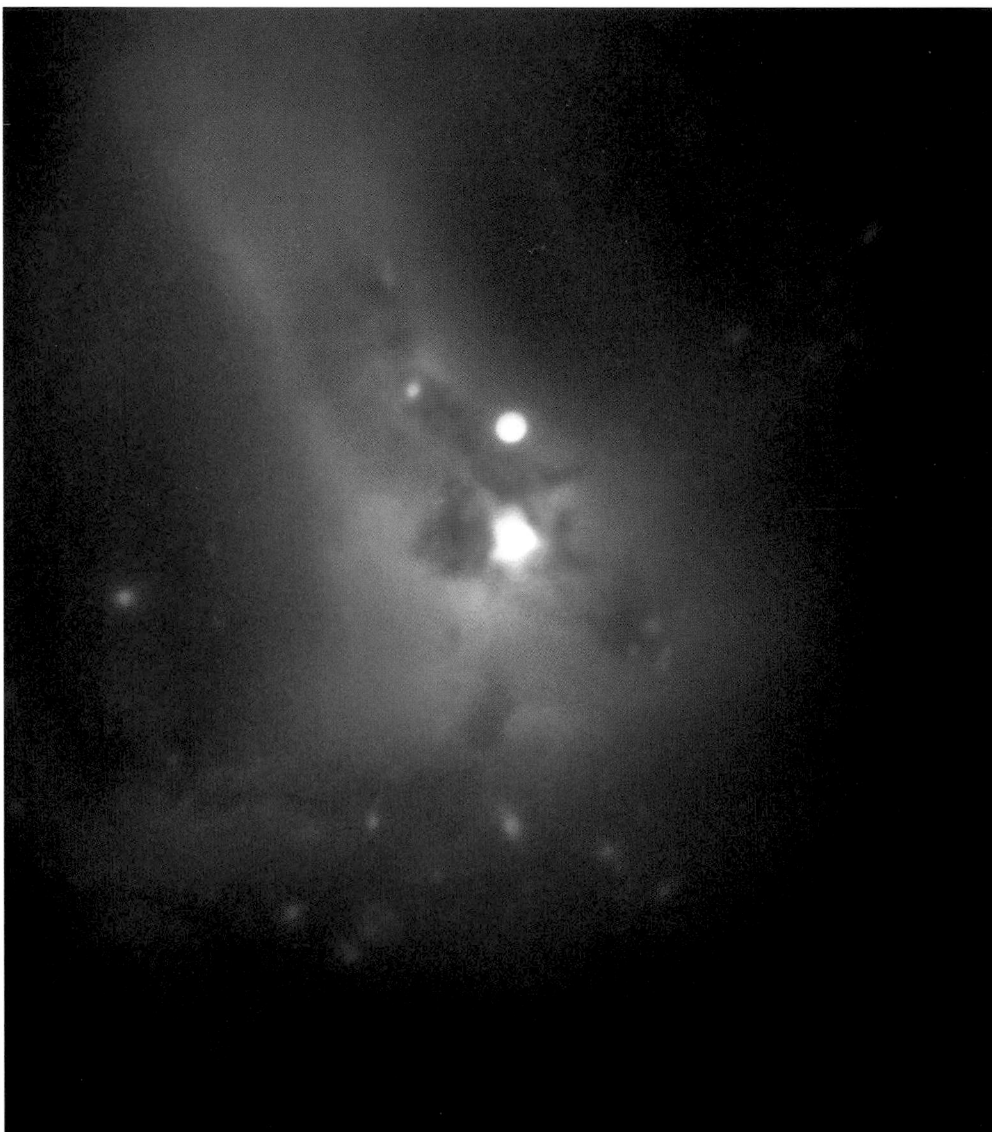

OBJECTS OF INTEREST

Like some of its adjoining constellations, such as Horologium and Pictor, Caelum is a celestial desert. Situated far from the Milky Way, it has no clusters or nebulae, but instead features just a few faint galaxies and some double stars.

Gamma (γ) Caeli is a tight, unequal pair consisting of a 5th magnitude yellow star and an 8th magnitude white star, which look attractive in a 4-inch (10 cm) telescope. There is another white-colored pair with less than one-third of the separation in the next telescope field to the south.

All of the galaxies in this constellation are faint, and generally of little interest. The best of these is the galaxy trio—**NGC 1595**, **NGC 1598**, and **ESO 202-23** —which are located in the far south of the constellation. However, even in an 8-inch (20 cm) telescope under a dark rural sky, three tiny smudges of haze with weakly brighter centers are all that can be made out by amateur sky-watchers.

Left This false-color composite image from the European Southern Observatory's Very Large Telescope (VLT) shows the center of the merging galaxy system known as ESO 202-23. This faint galaxy is located in the southern regions of the inconspicuous constellation of Caelum.

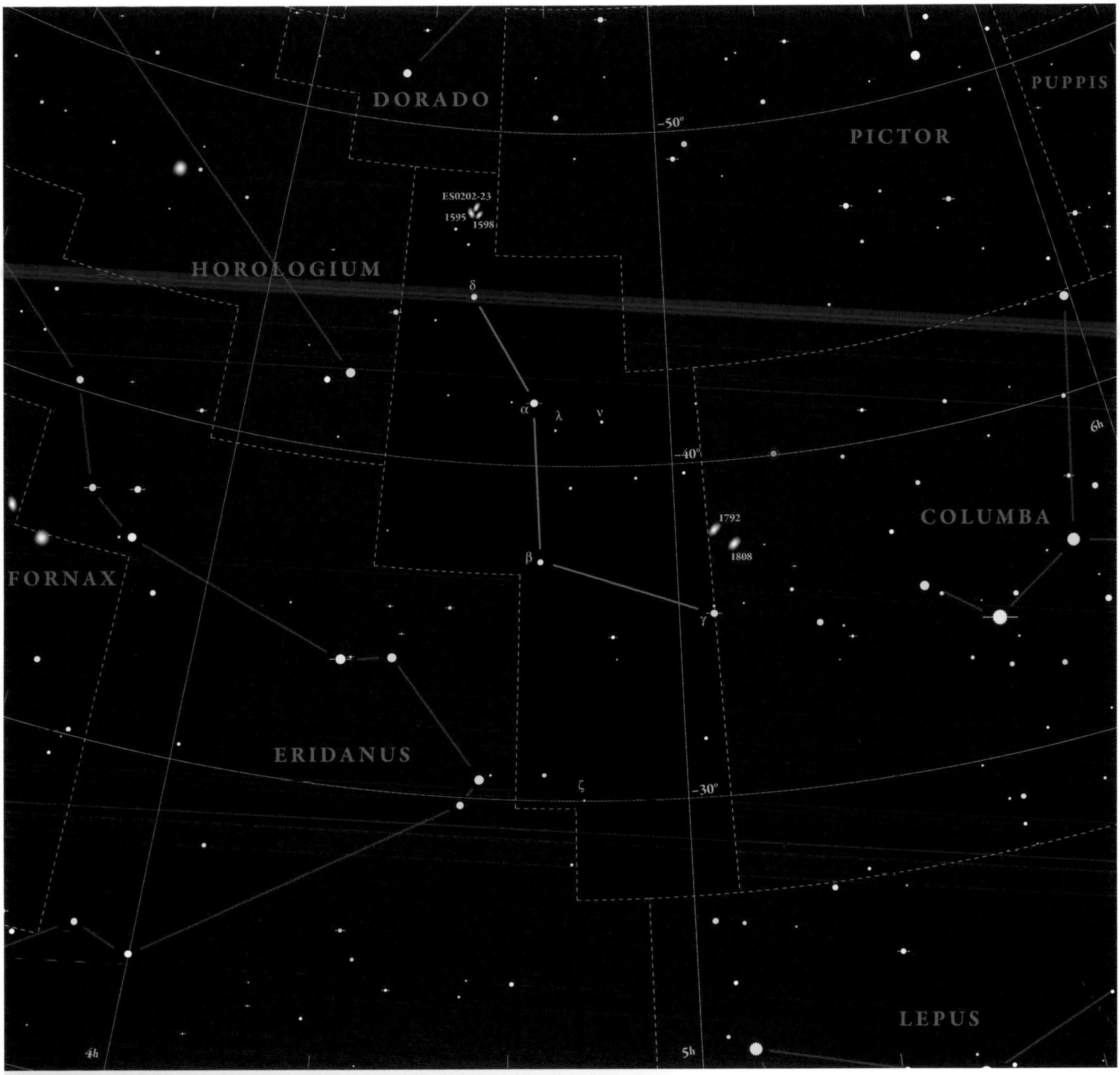

DORADO
PUPPIS
PICTOR
-50°
ESO202-23
1595
1598
HOROLOGIUM
δ
α λ ν
-40°
6h
1792
1808
COLUMBA
FORNAX
β
γ
ERIDANUS
ζ
-30°
LEPUS
4h
5h

FACT FILE

Caelum
SEE-lum
The Chisel

Genitive
Caeli

Abbreviation
Cae

Right Ascension
5 hours

Declination
–40°

Visibility
40°N to 90°S

Notable Features
NGC 1595
NGC 1598
ESO 202-23

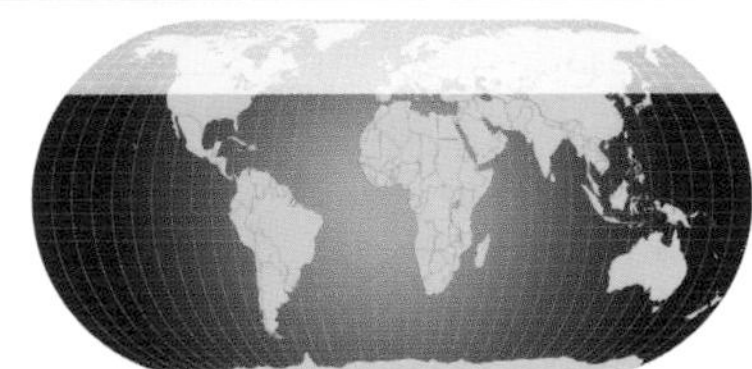

NORTHERN HEMISPHERE—as viewed from New York, USA, at 10 PM on the 15th of each month

JAN

FEB

MAR

APR

MAY

JUN

JUL

AUG
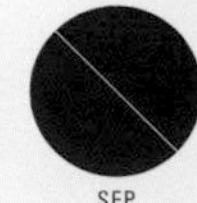
SEP

OCT
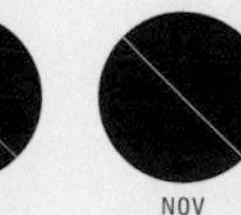
NOV
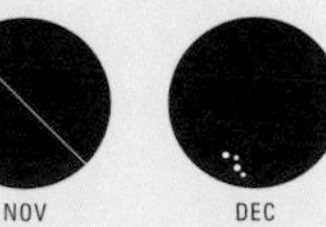
DEC

SOUTHERN HEMISPHERE—as viewed from Sydney, Australia, at 10 PM on the 15th of each month

JAN

FEB

MAR

APR

MAY

JUN
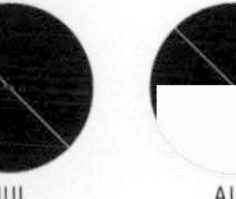
JUL
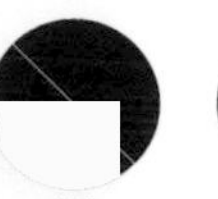
AUG

SEP

OCT

NOV

DEC

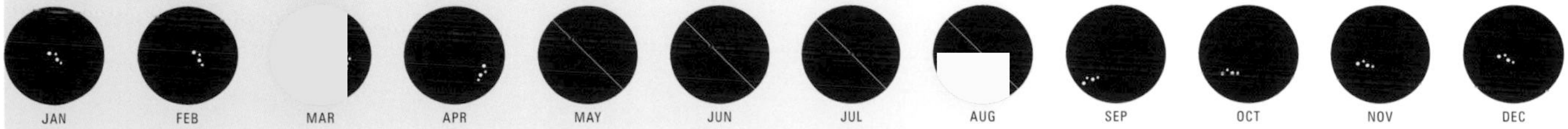

Camelopardalis

Camelopardalis, the Giraffe, is an obscure seventeenth-century constellation fashioned from a large but relatively dim swathe of the northern sky.

HEADS UP

German astronomer Jakob Bartsch created Camelopardalis in the early seventeenth century. He also created two other constellations honoring animals, Monoceros and Columba (the Unicorn and the Dove, respectively). Other references credit Petrus Kaerius of the Netherlands with first drawing Camelopardalis on his celestial globes. Until recently, this constellation was frequently called Camelopardus.

THE STARS OF CAMELOPARDALIS

Eight stars, ranging from magnitude 4.0 down to 4.8, outline a vaguely giraffe-shaped figure that few amateur astronomers have traced out more than once or twice. Finding the correct stars is made more difficult by the many only slightly fainter ones sprinkling this constellation, especially in its southern parts, which are within the Milky Way.

OBJECTS OF INTEREST

Most amateur astronomers begin their starhops from jumping-off points in one of the adjacent much brighter constellations. Through 7 × 50 binoculars the open cluster **NGC 1502** looks rather like a ledge in a waterfall because it lies between the delightful 2.5-degree-long chain of mainly 8th magnitude stars to the northwest known as **Kemble's Cascade**, and a short arc of four stars tumbling southward. In a telescope, bright and compact NGC 1502 features a band of four doubles across its center.

The magnitude 11.5 bluish planetary nebula **NGC 1501**'s central star is visible at high power with a 10-inch (25 cm) instrument at the darker center of the disk that is 52 arc-seconds in diameter.

NGC 2403 is one of the nearer spirals and is gravitationally bound to the showpiece galaxies M81 and M82 in Ursa Major. Magnitude 8.5 NGC 2403 has a fairly bright core, but its halo is somewhat overpowered by two bright foreground stars situated near either end of the 12 arc-minute oval. Large telescopes show some mottling.

An even closer spiral is **IC 342**. Once considered to be a member of the Local Group of galaxies, it is now thought to be located just outside our group. Unfortunately, IC 342 lies behind the Milky Way, which masks much of its light. All that can be seen of this large, face-on, magnitude 12 galaxy through an 8-inch (20 cm) or preferably larger telescope is a faint amorphous patch camouflaged by many foreground stars; its nucleus is hidden among these stars. However, this galaxy's interest lies not so much in the few details that are visible as in the knowledge that you are seeing light from what is probably the nearest spiral galaxy after M31 and M33.

The fine binary **Struve 1694** represents the giraffe's head in this constellation. Separated by nearly 22 arc-minutes (just over a half-moon diameter), it is pale yellow and lilac in small achromatic refractors. Other opinions on the colors range from bluish-white and greenish to simply both white.

Above In 2004, NGC 2403, located just beyond the thirty galaxies we know as the Local Group, was home to one of the brightest and closest supernovae observed in modern times.

Right Usually veiled by stars, dust clouds, and cosmic gases, the pink, star-forming spiral arms of the majestic IC 342 are clearly visible in this image taken under dark skies at Kitt Peak National Observatory, Arizona, USA.

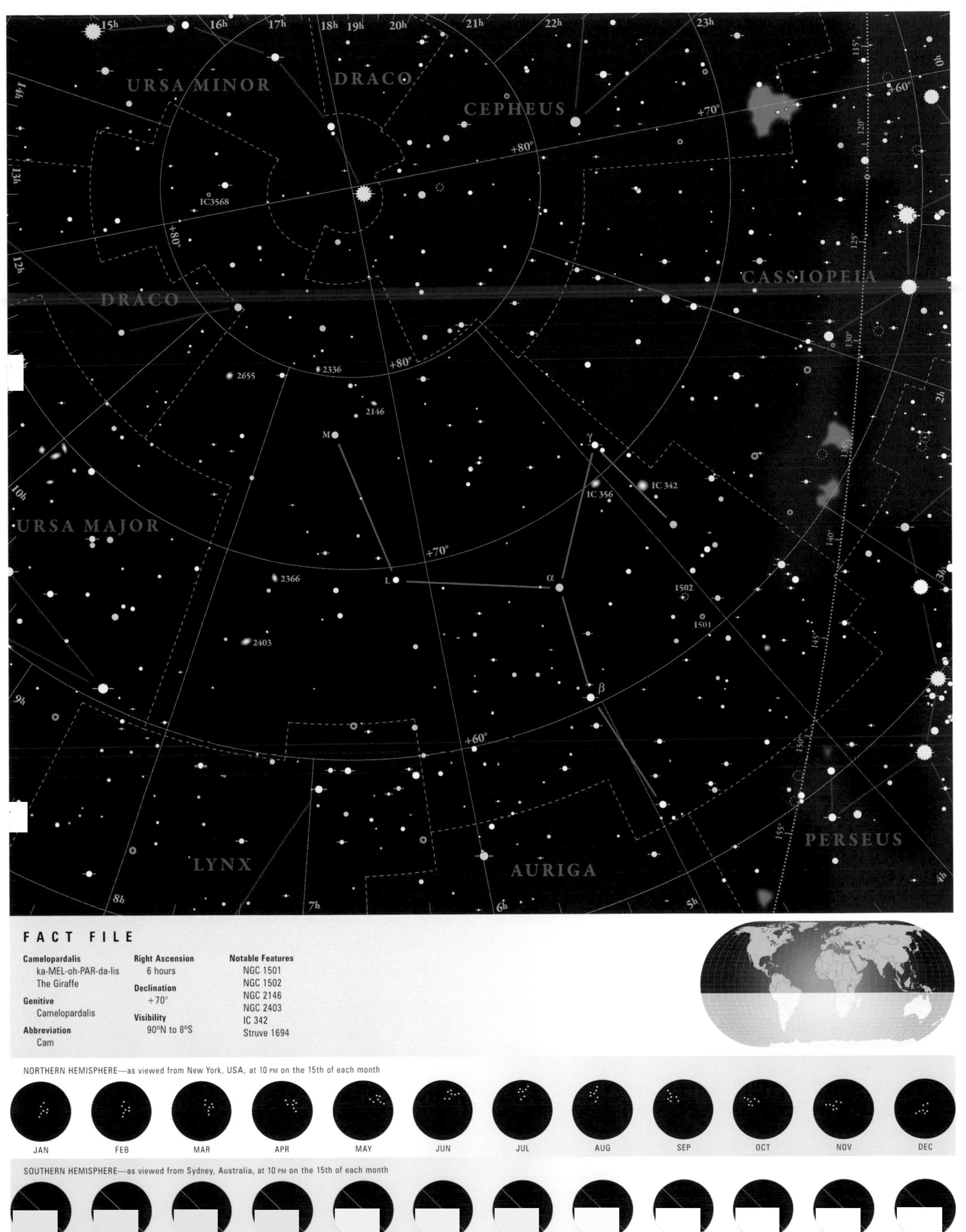

FACT FILE

Camelopardalis
ka-MEL-oh-PAR-da-lis
The Giraffe

Genitive
Camelopardalis

Abbreviation
Cam

Right Ascension
6 hours

Declination
+70°

Visibility
90°N to 8°S

Notable Features
NGC 1501
NGC 1502
NGC 2146
NGC 2403
IC 342
Struve 1694

NORTHERN HEMISPHERE—as viewed from New York, USA, at 10 PM on the 15th of each month

JAN FEB MAR APR MAY JUN JUL AUG SEP OCT NOV DEC

SOUTHERN HEMISPHERE—as viewed from Sydney, Australia, at 10 PM on the 15th of each month

JAN FEB MAR APR MAY JUN JUL AUG SEP OCT NOV DEC

Cancer

Overshadowed by the spectacular neighboring constellations of Gemini and Leo, the starry crustacean plays a mere cameo role in the night sky.

A CRUSHING BLOW

Most mythological figures faced unfathomable danger, performed the almost impossible, or slew an implacable beast to gain their place in the starry firmament. Cancer's role, however, was both short and unspectacular (see the constellation description for Hydra, the serpent). As Heracles (Latinized as Hercules) fought to subdue the serpent, Cancer emerged from the swamp and attacked his feet. Heracles crushed him with his foot, and Zeus's wife, Hera, an enemy of Heracles, promoted Cancer to the sky.

THE STARS OF CANCER

Cancer is the faintest of all the zodiacal constellations. From urban and many suburban locations, it simply appears as a blank space between the constellations of Gemini and Leo. It is inconspicuous even in a rural sky, and there is no resemblance between its mainly 5th and 6th magnitude stars and the outline of a crab.

Above right The best views of Praesepe (M44), the Beehive Cluster, are through a finderscope or binoculars. This open star cluster is about 600 light-years away.

Beta (β) Cancri is the brightest star in the constellation, with Alpha (α) being dimmer than at least three others as well. Beta Cancri is a K-type giant star about three times more massive than our Sun and 48 times brighter, living out the last parts of its life approximately 290 light-years away. More interesting are Gamma (γ) and Delta (δ) Cancri near the center of the figure, which are named in Latin "Asellus Borealis" and "Asellus Australis"—the northern and southern ass, respectively.

As told by Eratosthenes, after the Olympian gods had overthrown the Titans they had to defeat the giants. The two Olympian gods Dionysus and Hephaestus rode to the clash on donkeys. As they came near their donkeys were braying, and the giants, never having heard the bray of a donkey before, fled in terror. Dionysus elevated the humble donkeys to the heavens, placing them opposite each other on either side of a small cloudy patch that the Greeks called Phatne—"the manger." We now know that this cloudy patch is a star cluster popularly called the Beehive, but also known as Praesepe, a Latin name that can mean both "manger" and "hive."

OBJECTS OF INTEREST

Praesepe, or **M44**, is without doubt Cancer's showpiece object. A huge open cluster at more than a moon's diameter, it was known to the ancients and was easily visible to the naked eye in a dark sky almost between Gamma (γ) and Delta (δ) Cancri. Binoculars show 10 to 20 stars evenly spread over a slightly hazy background, and a small telescope at low power reveals more than 50 stars.

M67 is another excellent cluster near Alpha (α) Cancri. A medium-size telescope reveals a beautiful cloud of more than 50 medium-faint stars evenly spread over a moon's diameter of sky. Cancer also has an abundance of galaxies, of which **NGC 2775** is the best. An 8-inch (20 cm) telescope shows it as a small, round, slightly oval patch of mist with a much brighter center.

Left The M67 open star cluster is thought to be between four and five billion years old. It contains some 500 stars with magnitudes ranging from 10 to 16, although many more are fainter. They are most likely slightly younger than our Sun.

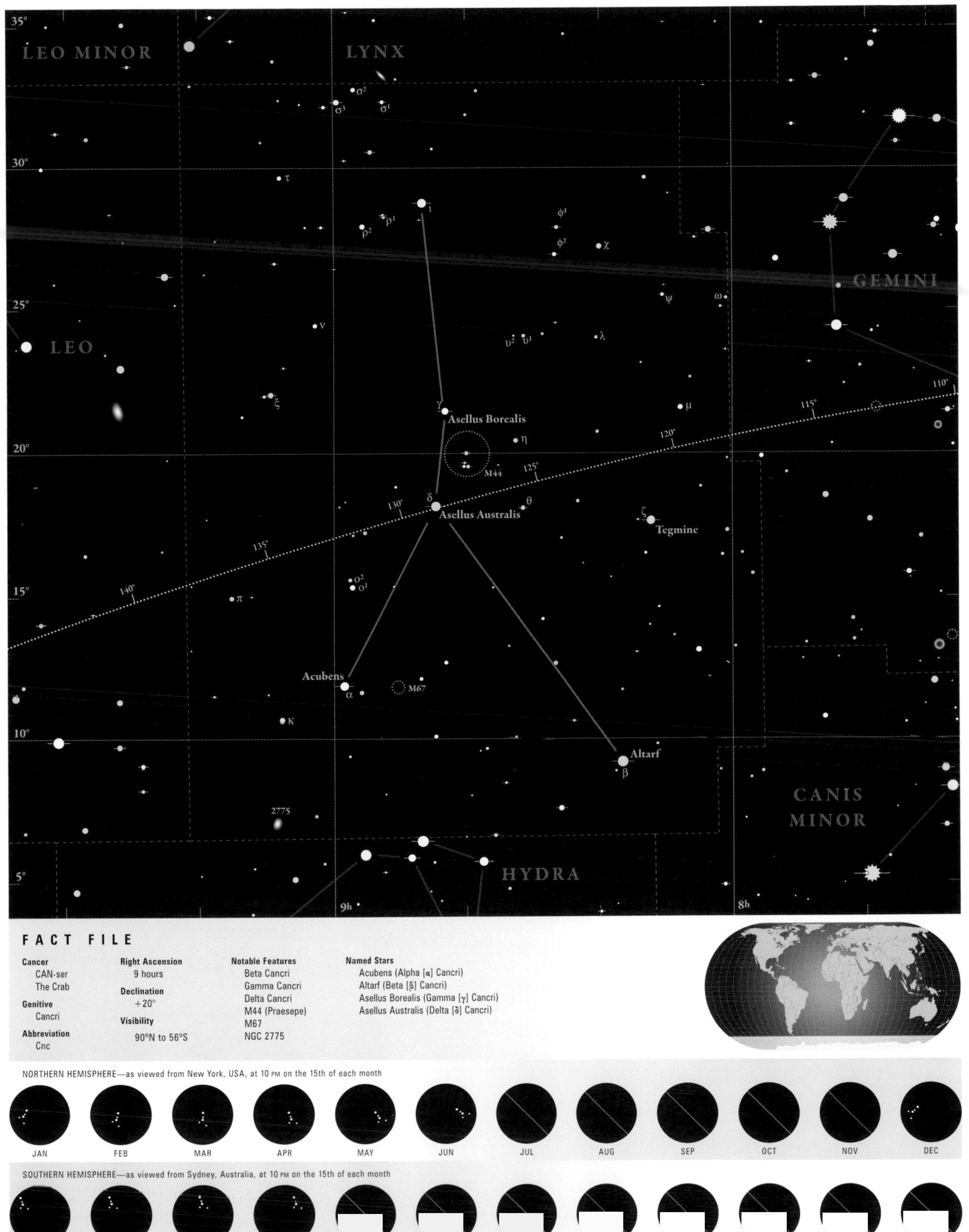

FACT FILE

Cancer
CAN-ser
The Crab

Genitive
Cancri

Abbreviation
Cnc

Right Ascension
9 hours

Declination
+20°

Visibility
90°N to 56°S

Notable Features
Beta Cancri
Gamma Cancri
Delta Cancri
M44 (Praesepe)
M67
NGC 2775

Named Stars
Acubens (Alpha [α] Cancri)
Altarf (Beta [β] Cancri)
Asellus Borealis (Gamma [γ] Cancri)
Asellus Australis (Delta [δ] Cancri)

NORTHERN HEMISPHERE—as viewed from New York, USA, at 10 PM on the 15th of each month

JAN FEB MAR APR MAY JUN JUL AUG SEP OCT NOV DEC

SOUTHERN HEMISPHERE—as viewed from Sydney, Australia, at 10 PM on the 15th of each month

JAN FEB MAR APR MAY JUN JUL AUG SEP OCT NOV DEC

Canes Venatici

This small northern constellation is one of three constellations representing dogs. It contains the famous Whirlpool Galaxy, as well as seven other showpieces.

LEADERS OF THE PACK

Canes Venatici, the Hunting Dogs of Boötes, chase the Great Bear endlessly around the pole. Although Boötes can be traced back to the Babylonians, the idea of providing him with hunting dogs is much more recent, dating from the seventeenth century.

THE STARS OF CANES VENATICI

The constellation has only two obvious stars, Alpha (α) and Beta (β) Canum Venaticorum. They represent the two dogs, named Asterion and Chara, but there is no agreement as to which star represents which dog—modern charts can be found that label Beta (β) with either name. In any case, Alpha (α) has been called Cor Caroli (meaning "Charles's Heart") since the restoration of the English monarchy, when Charles II ascended the vacant throne. Most references logically say that Cor Caroli refers to the beheaded King Charles I, but some say that it refers to his son, Charles II. Edmond Halley, who named the star, may have been intentionally ambiguous so as to reap the maximum reward as the second Astronomer Royal. Cor Caroli has remained a well-known name because it is a famous double star with widely separated components, tinted bluish and greenish.

OBJECTS OF INTEREST

M3 (NGC 5272) is a fine symmetrical globular cluster of magnitude 5.9 that shows some slight resolution of outliers with a 4-inch (10 cm) telescope; with a 16-inch (40 cm) it is glorious. It has numerous long star chains, especially on the western side, with a cloud of much fainter stars surrounding the chains. There is a rich inner halo, and the concentration rises to a wide bright core with countless stellar pinpoints.

The Whirlpool Galaxy, M51 (NGC 5194), is easy to spot with binoculars. It is magnitude 8.4 and less than one-third of a moon-diameter. Both M51 and its satellite galaxy, **NGC 5195**, have obvious bright nuclei in a 6-inch (15 cm) telescope. M51's high surface brightness makes it easier to discern the spiral structure in this face-on galaxy than in any other, except M33. An 8-inch (20 cm) telescope begins to show the arms as a splitring structure and also reveals the galaxy's brightest knot. With a 16-inch (40 cm) telescope, the two spiral arms are very prominent and there are five bright knots along them. This aperture also shows the companion galaxy's central bar.

Other bright spiral galaxies for binoculars are comet-like **M94 (NGC 4736)**, elongated **M63 (NGC 5055)**, and **M106 (NGC 4258)**. Telescopes show a bright core in M94 and a bright stellar nucleus in M106. Large scopes unveil M106's faint spiral arms with several knots in them. There are three long and mottled edge-on galaxies suitable for moderate apertures—the very thin **NGC 4244**, the aptly named **Humpback Whale Galaxy (NGC 4631)**, and the **Hockey Stick Galaxy (NGC 4656/7)**.

Left This Hubble Telescope image reveals the heart of the magnificent Whirlpool Galaxy (M51). Astronomers surmise that the spiral structure is due to M51's gravitational interactions with a smaller neighboring galaxy.

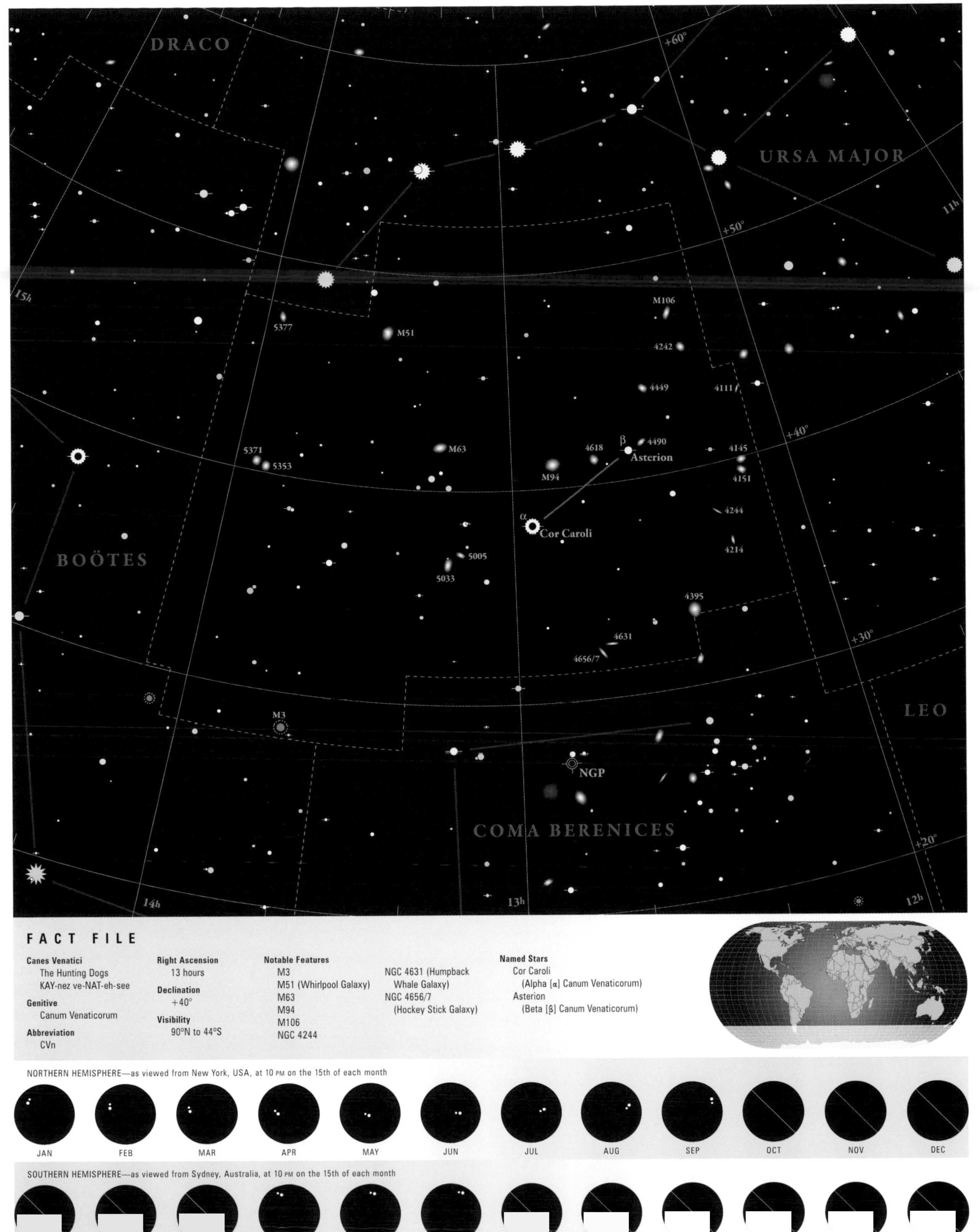

FACT FILE

Canes Venatici
The Hunting Dogs
KAY-nez ve-NAT-eh-see

Genitive
Canum Venaticorum

Abbreviation
CVn

Right Ascension
13 hours

Declination
+40°

Visibility
90°N to 44°S

Notable Features
M3
M51 (Whirlpool Galaxy)
M63
M94
M106
NGC 4244
NGC 4631 (Humpback Whale Galaxy)
NGC 4656/7 (Hockey Stick Galaxy)

Named Stars
Cor Caroli
(Alpha [α] Canum Venaticorum)
Asterion
(Beta [β] Canum Venaticorum)

NORTHERN HEMISPHERE—as viewed from New York, USA, at 10 PM on the 15th of each month

JAN FEB MAR APR MAY JUN JUL AUG SEP OCT NOV DEC

SOUTHERN HEMISPHERE—as viewed from Sydney, Australia, at 10 PM on the 15th of each month

JAN FEB MAR APR MAY JUN JUL AUG SEP OCT NOV DEC

Canis Major and Canis Minor

According to legend, Orion's hunting dogs—one large, one small—eternally pursued Lepus the Hare across the night sky.

OLD SKY DOGS

Canis Major and Minor were seen by Arab cultures as the faithful hunting dogs of Orion. This is reflected in an Arabic title for the constellation, Al Kalb al Jabbar, or the "Dog of the Giant." The Greek mythologists Eratosthenes and Hyginus thought Canis Major represented the hound Laelaps, a dog so swift no prey could outrun it.

THE STARS OF CANIS MAJOR AND CANIS MINOR

It takes relatively little imagination to discern a dog in the relatively bright stars of Canis Major. Brilliant Sirius marks its eye, with 2nd magnitude Beta (β) as a forepaw. The triangle formed by Omicron^2 (o^2), Delta (δ), and Epsilon (ε) marks the rump, and lines to Eta (η) and Zeta (ζ) make the tail and back leg, respectively.

Alpha (α) Canis Majoris is named Sirius, from the Greek word for "scorching" or "searing." Only 8.6 light-years distant, this A-class dwarf star is just a little more than twice as heavy as our Sun but 26 times as luminous—it is the most luminous star in our Sun's neighborhood and the brightest star in the night sky. Sirius is also a binary, having a close-in, dim, white-dwarf companion (Sirius B) almost exactly the mass of the Sun, but only 7,460 miles (12,000 km) in diameter. It is affectionately known as the "Pup," but seeing it is a severe test of both observer and telescope because of Sirius's overwhelming brilliance.

As for Canis Minor, the small dog, all the other stars are very faint, apart from Procyon and 3rd magnitude Beta (β). Alpha (α) Canis Minoris is known as Procyon, an ancient Greek name meaning "preceding the Dog." In the Northern hemisphere it rises just before Sirius. It is an F-class sub-giant, 11.4 light-years away, seven times more luminous than our Sun, and about 40 percent more massive. It too has a companion white-dwarf star half as massive as that of Sirius B.

OBJECTS OF INTEREST

The small dog has little to offer the observer, save a couple of small uninteresting clusters and some tiny faint galaxies. On the other hand, Canis Major has many interesting sights. **M41** is a bright open cluster more than a moon-diameter across and faintly visible to the unaided eye. A 10- to 15-inch (25–38 cm) telescope resolves this cluster into a cloud of faint points.

In a 6-inch (15 cm) telescope **NGC 2362** is one of the most attractive open clusters in the sky. A dense triangular throng of faint stars seems to spiral to the center, where the brilliant central star **Tau (τ) Canis Majoris** resides with two faint companions. Nearby is **NGC 2354**, another populous crowd of faint sparks nearly a half moon-diameter across. In the north of Canis Major is **NGC 2359** or **Thor's Helmet**, an interesting bright bubble of gas that is illuminated by a super-hot Wolf-Rayet star shown well by an 8-inch (20 cm) telescope.

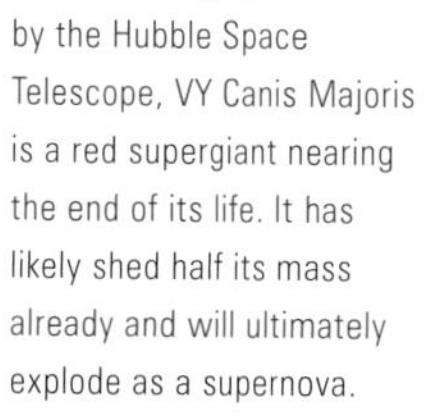

Below Photographed here by the Hubble Space Telescope, VY Canis Majoris is a red supergiant nearing the end of its life. It has likely shed half its mass already and will ultimately explode as a supernova.

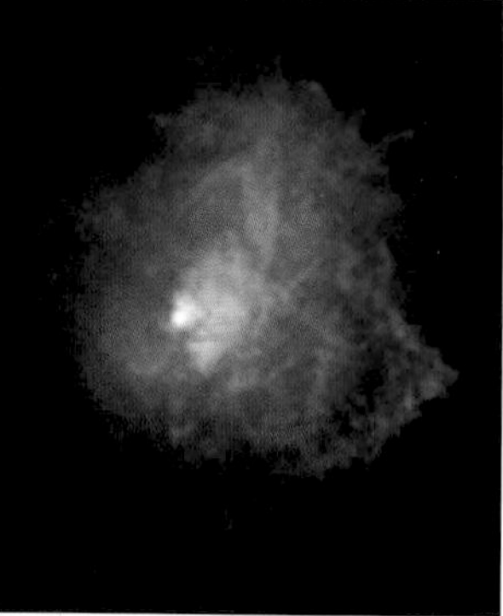

Right The dazzling Sirius (bottom, center) shines among the stellar company of red and blue supergiants Betelguese and Rigel in the constellation of Orion (center, right). Procyon (top, left) is the brightest star in Canis Minor.

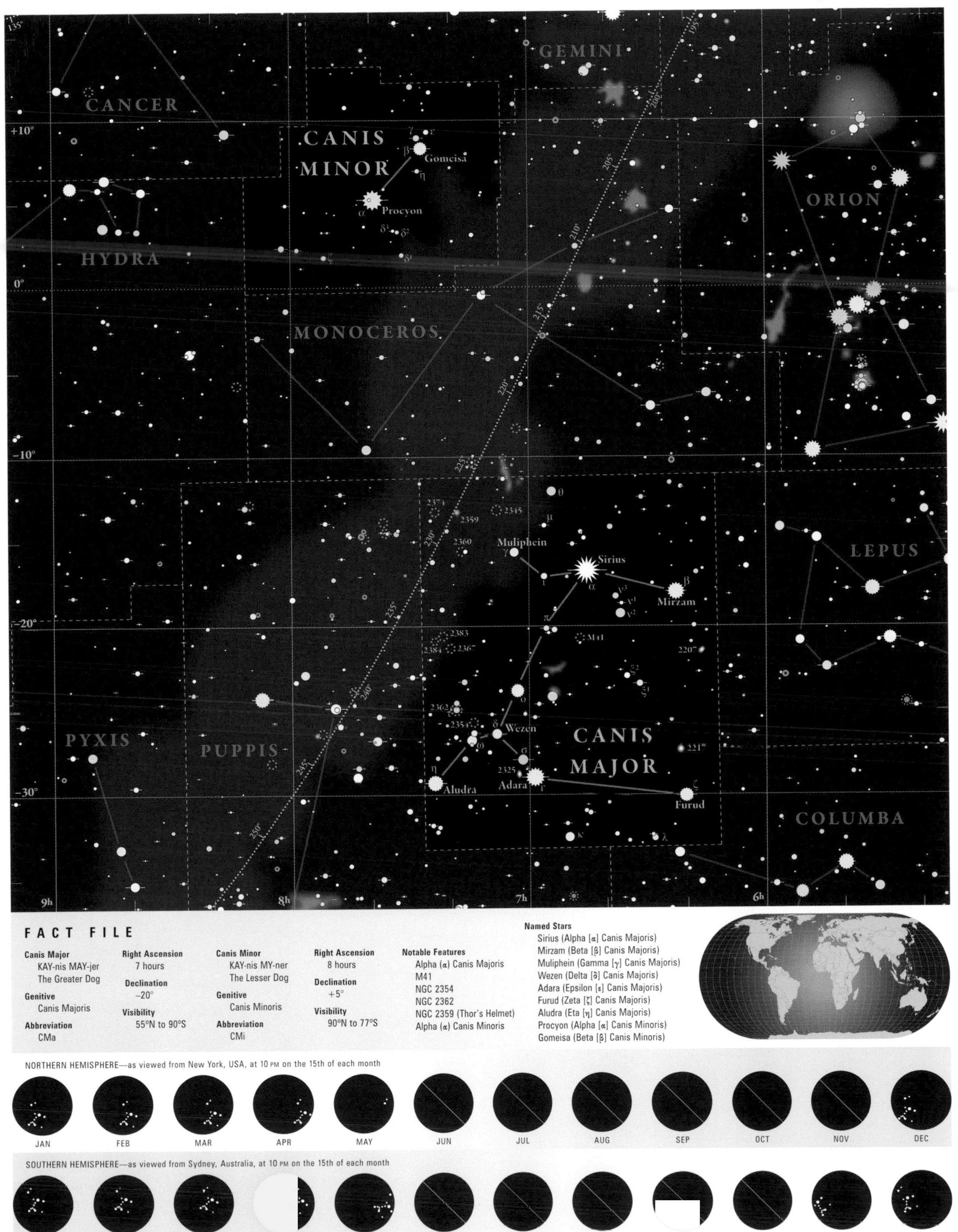
GEMINI
CANCER
CANIS MINOR
Gomeisa
Procyon
HYDRA
ORION
MONOCEROS
Muliphein
Sirius
Mirzam
M41
LEPUS
PYXIS
PUPPIS
Wezen
CANIS MAJOR
Aludra
Adara
Furud
COLUMBA
+10°
0°
–10°
–20°
–30°
9h
8h
7h
6h
FACT FILE
Canis Major
KAY-nis MAY-jer
The Greater Dog
Genitive
Canis Majoris
Abbreviation
CMa
Right Ascension
7 hours
Declination
–20°
Visibility
55°N to 90°S
Canis Minor
KAY-nis MY-ner
The Lesser Dog
Genitive
Canis Minoris
Abbreviation
CMi
Right Ascension
8 hours
Declination
+5°
Visibility
90°N to 77°S
Notable Features
Alpha (α) Canis Majoris
M41
NGC 2354
NGC 2362
NGC 2359 (Thor's Helmet)
Alpha (α) Canis Minoris
Named Stars
Sirius (Alpha [α] Canis Majoris)
Mirzam (Beta [β] Canis Majoris)
Muliphein (Gamma [γ] Canis Majoris)
Wezen (Delta [δ] Canis Majoris)
Adara (Epsilon [ε] Canis Majoris)
Furud (Zeta [ζ] Canis Majoris)
Aludra (Eta [η] Canis Majoris)
Procyon (Alpha [α] Canis Minoris)
Gomeisa (Beta [β] Canis Minoris)
NORTHERN HEMISPHERE—as viewed from New York, USA, at 10 PM on the 15th of each month
JAN
FEB
MAR
APR
MAY
JUN
JUL
AUG
SEP
OCT
NOV
DEC
SOUTHERN HEMISPHERE—as viewed from Sydney, Australia, at 10 PM on the 15th of each month
JAN
FEB
MAR
APR
MAY
JUN
JUL
AUG
SEP
OCT
NOV
DEC

Capricornus

Could Capricornus be the oldest of all constellations? The faint stars were recognized as early as the Sumerian and Chaldean civilizations, who knew it as Suhur-mash-ha, the Goat-fish.

PRIME POSITION

While it may seem odd that one of the faintest of the zodiacal constellations might be the oldest, its significance was not in the stars themselves but in their position. While the solstice today occurs when the Sun is in Sagittarius, thousands of years ago Capricornus marked the Sun's most southerly position in the sky.

Much later, the Ancient Greeks associated the stars of Capricornus with their god Pan. During the war between the Olympian gods and the Titans, Pan warned the Olympians about the approach of the terrible monster Typhon, sent by Gaia (Mother Earth) against them. The gods adopted various disguises to evade Typhon, while Pan partly submerged himself in the river Nile, transforming the lower part of his body into a fish.

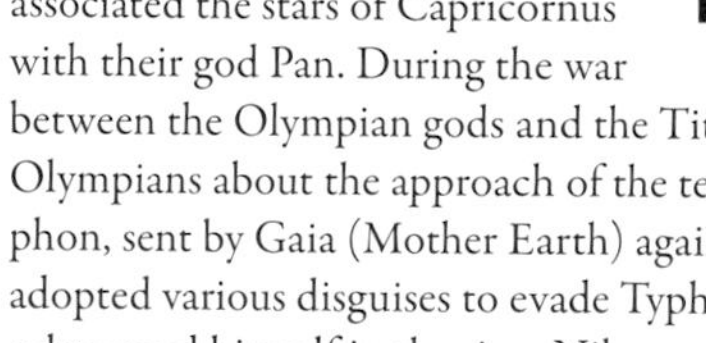

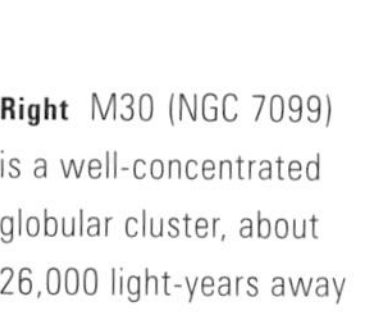

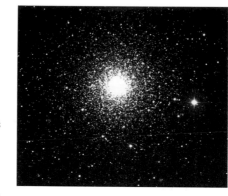

Right M30 (NGC 7099) is a well-concentrated globular cluster, about 26,000 light-years away and 75 light-years across.

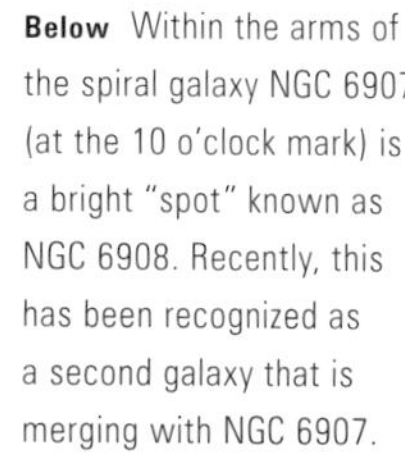

Below Within the arms of the spiral galaxy NGC 6907 (at the 10 o'clock mark) is a bright "spot" known as NGC 6908. Recently, this has been recognized as a second galaxy that is merging with NGC 6907.

THE STARS OF CAPRICORNUS

Although they hardly resemble a goat (even one with a fish tail), the stars of Capricornus are relatively easy to pick out. The major stars make a large triangle, with 4th magnitude Alpha (α) and Omega (ω), along with 3rd magnitude Delta (δ) making up the points. None of the stars is particularly bright, but they strongly resemble the triangular outline of a bikini bottom.

The brightest star in Capricornus, Delta (δ), is at the eastern point of the triangle and is otherwise known as Deneb Algedi meaning "the tail of the kid." It is a white A-class sub-giant star only 39 light-years away, and is almost nine times more luminous than the Sun.

At the eastern vertices are Alpha (α) and Beta (β), both of which are wide, naked-eye pairings. Alpha^1 (α^1) and Alpha^2 (α^2), also named Algedi ("the kid"), are a pair of yellow giant stars, much more massive than our Sun but otherwise unrelated to each other. The fainter Alpha^1 is 109 light-years away while Alpha^2 is more than 690 light-years distant. Beta^1 (β^1) and Beta^2 (β^2) form a rare naked-eye pair that is a true binary. Located 330 light-years distant, the two stars are separated by at least 21,000 times the Earth–Sun distance.

OBJECTS OF INTEREST

Capricornus has comparatively little to offer the amateur sky-watcher, although it does adjoin Sagittarius, one of the treasure troves of the sky. **M30** is a medium-bright, well-concentrated globular cluster that resolves into faint stars with an 8-inch (20 cm) telescope. On the northern and western sides of the halo are two prominent "arms" of stars that are quite distinctive.

A little south of **Beta Capricorni** is the small triangle of 5th and 6th magnitude **Omicron (ο), Pi (π), and Rho (ρ) Capricorni**. All three are lovely binary pairs of differing natures; they show nicely in a 6-inch (15 cm) telescope.

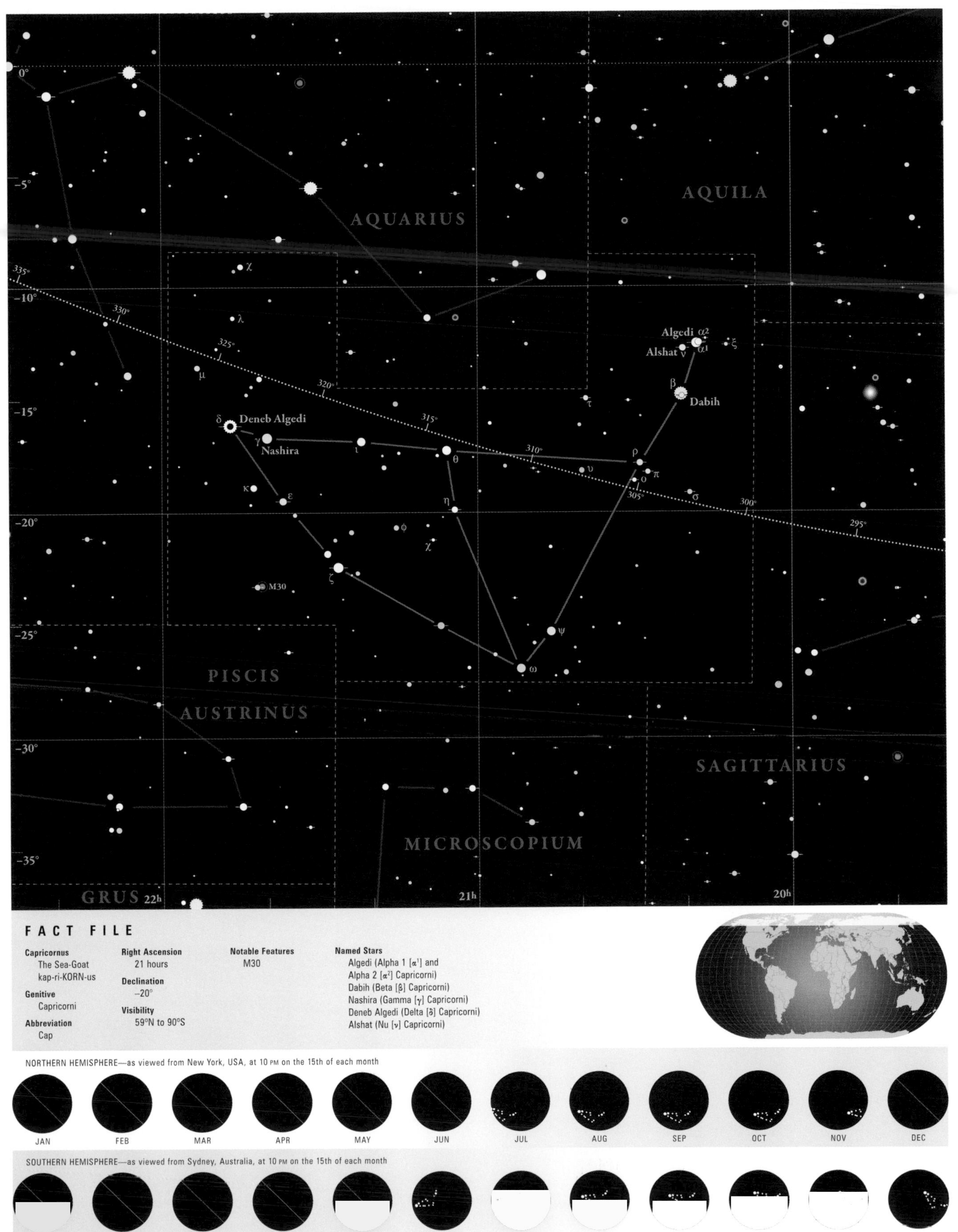

FACT FILE

Capricornus
The Sea-Goat
kap-ri-KORN-us

Genitive
Capricorni

Abbreviation
Cap

Right Ascension
21 hours

Declination
–20°

Visibility
59°N to 90°S

Notable Features
M30

Named Stars
Algedi (Alpha 1 [α¹] and Alpha 2 [α²] Capricorni)
Dabih (Beta [β] Capricorni)
Nashira (Gamma [γ] Capricorni)
Deneb Algedi (Delta [δ] Capricorni)
Alshat (Nu [ν] Capricorni)

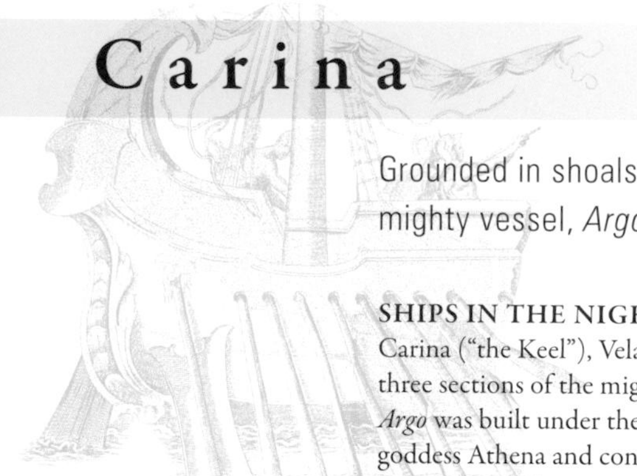

Carina

Grounded in shoals of stars, clusters, and nebulae is the keel of Jason's mighty vessel, *Argo*, with its crew of heroes and adventurers.

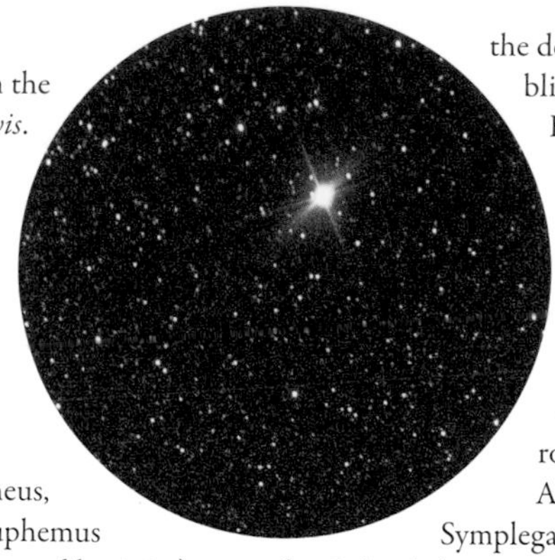

Above right The second-brightest star in the sky, Canopus (Alpha [α] Carinae) is a rare, class F, yellow-white supergiant star.

SHIPS IN THE NIGHT

Carina ("the Keel"), Vela, and Puppis form the three sections of the mighty ship *Argo Navis*. *Argo* was built under the guidance of the goddess Athena and consecrated by the sea-god, Poseidon. Minerva, the goddess of wisdom and craft, placed a plank from the Speaking Oak of Dodona in the prow of the ship, enabling *Argo* to counsel and guide its crew. Jason, his fifty Argonauts, and many Greek heroes, including Castor and Pollux (represented by Gemini), Orpheus, Heracles (Hercules), and the helmsman Euphemus sailed to recover the Golden Fleece (represented by Aries). Jason set out from Thessaly in Greece to search for Colchis, reputed to be on the eastern coast of the Black Sea, where King Aeëtes had custody of the fleece.

Their first port of call was Lemnos, an island inhabited solely by women. According to legend, the women living there had been deserted by their husbands for Thracian women. In revenge, the women murdered every man on the island. Jason and his crew (except Heracles) spent considerable time helping the women to "repopulate" Lemnos. After other adventures, they landed at the Thracian court of King Phineus, who was under punishment for revealing the deliberations of the gods to men. He was blinded by the gods and afflicted by the Harpies—large birds with women's faces—who prevented him from eating more than the barest amount necessary for survival. Jason took pity on Phineus and slew the Harpies. In return, Phineus revealed the exact location of Colchis and explained how to defeat the clashing rocks of the Symplegades.

After a successful encounter with the Symplegades (see the constellation description for Columba), Jason arrived in Colchis, but King Aeëtes demanded that Jason perform several seemingly impossible tasks before he handed over the fleece. Aeëtes's daughter Medea, a powerful sorceress, fell in love with Jason and decided to aid him against her father.

In the first task, Jason had to yoke two fire-breathing oxen and plow a field with them. Medea provided him with a balm that protected him from being burned. Aeëtes then ordered Jason to sow the field with dragon's teeth. The teeth sprouted into a throng of warriors, who began to attack him. Medea told Jason to throw a rock into their midst; in the confusion the warriors killed each other, and Jason was victorious.

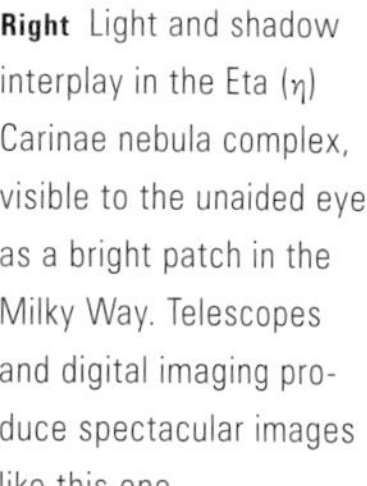

Right Light and shadow interplay in the Eta (η) Carinae nebula complex, visible to the unaided eye as a bright patch in the Milky Way. Telescopes and digital imaging produce spectacular images like this one.

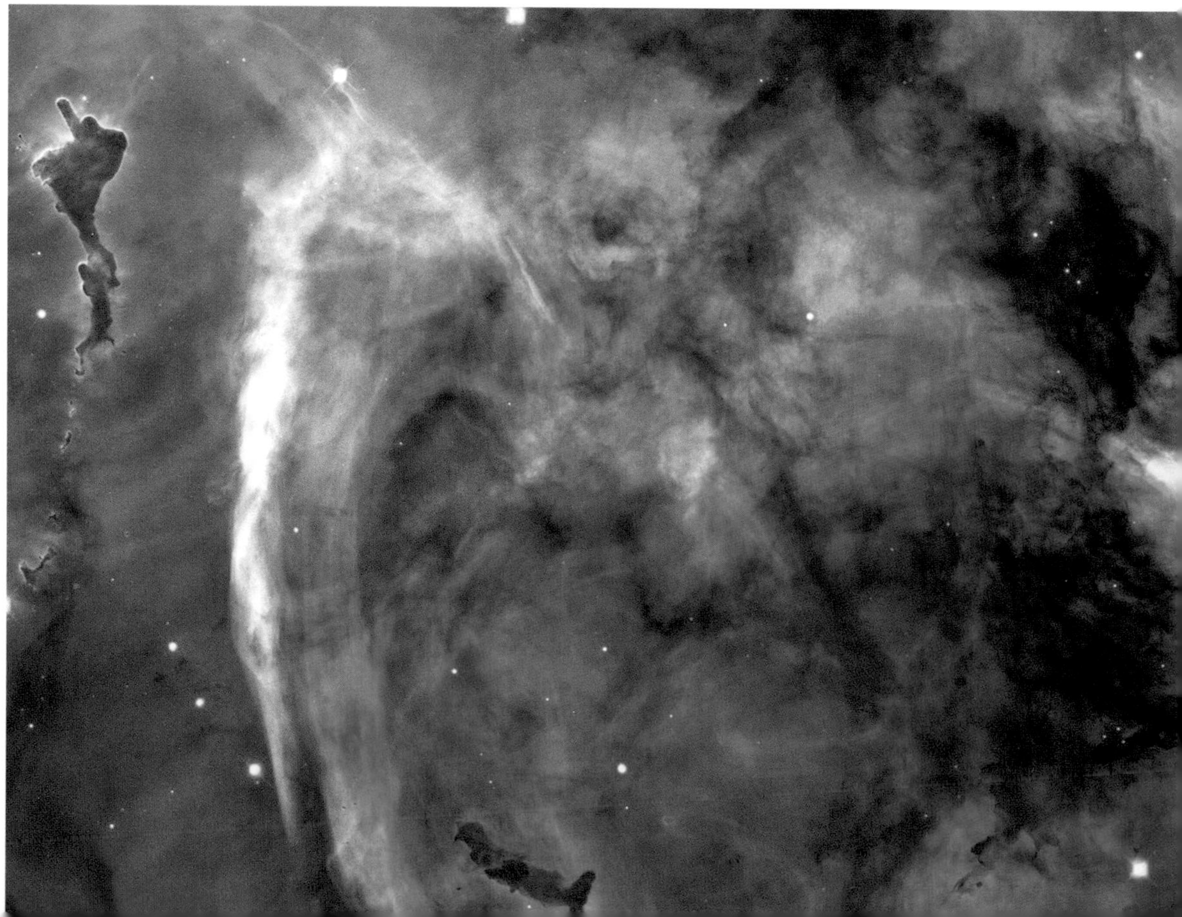

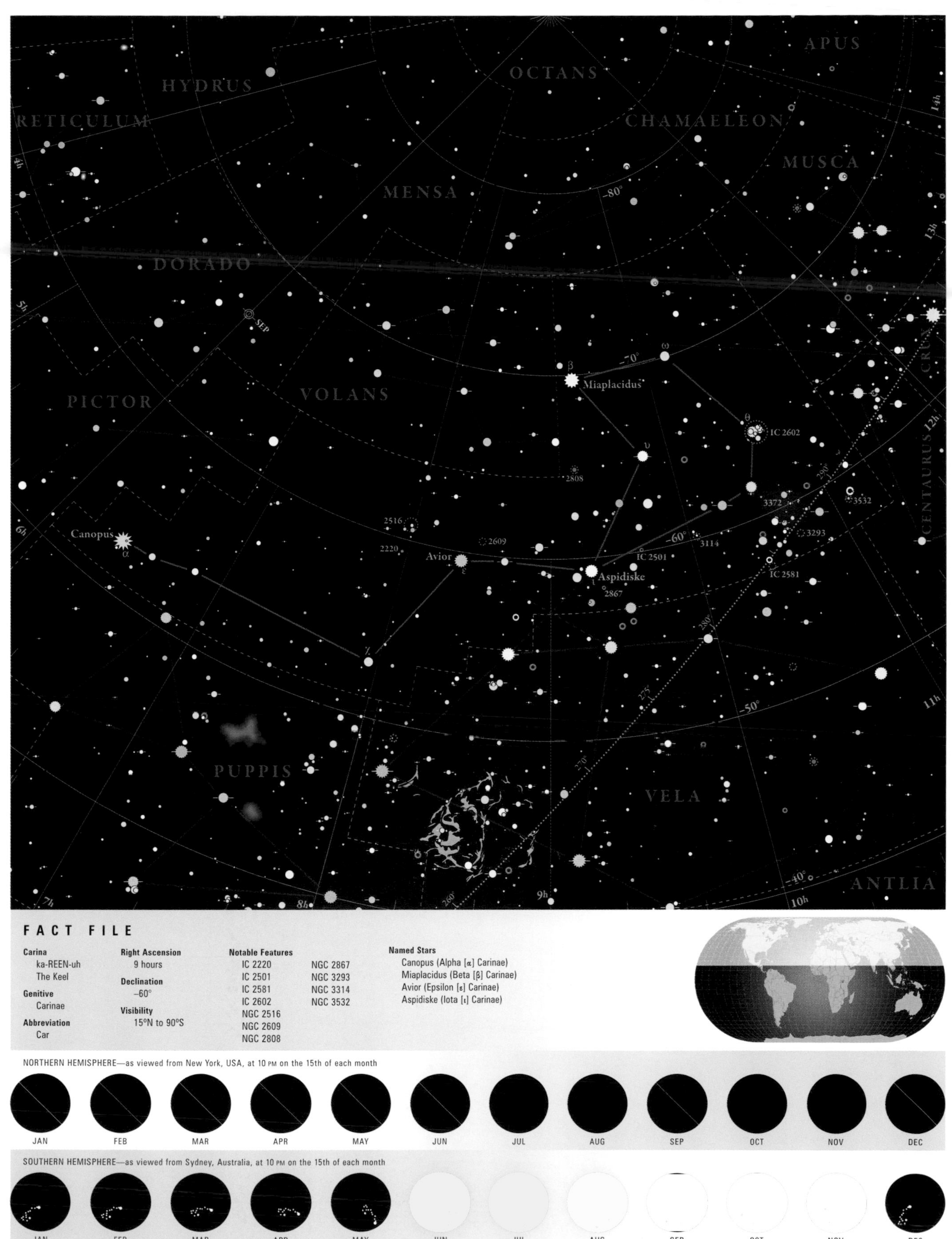

FACT FILE

Carina
ka-REEN-uh
The Keel

Genitive
Carinae

Abbreviation
Car

Right Ascension
9 hours

Declination
−60°

Visibility
15ºN to 90ºS

Notable Features
IC 2220
IC 2501
IC 2581
IC 2602
NGC 2516
NGC 2609
NGC 2808
NGC 2867
NGC 3293
NGC 3314
NGC 3532

Named Stars
Canopus (Alpha [α] Carinae)
Miaplacidus (Beta [β] Carinae)
Avior (Epsilon [ε] Carinae)
Aspidiske (Iota [ι] Carinae)

NORTHERN HEMISPHERE—as viewed from New York, USA, at 10 PM on the 15th of each month

JAN FEB MAR APR MAY JUN JUL AUG SEP OCT NOV DEC

SOUTHERN HEMISPHERE—as viewed from Sydney, Australia, at 10 PM on the 15th of each month

JAN FEB MAR APR MAY JUN JUL AUG SEP OCT NOV DEC

Above The massive star partly hidden in the center of the Carina Nebula is Eta (η) Carinae. This image is a good example of how very massive stars rip apart the molecular clouds that give birth to them, as the stellar winds ionize the surrounding gas and blow it away.

Aeëtes still refused to hand over the fleece, which was guarded by an unsleeping dragon (possibly represented by the far northern constellation of Draco), and began planning the destruction of the Argonauts. Medea and Jason secretly searched for and eventually found the fleece hanging on a tree. Jason stole the fleece after Medea had cast a sleeping spell on the dragon.

In their escape, Medea killed her brother Apsyrtus, and as *Argo* rowed away from Colchis—pursued by Aeëtes—she callously threw parts of Apsyrtus's corpse into the sea in order to delay her father, who was forced to stop and retrieve them.

This last act brought down the wrath of Zeus upon the Argonauts and they were battered by storm after storm until *Argo*'s oracle prow advised them to stop at the island of the nymph Circe and make offerings for purification. The Argonauts finally sailed past the island of the Sirens, the women whose beautiful songs so entranced sailors that their ship crashed into the islands. When Orpheus heard their voices, he produced his lyre (now represented in the sky by the constellation Lyra) and played a melody of unparalleled beauty, drowning out the Sirens' song.

The stars of Argo were uniformly associated with a ship by the ancient cultures. In Egypt it was believed to be the boat that carried Isis and Osiris over the deluge. The Hindus said that it performed the same function for mother and child Isi and Iswara; they called the ship *Argha*, similar to the Greek title. For the Hebrews, of course, it represented Noah's Ark.

Despite Argo's lofty pedigree, Edmund Halley tore a raft of stars from it (in particular Carina and Vela) during the seventeenth century to form a new constellation, Robor Carolinum—the Royal Oak. It commemorated a tree where Prince Charles (later Charles II) hid after his defeat by Cromwell in the English civil war. While Halley's invention was reproduced in several eighteenth- and nineteenth-century star-atlases, it never really caught on. Finally, Abbé Nicolas Louis de Lacaille, with no better excuse than to make Argo more manageable, divided it into the three current sections.

THE STARS OF CARINA

Argo was easily the largest of all constellations and Carina is a sprawling group bearing little resemblance to the hull of a Greek ship, with the eastern end of the constellation immersed in the bright Milky Way. Alpha (α) Carinae, or Canopus, at magnitude –0.7, is the second brightest star in the entire sky. It is a yellowish-white F-class supergiant about nine times more massive than the Sun, about 15,000 times more luminous, and 313 light-years distant.

Superlatives fail miserably to adequately describe Eta (η) Carinae. Of stars in our galaxy that have been studied in detail, it is among the four largest and most likely the champion. Eta weighs in at about 120 times as massive and a staggering 5,000,000 times more luminous than our Sun. Eta Carinae radiates as much energy in 6 seconds as our Sun does in a full year. At 8,000 light-years distant, it lies at the heart of the Eta Carinae nebula complex within a throng of stars and clusters. Two eruptions, in the 1840s and in 1890, produced the tiny Homunculus Nebula that surrounds it, and also saw it climb for a time to be the second brightest star in the sky. Now just visible to the naked eye, it is doomed to die quite soon (in astronomical time-frames) in a hypernova, eventually producing a black hole.

OBJECTS OF INTEREST

The eastern end of the constellation Carina is one of the richest treasure troves of open clusters and nebulae in the sky. The **Eta (η) Carinae nebula complex (NGC 3372)** is a huge, highly detailed, and bright emission nebula, visible to the unaided eye as a large bright patch of the Milky Way. Measuring several moon-diameters across, it is best appreciated at very low magnifications. The whole nebula is cut in two by a V-shaped swathe of dark nebula.

Near the center of the nebula complex is the star **Eta (η) Carinae** itself. At low magnifications it looks orange and a little bloated compared to other stars, but at high magnifications in a moderate-sized telescope, this orange spot is resolved into the extraordinary bi-polar **Homunculus Nebula**. There are several smaller star clusters scattered over the face of the nebula complex, and these contain some of the most massive stars in our Milky Way.

In the area around Eta Carinae, the rich and compact **NGC 3293** packs an emphatic punch of stars in a tiny area of the sky. **NGC 3532** and **NGC 3114** are both large and rich assemblages of stars ranked among the best in the sky. In a 6-inch (15 cm) telescope, NGC 3532 appears oval in shape and contains hundreds of stars, with a bright yellow star at one end. The scattered icy diamonds of **IC 2602**, the **Theta (θ) Carinae** cluster that is otherwise known as the **Southern Pleiades**, are judged by many sky-watchers as the second-best cluster in the sky for binoculars; this cluster shows at least 25 stars, many with bluish hues.

Near Epsilon (ε) Carinae—the most southwestern star in the "false cross" asterism—is the naked-eye star cluster **NGC 2516**. With a 6-inch (15 cm) telescope, it is a loose but rich cluster of medium-bright stars, many with various subtle color tints. Quite close to NGC 2516 is the small reflection nebula **IC 2220**. Photographs show it to have a striking resemblance to a Toby jug. An 8-inch (20 cm) telescope gives a good view of it.

Halfway between Beta (β) and Iota (ι) Carinae is the glorious globular cluster **NGC 2808**. At high magnifications under a rural sky, an 8-inch (20 cm) telescope will resolve it into a myriad faint specks with an intense concentration at the center.

Above The dark dusty Keyhole Nebula stands out against the background of glowing gas in the center of the Eta (η) Carinae Nebula.

Left This Hubble Space Telescope close-up shows a 3 light-year wide portion of the Eta (η) Carinae Nebula, which has a total diameter of more than 200 light-years. Dramatic dark dust knots are sculpted by the high-velocity stellar winds and high-energy radiation from Eta (η) Carinae, although the star is not included in the frame.

Cassiopeia

The distinctive constellation of Cassiopeia, named for the queen of the ancient kingdom of Æthiopia, is a useful starting point for learning about the night sky.

STAR QUALITY

Queen Cassiopeia boasted about her daughter Andromeda's beauty, which angered the god Nereus and caused disaster to befall her country. This myth is more fully detailed in the constellation description of Andromeda.

THE STARS OF CASSIOPEIA

This W-shaped group of 2nd and 3rd magnitude stars forms one of the most obvious constellations.

Gamma (γ) Cassiopeiae is an erratic young subgiant star of spectral class B0. Its color, diameter, and temperature can all fluctuate dramatically, and its brightness has varied unpredictably, the greatest range being from a maximum of magnitude 1.6 in 1937 down to 3.0 in 1940 after the star ejected an obscuring shell of gas and dust. Two nearby arrowhead-shaped nebulosities that point toward Gamma (γ)—IC 59 and IC 63—may have been ejected in earlier flare-ups. The nebulae are dusty and shine mainly by reflecting the light of the star. Under dark transparent skies a 10-inch (25 cm) telescope will show them.

OBJECTS OF INTEREST

Eta (η) Cassiopeiae is an unusual color-contrast double star. The primary is generally agreed to be yellow, but the comes has been variously described as purple, garnet, deep orange, orange–brown, purple–brown, and "perhaps brown."

Straddling the Milky Way, Cassiopeia offers a great many galactic clusters. The descriptions are of the view with an 8-inch (20 cm) reflector. The rich and young **M52 (NGC 7654)** has a bow-shaped group at the center.

NGC 7789 is marvelous, with a hundred faint stars in concentric arcs over the unresolved glow of hundreds more. (Astrophysical studies have found 583 members.)

The striking **NGC 457** is variously known as the **ET Cluster** (after the lovable movie character), the **Owl Cluster**, and the **Dragonfly Cluster**, due to its lines of stars. The yellow 5th magnitude supergiant star Phi (ϕ) Cassiopeiae and a blue 7th magnitude supergiant form the owl's eyes, while a 9th magnitude orange supergiant highlights one wing. NGC 457 is a crowd-pleaser on public observing nights.

M103 (NGC 581) is a small arrowhead of 20 stars. The three brightest, one of them yellow, mark the triangle's vertices.

Nearby **NGC 663** is flanked by **NGC 654** and **NGC 659** in the same low-power field. Large and bright NGC 663 has two ovals of stars, and a diamond-shaped group marks its center. Several star-streamers lead toward NGC 659. NGC 654 is an attractive little clump containing two binaries.

An 8-inch (20 cm) telescope shows two dwarf elliptical galaxies—**NGC 185** and the elongated and very faint **NGC 147**. They are interesting because they are satellites of the great Andromeda Galaxy, 7 degrees south.

NGC 281, the **Pac-Man Nebula**, appears when a UHC filter is used. An 8-inch (20 cm) telescope reveals a bright hemisphere with a dark intrusion from the flat side.

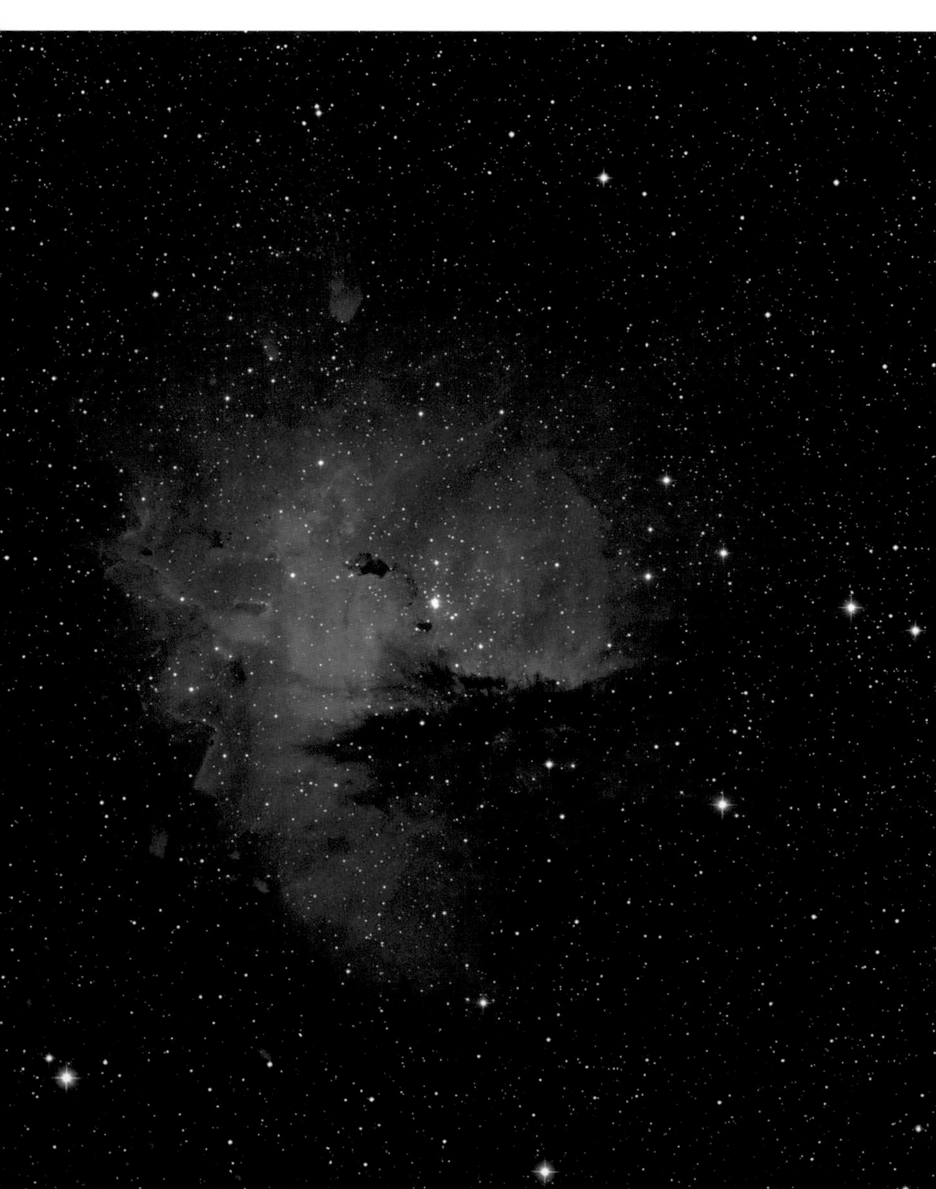

Left Named for its resemblance to the 1980s arcade game character, the Pac-Man Nebula (NGC 281) is a bustling star-making factory filled with small open clusters, a glowing red emission nebula, and gigantic lanes of gas and dust.

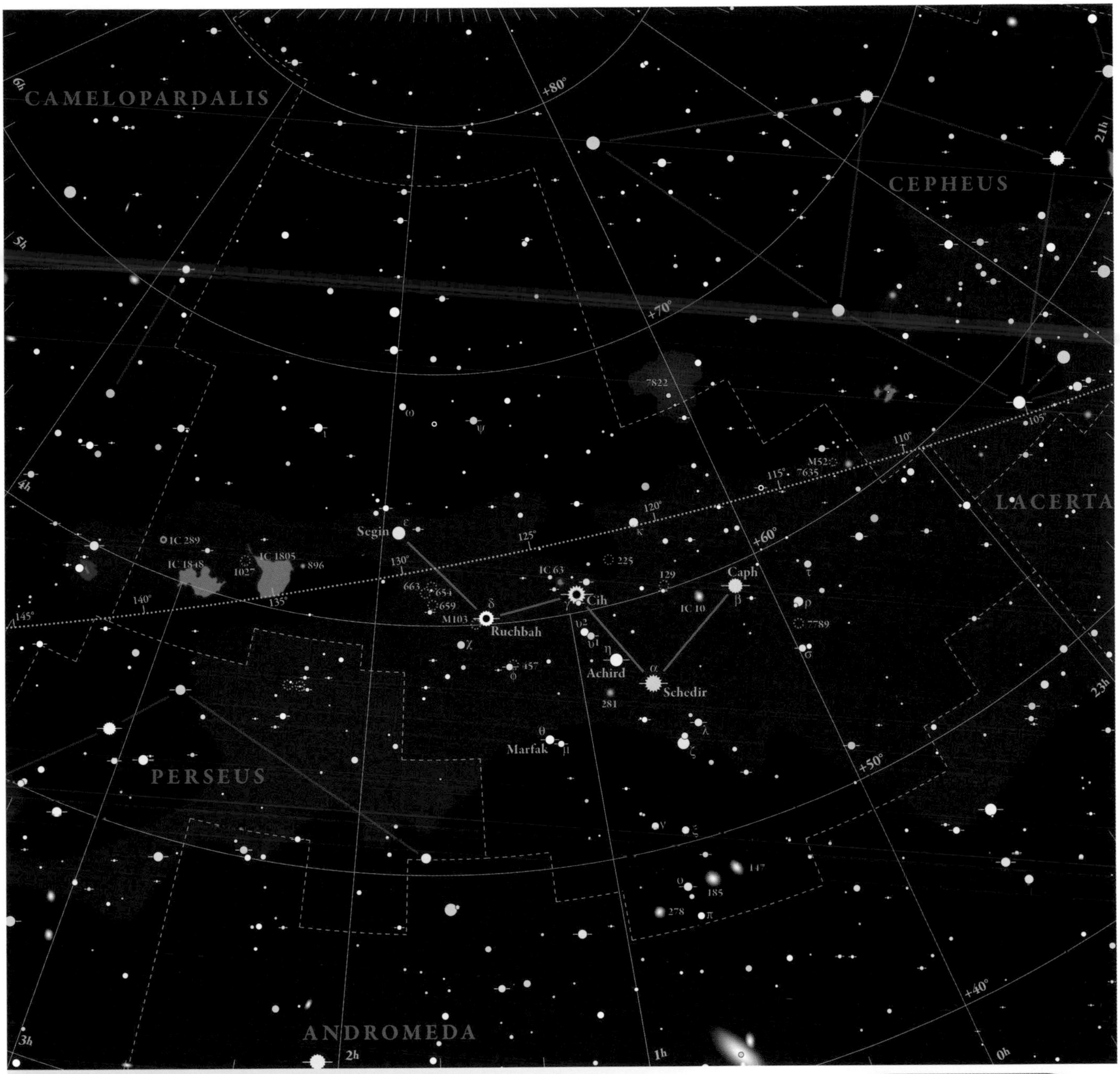

FACT FILE

Cassiopeia
kass-ee-oh-PEE-aah
Cassiopeia, the Queen

Abbreviation
Cas

Genitive
Cassiopeiae

Right Ascension
1 hour

Declination
+60°

Visibility
90°N to 21°S

Notable Features
M52
NGC 281 (Pac-Man Nebula)
NGC 7789
NGC 457 (ET Cluster)
NGC 663
NGC 654
NGC 659

Named Stars
Schedir (Alpha [α] Cassiopeiae)
Caph (Beta [β] Cassiopeiae)
Cih (Gamma [γ] Cassiopeiae)
Ruchbah (Delta [δ] Cassiopeiae)
Segin (Epsilon [ε] Cassiopeiae)
Achird (Eta [η] Cassiopeiae)
Marfak (Theta [θ] Cassiopeiae)
Marfak (Mu [μ] Cassiopeiae)

NORTHERN HEMISPHERE—as viewed from New York, USA, at 10 PM on the 15th of each month

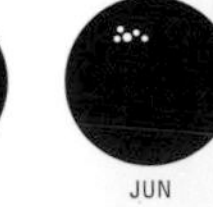
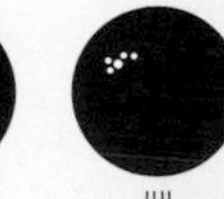

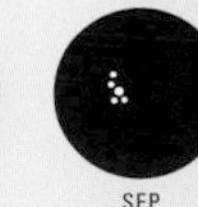

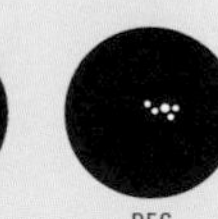

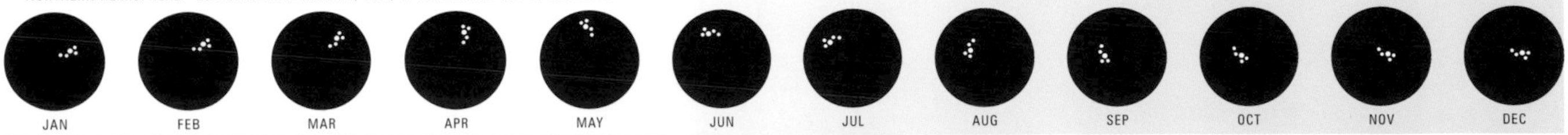

JAN FEB MAR APR MAY JUN JUL AUG SEP OCT NOV DEC

SOUTHFRN HEMISPHERE—as viewed from Sydney, Australia, at 10 PM on the 15th of each month

JAN FEB MAR APR MAY JUN JUL AUG SEP OCT NOV DEC

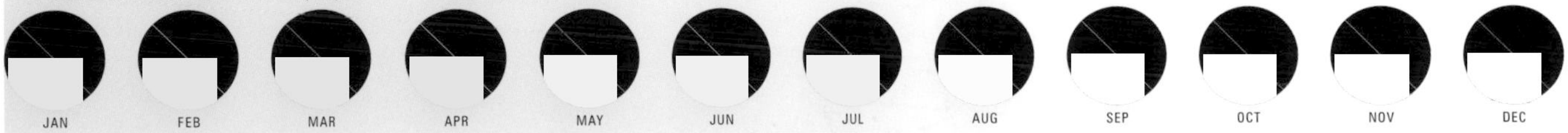

Centaurus

Centaurus is a constellation rich in both mythology and spectacular celestial sights. It also contains the closest stars to our own Sun.

HORSE PLAY

Ixion was a mortal whom Zeus invited to Mount Olympus out of pity for his circumstances, but instead of being grateful, Ixion lusted after Zeus's wife Hera (Juno). Learning of Ixion's intentions, Zeus formed a cloud into the likeness of Hera as a trap. Ixion seduced this phantom-goddess cloud and she bore him a son—Chiron—the first of all Centaurs. The mythical Centaurs possessed the body and legs of a horse with a human torso and head. While centaurs were generally savage and wild, Chiron was civilized and learned. He was skilled in archery, sciences, and medicine, and ultimately became the tutor of Heracles (Hercules), Achilles, Jason, and other mythological figures.

Because Chiron was the offspring of a goddess he was immortal—but this turned out to be his ruin. Heracles accidentally shot him with an arrow dipped in lethal poison and Chiron, unable to die from the venomous wound, was doomed to an eternity of unbearable pain. Zeus took pity on Chiron and transferred his immortality to Prometheus so that he could die at last and thus bring an end to his suffering. Zeus then placed Chiron among the stars as the constellation Centaurus.

Right An emission nebula in Centaurus, IC 2944 contains two of the brightest stars in the universe. The dense clumps of gas and dust are Thackery's Globules, which are the eroded remains of a dusty nebula, dissipated by ultraviolet radiation from the bright stars.

THE STARS OF CENTAURUS

Centaurus is a very large constellation whose stars outline a centaur quite convincingly, with front and hind legs wrapping around the Southern Cross. Rigel Kentaurus and Hadar, the 3rd and 11th brightest stars in the night sky, mark the two forehooves of the centaur and also act as brilliant pointers to the Southern Cross.

Rigel Kentaurus, or Alpha (α) Centauri, is the closest star system to our Sun, just over 4.3 light-years distant. It is a triple star consisting of two sun-like stars and a third member of the system, Proxima Centauri, which is a tiny red-dwarf star that is technically the closest star to the sun. Despite its proximity, you need a powerful telescope to see Proxima Centauri as a faint dot.

OBJECTS OF INTEREST

Omega (ω) Centauri (NGC 5139) is the largest, brightest, and most easily resolved globular star cluster in the entire sky. To find it, draw a line from Hadar (Beta [β] Centauri) northward to Epsilon (ε) Centauri and extend it once again. Containing several million stars, in an 8-inch (20 cm) telescope Omega Centauri resolves into myriad faint specks with a slight compression toward the center. Nearby is the enigmatic **Centaurus A (NGC 5128)**. This peculiar lenticular galaxy has recently merged with a dust-rich spiral, giving it a most unusual appearance. Its core harbors a super-massive black hole, one of the most powerful extra-galactic radio sources in the sky. A 6-inch (15 cm) telescope will show the narrow tapered form of the beautiful edge-on spiral galaxy **NGC 4945**, provided you have a dark sky.

Alpha (α) Centauri is perhaps the most spectacular multiple star in the sky. Viewed in a small telescope, its two brilliant sun-like stars look very much like approaching car-headlights.

Left The galactic cannibal Centaurus A is a lenticular galaxy that has recently merged with a dusty spiral galaxy. The arc of blue stars stretching above the galaxy in this image is a remnant of the merger.

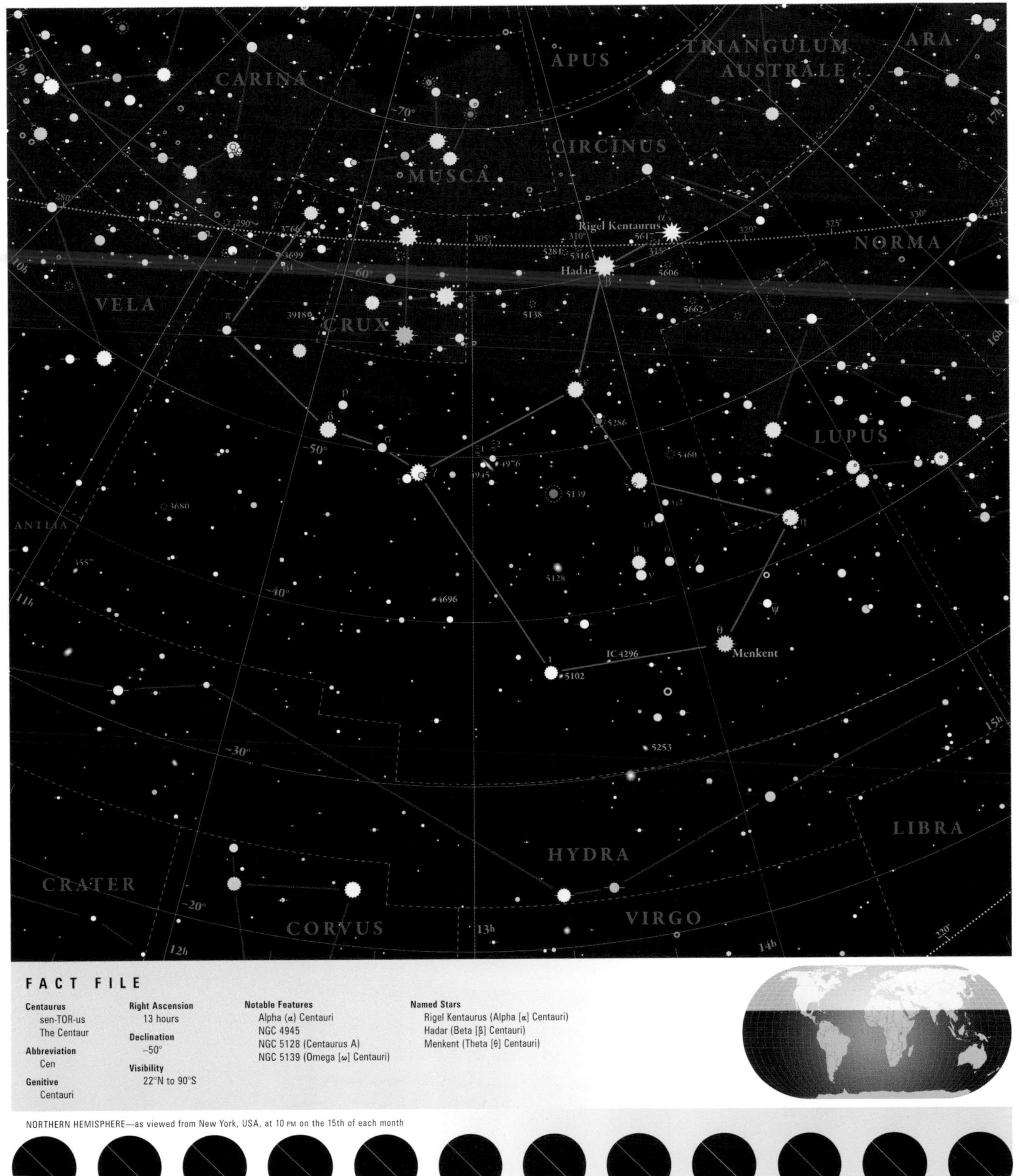

FACT FILE

Centaurus
sen-TOR-us
The Centaur

Abbreviation
Cen

Genitive
Centauri

Right Ascension
13 hours

Declination
–50°

Visibility
22°N to 90°S

Notable Features
Alpha (α) Centauri
NGC 4945
NGC 5128 (Centaurus A)
NGC 5139 (Omega [ω] Centauri)

Named Stars
Rigel Kentaurus (Alpha [α] Centauri)
Hadar (Beta [β] Centauri)
Menkent (Theta [θ] Centauri)

NORTHERN HEMISPHERE—as viewed from New York, USA, at 10 PM on the 15th of each month

JAN FEB MAR APR MAY JUN JUL AUG SEP OCT NOV DEC

SOUTHERN HEMISPHERE—as viewed from Sydney, Australia, at 10 PM on the 15th of each month

JAN FEB MAR APR MAY JUN JUL AUG SEP OCT NOV DEC

Cepheus

Cepheus, named after the King of ancient Æthiopia, is a rather dim constellation of the north polar regions. Its southern parts are in the Milky Way.

A FAMILY AFFAIR

King Cepheus played a minor role in the myth depicted under the constellation of Andromeda. He was the father of Andromeda and husband of Queen Cassiopeia.

THE STARS OF CEPHEUS

Cepheus is outlined by a thin pentagon that somewhat resembles a child's drawing of a house. Be aware that the four stars at the base of the pentagon can also be used to form a nearly identical figure that mistakenly uses a star in Draco.

The supergiant star Delta (δ) Cephei is the prototype of the pulsating Cepheid variable stars. Once it was learned that the intrinsic luminosity of Cepheids was related to their period, astronomers had a powerful tool for determining the distances of nearby galaxies. The changing brightness of Delta can be monitored with little or no optical equipment. Over 5.4 days its magnitude varies from 3.48 to 4.37.

The supergiant star Mu (μ) Cephei is famous for its deep orange color. The famous astronomer William Herschel named it the Garnet Star.

Right The Fireworks Galaxy (NGC 6946), located nearly 20 million light-years from Earth, is a stunning face-on spiral on the border of Cepheus and Cygnus.

OBJECTS OF INTEREST

Delta (δ) Cephei is also a popular double star, with yellow and blue components of magnitudes 4.1 (variable) and 6.3, 41 arc-seconds apart. **Struve 2873 (Σ2873)** is a matched yellow pair of 7th magnitude suns separated by 14 arc-seconds.

The very rich open cluster NGC 6939 and the face-on spiral galaxy NGC 6946 are in the same field of view. Their fame is probably due more to the contrast that these two very disparate objects create in the eyepiece than to their actual appearance. **NGC 6939**, magnitude 7.8 and a third moon-diameter, includes 300 stars, of which about 60, mostly 12th magnitude or fainter, can be seen with a 10-inch (25 cm) telescope against the background haze from the unresolved fainter stars.

Viewed in an 8-inch (20 cm) telescope, **NGC 6946**, magnitude 8.8, is a large amorphous patch with an unusual number of foreground stars because of this nearby galaxy's position, which is buried behind the Milky Way. A 16-inch (40 cm) telescope reveals two broad spiral arms and a weak central condensation.

The planetary nebula **NGC 40**, magnitude 12.4, is evenly illuminated across its 37 arc-second-round disk in an 8-inch (20 cm) telescope, with a dominatingly bright (magnitude 11.6) central star. With very large apertures the disk becomes an annulus, brighter on its outermost edge.

NGC 188, among the oldest known galactic clusters, has survived the galaxy's gravitational tugs for nine billion years—most open clusters dissipate much sooner. In small telescopes, this cluster's 550 stars appear as an unresolved hazy patch of magnitude 8.1, and a half-moon diameter. However, when viewed through a 10-inch (25 cm) telescope 50 starry points are resolved.

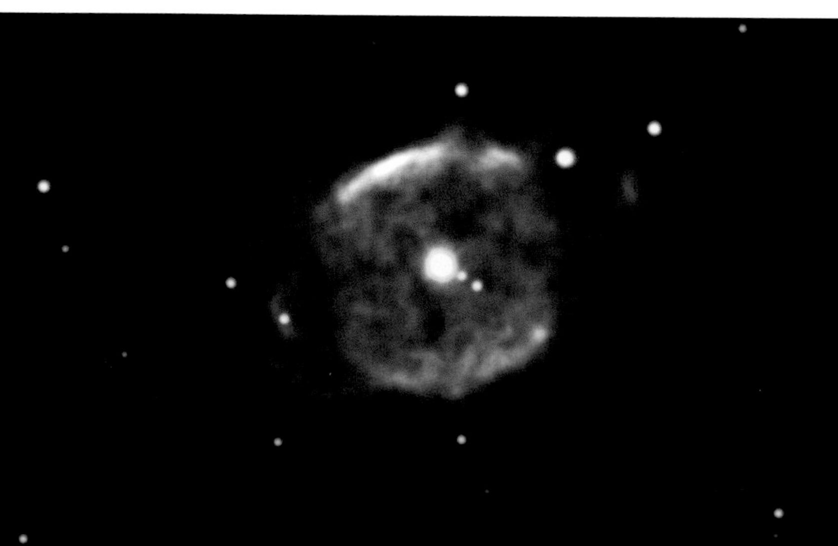

Left Discovered by William Herschel in 1788, the Bowtie Nebula (NGC 40) reflects a star in demise. Astronomers predict that in thirty thousand years this star will shrink to an Earth-size white dwarf.

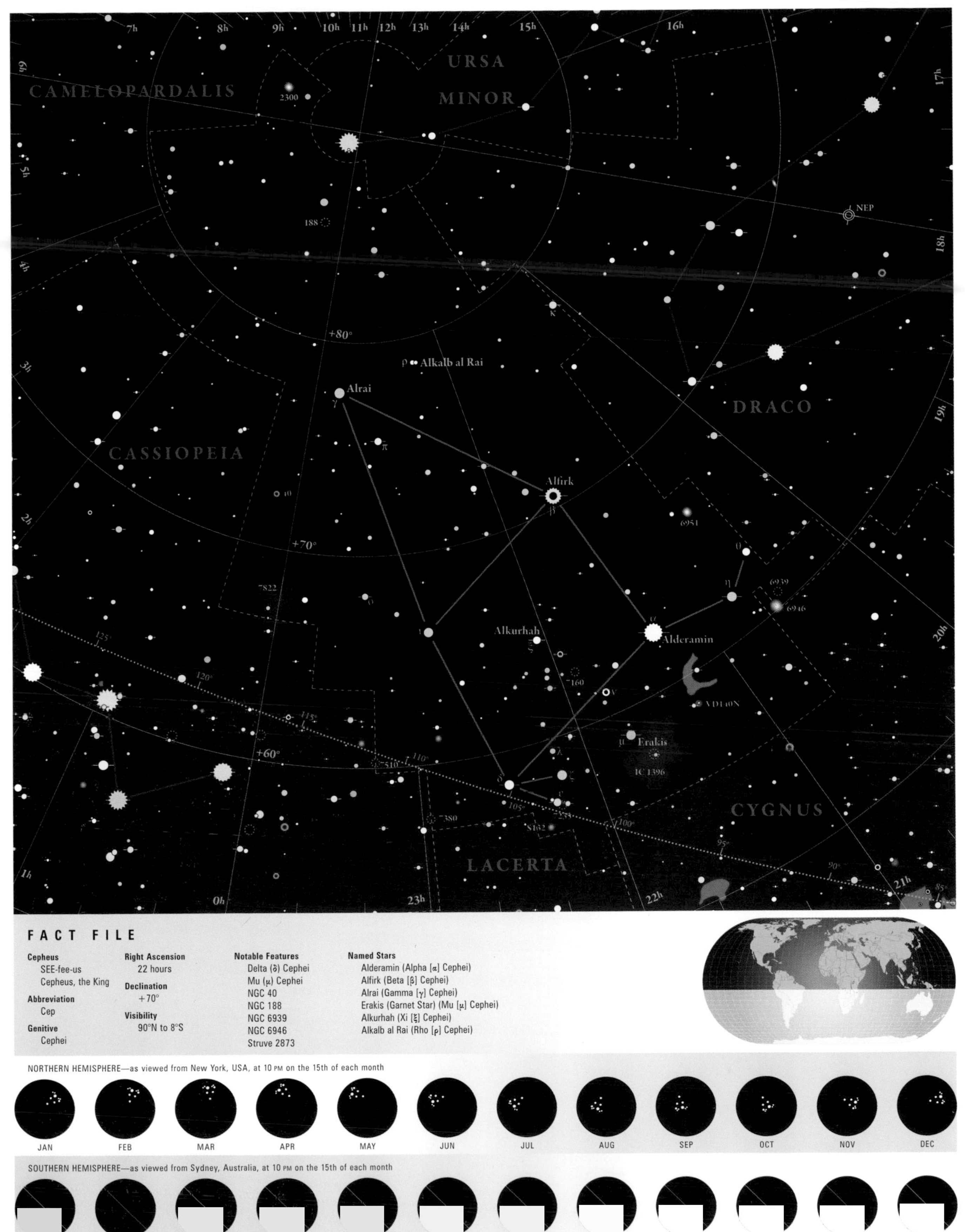

FACT FILE

Cepheus
SEE-fee-us
Cepheus, the King

Abbreviation
Cep

Genitive
Cephei

Right Ascension
22 hours

Declination
+70°

Visibility
90°N to 8°S

Notable Features
Delta (δ) Cephei
Mu (μ) Cephei
NGC 40
NGC 188
NGC 6939
NGC 6946
Struve 2873

Named Stars
Alderamin (Alpha [α] Cephei)
Alfirk (Beta [β] Cephei)
Alrai (Gamma [γ] Cephei)
Erakis (Garnet Star) (Mu [μ] Cephei)
Alkurhah (Xi [ξ] Cephei)
Alkalb al Rai (Rho [ρ] Cephei)

NORTHERN HEMISPHERE—as viewed from New York, USA, at 10 PM on the 15th of each month

JAN FEB MAR APR MAY JUN JUL AUG SEP OCT NOV DEC

SOUTHERN HEMISPHERE—as viewed from Sydney, Australia, at 10 PM on the 15th of each month

JAN FEB MAR APR MAY JUN JUL AUG SEP OCT NOV DEC

Cetus

The Arabs perceived the stars of Cetus as a whale—strange as their waters do not normally support whales. The Greeks associated it with the sea-monster in the tale of Perseus and Andromeda.

GIANT OF THE SKY

Sixteenth- and seventeenth-century cartographers such as Plancius and Bayer drew Cetus as a curious and somewhat comical mixture of whale and monster. Plancius and Schiller, the cartographers who devised the "Christian" star-atlas, interpreted it as the whale that swallowed Jonah. This is the likely source for the modern name Cetus—Latin for whale.

THE STARS OF CETUS

Although Cetus is huge, it has few bright stars; there are none of 1st magnitude and just one of 2nd. Alpha (α), Gamma (γ), and Delta (δ) form an easily identified triangle at the western end of the constellation, on an otherwise almost blank canvas. These three, with the many other faint stars, bear little resemblance to any sea creature. The lucida of Cetus is 2nd magnitude Beta (β) Ceti, otherwise known as Deneb Kaitos, from the Arabic meaning "whale's tail." It is a reasonably common K-type giant star, 96 light-years distant and 145 times more luminous than our Sun.

More interesting is Omicron (o) Ceti, also known as Mira, or the Wonder Star. The sixteenth-century astronomer David Fabricus was first to note that it varied in brightness, occasionally appearing as bright as 2nd magnitude, but at most other times was invisible. Mira is now recognized as the archetype of the class of variable star named after it. An extremely evolved low-mass star in its dotage, Mira rhythmically pulses in size, temperature, and brightness between 3rd and 10th magnitude, over a 330-day cycle.

OBJECTS OF INTEREST

The most interesting objects in Cetus are its abundant galaxies. The brightest and best is **M77**, close by to Gamma (γ) Ceti. Even the smallest of telescopes will show it looking like a faint star with some surrounding fuzz, and an 8-inch (20 cm) telescope reveals its brilliant stellar nucleus and bright, well-resolved halo. M77 is a Seyfert Galaxy—a type of galaxy sporting an extremely active nucleus containing a super-massive "feeding" black hole. The fainter edge-on spiral galaxy **NGC 1055** is in the same field.

Near Beta (β) Ceti, **NGC 247** is a member of the nearby Sculptor group of galaxies, just a few million light-years distant. It is a huge shy galaxy, nearly half a moon-diameter long, and can be glimpsed in binoculars under a rural sky. In pristine conditions, an 8-inch (20 cm) telescope shows a large, long, oval halo with a small faint nucleus.

NGC 246 is Cetus's only planetary nebula. An 8-inch (20 cm) telescope displays its small, round, faint, blotchy halo, with four stars in a Y-shaped asterism superimposed.

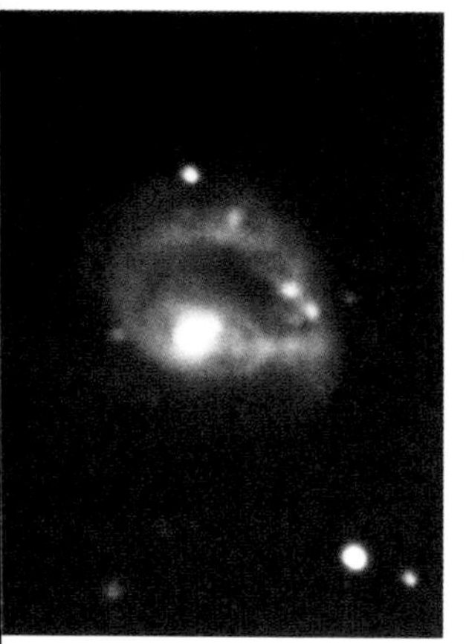

Above Taken by the Mayall Telescope, this is an image of NGC 985,a peculiar ring galaxy in the constellation Cetus. It has a violently active Seyfert nucleus.

Right Image of NGC 1068—the nearest and brightest example of a type II Seyfert galaxy—is located in the constellation Cetus.

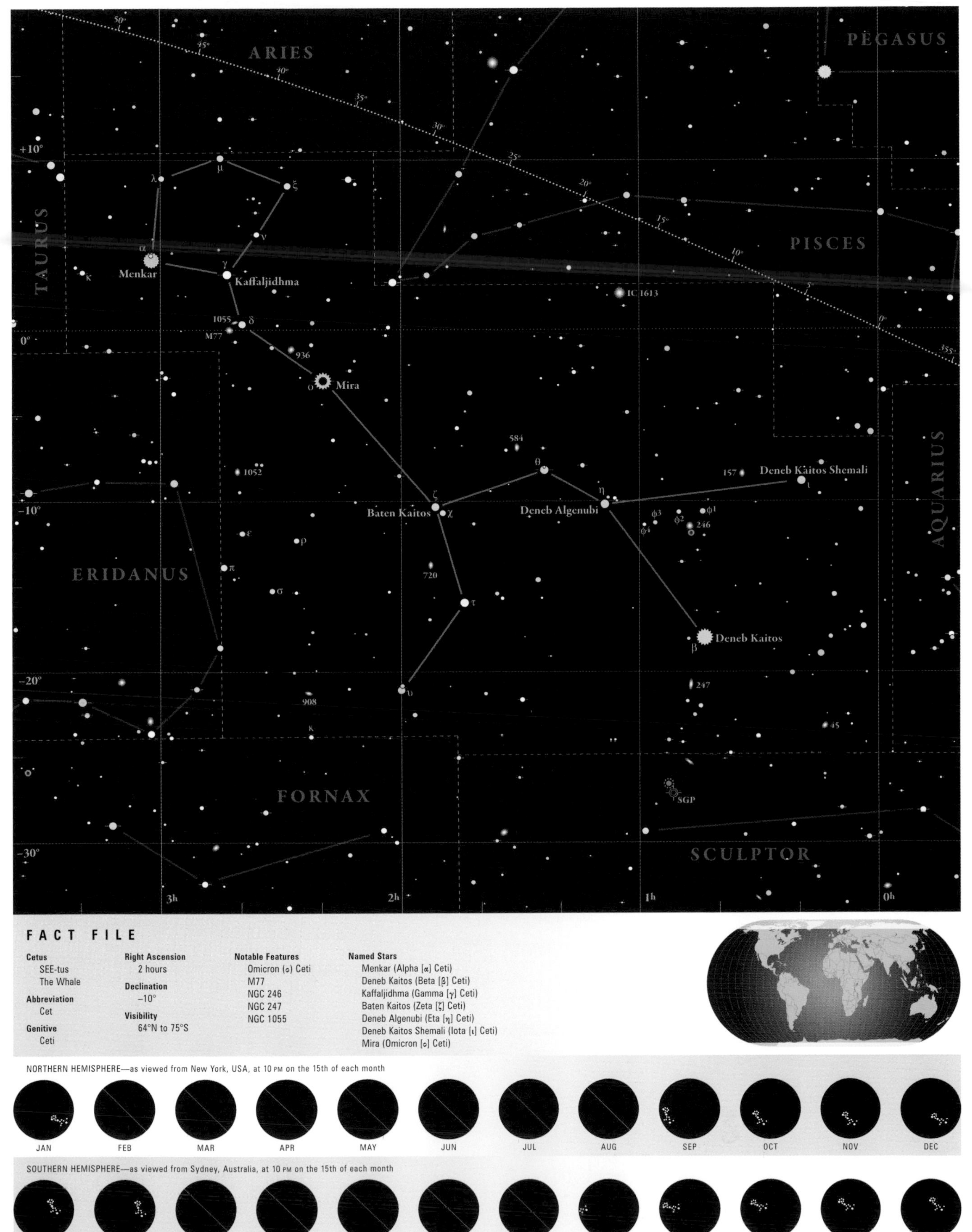

FACT FILE

Cetus
SEE-tus
The Whale

Abbreviation
Cet

Genitive
Ceti

Right Ascension
2 hours

Declination
–10°

Visibility
64°N to 75°S

Notable Features
Omicron (ο) Ceti
M77
NGC 246
NGC 247
NGC 1055

Named Stars
Menkar (Alpha [α] Ceti)
Deneb Kaitos (Beta [β] Ceti)
Kaffaljidhma (Gamma [γ] Ceti)
Baten Kaitos (Zeta [ζ] Ceti)
Deneb Algenubi (Eta [η] Ceti)
Deneb Kaitos Shemali (Iota [ι] Ceti)
Mira (Omicron [ο] Ceti)

NORTHERN HEMISPHERE—as viewed from New York, USA, at 10 PM on the 15th of each month

JAN FEB MAR APR MAY JUN JUL AUG SEP OCT NOV DEC

SOUTHERN HEMISPHERE—as viewed from Sydney, Australia, at 10 PM on the 15th of each month

JAN FEB MAR APR MAY JUN JUL AUG SEP OCT NOV DEC

Chamaeleon

Although it lies well away from the Milky Way, this group of faint stars still harbors a few delights for the telescope or binoculars.

CAMOUFLAGE STAR

As late as the seventeenth century, astronomers did not recognize the star group comprising Chamaeleon (the long-tongued lizard) as a formal constellation, which explains why there is no mythology associated with it.

The sixteenth-century Dutch explorers Pieter Dirkszoon Keyser and Frederick de Houtman devised the new constellation in the form of a lizard after their extensive exploration of the East Indies and the southern stars. Petrus Plancius introduced the constellation in the guise of a chamaeleon—the African lizard that can change its color—on a star-map in 1598. Bayer also adopted this portrayal in his *Uranometria*, published in 1603. Since Bayer's time, there has been no significant change to the form or stars of Chamaeleon.

Above right Traveling at a blistering 100 miles (160 km) per second, the shock wave from the glowing multicolored jet known as Herbig-Haro 49/50 is found in the Chamaeleon 1 star-forming region.

THE STARS OF CHAMAELEON

The principal stars of this constellation form a long thin diamond shape, with the lengthened ends pointing east–west. The center of the diamond is directly south of a point between the main stars of Musca and the bright 2nd magnitude star Miaplacidus in Carina.

None of the stars of Chamaeleon is particularly bright, but there is a remarkable consistency between the top five, all of which are between magnitudes 4.05 and 4.45. The shape can be a little difficult to make out in the bright skies of suburbia, but it is easy to see in a rural sky.

At the western end of the diamond, Alpha (α) Chamaeleontis is a star quite similar to our Sun. With a spectral type of F5, and a mass about 50 percent more than the Sun, it shines at magnitude 4.05 from 63.5 light-years away.

OBJECTS OF INTEREST

One of the most beautiful double stars in the heavens for binoculars is **Delta (δ) Chamaeleontis**. It has contrasting yellow and light blue components that are easy to discern as separate in handheld instruments. Although they are so close to each other, the two stars are an optical pair only; they are not related to each other.

Chamaeleon is well away from the plane of the Milky Way where most of the star-forming regions are located. Despite that, the closest nebula to the Sun that is currently spawning stars is located in Chamaeleon, just over 500 light-years distant. The small reflection nebula **IC 2631** that surrounds a 9th magnitude star is the brightest portion of a large yet faint nebula, and is visible in an 8-inch (20 cm) telescope in a dark sky.

The southernmost bright deep-sky object is also located in Chamaeleon. **NGC 3195** is a moderately bright planetary nebula that is seen as a small, grayish, oval disc in telescopes 4 inches (10 cm) and larger. **NGC 2915** is the brightest galaxy in Chamaeleon. Its small, oval, hazy form can be glimpsed in an 8-inch (20 cm) telescope.

Below Dark molecular clouds surround the bright planetary nebula NGC 3195, visible in the skies south of the Earth's equator.

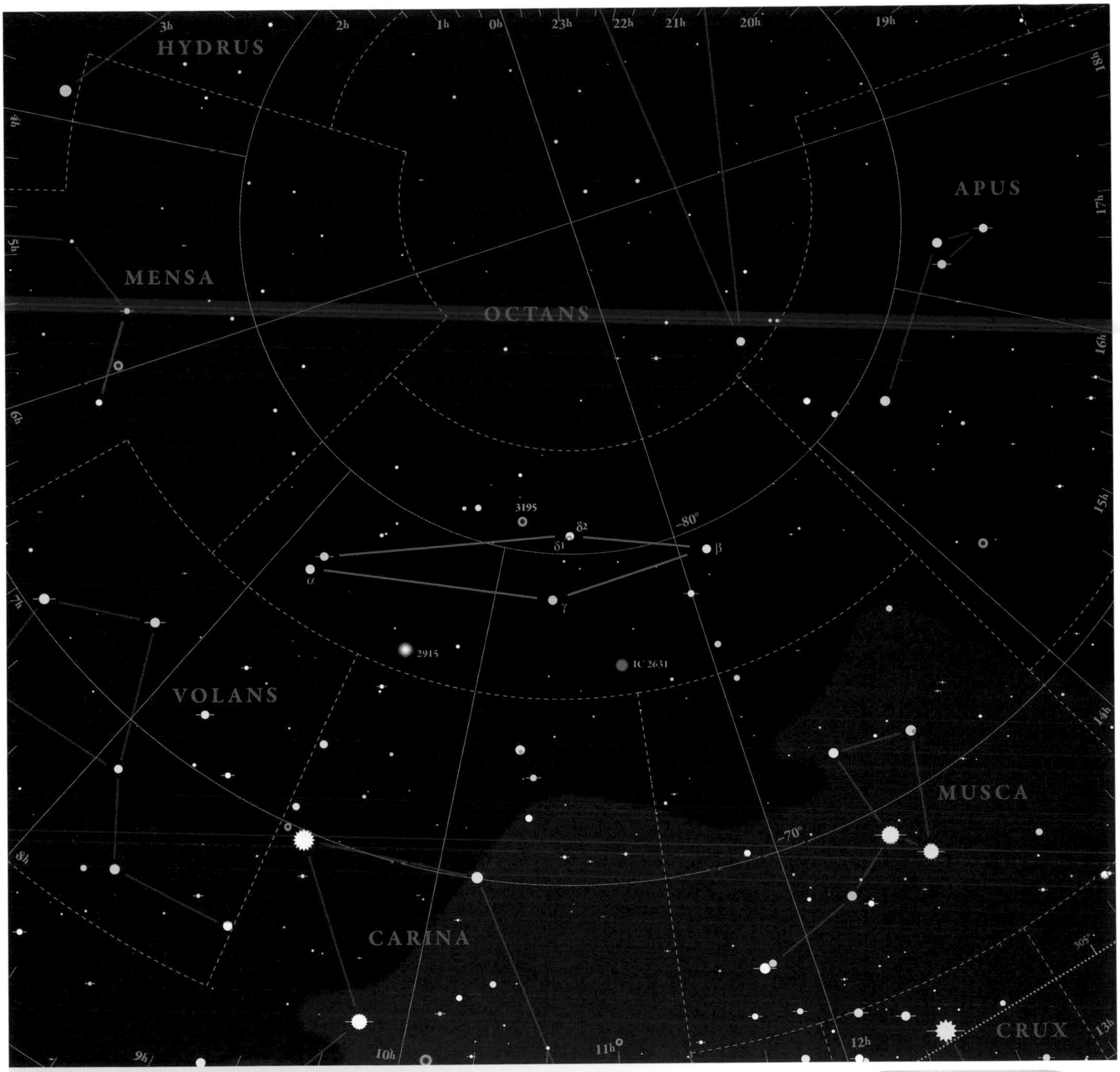

HYDRUS
MENSA
OCTANS
APUS
VOLANS
CARINA
MUSCA
CRUX
3195
2915
IC 2631
−80°
−70°

FACT FILE

Chameleon
ka-MEE-lee-un
The Long-tongued Lizard

Abbreviation
Cha

Genitive
Chamaeleontis

Right Ascension
11 hours

Declination
−80°

Visibility
5°N to 90°S

Notable Features
Delta (δ) Chamaeleontis
IC 2631
NGC 2915
NGC 3195

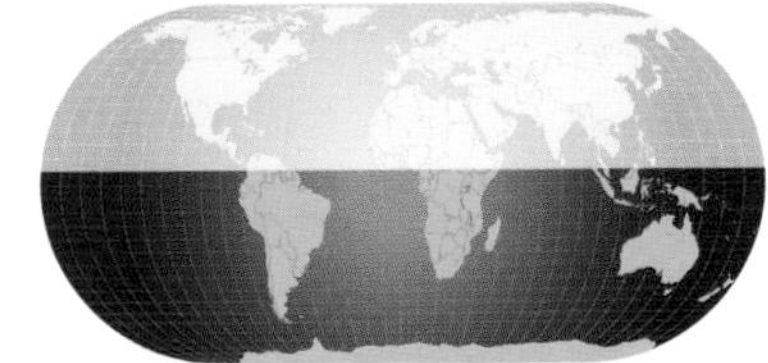

NORTHERN HEMISPHERE—as viewed from New York, USA, at 10 PM on the 15th of each month

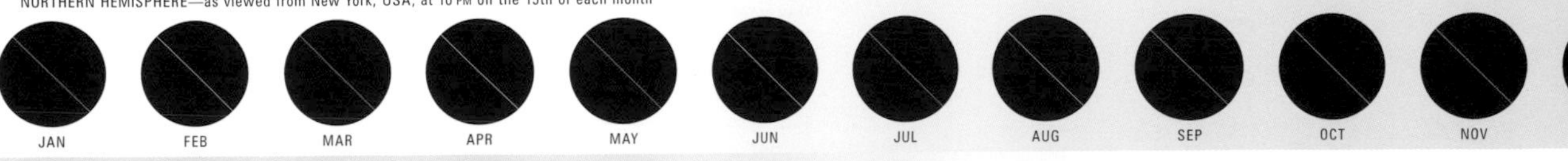

JAN FEB MAR APR MAY JUN JUL AUG SEP OCT NOV DEC

SOUTHERN HEMISPHERE—as viewed from Sydney, Australia, at 10 PM on the 15th of each month

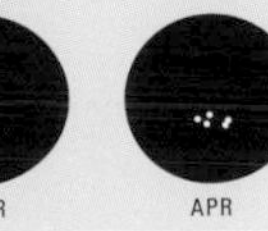

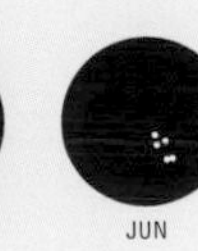

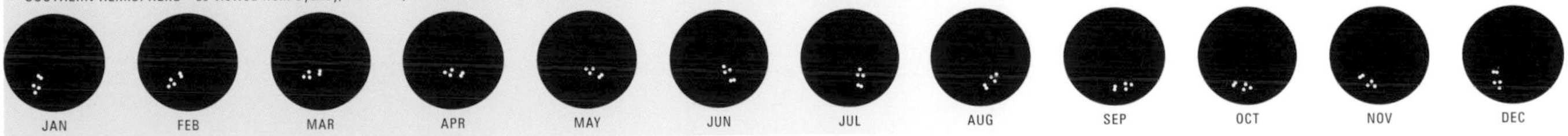

JAN FEB MAR APR MAY JUN JUL AUG SEP OCT NOV DEC

Circinus

Circinus is a very small faint constellation deeply immersed in the southern Milky Way. It was unknown until Lacaille named it during his visit to the Cape Observatory around 1751.

FOLLOW THE COMPASS

Like many of the constellations invented by Lacaille, it represents an instrument of science or engineering—in this case, a pair of compasses. It is found very close to Triangulum Australe, an older constellation devised by Keyser and de Houtman.

Right Hubble Space Telescope image of a type 2 Seyfert galaxy, a class of mostly spiral galaxies that have compact centers and are believed to contain massive black holes. The galaxy lies 13 million light-years away, in the southern constellation Circinus.

THE STARS OF CIRCINUS

The three main stars of Circinus form a long narrow isosceles triangle, looking much like folded compasses, a little eastward from the third brightest star, Alpha (α) Centauri. Alpha (α) Circini, at magnitude 3.19, is the brightest star in the constellation and has a very strange spectrum. Formally, its spectral type is ApSrEuCr where the "p" denotes peculiar. The spectrum is enhanced with absorption lines of heavy elements—strontium, europium, and chromium—highly unusual for an A-type dwarf star that is a little more massive and several times brighter than our Sun. Gamma (γ) Circini is a close binary star with components of magnitude 4.5 and 5.4, which in excellent conditions is resolvable with a 6 inch (15 cm) telescope at high power.

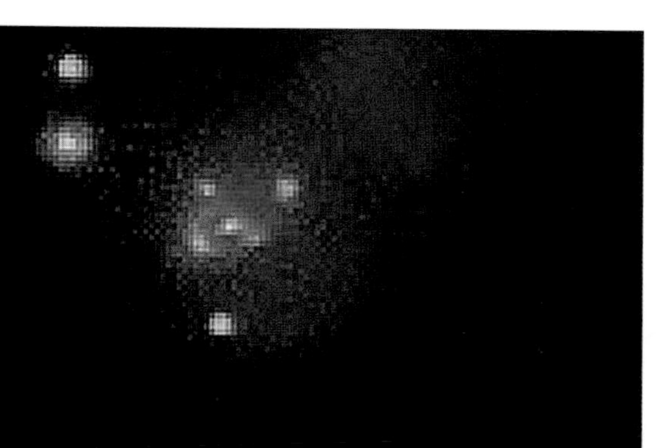

Below This Chandra X-ray image shows the inner portion of the Circinus Galaxy, with north at the top of the image and east to the left. A bright compact emission source is present at the center of the image. That nuclear source is surrounded by a diffuse X-ray halo that extends out several hundred light-years.

OBJECTS OF INTEREST

Circinus is immersed deeply within the Milky Way, but it has few good objects to see with a telescope. This is because thick clouds of gas and dust within our Galaxy screen off objects unless they are relatively close or exceptionally bright. Near the border with Lupus is a pair of open clusters, **NGC 5822** and **NGC 5823**, which are only two moon-diameters apart. The former is a very large scattered cluster of more than 100 stars from magnitudes 8 to 12, while NGC 5823 is a slightly fainter, more compact grouping of stars with a triangular outline. A 4-inch (10 cm) or 6-inch (15 cm) telescope will show both nicely.

About six moon diameters west of Alpha (α) Circini is the remarkable galaxy known as the **Circinus Dwarf,** or **ESO 97-13**, although the title dwarf is a misnomer. Originally thought to be a dwarf galaxy, this huge spiral is heavily obscured by gas and dust within the Milky Way. Studied at radio wavelengths that penetrate the gas and dust, it is almost three moon-diameters across!

Lying about 20 million light-years distant, Circinus Dwarf is an energetic galaxy with a star-bursting core and an ultra-luminous "Seyfert" nucleus. In excellent conditions an 8-inch (20 cm) telescope faintly shows just the very small hazy core of this enigmatic giant.

Of additional interest to observers is the bright planetary nebula **NGC 5315**, which is close to the western border with Musca. A 4-inch (10 cm) telescope at high magnification, on a clear night, will show it as a tiny bright blue disk suspended in a rich field of stars.

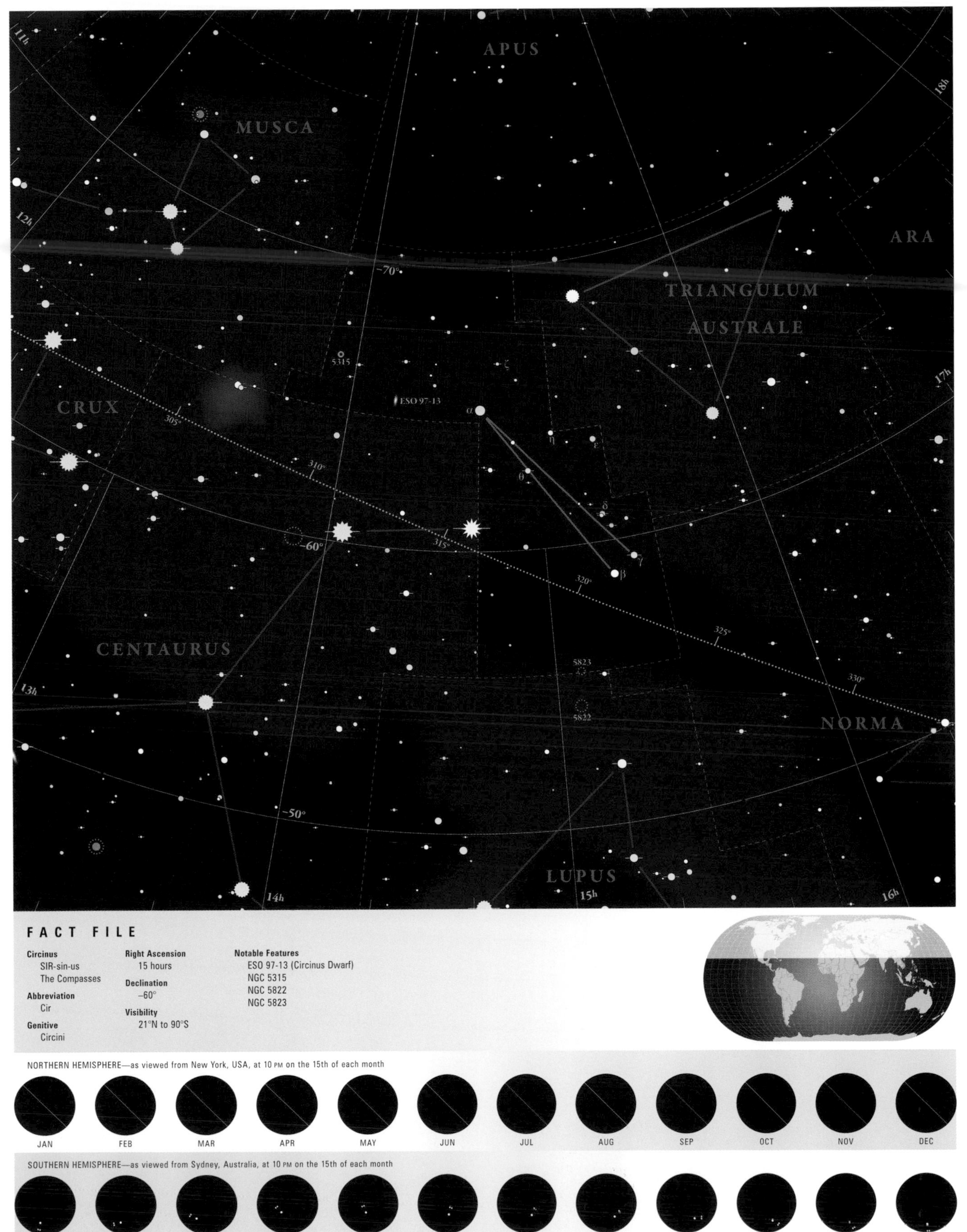
APUS
MUSCA
ARA
TRIANGULUM AUSTRALE
CRUX
CENTAURUS
NORMA
LUPUS
5315
ESO 97-13
5823
5822
-70°
-60°
-50°
305°
310°
315°
320°
325°
330°
11h
12h
13h
14h
15h
16h
17h
18h
FACT FILE
Circinus
SIR-sin-us
The Compasses
Abbreviation
Cir
Genitive
Circini
Right Ascension
15 hours
Declination
–60°
Visibility
21°N to 90°S
Notable Features
ESO 97-13 (Circinus Dwarf)
NGC 5315
NGC 5822
NGC 5823
NORTHERN HEMISPHERE—as viewed from New York, USA, at 10 PM on the 15th of each month
JAN FEB MAR APR MAY JUN JUL AUG SEP OCT NOV DEC
SOUTHERN HEMISPHERE—as viewed from Sydney, Australia, at 10 PM on the 15th of each month
JAN FEB MAR APR MAY JUN JUL AUG SEP OCT NOV DEC

Columba

Do the stars of this constellation represent Noah's dove or Jason's dove? It depends on who is telling the story.

PEACEFUL NIGHT FLIGHT

Petrus Plancius devised Columba in the sixteenth century using some ungrouped stars south of Canis Major. Plancius's inspiration came from his alternative view of the stars of Argo representing Noah's ark. Plancius imagined Columba as the dove that was released by Noah after the flood, which then returned with an olive twig, signifying the existence of dry land. Other celestial cartographers agreed and rapidly adopted Columba as a constellation.

Columba may also represent the dove that the Argonauts released in a narrow sea channel at the mouth of the Black Sea, called the Symplegades. As ships approached, the rocks on either side of the towering crags would rush together, crushing any vessel unfortunate enough to be sailing there. Before Jason attempted the passage, he released a dove that sped between the crags causing them to crash together; fortunately, the dove lost only a few tail feathers. *Argo* was rowed through at full speed behind the dove. The rocks crashed together again, grazing *Argo*'s stern but causing minimal damage. Ever since Jason and the *Argo* defeated them, the Symplegades have remained open and stationary.

THE STARS OF COLUMBA

Although the stars of Columba look very little like a dove, the crooked line of 3rd and 4th magnitude stars from Epsilon (ε) through Alpha (α), Beta (β), Gamma (γ), and Delta (δ) is an easily recognized pattern south of Sirius.

Alpha (α) Columbae, otherwise known as Phact, meaning "ring dove," is a blue B-class sub-giant star 270 light-years distant and a thousand times more luminous than our Sun.

Even more interesting is the 5th magnitude Mu (μ) Columbae, a runaway star. Mu is an O-type dwarf star that is careering across the galaxy at 73 miles (117 km) per second, about eight times faster than the Sun. Astronomers studying this star and a similar one in Auriga (AE Aurigae) discovered they are traveling directly away from each other at a combined velocity of nearly 125 miles (200 km) per second. Tracing their paths backward, they converge almost exactly on Iota (ι) Orionis, near M42. An extremely close encounter between a pair of binary stars ejected these two in opposite directions about 2.5 million years ago.

OBJECTS OF INTEREST

The best of Columba's features is the bright globular cluster **NGC 1851** (telescope). Almost a naked-eye object, it is a heavily concentrated swarm of faint stars with a dramatic, sharply brighter center. A 6-inch (15 cm) or 8-inch (20 cm) telescope shows a good view of it.

Lying well off the Milky Way, there are no open clusters or nebulae—it is galaxies that dominate Columba. Many are small and faint, but the spiral galaxies **NGC 1792** and **NGC 1808** (telescope) are just visible in a 4-inch (10 cm) telescope; they are bright, attractive, spindle-shaped patches of haze in an 8-inch (20 cm) telescope under a rural sky.

Above One of the most appealing objects in the constellation of Columba is the globular cluster NGC 1851. A faint sprinkling of stars surrounds its bright and striking center.

Right NGC 1792, on the Columba–Caelum border, is a so-called starburst spiral galaxy. It is rich in neutralized hydrogen gas, evidence of its rapid formation of new stars.

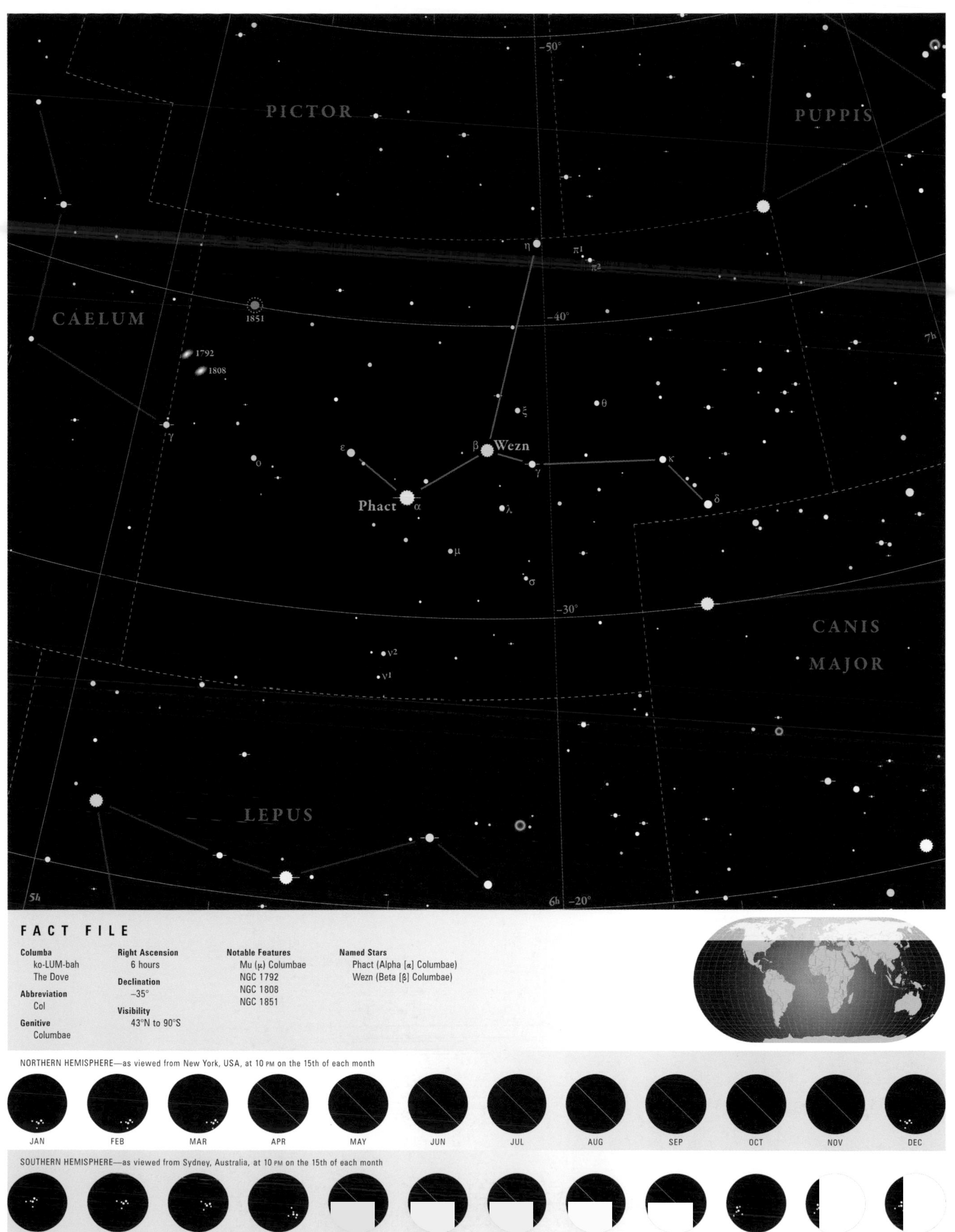

FACT FILE

Columba
ko-LUM-bah
The Dove

Abbreviation
Col

Genitive
Columbae

Right Ascension
6 hours

Declination
−35°

Visibility
43°N to 90°S

Notable Features
Mu (μ) Columbae
NGC 1792
NGC 1808
NGC 1851

Named Stars
Phact (Alpha [α] Columbae)
Wezn (Beta [β] Columbae)

Coma Berenices

Coma Berenices—usually called simply Coma—is a unique constellation: It is the only one that consists mainly of a single cluster of stars.

HAIR WARS

Queen Berenice of Egypt, wanting to ensure her husband's safe return from war, cut off her long golden tresses and took them to the temple as an offering to Aphrodite. Upon the king's return, she brought him to the temple to show him her role in his survival, but her offering was missing. The high priest, thinking with the clarity that mortal danger brings, told the royal couple that the gods had been so pleased with her gift that they had placed it in the sky. He motioned toward a fuzzy patch in the heavens, known ever after as Coma Berenices—"The Hair of Berenice." The small cloud also represents the tuft of hair at the end of Leo's tail.

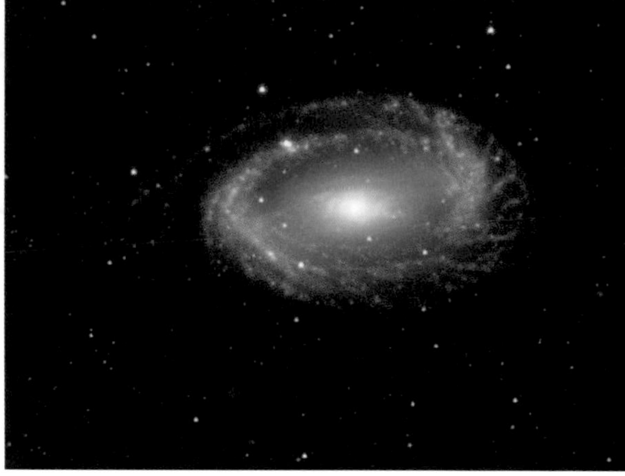

Right Most spiral galaxies have two or more arms, but the intriguing NGC 4725 in Coma Berenices has only one. Photographed in infrared by the Spitzer Space Telescope, this galaxy is 100,000 light-years across and lies 41 million light-years away.

THE STARS OF COMA BERENICES

None of the constellation's three brightest stars is a member of the Coma star cluster, but since the three are only 4th magnitude, it is the triangular open cluster that the naked-eye observer notices.

The cluster, Melotte 111, is ten moon-diameters wide, since it is one of the closest to us. Binoculars show it best. The unaided eye can discern about 20 of the cluster's 273 stars under excellent sky-watching conditions.

OBJECTS OF INTEREST

There are more than thirty galaxies in Coma Berenices that are suitable for moderate apertures. The spiral galaxy **M64 (NGC 4826)** is known as the **Black Eye Galaxy** because of the large dust cloud spanning its oval core. The dust is visible with an 8-inch (20 cm) telescope. The Virgo Galaxy Cluster spills into southern Coma Berenices, where there are six Messier galaxies present. The two best are **M99 (NGC 4254)** and **M100 (NGC 4321)**. Large amateur telescopes reveal the spiral arms in both these face-on galaxies. **NGC 4565** is the finest edge-on spiral in either hemisphere. At high power, the 14-arc-minute-long needle spans the field of view—an unforgettable sight. At its center there is a small round core, crossed by a delicate equatorial dust lane.

M53 (NGC 5024) is a globular cluster that begins to resolve into stellar pinpoints in a 6-inch (15 cm) telescope. M53's hundreds of thousands of stars collectively shine at magnitude 7.5. Only two moon-diameters southeast of M53 lies the globular **NGC 5053**. Although both are similar in size, NGC 5053 is only magnitude 9.8, as it has been stripped of most of its stars during its many orbits through the Milky Way's massive core.

Within a few more orbits NGC 5053 will dissipate entirely. Users of small telescopes enjoy the wide color-contrast binary **24 Comae Berenices**, which features a 5th magnitude light orange primary and a bluish 6th magnitude comes.

Left Recent studies of M64 suggest that the interstellar gas in its outer regions is spinning in the opposite direction to the gas and stars in the inner regions.

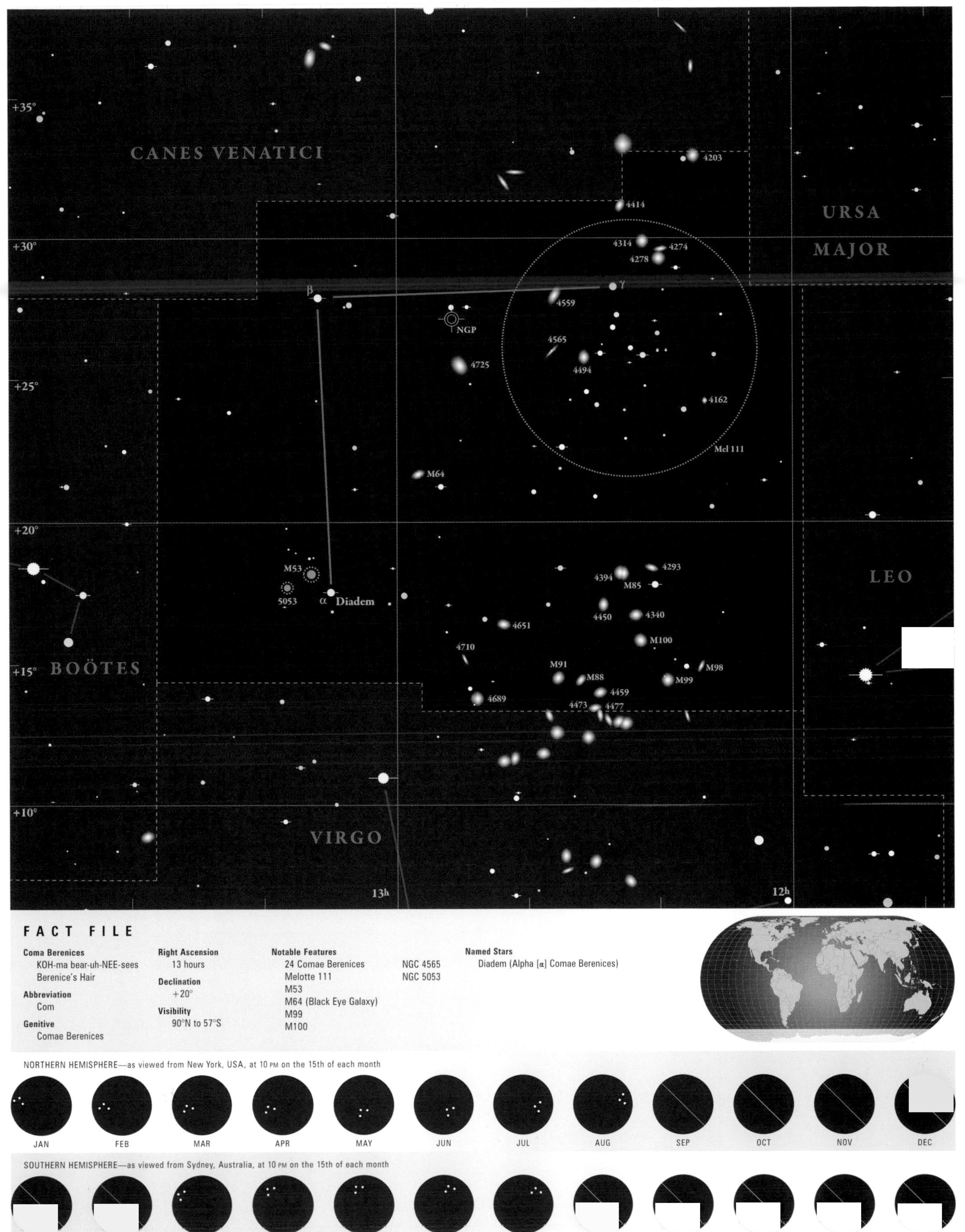

FACT FILE

Coma Berenices
KOH-ma bear-uh-NEE-sees
Berenice's Hair

Abbreviation
Com

Genitive
Comae Berenices

Right Ascension
13 hours

Declination
+20°

Visibility
90°N to 57°S

Notable Features
24 Comae Berenices
Melotte 111
M53
M64 (Black Eye Galaxy)
M99
M100
NGC 4565
NGC 5053

Named Stars
Diadem (Alpha [α] Comae Berenices)

NORTHERN HEMISPHERE—as viewed from New York, USA, at 10 PM on the 15th of each month

JAN FEB MAR APR MAY JUN JUL AUG SEP OCT NOV DEC

SOUTHERN HEMISPHERE—as viewed from Sydney, Australia, at 10 PM on the 15th of each month

JAN FEB MAR APR MAY JUN JUL AUG SEP OCT NOV DEC

Corona Australis

Corona Australis, the Southern Crown, was known to the Greeks and was one of Ptolemy's original 48 constellations, perceived as a wreath similar to those worn by Olympic athletes.

CROWN OF STARS

Another Greek astronomer, Aratus, recognized the group of stars as a wreath laid at the feet of Sagittarius. There are few, if any, legends associated with the group, although some may have perceived it as the crown placed in the sky by Dionysus after he retrieved his mother Semele from the underworld. This legend is also sometimes associated with the northern counterpart of Corona Australis—Corona Borealis.

Right This image of a star-forming region in Corona Australis shows bright white young stars and reddish protostars glowing in near-infrared wavelengths through the surrounding nebulosity.

THE STARS OF CORONA AUSTRALIS

None of the stars of Corona Australis is bright, although their consistent brightness and regular placement around the crescent makes the group easy to identify. If Theta (θ), Kappa (κ), and Lambda (λ) are added to the usual crescent, in modern "stick-figure" star maps, they almost make a complete circle.

Alfecca Meridiana, or Alpha (α) Coronae Australis, at magnitude 4.11, is the equal-brightest star in the constellation with Beta (β). It is similarly named to the Alpha star of the constellation's northern counterpart, though there the star's name is spelled Alphecca, from the Arabic for "break." It is an A-type dwarf star about 2.5 times as massive as our Sun, 31 times brighter, and 130 light-years distant. In most respects, it is a very similar star to the brightest star in the night sky, Sirius.

OBJECTS OF INTEREST

Corona Australis adjoins Sagittarius where the Milky Way is at its brightest and thickest. Despite that, it is not overly endowed with bright objects for the amateur sky-watcher. This is mostly due to the large patches of dark nebulae and obscuring material that block our view of more distant objects in that part of the sky.

A noteworthy globular cluster, barely below naked-eye visibility, is **NGC 6541**, located in the western end of the constellation. A 6-inch (15 cm) telescope displays a large, bright, hazy halo with many faint stars in the outer regions, steeply concentrated toward the center.

The only planetary nebula within the constellation is the bright **IC 1297**, which appears in an 8-inch (20 cm) telescope as a round azure-colored disk with many stars in the background.

The reflection/dark nebula complex of **NGC 6726, NGC 6729** and **IC 4812**, found in the same field as the bright globular cluster NGC 6723 (lying just over the border in Sagittarius), is very interesting in an 8-inch (20 cm) telescope. The well-resolved globular contrasts with several faint knots of nebulosity surrounding some stars, and with a very dark patch—the dark nebula **Bernes 157**—all in the one field.

As a bonus, the lovely, nearly equal magnitude 6 double star **Brisbane 14** is also visible.

Left The dust in these interstellar clouds glows in the reflected light of the stars of Corona Australis.

ARA
TELESCOPIUM
SCORPIUS
SAGITTARIUS
Alfecca Meridiana
17h
18h
19h
20h
-50°
-40°
-30°
6541
6726
6729
IC 1297

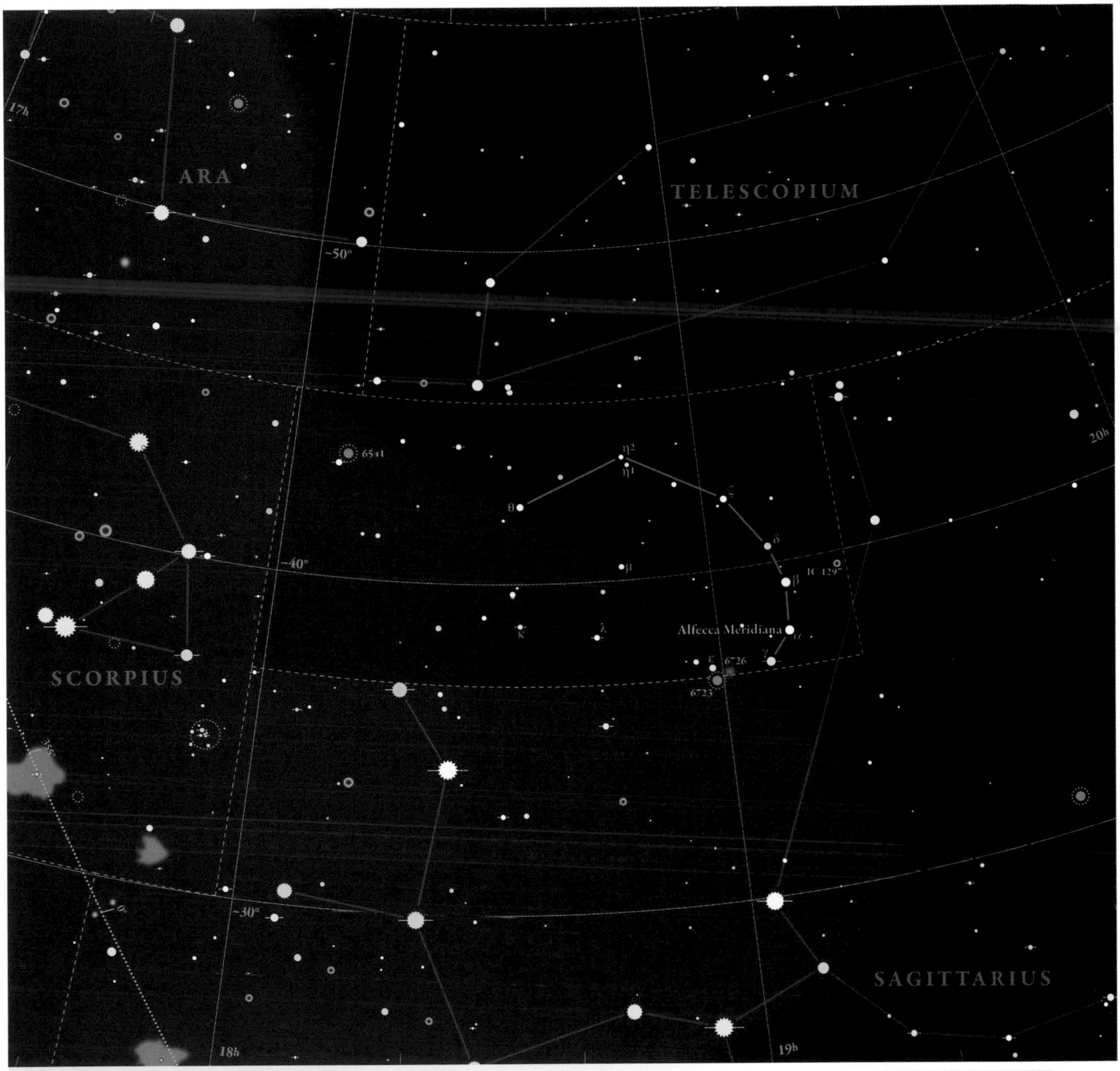

FACT FILE

Corona Australis
kor-OH-nah os-TRAH-lis
The Southern Crown

Abbreviation
CrA

Genitive
Coronae Australis

Right Ascension
19 hours

Declination
–40°

Visibility
42°N to 90°S

Notable Features
Bernes 157
Brisbane 14
IC 1297
IC 4812
NGC 6541
NGC 6726
NGC 6729

Named Stars
Alfecca Meridiana (Alpha [α] Coronae Australis)

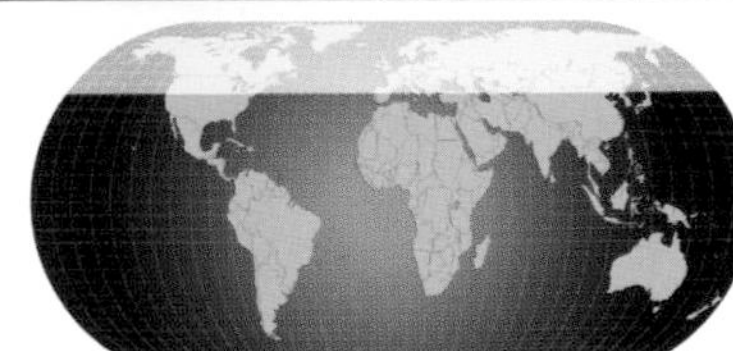

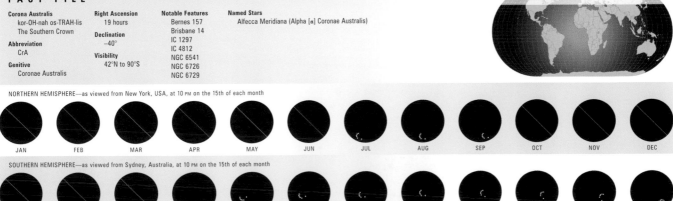

NORTHERN HEMISPHERE—as viewed from New York, USA, at 10 PM on the 15th of each month

JAN
FEB
MAR
APR
MAY
JUN
JUL
AUG
SEP
OCT
NOV
DEC

SOUTHERN HEMISPHERE—as viewed from Sydney, Australia, at 10 PM on the 15th of each month

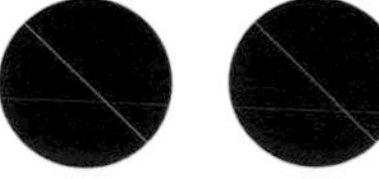

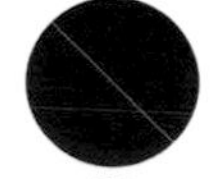

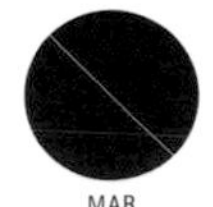

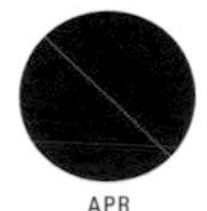

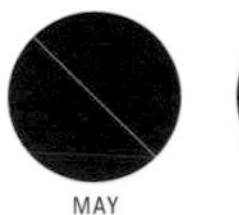

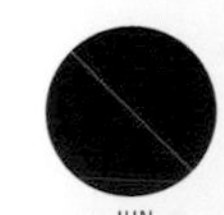

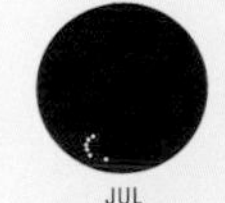

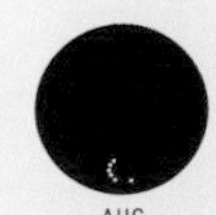

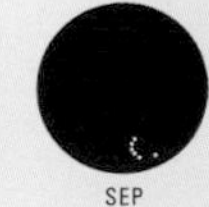

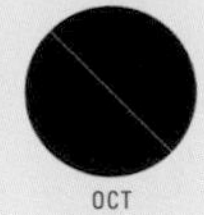

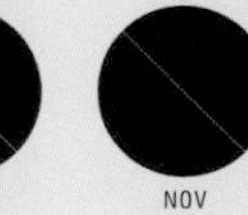

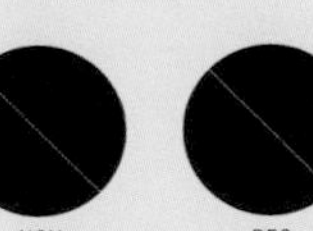

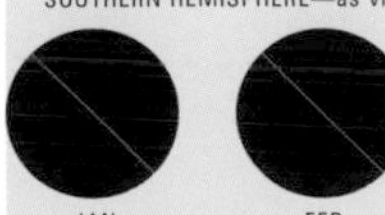

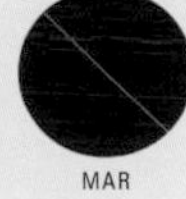

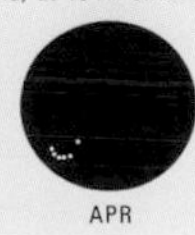

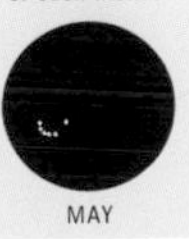

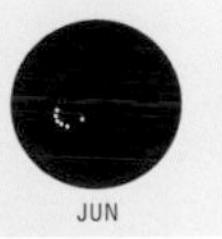

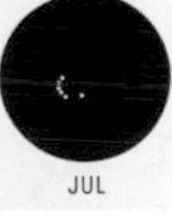

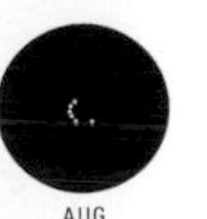

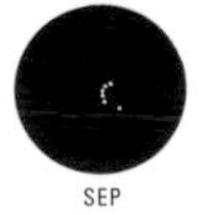

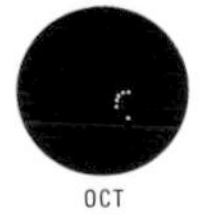

JAN
FEB
MAR
APR
MAY
JUN
JUL
AUG
SEP
OCT
NOV
DEC

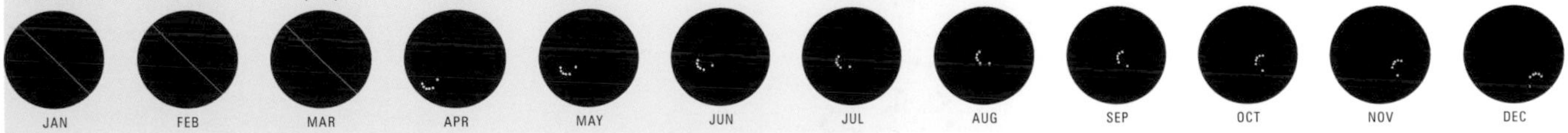

Corona Borealis

Corona Borealis, the Northern Crown, is one of the sky's constellations that actually resembles what it is meant to represent—a bejeweled crown.

Right Abell 2065 is a massive galaxy cluster 1.5 billion light-years from our Solar System, and is a challenging target for a large telescope.

CROWNING GLORY

Ariadne was the daughter of King Minos of Crete, the regent who built the Labyrinth under his palace at Knossos. The Minotaur, a flesh-eating monster with a man's body and a bull's head and tail, was held within the Labyrinth. Athens, as part of its tribute to powerful Crete, was obliged at regular intervals to send human sacrifices of seven youths and seven maidens, who were then imprisoned in the maze with the Minotaur. Theseus, son of the Athenian king, heroically volunteered to go to Crete with the hope of slaying the monster. Ariadne, having fallen in love with Theseus, gave him a sword and a ball of string to allow him to find his way out of the Labyrinth if he survived. He slew the Minotaur, rescued the Athenians sent with him, and fled Crete with Ariadne, who had prepared a ship. Theseus cruelly abandoned Ariadne on the island of Naxos. She was despondent, but soon she met and married the god Dionysus (Bacchus) instead. Upon her death, he placed in the sky the crown that he had given her at their wedding.

THE STARS OF CORONA BOREALIS

Alphecca, Alpha (α) Coronae Borealis, is a member of the Ursa Major Moving Cluster, Collinder 285, which also includes the five central stars in the Big Dipper asterism and various other stars in Ursa Major, Leo Minor, and Draco. This is the closest cluster to the Solar System, although there is some controversy as to whether this sparse, widely scattered group can be considered a true open cluster since its self-gravity could not hold the group together.

R Coronae Borealis is a variable star that is normally about magnitude 5.8, but at unpredictable and irregular intervals it fades to anywhere between magnitude 9 and 15 as carbon soot builds up in its atmosphere.

OBJECTS OF INTEREST

Struve 1932 is a 1.6 arc-seconds matched yellow binary of magnitudes 7.3 and 7.4. **Zeta (ζ) Coronae Borealis** has greenish-white and pale blue components of magnitudes 5.1 and 6.0, separated by 6.3 arc-seconds.

The constellation's only deep-sky objects are a few galaxies that are very faint and seldom observed, with the exception of the very challenging **Corona Borealis Galaxy Cluster, Abell 2065**. On a superb night, a 16-inch (40 cm) reflector reveals eight of the cluster's 400-plus swarming galaxies. This huge galaxy cluster is so distant—1.5 billion light-years away—that it covers only one moon-diameter of sky. These are the most distant normal galaxies visible in the eyepieces of amateur telescopes, although a gravitationally lensed quasar in Ursa Major can be seen halfway across the universe.

Right The arc of bright stars that form Corona Borealis show the various colors that make them the jewels in the crown.

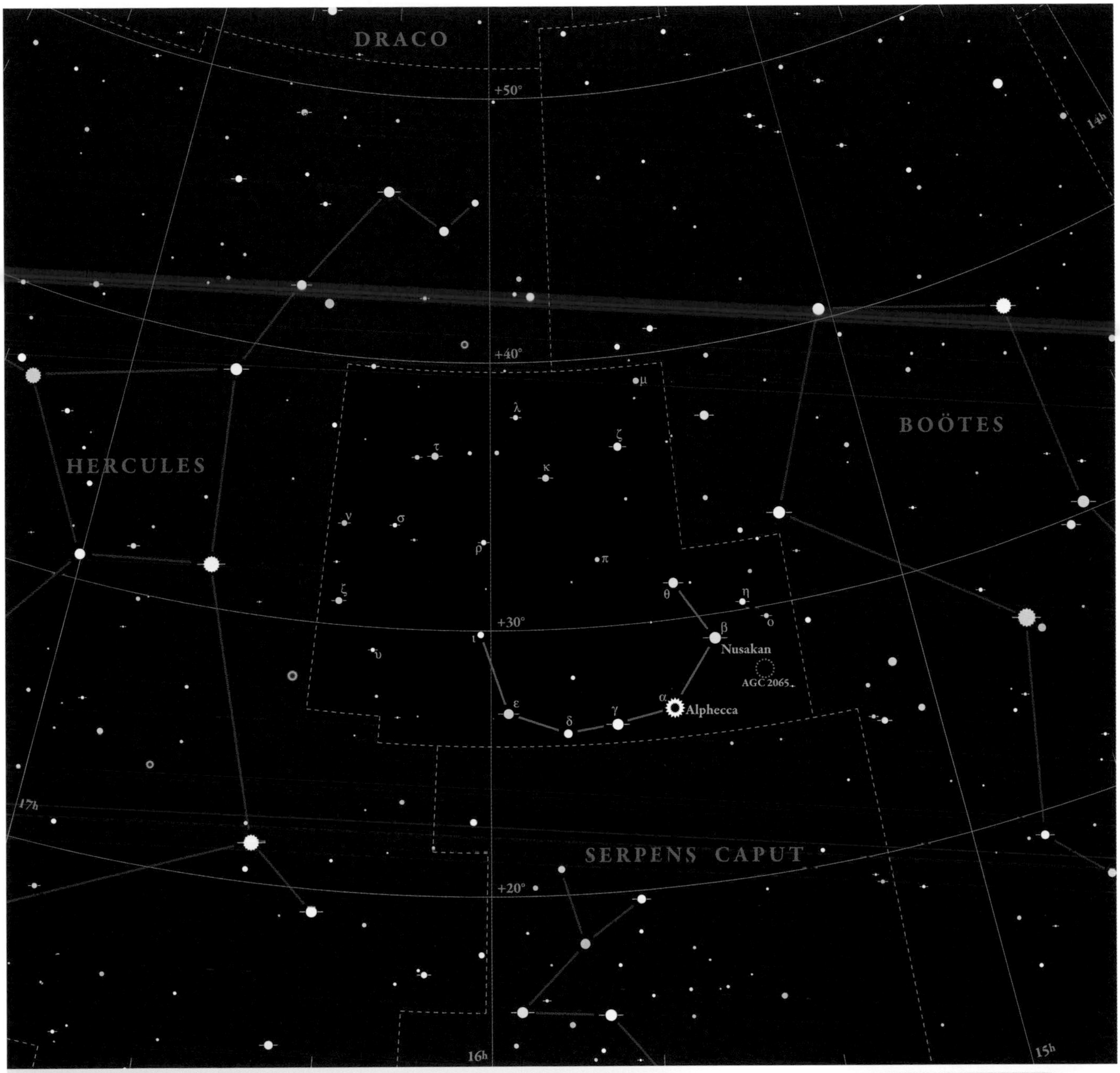

FACT FILE

Corona Borealis
kor-OH-nah bor-ee-AL-is
The Northern Crown

Abbreviation
CrB

Genitive
Coronae Borealis

Right Ascension
16 hours

Declination
+30°

Visibility
90°N to 54°S

Notable Features
Abell 2065 (Corona Borealis Galaxy Cluster)
R Coronae Borealis
Struve 1932
Zeta (ζ) Coronae Borealis

Named Stars
Alphecca (Alpha [α] Coronae Borealis)
Nusakan (Beta [β] Coronae Borealis)

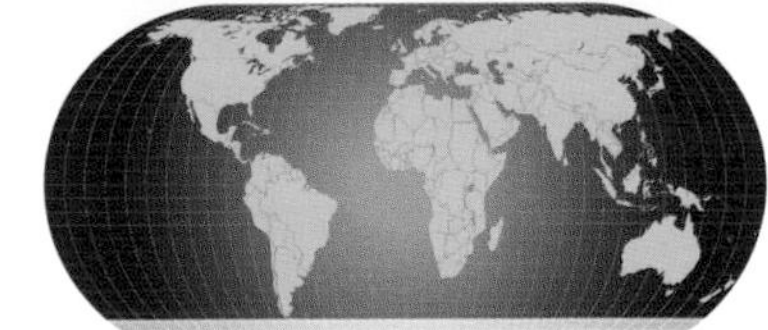

NORTHERN HEMISPHERE—as viewed from New York, USA, at 10 PM on the 15th of each month

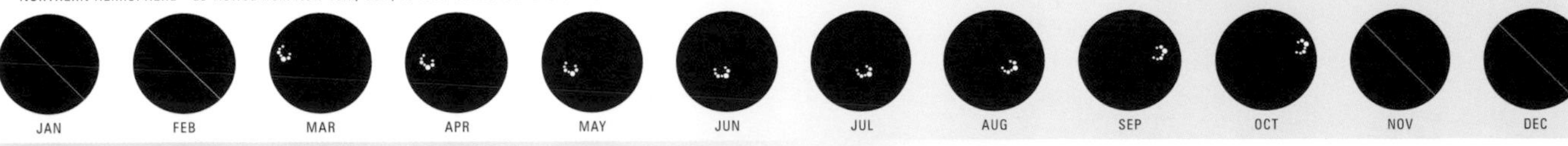

SOUTHERN HEMISPHERE—as viewed from Sydney, Australia, at 10 PM on the 15th of each month

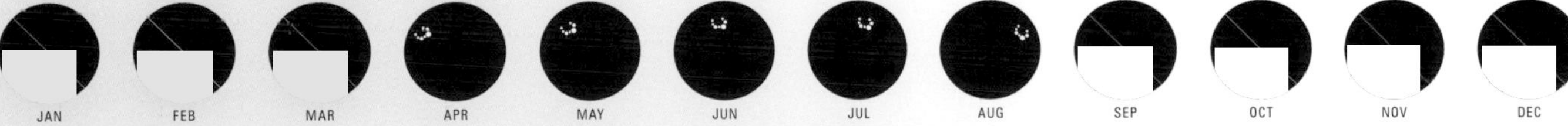

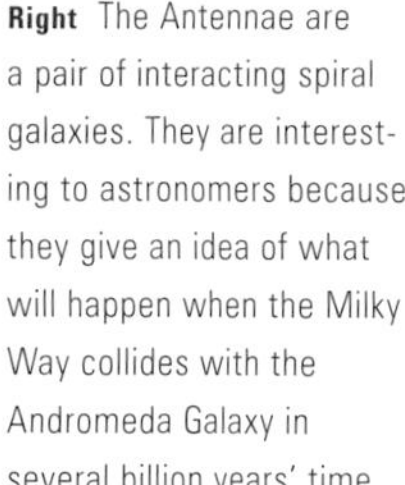

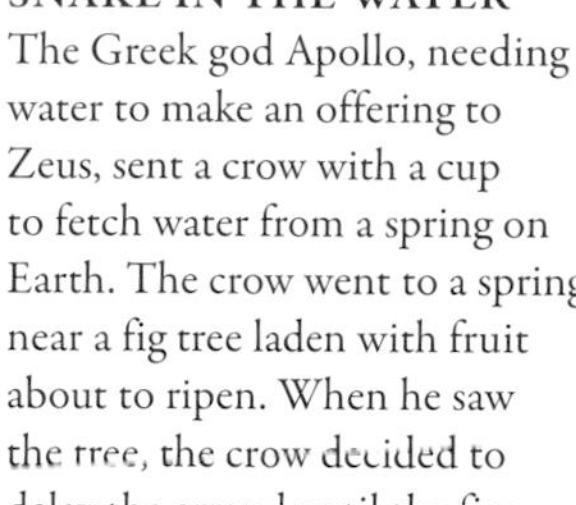

Corvus and Crater

The small southern constellations of Corvus the Crow and Crater the Cup share a myth involving a god, some lies, and a serpent.

SNAKE IN THE WATER

The Greek god Apollo, needing water to make an offering to Zeus, sent a crow with a cup to fetch water from a spring on Earth. The crow went to a spring near a fig tree laden with fruit about to ripen. When he saw the tree, the crow decided to delay the errand until the figs ripened and he had had his fill.

But Corvus tarried too long while gorging himself. Realizing that he would be punished, he filled the cup at the spring and picked up a water snake. On his return to Apollo, he presented the serpent and explained that it had prevented him from filling the cup until he could kill it. Apollo saw through the deception and in his anger flung the Cup (Crater), the Snake (Hydra), and the Crow (Corvus) into the sky. As further punishment, Apollo made the crow thirsty for eternity, causing it to have a harsh "caw-caw" call rather than a song.

Right The Antennae are a pair of interacting spiral galaxies. They are interesting to astronomers because they give an idea of what will happen when the Milky Way collides with the Andromeda Galaxy in several billion years' time.

THE STARS OF CORVUS AND CRATER

Although they look nothing like a bird, the main stars of Corvus stand out strongly against their fairly dim neighbors. The main part is a distinctive box of stars made by Beta (β), Epsilon (ε), Gamma (γ), and Delta (δ), which are all reasonably bright 3rd magnitude stars. The faint stars of Crater do little to excite the imagination of the sky-watcher. The open circle formed by Eta (η), Zeta (ζ), Gamma (γ), Delta (δ), Epsilon (ε), and Theta (θ) form the bowl of the cup, but as all are 4th and 5th magnitude, they require a dark sky to see them well.

Above This image shows a night-time view of the constellation of Corvus, in which the galaxies of NGC 4038 and NGC 4039—the Antennae—feature prominently.

Alpha (α) Corvi is another example of how Bayer did not always follow from brightest to faintest within a constellation. Only the fifth brightest star, it is a very youthful F-type dwarf star, about 48 light-years distant, and a little more massive, hotter, and more luminous than our Sun.

OBJECTS OF INTEREST

Far from the plane of the Milky Way, most of the objects for the amateur sky-watcher in Corvus and Crater are the small and faint galaxies. The most spectacular is the colliding galaxy pair **NGC 4038** and **NGC 4039** in Corvus—these are known as the **Antennae**. An 8-inch (20-cm) telescope under a dark sky shows the common halo, looking rather like a mottled and blotchy, curled-up, peeled shrimp, but the long faint "antenna" extensions are invisible in amateur telescopes.

In a 6-inch (15-cm) telescope, the bright planetary nebula **NGC 4361**, almost in the center of the "box" of Corvus, is a bright grayish haze with a prominent central star.

Of the many faint galaxies in Crater, the best is the spiral galaxy **NGC 3511**, which has a faint oval halo and a slightly brighter core.

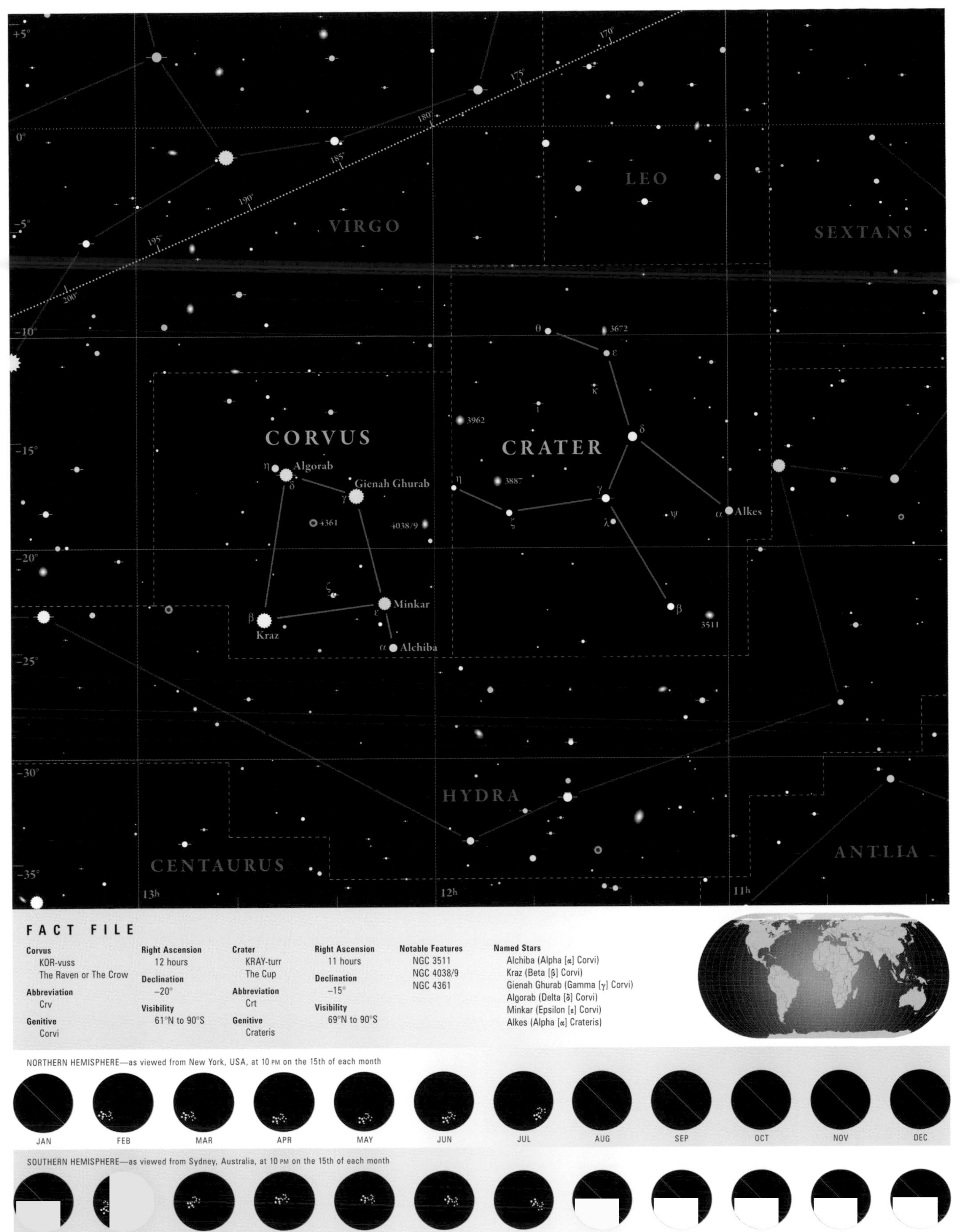

FACT FILE

Corvus
KOR-vuss
The Raven or The Crow

Abbreviation
Crv

Genitive
Corvi

Right Ascension
12 hours

Declination
−20°

Visibility
61°N to 90°S

Crater
KRAY-turr
The Cup

Abbreviation
Crt

Genitive
Crateris

Right Ascension
11 hours

Declination
−15°

Visibility
69°N to 90°S

Notable Features
NGC 3511
NGC 4038/9
NGC 4361

Named Stars
Alchiba (Alpha [α] Corvi)
Kraz (Beta [β] Corvi)
Gienah Ghurab (Gamma [γ] Corvi)
Algorab (Delta [δ] Corvi)
Minkar (Epsilon [ε] Corvi)
Alkes (Alpha [α] Crateris)

NORTHERN HEMISPHERE—as viewed from New York, USA, at 10 PM on the 15th of each month

JAN FEB MAR APR MAY JUN JUL AUG SEP OCT NOV DEC

SOUTHERN HEMISPHERE—as viewed from Sydney, Australia, at 10 PM on the 15th of each month

JAN FEB MAR APR MAY JUN JUL AUG SEP OCT NOV DEC

Crux

Crux is the impressive centerpiece of the heavens for sky-watchers in the Southern hemisphere. Surprisingly, it was not recognized as a constellation in its own right until the sixteenth century.

A CROSS POINT

Because of the effects of precession, Crux was visible low on the horizon for the Greek and Middle Eastern cultures at about the time of Christ's crucifixion. The Romans could also see Crux from Alexandria, and called it "Thronis Caesaris" to honor one of Rome's great emperors, Augustus. Although well known to civilizations for millennia, it was not depicted as an "official" constellation until Petrus Plancius placed it among the feet of the Centaur in his 1598 star-globe; at that time it was still seen as part of Centaurus.

The indigenous peoples of the Southern hemisphere have a multitude of mythologies and stories wound around the cross. To the Maori people of New Zealand it represents an anchor, with the pointer stars as the rope. Some tribes of Australian Aborigines see it as a stingray, with the pointer stars representing two pursuing sharks. In Hindu astrology, Crux is named "Trishanku," while the Incas commemorated it in stone at Machu Picchu.

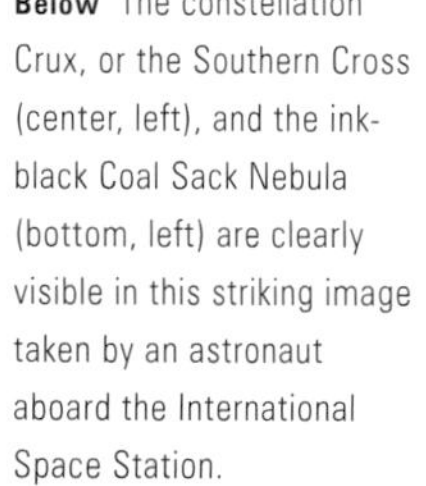

Below The constellation Crux, or the Southern Cross (center, left), and the ink-black Coal Sack Nebula (bottom, left) are clearly visible in this striking image taken by an astronaut aboard the International Space Station.

THE STARS OF CRUX

Crux is the smallest constellation in the sky, but perched on almost the brightest section of the southern Milky Way, it punches well above its weight in bright stars. Starting with brilliant Acrux, or Alpha (α) Crucis, the stars descend in brightness, running clockwise through Beta (β) (also called Mimosa), Gamma (γ), Delta (δ), and Epsilon (ε) Crucis.

Acrux, the twelfth brightest star in the sky, and Mimosa, the nineteenth brightest, are massive blue giants, 320 and 350 light-years distant, respectively. Gamma (γ) is the twenty-fourth brightest, and is an M-class red giant star somewhat closer to us at 88 light-years away. To the naked eye, it has an obvious orange tint.

OBJECTS OF INTEREST

Acrux is one of the most striking binary stars in the sky. Its two bluish-white diamonds of virtually equal brilliance are tight together, but can still be resolved in small telescopes. A somewhat fainter third star close by is not part of the system but is a little in the foreground.

Sapphire-like **Mimosa** also has an optical companion—a faint, blood-red carbon star, which is an incredible contrast. Between Acrux and Mimosa, and extending a little further southward, is the most easily seen dark nebula in the sky, the famous **Coal Sack**. This cold cloud of dark hydrogen is about 500 light-years away and looks like a hole in the brilliant Milky Way.

Also close to Mimosa is arguably the finest open star cluster in the sky, **NGC 4755**, which is known as the **Jewel Box Cluster**. About 7,000 light-years away, the brighter stars have a distinctive A-frame shape, with a host of fainter stars to the side and an ember-like red supergiant star near the center.

CENTAURUS

Gacrux

Mimosa

Acrux

Coal Sack

MUSCA

CARINA

CIRCINUS

FACT FILE

Crux
Kruks
The Southern Cross

Abbreviation
Cru

Genitive
Crucis

Right Ascension
12 hours

Declination
–60°

Visibility
22°N to 90°S

Notable Features
Alpha (α) Crucis
Beta (β) Crucis
NGC 4755 (Jewel Box)
Coal Sack

Named Stars
Acrux (Alpha [α] Crucis)
Mimosa (Beta [β] Crucis)
Gacrux (Gamma [γ] Crucis)

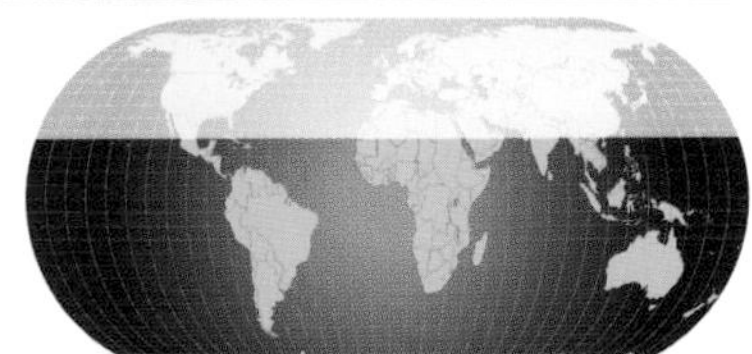

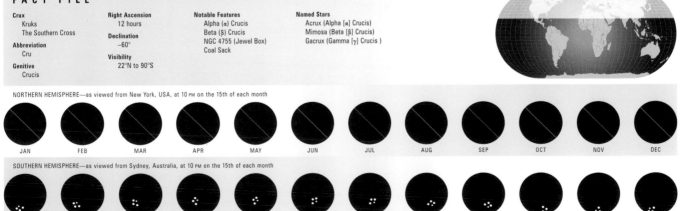

NORTHERN HEMISPHERE—as viewed from New York, USA, at 10 PM on the 15th of each month

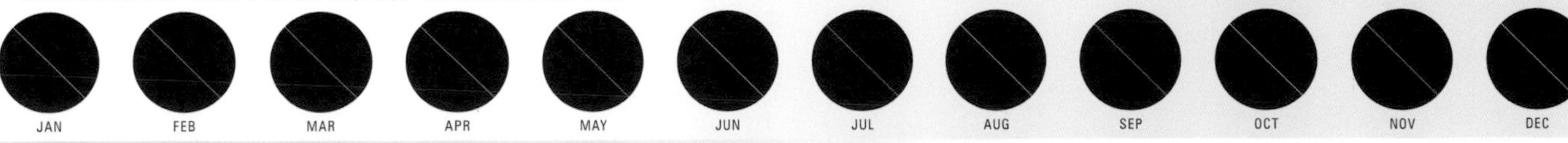

SOUTHERN HEMISPHERE—as viewed from Sydney, Australia, at 10 PM on the 15th of each month

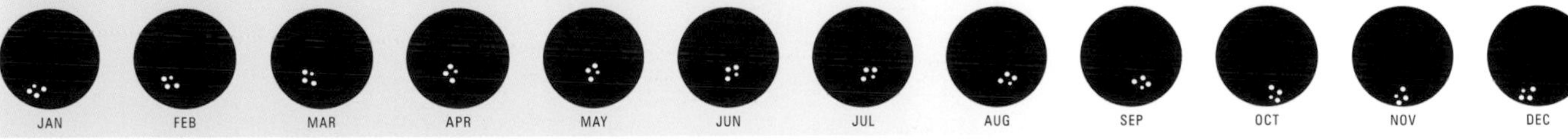

Cygnus

High in the northern summer sky, Cygnus, the Swan, offers the brightest part of the northern Milky Way—the magnificent Cygnus Starcloud.

SWAN SONG

Zeus once disguised himself as a swan to help him seduce Leda, the wife of the king of Sparta. The daughter resulting from their union became Helen of Troy.

THE STARS OF CYGNUS

The brighter stars of Cygnus constitute the Northern Cross, an asterism. The same stars, plus some fainter ones that fill out the wings, outline the swan.

Alpha (α) Cygni (Deneb) is 1st magnitude, despite being 1,500 light-years away, because it has sixty thousand times the luminosity of our Sun. Notably, the first star to have its distance (11.4 light-years) measured by its annual parallax was 61 Cygni, in 1840. This is a splendid, wide pair of deep-orange dwarfs.

OBJECTS OF INTEREST

Albireo, Beta (β) Cygni, is the sky's most famous color-contrast double star. Its widely separated yellow and deep-blue stars are magnitudes 3.1 and 5.1.

M39 (NGC 7092) is a magnitude 4.6 patch, known to Aristotle. In binoculars M39 becomes an attractive triangular open cluster of 25 stars, and the branching, six-moon-diameters long, dark nebula **B168** is in the same Milky Way field, 100 arc-minutes southeast. Cygnus has about 60 clusters suitable for amateur telescopes. (See Lacerta and Vulpecula, both carved out of this section of the Milky Way, for representative examples.)

Of the fourteen planetary nebulae visible in Cygnus, the best is **NGC 6826**. With averted (peripheral) vision, the bluish 25 arc-seconds disk of the Blinking Planetary (magnitude 8.8) is so bright that it hides the magnitude 10.6 central star. When direct vision is used, in turn, the star almost overpowers the disk.

Intense stellar winds from a massive young Wolf-Rayet star created **NGC 6888,** the **Crescent Nebula**. With an 8-inch (20 cm) telescope and an OIII filter, the crescent-shaped portion is revealed as just the brighter end of an 18 by 13 arc-minute oval.

The **Veil Nebula** is a magnificent supernova remnant. Ordinary binoculars reveal hook-shaped **NGC 6992/5**. A 3-inch (8 cm) aperture shows filamentary **NGC 6960** and **Pickering's Triangular Wisp** (uncataloged). Many of the strands and prongs that appear on images can be seen with an OIII filter and a 16-inch (40 cm) telescope, which also takes in **NGC 6979**.

NGC 7000 is the 3-degree-long **North America Nebula**. The unaided eye sees only a wedge-shaped star-cloud here, but a 4-inch (10 cm) telescope reveals both "Florida" and "Mexico," which consist of mostly emission nebulosity. Dark nebulae form the "oceans" surrounding the glowing "continent."

Another complex of bright and dark nebulae is the **Gamma (γ) Cygni Nebula, IC 1318**.

Cygnus's largest dark nebulae can be seen with the unaided eye. The **Northern Coalsack**, near Deneb, is the bulbous northern end of the **Great Rift**, a complex of dust clouds that bisects our galaxy for 130 degrees. The **Funnel Cloud Nebula (LG3)** runs 12 degrees from Eta (η) Cephei to just west of M39.

Above The North America Nebula is named because its shape appears to resemble that continent, when it is viewed with a small telescope.

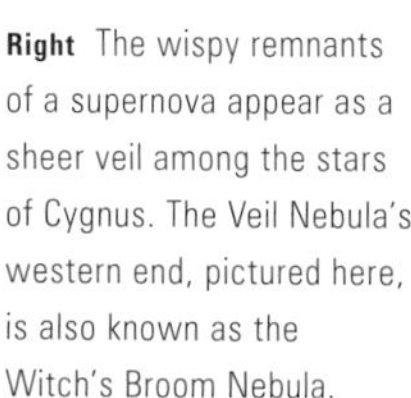

Right The wispy remnants of a supernova appear as a sheer veil among the stars of Cygnus. The Veil Nebula's western end, pictured here, is also known as the Witch's Broom Nebula.

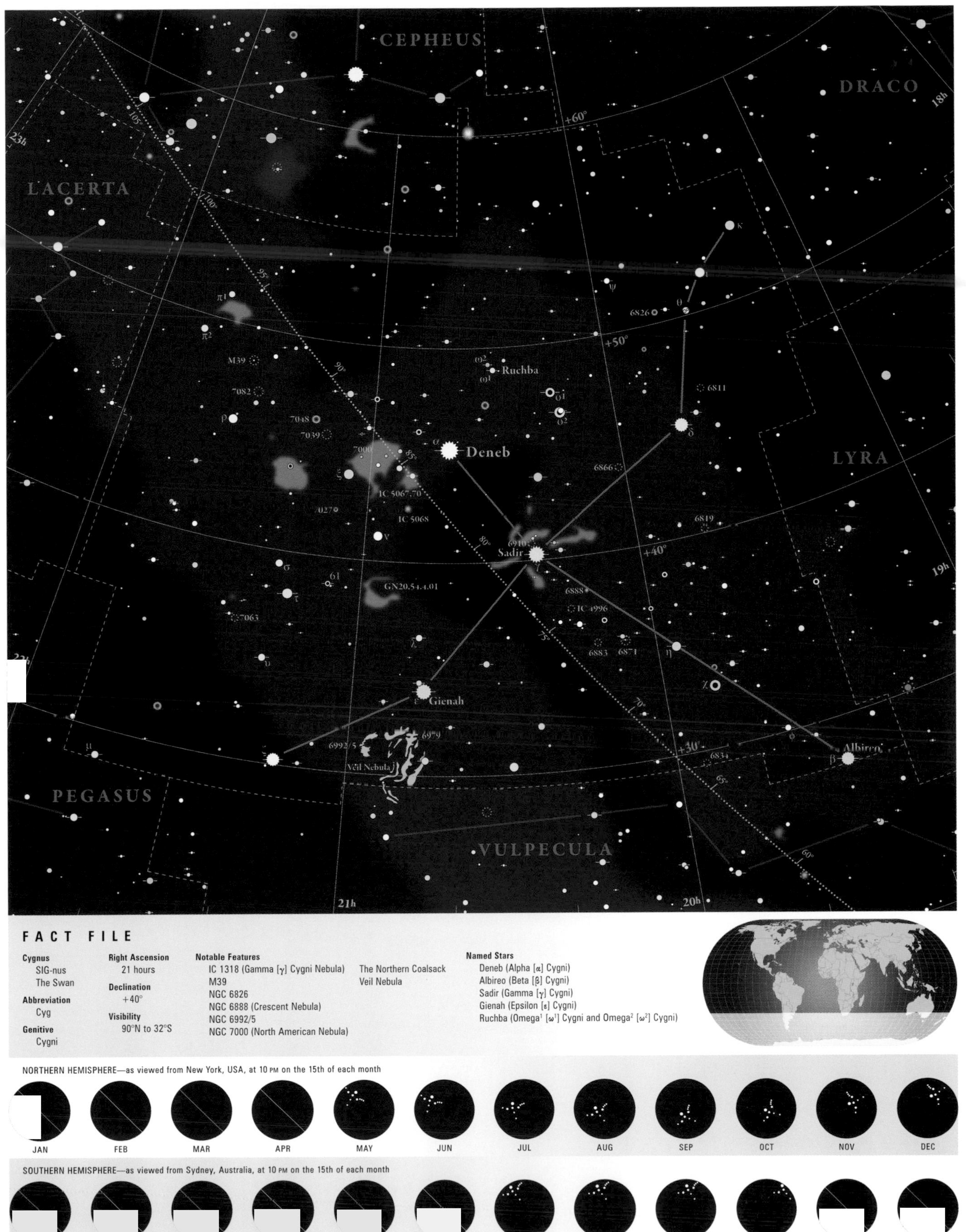

FACT FILE

Cygnus
SIG-nus
The Swan

Abbreviation
Cyg

Genitive
Cygni

Right Ascension
21 hours

Declination
+40°

Visibility
90°N to 32°S

Notable Features
IC 1318 (Gamma [γ] Cygni Nebula)
M39
NGC 6826
NGC 6888 (Crescent Nebula)
NGC 6992/5
NGC 7000 (North American Nebula)
The Northern Coalsack
Veil Nebula

Named Stars
Deneb (Alpha [α] Cygni)
Albireo (Beta [β] Cygni)
Sadir (Gamma [γ] Cygni)
Gienah (Epsilon [ε] Cygni)
Ruchba (Omega[1] [ω^1] Cygni and Omega[2] [ω^2] Cygni)

NORTHERN HEMISPHERE—as viewed from New York, USA, at 10 PM on the 15th of each month

JAN FEB MAR APR MAY JUN JUL AUG SEP OCT NOV DEC

SOUTHERN HEMISPHERE—as viewed from Sydney, Australia, at 10 PM on the 15th of each month

JAN FEB MAR APR MAY JUN JUL AUG SEP OCT NOV DEC

Delphinus

This distinctively shaped constellation is faint and far away, but the stars of the Dolphin are easy to identify.

DOLPHIN SWEET TALK

The Greeks devised the constellation of Delphinus, and it was most likely associated with the sea-god, Poseidon (Neptune). After the defeat of the Titans, Zeus, Hades, and Poseidon each set up his own realm. Poseidon's was the ocean, and after building his palace he began to look for a wife, and his thoughts turned to his fifty sea-nymphs—the Nereids.

He courted Amphitrite, but she resisted him and fled to the Atlas Mountains. A succession of messengers dispatched to persuade her returned unsuccessful, so Poseidon finally sent a dolphin. Moved by the dolphin's earnest pleas, Amphitrite relented and returned to the sea to become Poseidon's wife, and bore him many children. Grateful Poseidon elevated the dolphin to the sky in return. In more recent times, the constellation has become colloquially known as "Job's Coffin," but the origin and meaning of the appellation are now lost.

THE STARS OF DELPHINUS

Although Delphinus is a very small constellation with fairly faint stars, the compact diamond-plus-one shape formed by Alpha (α), Beta (β), Gamma (γ), Delta (δ), and Epsilon (ε) Delphini (between 1st magnitude Altair and the Great Square) is very easy to pick out. All five stars are 4th magnitude, and are almost equally bright.

Alpha (α) and Beta (β) Delphini were named for the first time in the Palermo star catalog of 1814 and were called Sualocin and Rotanev, with no obvious clues as to the origin of these curious names. In the late nineteenth century, however, the British astronomer Thomas Webb noticed that these names written backward spell "Nicolaus Venator"—the name of an assistant to the famous Italian astronomer Giuseppe Piazzi.

Beta (β) Delphini is the lucida of Delphinus; it is a neighboring pair of F-type sub-giants, 97 light-years away, that are, respectively, eight and eighteen times more luminous than our Sun.

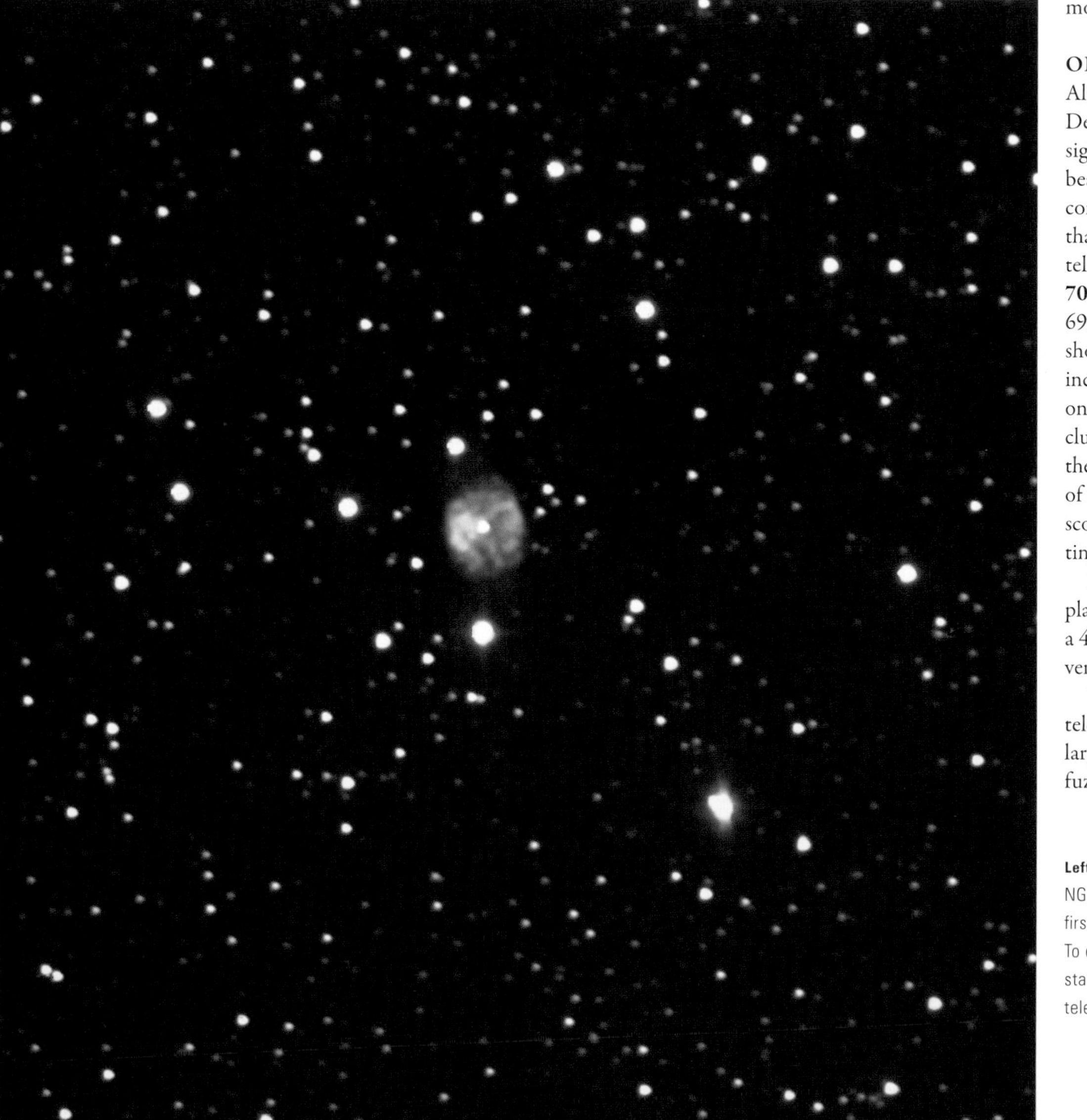

OBJECTS OF INTEREST

Although it is a small constellation, Delphinus contains several interesting sights. **Gamma (γ) Delphini** is a beautiful binary, with slightly uneven components of pale and golden yellow that can be resolved in virtually any telescope. Both **NGC 6934** and **NGGC 7006** are globular star clusters: NGC 6934 is small, moderately faint, and shows hints of weak resolution in an 8-inch (20 cm) telescope; NGGC 7006 is one of our Milky Way's most remote clusters—about 135,000 light-years from the Sun, and almost as far from the center of the Galaxy. An 8-inch (20 cm) telescope is needed to see this cluster as a tiny misty blur.

NGC 6891 is a tiny bright blue planetary nebula, which is just visible in a 4-inch (10 cm) telescope and looks very pretty in an 8-inch (20 cm).

When viewed in an 8-inch (20 cm) telescope, **NGC 6905** is a slightly larger, faintly blue, oval planetary with fuzzy edges.

Left The diminutive gemlike planetary nebula NGC 6905, also known as the Blue Flash, was first discovered by William Herschel in 1782. To observe the 14th magnitude bluish central star requires the use of a 10-inch (25 cm) telescope under very dark rural skies.

FACT FILE

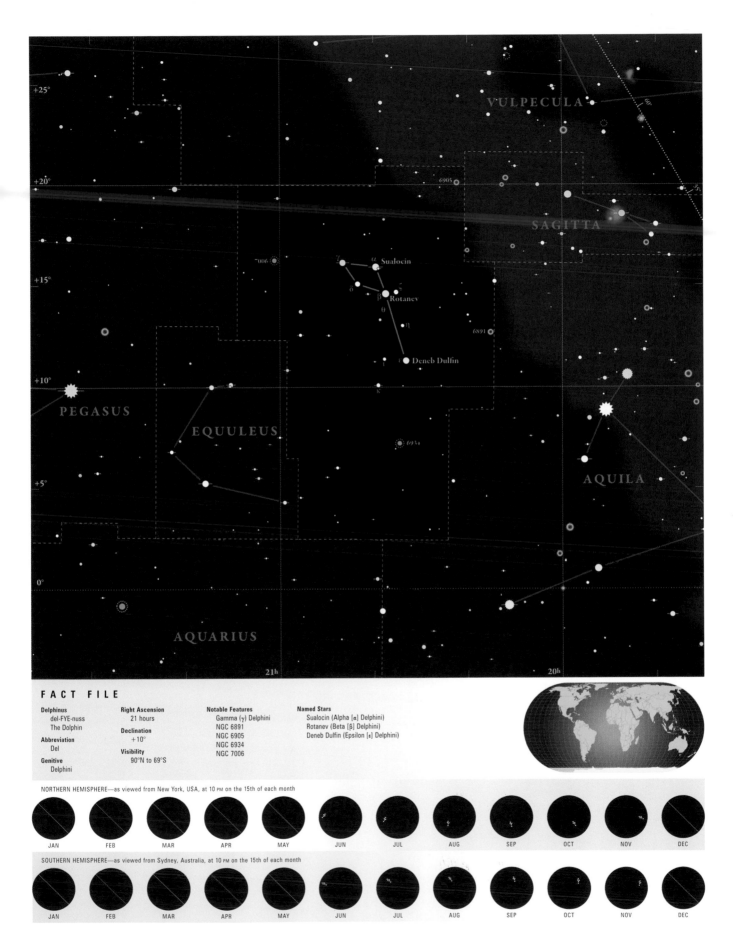

Delphinus
del-FYE-nuss
The Dolphin

Abbreviation
Del

Genitive
Delphini

Right Ascension
21 hours

Declination
+10°

Visibility
90°N to 69°S

Notable Features
Gamma (γ) Delphini
NGC 6891
NGC 6905
NGC 6934
NGC 7006

Named Stars
Sualocin (Alpha [α] Delphini)
Rotanev (Beta [β] Delphini)
Deneb Dulfin (Epsilon [ε] Delphini)

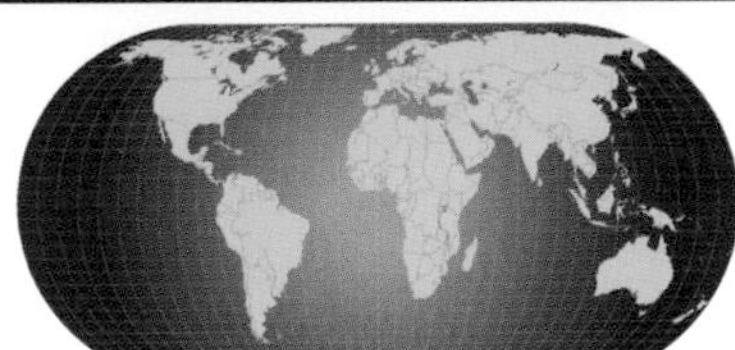

NORTHERN HEMISPHERE—as viewed from New York, USA, at 10 PM on the 15th of each month

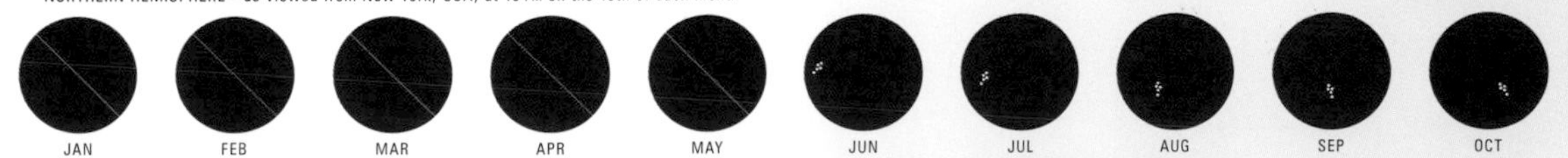

SOUTHERN HEMISPHERE—as viewed from Sydney, Australia, at 10 PM on the 15th of each month

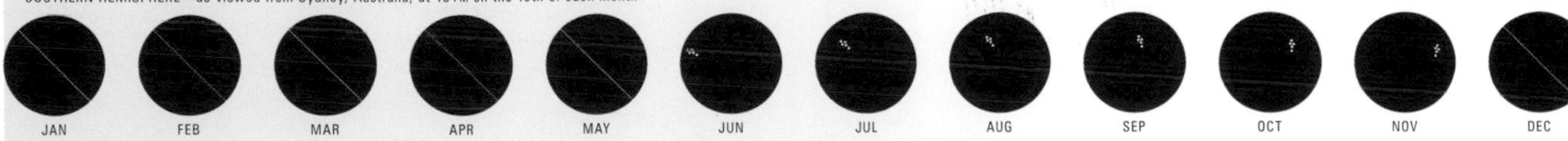

Dorado

Dorado is home to the famed Large Magellanic Cloud. The stars of Dorado are too far south to be seen in Europe, and accordingly no western mythology is attached to the constellation.

WHAT FISH IS THAT?

The Dutch explorers and cartographers Pieter Dirkszoon Keyser and Frederick de Houtman were the first to methodically observe Dorado's stars. Petrus Plancius introduced it on his star-globe in 1598, depicting it as the gold-colored dolphinfish found in tropical waters. Fittingly, it is found beside Volans, the flying fish, which is occasionally the dolphinfish's prey. In 1603, Johannes Bayer drew the constellation in *Uranometria* with its modern name Dorado, the goldfish. Later still, J. E. Bode and some other cartographers portrayed it as Xiphias, the swordfish. In modern times, it alternates between the names goldfish and swordfish, but its stars have remained largely unchanged.

Right The mammoth NGC 2070 (Tarantula Nebula) is the brightest object in the LMC. This giant is 30 times greater in size than the Great Nebula in Orion.

THE STARS OF DORADO

The principal stars of Dorado bear no real resemblance to any sea-creature. They form a long straggling line running from 4th magnitude Gamma (γ) in the north, through Alpha (α), Zeta (ζ), and Delta (δ) to 4th magnitude Beta (β) in the south, near the Large Magellanic Cloud.

Beta (β) Doradus is one of the brightest examples of a Cepheid variable star in the sky, varying rhythmically between magnitude 3.4 and 4.1 over a 9.9-day period. The Cepheid variables, named after the prototype Delta (δ) Cephei, are among the most important and useful pulsating variable stars. Using the Hubble Space Telescope, astronomers can identify them in other galaxies and use them as "standard candles" to determine the distance to that galaxy. Beta Doradus is more than 1,100 light-years away and 3,000 times brighter than our Sun.

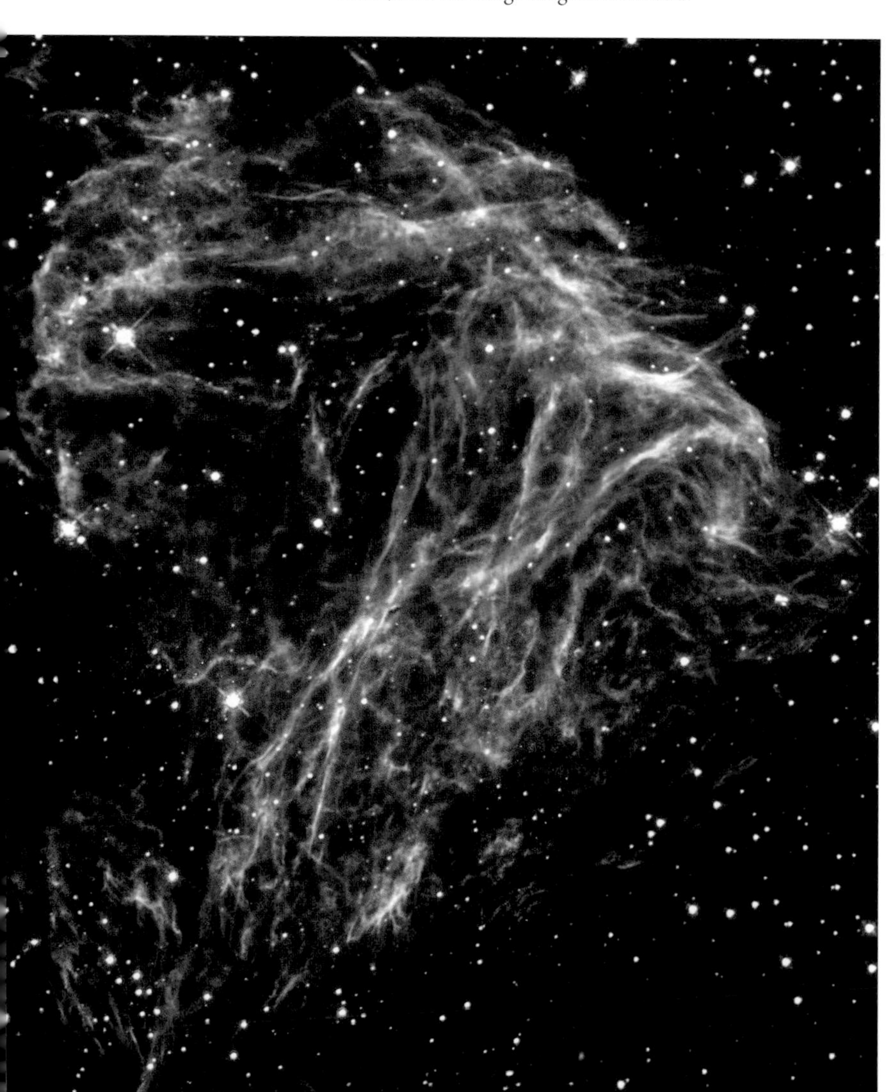

OBJECTS OF INTEREST

Dorado's main claim to fame is that it is the home of the **Large Magellanic Cloud (LMC)**. Visible to the naked eye, and looking like a large detached section of the Milky Way, it is in reality a satellite galaxy orbiting the Milky Way, about 165,000 light-years distant. With binoculars or a telescope, the cloud is alive with a vast collection of star clouds, clusters, knots, and thousands of faint points.

Most famous of all is the gigantic **Tarantula Nebula, NGC 2070**. Faintly visible to the naked eye as a hazy-looking spot at the eastern end of the cloud, it is a vast nebulous star factory in the process of assembling clusters of gargantuan stars. If it were 1,500 light-years away, like M42, it would cover most of Orion and be bright enough to cast prominent shadows in the "dark." The only naked-eye supernova since the invention of the telescope occurred near the Tarantula Nebula in 1987, peaking at 3rd magnitude.

Apart from the LMC, Dorado is home to some bright and prominent galaxies. **NGC 1566**, the **Spanish Dancer Galaxy**, is a lovely two-armed spiral for an 8-inch (20 cm) telescope, and the bright pair of lenticular galaxies, **NGC 1549** and **NGC 1553**, are found together in the same medium-power field. A 6-inch (15 cm) telescope shows a good view of them.

Left Spanning 30 light-years, the glowing debris from a supernova in the Large Magellanic Cloud (cataloged as N49) displays intricate filaments of gas and dust in this composite taken by the Hubble Space Telescope.

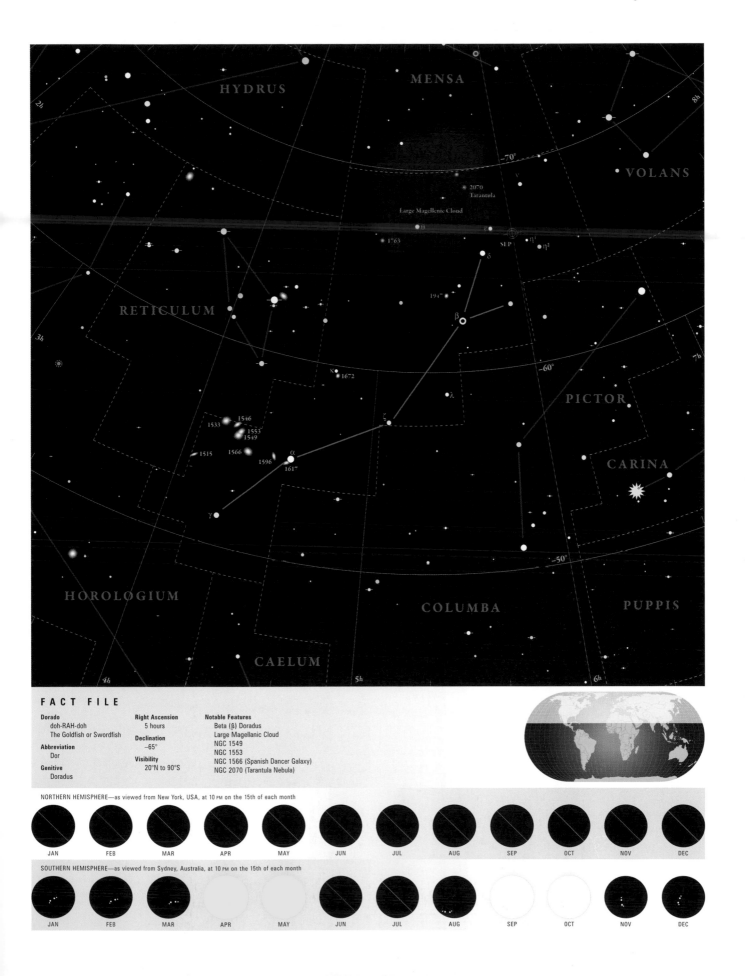

HYDRUS
MENSA
VOLANS
-70°
2070
Tarantula
Large Magellanic Cloud
θ
ε
SEP
η¹
η²
1763
δ
1947
β
RETICULUM
3h
2h
8h
7h
-60°
κ
1672
λ
PICTOR
ζ
1546
1533
1553
1549
1515
1566
1596
α
1617
CARINA
γ
-50°
HOROLOGIUM
COLUMBA
PUPPIS
CAELUM
4h
5h
6h

FACT FILE

Dorado
doh-RAH-doh
The Goldfish or Swordfish

Abbreviation
Dor

Genitive
Doradus

Right Ascension
5 hours

Declination
−65°

Visibility
20°N to 90°S

Notable Features
Beta (β) Doradus
Large Magellanic Cloud
NGC 1549
NGC 1553
NGC 1566 (Spanish Dancer Galaxy)
NGC 2070 (Tarantula Nebula)

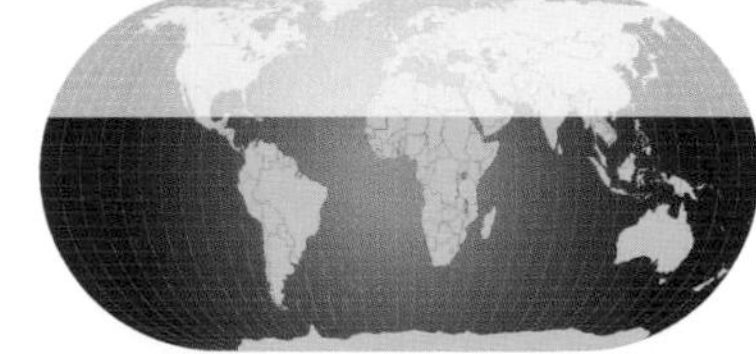

NORTHERN HEMISPHERE—as viewed from New York, USA, at 10 PM on the 15th of each month

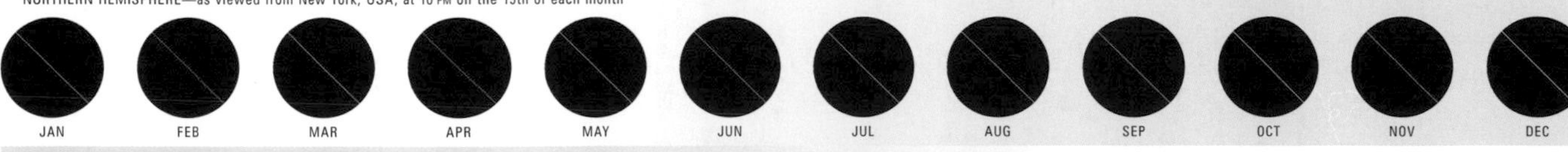

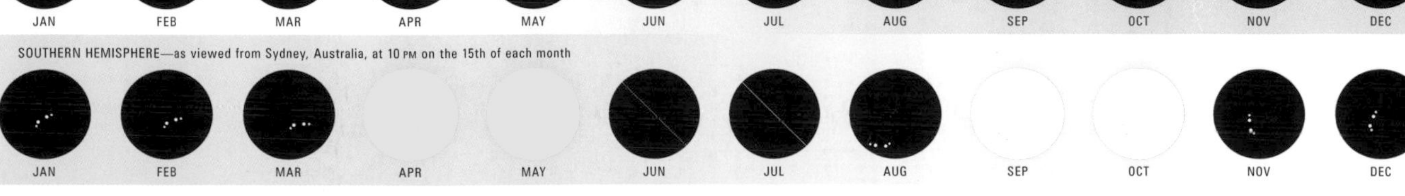

Draco

The Dragon is a far northern constellation and the eighth largest in area. It is best to view during the warmer months.

ENTER THE DRAGON

One of the twelve labors of Hercules was to obtain the golden apples of the Hesperides. In some versions of the myth, a dragon guarded the entrance to the Hesperides. Alternatively, an unsleeping dragon watched over the Golden Fleece; Jason managed to steal the treasure after putting the dragon to sleep.

THE STARS OF DRACO

Alpha (α) Draconis—Thuban—was the North Pole Star for a period of more than two thousand years, during which time the Egyptian Pyramids were constructed.

The mostly 3rd and 4th magnitude stars of Draco's tail are not too difficult to pick out as it winds around the Little Dipper at a fairly constant distance, but Draco does get a little confused where it turns back upon itself. The most obvious part of the constellation is the dragon's head, a small quadrangle of stars near Lyra.

OBJECTS OF INTEREST

Nu (ν) Draconis is a headlight double for binoculars with two white magnitude 4.9 stars situated about one arc-minute apart.

The stars **16 and 17 Draconis** look very similar (magnitudes 5.4 and 5.5), and are separated by 90 arc-seconds. A telescope transforms this white pair into a trio, revealing a yellowish magnitude 6.4 companion, just 3.4 arc-seconds from 17 Draconis.

Users of small telescopes can also enjoy two wide yellowish binaries **Psi (ψ) Draconis**, (magnitudes 4.9 and 6.1, at 30 arc-seconds apart), and **41 and 40 Draconis** (magnitudes 5.7 and 6.1, at 19 arc-seconds apart).

The lenticular galaxy **NGC 5866** is 6.6 by 3.2 arc-minutes and magnitude 9.9. An 8-inch (20 cm) reflector shows a small elongated spindle, brighter in the center, like a miniature of Virgo's M104. (NGC 5866 is sometimes identified as M102, but most authorities think that M102 was an error, a duplicate observation of M101.)

NGC 5907, the edge-on **Splinter Galaxy**, is 12 by 2 arc-minutes and magnitude 10.3. An 8-inch (20 cm) telescope shows an extremely thin needle of faint light with a slight central condensation.

The **Sampler** consists of **NGC 5981** (a mottled edge-on galaxy), **NGC 5982** (an elliptical), and **NGC 5985** (a face-on spiral), all in the same field of view.

Draco's showpiece is the planetary nebula **NGC 6543**, the **Cat's Eye Nebula**. Through an 8-inch (20 cm) telescope there is a bright (magnitude 8.1), blue, oval (23 by 17 arc-seconds) disk that is slightly fainter at the southern end. The magnitude 10.9 central star can be glimpsed with difficulty by using direct vision to reduce the brightness of the nebulosity. Through a 16-inch (40 cm) telescope the disk is strikingly blue–green at low power, the central star is obvious, and the 14th magnitude galaxy **NGC 6552** pops into view, only 9 arc-minutes to the east. The planetary becomes annular with an OIII filter, with a tiny darker center around the star. This filter also reveals an outer diffuse glow out to five disk-diameters width—this would be gas shed by the star before the planetary nebula formed.

Below The Hubble Space Telescope produced this clear image of the Draco galaxy NGC 5866, with its dark dust lane tilted edge-on to our line of sight.

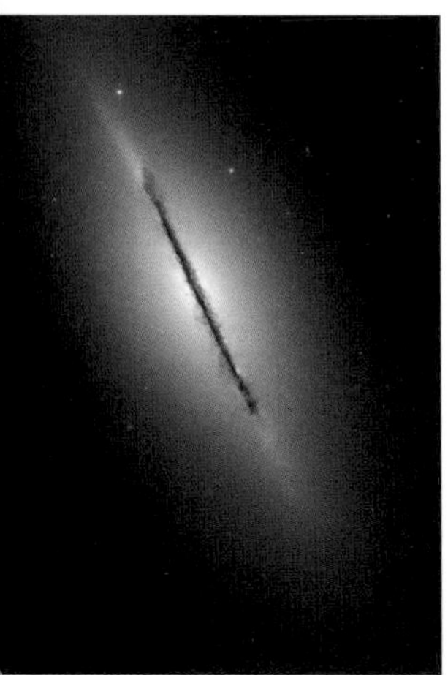

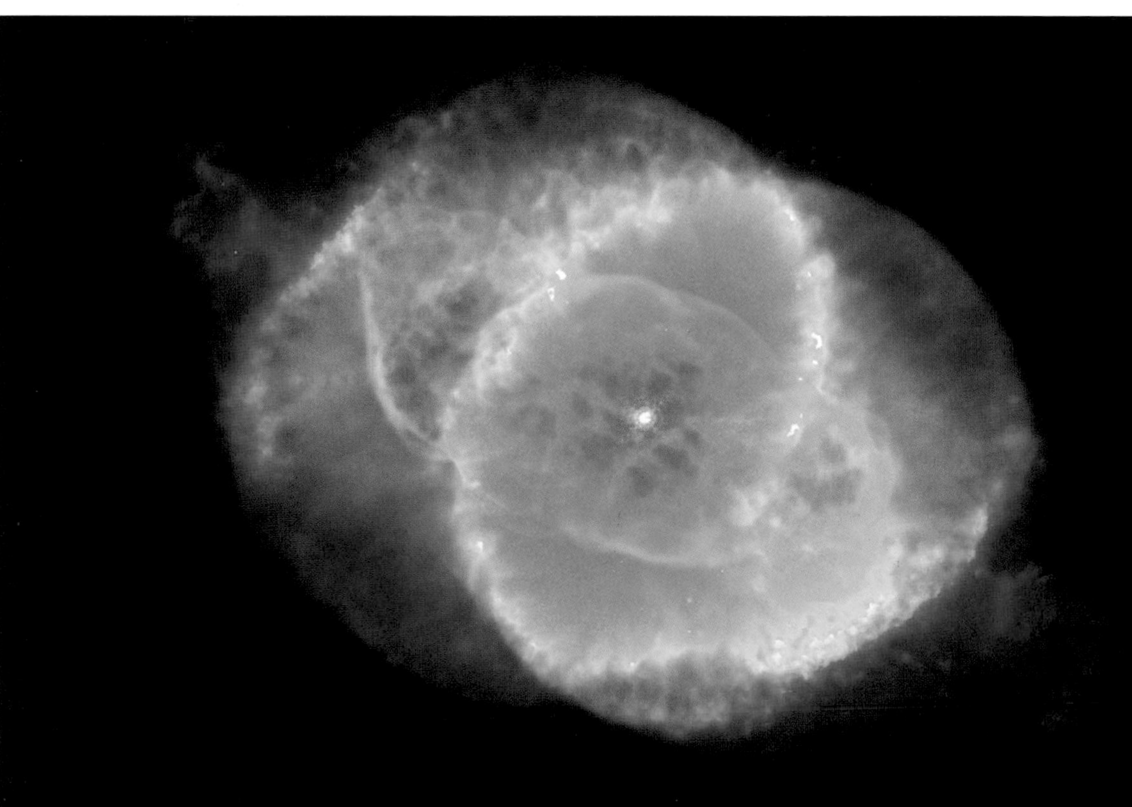

Right The Cat's Eye Nebula is a complex planetary nebula that reveals much about the dynamic processes that take place at the end of a star's life.

CEPHEUS
CAMELOPARDALIS
CYGNUS
URSA MINOR
LYRA
URSA MAJOR
HERCULES
BOÖTES
Nodus Secundus
Alsafi
Altais
Dsiban
NEP
Aldhibah
Grumium
Etamin
Kuma
Rastaban
Thuban
Eldsich
6643
6503
6340
6543
3147
4236
4125
6015
5985
5982
5981
5907
5866
+50°
+60°
+70°
+80°
+40°

F A C T   F I L E

Draco
DRAY-ko
The Dragon

Abbreviation
Dra

Genitive
Draconis

Right Ascension
17 hours

Declination
+65°

Visibility
90°N to 4°S

Notable Features
Alpha (α) Draconis
Nu (ν) Draconis
Psi (ψ) Draconis
NGC 5866
NGC 5907 (Splinter Galaxy)
The Sampler (NGC 5981, NGC 5982, NGC 5985)
NGC 6543 (Cat's Eye Nebula)

Named Stars
Thuban (Alpha [α] Draconis)
Rastaban (Beta [β] Draconis)
Etamin (Gamma [γ] Draconis)
Altais (Delta [δ] Draconis)
Nodus Secundus (Epsilon [ε] Draconis)
Aldhibah (Zeta [ζ] Draconis)
Eldsich (Iota [ι] Draconis)
Kuma (Nu [ν] Draconis)
Grumium (Xi [ξ] Draconis)
Alsafi (Sigma [σ] Draconis)
Dsiban (Psi [ψ] Draconis)

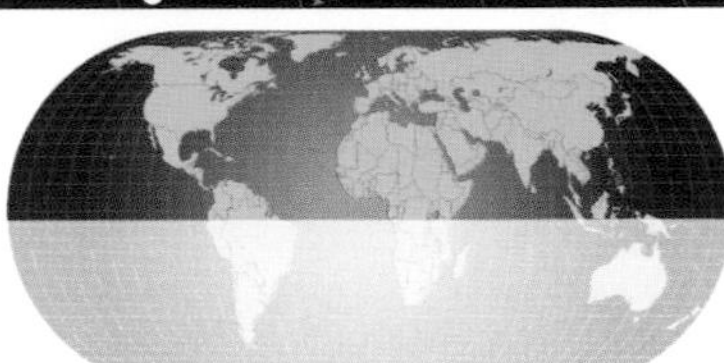

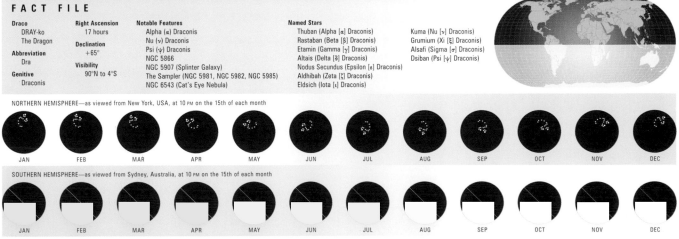

NORTHERN HEMISPHERE—as viewed from New York, USA, at 10 PM on the 15th of each month

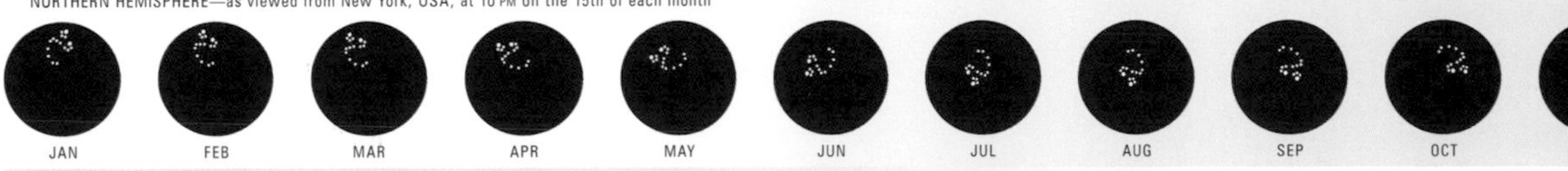

SOUTHERN HEMISPHERE—as viewed from Sydney, Australia, at 10 PM on the 15th of each month

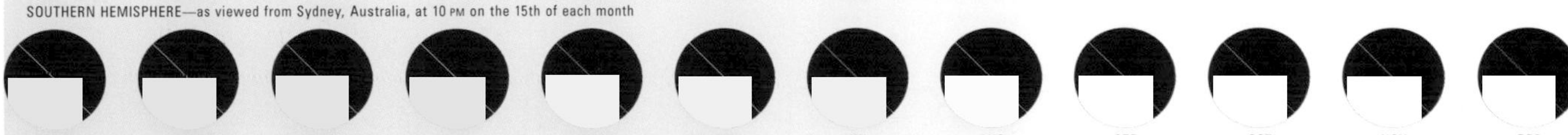

Equuleus

Equuleus is the second smallest constellation in the sky. Unlike the richly endowed and smallest constellation Crux, it contains only a few faint galaxies, plus a standard selection of double stars.

GALLOPING ACROSS THE SKIES

Equuleus, the Little Horse, has been connected to Celeris, the brother of the flying horse, Pegasus. Named by the Greek astronomer Hipparchus around 150 BCE, this constellation was one of the first created by a known individual rather than originating in the mists of time.

THE STARS OF EQUULEUS

A small trapezoid of 4th and 5th magnitude stars forms a distinctive figure, but best viewing is under suburban or rural skies. Alternatively, binoculars can be used because the trapezoid is only 5 degrees in its longest dimension.

OBJECTS OF INTEREST

There is enough to keep an observer occupied for part of an evening, with the help of the impressive globular cluster M15, which is in Pegasus, only 1 degree east of the constellation border. There is a good selection of double stars suitable for small telescopes. Pretty **Epsilon (ε) Equulei** has a magnitude 6 yellow star with a magnitude 7.1 blue secondary, separated by 10.7 arc-seconds.

Lambda (λ) Equulei is a 2.8 arc-seconds matched pair of magnitude 7.4 stars, both pale yellow. **Struve 2765 (Σ2765)** is a pair of white stars, magnitudes 8.4 and 8.6, 2.8 arc-seconds apart. Very similar is **Struve 2786 (Σ2786)**, a binary of magnitudes 7.2 and 8.3 that are 2.5 arc-seconds apart. **Struve 2793 (Σ2793)** is a wide (27 arc-seconds) color-contrast binary of magnitudes 7.8 and 8.5, the stars of which are yellow and bluish.

Observers wanting a bigger challenge can try the very unequal pair **Gamma (γ) Equulei** of magnitudes 4.7 and 11.5 at a separation of 1.9 arc-seconds in an 8-inch (20 cm) telescope; use high power on a steady night. The primary is yellow. The two stars have a common proper motion and so appear to be connected.

The only galaxy suitable for moderate apertures is the magnitude 11.5 spiral **NGC 7015** which is 1.6 arc-minutes in diameter. It is elongated and exhibits only a slight brightening at its center in a 12-inch (30 cm) reflector.

The galaxy **NGC 7040** is only magnitude 14.1 and quite small—1.0 by 0.7 arc-minutes. A 16-inch (40 cm) Newtonian reveals an elongated diffuse halo without any central brightening. A row of three barely visible stars lies on the southern edge of the halo. **NGC 7046** is a magnitude 13.1 barred spiral galaxy of 1.6 by 1.4 arc-minutes. In a 12-inch (30 cm) reflector it is just an amorphous glow, but an 18-inch (45 cm) telescope reveals a brighter center.

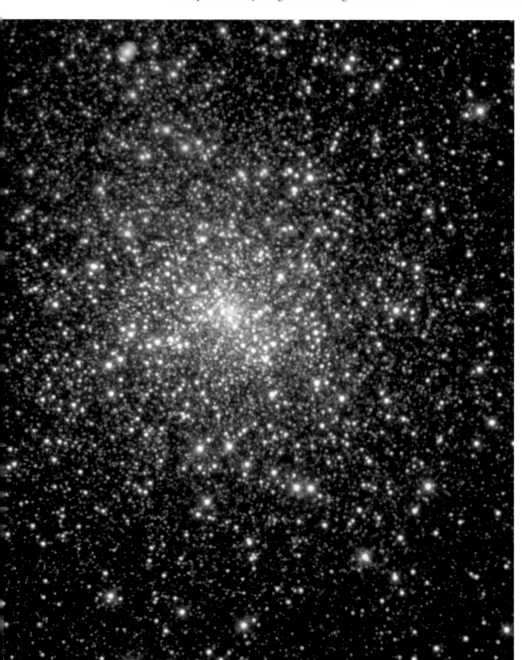

Left The splendid M15 globular cluster does not reside in Equuleus but is only a one-degree starhop away in the adjoining constellation of Pegasus. This Hubble Space Telescope image reveals thousands of individual stars and, like many globulars, M15 harbors ancient stars, some maybe 12 million years old.

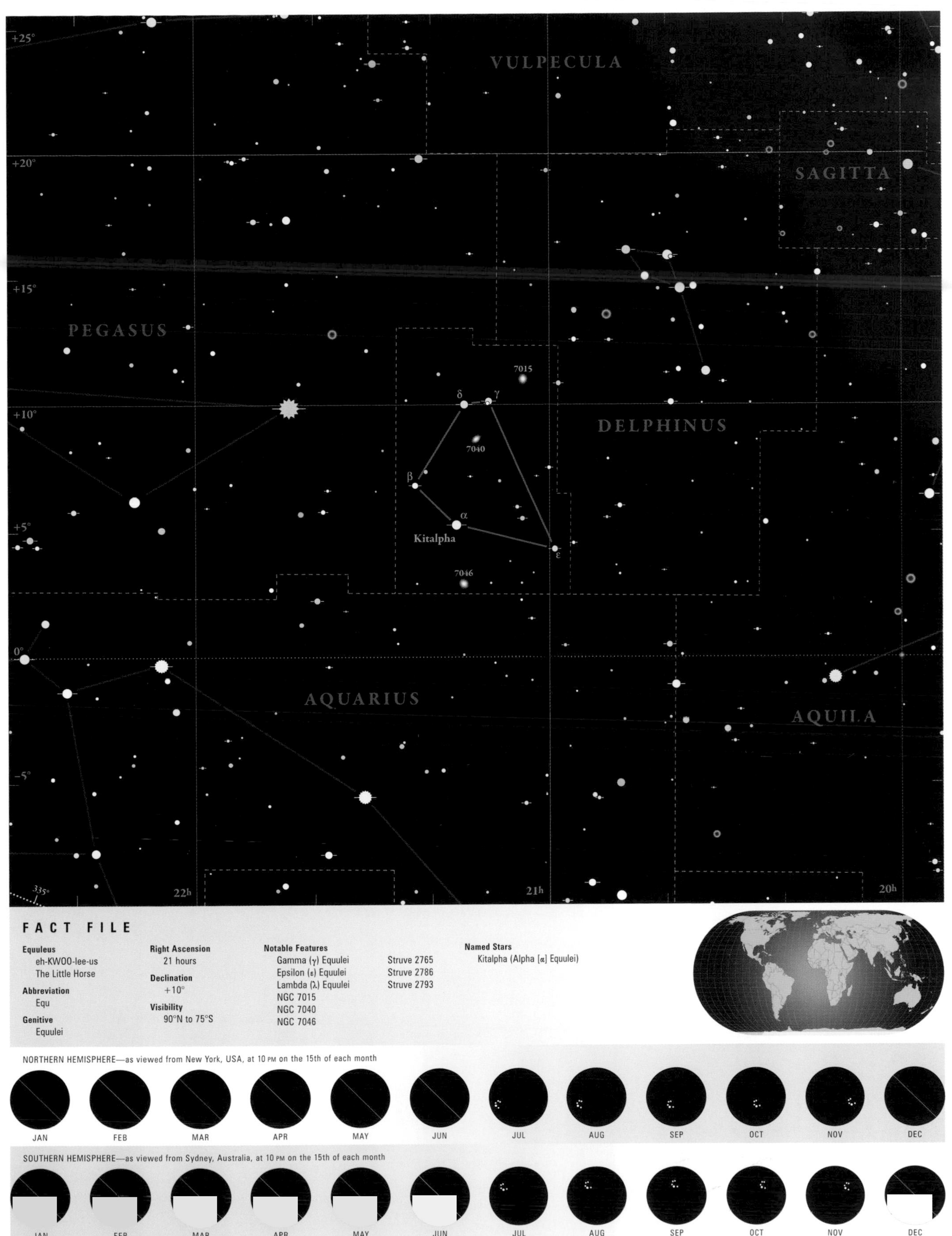

FACT FILE

Equuleus
eh-KWOO-lee-us
The Little Horse

Abbreviation
Equ

Genitive
Equulei

Right Ascension
21 hours

Declination
+10°

Visibility
90°N to 75°S

Notable Features
Gamma (γ) Equulei
Epsilon (ε) Equulei
Lambda (λ) Equulei
NGC 7015
NGC 7040
NGC 7046
Struve 2765
Struve 2786
Struve 2793

Named Stars
Kitalpha (Alpha [α] Equulei)

Eridanus

To ancient astronomers, Eridanus almost universally represented the River of the Night. For the Romans, it represented the Po; for others it was the Rhine, the Nile, the Tigris, or the Euphrates.

STAR STREAM

In Greek mythology, Eridanus is associated with Phaethon, the son of Helios, the Sun God. Phaethon persuaded his father to let him drive the Chariot of the Sun across the sky. Young Phaethon quickly discovered he could not control the powerful and wayward horses, and many places on Earth were either burned black or scorched. To end Earth's suffering, Zeus struck Phaethon with a thunderbolt, causing him to crash the horses and chariot into Eridanus, thus quenching their heat. Following European exploration of the southern stars, Eridanus found new tributaries and was extended far southward to reach the brilliant star Achenar.

THE STARS OF ERIDANUS

In modern times, one end of the river is marked by Alpha (α) Eridani, or Achenar, from the Arabic for "end of the river," while the other end is at 3rd magnitude Beta (β) Eridani, also named Cursa ("footstool"), which is next to Rigel in Orion. Achenar is the ninth brightest star in the sky—a hot, B-type blue giant about six times heavier and over five thousand times brighter than the Sun, which is 144 light-years away. Theta (φ) Eridani, or Acamar, has the same root and meaning as Achenar and once marked the end of the river. It is a showpiece binary pair of A-type stars 160 light-years distant.

In the north of Eridanus, Omicron^2 (o^2) Eridani is a triple star only about 16 light-years away. It consists of an orange dwarf circled by a binary pair comprising a tiny red-dwarf and a white-dwarf star. This is one of the brightest and the most easily seen white-dwarf stars in the sky.

OBJECTS OF INTEREST

Sitting well away from the plane of the Milky Way, Eridanus's most numerous objects for the amateur sky-watcher are galaxies, many of which are bright and interesting. **NGC 1291** is a face-on spiral galaxy that a 6-inch (15 cm) telescope shows as a large faint patch of haze with a small, brighter spot in the center. The edge-on spiral **NGC 1532** looks like a needle-like sliver of mist, almost half a moon diameter long in an 8-inch (20 cm) telescope; it has a small elliptical companion galaxy.

At the northern end of the constellation, **NGC 1232** is another bright face-on spiral galaxy that is well displayed by a 6-inch (15 cm) telescope. **NGC 1300** is an archetypical barred spiral system that requires a very large telescope to observe its two beautiful symmetrical arms. Not least is **NGC 1535**, one of the showpiece planetary nebulae of the sky. In an 8-inch (20 cm) telescope it shows as a small, bright azure disk with fuzzy edges.

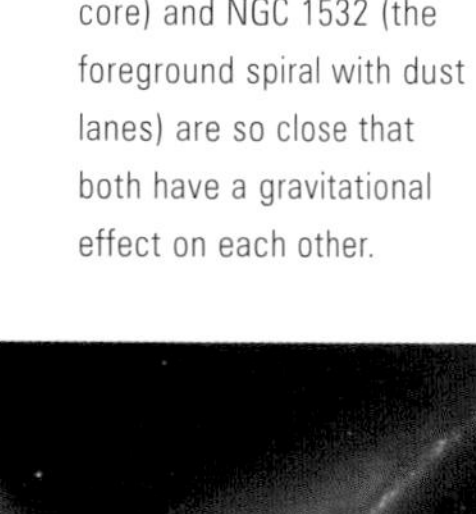

Below The interacting galaxies of NGC 1531 (on the left, with the bright core) and NGC 1532 (the foreground spiral with dust lanes) are so close that both have a gravitational effect on each other.

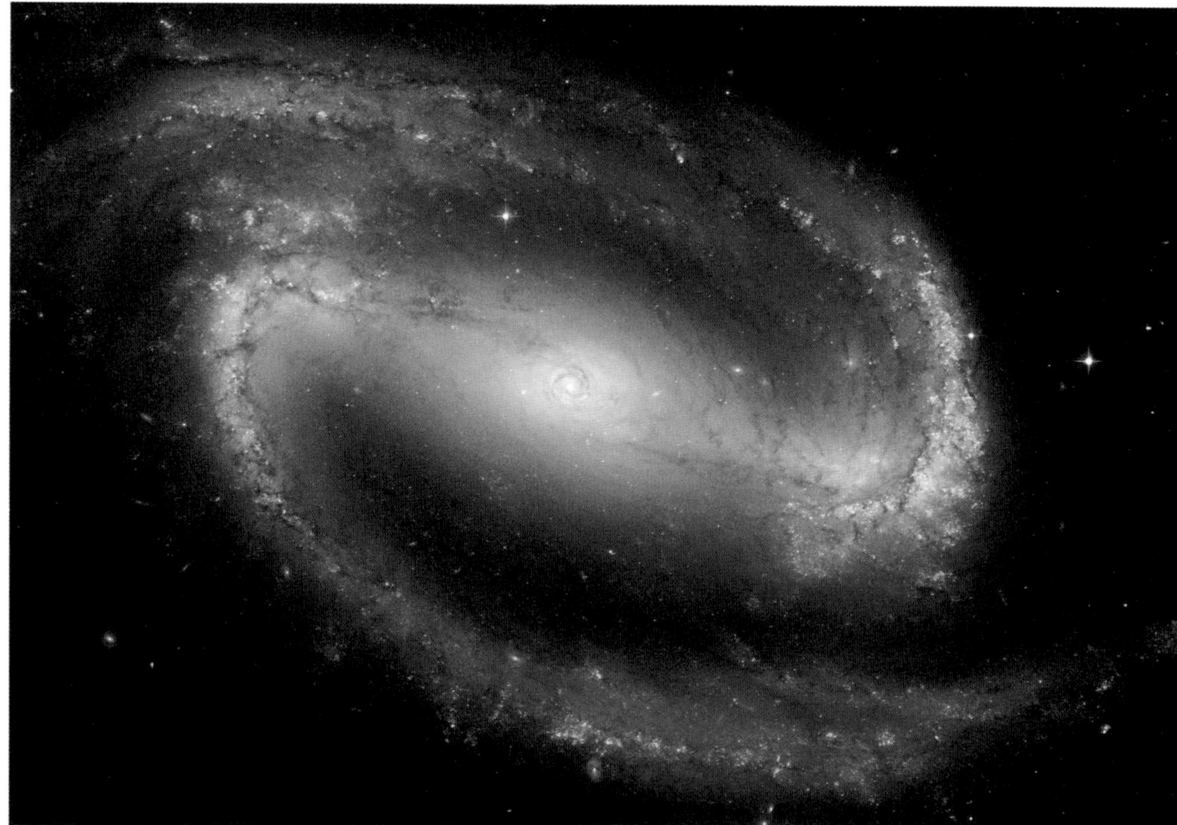

Right The 10th magnitude barred spiral galaxy NGC 1300 has a beguiling bright core and spans more than 100,000 light-years. Its dominant central bar and majestic spiral arms are revealed here by one of the largest Hubble Space Telescope images ever made of a complete galaxy.

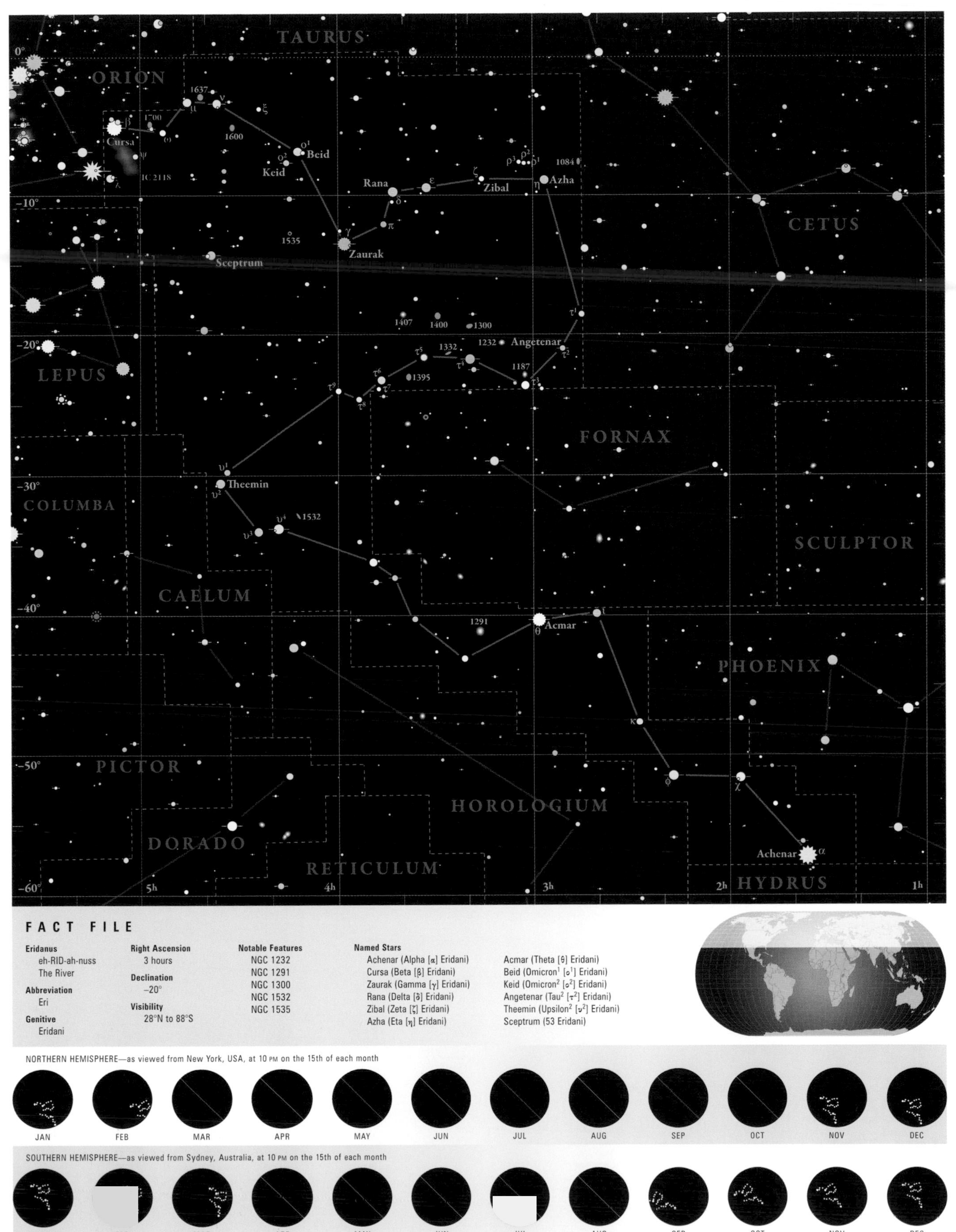

FACT FILE

Eridanus
eh-RID-ah-nuss
The River

Abbreviation
Eri

Genitive
Eridani

Right Ascension
3 hours

Declination
–20°

Visibility
28°N to 88°S

Notable Features
NGC 1232
NGC 1291
NGC 1300
NGC 1532
NGC 1535

Named Stars
Achenar (Alpha [α] Eridani)
Cursa (Beta [β] Eridani)
Zaurak (Gamma [γ] Eridani)
Rana (Delta [δ] Eridani)
Zibal (Zeta [ζ] Eridani)
Azha (Eta [η] Eridani)
Acmar (Theta [θ] Eridani)
Beid (Omicron^1 [o^1] Eridani)
Keid (Omicron^2 [o^2] Eridani)
Angetenar (Tau^2 [τ^2] Eridani)
Theemin (Upsilon^2 [υ^2] Eridani)
Sceptrum (53 Eridani)

NORTHERN HEMISPHERE—as viewed from New York, USA, at 10 PM on the 15th of each month

JAN FEB MAR APR MAY JUN JUL AUG SEP OCT NOV DEC

SOUTHERN HEMISPHERE—as viewed from Sydney, Australia, at 10 PM on the 15th of each month

JAN FEB MAR APR MAY JUN JUL AUG SEP OCT NOV DEC

Fornax

The fascinating galaxies of Fornax are far from the Milky Way. Its faint stars were visible to ancient astronomers, but were not grouped into a pattern.

FIRE IN THE SKY

Abbé Nicolas Louis de Lacaille invented Fornax during his visit to the Cape Observatory between 1751 and 1752. In doing so, he placed a large bend in the great river Eridanus, so that it that now flows around, instead of through, Fornax.

Lacaille initially named the constellation Fornax Chimiae, "the chemical furnace," in honor of the famous French chemist Antoine Lavoisier who was guillotined in 1794 at the height of the French Revolution. Johann Bode later called it Apparatus Chemicus in his *Uranographia* of 1801. In the same star-maps, Bode also "stole" some stars from Fornax and neighboring Sculptor to form another new constellation to honor the invention of the electrical generator: Machina Electrica. However, Bode's new group generated little interest, and with that exception there has been no significant change to the stars of Fornax since.

THE STARS OF FORNAX

Like most of Lacaille's constellations, the stars of Fornax do little to outline the intended figure or stimulate the sky-watcher. Alpha (α) and Beta (β) Fornacis are at the dim end of 4th magnitude, while most of Fornax's other stars require a pristine sky to be seen, even as faint motes.

At only 46 light-years from the Sun, Alpha (α) Fornacis is relatively close. To the unaided eye it looks single, but is in fact a close binary of sun-like stars. The brighter is an F-type sub-giant, about 25 percent more massive than our Sun, while the lesser G-type member is 25 percent less massive. They orbit around each other over a period of 246 years, with an average separation that is a little more than the distance from the Sun to Pluto.

OBJECTS OF INTEREST

Located far from the plane of the Milky Way, the main attraction of Fornax is its galaxies. At the Fornax–Eridanus border is the **Fornax cluster**, one of our closest major galaxy clusters at about 60 million light-years distant. Of its sixty-odd members, at least twenty are observable in an 8-inch (20 cm) telescope.

At the center of the cluster are the lenticular galaxies **NGC 1399** and **NGC 1404**, which stand side by side as small bright hazes with star-like nuclei. **NGC 1365** is probably the most elegant barred spiral galaxy in our sky. The central bar is visible in a 4-inch (10 cm) telescope as a small lengthened haze, while an 8-inch (20 cm) telescope under a dark rural sky will just reveal its slender arms, like threads of the faintest gossamer. Near the edge of the cluster is **NGC 1316**, a gargantuan elliptical galaxy criss-crossed with the dust lanes of a smaller galaxy with which it has recently merged.

One of the brighter planetary nebulae in the sky also calls Fornax home. The smallest of real telescopes show **NGC 1360** as a large gray haze that has an easily visible central star.

Above The unusual mélange of stars, dust, and gas in NGC 1316, also known as Fornax A, indicates that this giant elliptical galaxy may be the result of the past collision and merger of two gas-rich spiral galaxies.

Right Relatively nearby at 60 million light-years away, the barred spiral galaxy NGC 1365 is a premier member of the Fornax Galaxy Cluster. At the center of its bright core lies a massive black hole.

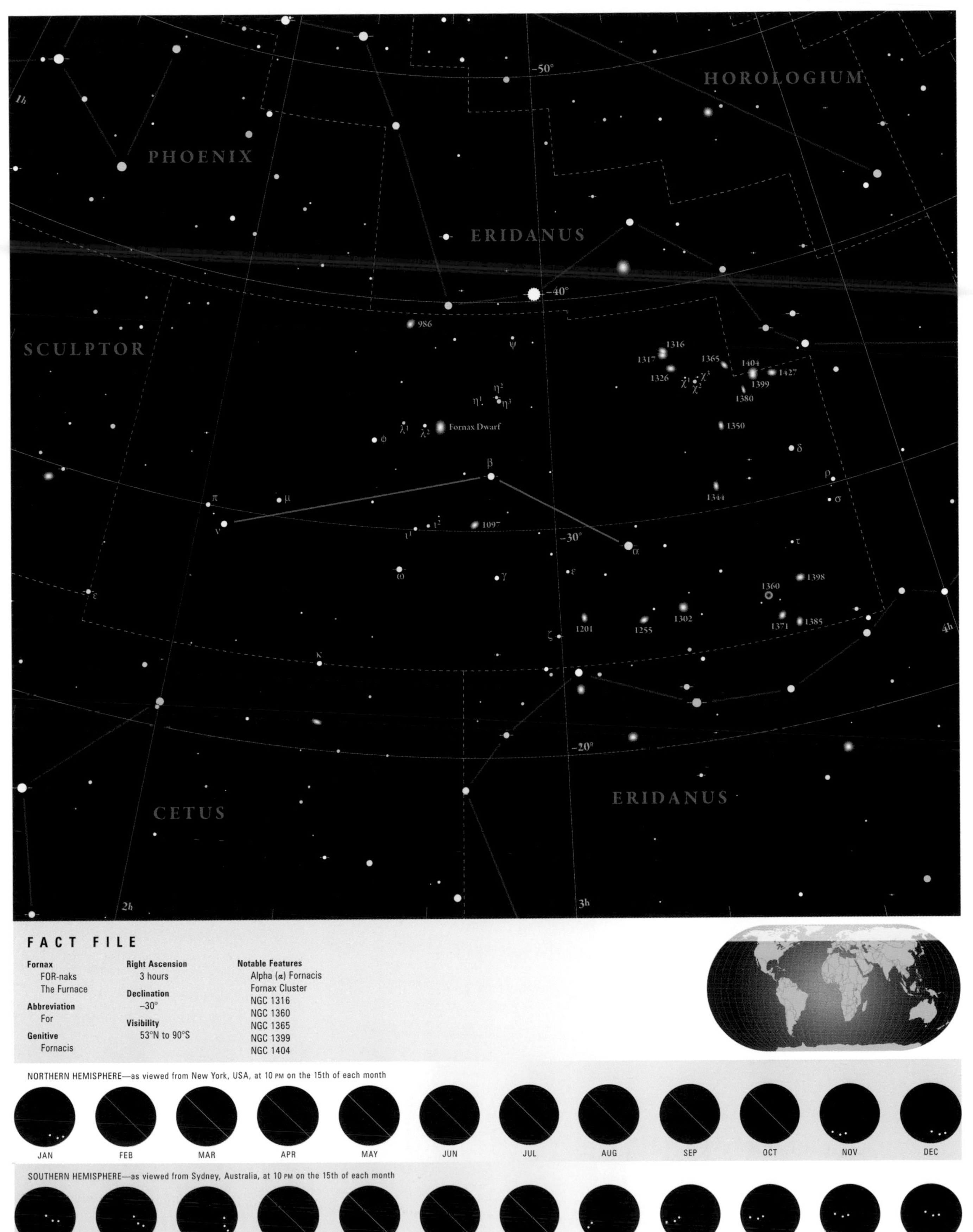

FACT FILE

Fornax
FOR-naks
The Furnace

Abbreviation
For

Genitive
Fornacis

Right Ascension
3 hours

Declination
–30°

Visibility
53°N to 90°S

Notable Features
Alpha (α) Fornacis
Fornax Cluster
NGC 1316
NGC 1360
NGC 1365
NGC 1399
NGC 1404

NORTHERN HEMISPHERE—as viewed from New York, USA, at 10 PM on the 15th of each month

JAN FEB MAR APR MAY JUN JUL AUG SEP OCT NOV DEC

SOUTHERN HEMISPHERE—as viewed from Sydney, Australia, at 10 PM on the 15th of each month

JAN FEB MAR APR MAY JUN JUL AUG SEP OCT NOV DEC

Gemini

The inseparable twins of Gemini stand with their feet in the Milky Way, making a well-known pattern in the sky.

TWIN ADVENTURES

Gemini represents Leda's sons, Castor and Pollux. Although twins, they had different fathers—Pollux was fathered by Zeus, and Castor by Tyndareus, the Spartan king. Their sister Helen became the queen of Sparta, and her abduction by Paris led to the Trojan War.

The twins shared a great many adventures, including the epic voyage of the *Argo*. Heracles (Hercules) learned his swordsmanship from Pollux. Castor and Pollux became embroiled in a fight with another pair of twins, Idas and Lynceus, over the affections of Phoebe and Hilaira. Lynceus slew Castor but Zeus took revenge and killed Lynceus with a thunderbolt. Immortal Pollux mourned ceaselessly for his brother and longed to follow him to Hades. Out of pity, Zeus allowed them to occupy Olympia and Hades on alternate days, and from there they found their way into the heavens.

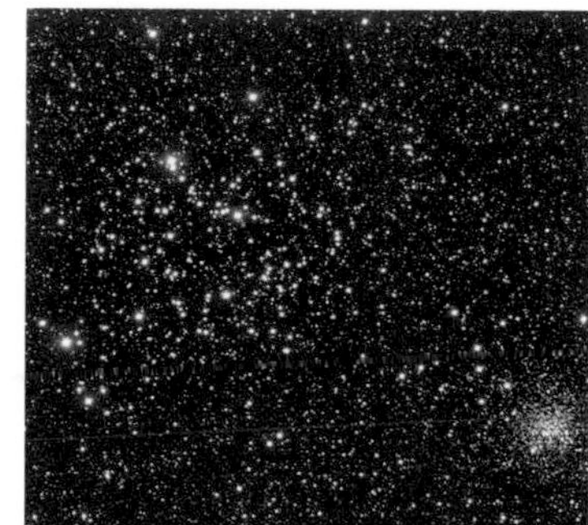

Right Bright blue stars often distinguish younger open clusters such as M35, seen here on the left of the image. The older, more compact, and distant cluster to the right is NGC 2158.

THE STARS OF GEMINI

Gemini is one of the easier constellations to find. Two almost equally bright stars, Castor and Pollux, mark the heads of the twins. An easily seen line of 2nd and 4th magnitude stars through Kappa (κ), Delta (δ), Zeta (ζ), and Gamma (γ) make up the body of Pollux. A parallel line of stars, from bright Castor through 3rd and 4th magnitude stars Tau (τ), Epsilon (ε), Mu (μ), and Eta (η) marks Castor's body.

Carrying the Beta (β) designation, 1st magnitude Pollux is the lucida of Gemini. It is an orange K-type giant star only 34 light-years distant, and is about 70 percent heavier than our Sun and 46 times brighter. Pollux is the closest "giant" star to our Sun. The 2nd magnitude Castor, or Alpha (α) Geminorum, is a complex sextuple system, 50 light-years distant, composed of four A-type stars and a pair of faint red-dwarfs.

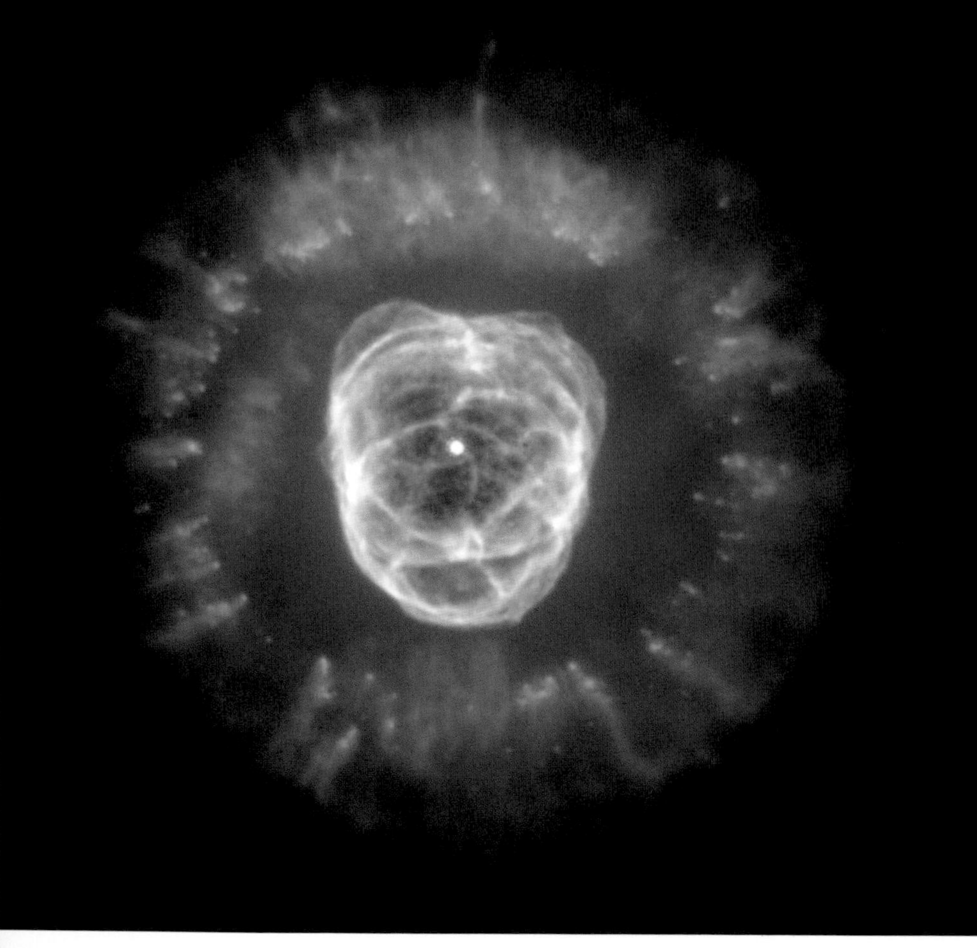

OBJECTS OF INTEREST

Castor is one of the most attractive double stars in the sky, with the two bright, almost equal components quite close to each other and beautifully displayed in a small telescope. In fact, each is a spectroscopic binary; another close-by dim orange star is also a spectroscopic binary that in turn orbits the pair of pairs.

The **Eskimo Nebula** or **NGC 2392** is a beautiful planetary nebula; in an 8-inch (20 cm) telescope, it shows as a small, round, pale blue disc with fuzzy edges. This nebula contains a brighter inner disk and an easily seen central star.

Near Eta (η) Geminorum is the splendid open cluster **M35**. In a dark sky, it is faintly visible to the naked eye; in a small telescope it shows dozens of stars in loops and sprays more than a moon-diameter in size.

Close by the outskirts of M35 is another open cluster, **NGC 2158**. An 8-inch (20 cm) telescope is needed to resolve this very distant cluster into a small patch of haze with a number of faint glimmers sewn in.

Left Resembling a head encircled by a fur-lined parka hood, the Eskimo Nebula (NGC 2392) contains complex and unusual orange filaments in its outer layer, some a light-year in length

FACT FILE

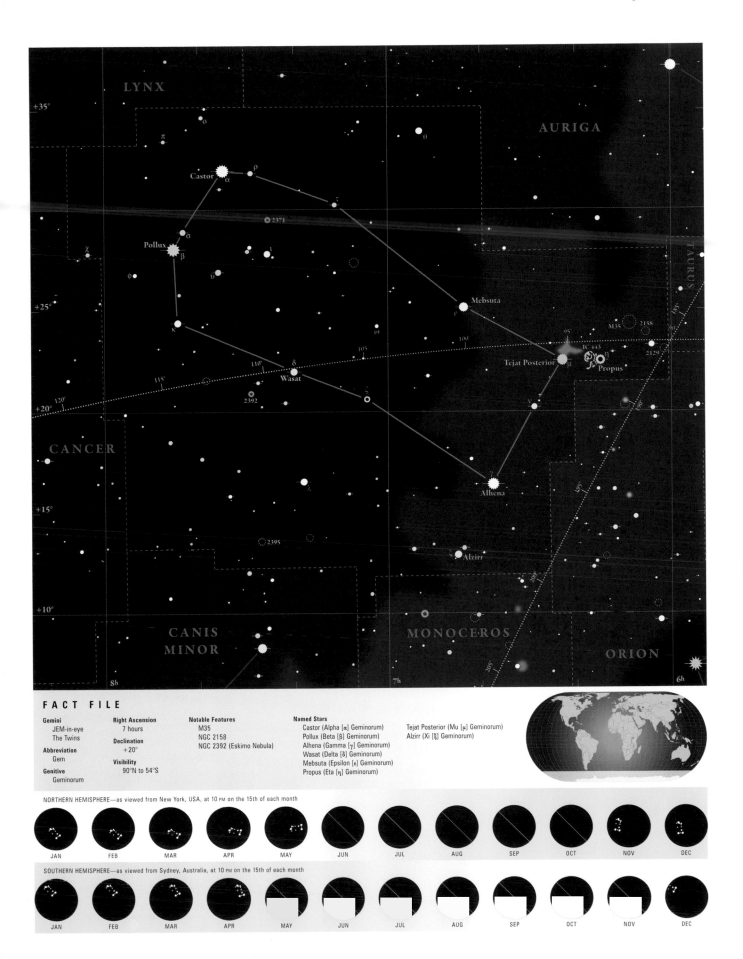

Gemini
JEM-in-eye
The Twins

Abbreviation
Gem

Genitive
Geminorum

Right Ascension
7 hours

Declination
+20°

Visibility
90°N to 54°S

Notable Features
M35
NGC 2158
NGC 2392 (Eskimo Nebula)

Named Stars
Castor (Alpha [α] Geminorum)
Pollux (Beta [β] Geminorum)
Alhena (Gamma [γ] Geminorum)
Wasat (Delta [δ] Geminorum)
Mebsuta (Epsilon [ε] Geminorum)
Propus (Eta [η] Geminorum)
Tejat Posterior (Mu [μ] Geminorum)
Alzirr (Xi [ξ] Geminorum)

NORTHERN HEMISPHERE—as viewed from New York, USA, at 10 PM on the 15th of each month

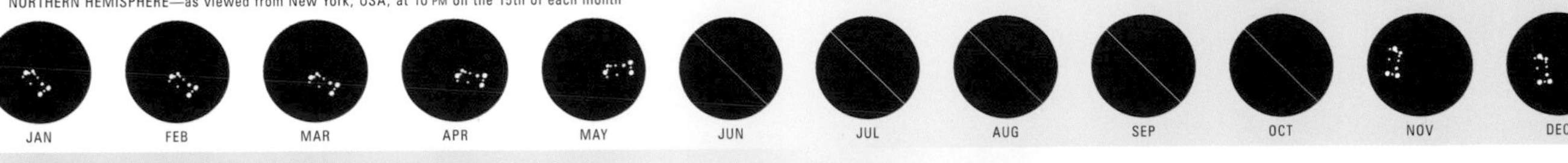

SOUTHERN HEMISPHERE—as viewed from Sydney, Australia, at 10 PM on the 15th of each month

JAN FEB MAR APR MAY JUN JUL AUG SEP OCT NOV DEC

Grus

The Crane is one of the sky's brighter birds. The stars making up the modern constellation of Grus have been associated with a bird of some description since ancient times.

HIGH FLYERS

The Egyptians certainly saw the constellation of Grus as a crane, a bird that also symbolized the office of Astronomer because of its high-altitude flight.

Before modern times, the star group making up Grus was usually considered part of the adjoining constellation Piscis Austrinus. Plancius first showed it on a celestial globe in 1598 as "Krane Grus"—both Dutch and Latin for crane. A later Plancius globe, however, showed it as "Phoenicopterus"—the Flamingo. The star catalog compiled by sixteenth-century Dutch explorers Pieter Dirkszoon Keyser and Frederick de Houtman named it the Heron. But in Bayer's *Uranometria* of 1603, Grus was unmistakably portrayed as a constellation in its own right, once again as a crane. Julius Schiller, a contemporary of Bayer, instead named it the Stork when compiling his alternative "Christian" star-atlas, but the name never caught on, and a crane it has remained.

THE STARS OF GRUS

Although many are faint, the stars of Grus fairly clearly outline a crane. The head is marked by Gamma (γ) in the north, and a slightly curved line of faint stars leading south through Lambda (λ) and Delta (δ) to Beta (β) forms the long neck and shoulders. Alpha (α) is one wing, while Iota (ι) and Theta (θ) outline the other. Epsilon (ε) and Zeta (ζ) at the southern end of the group form the tail.

Alpha (α) Gruis, also known as Alnair, is a blue B-class sub-giant, and is the 31st brightest star in the sky. Alnair is the Arabic for "the bright one." Located 101 light-years distant, it is more than 120 times brighter than our Sun.

OBJECTS OF INTEREST

Remote from the bright lights of the Milky Way, Grus has no clusters or nebulae to offer apart from one bright planetary. Galaxies abound, however, and several are nice examples for small telescopes.

NGC 7582, 7590, and **7599** is a trio of bright spiral galaxies about 70 million light-years distant. All three spindle-shaped objects, each with a brighter center, are located in the same low-power field. In an adjoining field is another bright, face-on spiral known as **NGC 7552**, which possesses a small but easily seen halo.

Along the neck of the crane, **Delta^1 (δ^1)** and **Delta^2 (δ^2)**, along with **Mu^1 (μ^1)** and **Mu^2 (μ^2) Gruis**, are a pair of optical doubles that can just be resolved as separate stars with the naked eye or binoculars. All are either yellow or orange giants. **IC 5150,** the **Spare Tire Nebula**, is one of the most attractive planetary nebulae in this part of the sky. An 8-inch (20 cm) telescope shows it as a thick-looking donut of translucent gas.

Left The serenely beautiful multi-armed spiral galaxy NGC 7424 is 40 million light-years away in Grus and is seen almost face-on.

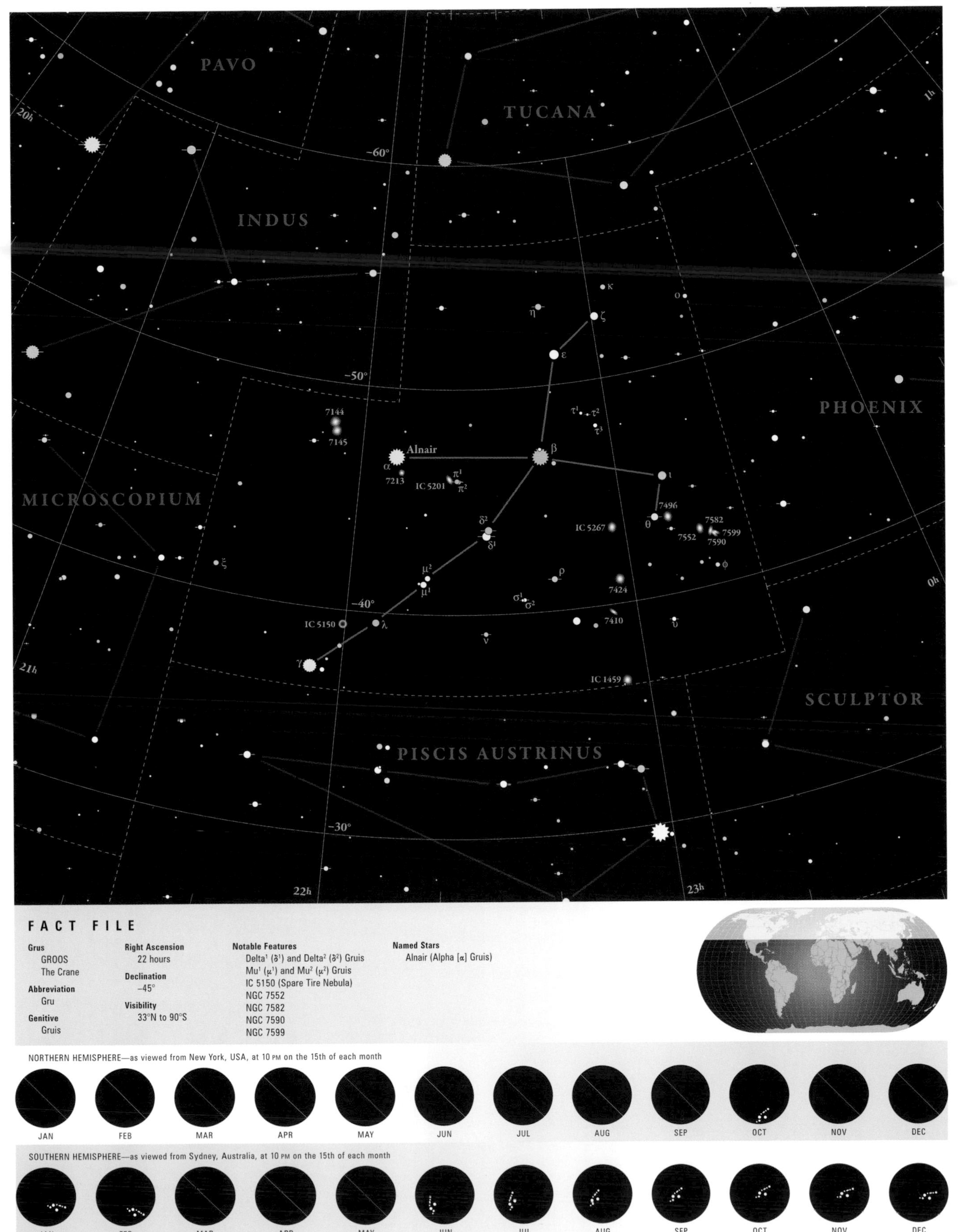

FACT FILE

Grus
GROOS
The Crane

Abbreviation
Gru

Genitive
Gruis

Right Ascension
22 hours

Declination
–45°

Visibility
33°N to 90°S

Notable Features
Delta[1] (δ^1) and Delta[2] (δ^2) Gruis
Mu[1] (μ^1) and Mu[2] (μ^2) Gruis
IC 5150 (Spare Tire Nebula)
NGC 7552
NGC 7582
NGC 7590
NGC 7599

Named Stars
Alnair (Alpha [α] Gruis)

Hercules

Hercules, the fifth-largest constellation, contains two showpiece globular clusters, one being the spectacular northern-sky Hercules Cluster.

TO THE EDGE OF THE WORLD

Hercules was the half-mortal son of Zeus. Hera, Zeus's jealous spouse, observed Hercules's super-human strength and guessed his parentage. Hera made Hercules temporarily mad, and in that state he killed his family. As penance, he had to undertake the Twelve Labors of Hercules. The first was killing the Nemean Lion, a beast that was impervious to all weapons. Hercules strangled the lion and wore its pelt ever afterward. Leo, Draco, Hydra, and Cancer all represent monsters that Hercules overcame. While engaged in his tenth labor Hercules reached the mouth of the Mediterranean, where he built the Pillars of Hercules as monuments to his voyage to the edge of the known world.

THE STARS OF HERCULES

Hercules lacks bright stars, but it does have a distinctive group—Epsilon (ε), Zeta (ζ), Eta (η) and Pi (π)—called the Keystone. Star-chains radiate out from each corner, marking the hero's limbs and head.

Right The turtle-shaped planetary nebula NGC 6210 lies about 6,500 light-years away in the constellation of Hercules. The central dying star violently expels jets of hot gas in a variety of unusual shapes.

OBJECTS OF INTEREST

The finest double stars are **Alpha (α) Herculis**, which is a beautiful orange and green binary of magnitudes 3.5 (variable) and 5.4, at about 5 arc-seconds. **Rho (ρ) Herculis** is a 4.1 arc-second white pair of magnitudes 4.6 and 5.6.

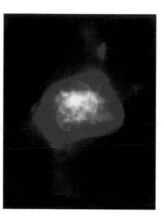

95 Herculis, a matched 5th magnitude pair at 6.3 arc-seconds, was called "apple green and cherry red" by nineteenth-century observers using achromatic refractors (which introduce some false color). In an 8-inch (20 cm) reflector the components are silver and gold.

M13 (NGC 6205), or the **Hercules Cluster**, is a naked-eye globular cluster. A 3-inch (8 cm) refractor makes it granular, with very few outliers resolved. Two unresolved star-chains are visible as tentacle-like glows. A 4-inch (10 cm) refractor resolves M13 right across its core. An 8-inch (20 cm) reveals six star-chains, some of which are backed by a fainter glow of unresolved stars. There are three connected dark lanes in the core, a feature called **The Propeller**. Look 0.5 degrees north-northeast of M13 for the small, lens-shaped 12th magnitude galaxy **NGC 6207**.

M92 (NGC 6341) is a distinctive globular when viewed with an 8-inch (20 cm) telescope. The brightest halo stars make the overall shape rectangular, in a 3 to 1 ratio. Within the rectangle, the inner halo and core are round. There are 35 fairly bright stars, plus as many intermittently visible faint stars. The underlying haze is splotchy. The brightness jumps at the edge of the 40 arc-second core, and then remains fairly constant across the star-studded core. The 16-inch (40 cm) view is markedly different: There are at least two hundred stars, masking the rectangular shape. The terms "outer halo," "inner halo," and "core" also no longer apply—the star density increases steadily from the outermost fringe to the resolved center.

NGC 6210 is a magnitude 8.8 planetary nebula. It appears when viewed with a 2-inch (6 cm) refractor and in an 8-inch (20 cm) telescope shows a blue "hazy star" only 14 arc-seconds in diameter.

Left A northern globular cluster of Herculean stature, M13 (NGC 6205) contains hundreds of thousands of stars. In 1974, it was the target for one of the first ever radio messages beamed to a possible extra-terrestrial audience.

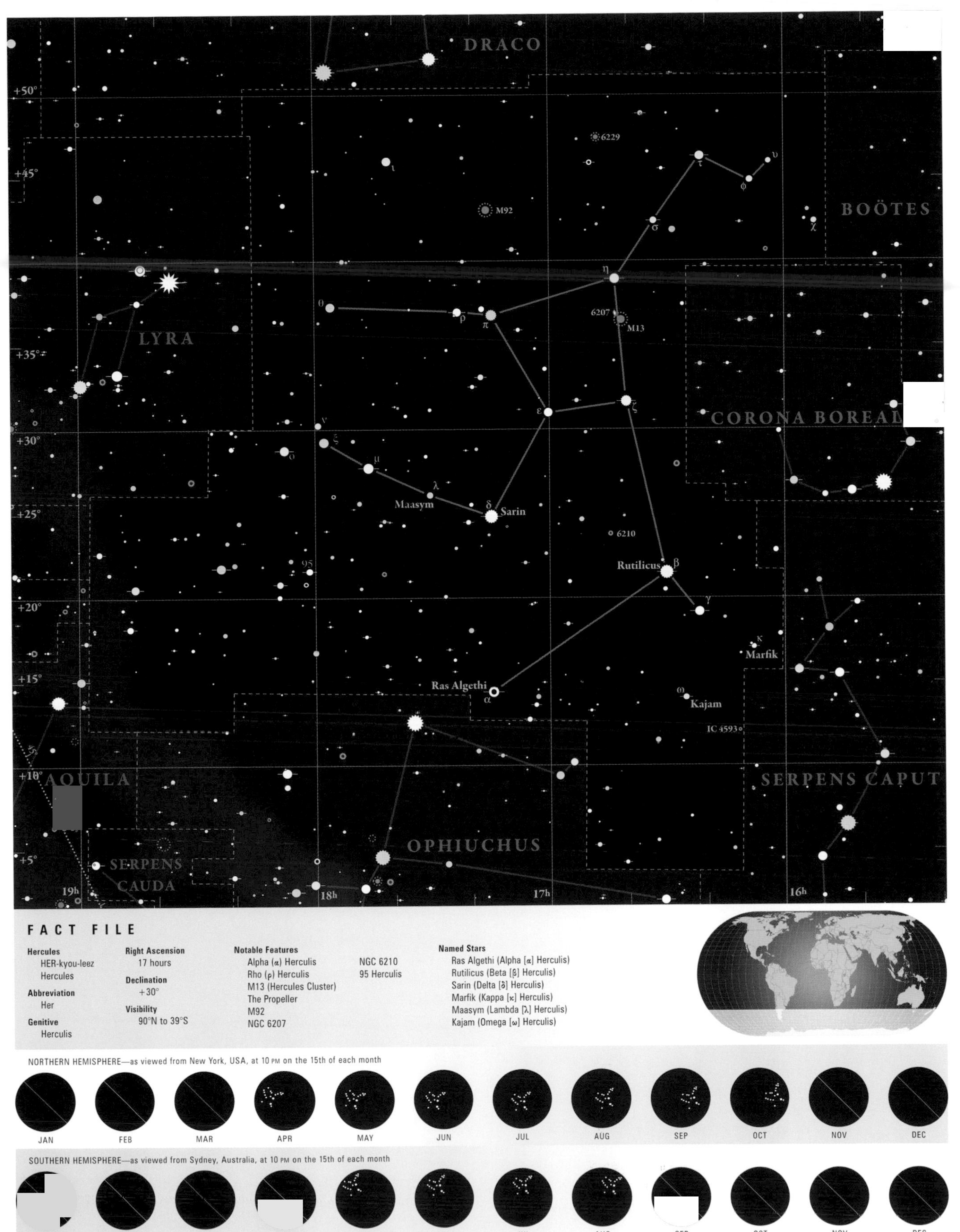

FACT FILE

Hercules
HER-kyou-leez
Hercules

Abbreviation
Her

Genitive
Herculis

Right Ascension
17 hours

Declination
+30°

Visibility
90°N to 39°S

Notable Features
Alpha (α) Herculis
Rho (ρ) Herculis
M13 (Hercules Cluster)
The Propeller
M92
NGC 6207
NGC 6210
95 Herculis

Named Stars
Ras Algethi (Alpha [α] Herculis)
Rutilicus (Beta [β] Herculis)
Sarin (Delta [δ] Herculis)
Marfik (Kappa [κ] Herculis)
Maasym (Lambda [λ] Herculis)
Kajam (Omega [ω] Herculis)

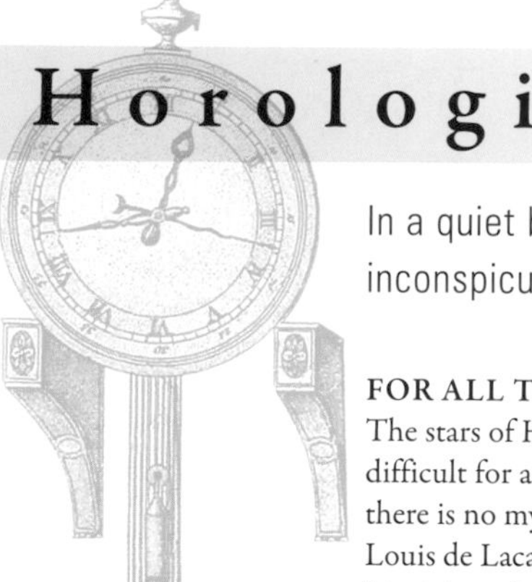

Horologium

In a quiet backwater by the celestial river of Eridanus, the inconspicuous stars of Horologium became the sky's timepiece.

FOR ALL TIME

The stars of Horologium are low in the southern sky and difficult for ancient astronomers to see. As a consequence, there is no mythology associated with them. Abbé Nicolas Louis de Lacaille invented Horologium Oscillitorium during his visit to the Cape Observatory between 1751 and 1752. Like many of Lacaille's constellations, it represents an instrument of science or engineering; Horologium commemorates the invention of the pendulum clock by Danish astronomer Christian Huygens in 1656. Since those times, there has been no significant change to its form or stars, although its name has been worn down to simple Horologium.

THE STARS OF HOROLOGIUM

The dim stars of Horologium form a long curved line following the banks of the great celestial river Eridanus and bear no apparent resemblance to a pendulum clock. The group starts a little west of the kite-shaped group of stars in Reticulum and winds its way north via several 5th and 6th magnitude stars to finish at the constellation's lucida, Alpha (α) Horologii.

As the only 4th magnitude star within the constellation, Alpha (α) Horologii is a "garden variety" K-type orange giant star well into the latter stages of its life. It began as an A-type dwarf star that probably looked a lot like Sirius, the brightest star in the night sky. Now about a billion years old, it subsists not only on hydrogen but on helium fusion for energy. Located about 117 light-years away, it is 47 times more luminous than our Sun, and will soon expand further to become a full-blown red-giant star.

R Horologii is a "Mira" type variable star—a pulsating red-giant in its dotage that fluctuates in brightness regularly although not frequently. Having one of the longest periods of stars in its class, it reaches its maximum every 407 days, briefly becoming one of the brighter stars within the constellation. It also has one of the greatest amplitudes in its class, swinging between naked-eye magnitude 4.7 and a very faint magnitude 14.3.

OBJECTS OF INTEREST

Well away from the riches of the Milky Way, Horologium comprises little to excite the sky-watcher. The bright globular cluster **NGC 1261** is easily the best. It is a quite distant globular, at about 54,000 light-years away. An 8-inch (20 cm) telescope shows some faint stars resolved over a well-concentrated hazy halo.

The many galaxies of Horologium are, like its stars, mainly faint and small. **NGC 1448** is just visible in a 6-inch (15 cm) telescope and appears as a beautiful, sharp-tipped, spindle-shaped galaxy with a brighter center.

Right The resplendent NGC 1512 barred spiral galaxy resides in the constellation of Horologium. At 30 million light-years distant, it is a neighbor to our own Milky Way and is bright enough to be seen in amateur scopes.

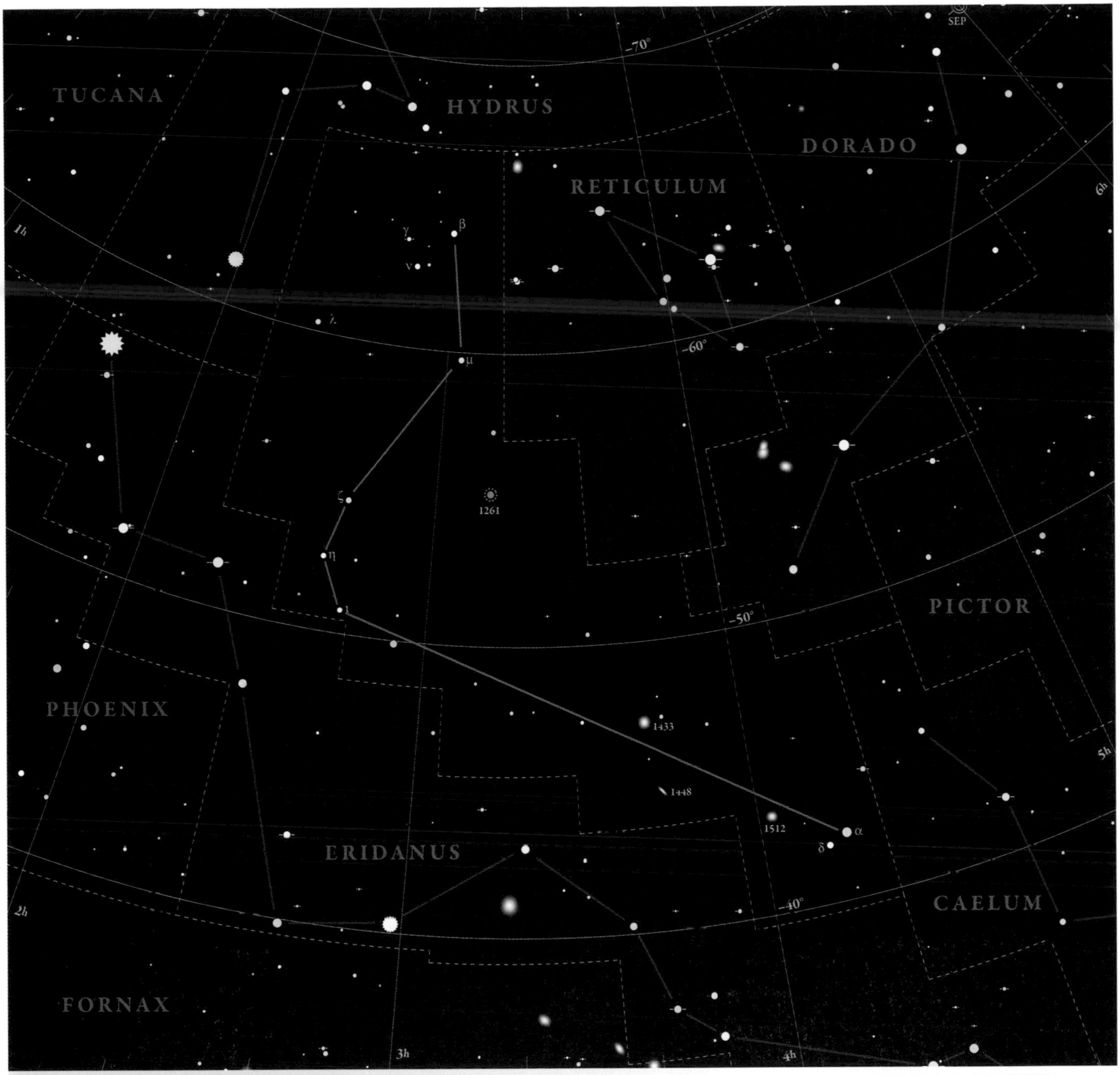

TUCANA
HYDRUS
DORADO
RETICULUM
PICTOR
PHOENIX
ERIDANUS
CAELUM
FORNAX
SEP
-70°
-60°
-50°
-40°
1h
2h
3h
4h
5h
6h
1261
1433
1448
1512

FACT FILE

Horologium
hor-oh-LOW-jee-um
The Clock

Abbreviation
Hor

Genitive
Horologii

Right Ascension
3 hours

Declination
−60°

Visibility
21°N to 90°S

Notable Features
NGC 1261
NGC 1448

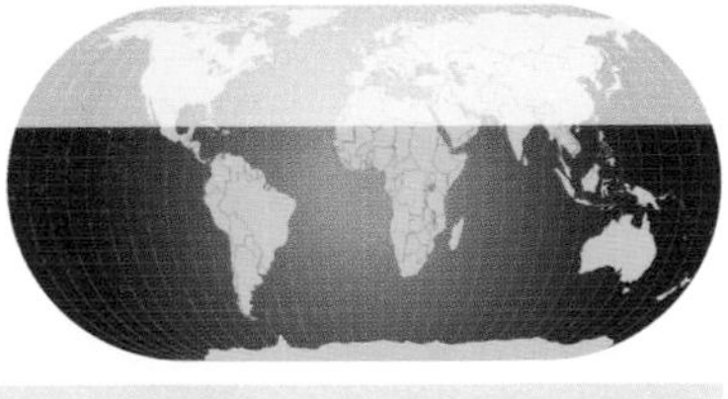

NORTHERN HEMISPHERE—as viewed from New York, USA, at 10 PM on the 15th of each month

JAN

FEB

MAR

APR

MAY

JUN
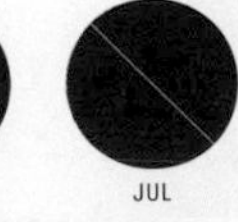
JUL
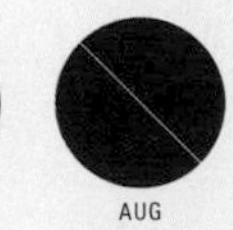
AUG

SEP

OCT

NOV
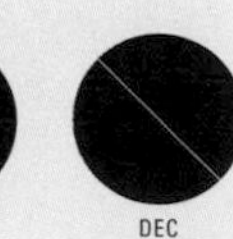
DEC

SOUTHERN HEMISPHERE—as viewed from Sydney, Australia, at 10 PM on the 15th of each month

JAN

FEB

MAR

APR

MAY

JUN

JUL

AUG

SEP

OCT
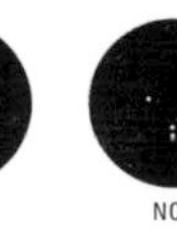
NOV

DEC

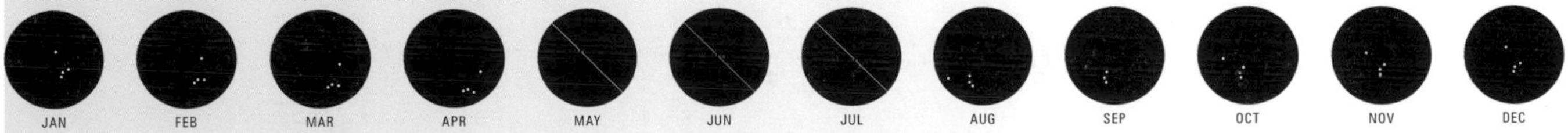

Hydra

The water snake is the largest of all modern constellations, and Ptolemy also listed it as one of his 48 constellations.

HEAD TO HEAD

Hydra was unknown before Hellenistic times. The Greeks saw Hydra as the serpent that Heracles (Hercules) slew as one of his twelve labors. After dispatching the Nemean Lion, Heracles went to the Peloponnese in search of the many-headed water snake with one immortal head. After wrestling it, Heracles clubbed one of its heads to death, but two new heads immediately grew in its place. A crab (commemorated by the constellation Cancer) emerged from the swamp during the battle to help Hydra by attacking Heracles's heels, but he smashed it underfoot.

Iolaus, Heracles's charioteer, came to his rescue by setting a grove of trees on fire and used a burning brand to cauterize the stump of each severed head as Heracles, one by one, chopped them off. Finally, Heracles removed the immortal head and buried it under a heavy stone. (An alternative mythology of Hydra is related in the constellation descriptions of Corvus and Crater.)

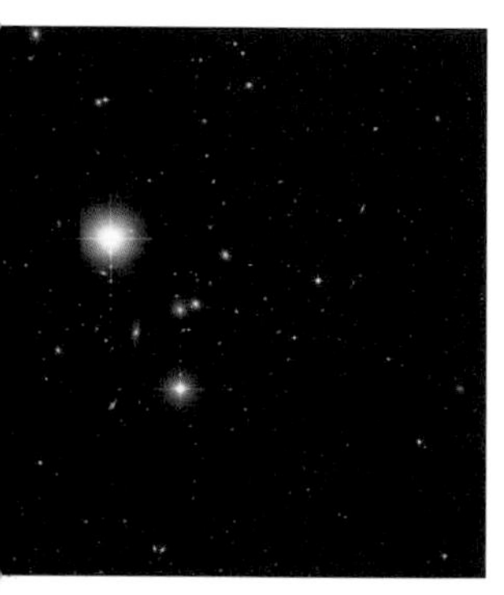

Above Abell 1060 is a large cluster of galaxies, of which the two brightest are NGC 3309 and NGC 3311, visible as fuzzy blobs in the center of the image.

THE STARS OF HYDRA

Winding across a quarter of the sky, from Cancer to Centaurus, Hydra is the largest of all constellations, but it is hardly conspicuous. The "head" asterism—just to the south of Cancer and composed of 3rd and 4th magnitude stars—is reasonably easy to see from suburbia, but the balance of the winding group is little more than a ribbon of faint dots. In 1799, the French astronomer Joseph Jérôme de Lalande formed a new constellation using several stars from the tail of Hydra and some from adjoining Antlia, naming it Felis—the Cat—but it was later abandoned.

The lucida of Hydra is the 2nd magnitude Alpha (α) Hydrae, known as Alphard, meaning "the lonely one." Easily the brightest star in a large blank area, it is a K-type giant star 175 light-years distant and 400 times more luminous than our Sun.

OBJECTS OF INTEREST

The most numerous objects in Hydra for the amateur sky-watcher are galaxies, but due to Hydra's extreme size, nearly all types of objects are represented. At the northwestern end of Hydra, near Monoceros, is **M48**, a very large, scattered, open cluster of medium-bright stars. The planetary nebulae are represented by **NGC 3242**, also known as the **"Ghost of Jupiter."** In a 6-inch (15 cm) telescope, it looks like a bright, small, azure disk about the apparent size of Jupiter; at higher magnifications it resembles an eye.

NGC 3309 and **3311** are the largest and brightest galaxies at the center of **Abell 1060**—a large cluster of galaxies, more than a dozen of which are visible through an 8-inch (20 cm) telescope as small hazes. Near Corvus, **M68** is a large, partly resolved, unconcentrated globular cluster, while in the tail is the remarkable galaxy **M83**. Nearly a half-moon in diameter, M83 is a barred spiral galaxy with a large faint halo containing the brighter bar near the center and a bright nucleus; in perfect conditions, the two-armed spiral pattern can be seen in an 8-inch (20 cm) telescope.

Below This Hubble Space Telescope image shows the warped disk of the spiral galaxy ESO 510–G13 in the constellation of Hydra.

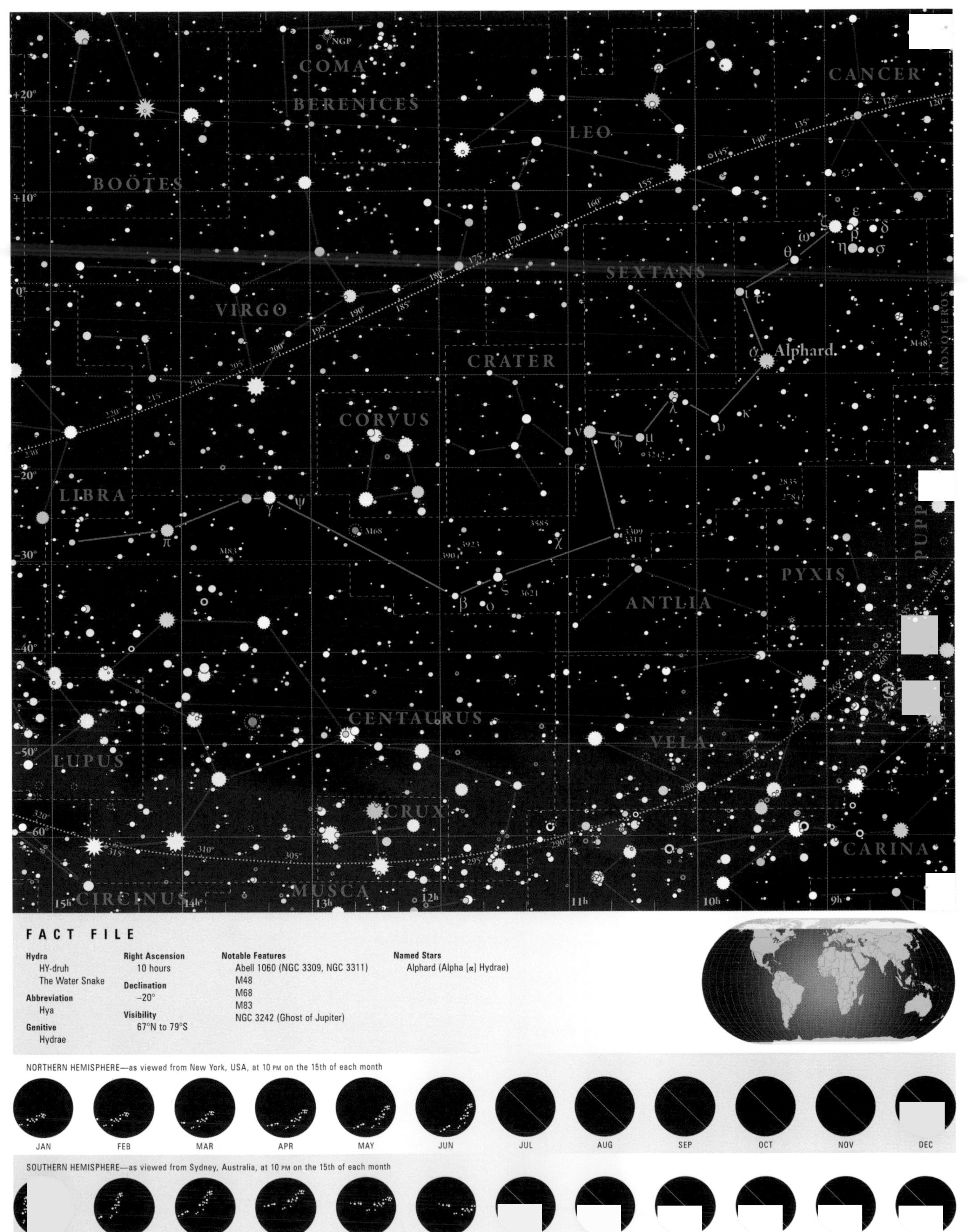

FACT FILE

Hydra
HY-druh
The Water Snake

Abbreviation
Hya

Genitive
Hydrae

Right Ascension
10 hours

Declination
–20°

Visibility
67°N to 79°S

Notable Features
Abell 1060 (NGC 3309, NGC 3311)
M48
M68
M83
NGC 3242 (Ghost of Jupiter)

Named Stars
Alphard (Alpha [α] Hydrae)

NORTHERN HEMISPHERE—as viewed from New York, USA, at 10 PM on the 15th of each month

JAN FEB MAR APR MAY JUN JUL AUG SEP OCT NOV DEC

SOUTHERN HEMISPHERE—as viewed from Sydney, Australia, at 10 PM on the 15th of each month

JAN FEB MAR APR MAY JUN JUL AUG SEP OCT NOV DEC

Hydrus

Tucked away in a quiet backwater, Hydrus, the lesser water snake, is the male counterpart of the largest constellation in the sky—Hydra.

SNAKING THROUGH THE SKY

The group of stars that make up Hydrus, located close to the South Celestial Pole, was unknown to astronomers in Europe until modern times, and there is no mythology attached to the constellation.

The sixteenth-century Dutch explorers and cartographers Pieter Dirkszoon Keyser and Frederick de Houtman invented the figure during their explorations. The first to introduce it on a star-globe was Petrus Plancius in 1598, and the figure was adopted by Johannes Bayer in his *Uranometria*, published in 1603. Since those times there has been little change to Hydrus, save that in the eighteenth century, Lacaille "stole" two of its stars when forming the neighboring constellation of Octans; they were never returned to Hydrus.

THE STARS OF HYDRUS

The outline of Hydrus is faint, but it is possible see the resemblance to a serpent. The long right-angled triangle formed by Gamma (γ), Beta (β), and Epsilon (ε) forms the coiled body, while a curved string of stars between Delta (δ) and Alpha (α) represent the upraised head poised to strike.

Alpha (α) Hydri, the second brightest star of Hydrus, is easy to find, 5 degrees south of the brilliant 1st magnitude Achenar in Eridanus. Only a fraction of a magnitude dimmer than the lucida Beta (β), it is an F-class dwarf star, more than 20 times brighter than the Sun, that we see from a distance of 71 light-years. Beta (β) Hydri is, in some ways, close to being a clone of our Sun. Only 24 light-years distant, it has an almost identical spectrum, although it is a little more massive, brighter, older, and more evolved than our luminary.

Third magnitude Gamma (γ) Hydri is a red-giant star over 200 light-years distant. Probably weighing in at about twice the mass of the Sun, its cool surface shows it to be an evolved star well on its way to ending up as a white dwarf.

OBJECTS OF INTEREST

Because Hydrus is far from the Milky Way, there are no bright objects from within our galaxy to view. Only a couple of degrees north of Alpha (α) Hydri is an attractive binary star, **h3475**. The similarly bright 6th magnitude component stars require steady seeing and a 4-inch (10 cm) telescope to resolve well. They look rather like snake eyes staring back out of the eyepiece, which is appropriate given the constellation in which they reside.

Hydrus contains many tiny faint galaxies, of which **NGC 1511** is easily the best. Its small, hazy, oval form can be glimpsed with an 8-inch (20 cm) telescope under a dark rural sky.

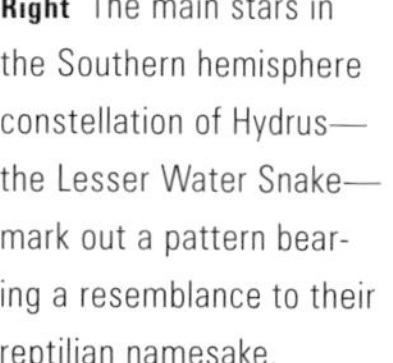

Right The main stars in the Southern hemisphere constellation of Hydrus—the Lesser Water Snake—mark out a pattern bearing a resemblance to their reptilian namesake.

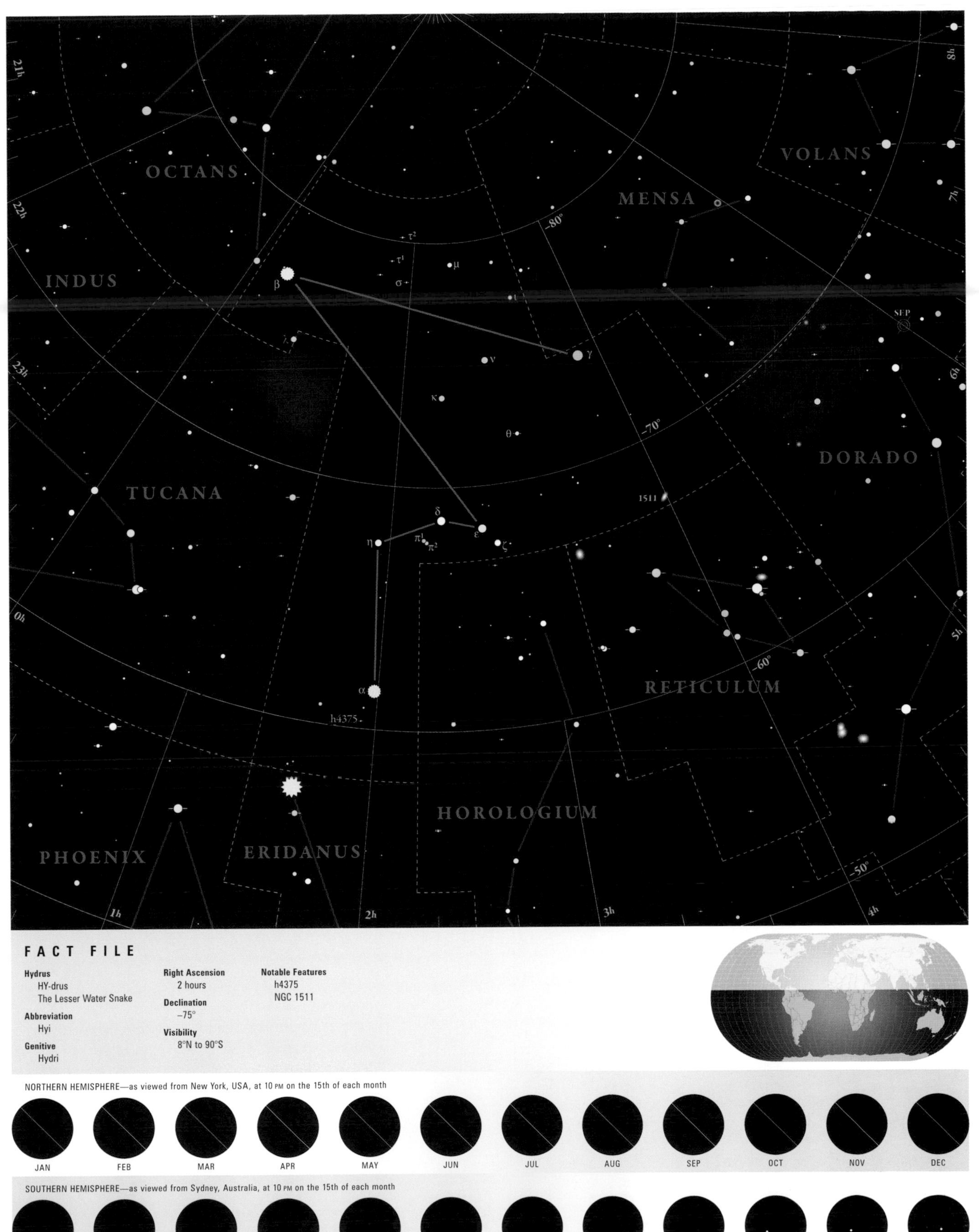

FACT FILE

Hydrus
HY-drus
The Lesser Water Snake

Abbreviation
Hyi

Genitive
Hydri

Right Ascension
2 hours

Declination
–75°

Visibility
8°N to 90°S

Notable Features
h4375
NGC 1511

NORTHERN HEMISPHERE—as viewed from New York, USA, at 10 PM on the 15th of each month

JAN FEB MAR APR MAY JUN JUL AUG SEP OCT NOV DEC

SOUTHERN HEMISPHERE—as viewed from Sydney, Australia, at 10 PM on the 15th of each month

JAN FEB MAR APR MAY JUN JUL AUG SEP OCT NOV DEC

Indus

The faint stars of Indus harbor a multitude of galaxies—and a star that is one of the closest to our planet.

GOING NATIVE

The star-group now recognized as Indus is too far south to have been visible to early European observers. The Dutch explorers and cartographers Pieter Dirkszoon Keyser and Frederick de Houtman were the first to methodically observe its stars, and saw it as a native man.

Following this, Petrus Plancius introduced the Indian—a Native American—on his star-globe in 1598, drawn as if hunting. In 1603, Johannes Bayer adopted the form of Indus in his *Uranometria*.

In his attempt to compile a "Christian" star atlas, a contemporary of Bayer, Julius Schiller, combined Indus and the adjoining constellation of Pavo to portray the long-suffering biblical figure Job. Unlike the real Job, Schiller's constellation did not stand the test of time. That apart, there has been no significant change to the stars of Indus since Bayer's time.

Below This is an image of the constellation Indus taken from New Zealand in July, 2007. Being a Southern hemisphere constellation, it was named in the seventeenth century after discovery and has no mythology attached to it.

THE STARS OF INDUS

The principal stars in Indus, Alpha (α), Beta (β), Delta (δ), and Theta (θ) make a long triangle that bears no real resemblance to the figure Bayer drew. It is very unusual for a far-southern star forming part of a "modern" constellation to have a common name, but Jesuit missionaries named Alpha (α) Indi ("The Persian") long before the stars became an Indian. It is a K-type orange giant star that is about 60 times brighter than our Sun and about 100 light-years distant.

One of our Solar System's closest neighbors is 5th magnitude Epsilon (ε) Indi, located only 11 light-years away. Epsilon is a K-class dwarf star—combined with the M-class dwarfs, the most common stars of all. It weighs in at just on 70 percent of our Sun's mass and is barely one fifth as luminous. Epsilon is intrinsically the least luminous star in the whole sky that is visible to the naked eye.

OBJECTS OF INTEREST

Not only are the stars of Indus few, they are also located well away from the plane of the Milky Way, with no star clusters or nebulae for the amateur observer. While the constellation of Indus is rich in galaxies, just about all of these are distant and small.

Theta (θ) Indi is a very attractive binary star with close, light yellow and orangish components that are easily resolved in a 6-inch (15 cm) telescope.

Of the galaxies, **NGC 7090** is easily the best. Lying virtually between Delta (δ) and Theta (θ), it is an edge-on barred spiral galaxy with a long thin halo when viewed with an 8-inch (20 cm) telescope.

There is a faint star superimposed near one end, which has tricked many amateur sky-watchers into thinking they may have discovered a supernova. Although it is very small, **NGC 7049** is fairly bright in an 8-inch (20 cm) telescope, displaying a conspicuous nucleus.

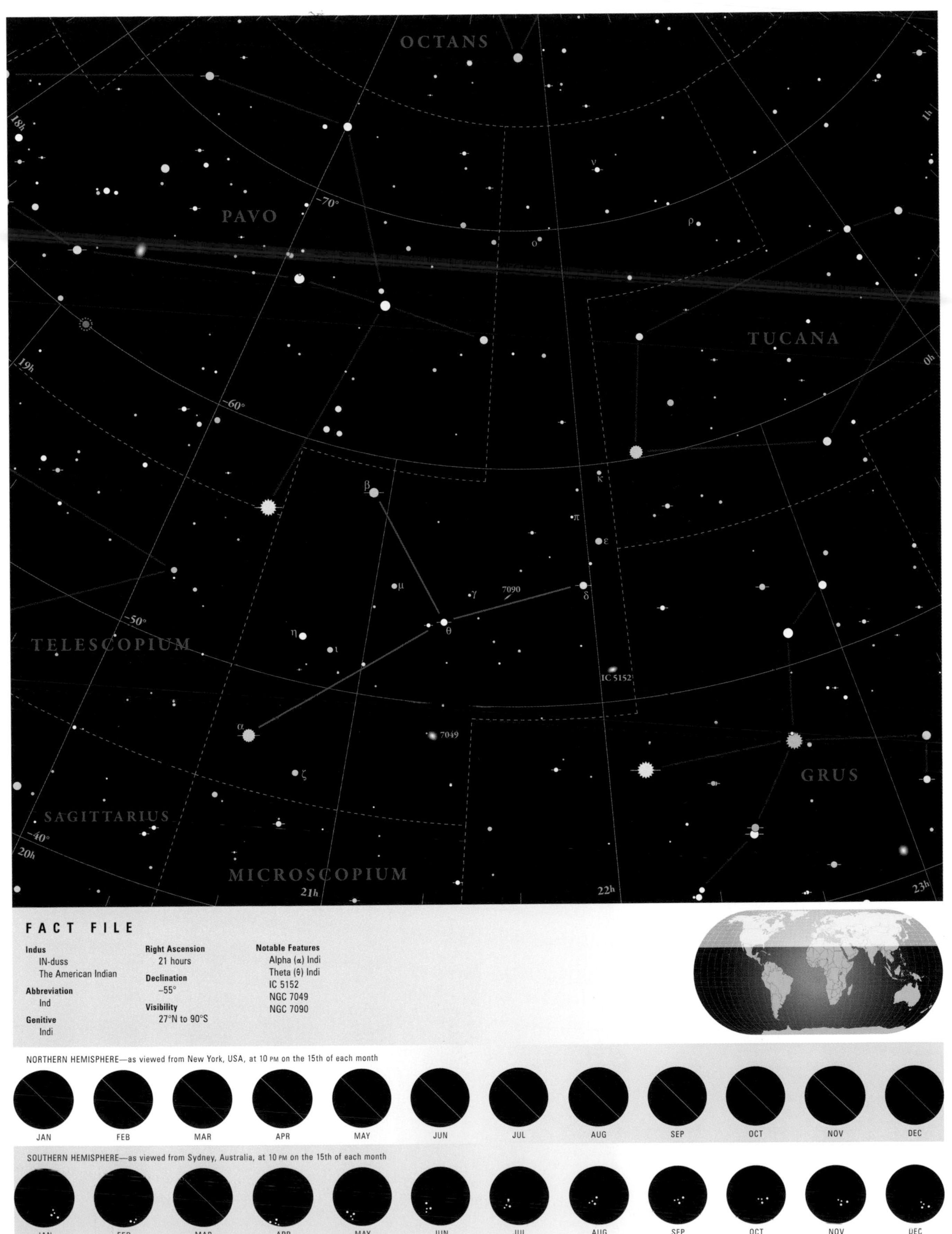

FACT FILE

Indus
IN-duss
The American Indian

Abbreviation
Ind

Genitive
Indi

Right Ascension
21 hours

Declination
−55°

Visibility
27°N to 90°S

Notable Features
Alpha (α) Indi
Theta (θ) Indi
IC 5152
NGC 7049
NGC 7090

NORTHERN HEMISPHERE—as viewed from New York, USA, at 10 PM on the 15th of each month

JAN FEB MAR APR MAY JUN JUL AUG SEP OCT NOV DEC

SOUTHERN HEMISPHERE—as viewed from Sydney, Australia, at 10 PM on the 15th of each month

JAN FEB MAR APR MAY JUN JUL AUG SEP OCT NOV DEC

Lacerta

Lacerta, the Lizard, is a small constellation carved out of the northern Milky Way. It has many minor open clusters, but not a single named star.

Right This is an image of the area determined by the European Space Agency and the Astronomical Observatory of Palermo that will be searched for habitable planets by Eddington, the European mission due for launch at the beginning of 2008. It will point in the direction of the constellation of Lacerta, which is positioned on the Milky Way, ensuring the rich population of stars essential to Eddington's success.

STELLAR SCALES

Located between the constellations of Cygnus, Cassiopeia, and Andromeda, Lacerta is one of the small northern constellations introduced by the seventeenth-century German astronomer Johann (Johannes) Hevelius. Some astronomers, including Johann Bode, suggested alternative names, such as "Frederick's Honors" in remembrance of King Frederick the Great of Prussia.

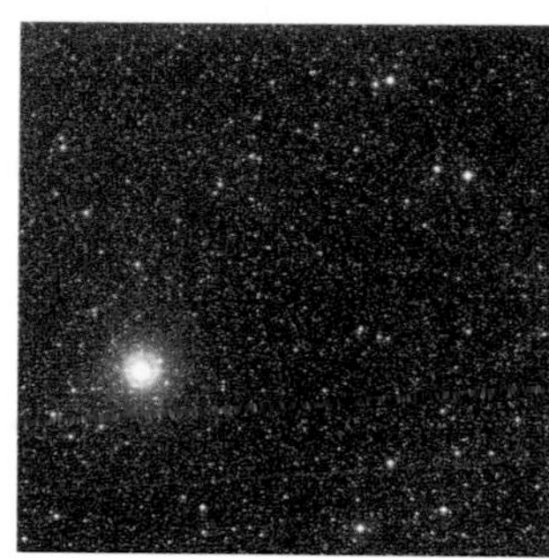

THE STARS OF LACERTA

Lacerta has no great splendors, but it does contain three open clusters that are of interest to the sky-watcher. The main stars form a zigzag, perhaps following the path of a darting lizard. These stars are magnitude 4 and 4.5, and stand out surprisingly well under dark skies. Binoculars will help under suburban skies, since the zigzag is only two binocular-fields long.

OBJECTS OF INTEREST

There are several fine double stars that are suitable for small telescopes. **Struve 2894 (Σ2894)** is just such a pair, at magnitudes 6.1 and 8.3, separated by 15.6 arc-seconds. Dawes gave the colors as "white and blue."

Struve 2902 (Σ2902) is a yellowish and white double of magnitudes 7.6 and 8.5, separated by 6.4 arc-seconds.

8 Lacertae is an all-white quadruple system. The main pair is magnitudes 5.7 and 6.5, and 22 arc-seconds apart. At about twice and four times this separation there are components of magnitudes 10.5 and 9.3, respectively.

Struve 2942 (Σ2942) is a tighter, 2.8 arc-second pair of magnitudes 6.1 and 8.3 that Struve called "reddish gold and ashen." Three of the best open clusters in the rich Lacerta Milky Way are **NGC 7209**, with a magnitude of 7.7 and a half moon-diameter width. With an 8-inch (20 cm) reflector, it is a fine but unconcentrated group. There are many 10th and 11th magnitude stars, including two red ones. Three trios lie along the following side. A 6th magnitude orange star sits just beyond the northern edge of the cluster, and the adjoining Milky Way fields feature additional colorful stars. **NGC 7243** is a brighter, larger, and looser cluster of 40 stars. It is a scattered irregular group, centered on the double star Struve 2890 (Σ2890)—magnitudes 8.5 and 8.5 at 9.4 arc-seconds.

NGC 7245 is the smallest and faintest of the three clusters, but that is due to its great distance of 6,000 light-years. Its brightest star is magnitude 13, so an 8-inch (20 cm) telescope sweeping at low power sees just a faint, unresolved, elongated glow when it happens upon NGC 7245. High power resolves about 35 of its 169 stars.

The tiny planetary nebula **IC 5217** at 6.6 arc-seconds, is barely non-stellar at 175 times magnification with a 16-inch (40 cm) telescope. This magnitude 12.6 target is fainter than two adjacent stars without a filter, but it is the brightest object in the 260 times field with an OIII filter—the filter that revolutionized planetary nebula hunting. IC 5217 finally shows a little round disk at 520 times. A perfect arc of nine stars lies 7 arc-minutes to the northwest.

Left Sharpless 126 is a nebula named after the astronomer Stewart Sharpless who published the second and final version of his famous nebula catalog in 1959. Sharpless 126 is found in the constellation Lacerta.

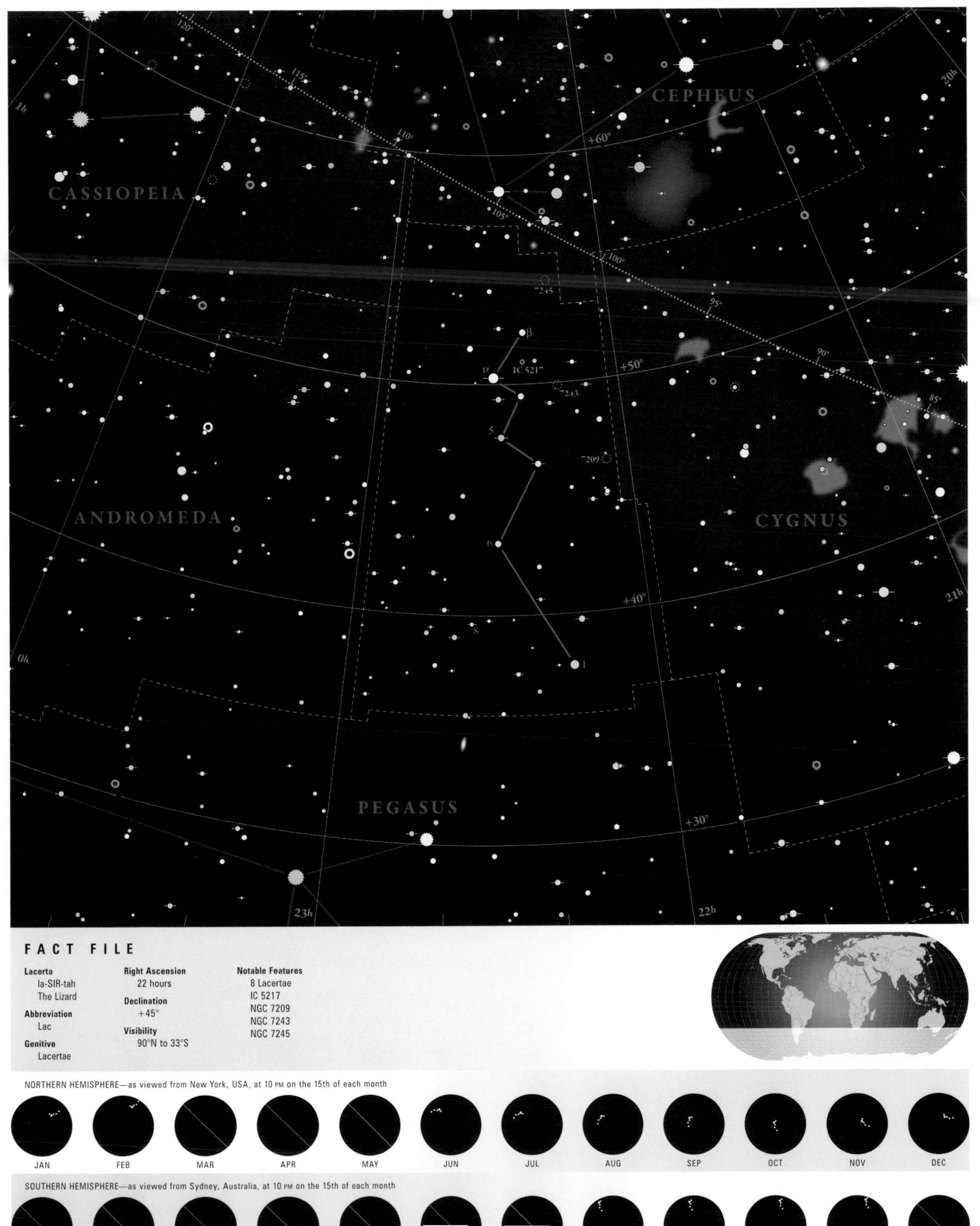

FACT FILE

Lacerta
la-SIR-tah
The Lizard

Abbreviation
Lac

Genitive
Lacertae

Right Ascension
22 hours

Declination
+45°

Visibility
90°N to 33°S

Notable Features
8 Lacertae
IC 5217
NGC 7209
NGC 7243
NGC 7245

NORTHERN HEMISPHERE—as viewed from New York, USA, at 10 PM on the 15th of each month

JAN FEB MAR APR MAY JUN JUL AUG SEP OCT NOV DEC

SOUTHERN HEMISPHERE—as viewed from Sydney, Australia, at 10 PM on the 15th of each month

JAN FEB MAR APR MAY JUN JUL AUG SEP OCT NOV DEC

Leo and Leo Minor

Leo, the Lion, is a prominent zodiacal constellation; it is the key to learning all of its surrounding constellations, including Leo Minor, the Little Lion.

THE STAR KING

Leo represents the Nemean Lion that Hercules (Heracles) killed with his bare hands (see Hercules). Leo's brightest star, Regulus, was an ancient symbol of monarchy. Johannes Hevelius formed Leo Minor (see Lacerta).

THE STARS OF LEO AND LEO MINOR

Unlike many of the constellations, Leo really does resemble its name. The lion's head is represented by a backward facing question mark—an asterism called the Sickle. A prominent right triangle makes the hindquarters, and strings of stars outline the legs of the reclining beast. The Coma Berenices Star Cluster represents the tuft at the end of a lion's tail.

Leo Minor consists primarily of a diamond shape, elongated east to west. It contains mainly 4th magnitude stars.

OBJECTS OF INTEREST

Gamma (γ) Leonis—or **Algieba**, meaning "the lion's mane"—is a famous golden binary pair of magnitudes 2.6 and 3.8. The secondary sometimes appears greenish-yellow. **Tau (τ) Leonis** is a wide 91-arc-second pair with 5th and 8th magnitude lemon yellow and pale blue components; **83 Leonis** (magnitudes 6.2 and 7.9, at 29 arc-seconds) is another double star in the same field.

Leo has over 80 galaxies visible with a 10-inch (25 cm) reflector. **NGC 2903** yields to 2-inch (6 cm) refractors; through an 8-inch (20 cm) telescope, this elongated spiral has a star-like nucleus. A 16-inch (40 cm) telescope adds an elongated core, a darker band west of the core, and a bright knot, cataloged as **NGC 2905**, at the northern tip of the galaxy.

Spirals **M95 (NGC 3351)** and **M96 (NGC 3368)** lead a group of gravitationally connected galaxies. The trio of elliptical **M105 (NGC 3379)**, lenticular **NGC 3384**, and faint spiral **NGC 3389**—an elongated blur in an 8-inch (20 cm) telescope—are part of this group. The five show little detail beyond brighter centers, and bright nuclei for M96, M105 and NGC 3384. Large apertures hint at M95's central bar, and an 8-inch (20 cm) telescope reveals four more members of M96. The oval spiral NGC 3521 is as large and bright as NGC 2903.

The **"Trio in Leo"** consists of three spirals: elongated **M65 (NGC 3623)**, **M66 (NGC 3627)**, and edge-on **NGC 3628**. A 2-inch (6 cm) refractor shows the two Messiers. An 8-inch (20 cm) adds a central condensation for M65, a star-like nucleus for M66, and mottling in NGC 3628. A 16-inch (40 cm) reveals a nucleus for M65, plus a knot at that galaxy's northern tip, M66's longer spiral arm and the root of the second, and NGC 3628's impressive long dust lane.

None of the galaxies in Leo Minor compare with Leo's deservedly better known ones, but the small constellation has 17 galaxies suitable for moderate apertures, including the edge-on streak of **NGC 3432**.

Above This ultraviolet image of the galaxy M95 (NGC 3351) reveals the underlying structure of this barred and ringed spiral.

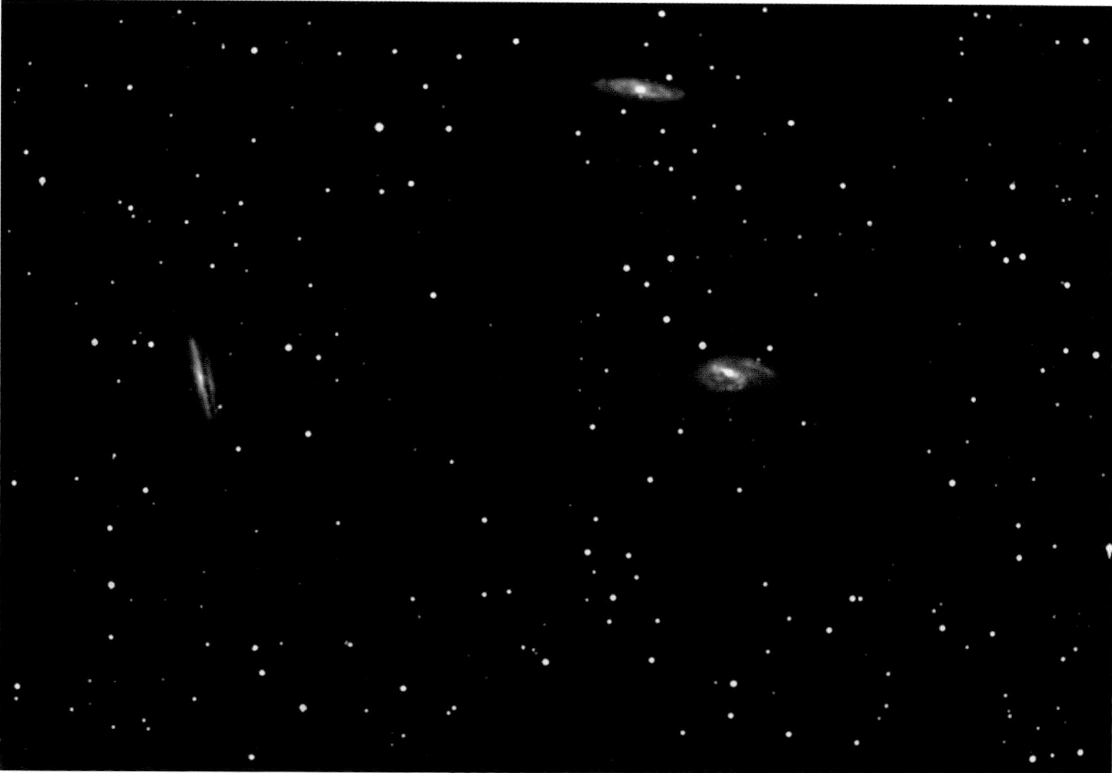

Right The "Trio in Leo": Galaxies M65 (lower right), M66 (upper right), and NGC 3628 (left) are all 65 million light-years from Earth.

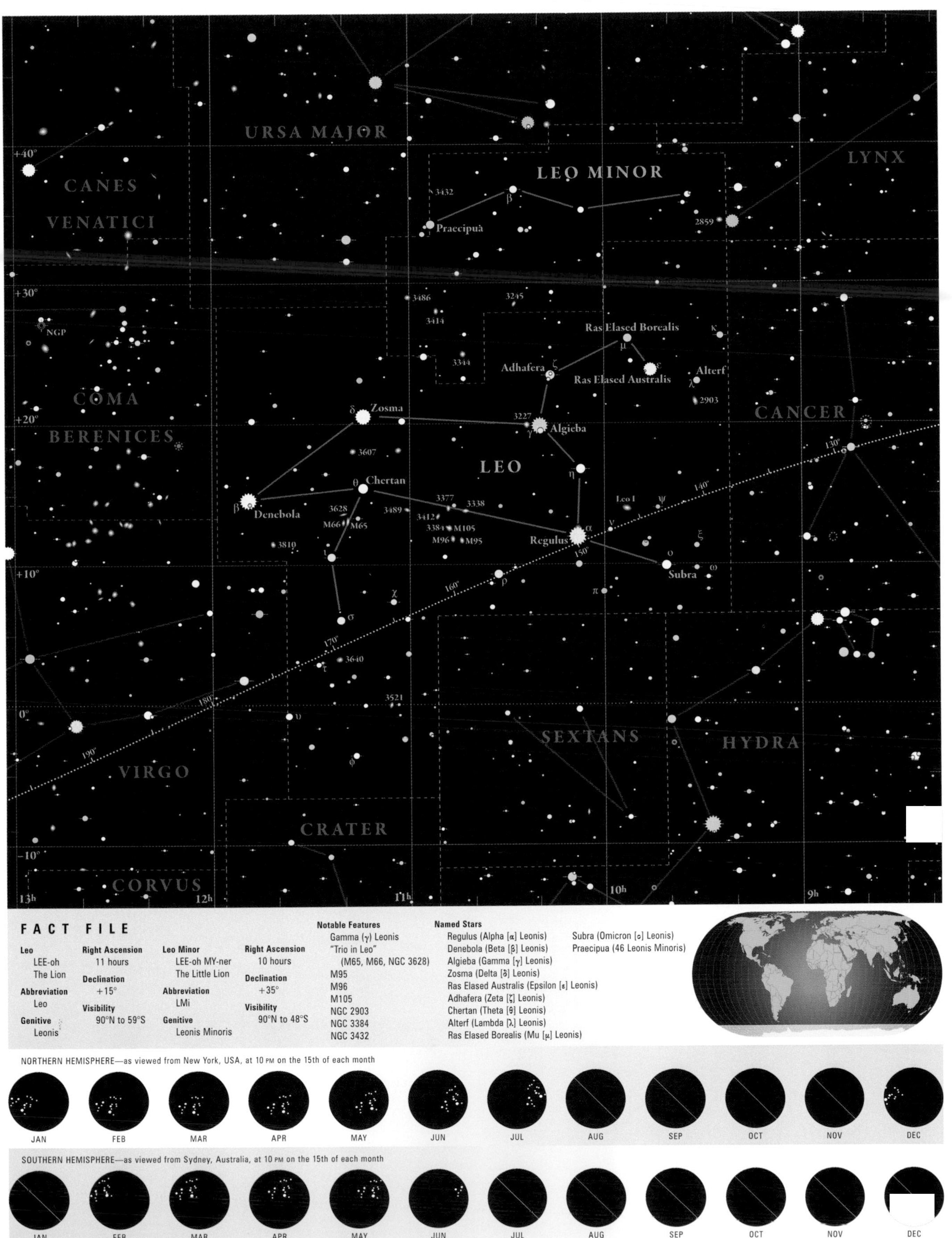
URSA MAJOR
LEO MINOR
LYNX
CANES VENATICI
COMA BERENICES
CANCER
LEO
SEXTANS
HYDRA
VIRGO
CRATER
CORVUS
Praecipua
Ras Elased Borealis
Adhafera
Ras Elased Australis
Alterf
Zosma
Algieba
Chertan
Denebola
Regulus
Subra
Leo I
NGP
+40°
+30°
+20°
+10°
0°
-10°
13h
12h
11h
10h
9h
FACT FILE
Leo
LEE-oh
The Lion
Abbreviation
Leo
Genitive
Leonis
Right Ascension
11 hours
Declination
+15°
Visibility
90°N to 59°S
Leo Minor
LEE-oh MY-ner
The Little Lion
Abbreviation
LMi
Genitive
Leonis Minoris
Right Ascension
10 hours
Declination
+35°
Visibility
90°N to 48°S
Notable Features
Gamma (γ) Leonis
"Trio in Leo"
(M65, M66, NGC 3628)
M95
M96
M105
NGC 2903
NGC 3384
NGC 3432
Named Stars
Regulus (Alpha [α] Leonis)
Denebola (Beta [β] Leonis)
Algieba (Gamma [γ] Leonis)
Zosma (Delta [δ] Leonis)
Ras Elased Australis (Epsilon [ε] Leonis)
Adhafera (Zeta [ζ] Leonis)
Chertan (Theta [θ] Leonis)
Alterf (Lambda [λ] Leonis)
Ras Elased Borealis (Mu [μ] Leonis)
Subra (Omicron [ο] Leonis)
Praecipua (46 Leonis Minoris)
NORTHERN HEMISPHERE—as viewed from New York, USA, at 10 PM on the 15th of each month
JAN FEB MAR APR MAY JUN JUL AUG SEP OCT NOV DEC
SOUTHERN HEMISPHERE—as viewed from Sydney, Australia, at 10 PM on the 15th of each month
JAN FEB MAR APR MAY JUN JUL AUG SEP OCT NOV DEC

Lepus

Crouched warily at the feet of Orion the Hunter is the constellation of Lepus, the Hare, one of Ptolemy's original 48 constellations.

NEED FOR SPEED

Ancient cultures mostly associated the stars of Lepus with a hare; however, some Arabs saw it as the throne of Orion, while the Egyptians, perceiving Orion as their god Osiris, believed the stars of Lepus to be his boat.

Strangely, modern celestial cartographers usually drew the constellation crouching or sleeping, although the ancients did not always see it that way. Canis Major, the swiftest of Orion's hounds, is found immediately eastward, eternally chasing Lepus, one of the fastest of animals, across the sky. The Macedonian Greek poet Aratus wrote of the pursuit: "Close behind he rises, and as he sets he eyes the setting hare."

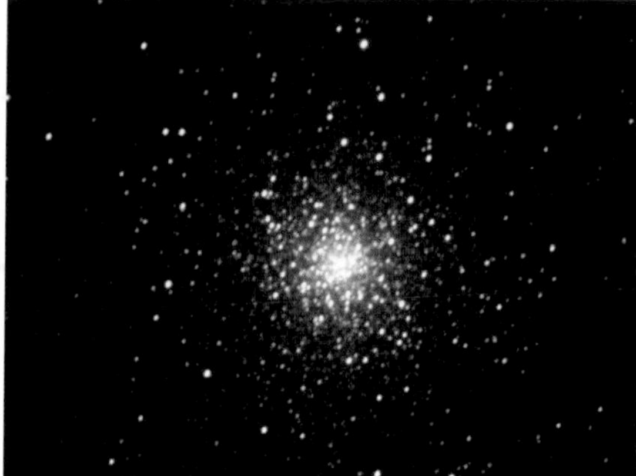

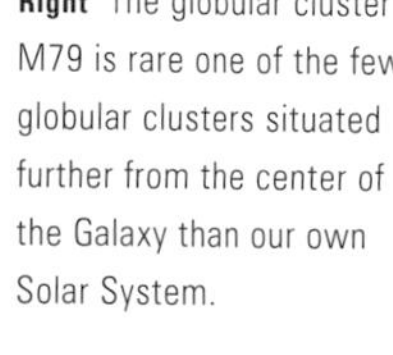

Right The globular cluster M79 is rare one of the few globular clusters situated further from the center of the Galaxy than our own Solar System.

THE STARS OF LEPUS

Although Lepus sits quietly under the feet of Orion, its stars are not completely overshadowed and are relatively easy to pick out. Two diverging curved lines of three stars—Delta (δ), Alpha (α), and Mu (μ) on the northern side, with Gamma (γ), Beta (β), and Epsilon (ε) on the southern side—create the main pattern. All are 3rd or 4th magnitude, and the group resembles not so much a hare as a cornucopia.

Alpha (α) Leporis, a bright 3rd magnitude star, looks faint only because it is so far away. About 1,300 light-years distant, it is an evolved F-type supergiant star that pours out 13,000 times the energy of our Sun; were it in our Solar System, it would engulf the planet Mercury.

R Leporis is one of the brightest "carbon stars" in the sky—you will need binoculars or a telescope to see it. It is a very evolved, pulsating, bloated giant star with clouds of carbon, forged in the nuclear furnaces of the star, enveloping it and blocking out what little bluish light it has, resulting in a star that looks like a deep red ember.

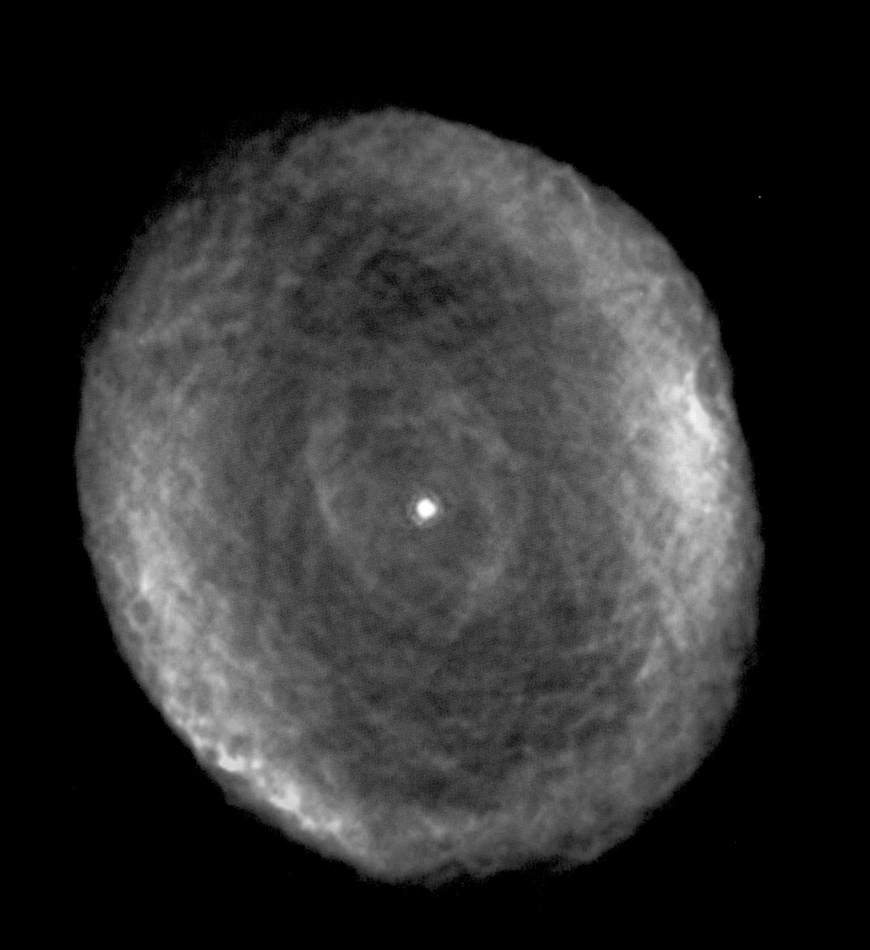

OBJECTS OF INTEREST

Lepus is home to several interesting objects for the amateur sky-watcher, the chief of which is the globular cluster **M79** . A 6-inch (15 cm) telescope shows it as a round, bright-centered haze with some very faint stars around the edges; while larger telescopes will resolve it almost to the center.

Not far from brilliant Rigel in Orion is the interesting planetary nebula **IC 418**, otherwise known as the **Spirograph Nebula** . Larger amateur telescopes reveal a small bright aqua disk with a well-defined edge, with the star that formed the nebula resting in the center. **NGC 2017** is found near Alpha (α) Leporis, and is often classified as an open cluster of stars. It is, in fact, a very complex multiple star system that a 4-inch (10 cm) telescope shows as a tight bright family of six stars. Larger telescopes reveal that some of them are close binaries.

Left The star at the center of IC 418 was a red giant a few thousand years ago, but then ejected its outer layers into space to form the Spirograph Nebula, which has now expanded to a diameter of about 0.1 light-years.

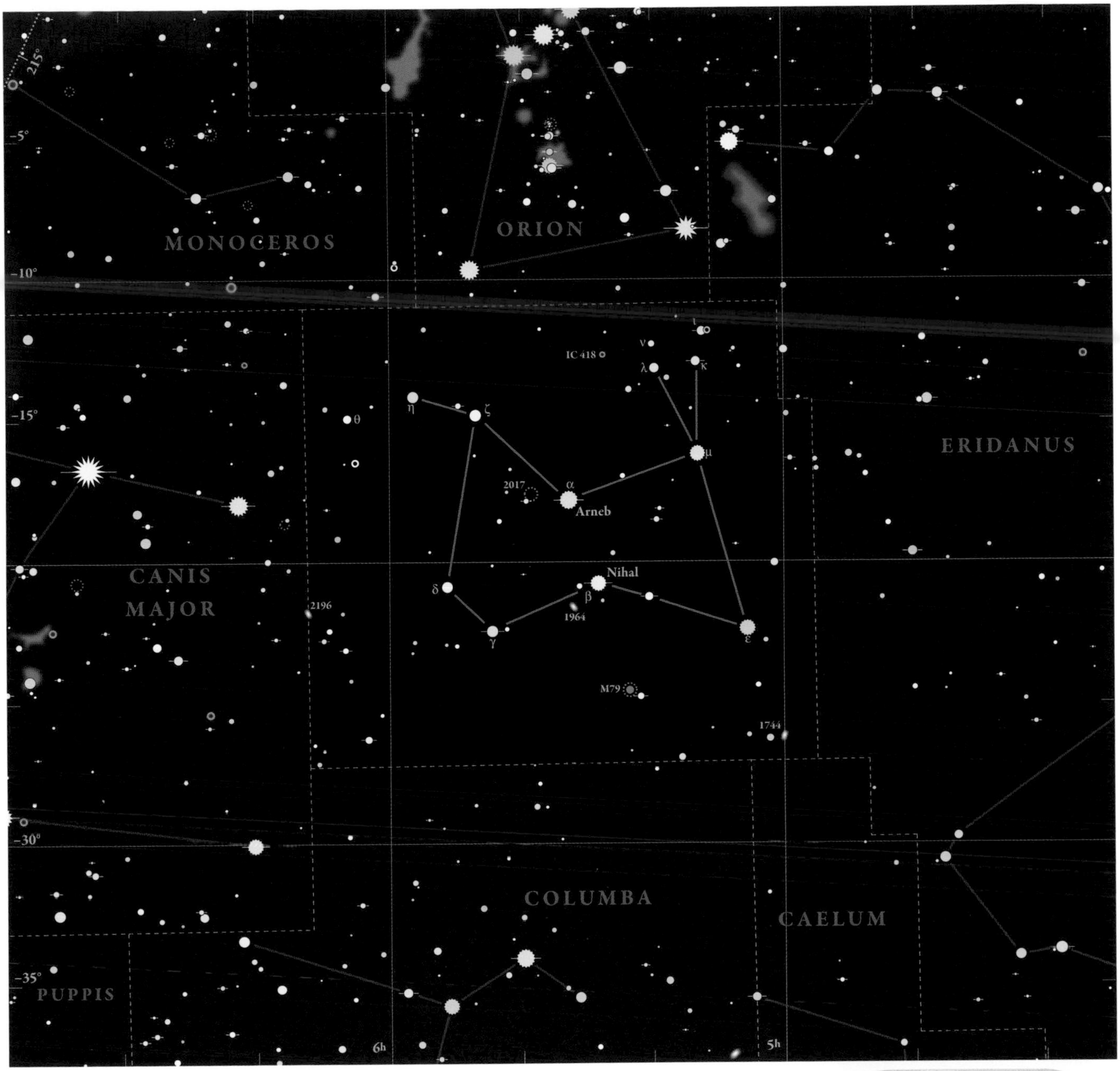

FACT FILE

Lepus
LEE-pus
The Hare

Abbreviation
Lep

Genitive
Leporis

Right Ascension
6 hours

Declination
–20°

Visibility
63°N to 90°S

Notable Features
IC 418 (Spirograph Nebula)
M79
NGC 2017
R Leporis

Named Stars
Arneb (Alpha [α] Leporis)
Nihal (Beta [β] Leporis)

NORTHERN HEMISPHERE—as viewed from New York, USA, at 10 PM on the 15th of each month

SOUTHERN HEMISPHERE—as viewed from Sydney, Australia, at 10 PM on the 15th of each month

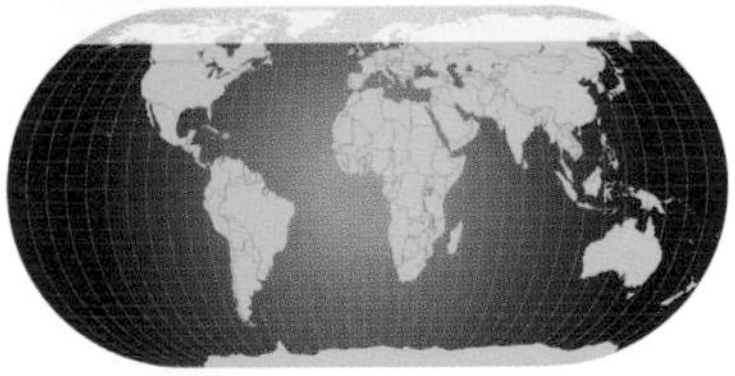

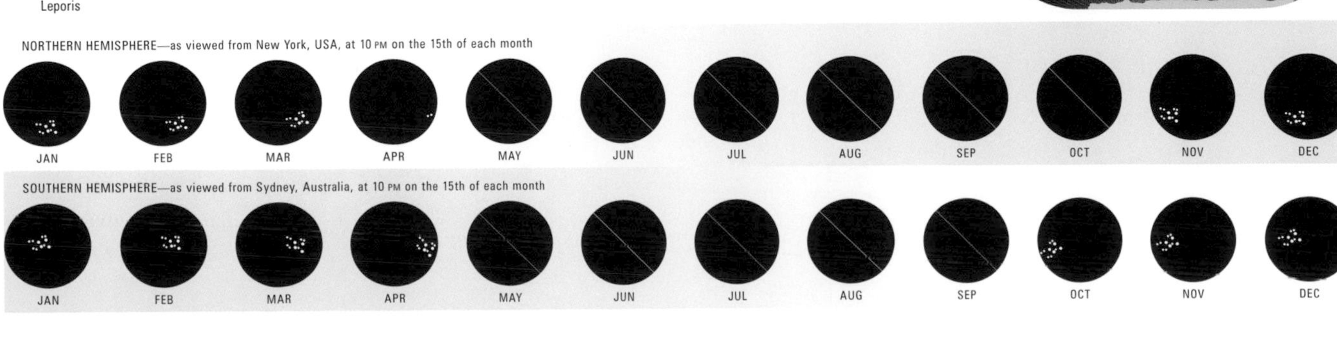

Libra

Located between Virgo to the west and Scorpius to the east, Libra is the youngest constellation of the astrological zodiac.

KEEPING A BALANCE

The Sumerians saw the stars of Libra as Zib-ba an-na—the "balance" of the heavens; at that time, the northern autumnal equinox occurred when the Sun was in the constellation of Libra. They also associated the group with the judgment of the living and the dead, but this interpretation seems to have skipped intervening civilizations until the Romans revived it in the first century BCE.

The Romans formed the modern constellation of Libra from stars that the Greeks perceived as the extended claws of the Scorpion. They believed that the Moon was in Libra when the city of Rome was founded, and saw Libra as symbolic of balance and order, as well as being a modernizing influence of the Empire itself.

Libra is also the only constellation of the zodiac that represents an inanimate object—the scales; the others are all mythological figures or animals.

THE STARS OF LIBRA

Most of the stars in Libra are faint, but the square shape of the main stars is quite easy to see. A line from Alpha (α) to Beta (β) makes the beam of the balance, while Gamma (γ) and Sigma (σ) represent the two weighing pans. Alpha (α) and Beta (β) Librae have common names from Arabic, disclosing their heritage as marking the claws of Scorpius—Zubenelgenubi and Zubenelschemali, meaning, respectively, the southern and northern claw of the scorpion.

The lucida of Libra is 3rd magnitude Beta (β) Librae. It is a blue B-type dwarf star that is about 160 light-years distant and 130 times more luminous than our Sun. While Alpha and Beta Librae were removed from Scorpius in Roman times, Libra's third brightest star, Sigma (σ) Librae wasn't re-assigned by astronomers until the nineteenth century. Also known as Zubenhakrabi, from the Arabic meaning "scorpion claw," it is a bloated M-type red giant star, 290 light-years distant. If it were in our Solar System, it would nearly reach Venus.

OBJECTS OF INTEREST

Although Libra is adjacent to the spectacular sights of Scorpius and Sagittarius, it has little to offer the amateur observer. Third magnitude **Alpha (α) Librae**, or **Zubenelgenubi**, is a naked-eye binary star with uneven components, which the keen-eyed can resolve as separate 3rd and 5th magnitude stars.

The globular cluster **NGC 5897** is Libra's chief adornment. In an 8-inch (20 cm) telescope it is a large hazy cloud with little concentration to the center and showing several resolved stars. Nearby to this globular is the small, fairly faint planetary nebula **Merrill 2-1**, which displays an evenly illuminated bluish disk in an 8-inch (20 cm) telescope.

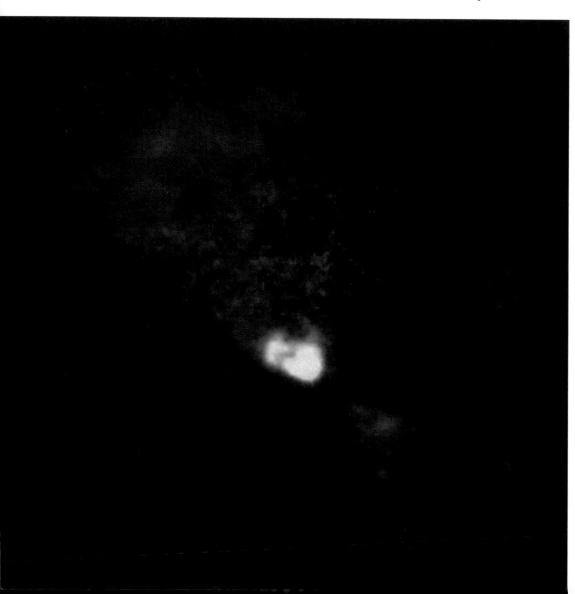

Left The core of the barred spiral galaxy, NGC 5728 is probably a supermassive black hole surrounded by a central region of ionized gas. The interaction between the visible and ultraviolet light give it a "lighthouse beacon" shape.

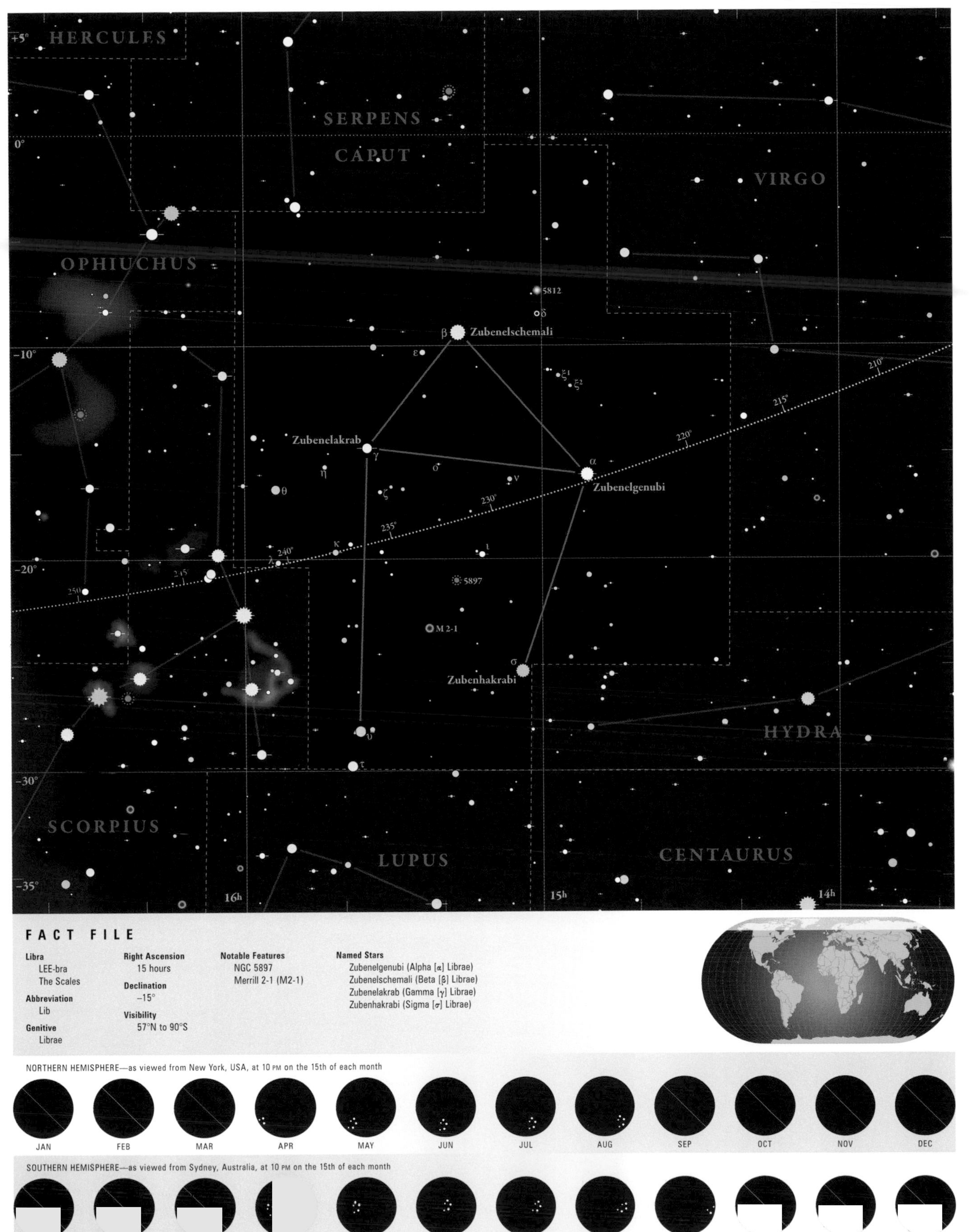

FACT FILE

Libra
LEE-bra
The Scales

Abbreviation
Lib

Genitive
Librae

Right Ascension
15 hours

Declination
−15°

Visibility
57°N to 90°S

Notable Features
NGC 5897
Merrill 2-1 (M2-1)

Named Stars
Zubenelgenubi (Alpha [α] Librae)
Zubenelschemali (Beta [β] Librae)
Zubenelakrab (Gamma [γ] Librae)
Zubenhakrabi (Sigma [σ] Librae)

NORTHERN HEMISPHERE—as viewed from New York, USA, at 10 PM on the 15th of each month

JAN FEB MAR APR MAY JUN JUL AUG SEP OCT NOV DEC

SOUTHERN HEMISPHERE—as viewed from Sydney, Australia, at 10 PM on the 15th of each month

JAN FEB MAR APR MAY JUN JUL AUG SEP OCT NOV DEC

Lupus

Lupus, the wild wolf, prowls around the feet of Scorpius. This small southern constellation has few bright stars.

WOLF AT THE DOOR

Since antiquity, the stars that make up the modern constellation of Lupus have been associated with a wild beast of some kind. The Greeks called it Therium—an unspecified wild animal—while the Romans simply called it "the beast," and imagined it as the carcass of a wild animal that is carried on a long spear or pole by the adjacent Centaurus to be sacrificed on the nearby altar, Ara.

In other ancient cultures, the people of the Euphrates Valley associated the asterism with Zibu, the Beast, or Urbat, the Beast of Death. In Babylon, it was named Ur-Idim, the wild dog, and to the Arabs it was Al-Asadah, the lioness. The stars finally became specifically associated with a wolf during the Renaissance, and Johannes Bayer was the first to map them in that form in *Uranometria*, published in 1603.

THE STARS OF LUPUS

Lupus has no 1st magnitude stars, but the constellation has an abundance of 2nd and 3rd magnitude stars. Although the stars are not hard to pick out between Centaurus and Scorpius, there is very little resemblance to the form of a wolf.

Alpha (α) Lupi is sometimes known as Kakkab, a name of Arabic origin meaning "the star left of the horned bull," which the Arabs associated with the group that we know as Centaurus. Located 515 light-years distant, it is a blue B-class subgiant star of about 11 solar masses, and is 20,000 times more luminous than our Sun.

In the year 1006 CE, the brightest supernova in recorded history occurred in Lupus, peaking at approximately magnitude -9. It was visible even in daylight for several weeks.

OBJECTS OF INTEREST

Lupus contains a measure of interesting objects for the amateur sky-watcher, although not as many as its position near the Milky Way might suggest. Part of the reason is the many dark clouds of gas and dust within our galaxy near Lupus, which tend to blot out the more distant objects. There are three globular star clusters in Lupus, the best being **NGC 5986**. In an 8-inch (20 cm) telescope at high powers it is a fairly well concentrated patch of haze peppered with very faint stars. **Xi (ξ) Lupi** is a lovely pair of white stars in a close binary that differ little in brightness.

The most interesting object in Lupus is the bright planetary nebula **IC 4406**, found near the border with Centaurus. It is faintly visible in a 4-inch (10 cm) telescope, and an 8-inch (20 cm) telescope shows it as a bright, steel-gray, slightly oval disk with diffuse edges. Nearby is the reasonably bright galaxy **NGC 5643**, which a 6-inch (15 cm) telescope shows as a faint, round, hazy patch with a hint of a bright spot in the center.

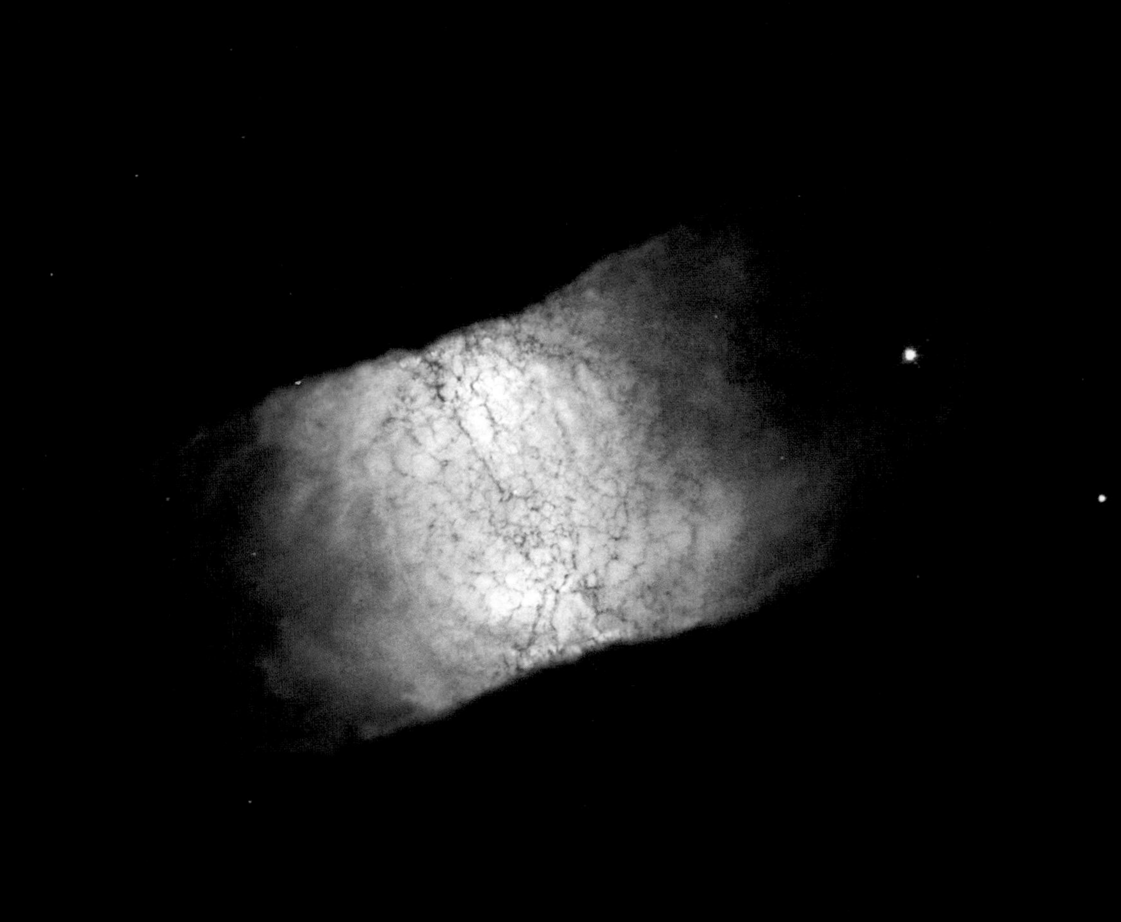

Left This Hubble Space Telescope image reveals the symmetry of the planetary nebula IC 4406, which is shaped like a donut that we are looking at from the side. The tendrils of dark dust give the Retina Nebula its name, as they look like blood vessels in the human eye.

CIRCINUS
ARA
NORMA
CENTAURUS
SCORPIUS
HYDRA
LIBRA

FACT FILE

Lupus
LEW-pus
The Wolf

Abbreviation
Lup

Genitive
Lupi

Right Ascension
15 hours

Declination
–45°

Visibility
33°N to 90°S

Notable Features
Xi (ξ) Lupi
IC 4406
NGC 5643
NGC 5986

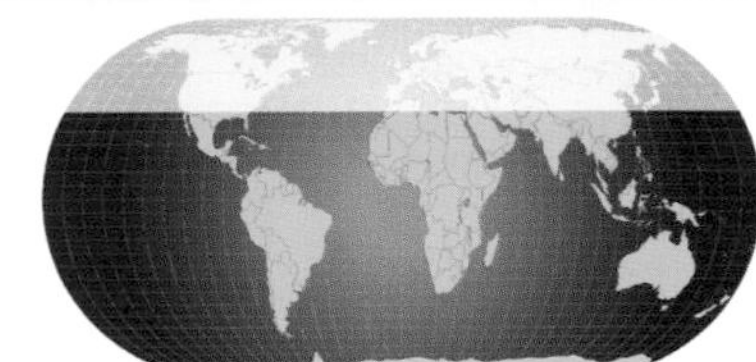

NORTHERN HEMISPHERE—as viewed from New York, USA, at 10 PM on the 15th of each month

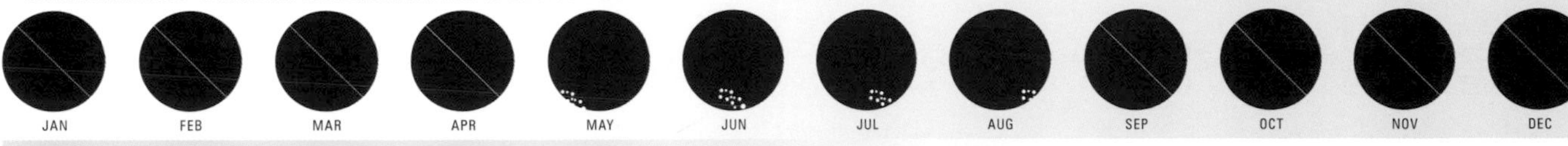

SOUTHERN HEMISPHERE—as viewed from Sydney, Australia, at 10 PM on the 15th of each month

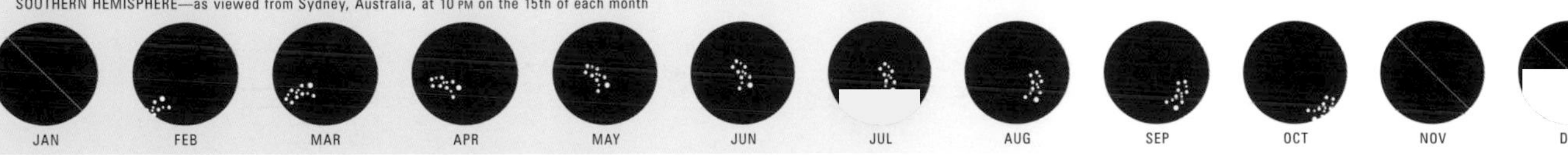

Lynx

Lynx is fashioned from a fairly large yet dim swathe of the northern sky. It runs between two other faint constellations, Leo Minor and Camelopardalis.

TAKE A CLOSER LOOK

Lynx was introduced by the seventeenth-century astronomer Johann (Johannes) Hevelius and it is one of the hardest constellations to find in the sky. Hevelius is said to have joked that he called his new constellation Lynx because only the lynx-eyed would be able to see it!

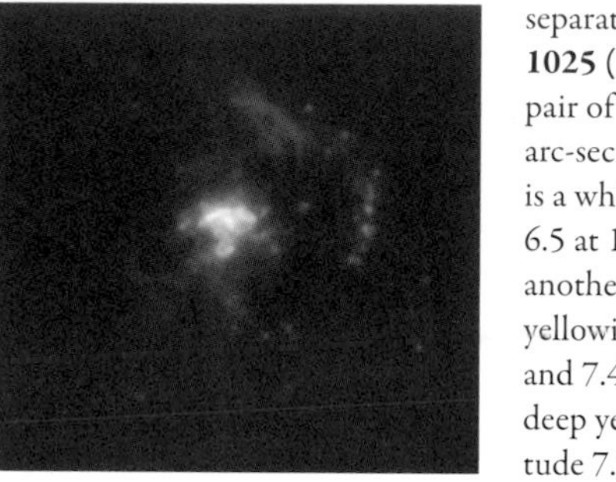

THE STARS OF LYNX

Lynx does possess one 3rd magnitude star, Alpha (α) Lyncis, and five 4th magnitude ones, but they are scattered over 545 square degrees. When observing in this constellation, amateur astronomers usually begin their starhops from jumping-off points in one of the adjacent brighter constellations of Ursa Major, Gemini, or Auriga. For example, the globular cluster NGC 2419 is an easy starhop from the bright star Castor in Gemini.

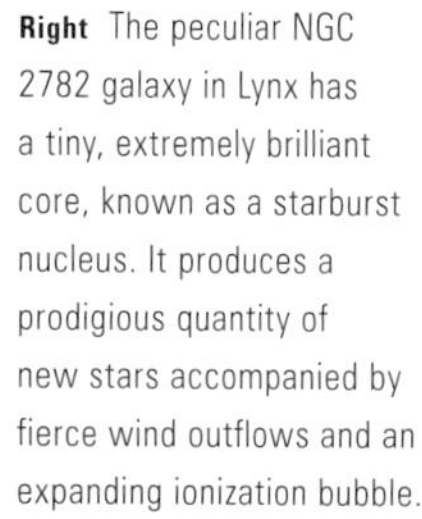

Right The peculiar NGC 2782 galaxy in Lynx has a tiny, extremely brilliant core, known as a starburst nucleus. It produces a prodigious quantity of new stars accompanied by fierce wind outflows and an expanding ionization bubble.

OBJECTS OF INTEREST

Struve 958 (Σ958) is a matched pale yellow pair of magnitude 6.3 stars at 4.8 arc-seconds separation. **Struve 1009 (Σ1009)** is a white binary of magnitudes 6.9 and 7.0, separated by 4.1 arc-seconds. **Struve 1025 (Σ1025)** is a light orange pair of magnitudes 8.3 and 8.6 at 26 arc-seconds separation. **19 Lyncis** is a white double of magnitudes 5.6 and 6.5 at 15 arc-seconds. **20 Lyncis** is another 15 arc-second double, a matched yellowish-white pair of magnitudes 7.3 and 7.4. **Struve 1282 (Σ1282)** is a deep yellow head-light binary of magnitude 7.5 stars at 3.6 arc-seconds. **Struve 1333 (Σ1333)** is a closer binary (1.6 arc-seconds, so good seeing is required) of magnitudes 6.4 and 6.7, both white. **Struve 1369 (Σ1369)** is a yellowish triple. The AB pair is magnitudes 7.0 and 8.0 at 25 arc-seconds. There is a magnitude 8.7 C component at 118 arc-seconds distance.

NGC 2419 was once commonly called the **Intergalactic Wanderer** because it is 275,000 light-years away, farther than the Magellanic Clouds. However, with the discovery of dark matter, we now know that NGC 2419 must still be gravitationally bound to the Milky Way. This is one of the most luminous of the more than 150 known globular clusters orbiting our galaxy, since only Omega (ω) Centauri, NGC 6388, and M54 have greater absolute magnitudes.

Despite its distance and consequent faintness (at magnitude 10.3), NGC 2419 is immediately obvious as an unresolved globular cluster with an 8-inch (20 cm) reflector. A 16-inch (40 cm) Newtonian shows a flat brightness profile across its core, with some fading in the halo. In recent years a few amateurs, who have obtained observing time on elderly 3-foot to 6 ½-foot (1–2 m) reflectors at professional observatories, have become the first observers ever to visually resolve a few of NGC 2419's stars.

The nearly edge-on spiral **NGC 2683** is easy with an 8-inch (20 cm). A 16-inch (40 cm) shows a large, elongated galaxy with hints of a nucleus and a dust lane. In an 8-inch (20 cm), **NGC 2782** is a small, round, magnitude 11.6 spiral galaxy with a star-like nucleus.

Left An artist's impression of the distant Lynx Arc, a gargantuan cluster of extremely hot young stars about 12 billion light-years away. One million times brighter than the Orion Nebula, the arc contains a million blue stars that are twice as hot as similar denizens of the Milky Way.

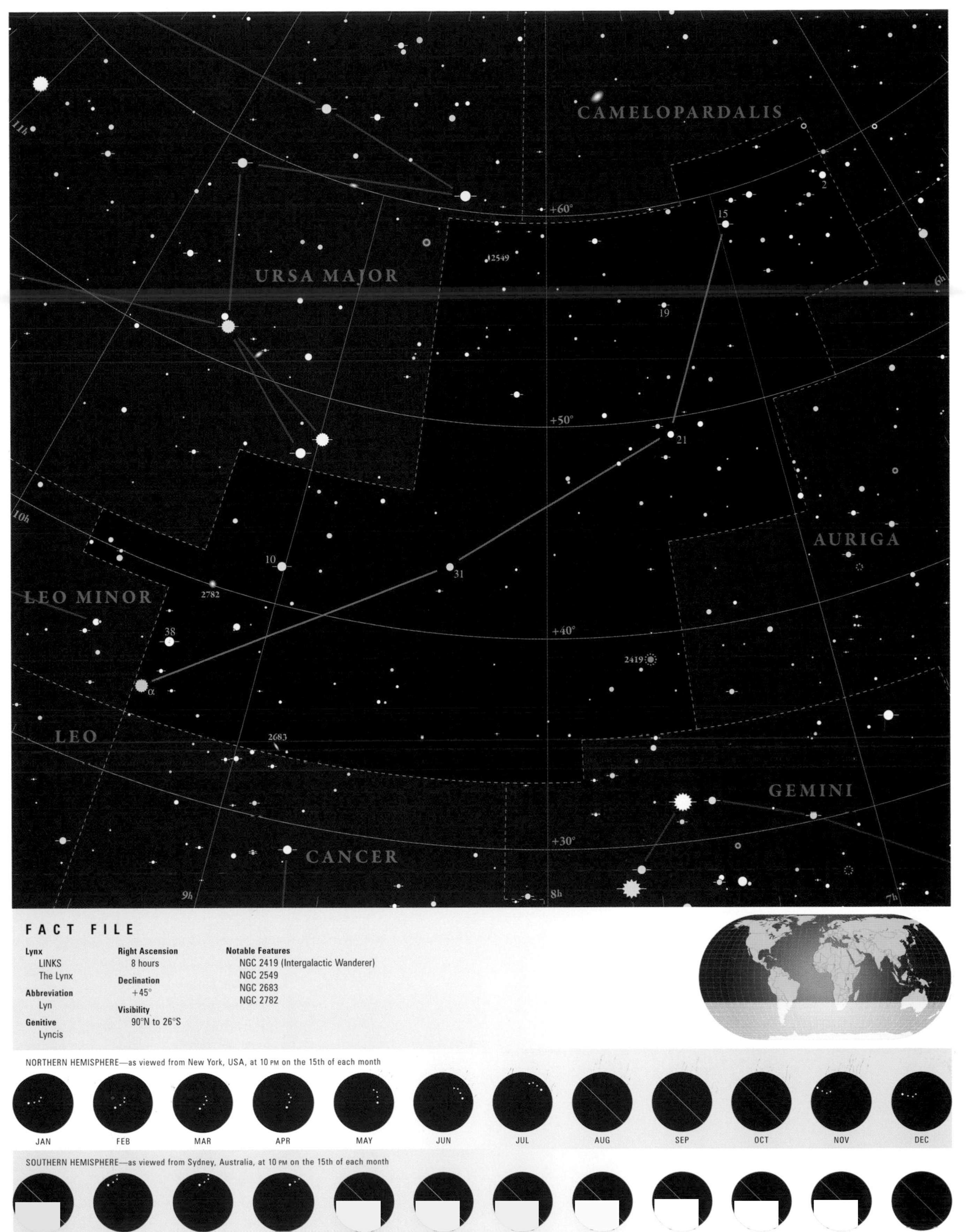

FACT FILE

Lynx
LINKS
The Lynx

Abbreviation
Lyn

Genitive
Lyncis

Right Ascension
8 hours

Declination
+45°

Visibility
90°N to 26°S

Notable Features
NGC 2419 (Intergalactic Wanderer)
NGC 2549
NGC 2683
NGC 2782

NORTHERN HEMISPHERE—as viewed from New York, USA, at 10 PM on the 15th of each month

SOUTHERN HEMISPHERE—as viewed from Sydney, Australia, at 10 PM on the 15th of each month

Lyra

Lyra is a small yet bright constellation that many sky-watchers visit regularly to see the feature called the Double-Double. On dark nights, most also take the opportunity to view the Ring Nebula.

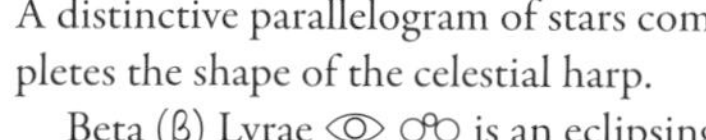

MUSICAL CHARMS

Lyra represents the first lyre, created by Hermes from a tortoise shell and later given to Orpheus. Orpheus's music charmed every creature. He used it to pass the guardians of Hades in order to plead for his young bride Eurydice, who was bitten by a viper. Pluto agreed to release Eurydice from the Underworld, but warned Orpheus not to look back at her as she followed him on the climb to the upper world. When Orpheus reached daylight he joyously turned to her, only to see her still within the cavern. He had looked too soon and lost her forever, since he was banned from another trip to the Underworld while he was alive.

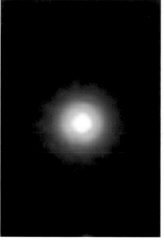

Right Vega, or Alpha (α) Lyrae, captured here by the Spitzer Space Telescope, is the fifth brightest star in the sky. Its diameter is almost three times that of our Sun.

THE STARS OF LYRA

Of the stars that in turn become the North Pole Star, the magnitude 0.0 Vega, Alpha (α) Lyrae, is by far the brightest. In ten thousand years, Vega will become the Pole Star. A distinctive parallelogram of stars completes the shape of the celestial harp.

Beta (β) Lyrae is an eclipsing variable ranging from magnitude 3.4 to 4.3 over 12.94 days. Both members of this contact binary are within the same gaseous envelope. Gravitational attraction and rapid rotation make both stars ellipsoidal. Beta's magnitude varies continually, depending on the changing areas of the stars presented toward us, plus the mutual eclipses. Beta is also a visual triple star. The main pair is magnitudes 3.4 and 8.6 at 46 arc-seconds; a third member is magnitude 9.9 at 86 arc-seconds.

OBJECTS OF INTEREST

Delta (δ) Lyrae makes a striking orange and blue binocular double of magnitudes 4.5 and 5.6, separated by one third of a moon diameter. They are the brightest gems in the sparse open cluster **Stephenson 1**. Large telescopes show several other bluish stars as well as another orange one.

Epsilon (ε) Lyrae—known as the **Double-Double**—is a famous quadruple star. Binoculars resolve two 5th magnitude stars that are 3.5 arc-minutes apart. Some people can succeed in viewing these stars with the unaided eye. On steady nights, a 3-inch (8 cm) refractor splits each star again. The AB pair is magnitudes 5.0 and 6.1, at 2.4 arc-seconds separation. The CD pair is magnitudes 5.2 and 5.5, also at 2.4 arc-seconds separation.

The Ring Nebula, also known as the planetary nebula **M57 (NGC 6720)**, is a little smoke-ring in small telescopes. In an 8-inch (20 cm) telescope, the oval fades at the ends, and it is brighter inside the annulus than outside. The magnitude 15 central star can be glimpsed with a 16-inch (40 cm) at 400 times on superb viewing nights. A 36-inch (0.9 m) reflector gives a view comparable to photographs, showing the star beside the central star, a star on the ring, and parallel bands of nebulosity inside the ring.

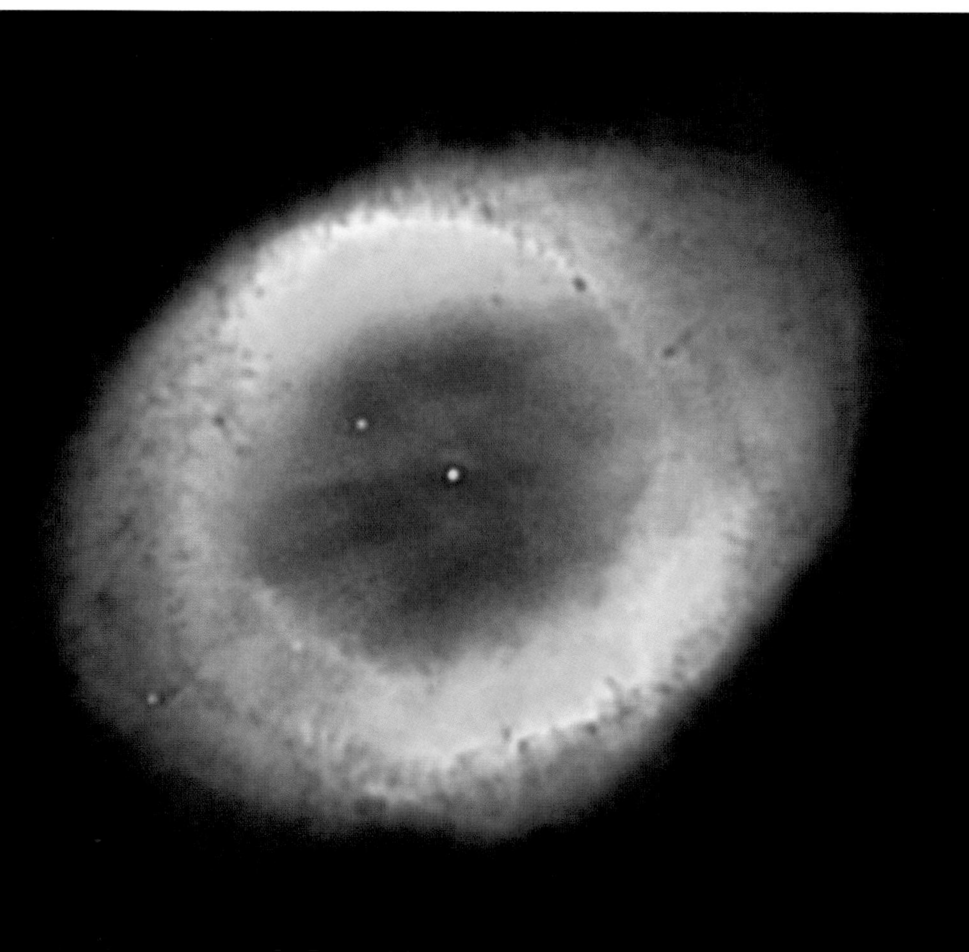

Left Astronomers of the past noticed the unusual shape of what we now popularly call the Ring Nebula, M57 (NGC 6720). Recent photographs indicate that this planetary nebula may be cylindrical rather than spherical in shape.

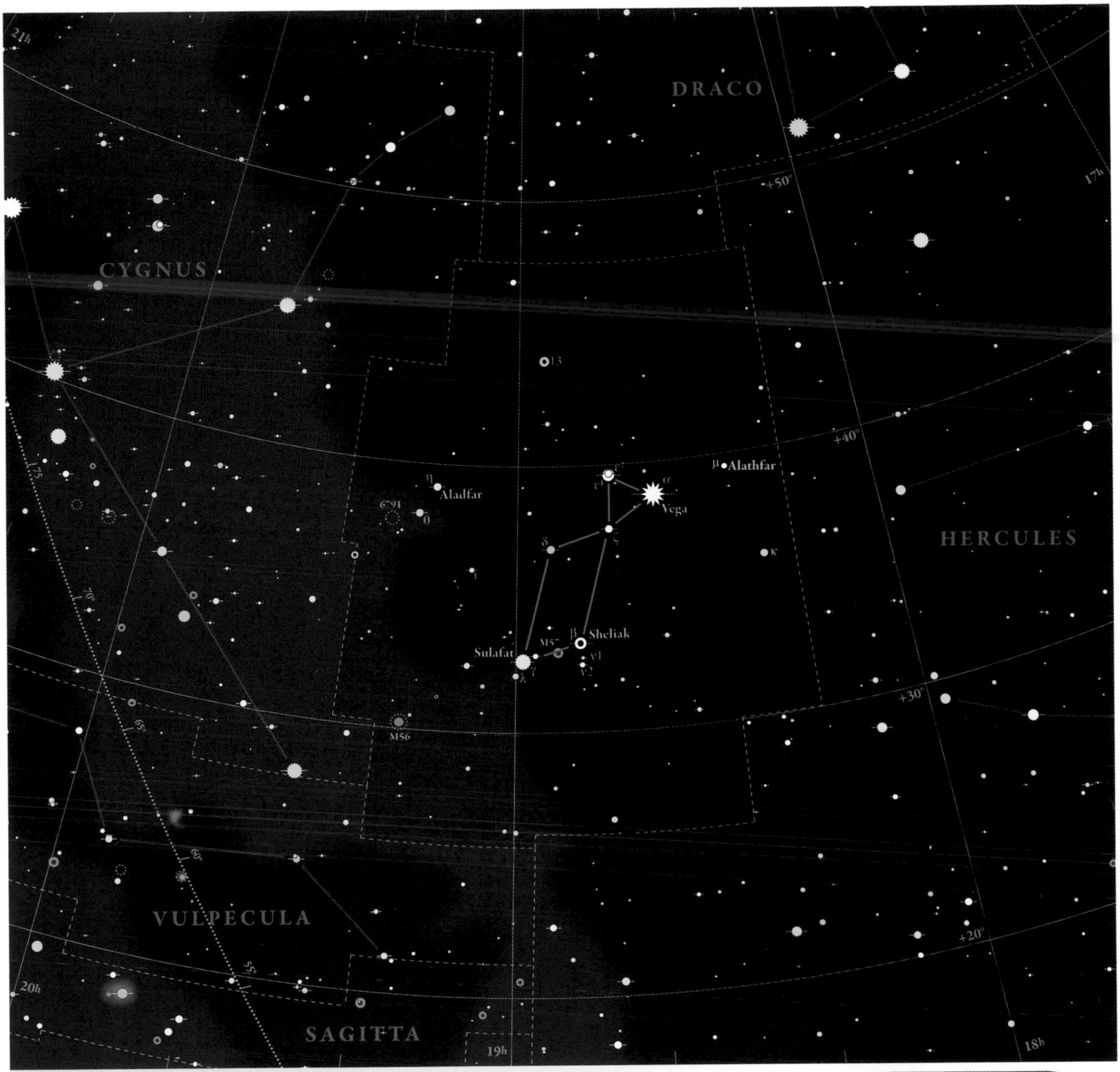

DRACO
CYGNUS
HERCULES
VULPECULA
SAGITTA
Vega
Alathfar
Aladfar
Sheliak
Sulafat
M57
M56
+50°
+40°
+30°
+20°
21h
20h
19h
18h
17h

FACT FILE

Lyra
LYE-rah
The Lyre

Abbreviation
Lyr

Genitive
Lyrae

Right Ascension
19 hours

Declination
+40°

Visibility
90°N to 45°S

Notable Features
Alpha (α) Lyrae
Delta (δ) Lyrae
Epsilon (ε) Lyrae (Double-Double)
M57 (Ring Nebula)
Stephenson 1

Named Stars
Vega (Alpha [α] Lyrae)
Sheliak (Beta [β] Lyrae)
Sulafat (Gamma [γ] Lyrae)
Aladfar (Eta [η] Lyrae)
Alathfar (Mu [μ] Lyrae)

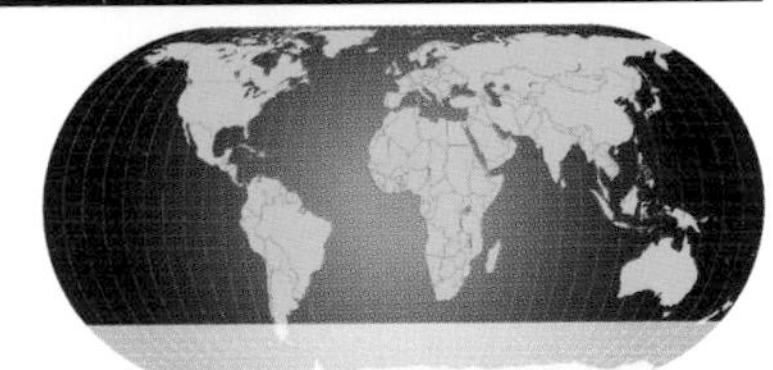

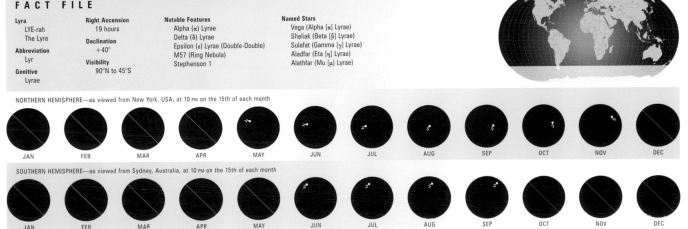

NORTHERN HEMISPHERE—as viewed from New York, USA, at 10 PM on the 15th of each month

JAN FEB MAR APR MAY JUN JUL AUG SEP OCT NOV DEC

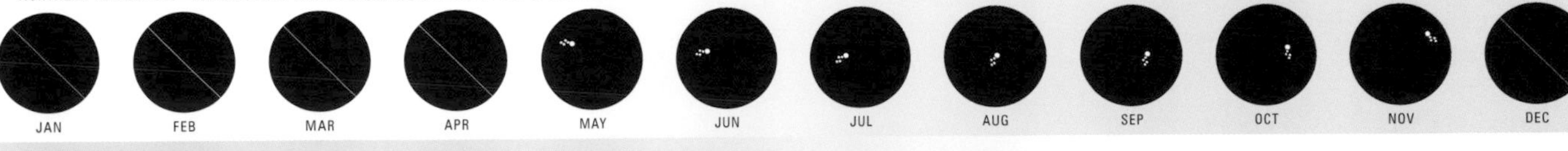

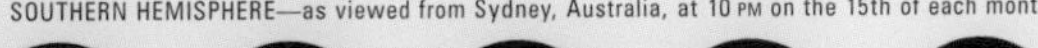

SOUTHERN HEMISPHERE—as viewed from Sydney, Australia, at 10 PM on the 15th of each month

JAN FEB MAR APR MAY JUN JUL AUG SEP OCT NOV DEC

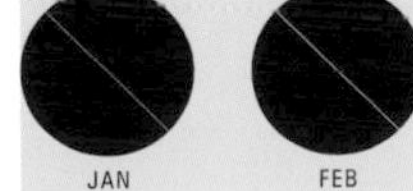

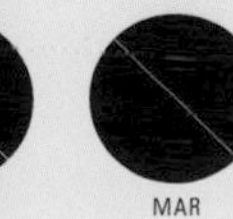

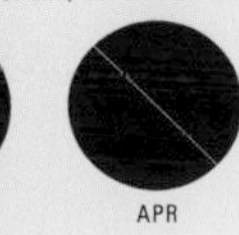

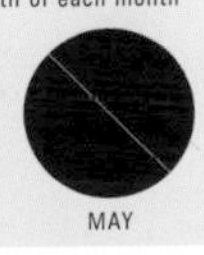

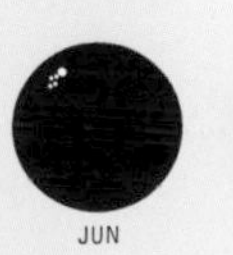

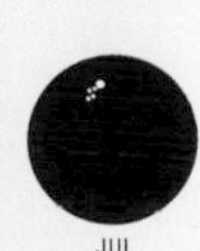

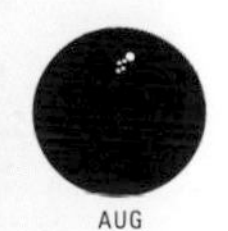

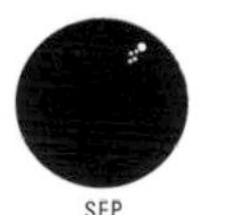

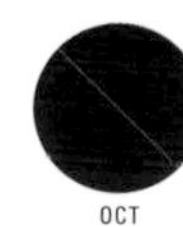

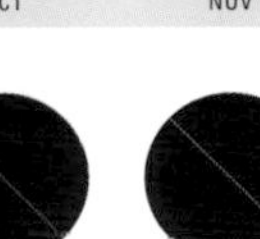

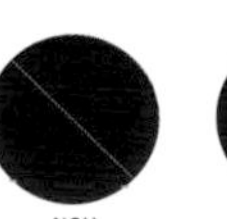

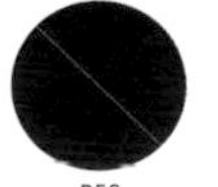

Mensa

One of the faintest and least distinguished of the constellations, Mensa is home to the southern half of the Large Magellanic Cloud.

MOUNTAIN GLIMMER

The dim stars of Mensa are too far south to have been seen by European astronomers, and therefore no mythology is associated with them. Even Bayer's *Uranometria* left it as a blank area of sky. Mons Mensae, the Table Mountain, is one of only two constellations in the sky that commemorate a landform.

Following his visit to the Cape of Good Hope during 1750 to 1751, Abbé Nicolas Louis de Lacaille introduced this constellation to commemorate one of Earth's most spectacular mountains—Table Mountain, situated above Cape Town in South Africa. Since that time, its stars and form have undergone no significant change, save that the name has been worn down to simple Mensa.

THE STARS OF MENSA

Even the brightest stars of Mensa are tiny 5th magnitude specks, and a rural sky is needed to make them out. The four principal stars in the figure, Alpha (α), Gamma (γ), Eta (η), and Beta (β) Mensae make up an inverted, block-shaped group beneath the Large Magellanic Cloud.

Fifth magnitude Alpha (α) Mensae is intrinsically the dimmest "Alpha" star in the sky and one of the closest solar analogs, at just 33 light-years distant. Alpha Mensae possesses 90 percent of the sun's mass and 80 percent of its luminosity. Although nowadays it is only a dim mote, 250,000 years ago it was at its closest point to the Sun, only 11 light-years distant, and almost shone at 2nd magnitude.

Pi (π) Mensae, about 59 light-years away, is another G-type star very similar to the Sun; we see it as a 6th magnitude dot. Astronomers have detected a gas-giant companion more than 10 times the mass of Jupiter orbiting Pi Mensae.

OBJECTS OF INTEREST

As Mensa is located well off the plane of the Milky Way, no star clusters or nebulae from our own galaxy are present. All of its galaxies are either too faint or small (usually both) to be interesting objects in a telescope.

The southern reaches of the **Large Magellanic Cloud** hang over the northern end of Mensa in similar fashion to the cloud and mist that forms the "tablecloth" over the real Table Mountain. Sweeping the southern edge of the cloud with a telescope or large binoculars from a rural sky will reveal the multitude of faint stars, starry knots, and hazy patches that makes up our closest extra-galactic neighbor.

Above Glimmering stars interspersed with wisps of gas provide dramatic testament to supernova 1987A that took place in the Large Magellanic Cloud (LMC) in 1987. Many of the nearby bright blue stars are massive, at more than six times the mass of our Sun.

Right The Large Magellanic Cloud (LMC), partly residing in Mensa, was once most likely a barred spiral galaxy disrupted by our own Milky Way. Stellar debris, glowing pink and purple, is photographed here by the Hubble Space Telescope.

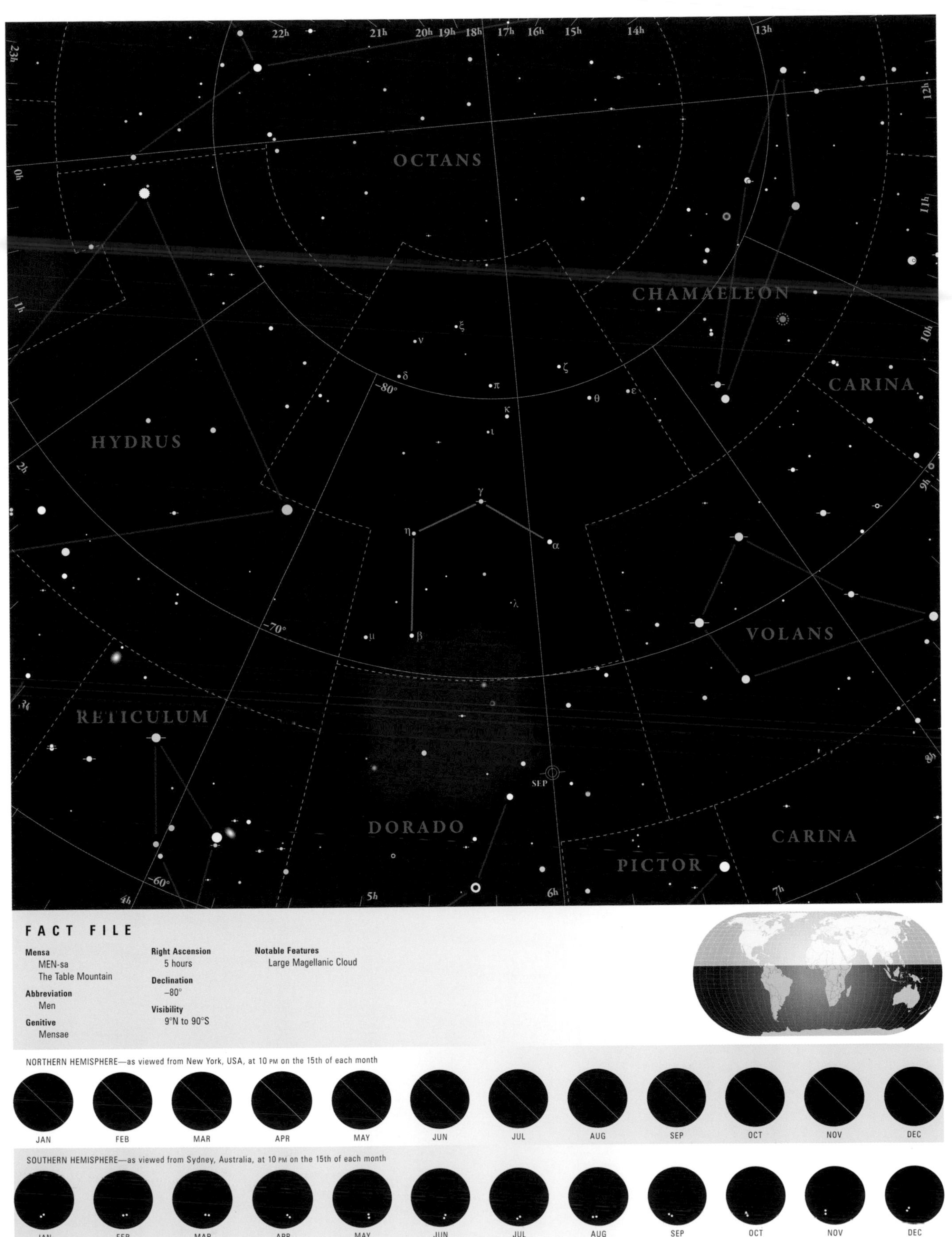

FACT FILE

Mensa
MEN-sa
The Table Mountain

Abbreviation
Men

Genitive
Mensae

Right Ascension
5 hours

Declination
−80°

Visibility
9°N to 90°S

Notable Features
Large Magellanic Cloud

NORTHERN HEMISPHERE—as viewed from New York, USA, at 10 PM on the 15th of each month

JAN FEB MAR APR MAY JUN JUL AUG SEP OCT NOV DEC

SOUTHERN HEMISPHERE—as viewed from Sydney, Australia, at 10 PM on the 15th of each month

JAN FEB MAR APR MAY JUN JUL AUG SEP OCT NOV DEC

Microscopium

Microscopium is a small constellation in the southern sky. Its common name, the Microscope, reflects its population of inconspicuous hard-to-see objects.

A DIM SPECTACLE

Abbé Nicolas Louis de Lacaille was the first to group the faint stars of Microscopium into a constellation of their own during his visit to the Cape Observatory between 1751 and 1752. As its name implies, the constellation commemorates the invention of the compound microscope by Zacharias Janssen, a Dutch spectacle-maker, in the late sixteenth century. To create the group, Lacaille plundered an adjoining constellation, Piscis Austrinus, and removed several stars.

In turn, Johannes Bode later "stole" several stars from Microscopium to create the constellation of Globus Aerostaticus, the hot-air balloon, in his *Uranographia* of 1803, but that star-group never gained acceptance.

Many amateur sky-watchers have remarked that the Microscope is the best of all tributes to Lacaille's apparent affection for the less exciting stars. Dim, small, and inconspicuous, there are few more obscure star groups than Microscopium, even among the constellations that Lacaille himself devised. The Microscope is a comparatively modern invention, so it has no associated legends or mythologies.

THE STARS OF MICROSCOPIUM

The main stars of Microscopium form a slightly crooked, inverted L-shaped asterism to the south of Capricornus. All are 5th magnitude or fainter, and the group bears little resemblance to a microscope. Alpha (α) Microscopii is the third brightest in the group, beaten to that honor by both Beta (β) and Gamma (γ). The lucida of Microscopium is Gamma (γ)—one of the stars "borrowed" by Lacaille from Piscis Austrinus. It is a G-class giant star that must have dazzled early humans. At about 2.5 times the mass of the Sun and 64 times its luminosity, it is nearing the end of its life, and now subsists on helium fusion in its core rather than hydrogen, as it did back when it was born a B-class dwarf.

The proper motion of Gamma, at 9 miles (15 km) per second almost directly away from us, reveals something very interesting: It is on an outward trajectory from a close encounter with our Sun. It is now 5th magnitude, but some 3.5 million years ago it was only 6 light-years distant and shone at magnitude −3.

OBJECTS OF INTEREST

Microscopium is some distance from the Milky Way and offers little to the amateur sky-watcher apart from some galaxies, most of which are faint and tiny. **NGC 6925** is the brightest, and thus is the easiest to see in an 8-inch (20 cm) telescope. It appears as a small hazy oval with a much brighter center.

Alpha (α) Microscopii is the constellation's most interesting double star. The 5th magnitude yellow star has a faint nearby companion that is easy to see in a 6-inch (15 cm) telescope.

Right Artist's impression of the view from the vicinity of a hypothetical terrestrial planet and moon orbiting the red dwarf star AU Microscopii. The relatively newborn 12-million-year-old star is surrounded by a very dusty disk of debris from the collision of comets and asteroids swirling around the young star.

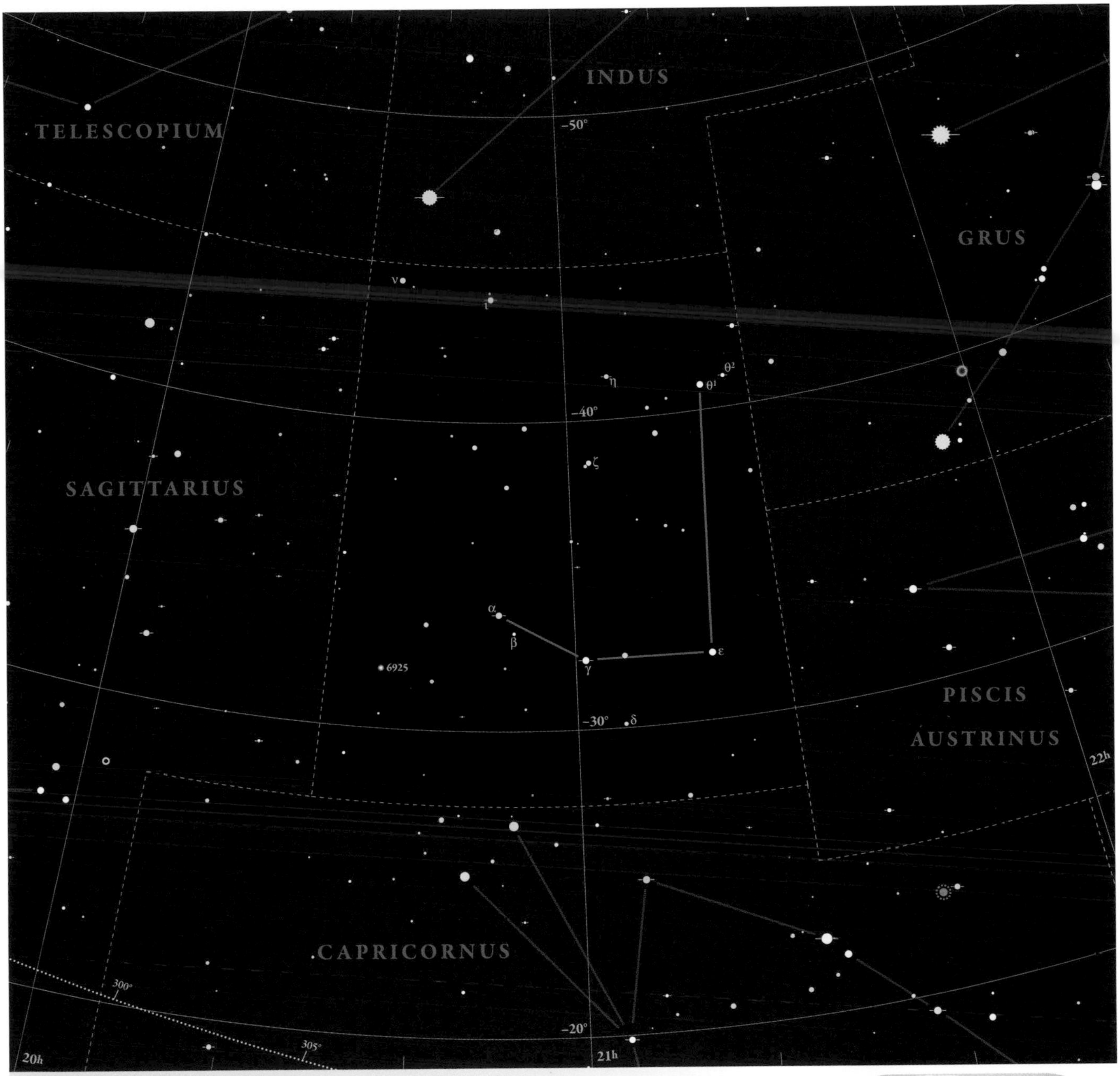

INDUS
TELESCOPIUM
GRUS
SAGITTARIUS
PISCIS AUSTRINUS
CAPRICORNUS
-50°
-40°
-30°
-20°
20h
21h
22h
300°
305°
6925
α β γ δ ε ζ η θ¹ θ² ι ν

FACT FILE

Microscopium
my-kro-SKO-pee-em
The Microscope

Abbreviation
Mic

Genitive
Microscopii

Right Ascension
21 hours

Declination
–35°

Visibility
44°N to 90°S

Notable Features
Alpha (α) Microscopii
NGC 6925

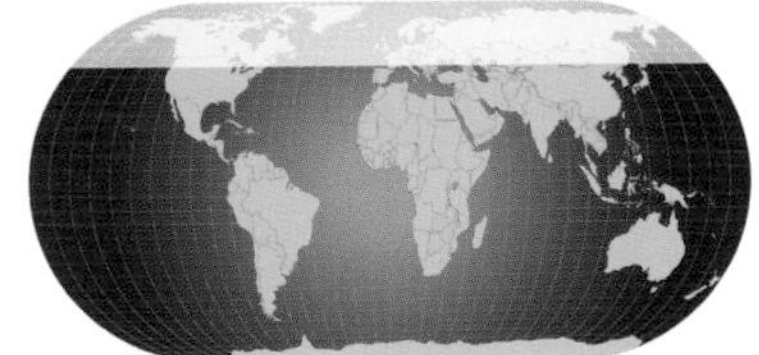

NORTHERN HEMISPHERE—as viewed from New York, USA, at 10 PM on the 15th of each month

JAN

FEB

MAR

APR

MAY

JUN
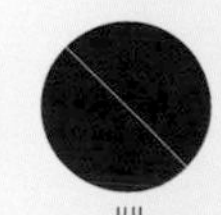
JUL

AUG

SEP

OCT

NOV
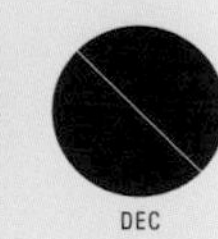
DEC

SOUTHERN HEMISPHERE—as viewed from Sydney, Australia, at 10 PM on the 15th of each month

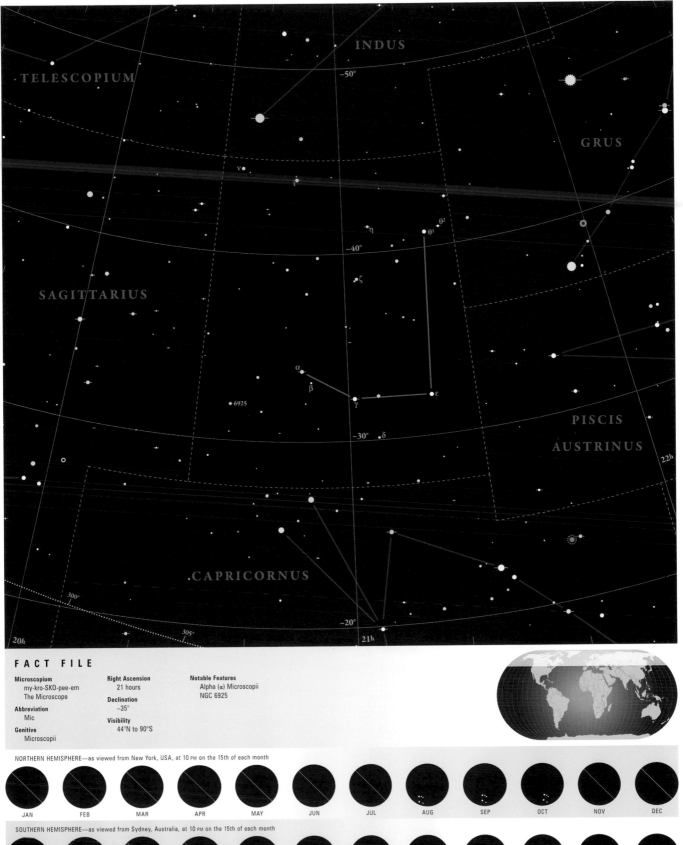

JAN

FEB
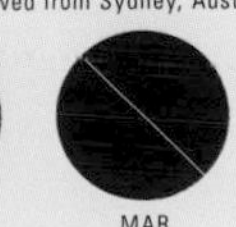
MAR
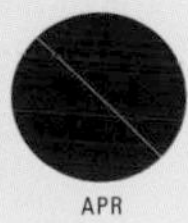
APR

MAY

JUN

JUL

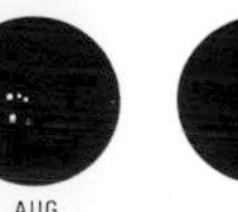
AUG
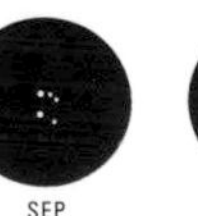
SEP

OCT

NOV

DEC

Monoceros

Hide-and-seek stars mark Monoceros, the shy unicorn. The constellation occupies quite a large area of the sky, but as no ancient cultures associated a figure with it, there is no mythology.

ONE-HORNED MYTH

Petrus Plancius is usually credited with the invention of Monoceros after he placed it in a star-globe in 1624. He probably took his inspiration from scripture, where a creature interpreted as a unicorn is occasionally mentioned. The figure was quickly adopted by other astronomers, and has remained in use until modern times.

THE STARS OF MONOCEROS

Although Monoceros flows directly through the Milky Way, it has no bright stars. Alpha (α) and Beta (β) Monocerotis are of mid-4th magnitude, and all other Bayer stars are 5th magnitude or fainter. Even in a dark sky it is impossible to see any resemblance to the fabled unicorn among the stars. The Milky Way background of Monoceros is much fainter and broader than most other areas, because we are looking outward toward the edge of our Galaxy.

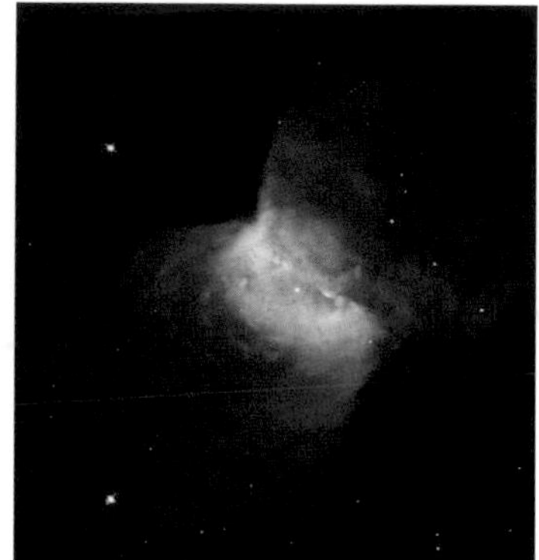

Right Emerging from the darkness like a giant moth, NGC 2346's butterfly-like shape represents the death throes of the binary star at the nebula's core.

Monoceros's brightest star is Beta (β), barely edging out Alpha for the title. One of the finest showpiece triple stars, it consists of two 5th and one 6th magnitude stars, with the two fainter stars close to each other, and the brightest orbiting that pair. All three are B-type stars, the most luminous of the three a giant, with the whole system 690 light-years distant. Near the famous Rosette Nebula is Plaskett's Star, one of the most massive binary star-systems known: Its components are O-type hypergiant stars, 6,600 light-years away, 51 and 43 times heavier than our Sun, and orbiting each other in just 14 days. Between them, they outshine the Sun more than a million-fold.

OBJECTS OF INTEREST

Monoceros abounds in open clusters and nebulae, and at least ten clusters are excellent viewing objects. The **Rosette Nebula**, or **NGC 2244** and **NGC 2237**, is the best of all; a 4-inch (10 cm) telescope shows a long rectangle formed by six stars surrounded by many fainter ones. Larger telescopes at ultra-low magnifications reveal a huge wreath of extremely faint nebulosity surrounding the group.

The **Christmas Tree Cluster**, or **NGC 2264**, is another bright favorite, the main stars having a slim, pine-tree-like outline. The famous **Cone Nebula** is in the same field, and sits just to the south of the cluster, but it is not really visible in amateur-class telescopes.

NGC 2232 is a triangular open cluster that is best seen at low powers, while **M50** is a medium-sized cloud of evenly distributed stars about half a moon in diameter.

The planetary nebula **NGC 2346** is a lovely, tiny, grayish-aqua disk, with a prominent central star that shows very well in an 8-inch (20 cm) telescope in a field of many stars.

Left The illuminating heart of the florid Rosette Nebula (NGC 2244) is actually the open star cluster known as NGC 2237. The ultraviolet light from the cluster causes the nebula's gas and dust to luminesce.

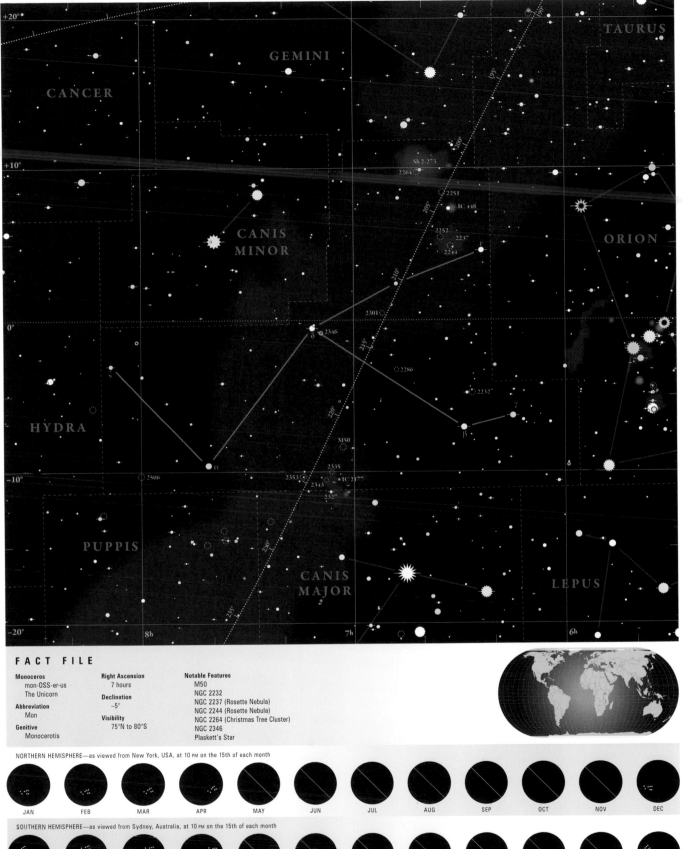

FACT FILE

Monoceros
mon-OSS-er-us
The Unicorn

Abbreviation
Mon

Genitive
Monocerotis

Right Ascension
7 hours

Declination
–5°

Visibility
75°N to 80°S

Notable Features
M50
NGC 2232
NGC 2237 (Rosette Nebula)
NGC 2244 (Rosette Nebula)
NGC 2264 (Christmas Tree Cluster)
NGC 2346
Plaskett's Star

NORTHERN HEMISPHERE—as viewed from New York, USA, at 10 PM on the 15th of each month

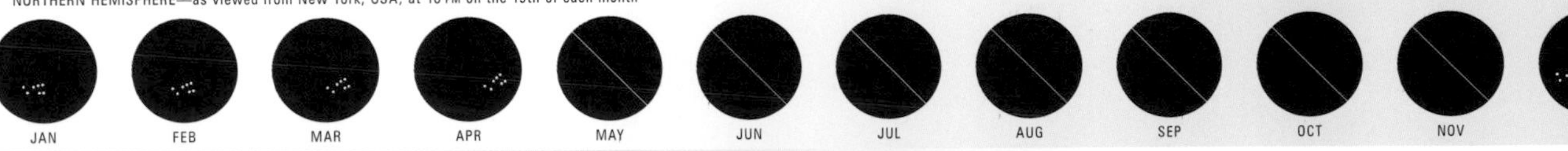

SOUTHERN HEMISPHERE—as viewed from Sydney, Australia, at 10 PM on the 15th of each month

Musca

Musca, the Fly, is a distinctive, easy-to-find group of stars to the south of Crux, the Cross, known to observers in the Southern hemisphere as the Southern Cross.

FLY-BEE-NIGHT

Early astronomers did not recognize the star group comprising Musca as a constellation because it lay too far south to be observed. The sixteenth-century Dutch explorers and cartographers Pieter Dirkszoon Keyser and Frederick de Houtman were the first to record its stars when they visited the East Indies, and the first to introduce it on a star-globe was Petrus Plancius in 1598. The stars were also recognized in Bayer's *Uranometria* published in 1603—not as a fly, but as Apis, the Bee. Over the course of the next couple of centuries, the constellation swapped identity several times.

Edmund Halley re-christened the pattern Musca-Apis—the Fly-Bee. In 1752, Abbé Nicolas Louis de Lacaille again recognized it as a fly, but in Bode's 1801 star-atlas, *Uranographia*, he restored it to Apis. However, by the end of the nineteenth century, Musca Australis—the Southern Fly—became the dominant appellation, and when the 88 modern recognized constellations were made permanent by the International Astronomical Union in 1929, it became known simply as Musca.

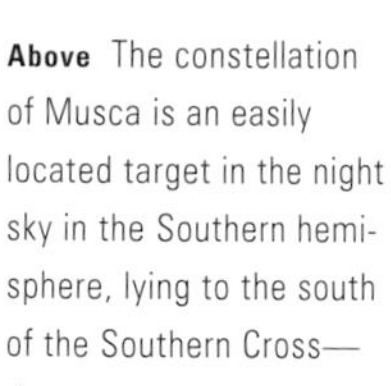

Above The constellation of Musca is an easily located target in the night sky in the Southern hemisphere, lying to the south of the Southern Cross—the constellation of Crux.

THE STARS OF MUSCA

Five 3rd and 4th magnitude stars compose an easily recognized "five-of-diamonds" asterism that makes up the shape of the fly. Four of the five stars in the group—Delta (δ) is the exception—all lie approximately 300 light-years away, although they are not physically related or gravitationally bound together in a cluster. Alpha (α) Muscae, in the center, is a hot B-class giant star, about eight times more massive than our Sun. In contrast, Epsilon (ε) is a cool red-giant star that, if placed in our Solar System, would nearly engulf the planet Venus.

OBJECTS OF INTEREST

Musca is home to two bright globular clusters: **NGC 4372** and **NGC 4833**, which are located at 19,000 and 21,000 light-years distant, respectively. NGC 4833 is located less than a degree north of magnitude 3.6 Delta (δ) Muscae, and in an 8-inch (20 cm) telescope is resolved into tiny stars with a hazy background. NGC 4372 is a slightly smaller and fainter cluster that looks very similar. Both show only mild compression toward their cores.

Beta (β) Muscae is a beautiful binary star that a 6-inch (15 cm) telescope will resolve into two close, slightly uneven bluish-white points. Planetary nebulae abound within Musca, but the best of them all is **NGC 5189**. This strange-looking object does not at first resemble a planetary, but an 8-inch (20 cm) telescope shows a fairly bright, bluish, lumpy-looking cloud of gas, dominated by a brighter streak through the center. Southeast of NGC 5189 (by 1½ degrees) is the famous planetary nebula **MyCn18**. Visible in an 8-inch (20 cm) telescope only as a faint star-like object, it is one of the most well-known and popular images captured by the Hubble Space Telescope.

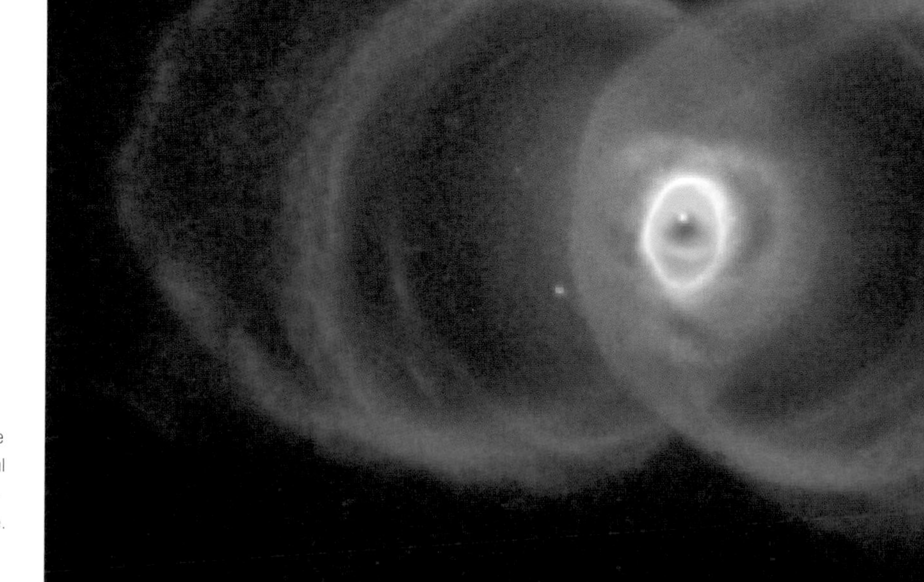

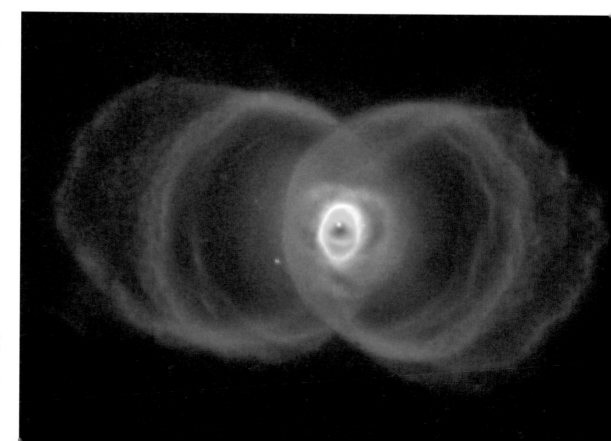

Right The central star of planetary nebula MyCn18 is in stellar demise. With its nuclear fuel depleted and its outer layers expelled, its core is now cooling to become a white dwarf. The rings of colorful glowing gas mark the distinctive "hourglass" shape.

FACT FILE

Musca
MUSS-kah
The Fly

Abbreviation
Mus

Genitive
Muscae

Right Ascension
12 hours

Declination
–70°

Visibility
13°N to 90°S

Notable Features
Beta (β) Muscae
IC 4191
MyCn18
NGC 4372
NGC 4833
NGC 5189

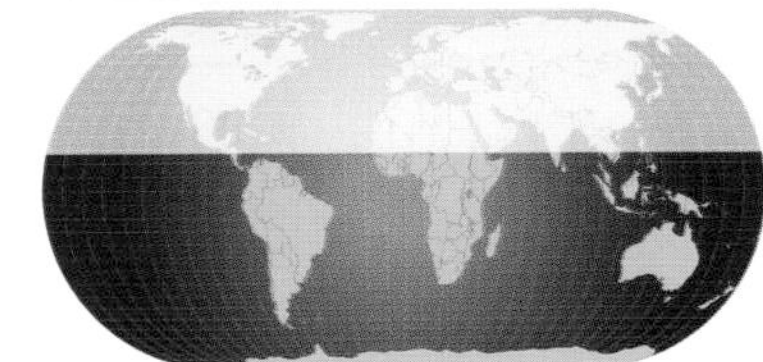

NORTHERN HEMISPHERE—as viewed from New York, USA, at 10 PM on the 15th of each month

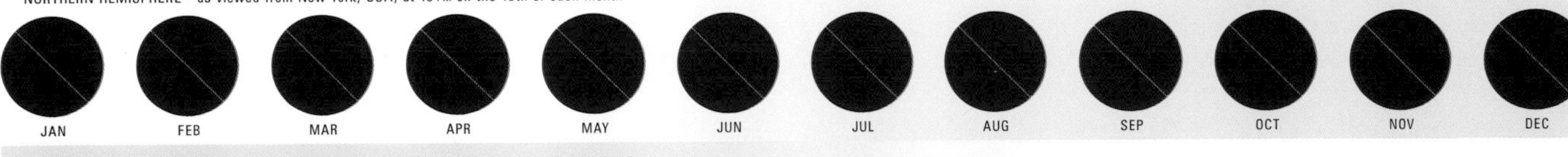

SOUTHERN HEMISPHERE—as viewed from Sydney, Australia, at 10 PM on the 15th of each month

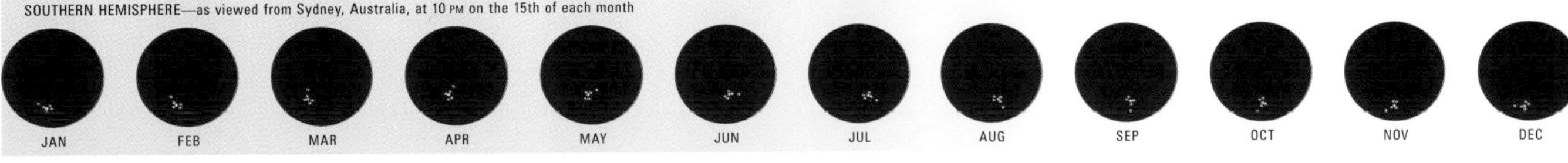

Norma

The magnificent backdrop provided by the Milky Way makes Norma, the Square, a fascinating star-scape.

GETTING IT STRAIGHT

Wedged between Lupus, Ara, and Scorpius, the stars of Norma have no mythology attached to them. The constellation has occasionally been called Triangulum Australe (no relation to the genuine Triangulum Australe, located further south), and Quadrans Euclidis, or Euclid's Square.

Abbé Nicolas Louis de Lacaille bestowed the final name change, Norma et Regula—the Square and the Rule—while at the Cape Observatory between 1751 and 1752. To form the constellation, Lacaille borrowed two stars from adjoining Scorpius, which subsequently became Alpha (α) and Beta (β) Normae. These stars have since been returned to Scorpius, and Norma's brightest star is now Gamma (γ). Over the subsequent years its name has shrunk to Norma.

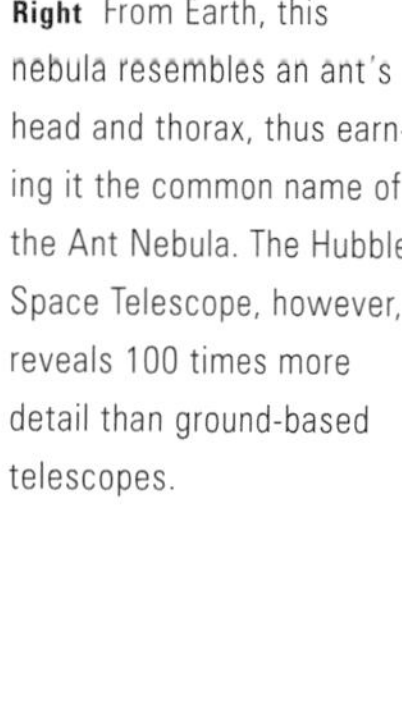

Right From Earth, this nebula resembles an ant's head and thorax, thus earning it the common name of the Ant Nebula. The Hubble Space Telescope, however, reveals 100 times more detail than ground-based telescopes.

THE STARS OF NORMA

The main stars of Norma, 4th magnitude Gamma² (γ^2) and 5th magnitude Gamma¹ (γ^1), along with Epsilon (ε) and Eta (η), make up a simple right-angled triangle amid many faint stars further in the background. To the naked eye, the Milky Way background throughout the Norma starcloud is a little different than it is elsewhere; it looks "gritty" by comparison. There are several naked-eye brighter knots, and a binocular sweep shows a blotchy, almost marbled texture.

Having to make do without an Alpha star, Gamma² Normae is the lucida. A yellow giant star 128 light-years distant, it has twice the Sun's mass and 45 times its luminosity. In contrast, its apparent companion, the slightly fainter Gamma¹, is an F-class supergiant nearly 1,500 light-years away.

OBJECTS OF INTEREST

NGC 6087 is the brightest open star cluster in Norma. It is visible to the unaided eye as a small milky knot, and binoculars will resolve a few faint stars within the strong haze. A 6-inch (15 cm) telescope resolves it into a thicket of about 30 stars. Only slightly fainter is the similar looking **NGC 6067**, also spectacular in a field resplendent with a thick carpeting of faint stars. Planetary nebulae abound in Norma, but most are either faint or small. The best is **SP-3**, which in an 8-inch (20 cm) telescope looks like a hazy, faint, donut-shaped ring of translucent haze.

NGC 6164-65 is one of the rarest types of nebulae in the sky—a wind-blown bipolar shell. Only two nebulae of this type are bright enough to be seen in amateur telescopes; the other is the Bubble Nebula NGC 7635 in Cassiopeia. NGC 6164–65 surrounds a brilliant O7 class supergiant star weighing in at about 40 solar-masses, which is releasing violent high-speed winds that are ionizing and sculpting the gas into a symmetrical S-shaped form. An 8-inch (20 cm) telescope used under a dark rural sky will barely show this rare sight.

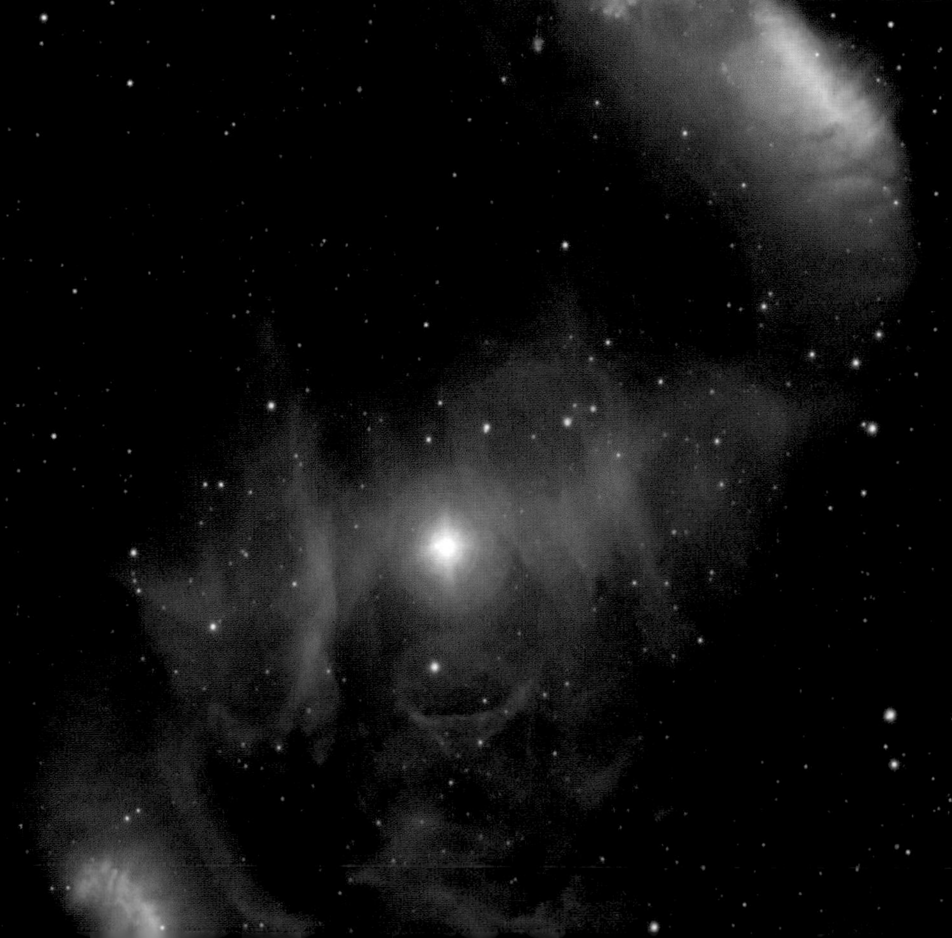

Left In the middle of the emission nebula NGC 6164–65 is an unusually massive star nearing the end of its life. The star, visible in the center of the image, is so hot that the ultraviolet light it emits heats up the gas that surrounds the star and blasts it into this twisted S-shape.

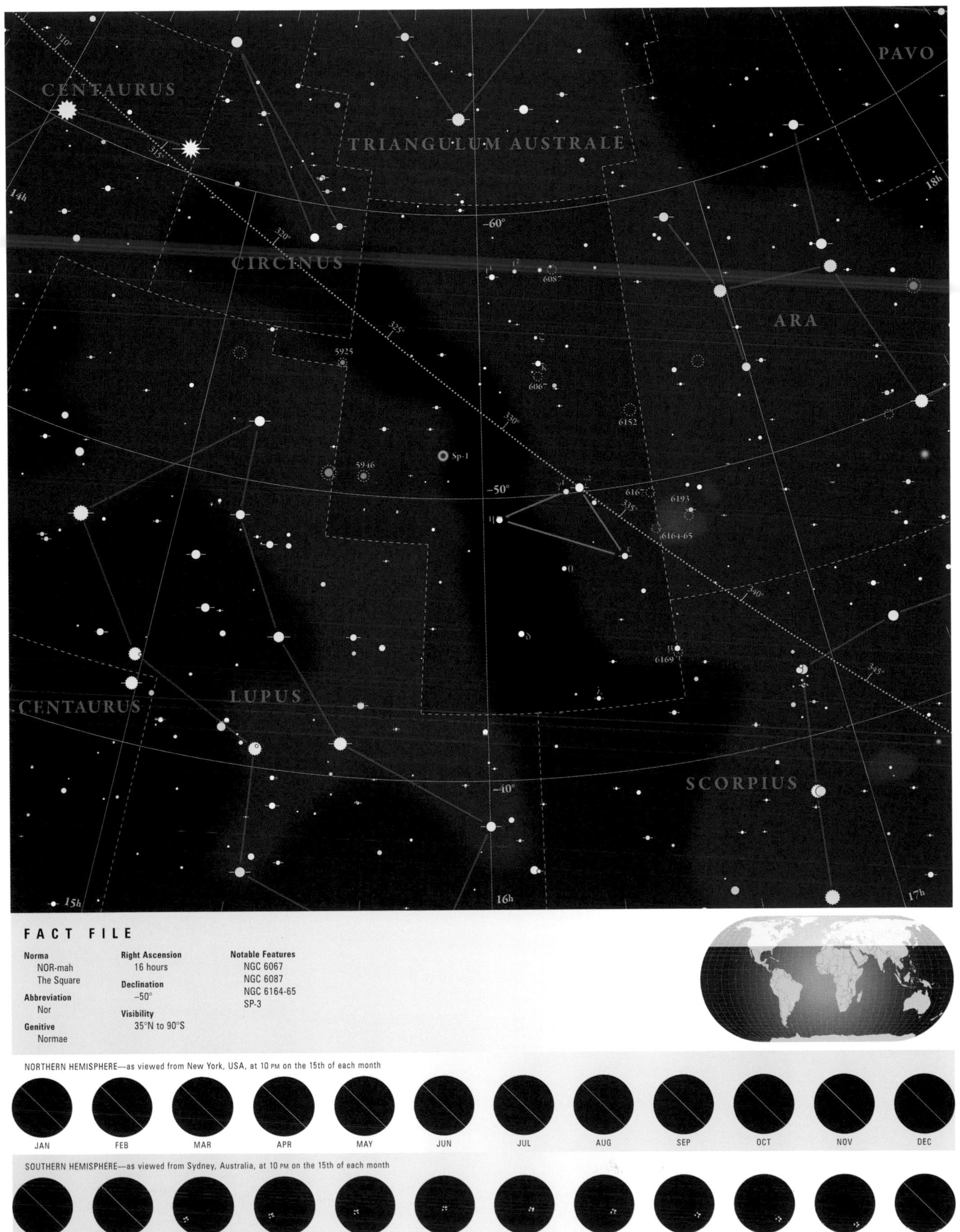

FACT FILE

Norma
NOR-mah
The Square

Abbreviation
Nor

Genitive
Normae

Right Ascension
16 hours

Declination
–50°

Visibility
35°N to 90°S

Notable Features
NGC 6067
NGC 6087
NGC 6164-65
SP-3

NORTHERN HEMISPHERE—as viewed from New York, USA, at 10 PM on the 15th of each month

JAN FEB MAR APR MAY JUN JUL AUG SEP OCT NOV DEC

SOUTHERN HEMISPHERE—as viewed from Sydney, Australia, at 10 PM on the 15th of each month

JAN FEB MAR APR MAY JUN JUL AUG SEP OCT NOV DEC

Octans

Dim and obscure, Octans is home to the anchor of the southern sky—the South Celestial Pole.

NAVIGATING THE STARS

Octans was unknown to ancient astronomers; Abbé Nicolas Louis de Lacaille devised the constellation and called it Octans Hadleianus during his visit to the Cape Observatory between 1751 and 1752. Although the South Celestial Pole resides there, it is difficult to see why the dim stars of Octans are collected into a constellation of their own. In fact, Lacaille had to steal two stars from the adjoining constellation of Hydrus, now Nu (ν) and Beta (β) Octantis (plus another fainter one), to complete the figure.

Like many constellations invented by Lacaille, Octans represents an instrument of science or engineering—in this case the astronomical and navigational octant, invented by John Hadley in 1730, and the forerunner of the sextant. There has been no significant change in its stars or form, save that its name has been worn down to simple Octans.

THE STARS OF OCTANS

The principal stars of Octans form a long scalene triangle beside the South Celestial Pole. Its lucida, rather surprisingly, is 4th magnitude Nu (ν) Octantis, which beats not only Alpha but also Beta to the title.

If one were looking for a snapshot of what our own Sun will look like in about seven billion years, Nu Octantis fits the bill exactly. It has an almost identical mass to the Sun, and probably spent most of its life looking much like our own luminary. About 12 billion years old, it has consumed its hydrogen supply. At present a slightly swollen and cooler K-type sub-giant, it is preparing to commence helium fusion in a hundred million years or so, and will become a full-blown red giant star some 60 times brighter again, before dying as a white-dwarf.

While the northern sky has bright Polaris to mark the pole, the nearest star to the South Celestial Pole is unassuming 5th magnitude Sigma (σ) Octantis.

Although they are poles apart in brightness (and distance), these two stars share a common trait: Both are pulsating F-class variable stars. While Polaris is a Cepheid variable, Sigma is of the Delta Scuti type, and varies by less than one-tenth of a magnitude over just a couple of hours.

OBJECTS OF INTEREST

The sky of Octans, like that of some of its neighboring constellations, is a virtual desert, and there is little to attract the amateur observer. **Collinder 411** is a sprawling open cluster of 7th to 10th magnitude stars, spread over more than one degree, and is best appreciated with large binoculars or a wide-field telescope at low power.

There are many faint and small galaxies in Octans. The only interesting thing about **NGC 2573** (Polarissima Australis) is that it is the closest galaxy to the South Celestial Pole, at less than 1° away.

Right This image shows the sky's rotation as stars arc around the South Pole, with Sigma (σ) Octans near the center of the rotation. Sigma (σ) Octans is off the pole by about a degree.

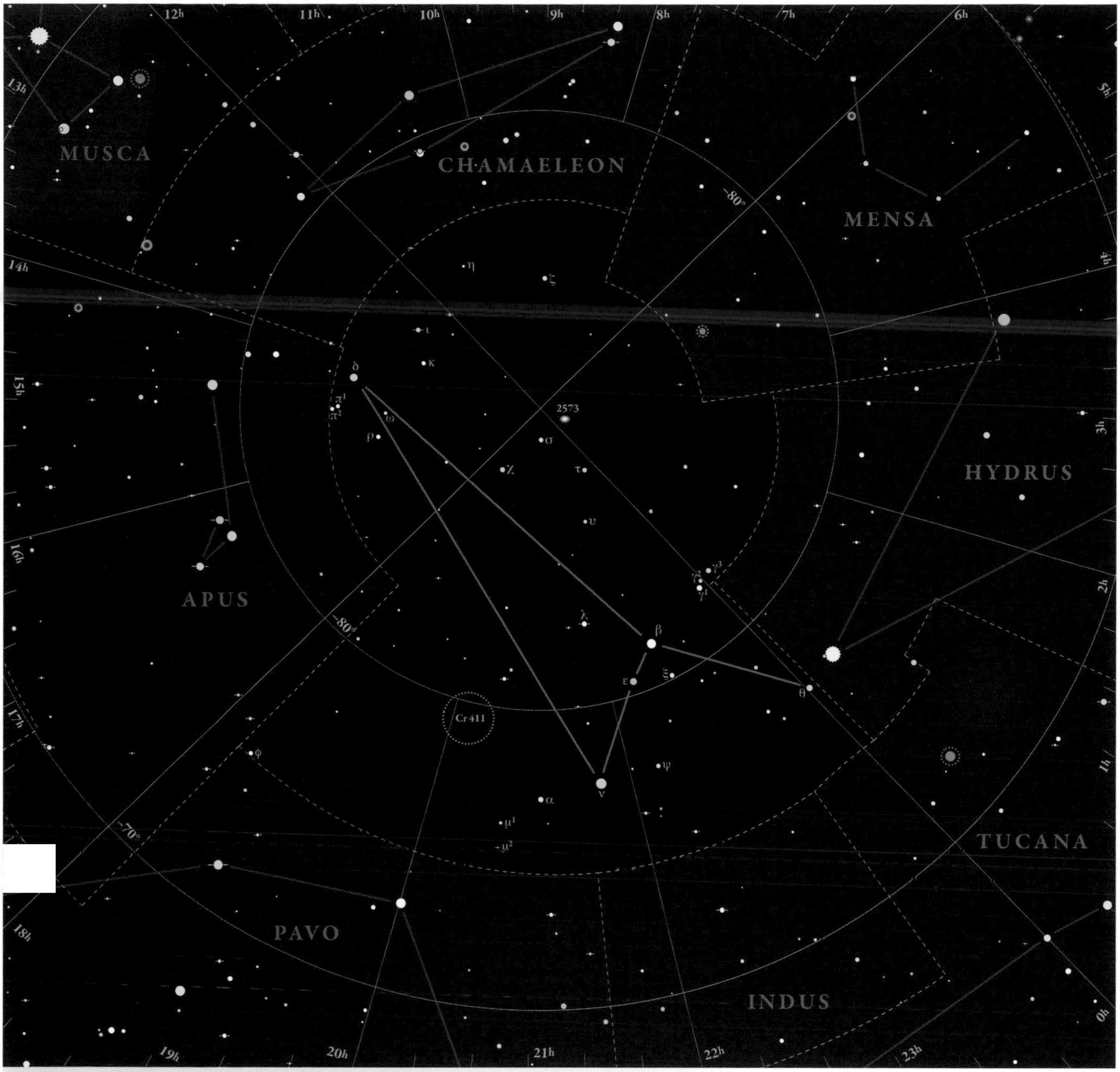

FACT FILE

Octans
OCK-tanz
The Octant

Abbreviation
Oct

Genitive
Octantis

Right Ascension
22 hours

Declination
–85°

Visibility
1°N to 90°S

Notable Features
Collinder 411 (Cr411)
NGC 2573

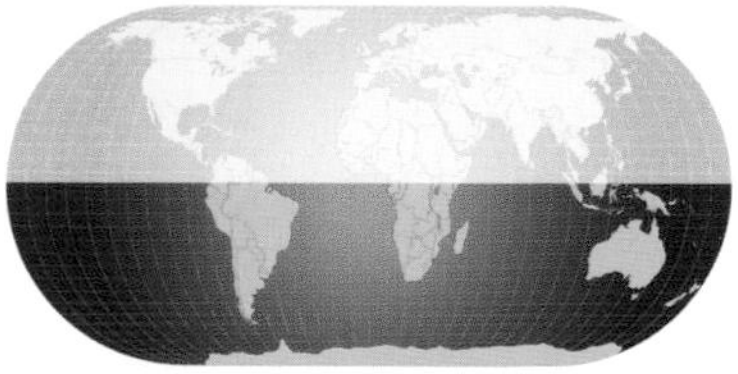

NORTHERN HEMISPHERE—as viewed from New York, USA, at 10 PM on the 15th of each month

JAN FEB MAR APR MAY JUN JUL AUG SEP OCT NOV DEC

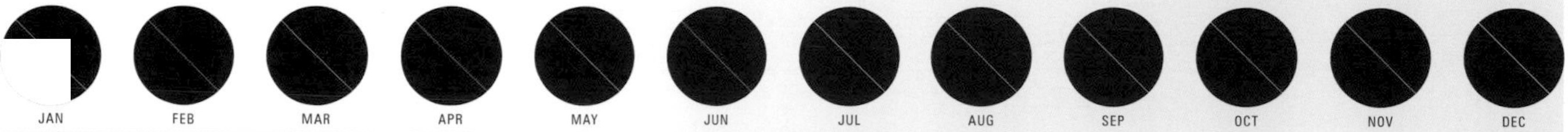

SOUTHERN HEMISPHERE—as viewed from Sydney, Australia, at 10 PM on the 15th of each month

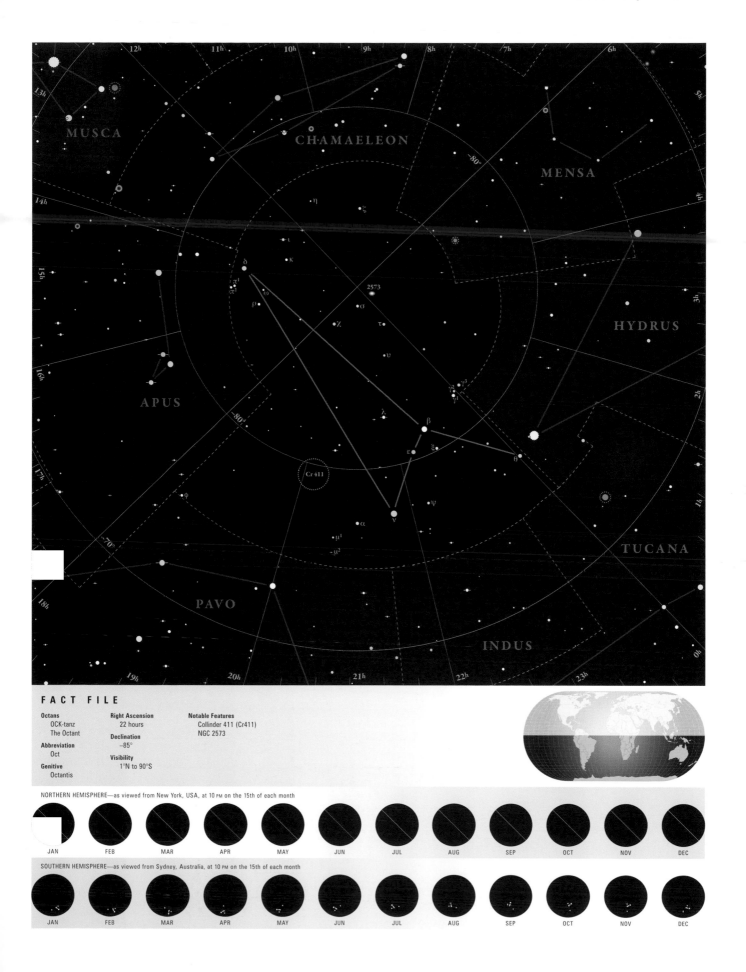

JAN FEB MAR APR MAY JUN JUL AUG SEP OCT NOV DEC

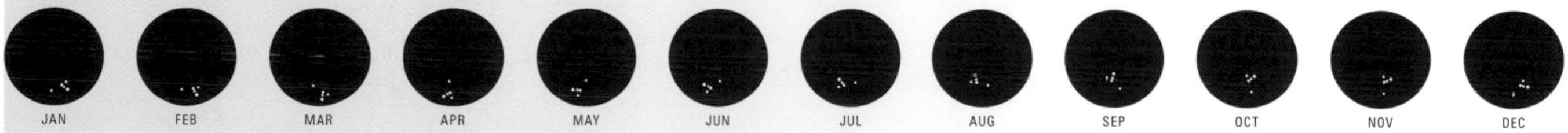

Ophiuchus

The stars of Ophiuchus represent the Serpent Bearer. This large Southern-hemisphere constellation is located near the center of the Milky Way.

HEALER OR SNAKE-OIL MERCHANT?

Ophiuchus literally means "the serpent bearer" and represents Asclepius, the son of the god Apollo and the nymph Coronis. Asclepius was raised by the centaur Chiron who taught him how to heal. According to legend, Asclepius once strangled a snake, but another snake slid up and administered a herb, thus resurrecting it. Before the second snake could escape, Asclepius snatched the herb out of its mouth and used it to revive the dead.

Eventually Asclepius became surgeon on the great ship *Argo* (see Carina) and revived several dead individuals, including Minos, the king of Crete. After Asclepius tried to revive Orion, Hades complained to Zeus that he would soon be left with no one in the Underworld. Zeus agreed, and killed Asclepius by striking him with a thunderbolt and placed him among the stars.

THE STARS OF OPHIUCHUS

Ophiuchus is a huge constellation with two 2nd magnitude stars, Alpha (α) and Eta (η), along with several of 3rd magnitude requiring much imagination to see as an older man holding a serpent across his body. Alpha (α) Ophiuchi, otherwise known as Rasalhague, meaning the "head of the serpent bearer," is only 47 light-years distant. An A-class giant star twice as massive as our Sun and 25 times brighter, it has a tiny faint companion that is too close in to be visible in amateur telescopes.

A small telescope is needed to see the most famous star in Ophiuchus—Barnard's Star. A 10th magnitude mote, Barnard's Star is an M-class red-dwarf star with only four thousandths of the Sun's luminosity, but is the second closest star system to our Sun, at just 6 light-years away. Interesting to astronomers, it has the highest proper motion of any star in the sky, racing across half a moon-diameter in one human lifetime.

OBJECTS OF INTEREST

Ophiuchus lies along the northern edge of the Milky Way. This constellation contains many interesting objects for the amateur sky-watcher, and is especially noted for its numerous bright globular clusters. The best in Ophiuchus are **M10** and **M12**, which are both excellent examples and can be seen in the same binocular field. Both are about one-third of a moon-diameter across, and in an 8-inch (20 cm) telescope are partly resolved into many stars with a hazy background. **M14** is also a large, weakly concentrated globular, but is not easy to resolve in small telescopes because of thick clouds of gas and dust between the cluster and our Sun. **M62**, near Antares in Scorpius, is another excellent cluster, barely below naked-eye visibility; an 8-inch (20 cm) telescope shows a large, evenly concentrated, well-resolved halo with hundreds of stars. **M19**'s peculiarity is its oval halo containing many resolved stars, while **M9** and **M107** are also bright, easily observed clusters.

An 8-inch (20 cm) telescope also provides a good view of **NGC 6369**, known as the **Little Ghost Nebula**. It is a lovely example of the annular form planetary, looking like a small, round, thick, grayish donut.

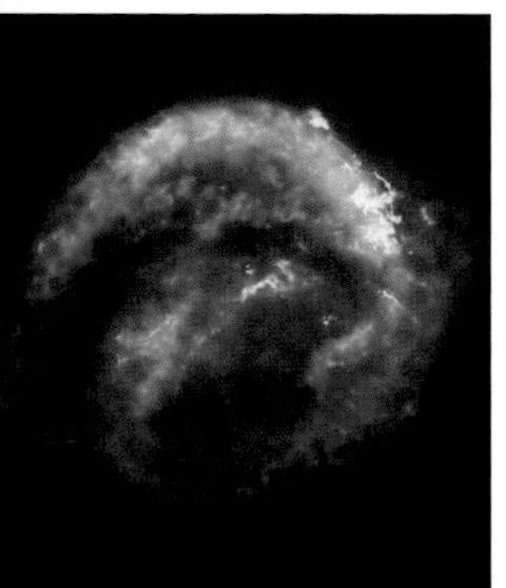

Above First seen by Johannes Kepler and other sky-watchers 400 years ago, this bubble of gas and dust is Kepler's supernova remnant found in the constellation of Ophiuchus.

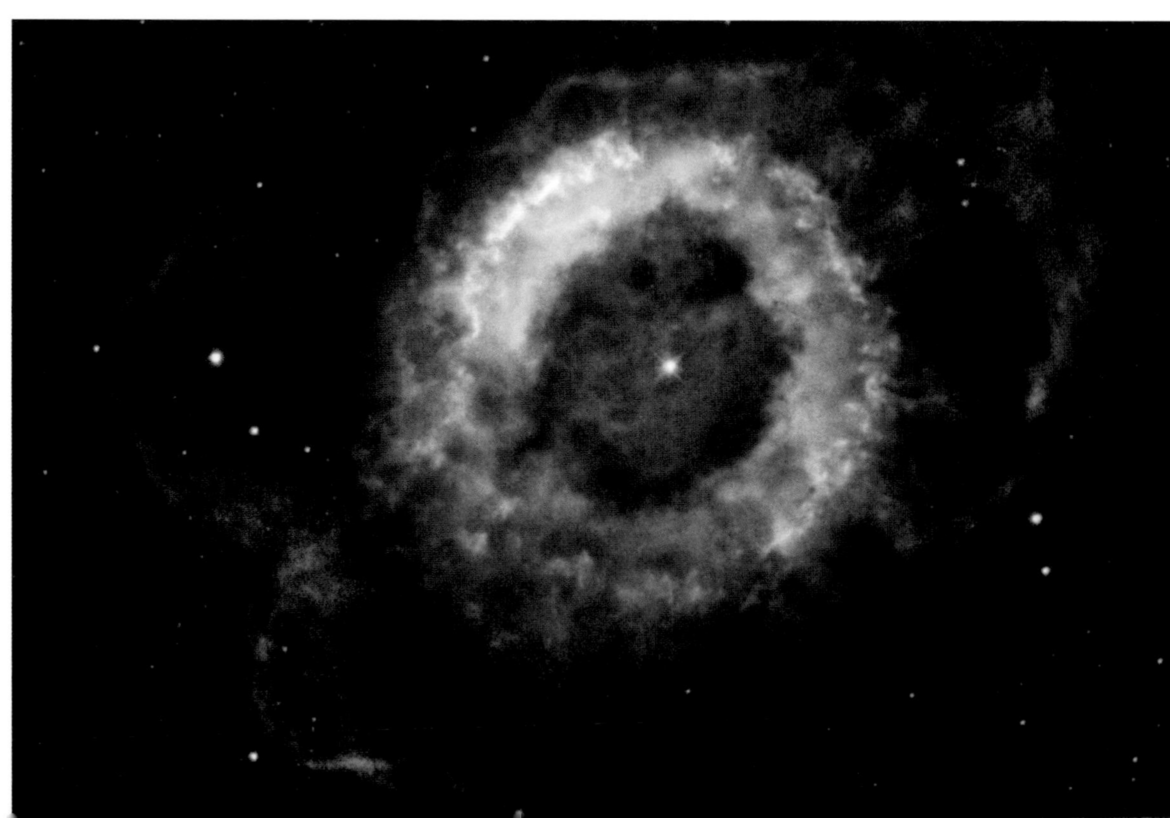

Right The ethereal Little Ghost Nebula (NGC 6369) affords us a glimpse of our own Sun's fate about five billion years from now. The blue–green ring, a light-year in diameter, marks the area where ultraviolet light has ionized oxygen atoms.

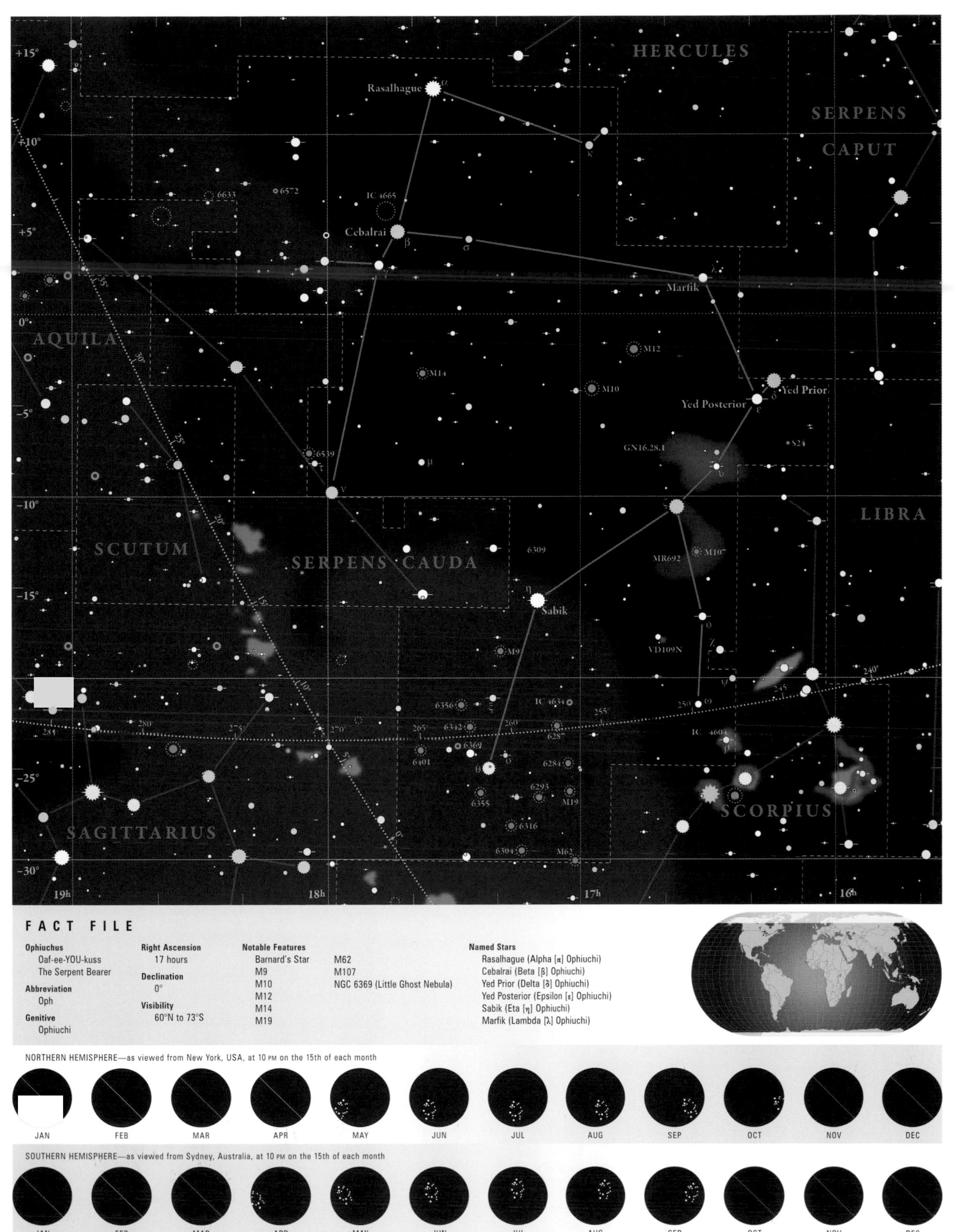

FACT FILE

Ophiuchus
Oaf-ee-YOU-kuss
The Serpent Bearer

Abbreviation
Oph

Genitive
Ophiuchi

Right Ascension
17 hours

Declination
0°

Visibility
60°N to 73°S

Notable Features
Barnard's Star
M9
M10
M12
M14
M19
M62
M107
NGC 6369 (Little Ghost Nebula)

Named Stars
Rasalhague (Alpha [α] Ophiuchi)
Cebalrai (Beta [β] Ophiuchi)
Yed Prior (Delta [δ] Ophiuchi)
Yed Posterior (Epsilon [ε] Ophiuchi)
Sabik (Eta [η] Ophiuchi)
Marfik (Lambda [λ] Ophiuchi)

Orion

Standing beside the Milky Way, Orion is one of the most prominent and best known constellations in the sky.

HUNTING THE HEAVENS

Orion has been widely recognized across the world as a distinctive group of stars, sitting astride the celestial equator. To the Chaldeans, these stars were known as Tammuz. To the Syrians, they traced out the giant Al Jabbar. The Egyptians saw Sahu, the soul of Osiris and, it has been suggested, aligned their massive burial pyramids to the positions of these stars in the sky.

The description of Orion as a giant and a hunter comes from Greek mythology. In one version of the myth, Artemis, the goddess of the Moon and the hunt, fell in love with him. Her twin brother Apollo challenged his sister to fire an arrow at a distant figure swimming in the sea. Unknown to her, the swimmer was Orion. Her arrow killed him and only then did she realize what she had done. Her grief is said to explain the cold light of the Moon. She placed his body among the stars, along with his hunting dogs, Canis Major and Canis Minor, that lie to Orion's east in the sky.

THE STARS OF ORION

The constellation's bright stars trace out a rectangle, crossed by the three stars of the belt. Outlying stars are often depicted to mark his upraised club, with the other outstretched arm carrying a shield or animal skin.

On the northeast corner of the rectangle is Betelgeuse, Alpha (α) Orionis, a red supergiant star that would encompass the orbit of Mars if it replaced the Sun in our Solar System. It appears orange rather than red to the eye. In stark contrast, the opposite corner of the rectangle is marked by the blue-white supergiant Rigel—a smaller star but several times brighter.

Above The surface of Alpha (α) Orionis (Betelgeuse), a red supergiant star, has a huge bright spot that is ten times bigger than the diameter of Earth.

Right This infrared image of the Trapezium in the famous Orion Nebula successfully spotlights the stars of the formation.

OBJECTS OF INTEREST

Orion's most famous feature is **M42 (NGC 1976)**, the **Great Nebula** in Orion. It lies amid the stars of Orion's sword, hanging from his belt. It is a glowing emission nebula marking the current location of star formation in the Orion giant molecular cloud, a star-forming factory that covers a large part of the sky within the boundaries of the constellation, but lying around 1,500 light-years away. M42 is readily visible to the unaided eye, but viewed with binoculars or a telescope it is one of the finest sights in the night sky. The larger the instrument used to view the glowing gas, the more detail will become apparent. The vibrant red color so apparent in many photographs of the Orion Nebula is too faint to be seen visually in a telescope, but many observers report a pale green tint in the view through telescopes larger than about 8 inches (20 cm). Four hot young stars in the heart of the nebula power its glow. These stars are known as the **Trapezium**, although many photographs "burn out" the central portion of the nebula and hide this formation. A newer generation of stars is lurking nearby, visible only by their infrared glow penetrating the dusty cocoon of their birth.

Just to the northeast of the Trapezium is another diffuse nebula, **M43 (NGC 1982)**. This almost circular patch of nebulosity seems connected to M42, although it is actually distinct and lit by a different star.

Another well-known nebula lurking among the gas and dust of Orion is the **Horsehead Nebula (IC 434)**, also known as **Barnard 33**. The distinctive horsehead shape is due to a dark cloud of dust superimposed on a bright background of emission nebula. Although spectacular in photographs, it is a very difficult visual target. Elsewhere in Orion, small knots of bright and dark nebulae can be tracked down by diligent observers.

Compared to the glories of Orion's nebulae, the rest of the constellation is a little lacking in exciting telescopic targets. Several of Orion's stars such as **Zeta (ζ) Orionis**, one of the belt stars, are double stars.

NGC 1981 is one of only a few star clusters within the constellation boundaries.

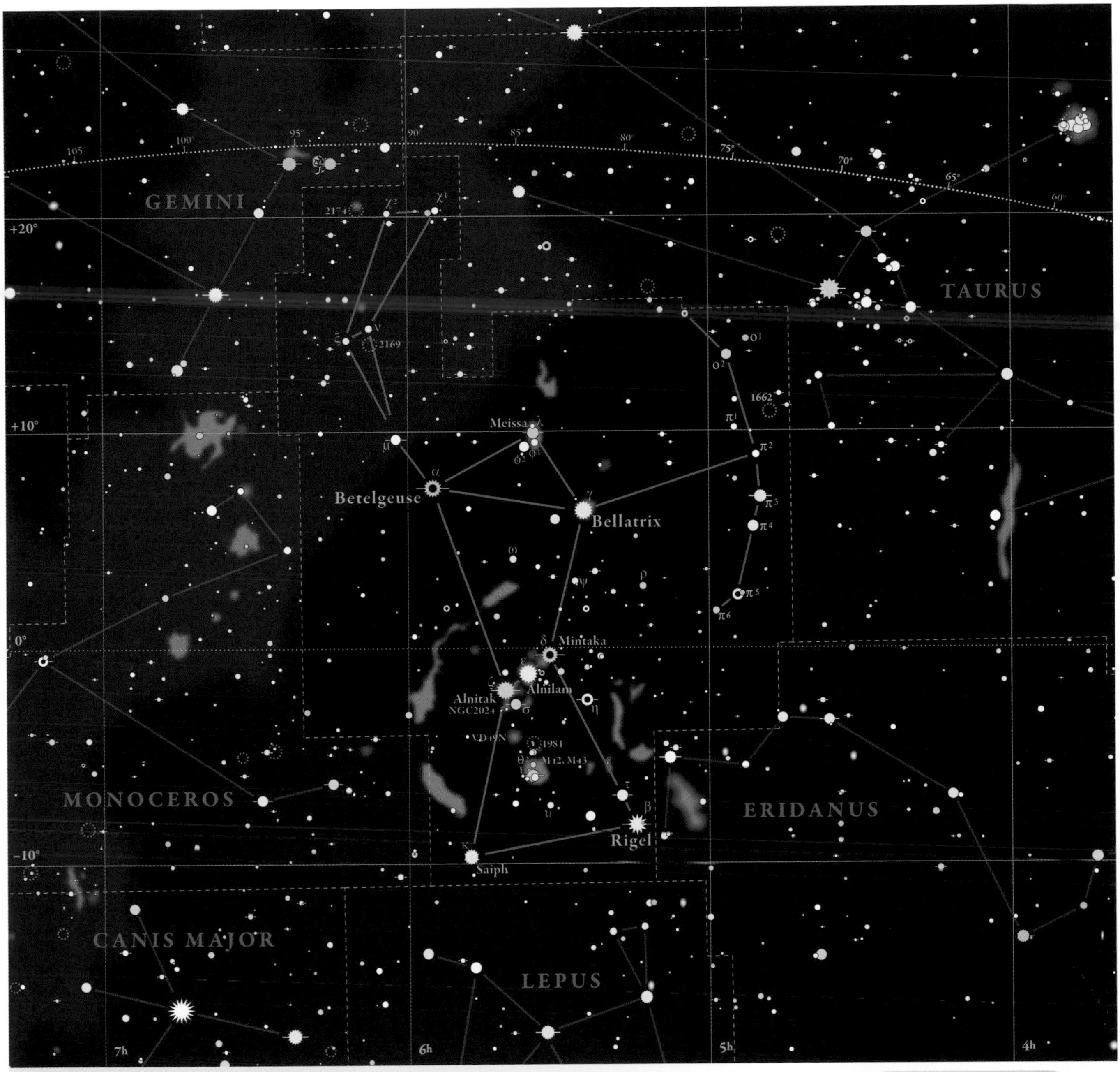

FACT FILE

Orion
Oh-RYE-un
The Hunter

Abbreviation
Ori

Genitive
Orionis

Right Ascension
5 hours

Declination
+5°

Visibility
75°N to 65°S

Notable Features
Alpha (α) Orionis
IC 434 (Horsehead Nebula, Barnard 33)
M42 (Great Orion)
NGC 1981
Trapezium
Zeta (ζ) Orionis

Named Stars
Betelgeuse (Alpha [α] Orionis)
Rigel (Beta [β] Orionis)
Bellatrix (Gamma [γ] Orionis)
Mintaka (Delta [δ] Orionis)
Alnilam (Epsilon [ε] Orionis)
Alnitak (Zeta [ζ] Orionis)
Saiph (Kappa [κ] Orionis)
Meissa (Lambda [λ] Orionis)

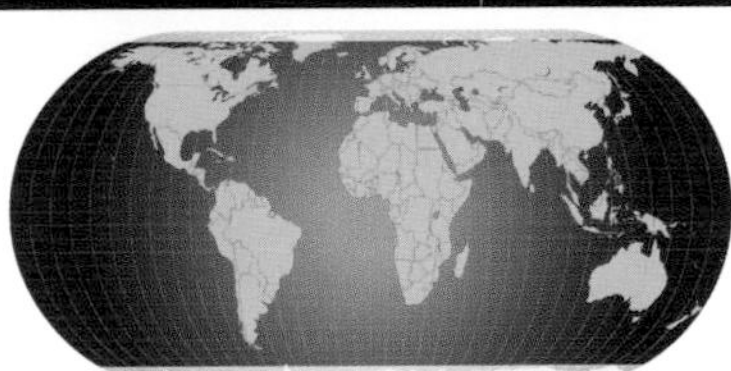

NORTHERN HEMISPHERE—as viewed from New York, USA, at 10 PM on the 15th of each month

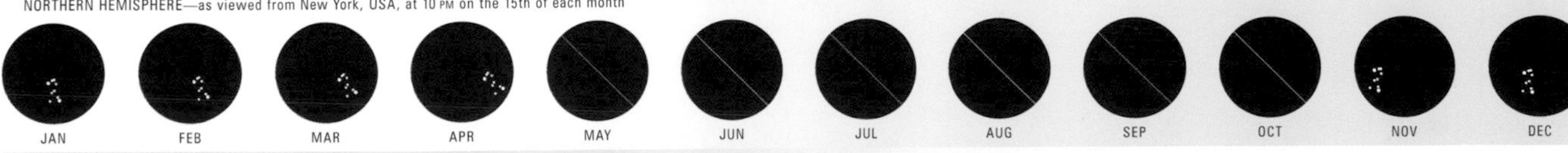

SOUTHERN HEMISPHERE—as viewed from Sydney, Australia, at 10 PM on the 15th of each month

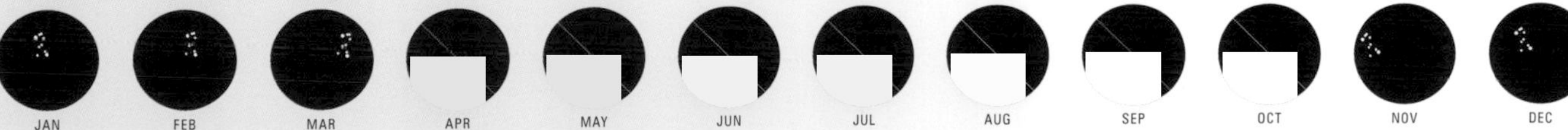

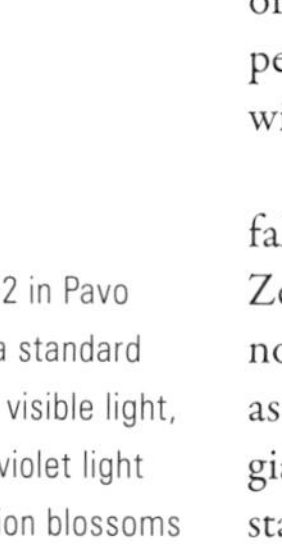

Pavo

The abundant faint stars of Pavo—the Peacock—reflect the legend of its magnificent tail.

AN EYE FOR AN EYE

First introduced onto a star-globe by Petrus Plancius in 1598, the stars of Pavo were methodically observed for the first time by Pieter Dirkszoon Keyser and Frederick de Houtman during their explorations to map the southern skies. Bayer followed the lead of Plancius in 1603, drawing the constellation as a peacock. However, Plancius and Bayer's inspiration might also have originated from the mythological peacock that was sacred to Zeus's wife, Hera (Juno).

Hera suspected that Zeus had fallen in love with the nymph Io. To deceive his wife, Zeus disguised Io as a heifer to hide her, but Hera was not easily fooled and demanded the heifer be given to her as a gift. Hera then asked one of her servants, an all-seeing giant with one hundred eyes called Argus Panoptes, to stand guard over it. When Zeus discovered his darling was guarded, he sent his son Hermes to slay Argus. Upon learning of Argus's death, Hera spread his manifold eyes over the tail of the peacock.

Right NGC 6782 in Pavo appears to be a standard spiral galaxy in visible light, but under ultraviolet light the central region blossoms into a beautiful and intricate structure with a bright ring surrounding the nucleus.

THE STARS OF PAVO

Alpha (α) Pavonis is also named "Peacock," and because this is an English name, it would seem to be a recent invention. Peacock is a magnitude 2 star and the forty-fifth brightest in the sky, residing 180 light years away. A blue sub-giant star about five times heavier than our Sun, it is over 200 times more luminous.

Alpha, together with the right-angled triangle of stars made up by 3rd and 4th magnitude Beta (β), Delta (δ), and Epsilon (ε) Pavonis directly to its south, is the most recognizable asterism in the constellation. Seen together with the many other faint stars, it is still difficult to imagine them making a peacock's outline.

OBJECTS OF INTEREST

Lying well off the plane of the Milky Way, the constellation of Pavo boasts just one star cluster—the magnificent globular **NGC 6752**. Nearly a moon-diameter across and barely visible to the naked eye, it is one of the largest and brightest globular clusters in the sky. A 6-inch (15 cm) telescope provides a magnificent view of its numerous faint stars, arranged in tentacles set amid a somewhat spartan background.

Quite close to NGC 6752 is **NGC 6744**. To get the best out of this large spiral galaxy, view it under a dark rural sky because its light is spread over a large area. An 8-inch (20 cm) telescope will show a sizable, faint-looking, milky halo with a brighter core.

Pavo has many other faint galaxies to offer, of which the long thin halo of **IC 5052** is the most interesting; an 8-inch (20 cm) telescope is the minimum required to show it well.

Left The face-on barred spiral galaxy NGC 6744 is similar in overall size, shape, and appearance to our own Milky Way. It lies approximately 25 million light-years away in the constellation of Pavo and contains more than a hundred thousand million stars.

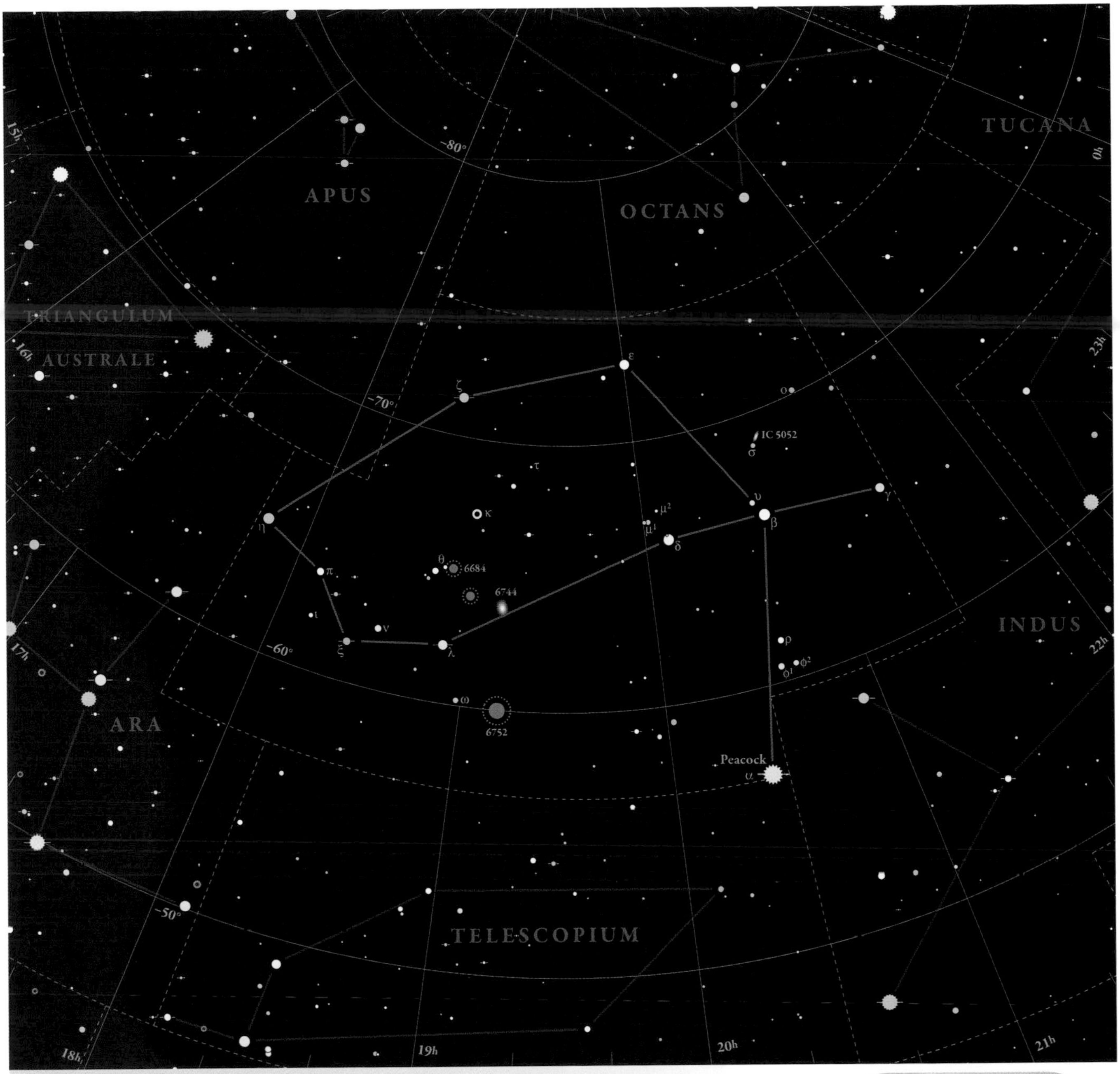

FACT FILE

Pavo
PAH-vo
The Peacock

Abbreviation
Pav

Genitive
Pavonis

Right Ascension
20 hours

Declination
−65°

Visibility
13°N to 90°S

Notable Features
Alpha (α) Pavonis
IC 5052
NGC 6744
NGC 6752

Named Stars
Peacock (Alpha [α] Pavonis)

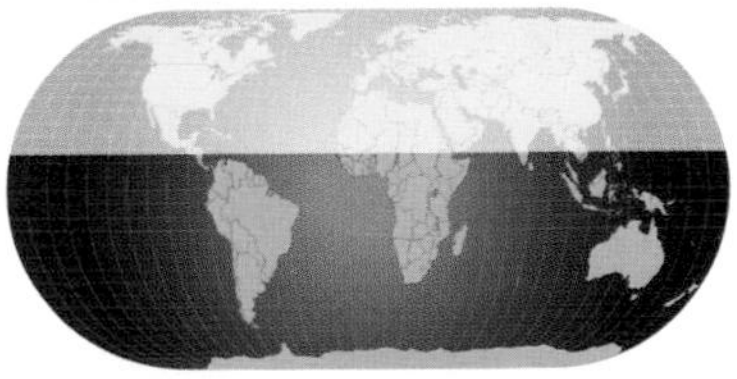

NORTHERN HEMISPHERE—as viewed from New York, USA, at 10 PM on the 15th of each month

JAN FEB MAR APR MAY JUN JUL AUG SEP OCT NOV DEC

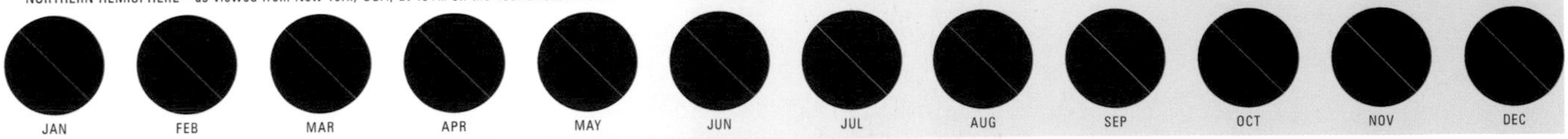

SOUTHERN HEMISPHERE—as viewed from Sydney, Australia, at 10 PM on the 15th of each month

JAN FEB MAR APR MAY JUN JUL AUG SEP OCT NOV DEC

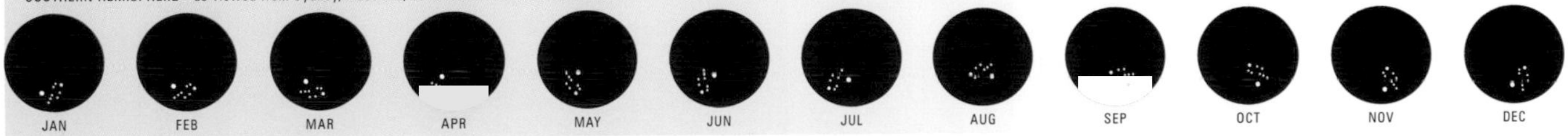

Pegasus

Pegasus, the Winged Horse, is the seventh-largest constellation. It has one splendor—the globular cluster M15.

FLYING RIDE

When Perseus slew the Gorgon Medusa, its blood dripped into the earth and spawned the magical winged horse called Pegasus. The goddess Athena delivered Pegasus to Bellerophon, who rode him on many adventures. Bellerophon was the first to demonstrate the worth of air power when he killed the fire-breathing Chimaera with arrows shot from aloft, beyond its horrible reach. Pegasus eventually flew to Mount Olympus—after throwing Bellerophon off—and was welcomed into Zeus's stables.

Right This Hubble Space Telescope image shows a close-up view of Stephan's Quintet, which is a group of five galaxies in the constellation Pegasus.

THE STARS OF PEGASUS

The body of the horse is outlined by the Great Square of Pegasus, a prominent 16- by 14-degree asterism. The star at the northeastern corner was long shared with Andromeda and is now officially labeled Alpha (α) Andromedae, but the concept of the Great Square has not suffered. The orange supergiant star Enif, Epsilon (ε) Pegasi, marks the nose. Other stars outline the horse's neck and forelegs. The constellation Andromeda doubles as an admirable pair of hind legs.

The magnitude 5.5 naked-eye star 51 Pegasi was the first star proven to have a planet orbiting it. IK Pegasi, which is 150 light-years distant, is the closest known supernova progenitor. This binary star is expected to produce a Type 1a supernova explosion sometime over the next few million years.

OBJECTS OF INTEREST

The magnificent globular cluster **M15 (NGC 7078)** is magnitude 6.3 and 18 arc-minutes in diameter. An 8-inch (20 cm) reflector reveals five long, ragged star-chains in the halo. Its core swells in brightness, rising to a brilliant central pip. M15's core has the second greatest surface brightness of all globulars, exceeded only by NGC 1851. A 16-inch (40 cm) telescope with an OIII-filter unveils the 3-arc-second, 13th magnitude planetary nebula **Pease 1**, only 25 arc-seconds north–northeast of M15's center, and one of only four planetaries known in globulars.

Struve 2841 (Σ2841) is a 22-arc-second yellow and greenish double star of magnitudes 6.4 and 7.9.

Spiral galaxy **NGC 7331** is magnitude 9.5 and 10 by 4 arc-minutes. An 8-inch (20 cm) telescope shows it as elongated with a bright center, a very bright nucleus, and a knot just north of the nucleus. A 16-inch (40 cm) telescope adds long faint spiral arms. A sharper edge—indicating a dust lane like the Andromeda Galaxy's—runs along the western side. Four small NGC companion galaxies can be seen, all to the east of NGC 7331.

Stephan's Quintet, NGCs 7317–20, is a clump of five galaxies, magnitude 13 to 14, located 0.5° south–southwest of NGC 7331. The largest, NGC 7320, is a foreground galaxy, but the other four galaxies are gravitationally connected. A 10-inch (25 cm) telescope shows four, failing to split the close pair, NGC 7318A/B, but a 12-inch (30 cm) reveals all five. Single faint galaxies are ignored, but groups of them delight amateur observers.

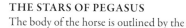

Left Image of NGC 7331 which was discovered by William Herschel in 1784. It is one of the brightest galaxies that is not included in Messier's catalog. It shows a fine spiral structure despite its small inclination from the edge-on position. Several background galaxies are also visible.

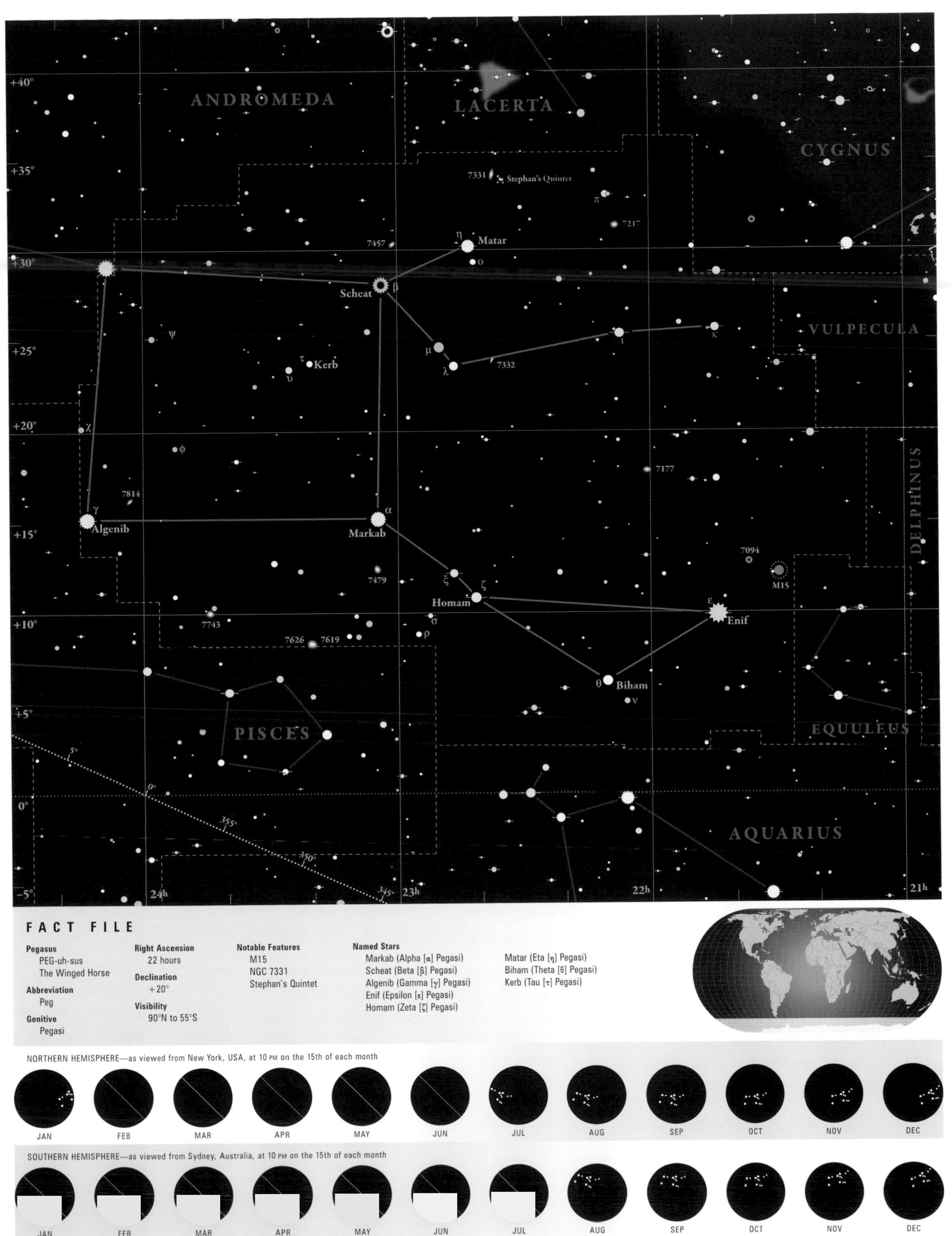

FACT FILE

Pegasus
PEG-uh-sus
The Winged Horse

Abbreviation
Peg

Genitive
Pegasi

Right Ascension
22 hours

Declination
+20°

Visibility
90°N to 55°S

Notable Features
M15
NGC 7331
Stephan's Quintet

Named Stars
Markab (Alpha [α] Pegasi)
Scheat (Beta [β] Pegasi)
Algenib (Gamma [γ] Pegasi)
Enif (Epsilon [ε] Pegasi)
Homam (Zeta [ζ] Pegasi)
Matar (Eta [η] Pegasi)
Biham (Theta [θ] Pegasi)
Kerb (Tau [τ] Pegasi)

Perseus

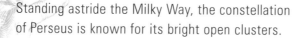

Standing astride the Milky Way, the constellation of Perseus is known for its bright open clusters.

BLOODSHED IN SANDALS

Perseus was the mortal son of Zeus. He was sent on a quest by King Polydectes to bring back the head of the snake-haired Gorgon Medusa, whose glance could turn people to stone. Knowing the difficulty of this task, Zeus dispatched Athena and Hermes to help. Athena lent her polished bronze shield to Perseus and instructed him to look only at the monster's reflection in the shield. Hermes gave him a scimitar and took him to nymphs who provided winged sandals, a cap of invisibility, and a pouch to hold the severed head.

Perseus flew to the island of the three Gorgons and, wearing his invisibility cap, beheaded Medusa, placing the head in the pouch. Perseus fled from the two remaining Gorgons with his winged sandals. The famed winged horse Pegasus arose from Medusa's spilled blood.

As he flew home Perseus saw Andromeda in peril, rescued her, and then married her. He used Medusa's head to turn his various enemies to stone.

THE STARS OF PERSEUS

This constellation bears a strong resemblance to the Greek letter Pi (π). The Double Cluster, one of the "little clouds" known in antiquity, represents Perseus's upraised scimitar.

Algol, or Beta (β) Persei, is an eclipsing variable star that falls from magnitude 2.1 to 3.3 once every 2.867321 days. The variability of Algol—the Demon's Head—must have been well known to the ancients. It strains credulity that this most obvious of naked-eye variable stars would represent Medusa's head by chance.

OBJECTS OF INTEREST

M76 (NGC 650/51) is a magnitude 10.1 planetary nebula. Also known as the **Cork Nebula**, it has two NGC numbers owing to its double-lobed structure. The south-western lobe is the brighter of the two.

The rich **Double Cluster** **(NGC 869** and **884)**, each 18 arc-minutes wide, is the constellation's show-piece, with over 200 stars visible. There is some resolution with 7 × 50 binoculars. An 8-inch (20 cm) telescope shows two trios and four orange stars within NGC 884. An oval of stars is prominent in NGC 869.

M34 (NGC 1039), at magnitude 5.2 and nearly a moon-diameter wide, is a fairly easy naked-eye open cluster. An 8-inch (200 mm) telescope shows star-chains and several doubles among M34's 60 stars. In a 16-inch (40 cm) telescope, the cluster's heart resembles Perseus, and there are three orange gems.

The splendid **Alpha (α) Persei Moving Group** is a 5-degree-long binocular cluster of bright stars.

With binoculars, the magnitude 6.4 cluster of **NGC 1528** is a prominent, unresolved fuzzy patch. An 8-inch (20 cm) at low power shows about 50 fairly equal-brightness stars, arranged in arcs and ovals.

The **California Nebula (NGC 1499)**, so named for its vague resemblance to the US state, is a large 145 by 40 arc-minute emission nebula , faintly illuminated by the 4th magnitude **Xi (ξ) Persei**. Telescopes show only the "Sierra Nevada" and the "California coastline." Viewing may be assisted by using a H-beta filter.

Above NGC 1023 is a SBO galaxy, meaning it has an overall disk shape with a central prominence of stars (similar to our Milky Way), but lacks spiral arms. Its rapidly rotating core may suggest a massive black hole.

Right The carnival of colors in this image represents the "beautiful chaos" of star-birth taking place in the reflection nebula NGC 1333, which is 1,000 light-years from Earth.

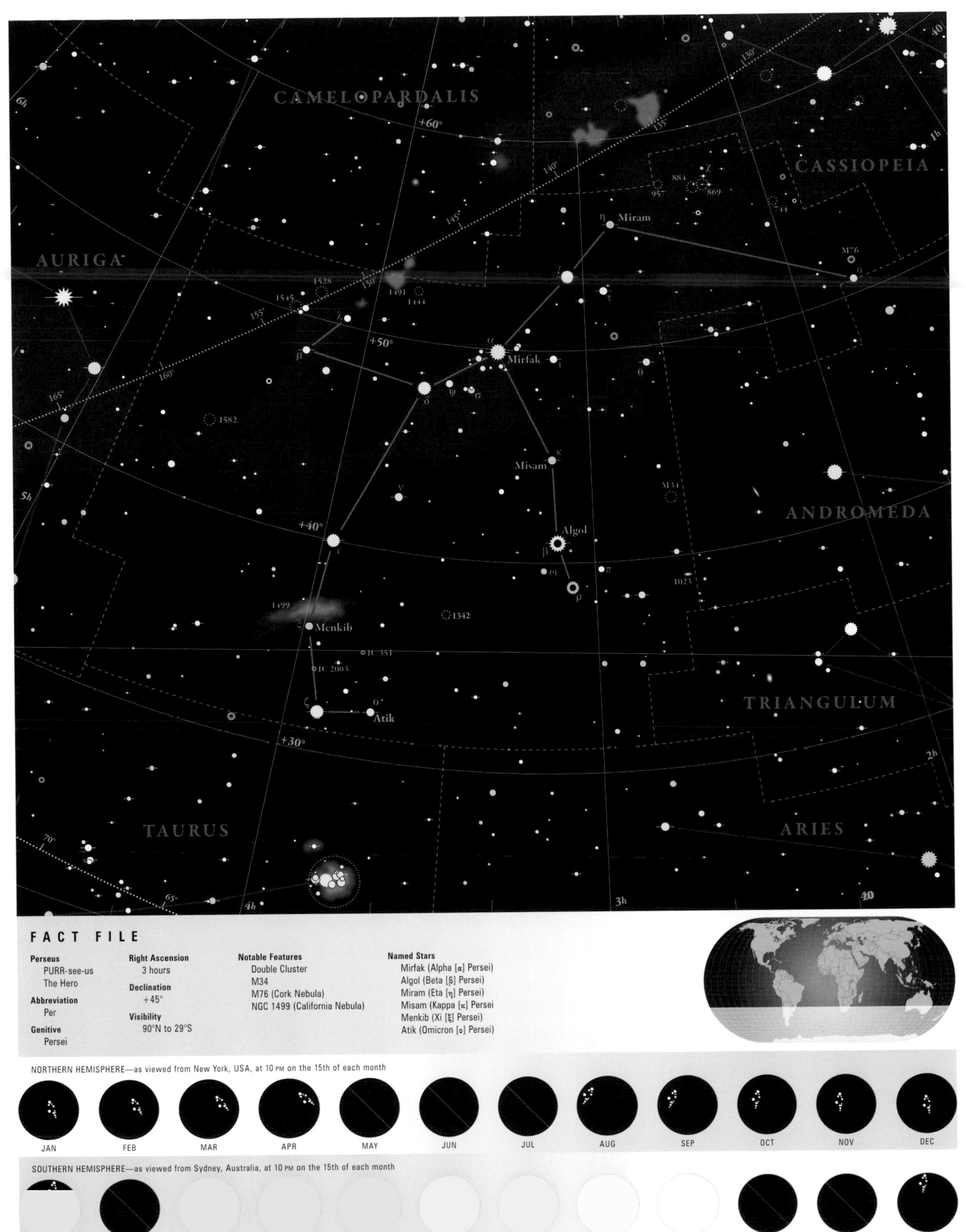

FACT FILE

Perseus
PURR-see-us
The Hero

Abbreviation
Per

Genitive
Persei

Right Ascension
3 hours

Declination
+45°

Visibility
90°N to 29°S

Notable Features
Double Cluster
M34
M76 (Cork Nebula)
NGC 1499 (California Nebula)

Named Stars
Mirfak (Alpha [α] Persei)
Algol (Beta [β] Persei)
Miram (Eta [η] Persei)
Misam (Kappa [κ] Persei
Menkib (Xi [ξ] Persei)
Atik (Omicron [ο] Persei)

NORTHERN HEMISPHERE—as viewed from New York, USA, at 10 PM on the 15th of each month

JAN FEB MAR APR MAY JUN JUL AUG SEP OCT NOV DEC

SOUTHERN HEMISPHERE—as viewed from Sydney, Australia, at 10 PM on the 15th of each month

JAN FEB MAR APR MAY JUN JUL AUG SEP OCT NOV DEC

Phoenix

The glorious Phoenix of legend has a wealth of mythology—but unfortunately a paucity of stars.

OUT OF THE FIRE

The Phoenix was a mythical bird of incredible beauty that reputedly lived for 500 years. As the end of its life approached, the Phoenix would build a fragrant nest of leaves and cinnamon bark that was set on fire by the hot noon sun. The bird was consumed by the fire, but left behind a tiny worm that would emerge from the cinders and grow into a new Phoenix.

The group of stars that make up Phoenix was recognized as a bird by many cultures: The constellation has been called the Griffin, the Eagle, the Young Ostriches (Arabia), and the Fire Bird (China).

The sixteenth-century Dutch explorers Pieter Dirkszoon Keyser and Frederick de Houtman devised Phoenix, mapping its stars during their explorations of the East Indies. Johannes Bayer included it in *Uranometria*, published shortly afterward in 1603. The stars of Phoenix have changed little since.

THE STARS OF PHOENIX

The stars of Phoenix, while generally dim, do outline a bird rising to flight as seen from the front. With bright Alpha (α) at the head, Beta (β), Zeta (ζ), Eta (η), and Epsilon (ε) form the rest of the body. Gamma (γ) and Delta (δ) make one wing, while dim Iota (ι) and Theta (θ) compose the other. The golf-club-shaped trio of Alpha (α), Kappa (κ), and Epsilon (ε) make up the most distinctive asterism.

Alpha (α) Phoenicis is also known as Ankaa; the name is a relatively modern invention derived from the Arabic name for the Phoenix bird. It is a nearby K-class giant star, 88 light-years distant, about 2.5 times the mass of the Sun and more than 80 times more luminous.

OBJECTS OF INTEREST

Phoenix is far from the Milky Way, and the only objects of interest to amateur sky-watchers are some double stars and galaxies.

Theta (θ) Phoenicis is an attractive pair of nearly equal white stars that are tight but that can be resolved in a 4-inch (10 cm) telescope. In the far south of the constellation, a slightly wider, unequal pair of 4th and 7th magnitude white gems makes up **Zeta (ζ) Phoenicis**.

Of the galaxies in this constellation, **NGC 625** is by far the best of a poor bunch. An oval, nearly edge-on spiral, this galaxy is quite easy to see in an 8-inch (20 cm) telescope.

The lenticular galaxy **NGC 7702** can just be glimpsed in an 8-inch (20 cm) telescope as a small haze, and photographs show this galaxy as strikingly similar in appearance to a botched picture of the planet Saturn.

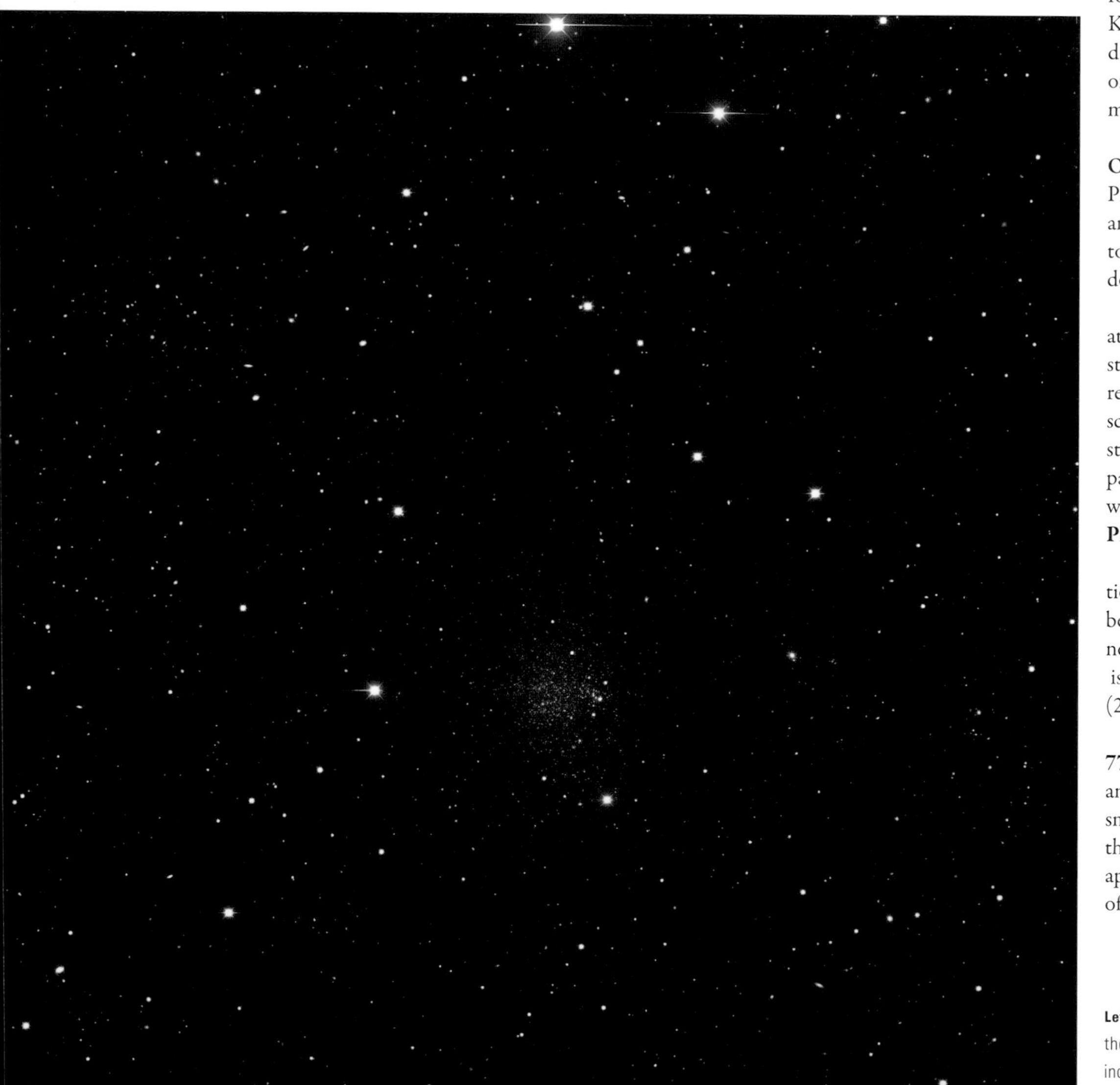

Left The Phoenix Dwarf Galaxy is one of the Local Group of nearby galaxies that includes the Milky Way and Andromeda galaxies, among others.

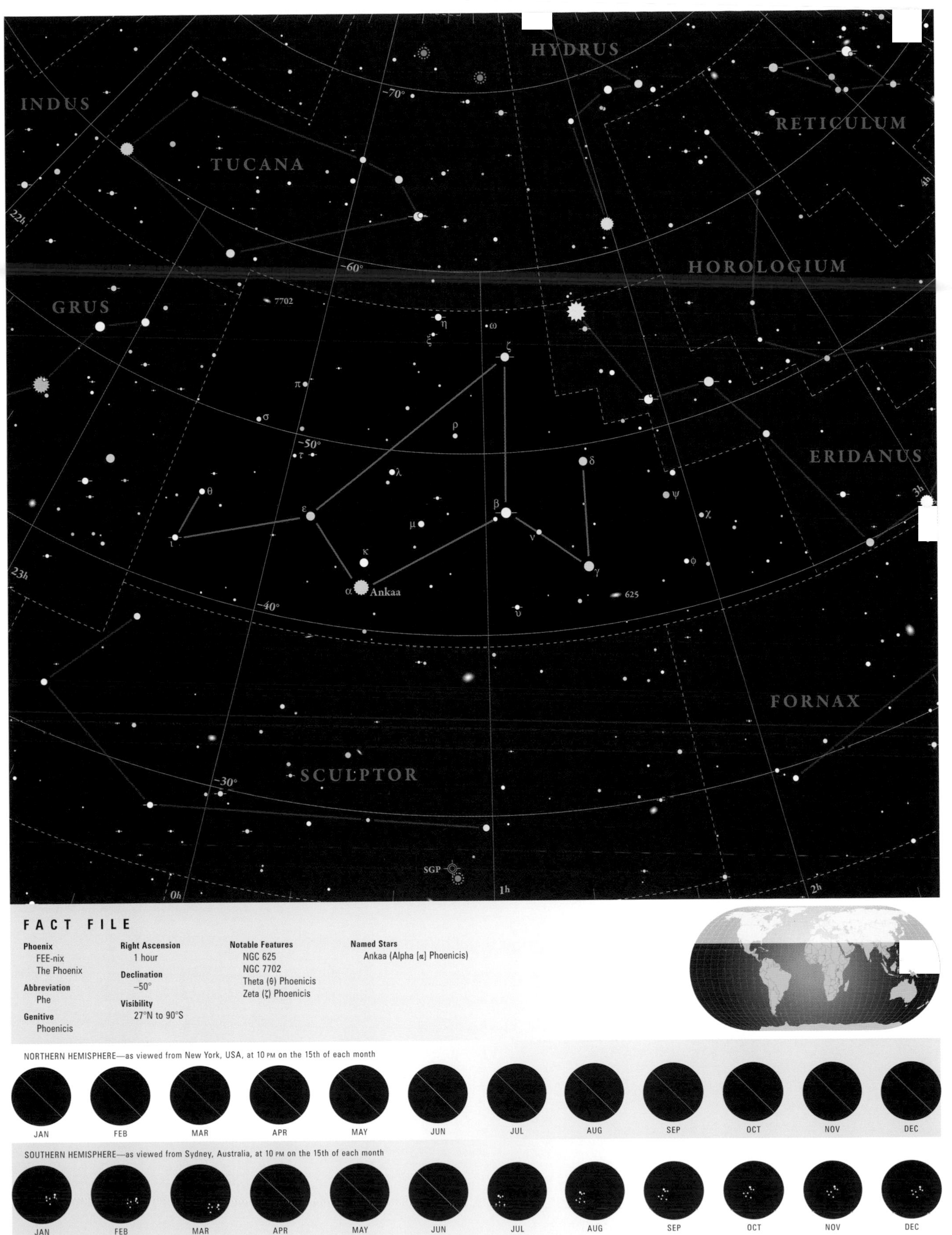
HYDRUS
INDUS
TUCANA
RETICULUM
HOROLOGIUM
GRUS
ERIDANUS
FORNAX
SCULPTOR
7702
625
Ankaa
SGP
–70°
–60°
–50°
–40°
–30°
22h
23h
0h
1h
2h
3h
4h
FACT FILE
Phoenix
FEE-nix
The Phoenix
Abbreviation
Phe
Genitive
Phoenicis
Right Ascension
1 hour
Declination
–50°
Visibility
27°N to 90°S
Notable Features
NGC 625
NGC 7702
Theta (θ) Phoenicis
Zeta (ζ) Phoenicis
Named Stars
Ankaa (Alpha [α] Phoenicis)
NORTHERN HEMISPHERE—as viewed from New York, USA, at 10 PM on the 15th of each month
JAN FEB MAR APR MAY JUN JUL AUG SEP OCT NOV DEC
SOUTHERN HEMISPHERE—as viewed from Sydney, Australia, at 10 PM on the 15th of each month
JAN FEB MAR APR MAY JUN JUL AUG SEP OCT NOV DEC

Pictor

Pictor—the painter—is supposed to represent an artist's easel and palette, but for the sky-watcher it is more reminiscent of a blank canvas.

STAR CANVAS

Ancient astronomers never used the group of stars now known as Pictor—southwest of *Argo*—to form a figure. Abbé Nicolas Louis de Lacaille devised Pictor as Le Chevalet et la Palette—the Easel and Palette—during his visit to the Cape Observatory between 1751 and 1752. Over the years since its invention, the constellation's name has evolved into the simpler modern form of Pictor, while its stars and shape have not changed.

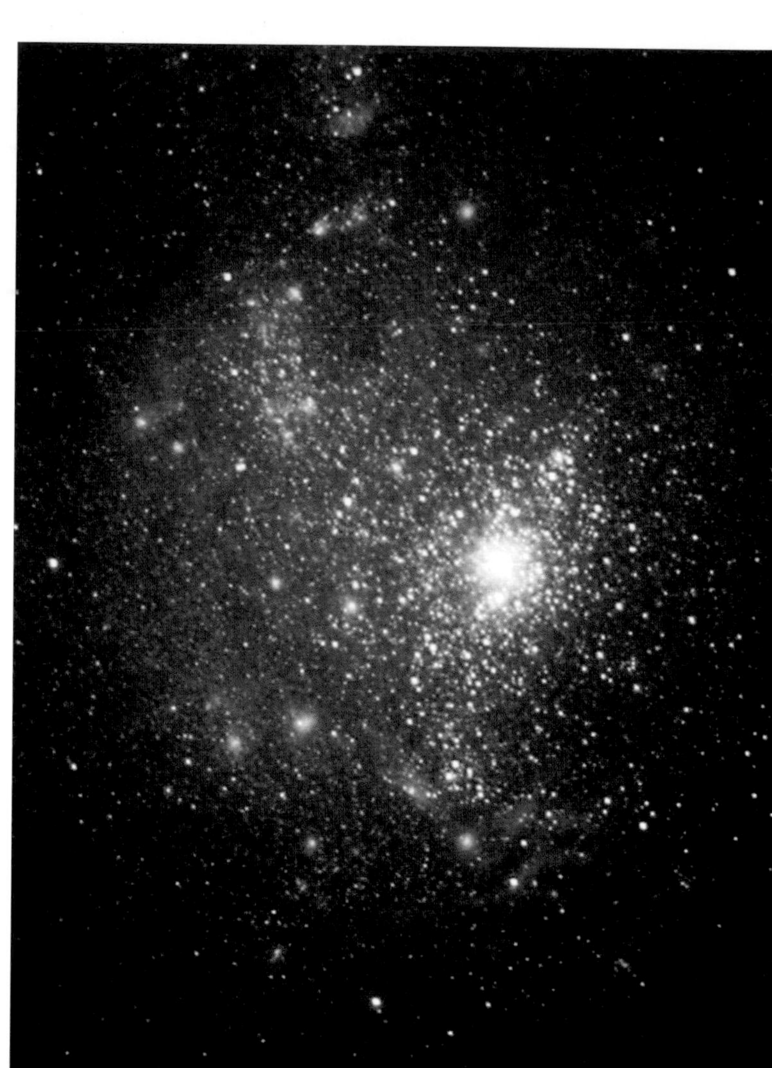

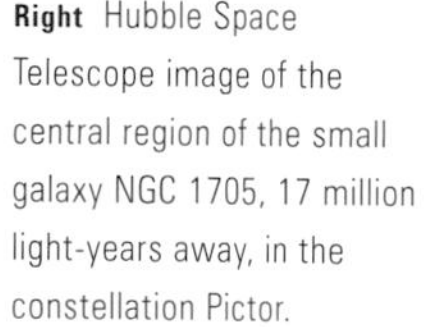

Right Hubble Space Telescope image of the central region of the small galaxy NGC 1705, 17 million light-years away, in the constellation Pictor.

THE STARS OF PICTOR

The principal stars of Pictor form a long, thin, scalene triangle to the west of brilliant Canopus, which is not too hard to pick out even from suburbia.

Pictor's lucida, 3rd magnitude Alpha (α) Pictoris, is an A-class subgiant star that we see from a distance of 100 light-years. Approximately 35 times more luminous than our Sun, this star is also about twice as massive.

Beta (β) Pictoris is probably the most famous of Pictor's stars. Nowadays, the detection of planets orbiting other stars hardly raises an eyebrow, but the discovery in 1983 of a warm dusty disk of planet-building material surrounding Beta (β) Pictoris, and only 63 light-years away, was a sensation.

In the following years, investigations revealed that the massive disk of dust extends to a distance greater than a thousand astronomical units from the star.

Recent findings show that the outer regions of the disk are warped, suggesting the presence of at least two large orbiting planets that might be still expanding. Beta (β) Pictoris is almost certainly another solar system in the making.

The overwhelming majority of Pictor's stars orbit our Milky Way Galaxy in one direction, like horses on a racetrack. Kapteyn's Star, a tiny, faint M-class red-dwarf star only 12 light-years away, was discovered by Jacobus Kapteyn and is somewhat of an individual—it orbits the Milky Way backward. Because it is going in the opposite direction to all the stars in the Sun's neighborhood, it moves quite quickly across our sky, covering a moon-diameter in just over 200 years—the second highest proper motion of any star.

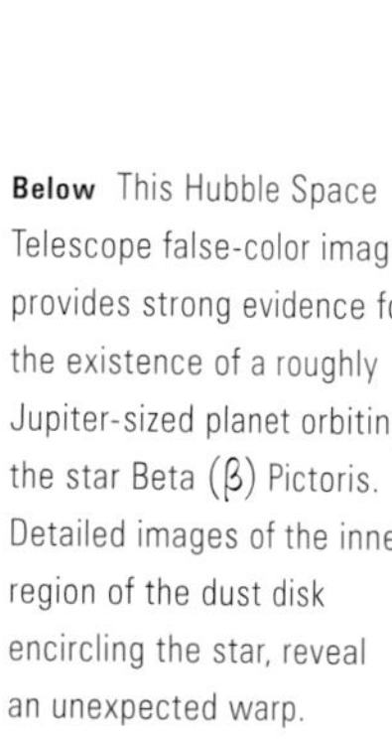

Below This Hubble Space Telescope false-color image provides strong evidence for the existence of a roughly Jupiter-sized planet orbiting the star Beta (β) Pictoris. Detailed images of the inner region of the dust disk encircling the star, reveal an unexpected warp.

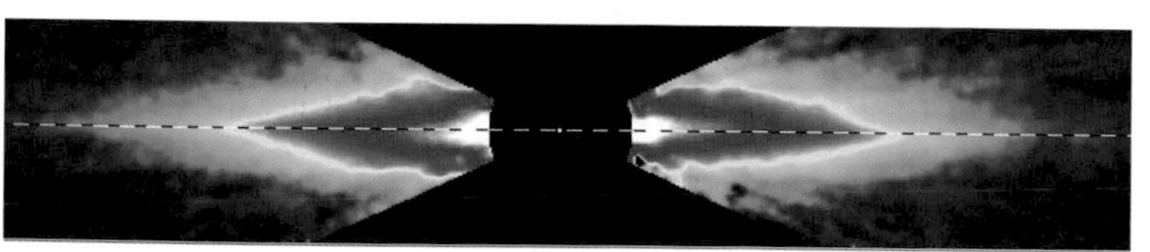

OBJECTS OF INTEREST

For sky-watchers, Pictor is the most arid of deserts, offering no clusters or nebulae at all, only very faint and tiny galaxies.

Of these, **NGC 1705** and **NGC 1803** are the best of a poor bunch. Both are very small and faint patches of haze, barely visible with an 8-inch (20 cm) telescope.

Dunlop 21 is a wide pair of 5th and 6th magnitude stars. The exceptionally keen-eyed observer can barely resolve it with the naked eye, on a clear dark night away from lights, but it is much easier to find in binoculars and even easier to resolve with a telescope.

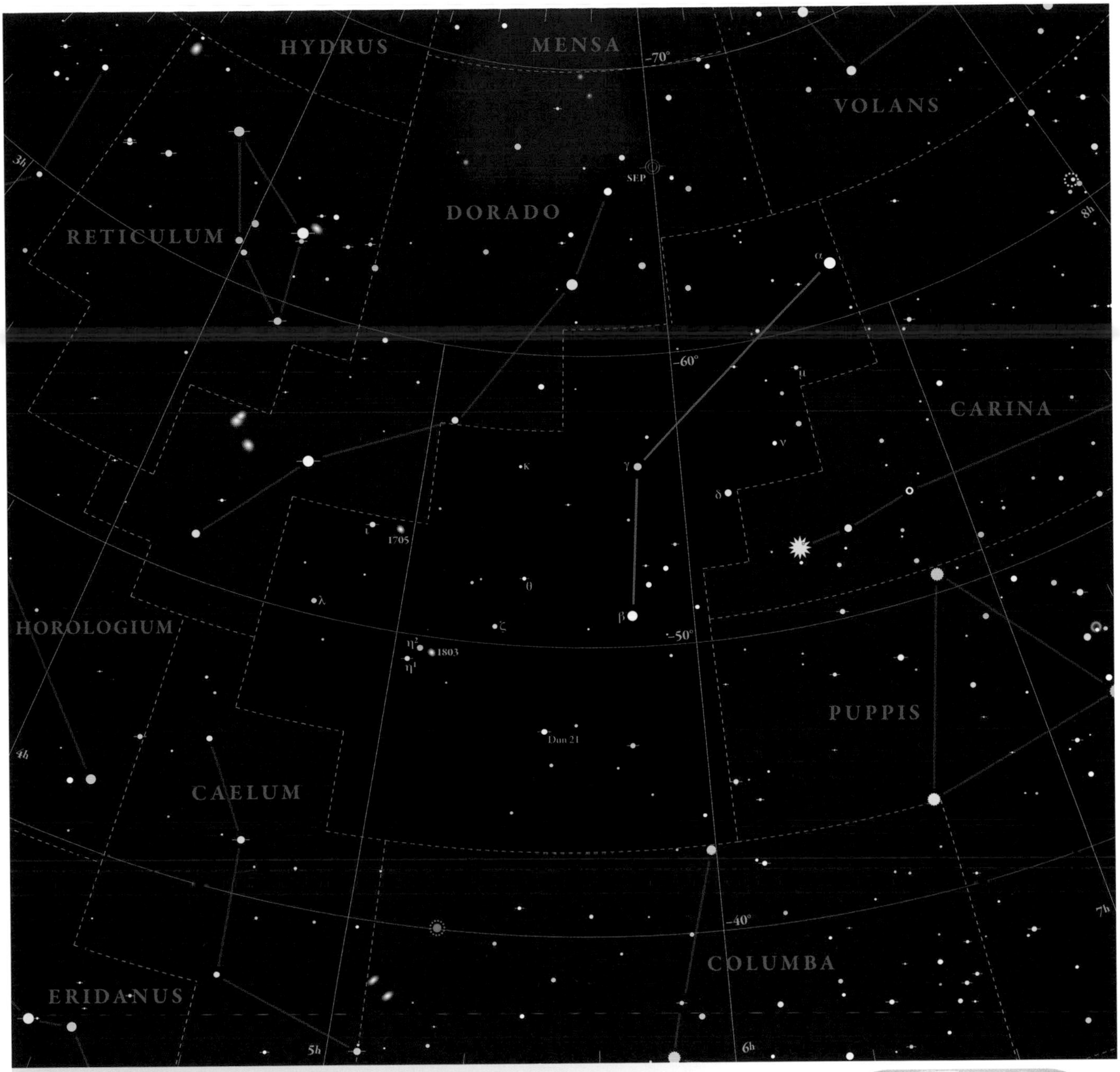

FACT FILE

Pictor
PICK-tor
The Painter's Easel

Abbreviation
Pic

Genitive
Pictoris

Right Ascension
6 hours

Declination
–55°

Visibility
25°N to 90°S

Notable Features
Dunlop 21
Kapteyn's Star
NGC 1705
NGC 1803

NORTHERN HEMISPHERE—as viewed from New York, USA, at 10 PM on the 15th of each month

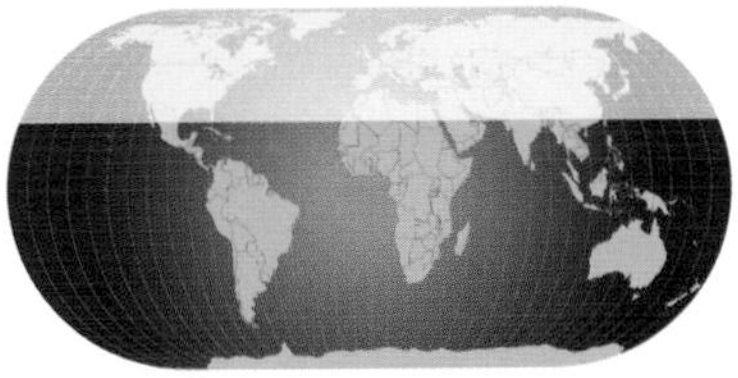

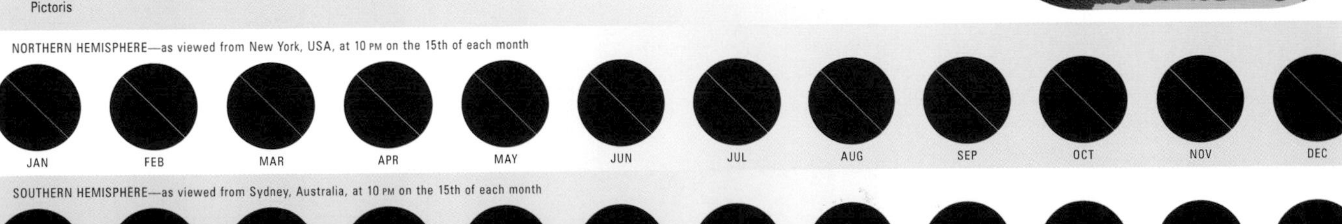

SOUTHERN HEMISPHERE—as viewed from Sydney, Australia, at 10 PM on the 15th of each month

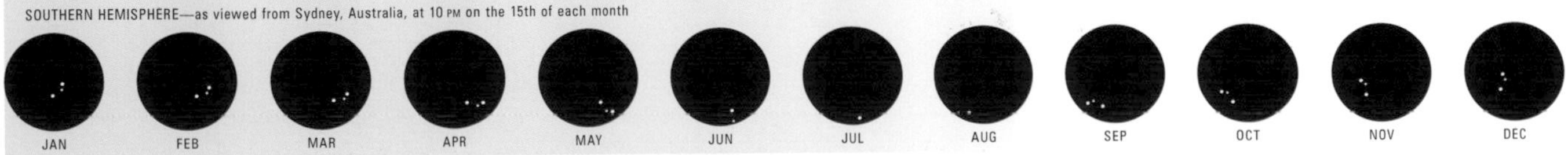

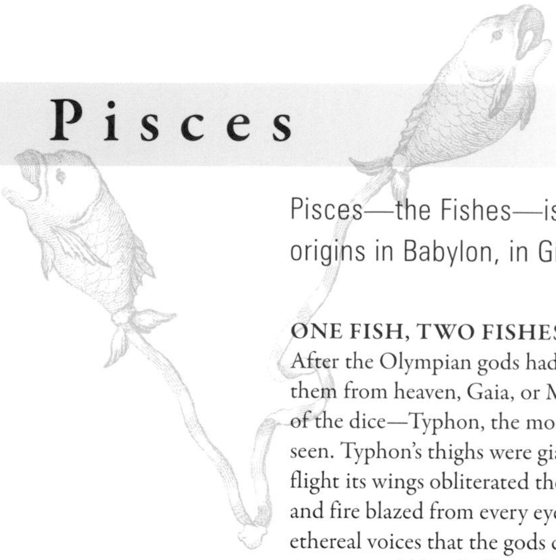

Pisces

Pisces—the Fishes—is one of the faintest figures of the zodiac. Although the tale probably had its origins in Babylon, in Greek myth the two fishes of Pisces represent Aphrodite and her son, Eros.

ONE FISH, TWO FISHES

After the Olympian gods had deposed the Titans and driven them from heaven, Gaia, or Mother Earth, cast her last throw of the dice—Typhon, the most awful monster the world had seen. Typhon's thighs were giant serpents, and when it took flight its wings obliterated the sun. It had 100 dragons' heads and fire blazed from every eye. Typhon sometimes spoke with ethereal voices that the gods could understand, but at other times roared like a bull or a lion, or hissed like a snake.

The terrified Olympians fled, and Aphrodite and Eros transformed themselves into fishes and vanished into the sea. In order not to lose each other in the dark waters of the Euphrates (some versions say the Nile), they tied their tails together with a cord.

THE STARS OF PISCES

Very few constellations, if any, are less conspicuous than Pisces. Only two 4th magnitude stars adorn the group, in the shape of a very elongated V to the southeast of the Great Square of Pegasus. The asterism known as the "circlet"—a rough pentagon of 4th and 5th magnitude stars made up by Gamma (γ), Theta (θ), Iota (ι), Kappa (κ), and Lambda (λ)—is the most recognizable part of the constellation, directly south of the Great Square of Pegasus.

Eta (η) Piscium, east of the Great Square, is the brightest star in Pisces, outshining Alpha (α) by nearly a half magnitude. It is a G-type giant star, about four times heavier than our Sun and 316 times more luminous, that we see from 300 light-years distance.

OBJECTS OF INTEREST

Pisces is some way from the Milky Way and, apart from several interesting double stars, the only objects it holds are galaxies, which are profuse. The gigantic Pisces–Perseus supercluster of galaxies lies far in the background to the stars of Pisces, and contains literally thousands of observable galaxies for professional-class telescopes.

In amateur telescopes, the best is **M74**. The key to getting the best from this very faint face-on spiral galaxy is viewing in a dark sky. It is visible in a 4-inch (10 cm) telescope, and an 8-inch (20 cm) shows a large, almost quarter-moon-diameter patch of gossamer with a large, slightly brighter core. Of the others, **NGC 676** is an attractive spindle, though small and faint, with a 9th magnitude star superimposed virtually dead center.

Alpha (α) Piscium is a lovely binary of slightly unequal stars that are quite close together but easily resolved in a 4-inch (10 cm) telescope.

Above Looking halfway across the universe, the Hubble Space Telescope captures a primeval galaxy using the effect of gravitational lensing from this massive galaxy cluster in the constellation Pisces.

Right M74 is a magnificent face-on spiral galaxy that is faint, but can be resolved in a medium-sized telescope as a patch of gossamer with a brighter core.

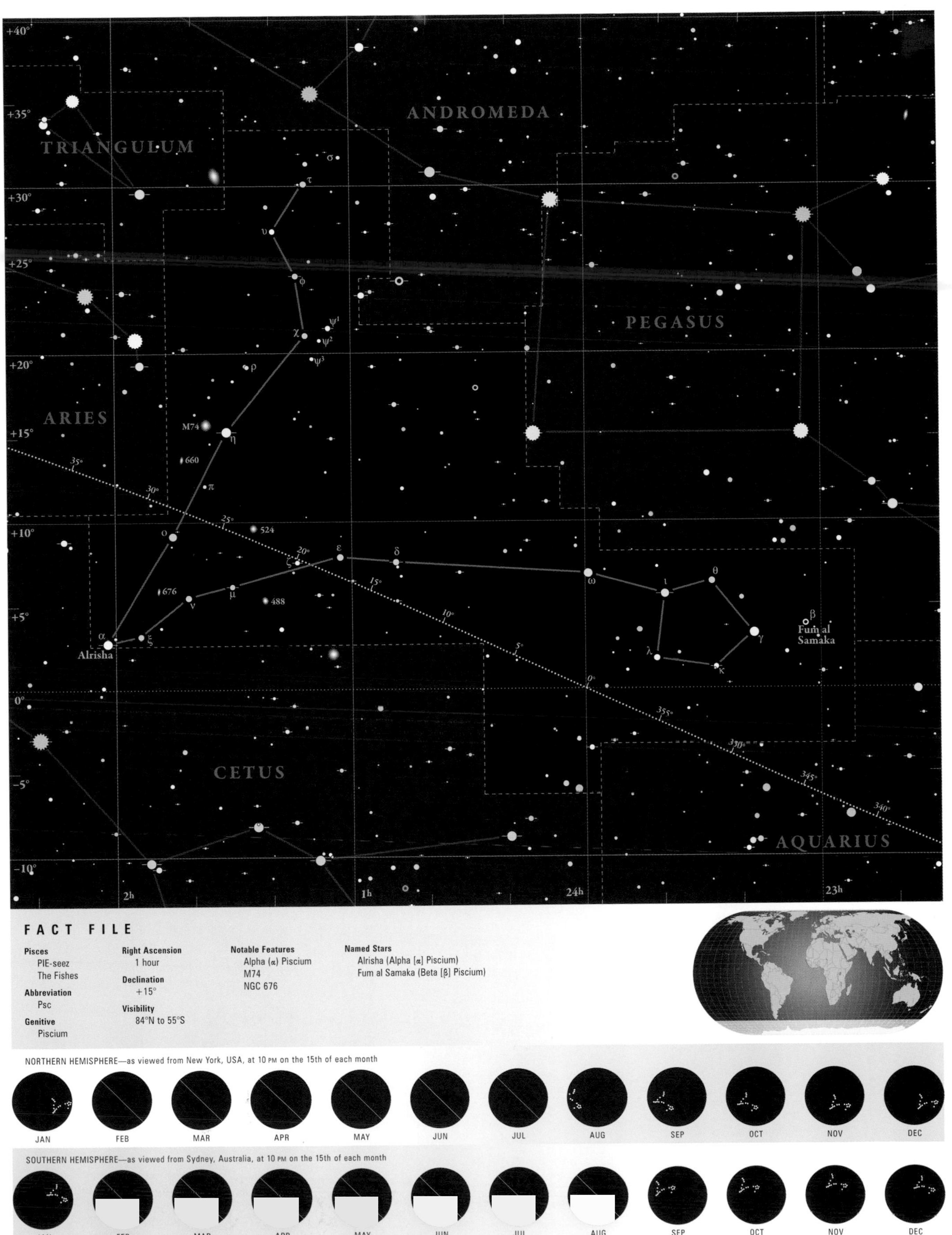

FACT FILE

Pisces
PIE-seez
The Fishes

Abbreviation
Psc

Genitive
Piscium

Right Ascension
1 hour

Declination
+15°

Visibility
84°N to 55°S

Notable Features
Alpha (α) Piscium
M74
NGC 676

Named Stars
Alrisha (Alpha [α] Piscium)
Fum al Samaka (Beta [β] Piscium)

NORTHERN HEMISPHERE—as viewed from New York, USA, at 10 PM on the 15th of each month

JAN FEB MAR APR MAY JUN JUL AUG SEP OCT NOV DEC

SOUTHERN HEMISPHERE—as viewed from Sydney, Australia, at 10 PM on the 15th of each month

JAN FEB MAR APR MAY JUN JUL AUG SEP OCT NOV DEC

Piscis Austrinus

One of the oldest constellations, Piscis Austrinus and its stars were uniformly associated with a fish by ancient cultures.

LESSER-KNOWN COUSIN

The origin of the mythology associated with the group is at best vague; it is most probably Babylonian. Piscis Austrinus was believed to be a parent of the fishes that form the better known zodiacal constellation of Pisces. Piscis Austrinus is also sometimes linked to the tale, recounted by Eratosthenes, that the Syrian fertility goddess Atargatis (also known in Greek as Derceto), having fallen into a lake near the Euphrates River in Syria, was saved from drowning by a huge fish.

Piscis Austrinus was originally a large constellation. During the sixteenth century a swathe of stars that now make up Grus was carved out, and in 1752 Lacaille removed several more to make Microscopium. In *Uranographia*, published in 1803, Johannes Bode borrowed more stars from Piscis Austrinus, plus a few from Microscopium, to form the now obsolete constellation Globus Aerostaticus, celebrating the invention of the hot-air balloon. Bode's balloon was later reproduced in a couple of other star-atlases, but it never really took off.

THE STARS OF PISCIS AUSTRINUS

Except for 1st magnitude Alpha (α) Piscis Austrini, which has always marked the mouth of the southern fish, the balance of the stars is quite faint. The oval loop made up by Alpha (α), Epsilon (ε), Lambda (λ), Theta (θ), Iota (ι), Beta (β), Gamma (γ), and Delta (δ) approximates the shape of a fish minus the tail, and is easy to make out to the west of Alpha Piscis Austrini, otherwise known as Fomalhaut.

From the Arabic meaning "mouth of the fish," Fomalhaut is the seventeenth brightest star in the night sky. A local star just 25 light-years distant, it is an A-type dwarf star that is about twice our Sun's mass and 16 times more luminous.

OBJECTS OF INTEREST

Well away from the plane of the Milky Way, Piscis Austrinus has very little to offer the amateur sky-watcher save some galaxies and some interesting double stars. The spiral galaxy **NGC 7314** is pretty easy to see in an 8-inch (20 cm) telescope, having an oval-shaped outline with a brighter center. Elliptical shaped **IC 5271**, near Gamma (γ) and Delta (δ), is another galaxy just bright enough to see in a 6-inch (15 cm) telescope.

Of the double stars, **Beta (β)** is the best. The two white 4th and 6th magnitude stars are easy to resolve in a 4-inch (10 cm) telescope. **Gamma (γ)** and **Delta (δ) Piscis Austrini** is a similar-looking binary, a bright star with a wide, much fainter companion.

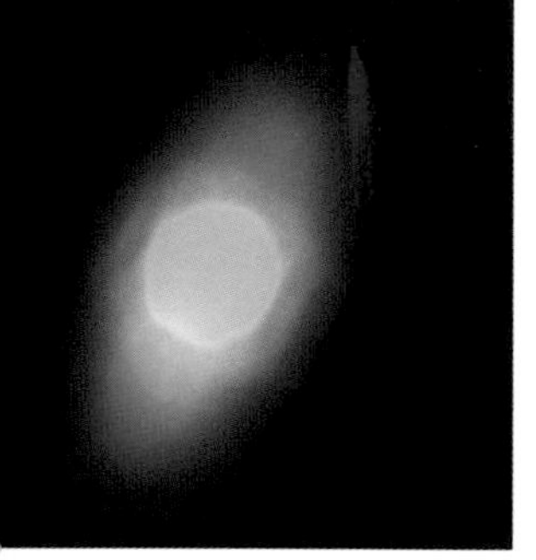

Above The mouth of the fish, Fomalhaut, is the seventeenth brightest star in the sky. This infrared image captured by the Spitzer Space Telescope shows a flattened disk of gas and dust that could be forming planets like those in our own Solar System.

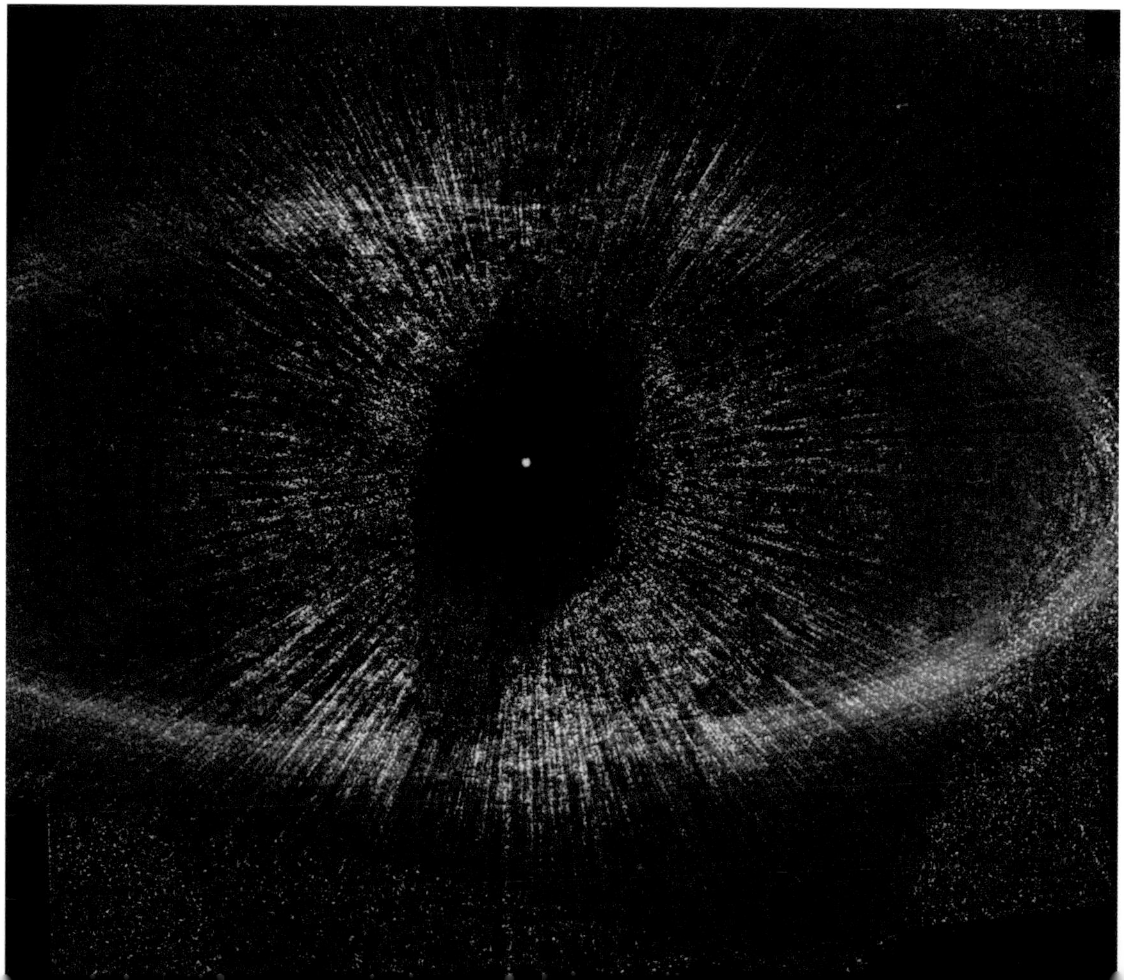

Right The Hubble Space Telescope's view of Fomalhaut in visible wavelengths reveals a distinct ring of dust and debris that is slightly off-center from the star. This could be a clue that there are one or more planets orbiting the star.

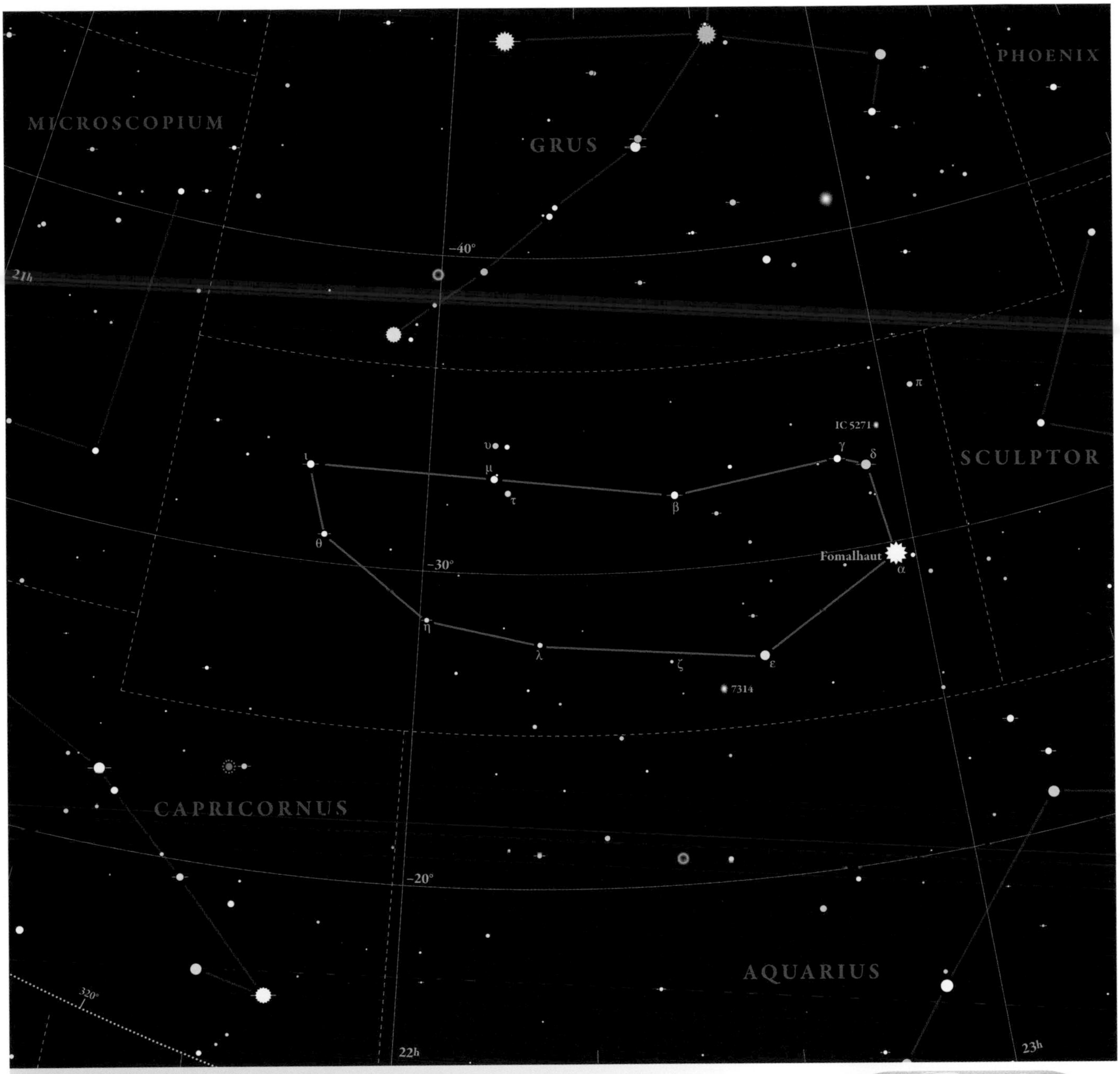

PHOENIX
MICROSCOPIUM
GRUS
SCULPTOR
CAPRICORNUS
AQUARIUS
Fomalhaut
IC 5271
7314

FACT FILE

Piscis Austrinus
PIE-siss oss-TRY-nuss
The Southern Fish

Abbreviation
PsA

Genitive
Piscis Austrini

Right Ascension
22 hours

Declination
–30°

Visibility
52°N to 90°S

Notable Features
Alpha (α) Piscis Austrini
Beta (β) Piscis Austrini
Delta (δ) Piscis Austrini
Gamma (γ) Piscis Austrini
IC 5271
NGC 7314

Named Stars
Fomalhaut (Alpha [α] Piscis Austrini)

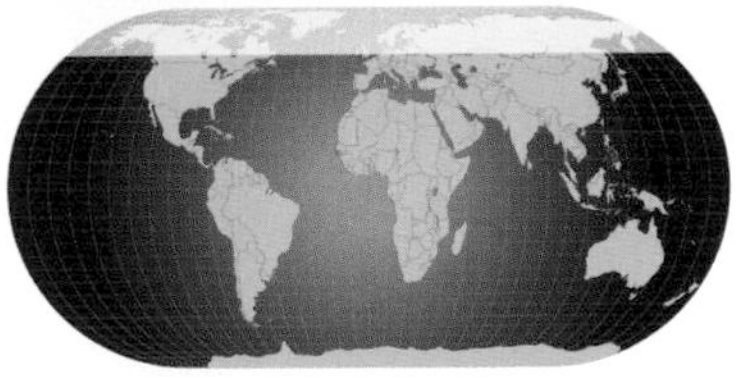

NORTHERN HEMISPHERE—as viewed from New York, USA, at 10 PM on the 15th of each month

JAN FEB MAR APR MAY JUN JUL AUG SEP OCT NOV DEC

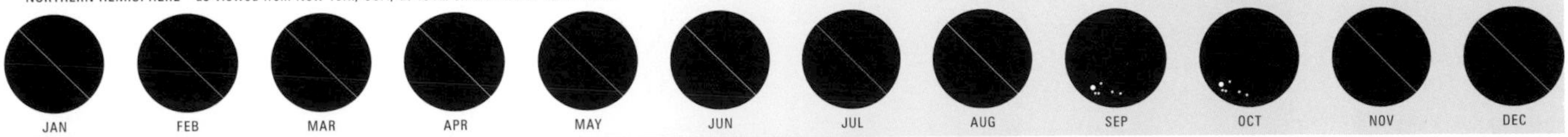

SOUTHERN HEMISPHERE—as viewed from Sydney, Australia, at 10 PM on the 15th of each month

JAN FEB MAR APR MAY JUN JUL AUG SEP OCT NOV DEC

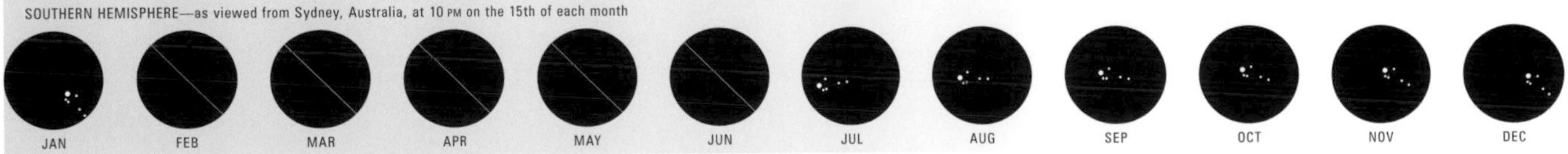

Puppis

With Carina and Vela, Puppis forms the three sections of the ancient superconstellation Argo Navis. Its spray of star clusters adorns the stern of Jason's ship, *Argo*.

NIGHT VOYAGE

Abbé Nicolas Louis de Lacaille dismembered the constellation of Argo in the seventeenth century, with no better excuse than to make it more manageable. In various modern texts, Puppis is sometimes referred to as the prow of *Argo*, but Puppis actually represents the stern or the poop deck. There is some variance in the star-atlases published during the sixteenth to nineteenth centuries as to which way *Argo* is sailing—northward or southward—and this is a possible cause of the confusion. (For the mythology related to Jason and *Argo*, see the constellation description for Carina.)

THE STARS OF PUPPIS

Puppis is immersed in a rich portion of the Milky Way. Its many medium-bright stars very roughly outline a side elevation of the rear of a Mediterranean-style carrack-type ship. The mid-ships are in the south, while the stern is at the north. When Argo was broken up by Lacaille, the Bayer designations of its stars remained with the individual, wherever that star fell, so Puppis has no Alpha or Beta star.

The brightest star of Puppis is a true superstar of the heavens. Second magnitude Zeta (ζ) Puppis, otherwise known as Naos—a name derived from the Greek word for "ship"—is one of the brightest O-type supergiant stars in the sky. Located 1,400 light-years away, it is 60 times heavier than our Sun. Much of the energy it emits is in the ultraviolet spectrum, owing to its extremely high surface temperature; it is more than 750,000 times brighter than our luminary. If humans were capable of seeing ultraviolet light, Naos would rival Jupiter in brilliance.

OBJECTS OF INTEREST

Lying on the Milky Way, Puppis is awash with bright star clusters. In the north of Puppis, below bright Sirius, are **M46** and **M47** ☍ ⌭. Both are large, bright, scattered clusters with different characters: M47 has medium-bright stars, while M46 is a throng of faint ones with a tiny doughnut-shaped planetary nebula **NGC 2438** ⌭ on the edges of the group. Not far south of these clusters is the bright planetary nebula **NGC 2440** ⌭, a tiny, oval, bluish-white spot with two tiny, nearly star-like points at its center.

Further south, near Naos, is the naked-eye cluster **NGC 2451** ◎ ☍ ⌭—an excellent bright cluster for binoculars, with the beautiful **NGC 2477** ⌭ just a little further west. It looks like a vast cloud of thousands of faint dots. On the western side of Naos, **NGC 2546** ⌭ is another wonderful cluster with a mixture of bright and faint stars. There are literally dozens more attractive clusters in Puppis, and scanning with binoculars on a dark night is a rewarding experience for the sky-watcher.

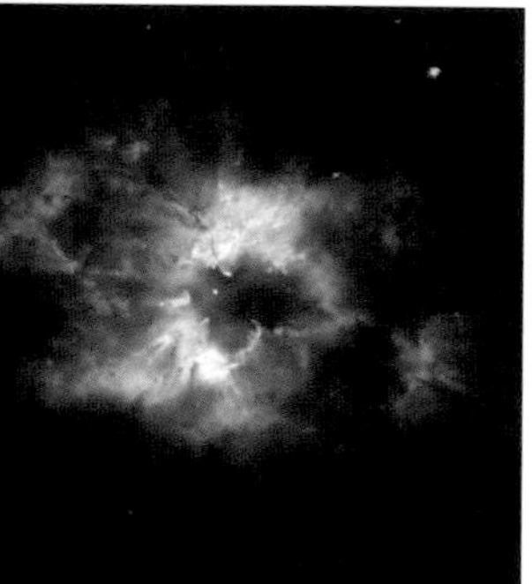

Above The central star of NGC 2440 in Puppis is one of the hottest white-dwarf stars known. It is seen here as the bright spot near the center of the nebula.

Right NGC 2467 is an active stellar nursery. The stars to the far left of the image have formed, and their birth nebulae have dissipated. Toward the center, dark dust lanes obscure parts of the nebula that are most likely forming new stars.

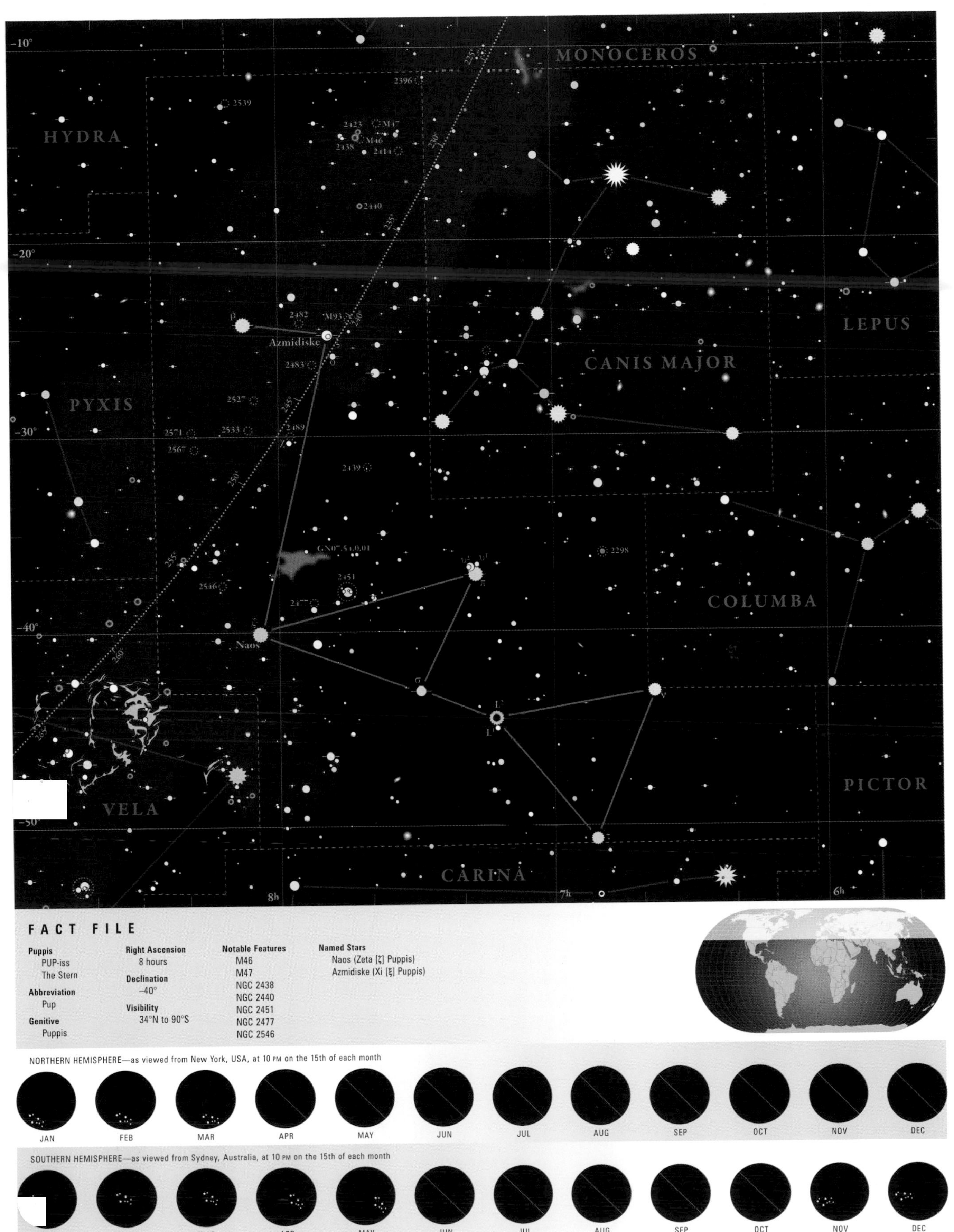

FACT FILE

Puppis
PUP-iss
The Stern

Abbreviation
Pup

Genitive
Puppis

Right Ascension
8 hours

Declination
–40°

Visibility
34°N to 90°S

Notable Features
M46
M47
NGC 2438
NGC 2440
NGC 2451
NGC 2477
NGC 2546

Named Stars
Naos (Zeta [ζ] Puppis)
Azmidiske (Xi [ξ] Puppis)

NORTHERN HEMISPHERE—as viewed from New York, USA, at 10 PM on the 15th of each month

JAN FEB MAR APR MAY JUN JUL AUG SEP OCT NOV DEC

SOUTHERN HEMISPHERE—as viewed from Sydney, Australia, at 10 PM on the 15th of each month

JAN FEB MAR APR MAY JUN JUL AUG SEP OCT NOV DEC

Pyxis

This constellation's full name used to be Pyxis Nautica. Does it represent a mariner's compass? A mast? Neither fits well with the great *Argo Navis*.

SETTING A COURSE

Pyxis was unknown to ancient astronomers: Abbé Nicolas Louis de Lacaille devised the constellation during his visit to the Cape Observatory between 1751 and 1752, using several stars near the stern of *Argo*, some in Vela, and other ungrouped stars. Like many of the constellations invented by Lacaille, Pyxis represents an instrument of science or engineering, in this case a nautical magnetic compass—a strange piece of equipment for an ancient Greek vessel.

Apparently because of this anomaly, the great nineteenth-century English astronomer Sir John Herschel proposed that Pyxis be abolished and the group renamed Malus, the Mast. In the sixteenth- to eighteenth-century star-atlases that preceded Lacaille, these stars were often depicted as forming a mast or spar of *Argo*. In any event, it is quite possible that the "real" *Argo* had no mast at all—it was usually described in Greek mythology as a galley rowed by 50 men. Herschel's proposal did not receive wide support, and Lacaille's invention remained. There has been no significant change to the group or its stars since this time, except that the name has been worn down to Pyxis.

THE STARS OF PYXIS

The main stars of Pyxis are 4th magnitude and fainter, bearing no resemblance to the depiction of the group as a wooden box containing a magnetic compass with a round face. Compared to the many bright stars of adjoining Vela and Puppis, it is a fairly dull group on the edge of the Milky Way.

Fourth magnitude—nearly 3rd—Alpha (α) Pyxidis is the lucida of Pyxis. Faint because it is very distant, it is a B-class giant star about 11 times more massive than and 18,000 times as luminous as our Sun. Its light reaches us from 830 light-years away, behind a dust cloud in our Milky Way that makes it a little dimmer than it might otherwise appear.

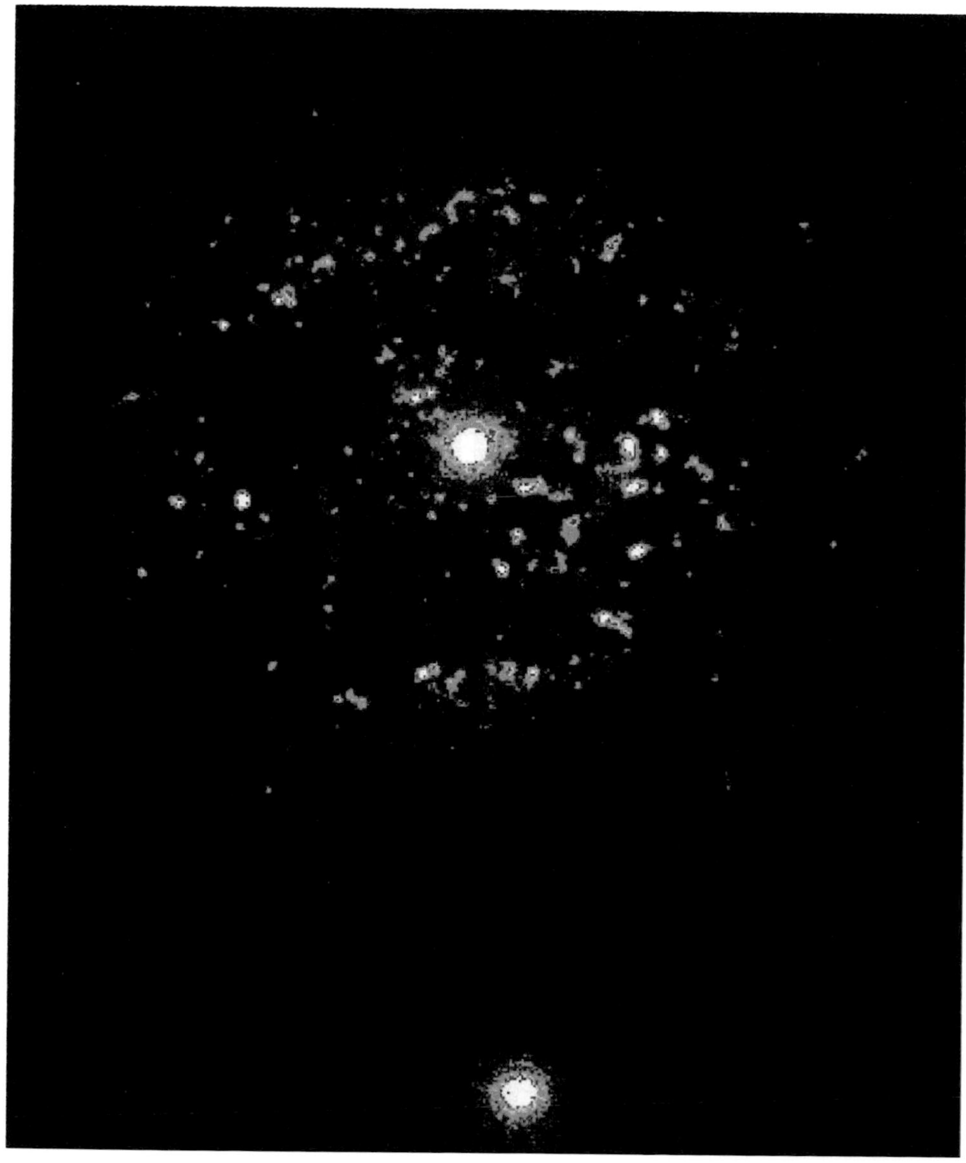

OBJECTS OF INTEREST

Being near the Milky Way, Pyxis has several open star clusters. The most interesting is **NGC 2818**, a scattered group of moderately faint stars made much more appealing by having a fairly bright planetary nebula in its midst. **NGC 2818A** is a lovely, irregularly-shaped, light blue patch on the outskirts of the cluster, and is no chance alignment—this planetary is almost certainly a bona-fide member of the star cluster.

There are also many galaxies in Pyxis, of which the brightest is **NGC 2613**. In an 8-inch (20 cm) telescope it looks like a bright hazy spindle, with many faint stars in the background.

IC 2469 is another very similar galaxy but it is slightly fainter and in the south of Pyxis.

Left This image of eruptions by the prolific and recurrent nova T Pyxidis has attracted the attention of many professional and amateur observers. T Pyxidis is 6,000 light-years away, in the dim southern constellation Pyxis, the Mariner's Compass.

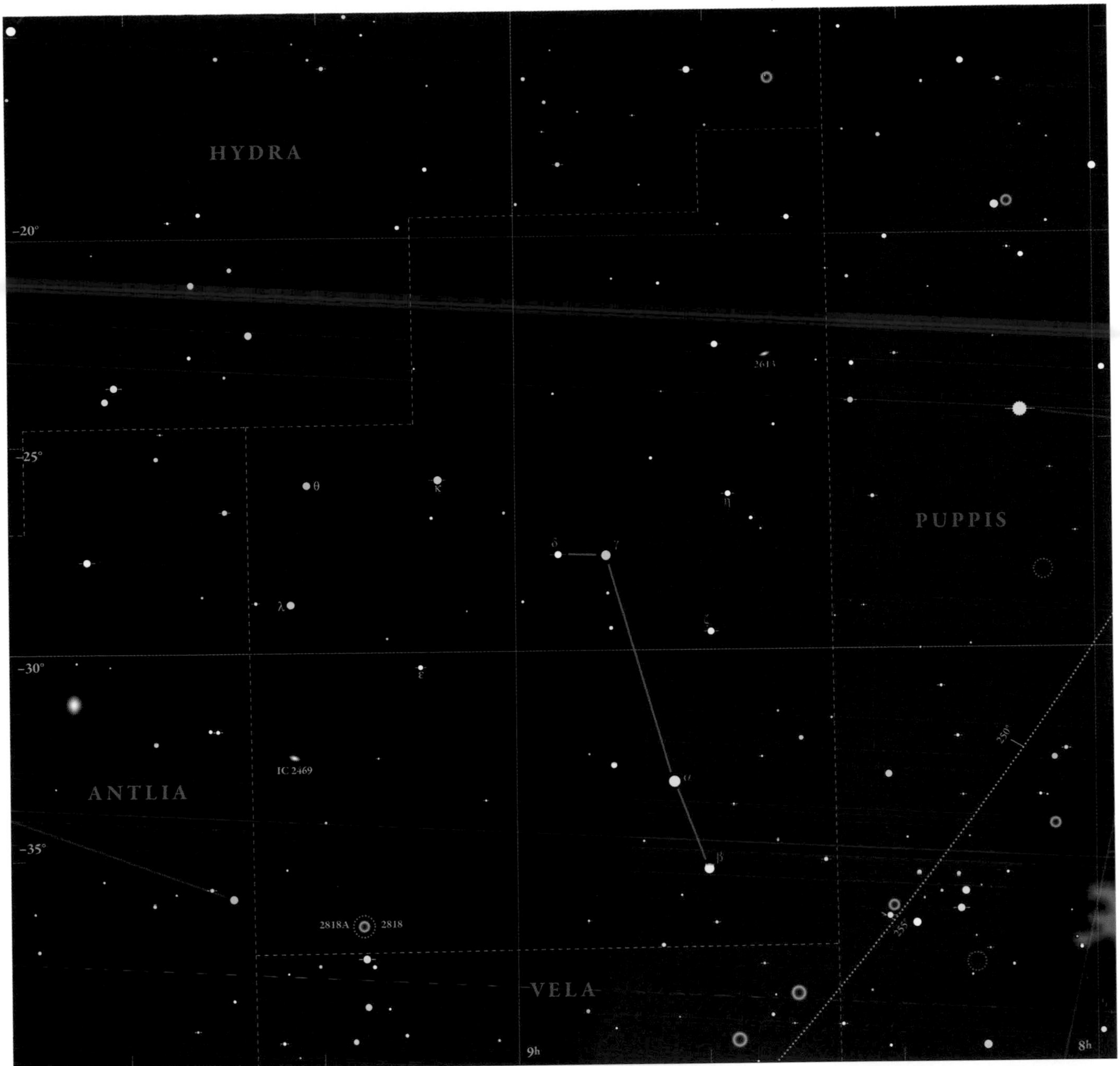

HYDRA

PUPPIS

ANTLIA

VELA

-20°

-25°

-30°

-35°

2613

IC 2469

2818A 2818

9h

8h

250°

255°

FACT FILE

Pyxis
PIK-sis
The Magnetic Compass

Abbreviation
Pyx

Genitive
Pyxidis

Right Ascension
9 hours

Declination
–30°

Visibility
50°N to 90°S

Notable Features
IC 2469
NGC 2613
NGC 2818
NGC 2818A

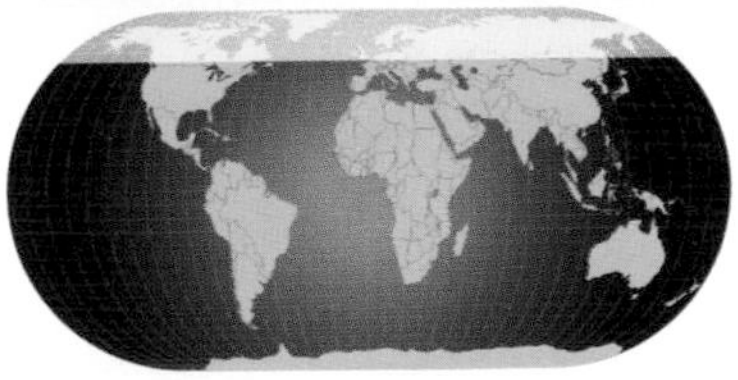

NORTHERN HEMISPHERE—as viewed from New York, USA, at 10 PM on the 15th of each month

JAN FEB MAR APR MAY JUN JUL AUG SEP OCT NOV DEC

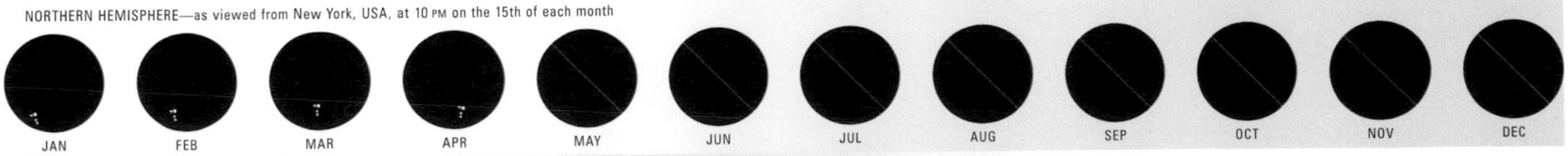

SOUTHERN HEMISPHERE—as viewed from Sydney, Australia, at 10 PM on the 15th of each month

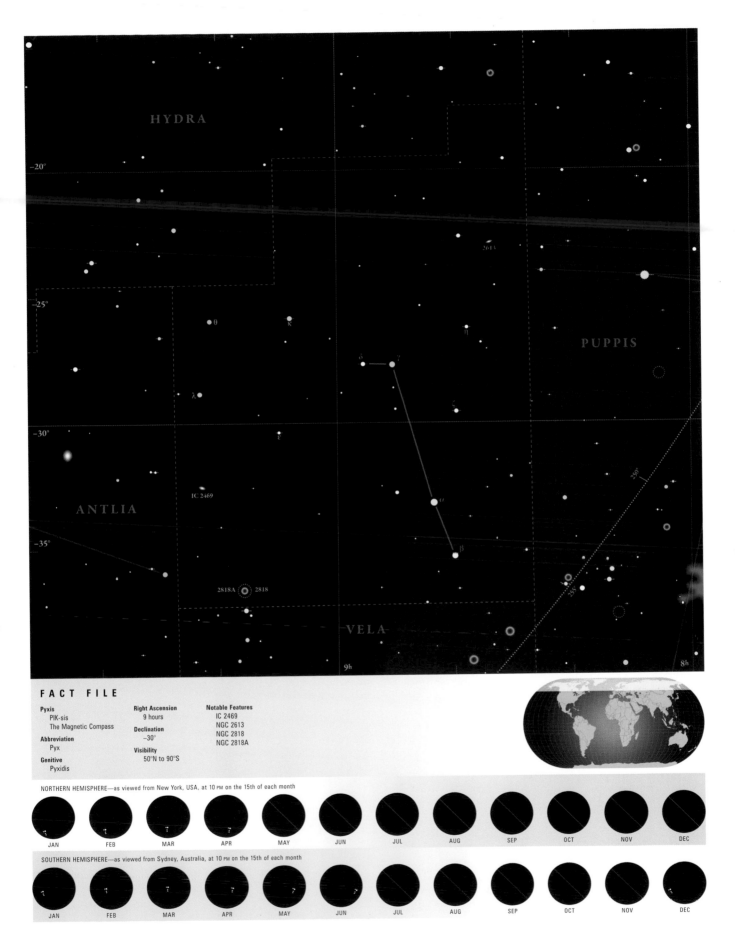

JAN FEB MAR APR MAY JUN JUL AUG SEP OCT NOV DEC

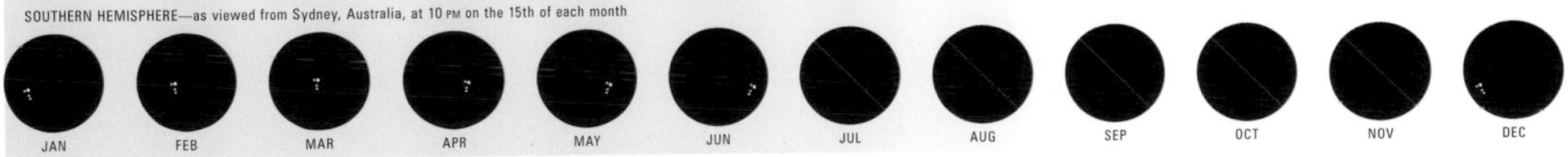

Reticulum

Could Reticulum, a small constellation in the southern skies, be the home of Unidentified Flying Objects?

CROSSHAIR OR NET?

Reticulum was unknown to ancient astronomers: It was devised by Abbé Nicolas Louis de Lacaille during his visit to the Cape Observatory between 1751 and 1752. However, Lacaille's interpretation may have been preceded by Issaac Habrecht of Strasbourg, who drew it as the Rhombus. Like many of the constellations invented by Lacaille, it represents an instrument of science or engineering—in this case, the crosshairs, or reticle—"little net"—used to measure the positions of stars in a telescope. Reticulum is also sometimes known as the Net.

THE STARS OF RETICULUM

The main stars of Reticulum, although fairly faint, form a distinctive, small, kite-shaped asterism west of the Large Magellanic Cloud. Either forming part of the diamond or falling within it, Alpha (α), Beta (β), Gamma (γ), Delta (δ), and Epsilon (ε) are all 3rd or 4th magnitude yellow or orange giant-type stars.

Alpha (α) Reticuli, shining at magnitude 3.35 from 163 light-years away, is a yellow G-class giant about 3.5 times more massive and 240 times more luminous than our Sun. It has a tiny companion star—an M-class red-dwarf that is a mere fraction of its big brother's brightness, and treads a lonely and distant orbit at least 60,000 years long.

A keen-eyed sky-watcher might just pick out the duplicity in 5th magnitude Zeta (ζ) Reticuli. A very wide binary pair of sun-clones, they are only 39 light-years away and a familiar topic for UFO enthusiasts. In the 1960s, Zeta (ζ) Reticuli shot to fame when Betty and Barney Hill claimed to have been abducted by aliens while driving their car in New Hampshire, USA. Upon their return, Betty drew a sketchy star map to indicate where the aliens were from. Amateur astronomer and UFO researcher, Marjorie Fish examined the map a few years later and asserted that it clearly identified Zeta (ζ) Reticuli. A number of claims in relation to UFOs and aliens from Zeta (ζ) Reticuli have subsequently been made—but, as always, offering little or no proof.

OBJECTS OF INTEREST

Located well away from the plane of the Milky Way, this constellation holds little of interest for the amateur sky-watcher, except for several galaxies.

NGC 1313 is easily the best, and is visible in a 4-inch (10 cm) telescope under a rural sky. Located a mere 15 million light-years away, NGC 1313 is a relatively nearby galaxy. In an 8-inch (20 cm) telescope, this galaxy has a quite large, faint, slightly blotchy halo with a slightly brighter core.

The spiral galaxy **NGC 1559** is easy to locate, only one low-power field southeast of Alpha (α) Reticuli. In an 8-inch (20 cm) telescope, it is a moderately bright, spindle-shaped galaxy with a slightly brighter center.

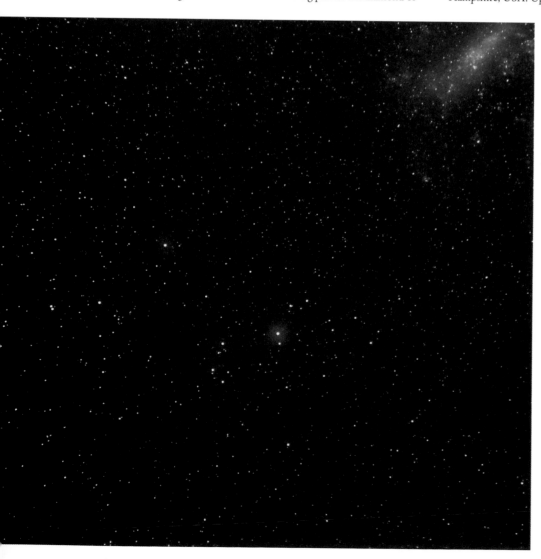

Left This is an image of the constellation Reticulum, represented by a series of four stars forming a diamond in the center of the image. It was named after the reticle, an instrument used to measure alignment.

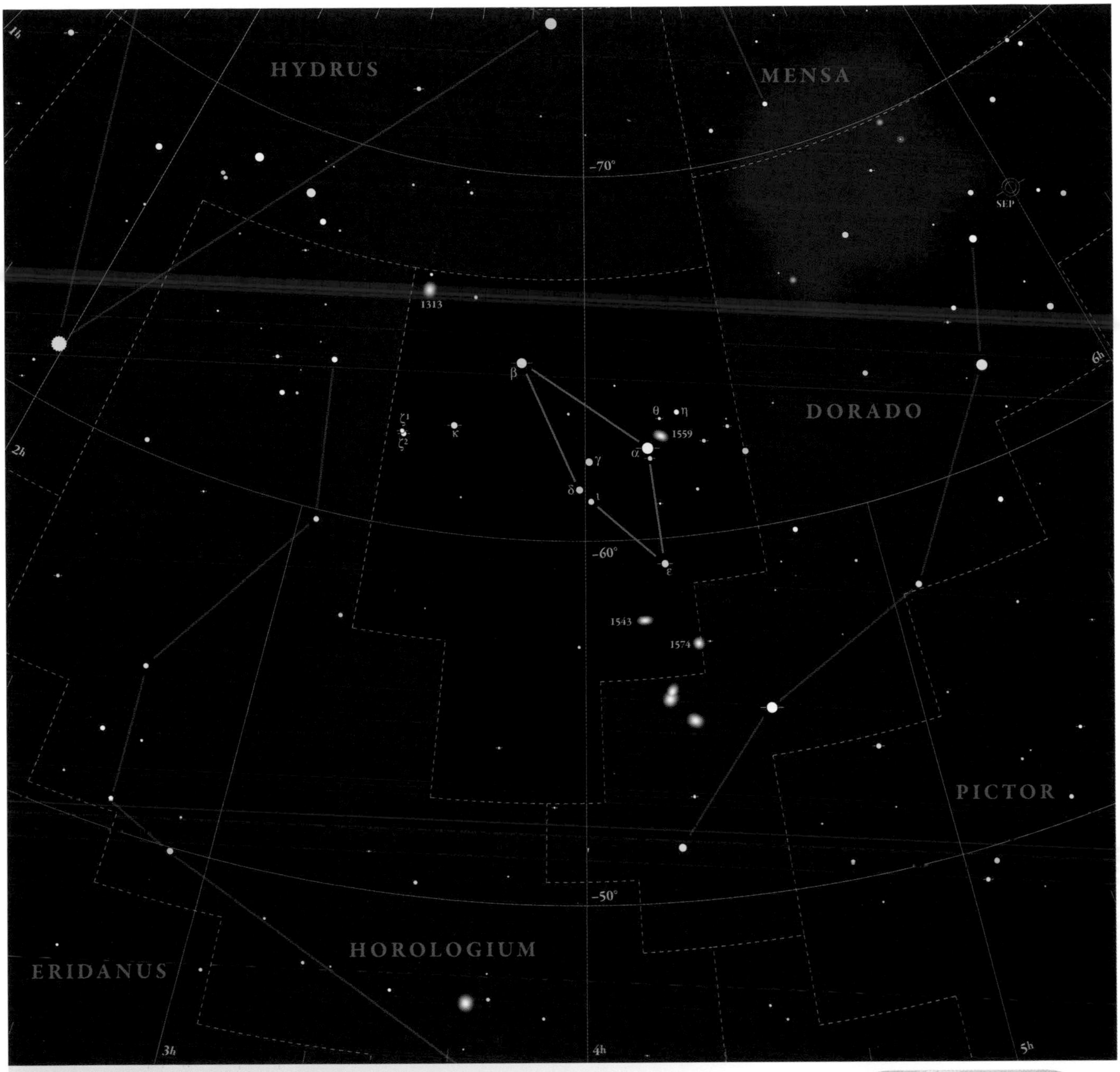
HYDRUS
MENSA
DORADO
PICTOR
HOROLOGIUM
ERIDANUS
SEP
1313
1559
1543
1574
β
θ
η
ζ¹
ζ²
κ
α
γ
δ
ι
ε
-70°
-60°
-50°
1h
2h
3h
4h
5h
6h

FACT FILE

Reticulum
reh-TIK-yuh-lum
The Reticle or The Net

Abbreviation
Ret

Genitive
Reticuli

Right Ascension
4 hours

Declination
–60°

Visibility
20°N to 90°S

Notable Features
NGC 1313
NGC 1559

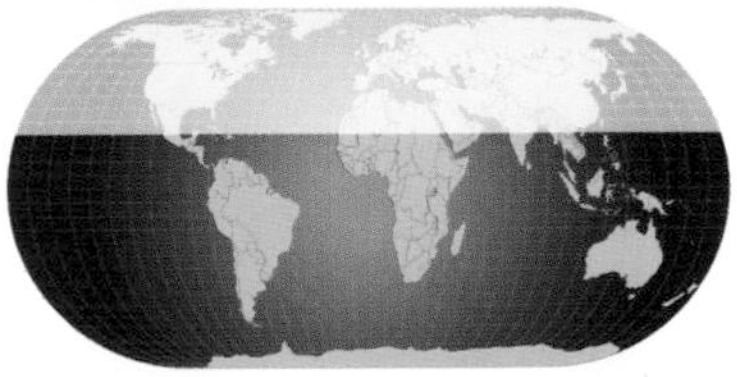

NORTHERN HEMISPHERE—as viewed from New York, USA, at 10 PM on the 15th of each month

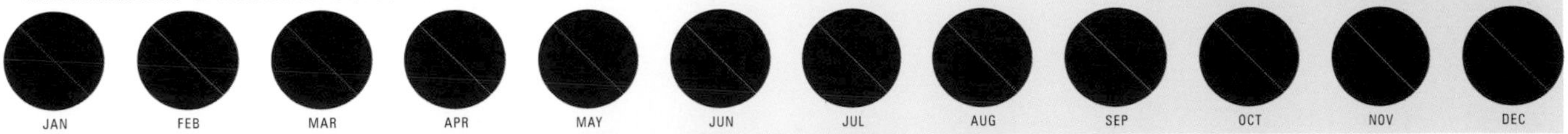
JAN
FEB
MAR
APR
MAY
JUN
JUL
AUG
SEP
OCT
NOV
DEC

SOUTHERN HEMISPHERE—as viewed from Sydney, Australia, at 10 PM on the 15th of each month

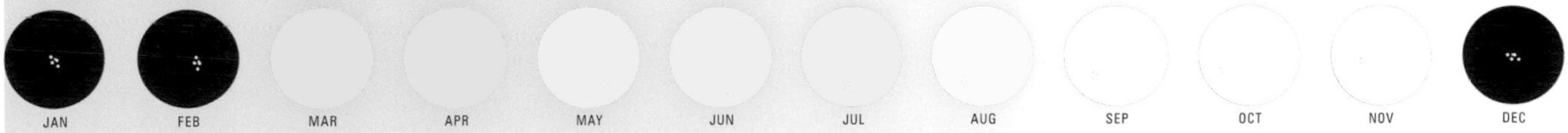
JAN
FEB
MAR
APR
MAY
JUN
JUL
AUG
SEP
OCT
NOV
DEC

Sagitta

As its Latin name suggests, the constellation of Sagitta certainly represents an arrow. But the tales of who shot this arrow are many and varied.

MYSTERIOUS TARGETS

Sagitta is an arrow, but who shot it? In Greek mythology, some believed that it was an arrow of Eros—Cupid. Others held it to be an arrow shot by the archer–centaur, Sagittarius, but it seems to be flying at a strange angle to be his.

Eratosthenes thought it was the arrow used by Apollo to slay the Cyclopes in retaliation against Zeus, because the Cyclopes manufactured the thunderbolts of Zeus that struck down Apollo's son Aesculapius, represented in the sky nearby by Ophiuchus the healer. Hyginus believed Sagitta was one of the arrows Heracles—Hercules—shot to kill the eagle that perpetually devoured the regenerating liver of Prometheus. Still others saw it as an arrow shot by Heracles to slay the Stymphalian birds as part of his labors. There are other suspects—but the mystery remains as to who shot Sagitta.

THE STARS OF SAGITTA

The faint stars of Sagitta convincingly outline an arrow—indeed, it is the only shape they could make. Side-by-side Alpha (α) and Beta (β) form the tail feathers, while the near-straight line formed by Delta (δ) and Gamma (γ) makes the shaft, with Eta (η) at the tip of the arrow.

Alpha (α) Sagittae is named Sham, from the Arabic meaning "arrow." It is one of the dimmest alpha stars in the whole sky, shining at 4th, almost 5th magnitude and a little dimmer than both Delta (δ) and Gamma (γ)—the lucida. Sham is faint only because it lies at a very distant 475 light-years. Four times heavier than our Sun, it has a very similar spectral type, but is a giant star outshining our luminary 350 times over.

OBJECTS OF INTEREST

Although it is immersed in the Milky Way, Sagitta has little to offer the amateur sky-watcher, mainly owing to its size—it is the third smallest of all constellations.

M71 is easily the most interesting object in Sagitta. This fascinating globular cluster was, until the 1970s, thought to be an extremely rich and compressed open cluster, but its spectrum reveals it to contain RR Lyrae-type variable stars, and to contain truly ancient stars (like all true globulars). Only 12,000 light-years distant, quite close for a globular, in an 8-inch (20 cm) telescope, M71 shows as a large haze with some faint resolved specks in a very rich starry field.

NGC 6886, **NGC 6879**, and **IC 4997** are all planetary nebulae with very similar characters. Each is a very tiny, well-resolved, azure disk that appears obviously different to a star only at high magnifications.

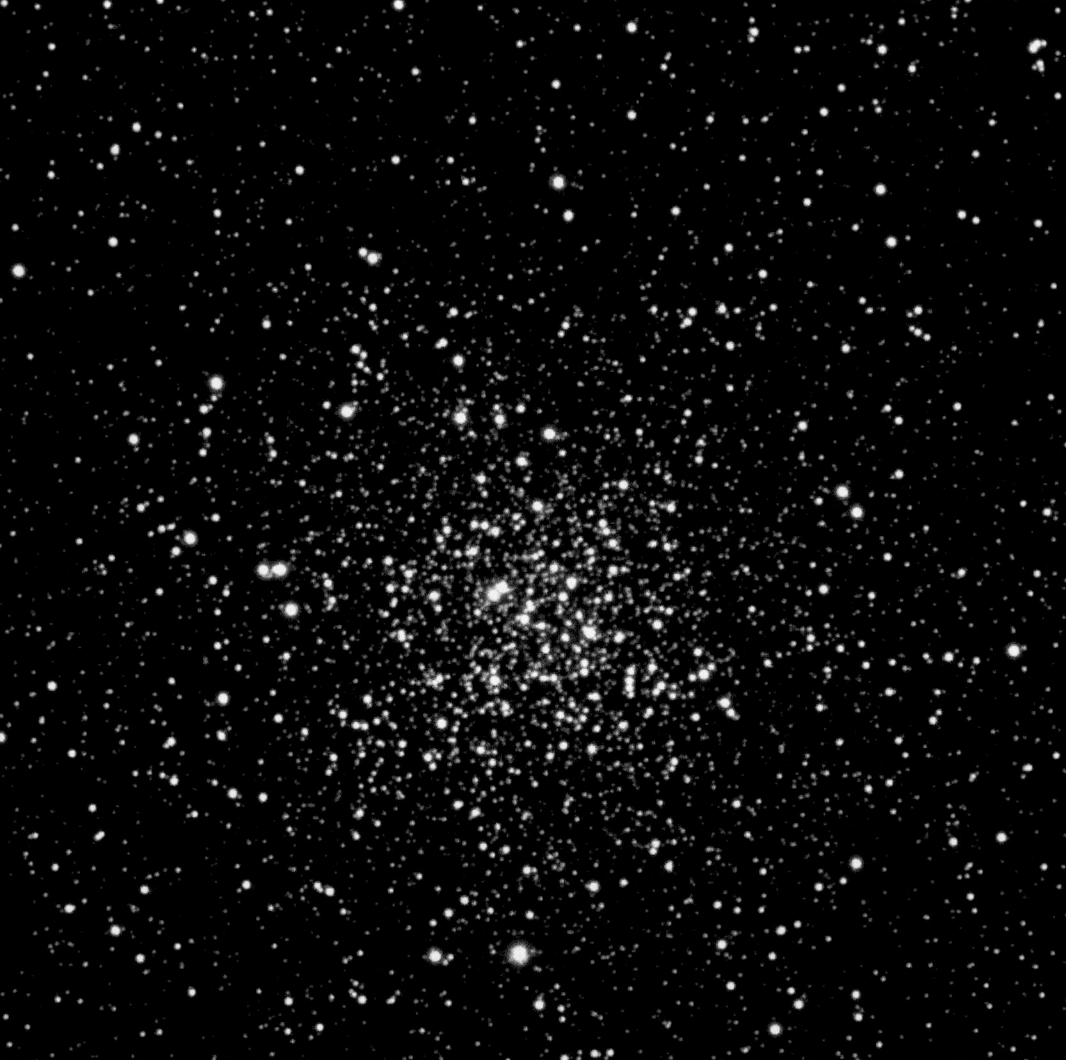

Left Satellite image of M71, a condensed open star cluster in the constellation Sagitta. At about 12,000 light-years from Earth, M71 is quite small, probably only about 30 light-years across.

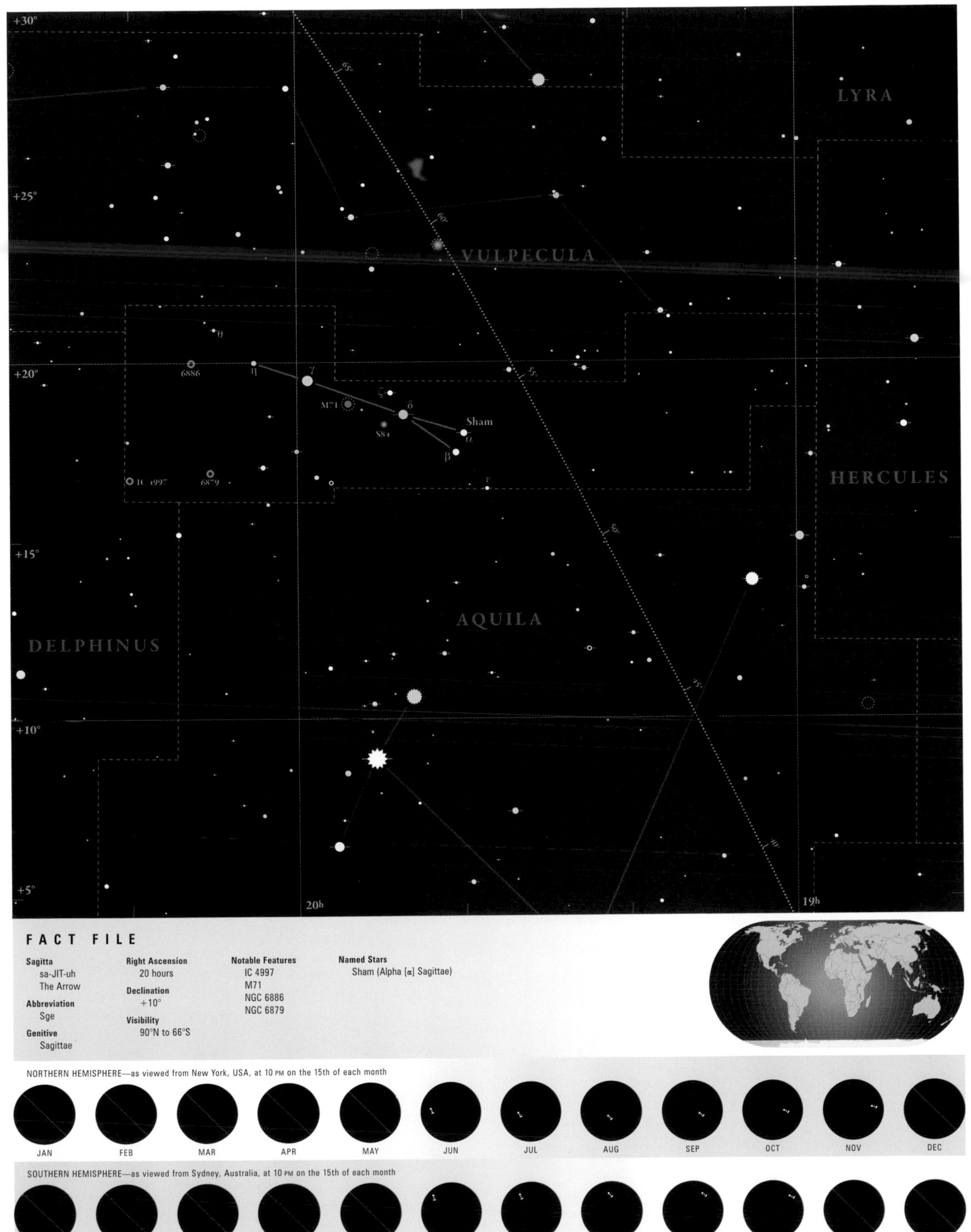

FACT FILE

Sagitta
sa-JIT-uh
The Arrow

Abbreviation
Sge

Genitive
Sagittae

Right Ascension
20 hours

Declination
+10°

Visibility
90°N to 66°S

Notable Features
IC 4997
M71
NGC 6886
NGC 6879

Named Stars
Sham (Alpha [α] Sagittae)

NORTHERN HEMISPHERE—as viewed from New York, USA, at 10 PM on the 15th of each month

JAN FEB MAR APR MAY JUN JUL AUG SEP OCT NOV DEC

SOUTHERN HEMISPHERE—as viewed from Sydney, Australia, at 10 PM on the 15th of each month

JAN FEB MAR APR MAY JUN JUL AUG SEP OCT NOV DEC

Sagittarius

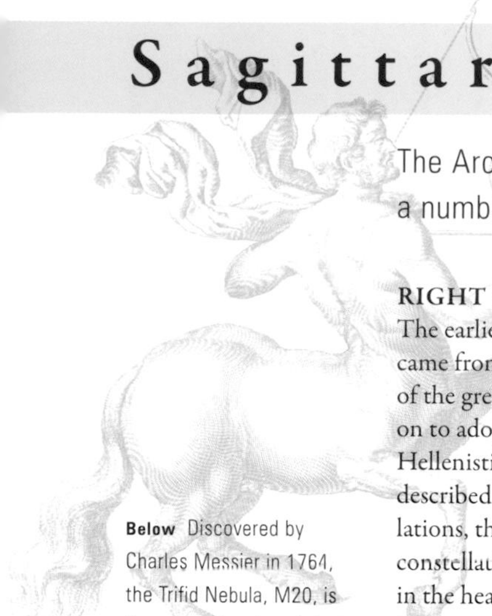

The Archer, which hosts the center of the Milky Way, has undergone a number of incarnations since its inception in ancient times.

Below Discovered by Charles Messier in 1764, the Trifid Nebula, M20, is famous for its three-lobed appearance. The red emission nebula with its young star cluster near its center is surrounded by a blue reflection nebula that is particularly conspicuous at the northern end.

RIGHT ON TARGET

The earliest recordings of the group known as Sagittarius came from the Sumerians, who identified it as Nergal, god of the great abode, or the underworld. The Greeks then went on to adopt the pattern, but there was controversy in the Hellenistic period about what the stars represented. Aratus described the pattern as though it were two separate constellations, the Bow and the Archer. Other Greeks associated the constellation with a figure of a Centaur that Chiron placed in the heavens to guide the Argonauts to Colchis. This interpretation appears to have led to the misidentification of Sagittarius as Chiron himself—but Chiron was already in the sky as Centaurus. Eratosthenes argued strongly that the stars of Sagittarius did not represent a Centaur, mainly because Centaurs did not use a bow, and instead identified it as the satyr Crotus. Crotus was the son of the god Pan and Eupheme, a nurse to the Muses—the nine daughters of Zeus. The Greeks imagined Crotus as a two-footed creature, like Pan but with a tail like a horse. Crotus is reputed to have invented archery, often went hunting on horseback, and lived on Mount Helicon among the Muses. Ultimately, the Romans seem to have settled on the interpretation that it was a Centaur, but one shooting a bow. The modern name has its roots in ancient Roman times: *Sagitta* is Latin for "arrow."

THE STARS OF SAGITTARIUS

Bayer's system of assigning Greek letters to designate individual stars within constellations did not always follow their diminishing brightness, sometimes relying instead on the importance of the star's position within the figure. Sagittarius is an excellent example of this type of ordering.

The Alpha (α) and Beta (β) stars of Sagittarius are both insignificant 4th magnitude points in the south of the constellation, far from the main pattern, and are 14th and 15th brightest, respectively. Instead, the two brightest stars are designated Epsilon (ε) and Sigma (σ) Sagittarii.

AQUILA
OPHIUCHUS
SERPENS CAUDA
SCUTUM
CAPRICORNUS
MICROSCOPIUM
CORONA AUSTRALIS
SCORPIUS
TELESCOPIUM
ARA
Albaldah
Ain al Rami
Nunki
Kaus Borealis
Terebellum
Ascella
Kaus Meridiani
Alnasi
Kaus Australis
Rukbat
Arkab Prior
Arkab Posterior
–10°
–20°
–30°
–40°
20h
19h
18h

FACT FILE

Sagittarius
sah-ji-TAIR-ee-us
The Archer

Abbreviation
Sgr

Genitive
Sagittarii

Right Ascension
19 hours

Declination
–25°

Visibility
40°N to 90°S

Notable Features
M8 (Lagoon Nebula)
M17 (Omega Nebula or Swan Nebula)
M18
M20 (Trifid Nebula)
M22
M23
M54
M55

Named Stars
Rukbat (Alpha [α] Sagittarii)
Arkab Prior (Beta[1] [β[1]] Sagittarii)
Arkab Posterior (Beta[2] [β[2]] Sagittarii)
Alnasi (Gamma [γ] Sagittarii)
Kaus Meridiani (Delta [δ] Sagittarii)
Kaus Australis (Epsilon [ε] Sagittarii)
Ascella (Zeta [ζ] Sagittarii)
Kaus Borealis (Lambda [λ] Sagittarii)
Ain al Rami (Nu[1] [ν[1]] Sagittarii)
Albaldah (Pi [π] Sagittarii)
Nunki (Sigma [σ] Sagittarii)
Terebellum (Omega [ω] Sagittarii)

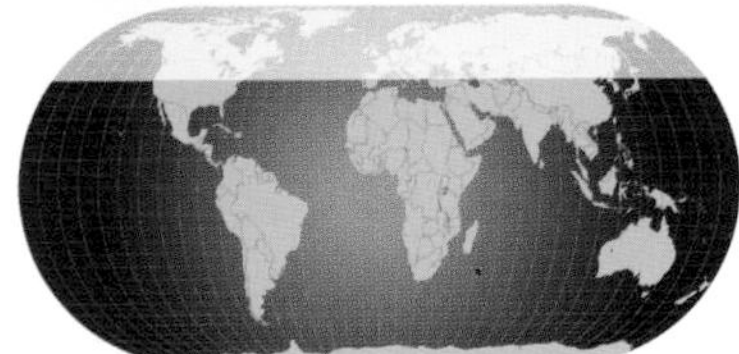

NORTHERN HEMISPHERE—as viewed from New York, USA, at 10 PM on the 15th of each month

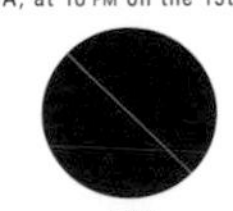

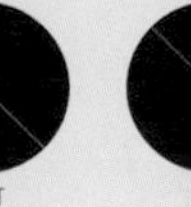
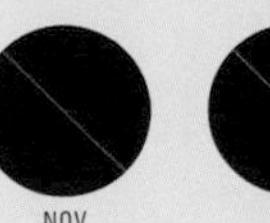

JAN FEB MAR APR MAY JUN JUL AUG SEP OCT NOV DEC

SOUTHERN HEMISPHERE—as viewed from Sydney, Australia, at 10 PM on the 15th of each month

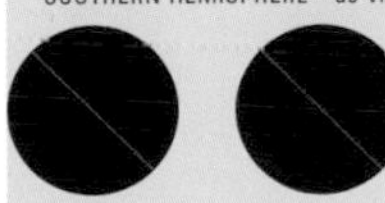

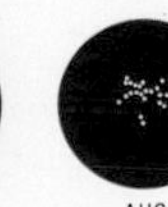

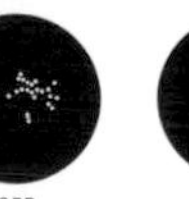

JAN FEB MAR APR MAY JUN JUL AUG SEP OCT NOV DEC

The main stars of Sagittarius are difficult to imagine as the outline of a Centaur pointing a bow and arrow toward Scorpius. However, the three main stars of the bow, Epsilon (ε), Delta (δ), and Lambda (λ) are easy to see, forming a curved line. These three stars have similar names, which originated in Arabic: Kaus Australis, Kaus Meridiani, and Kaus Borealis, meaning south, middle, and north of the bow, respectively. Gamma (γ) Sagittarii is named Alnasi ("the tip"), referring to the tip of the arrow.

Much more easily seen, the brighter stars convincingly outline a teapot. As the stars are of roughly equal brightness, this unofficial constellation is easy to make out even in suburban skies. The handle of the teapot is made by Tau (τ), Phi (ϕ), Zeta (ζ), and Sigma (σ) Sagittarii, while the lid of the teapot is formed by Phi (ϕ), Lambda (λ), and Delta (δ); the base is Zeta (ζ) and Epsilon (ε), and the spout is Epsilon (ε) and Gamma (γ). Epsilon (ε) Sagittarii, or Kaus Australis, is the brightest star in Sagittarius. It is a blue, B-class, bright giant star that is nearly 400 times more luminous than our Sun and located approximately 145 light-years distant.

Lying mid-way between the Lagoon Nebula and the Small Sagittarius Star Cloud is 4th magnitude Mu (μ) Sagittarii, otherwise known as Polis—from the Coptic for "foal." Polis is one of the true superstars of our sky; we see it from the immense estimated distance of 3,600 light-years. It is a brilliant blue, B-class supergiant, which if placed in our Solar System would nearly engulf Venus's orbit and even outshine the Sun 180,000 times over.

The Milky Way is at its very brightest in Sagittarius, for here we are looking toward the core of our Galaxy. The Large and Small Sagittarius star clouds form the most easily visible parts, and they have a strongly hazy texture when seen in a truly dark sky. They are even bright enough to be faintly visible in some urban areas. Sweeping these two adjoining regions with binoculars or a low-power telescope shows myriad knots and many darker patches.

Below Omega Nebula, M17 or NGC 6618, in the constellation Sagittarius. M17 is a bright emission nebula filled with young stars and vast lanes of opaque dust.

OBJECTS OF INTEREST

No constellation is richer in interesting objects for the amateur sky-watcher than Sagittarius; the constellation's wealth of open and globular clusters and nebulae of showpiece quality is unequaled.

The open cluster nebula complex **M8**, better known as the **Lagoon Nebula**, is located over 4,000 light-years away, yet is visible to the naked eye, even in suburbia, as a small hazy patch near the lid of the "teapot." M8 is a very large, bright, sprawling complex of emission nebulosity, with a swathe of superimposed dark nebulae and a scattering of diamond-like stars in an associated star cluster. Very low powers and a wide field are necessary to see the whole Lagoon, as it is nearly two moon-diameters across; higher powers show more of the streaky detail in its brighter parts.

A little over two moon-diameters north from the center of the Lagoon is the **Trifid Nebula**, or **M20**. Perhaps one of the most photogenic nebulae in the sky, the half moon-diameter patch of haze is cut into the three sections that give the Trifid its very distinctive appearance. Just north of the Trifid is the compact and bright open cluster **M21**, which a 6-inch (15 cm) telescope shows beautifully.

Two of Messier's Sagittarius globular clusters are close to Lambda (λ) Sagittarii—M22 and M28. **M22** is probably the third best of all the globular clusters. In binoculars it looks like a small misty cloud, but small telescopes show it nearly a half moon-diameter across, with more than a hundred faint resolved stars scattered over a halo showing slight compression to the center.

M28 has a different character: Much further away, it is a small, bright, misty ball of haze, difficult to resolve into stars, with a highly condensed center.

Three excellent open clusters that are shown well by small instruments surround the Small Sagittarius Star Cloud to the east, west, and north. **M23** to the west is a medium-sized, bright, and well-resolved cluster with an immensely rich backdrop. **M18** to the north of the cloud is a more compact grouping with a distinctive V-shaped pattern near the center that seems to have a darker ring around it. To the east of the cloud is **M25**, a very bright and scattered splash of stars, a moon diameter in size, that looks best in small low-power telescopes. The globular cluster **M54** is found close to Zeta (ζ) Sagittarii and looks similar in most respects to the nearby M28. As one of our Milky Way's largest and intrinsically brightest globulars, it has been harboring a secret only recently unlocked by astronomers. M54 is an "adopted" child—it was once part of a small elliptical galaxy called the Sagittarius Dwarf Elliptical Galaxy that our mighty Milky Way is in the process of dismembering and devouring. In the southeast of Sagittarius is the lovely globular cluster **M55**. In binoculars it has a faint cloud-like appearance, but an 8-inch (20 cm) telescope resolves it into a patch of tiny stars of nearly half a moon-diameter, which shows virtually no concentration to the center. There are several other bright and interesting globular clusters in Sagittarius, along with many more fainter ones.

Last, but by no means least, is the spectacular emission nebula **M17**, also known as the **Omega Nebula** or the **Swan Nebula**. An 8-inch (20 cm) telescope at low power shows a medium-size, bright patch of haze that looks much like the numeral 2, but with a greatly extended baseline. Moderate magnifications show blotchy stringy detail, particularly in the hook of the 2, where there is a very dark zone surrounded by brighter material.

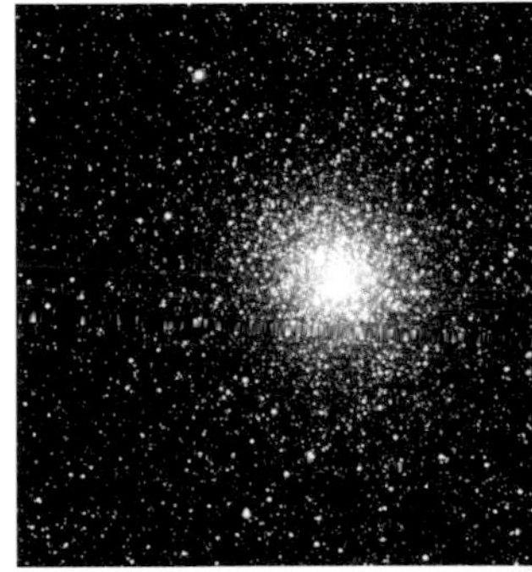

Above M22, or NGC 6656, a globular cluster in the constellation Sagittarius. A conspicuous naked-eye object, M22 is the brightest globular cluster visible from the Northern hemisphere.

Top Hubble Space Telescope image of the Omega Nebula, a hotbed of newly born stars wrapped in colorful blankets. The nebula resides 5,500 light-years away, in the constellation Sagittarius.

Scorpius

A beguiling associate of the zodiac, Scorpius is one of the most easily recognized constellations. It needs little or no imagination to make out the figure among its bright stars.

STING IN THE TAIL

Scorpius is one of the oldest constellations in the sky and was known as far back as the Sumerian civilizations, over 5,000 years ago, who knew it as Gir-Tab—the Scorpion.

In Greek mythologies, Scorpius is almost universally associated with Orion. Orion was a mighty hunter, who became so full of himself that he finally boasted that he could, and would, slay any and all animals on the face of the Earth. To counter Orion, Gaia sent a scorpion to contend with him. Long they battled, and when Orion finally became tired and slept, the scorpion stung and killed him.

An alternative version sees Orion stung to death by the scorpion sent by Gaia after he attempted to ravish Artemis, the Greek goddess of the hunt. Either way, pride was Orion's downfall and he was positioned in the sky almost exactly opposite his enemy, the scorpion.

Ever since, Orion rises only when Scorpius sets, and as the scorpion rises Orion flees over the western horizon. The Greeks also saw Scorpius as a constellation of two parts: The claws and the tail/body. Later, the Romans removed what the Greeks saw as the extended claws of Scorpius and turned it into a new constellation—Libra.

THE STARS OF SCORPIUS

Like the constellation of Sagittarius, the Bayer designations of the brighter stars of Scorpius do not follow their order of brightness. For example, after Alpha (α) Scorpii, the next brightest star is Lambda (λ).

Starting in the middle of Scorpius, the ruddy glow of 1st magnitude Alpha (α) Scorpii, or Antares, is the heart of the scorpion. The tail is marked out by the long curling string of stars through Tau (τ), Epsilon (ε), Mu (μ), Zeta (ζ), Eta (η), Theta (θ), Iota (ι), and Kappa (κ), with Lambda (λ) and Upsilon (υ) as the stinger at the end.

In the more modern version, the claws drawn inward toward the scorpion's head are formed by the slightly curved line made by Nu (ν), Beta (β), and Delta (δ) on one side, with Pi (π) and Rho (ρ) opposing. Alternatively, you could imagine the constellation as the Greeks did, and use Alpha (α) and Beta (β) Librae to mark the fully extended claws. These two stars have common names from Arabic that disclose their true heritage—Zubenelgenubi and Zubenelschemali, meaning, respectively the southern and northern claw of the scorpion.

Alpha (α) Scorpii is called Antares, from the Greek meaning the "rival of Mars"—Ares was the Greek god of war; to the Romans he was Mars. The fifteenth brightest star in the night sky, and shining at us from a distance of 600 light-years, it is the second closest red supergiant star to our Sun. More than 60,000 times more luminous than the Sun, Antares weighs in 15 times heavier than our luminary, and is a near certainty to explode as a supernova some time in the next million years. Antares is so huge that were it placed in our Solar System, its surface would reach mid-way between the orbits of Mars and Jupiter.

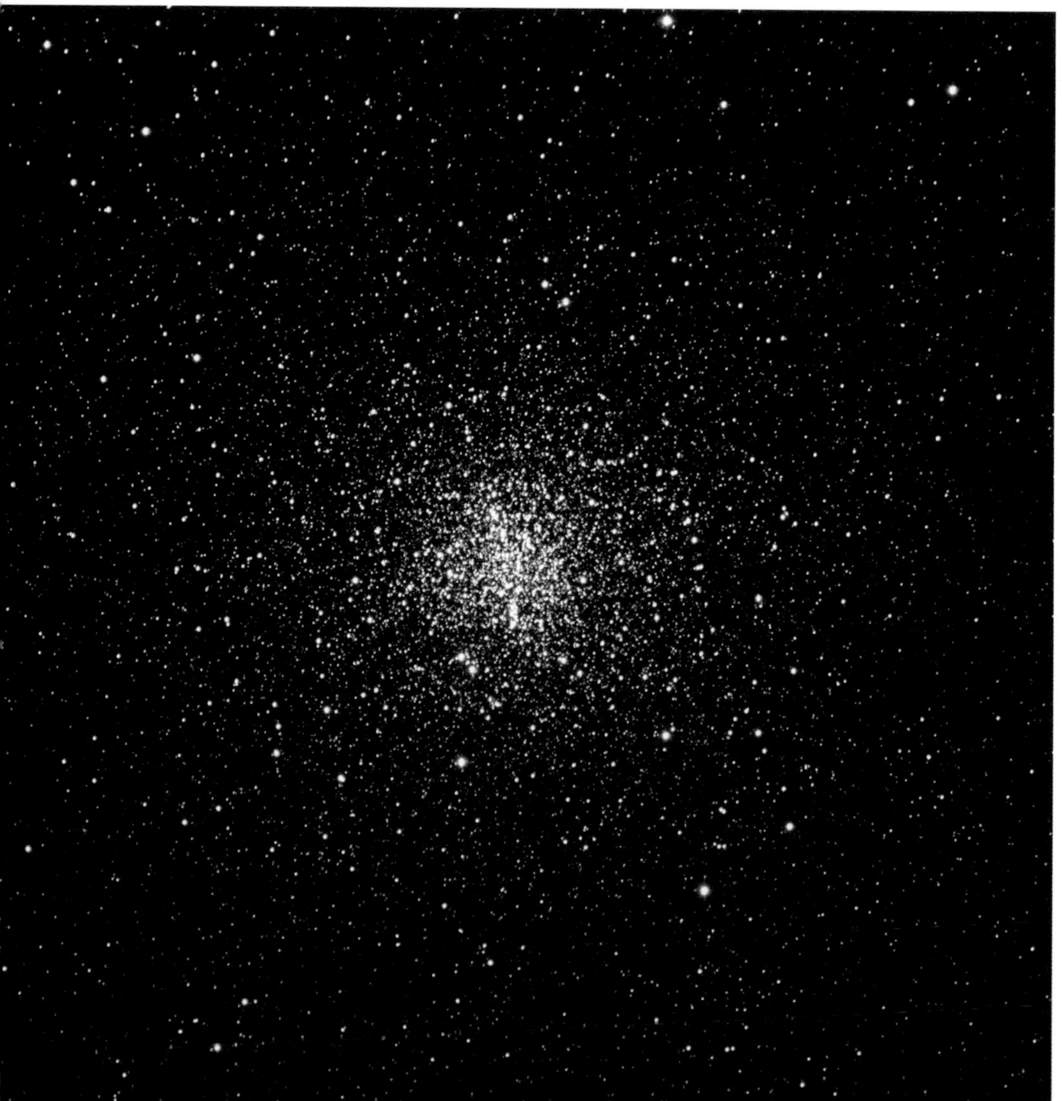

Left M4, a globular cluster in the constellation Scorpius, is about 7,000 light-years away from us, making it one of our nearest globular clusters. M4 shows an unusual central bar structure, and would be much brighter if not for considerable intervening interstellar dust, which also accounts for its slightly brownish color.

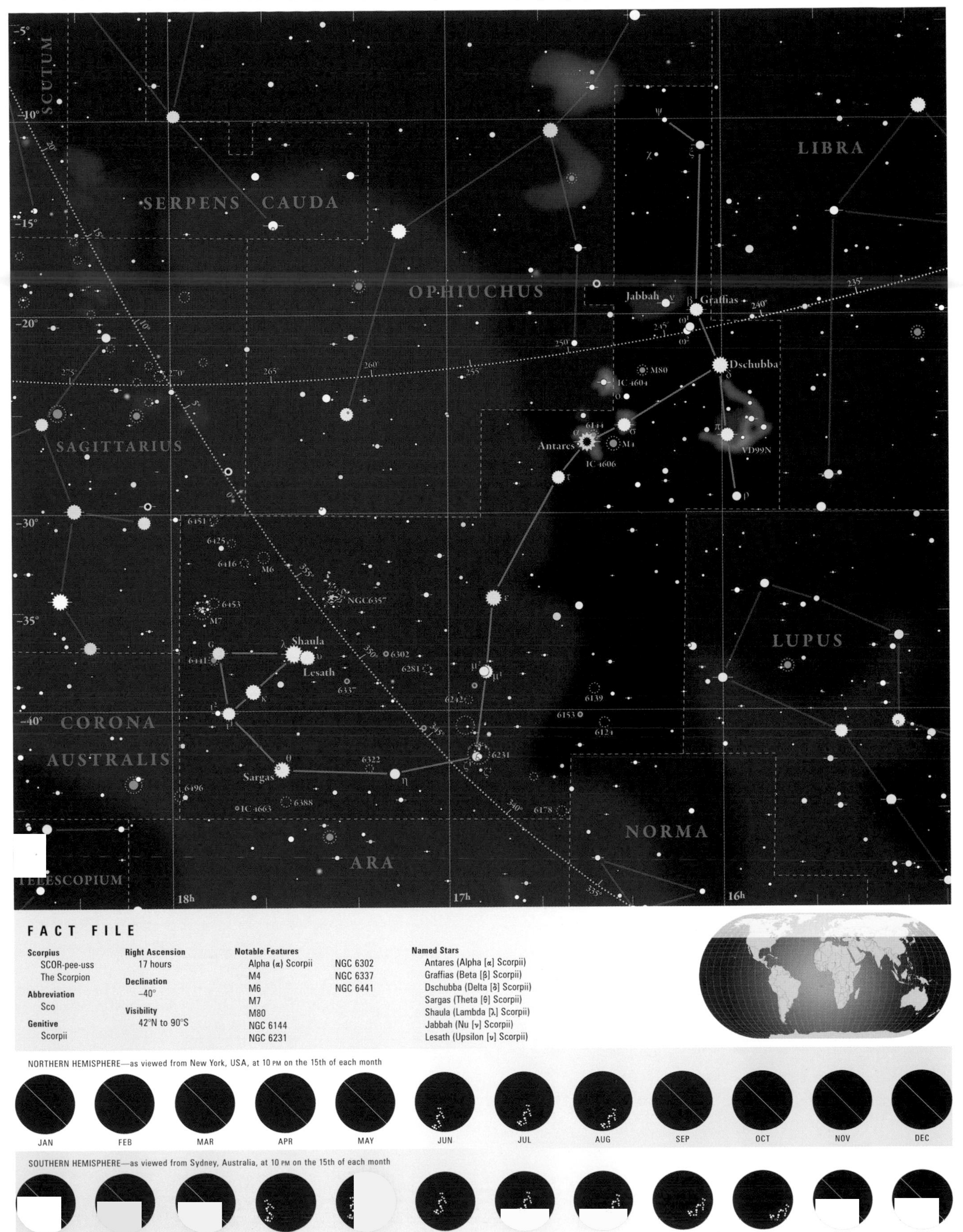

FACT FILE

Scorpius
SCOR-pee-uss
The Scorpion

Abbreviation
Sco

Genitive
Scorpii

Right Ascension
17 hours

Declination
–40°

Visibility
42°N to 90°S

Notable Features
Alpha (α) Scorpii
M4
M6
M7
M80
NGC 6144
NGC 6231
NGC 6302
NGC 6337
NGC 6441

Named Stars
Antares (Alpha [α] Scorpii)
Graffias (Beta [β] Scorpii)
Dschubba (Delta [δ] Scorpii)
Sargas (Theta [θ] Scorpii)
Shaula (Lambda [λ] Scorpii)
Jabbah (Nu [ν] Scorpii)
Lesath (Upsilon [υ] Scorpii)

Delta (δ) Scorpii, also named Dschubba, is currently behaving in an unusual manner. Known for many years to be a multiple star with four recognized components, it was supposed to be just one more garden-variety B-class dwarf star about 12 times heavier and 14,000 times as luminous as our Sun—and decidedly constant in brightness. In the year 2000, without warning, it began to slowly brighten. Normally the fifth brightest in Scorpius, within a few months it had risen to surpass all except Antares, and was approaching 1st magnitude. Its spectrum also underwent remarkable change, with the star transforming itself into a "Be" star with emission lines. Since then it has very slowly varied in brightness—that is stellar evolution happening before your very eyes!

OBJECTS OF INTEREST

There is a great deal for the amateur sky-watcher to see in Scorpius. Like the adjoining constellation Sagittarius, Scorpius is blessed with a fine array of showpiece objects, particularly open and globular clusters. The tail end of Scorpius is particularly rich, being deeply immersed in some of the most brilliant portions of the Milky Way.

Near Antares are two bright globular clusters, **M4** and **M80**, with vastly different characters. Located just over 10,000 light-years distant, M4 is probably the nearest globular cluster to our Solar System. In a 6-inch (15 cm) telescope, its large halo is well resolved into many faint stars with only slight compression toward the center. A distinctive feature of M4 is a line of stars of nearly equal spacing and brightness that cross its core. Even closer to Antares, and nearly buried in its glow, is the misty, almost ghostly form of the faint uncompressed globular cluster **NGC 6144**.

Below Known as the Bug Nebula, NGC 6302, in the constellation of Scorpius, is particularly interesting because it has a very hot central star.

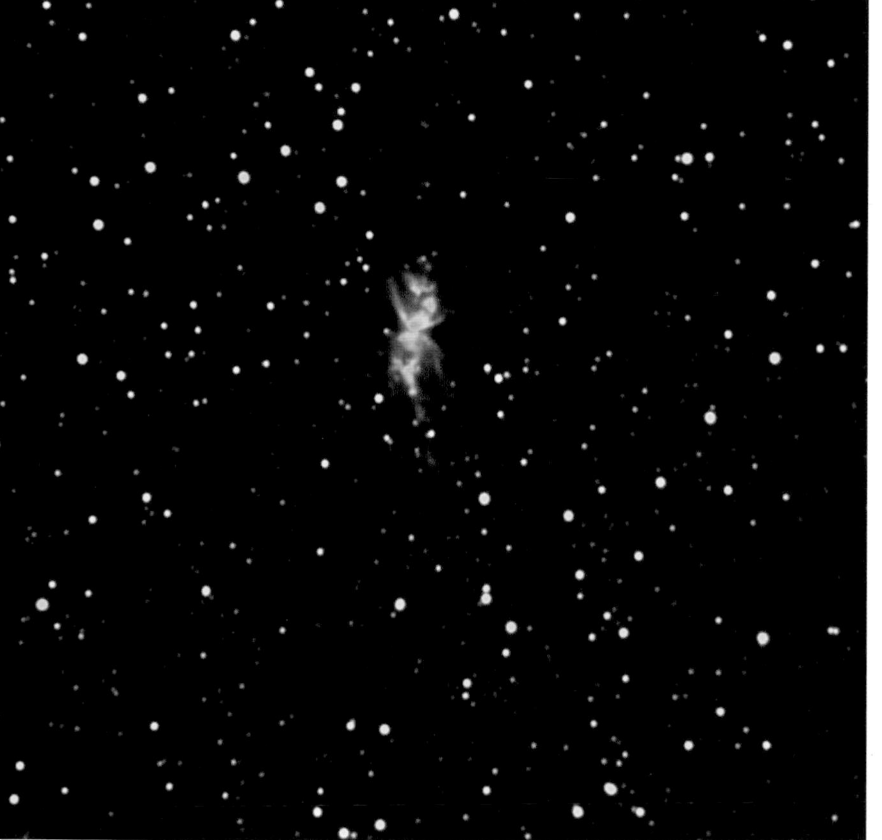

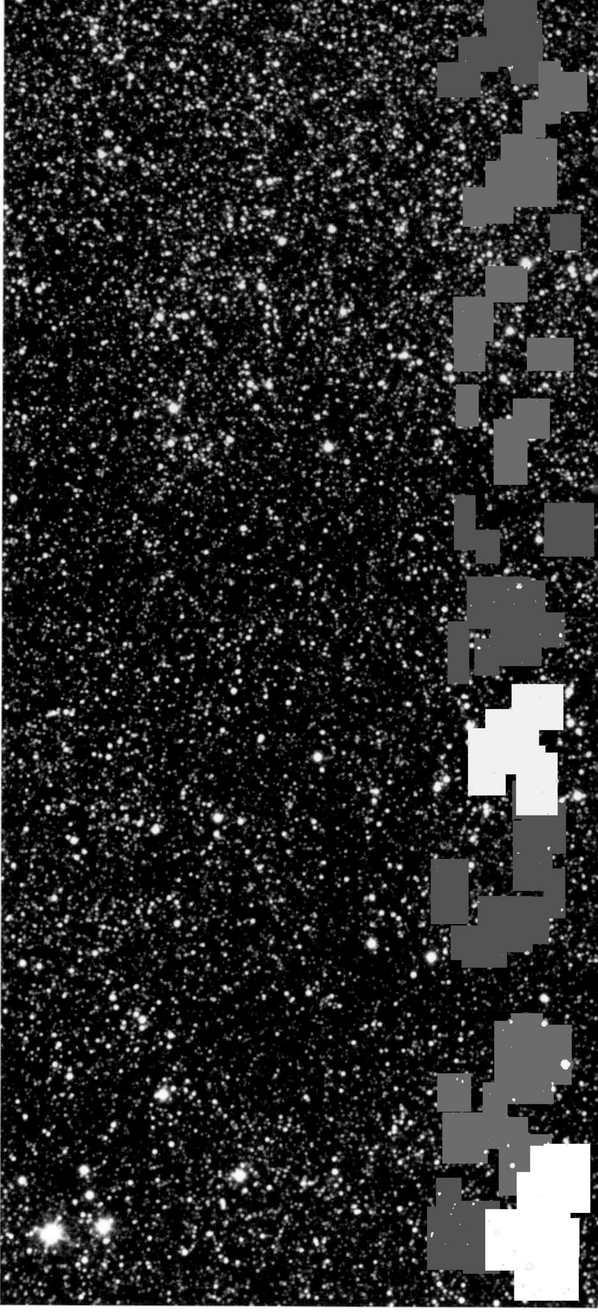

M80 lies about mid-way between Antares and Beta (β) Scorpii. Much smaller and somewhat fainter than M4, M80 is sharply compressed toward its center and is difficult to resolve in telescopes less than 10 inches (25 cm).

Antares itself is an interesting binary star that is notoriously difficult to resolve. The companion is a considerably fainter blue B-class dwarf star that sits close within Antares's glow and is often reported as looking green. This apparent coloration of the companion is due to Antares's strongly reddish hue. Chances of seeing this timid star are improved by observing in twilight and using high magnifications. **Nu (ν) Scorpii**, near Beta, is one of the showpiece multiple stars of the southern skies. Much like the famed Double-Double Epsilon (ε) Lyrae, it consists of a pair of pairs —the brighter ones are very close and difficult to resolve while the fainter pair is a little easier.

Mu¹ (μ^1) and **Mu²** (μ^2) **Scorpii** is the famous naked-eye pair along the back of the scorpion. While the stars seem so beautifully matched in brightness, they are only a chance

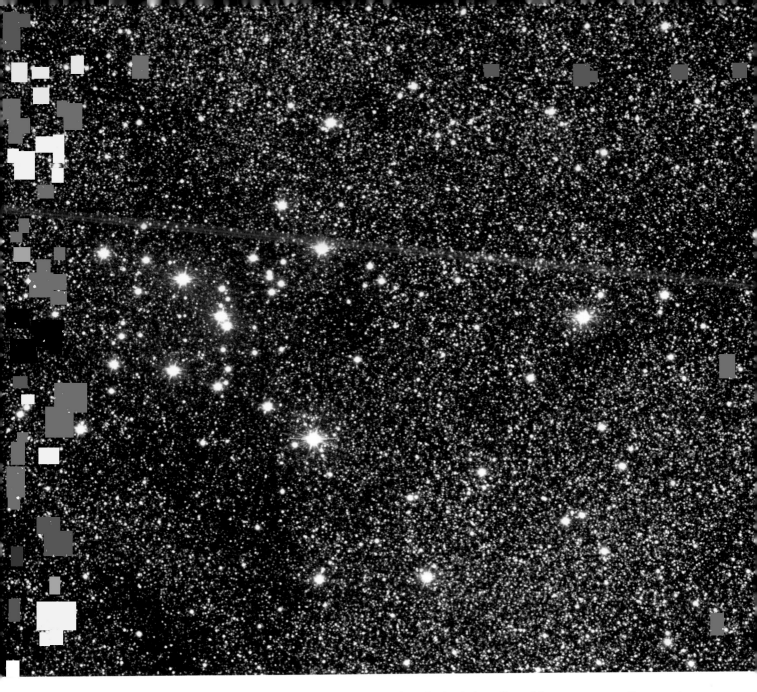

Above Known as early as the year 130 CE, when it was mentioned by Ptolemy, M7 is an open star cluster in the constellation Scorpius. Its 80-odd stars spread over some 20 light-years and are in excess of 200 million years old.

alignment and not a true binary. One star further along the tail from Mu is **Zeta (ζ) Scorpii** —itself also a naked-eye double star—and the remarkable open cluster **NGC 6231** . It is difficult to understand how the famous comet hunter Charles Messier could find M4, M80, M6, and M7 in Scorpius seemingly without observing this incredibly compact, bright, well-resolved cluster bursting with O- and B-class supergiant stars that are so beautifully displayed in a 4- to 6-inch (10 to 15 cm) telescope. Nearby, among many other open clusters, is **NGC 6281** , a small, bright, open cluster that strongly resembles the outline of a Christmas tree replete with fairy lights.

Near the stinger stars of the scorpion is the famous open star cluster pairing of **M6** and **M7** . M7 is a fairly conspicuous, small, hazy patch known since antiquity—even the Greek astronomer Ptolemy cataloged it. A few stars are resolved in binoculars, but in a small telescope at low power it is a huge dazzling splash of stars about two moon-diameters across.

M6 is very nearby, and only a little smaller and fainter; it is known as the **Butterfly Cluster** because its brighter stars clearly mark the outline of a butterfly at rest with outstretched wings. It is a beautiful object in a 6-inch (15 cm) telescope. Nearby, almost sitting atop of the 4th magnitude star G Scorpii, is the diminutive but strongly concentrated globular cluster **NGC 6441** , which is shown well—but not resolved—in a 6-inch (15 cm) telescope.

Scorpius is also home to more than a few beautiful planetary nebulae, the best of which are **NGC 6302** and **NGC 6337** , both located near the stinger stars of Scorpius—Lambda (λ) and Upsilon (υ). NGC 6302 is also called the **Bug Nebula**, and large amateur telescopes show it looking somewhat like a small bluish ant. An 8-inch (20 cm) telescope will show the small elongated body of the bug, and some feathery detail at the edges hints at its legs. NGC 6337 is a very thin and picturesque perfect ring of faint mist that is sometimes called the **Cheerio Nebula** because it resembles a piece of that famous breakfast cereal.

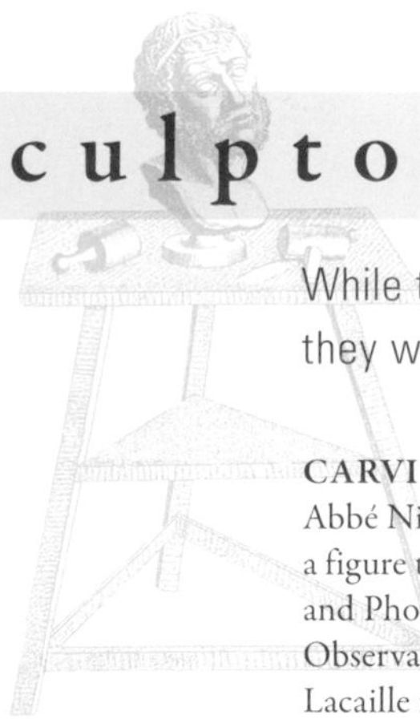

Sculptor

While the faint stars of Sculptor were visible to ancient astronomers, they were not collected into a constellation of their own.

CARVING A NICHE

Abbé Nicolas Louis de Lacaille devised a figure to fill the space between Cetus and Phoenix during his visit to the Cape Observatory between 1751 and 1752. Lacaille initially named the constellation in French as "l'Atelier du Sculpteur"—the Sculptor's Studio—and drew it as a tripod table with a stone bust and tools. He later Latinized it to "Apparatus Sculptoris," which over time has been worn down to simply Sculptor. Since the time of Lacaille, there has been no significant change to its stars or form.

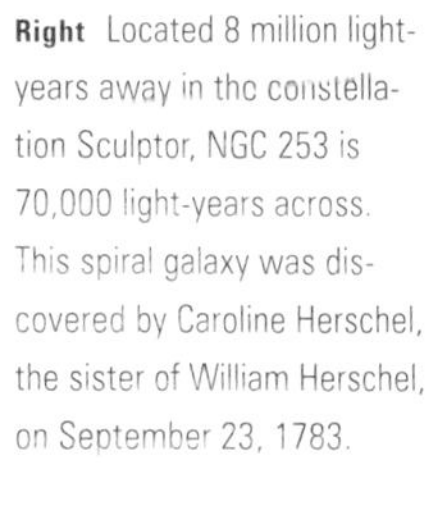

Right Located 8 million light-years away in the constellation Sculptor, NGC 253 is 70,000 light-years across. This spiral galaxy was discovered by Caroline Herschel, the sister of William Herschel, on September 23, 1783.

THE STARS OF SCULPTOR

Found directly north of Phoenix, the stars of Sculptor look like a random collection of faint stars, bearing no resemblance to any figure. Alpha (α), Beta (β), and Gamma (γ) are all at the faint end of 4th magnitude, while Delta (δ) is 5th magnitude—just a fraction dimmer than the other three. The lucida of the constellation, Alpha (α) Sculptoris, is only faint because it is far away. Fully 670 light-years distant from our Sun, Alpha is a B-class blue giant star that is 1,700 times more luminous and over 5 times more massive.

OBJECTS OF INTEREST

What Sculptor lacks in bright stars or a familiar shape, it more than makes up for in interesting objects for the amateur sky-watcher. Located at almost right angles to the plane of the Milky Way, Sculptor is rich in bright galaxies and home to our Local Group of Galaxies' nearest neighbors in space, the **Sculptor Group**.

The Sculptor Group is a minor cluster of mainly spiral galaxies centered 6 million light-years from the Milky Way. The largest and brightest is **NGC 253**, a huge spiral galaxy that in a 6-inch (15 cm) telescope appears as a long ellipse of haze approaching a moon-diameter in length, with a weakly brighter center. Further south is **NGC 300** another spiral that appears faint only because its light is spread over such a large area. A dark rural sky is needed to show it well, but in an 8-inch (20 cm) telescope it appears as an oval patch about half a moon-diameter in size.

NGC 55 is probably the closest galaxy to us in the group—little more than 3.5 million light-years away. It is a similar type of galaxy to the Large Magellanic Cloud, but is seen edge-on.

An 8-inch (20 cm) telescope shows a faint sliver almost a moon-diameter long, with some patchy brightenings off-center. In the same group, **NGC 7793** is a slightly smaller spiral.

Seeming a little out of place, the globular star cluster **NGC 288** is quite close to NGC 253. A 6-inch (15 cm) telescope shows a medium-size patch of haze with many stars sewn in, but relatively little concentration to the center of the cluster.

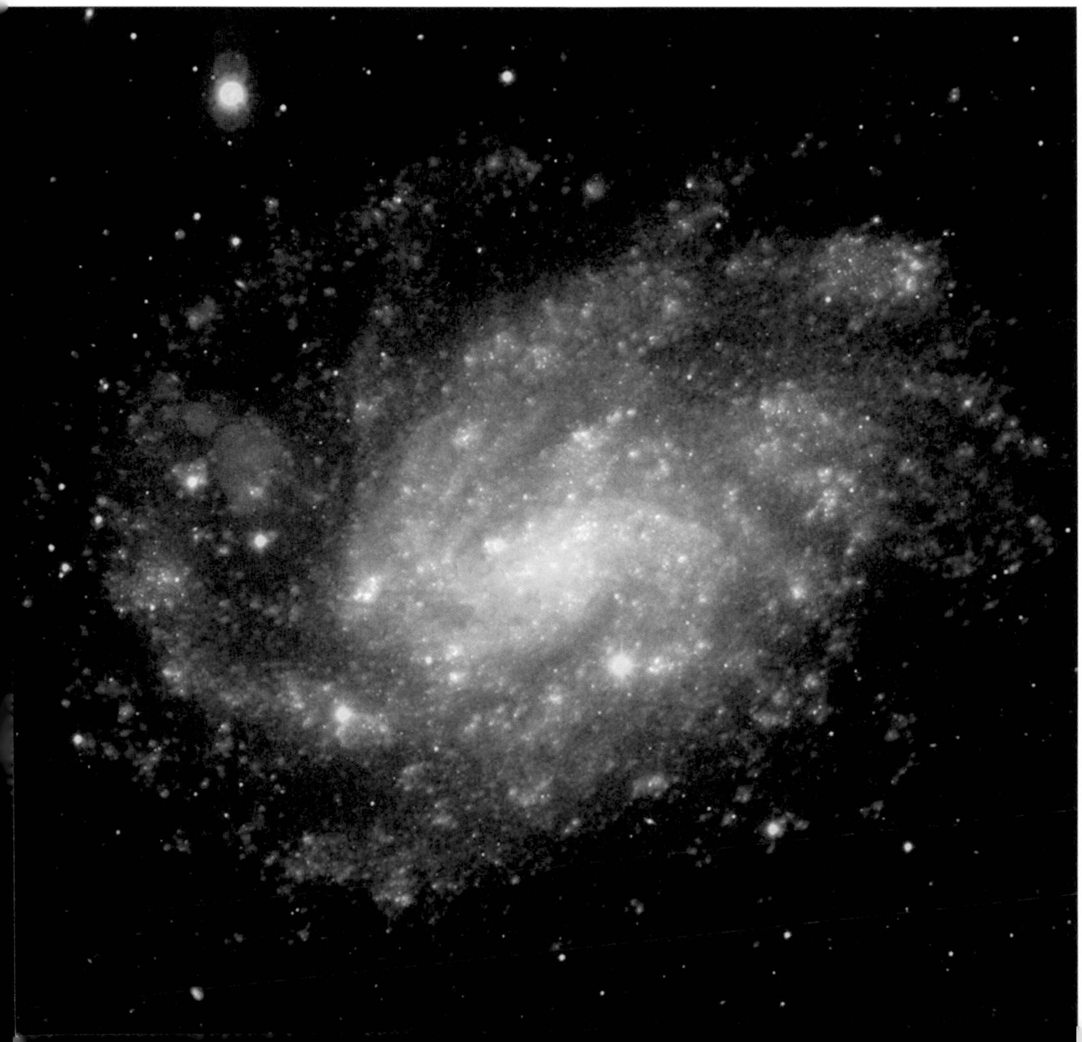

Left This composite image of NGC 300 shows star formation. Young hot blue stars dominate the outer spiral arms, while older stars congregate in the nuclear regions that appear yellow-green. Gases heated by hot young stars, and shocks due to winds from massive stars and supernova explosions, appear in pink Located nearly 7 million light-years away, NGC 300 is a member of the Sculptor Group.

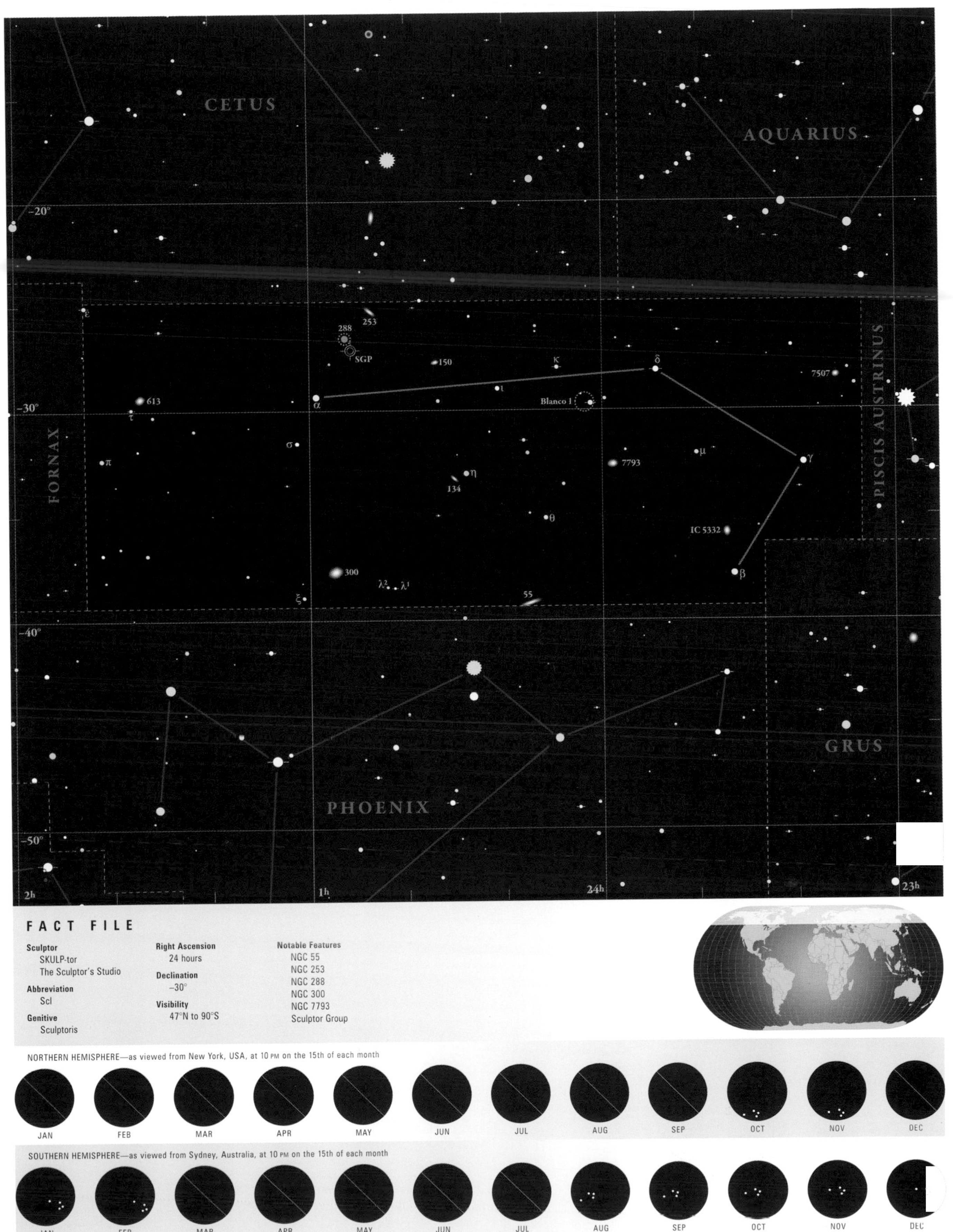

FACT FILE

Sculptor
SKULP-tor
The Sculptor's Studio

Abbreviation
Scl

Genitive
Sculptoris

Right Ascension
24 hours

Declination
−30°

Visibility
47°N to 90°S

Notable Features
NGC 55
NGC 253
NGC 288
NGC 300
NGC 7793
Sculptor Group

NORTHERN HEMISPHERE—as viewed from New York, USA, at 10 PM on the 15th of each month

JAN FEB MAR APR MAY JUN JUL AUG SEP OCT NOV DEC

SOUTHERN HEMISPHERE—as viewed from Sydney, Australia, at 10 PM on the 15th of each month

JAN FEB MAR APR MAY JUN JUL AUG SEP OCT NOV DEC

Scutum

The stars of the tiny constellation Scutum were easily visible to ancient cultures, but no figure was associated with the group except in China, where it was called "Pien"—the Heavenly Casque.

HEAVENLY SHIELD

Johannes Hevelius invented "Scutum Sobiescianum"—Sobieski's Shield—in 1684, when a chart of the constellation first appeared in *Acta Eruditorum*, a leading scientific journal of the day. Scutum is the only surviving constellation with a "political" background. It commemorates King John Sobieski III of Poland, who distinguished himself as a great military strategist in Poland's wars with the Turks.

The fact that King John also provided considerable assistance to Hevelius in rebuilding his observatory after it was razed by fire in 1679 is purely coincidental. Since that time, there has been no significant change to the constellation, save that the name has been shortened to simple Scutum.

THE STARS OF SCUTUM

The stars comprising Scutum are a faint group and are sometimes difficult to make out because of the bright Milky Way background. Alpha (α) and Beta (β) are 4th magnitude and make a Y-shaped asterism with 5th magnitude Gamma (γ) and Zeta (ζ); all others are fainter than 5th magnitude.

Scutum is also home to the Scutum Star-cloud—a very bright patch of Milky Way we can view well because of the lack of intervening dust and gas.

Alpha (α) Scuti is a cool K-class orange giant star about 1.7 times more massive than our Sun. It is probably in the last stages of its evolution, brightening and expanding into a full-blown red giant for the last time before it dies as a white dwarf. Located 175 light-years distant, it is 21 times bigger than our Sun and at least 130 times more luminous.

OBJECTS OF INTEREST

Deeply immersed in the Milky Way, Scutum is awash with open star clusters. The most interesting of these is **M11**, also known as the **Wild Duck Cluster**. A 6-inch (15 cm) telescope gives a good view of its compact, strong, V-shaped outline, with one bright leader at the apex of the V. Except for that one bright star, the remaining multitude are homogenous in brightness.

Close to M11 is the fascinating variable star **R Scuti**. Similar in mass to our Sun, R Scuti is very much in its dotage—a star that is probably transforming itself from red giant to white dwarf. It varies in brightness between 5th and 8th magnitude over a period of 71 days, with alternating "bright" and "faint" minima.

M26 is another compact cluster of faint stars, with a diamond-shaped asterism of four brighter ones near the center. This cluster looks particularly good in a 6-inch (15 cm) telescope.

NGC 6712 and **IC 1295** is an unusual pairing of a medium-bright globular star cluster with a fainter planetary nebula in the same field of view, which makes an interesting observation in an 8-inch (20 cm) telescope.

Above M11 is one of the richest and most compact open clusters, with nearly 3,000 stars packed into about 20 light-years diameter. Located in the constellation of Scutum, it is also called the Wild Duck Cluster.

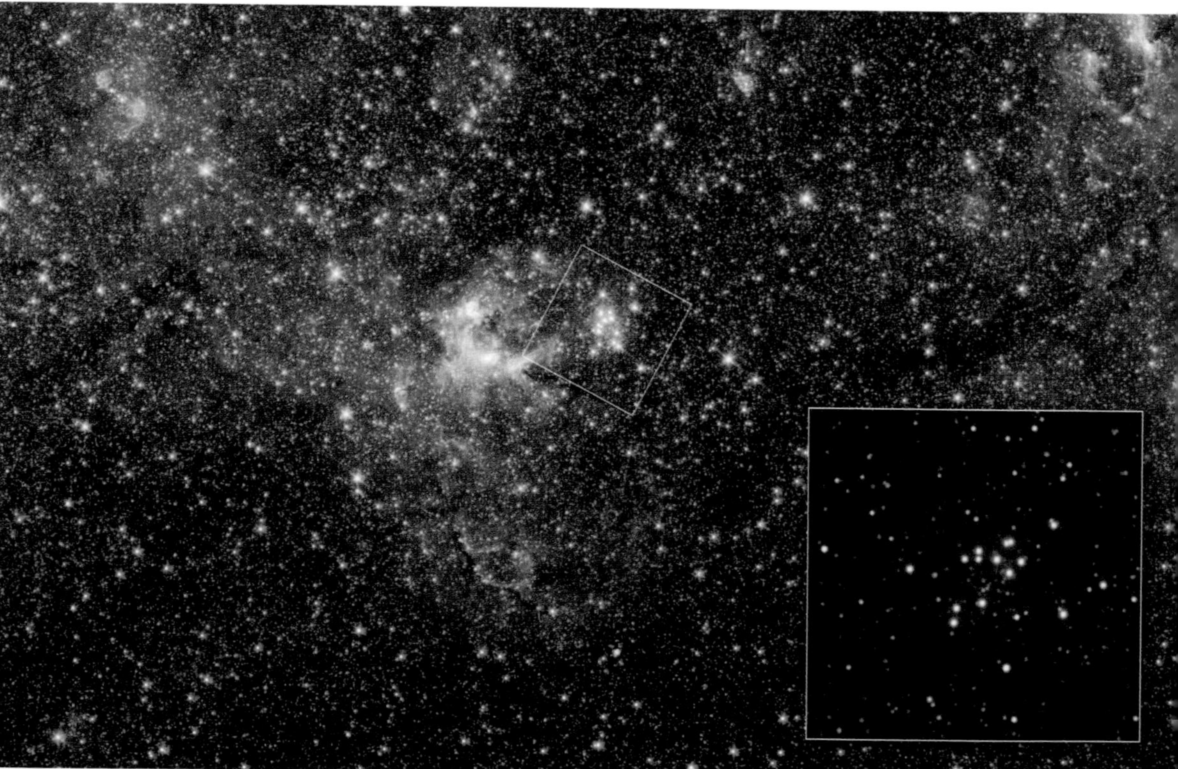

Right This Hubble Space Telescope image captures a galaxy cluster lying about 18,900 light-years away, in the direction of the constellation Scutum. The inset image, showing a close-up of the cluster, was captured by the Two Micron All Sky Survey (2MASS).

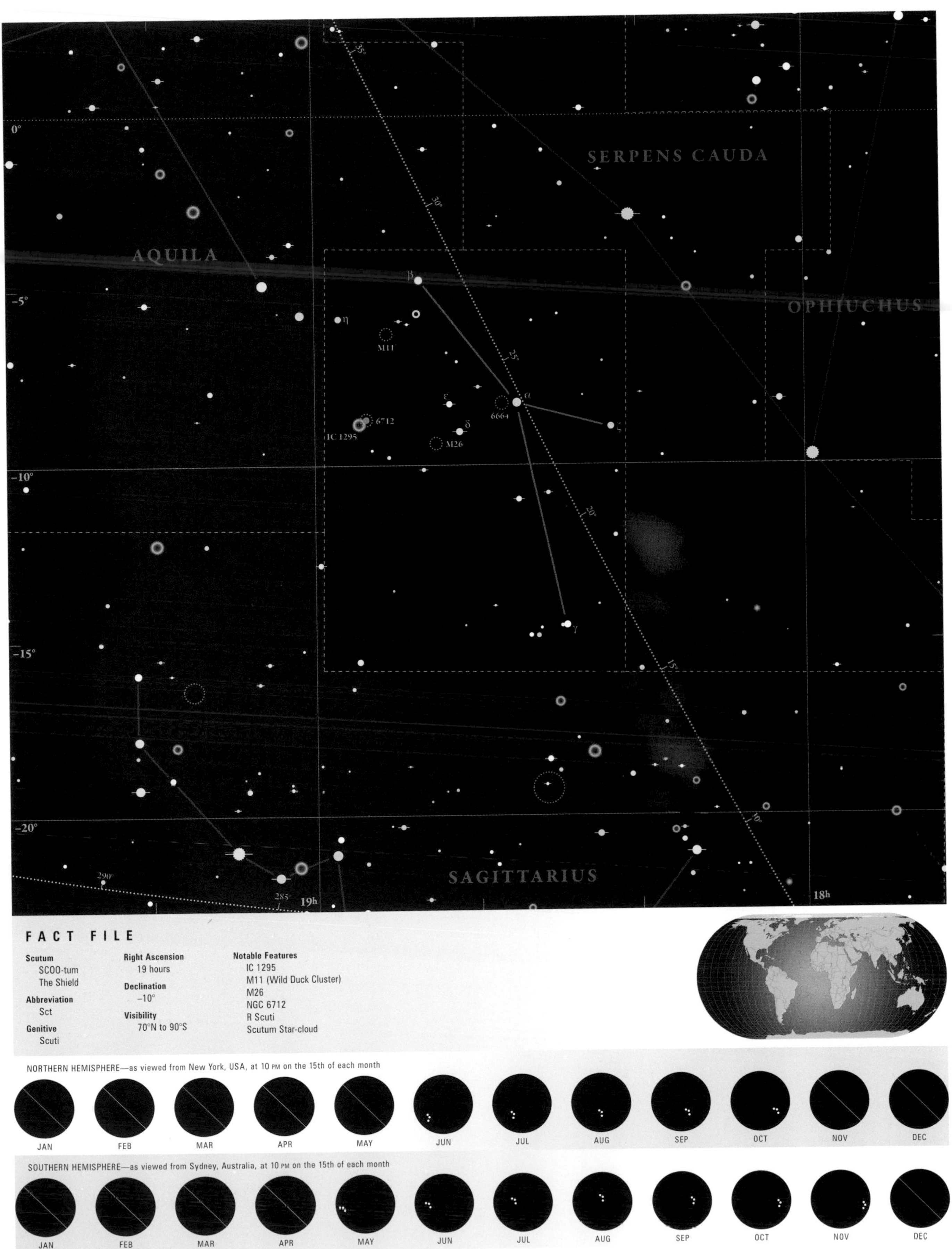

FACT FILE

Scutum
SCOO-tum
The Shield

Abbreviation
Sct

Genitive
Scuti

Right Ascension
19 hours

Declination
−10°

Visibility
70°N to 90°S

Notable Features
IC 1295
M11 (Wild Duck Cluster)
M26
NGC 6712
R Scuti
Scutum Star-cloud

NORTHERN HEMISPHERE—as viewed from New York, USA, at 10 PM on the 15th of each month

JAN FEB MAR APR MAY JUN JUL AUG SEP OCT NOV DEC

SOUTHERN HEMISPHERE—as viewed from Sydney, Australia, at 10 PM on the 15th of each month

JAN FEB MAR APR MAY JUN JUL AUG SEP OCT NOV DEC

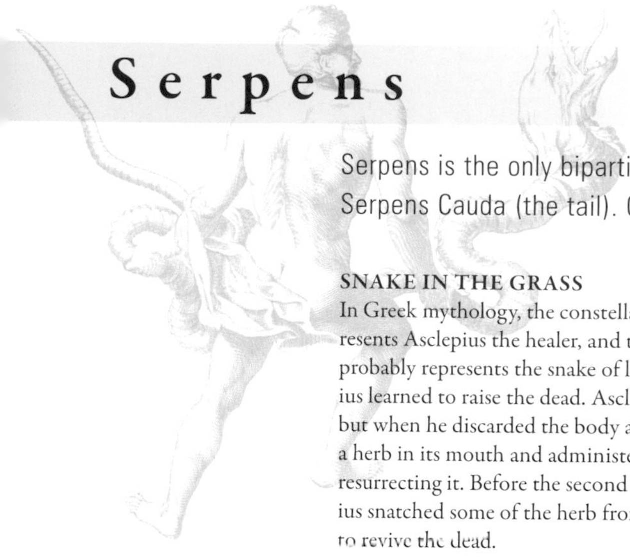

Serpens

Serpens is the only bipartite constellation, consisting of Serpens Caput (the head), and Serpens Cauda (the tail). Ophiuchus holds its head in one hand and its tail in the other.

SNAKE IN THE GRASS

In Greek mythology, the constellation of Ophiuchus represents Asclepius the healer, and the serpent he is holding probably represents the snake of legend from whom Asclepius learned to raise the dead. Asclepius strangled a snake, but when he discarded the body another snake slid up with a herb in its mouth and administered it to the dead snake, resurrecting it. Before the second snake could escape, Asclepius snatched some of the herb from its mouth and used it to revive the dead.

THE STARS OF SERPENS

The stars of Serpens are uniformly moderately faint, but the small triangle made by 4th magnitude Beta (β), Gamma (γ), and Kappa (κ) composing its head, with Delta (δ), Alpha (α), and Epsilon (ε) forming its neck, are easy to pick out in a dark sky. The remainder of the figure is difficult to distinguish from adjoining constellations.

Alpha (α) Serpentis, or Unukalhai—the serpent's neck—is a bright 3rd magnitude, K-type, orange giant star that lies just 76 light-years distant. Its diameter is 15 times that of our Sun, and it is 70 times more luminous.

Very similar to our Sun is 4th magnitude Gamma (γ) Serpentis—an F-type dwarf star 25 percent more massive but nearly three times brighter than our luminary. The extra mass makes a big difference to its total life expectancy; at only 3 billion years old, it has already lived about two-thirds of its main-sequence life.

OBJECTS OF INTEREST

Serpens contains a variety of interesting objects for the amateur sky-watcher. Numerous faint small galaxies dominate the northern end of the constellation, and the southern end, near Sagittarius, has several open clusters.

The brilliant globular cluster **M5** is a showpiece, magnificently set beside the double star 5 Serpentis. In a 6-inch (15 cm) telescope it is a large object about a half moon-diameter across, with an even concentration of its hundreds of resolved stars to the center, overlaying a strongly hazy halo.

In Serpens Cauda, **M16** is a large scattered cluster of medium-bright stars wrapped in the famous **Eagle Nebula**. Though the cluster is easy to see in a 4-inch (10 cm) telescope, the nebula is elusive in smaller instruments and requires a dark sky and a very large telescope.

IC 4756 is a huge scattered cluster of stars that looks good in a low-power telescope or powerful binoculars, while nearby **Theta (θ) Serpentis** is a lovely binary star with bright, white, twin A-type stars.

Above This eerie dark pillar-like structure is actually a column of cool interstellar hydrogen gas and dust that incubates stars. It is part of the Eagle Nebula (M16)—6,500 light-years away in the constellation Serpens.

Right Spitzer Space Telescope infrared image of the star-forming region, located approximately 848 light-years away. The reddish-pink dots are baby stars deeply embedded in the cosmic cloud of gas and dust.

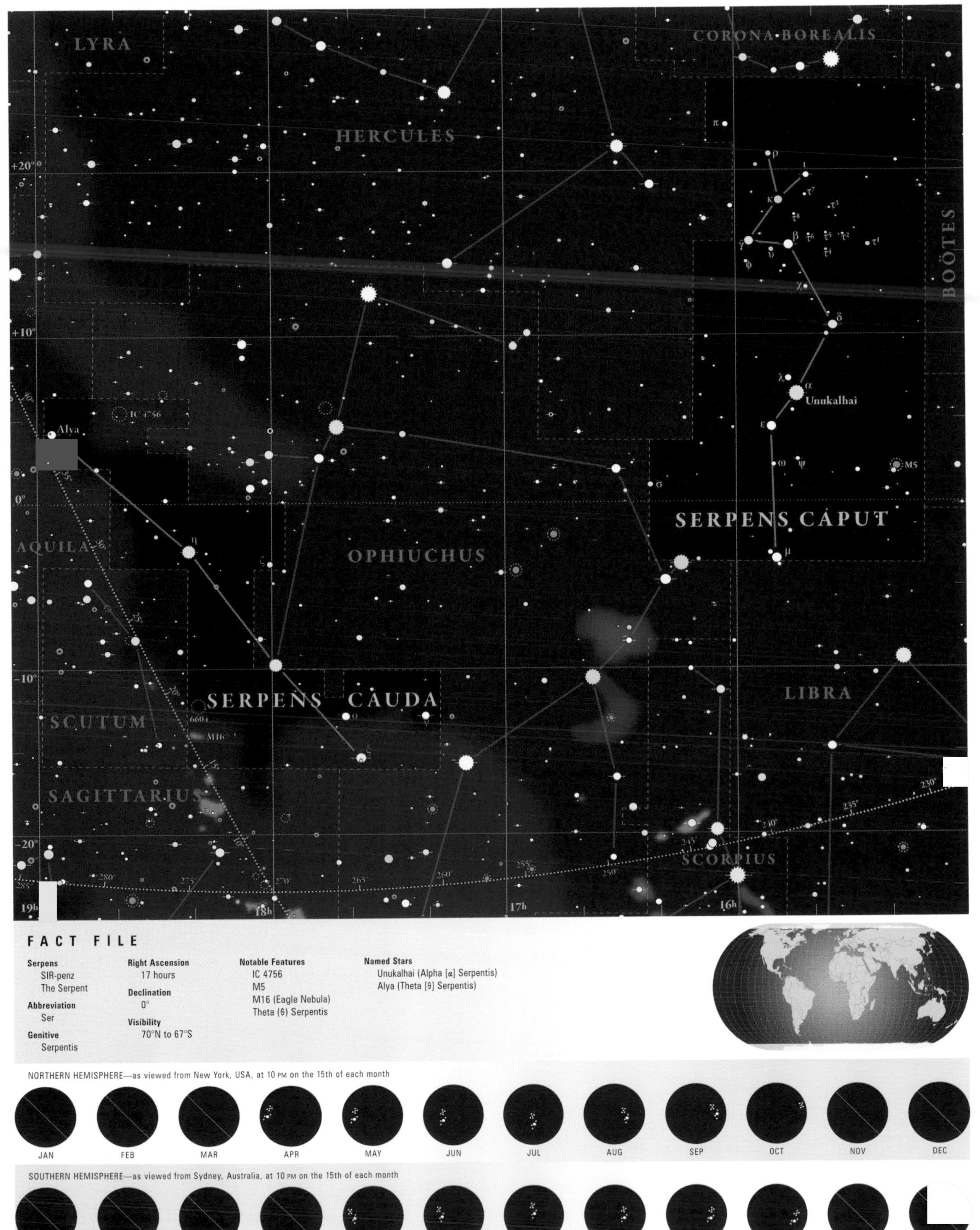

FACT FILE

Serpens
SIR-penz
The Serpent

Abbreviation
Ser

Genitive
Serpentis

Right Ascension
17 hours

Declination
0°

Visibility
70°N to 67°S

Notable Features
IC 4756
M5
M16 (Eagle Nebula)
Theta (θ) Serpentis

Named Stars
Unukalhai (Alpha [α] Serpentis)
Alya (Theta [θ] Serpentis)

NORTHERN HEMISPHERE—as viewed from New York, USA, at 10 PM on the 15th of each month

JAN FEB MAR APR MAY JUN JUL AUG SEP OCT NOV DEC

SOUTHERN HEMISPHERE—as viewed from Sydney, Australia, at 10 PM on the 15th of each month

JAN FEB MAR APR MAY JUN JUL AUG SEP OCT NOV DEC

Sextans

The Sextant marks the invention of the astronomical sextant. The Quadrant and the Octant were also commemorated in the sky, but only Octans and Sextans stood the test of time.

MAPPING THE STARS

The faint stars of Sextans, wedged between Leo and Hydra, were visible to ancient cultures but had no associated figure, so no mythology is attached to them. The famous Polish astronomer Johannes Hevelius invented the constellation in 1687 and called it Sextans Uraniae to commemorate the astronomical sextant. One of his favorite instruments, Hevelius used the sextant in preference to the telescope for the accurate measurement of star positions. The group has undergone no significant change since Hevelius, except that the name has been shortened to simple Sextans.

THE STARS OF SEXTANS

The faint stars of Sextans are difficult to see, except from a dark-sky site. Alpha (α), barely 4th magnitude, and 5th magnitude Beta (β) and Gamma (γ) Sextantis form a long triangle. Several other dim 5th and 6th magnitude motes do little to add to the figure.

Below Four-and-a-half million light-years from Earth, the irregular galaxy Sextans B is one of the most distant members of our Local Group of galaxies.

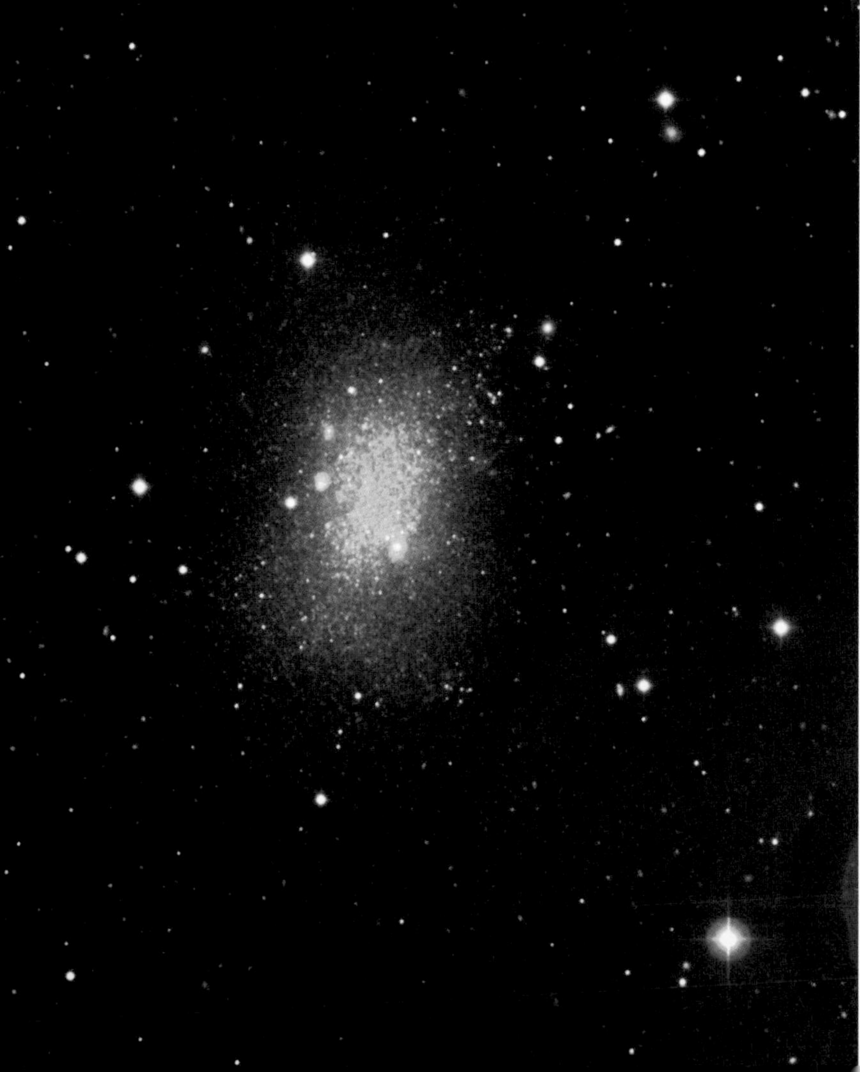

Alpha (α) Sextantis is a fairly typical A-type giant star, which we see from a distance of 286 light-years. About three times the mass of our Sun, this white giant is more than 120 times as luminous.

OBJECTS OF INTEREST

Located far from the bright lights of the Milky Way, Sextans has little to interest the amateur observer; nevertheless, a few of its many galaxies are bright and interesting. The best is **NGC 3115**, a very bright lenticular galaxy that we see almost edge-on. In a 6-inch (15 cm) telescope, it has a small, slim, spindle-shaped outline with a very bright nucleus.

The galaxy pair **NGC 3166** and **NGC 3169** is interesting in an 8-inch (20 cm) telescope. These two almost face-on spiral galaxies are very similar in appearance—small and round, with conspicuous nuclei, they are very close to each other with a faint sprinkling of stars in the background. The faint thin outline of the edge-on spiral galaxy **NGC 3044** is just visible in an 8-inch (20 cm) telescope as a gossamer-like thread.

Gamma (γ) Sextantis is an interesting binary star, only rarely resolvable in amateur-class telescopes. It is an equally matched pair of white A-type stars, orbiting each other every 77 years. Because the pair is about 260 light-years away, it is a severe test of telescope, observer, and conditions to see the two as distinct, even when at their widest separation. A 10-inch (25 cm) telescope barely shows the pair at maximum separation, which most recently occurred in 1996, and happens only two or three times each century.

Below Sextans A is a dwarf irregular galaxy that is 5 million light-years away—part of our Local Group. Young blue stars highlight areas of star formation. The bright orange star is in our own Milky Way Galaxy.

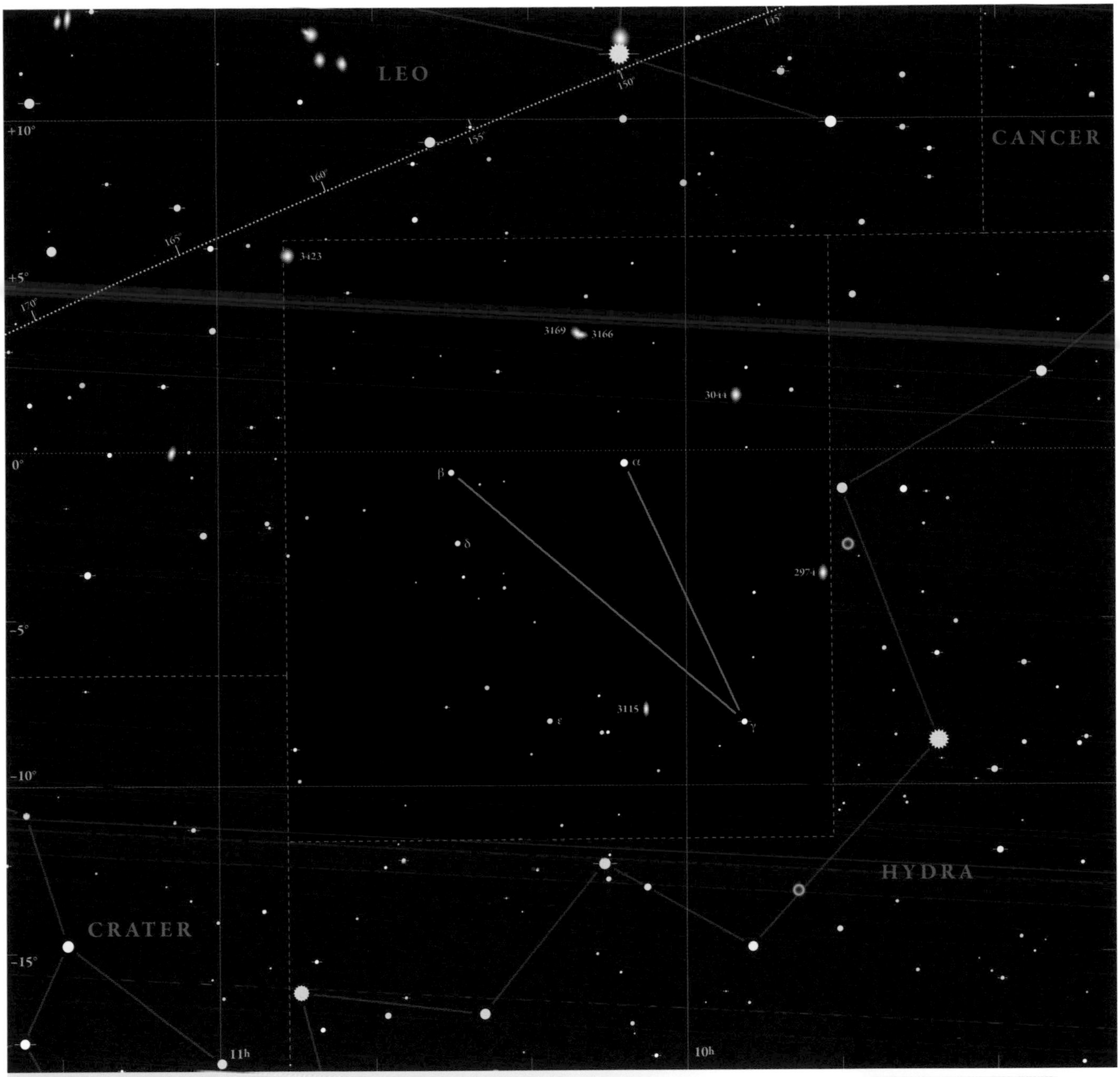

FACT FILE

Sextans
SEX-tanz
The Sextant

Abbreviation
Sex

Genitive
Sextantis

Right Ascension
10 hours

Declination
0°

Visibility
77°N to 85°S

Notable Features
Gamma (γ) Sextantis
NGC 3115
NGC 3166
NGC 3169
NGC 3044

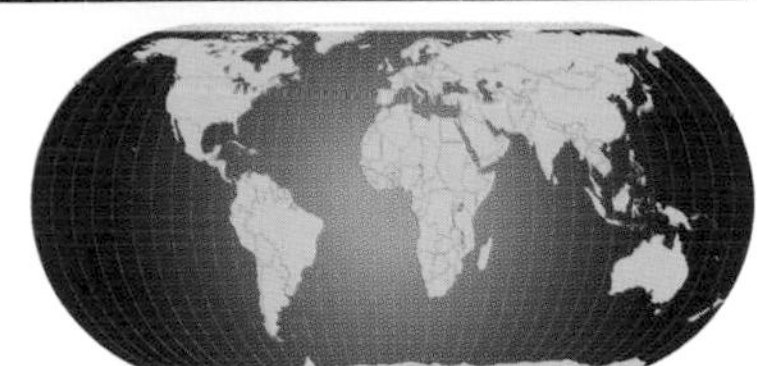

NORTHERN HEMISPHERE—as viewed from New York, USA, at 10 PM on the 15th of each month

JAN FEB MAR APR MAY JUN JUL AUG SEP OCT NOV DEC

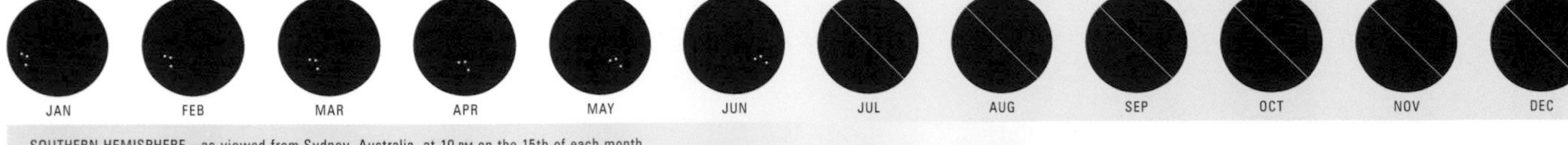

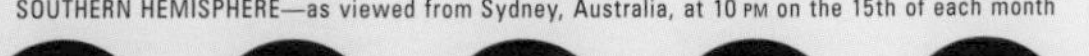

SOUTHERN HEMISPHERE—as viewed from Sydney, Australia, at 10 PM on the 15th of each month

JAN FEB MAR APR MAY JUN JUL AUG SEP OCT NOV DEC

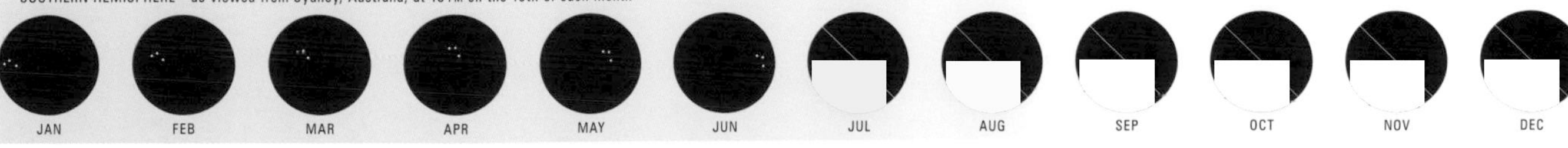

Taurus

The Bull is yet another disguise of Zeus, the seducer god who had his way with mortal women by metamorphosing into irresistible objects of desire.

GAMES ON THE BEACH

The Sumerians called Taurus the Bull of Light. The Egyptians worshipped it as Osiris–Apis. The Greeks linked it with Zeus's seduction of Europa, the daughter of King Agenor of Phoenicia. This story tells of a handsome bull approaching Europa while she was playing at the beach. Spellbound by the beautiful beast, she sat on its back. It swam to Crete, where Zeus revealed his true identity and seduced her.

Another myth associated with Taurus concerns the Pleiades, a naked-eye star cluster within the constellation. The Pleiades were seven daughters of Atlas, who was condemned to support the sky for eternity as a punishment for opposing the Olympians in their war against the Titans. In sorrow at the harshness of the sentence, the Pleiades committed suicide, whereupon Zeus, in pity, placed them in the sky. The Hyades, another naked-eye star cluster forming the bull's head, were also Atlas's daughters. When their brother Hyas died the Hyades wept endlessly. They too were housed in the sky, and the Greeks believed their tears were a sign of coming rain.

THE STARS OF TAURUS

Most of Taurus's stars are faint. Except for the V-shaped Hyades that make his prominent face, and the two horns marked by Beta (β) and Zeta (ζ), the remainder looks rather like a bull charging toward Orion.

Aldebaran, or Alpha (α) Tauri, is the brightest star in Taurus, and the thirteenth brightest in the night sky. Looking like part of the Hyades star-cluster, it is in reality a mere 60 light-years away. Aldebaran is a K-class orange giant star about 2.5 times more massive than our Sun, 40 times larger, and 350 times brighter.

OBJECTS OF INTEREST

The two outstanding objects in Taurus are the star clusters **Pleiades (M45)** and **Hyades** 👁 ∞ 🔭. The Pleiades are more than three moon-diameters across. In suburbia, keen-eyed observers can see seven Pleiades; perhaps a dozen are visible under rural skies, binoculars show at least 40, and a telescope with an ultra-wide field even more. Enmeshed in a hard-to-see reflection nebula, the Pleiades are only 440 light-years distant. The V-shaped Hyades that form the head of Taurus are even closer, only 150 light-years away. The group is so large that binoculars or naked-eye viewing is best.

NGC 1746 and **NGC 1647** 🔭 are large, scattered open clusters easily seen in small telescopes. Next to Zeta (ζ) Tauri is **M1**, the **Crab Nebula** 🔭, a cloud of debris discharged from a gaseous supernova observed by Chinese astronomers in 1054. The explosion was so immense that the star was visible during daylight for 23 days. In a 6-inch (15 cm) telescope it is a large, oval, misty cloud in a starry field.

Above The Pleiades (M45), or the Seven Sisters, is one of the highlights of Taurus. It is the best known and brightest open cluster in the sky, captured here in a Spitzer Telescope infrared image as a tangled mass of dust and young blue stars.

Right This Hubble Space Telescope mosaic image reveals the ethereal blue glow of electrons in the Crab Nebula's central pulsar. This dazzling supernova remnant spans 12 light-years.

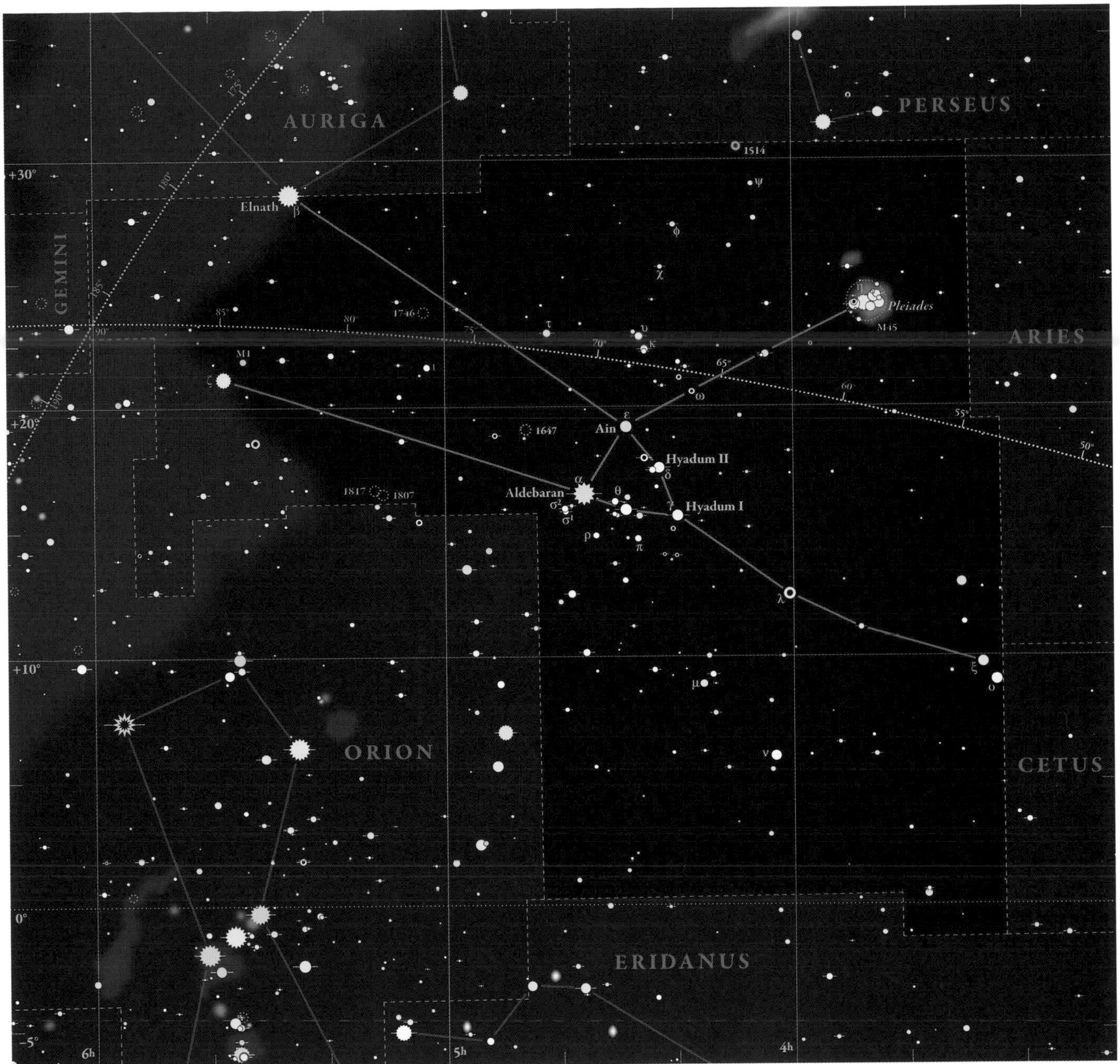

FACT FILE

Taurus
TOR-uss
The Bull

Abbreviation
Tau

Genitive
Tauri

Right Ascension
4 hours

Declination
+15°

Visibility
90°N to 56°S

Notable Features
M1 (Crab Nebula)
M45 (Pleiades)
NGC 1647
NGC 1746

Named Stars
Aldebaran (Alpha [α] Tauri)
Elnath (Beta [β] Tauri)
Hyadum I (Gamma [γ] Tauri)
Hyadum II (Delta [δ] Tauri)
Ain (Epsilon [ε] Tauri)

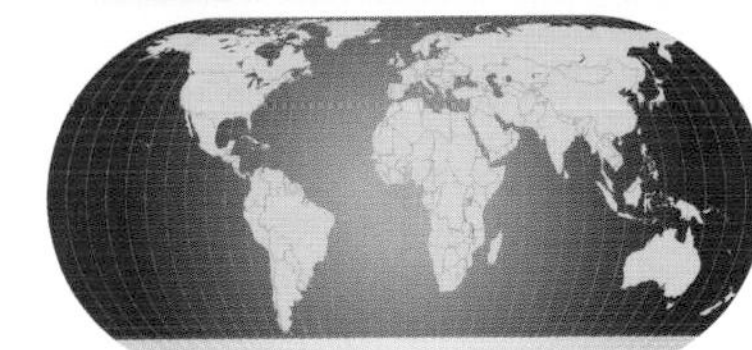

NORTHERN HEMISPHERE—as viewed from New York, USA, at 10 PM on the 15th of each month

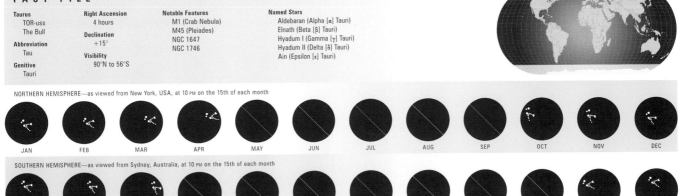

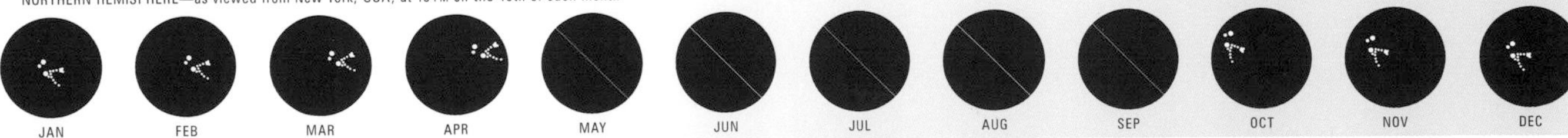

SOUTHERN HEMISPHERE—as viewed from Sydney, Australia, at 10 PM on the 15th of each month

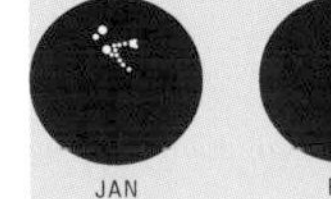

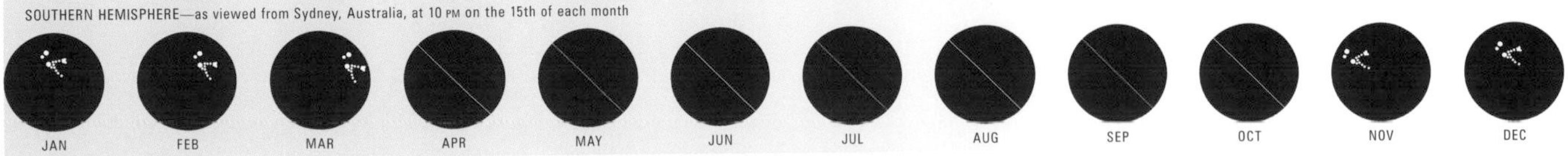

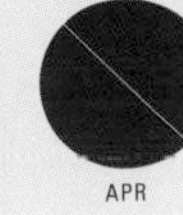

Telescopium

The Telescope's name, which is derived from the Greek meaning "far-seeing," is well chosen. An observer must be far-sighted indeed to distinguish its insignificant stars and tiny distant galaxies.

EYE IN THE SKY

The small, almost blank space occupied by the Telescope was visible—albeit low on the horizon—to early European astronomers, but the constellation was not named until Abbé Nicolas Louis de Lacaille created it when he visited the Cape Observatory in 1751 and 1752. Like many of Lacaille's constellations, it represents a scientific instrument, but it is hard to perceive its three brightest stars as a telescope, even with the multitude of 6th magnitude stars in the group.

THE STARS OF TELESCOPIUM

The principal stars are 4th magnitude Alpha (α) and Zeta (ζ) Telescopii, and 5th magnitude Epsilon (ε) Telescopii. They form a small, nearly right-angled triangle to the south of Scorpius, which points at it with the bright line of stars in its tail—Lambda (λ), Kappa (κ), and Iota (ι) Scorpii.

Alpha (α) Telescopii is a blue B-class star that we see from a distance of 250 light-years. A subgiant with 6 times the mass of our Sun, it is nearly 1,000 times as luminous. Its slightly strange spectrum shows that it is oddly deficient in helium for a star of its spectral class.

OBJECTS OF INTEREST

For a constellation not far from the plane of the Milky Way, Telescopium is not rich in interesting sights to observe. By far the best, at a distance of 43,000 light-years, is the globular cluster **NGC 6584**. In a field with many faint, scattered stars, an 8-inch (20 cm) telescope will begin to resolve some of its stars; it shows only slight compression to the center of its small halo.

Of the Telescope's many faint galaxies, the easiest to see is the elliptical galaxy **NGC 6868**. In a 6-inch (15 cm) telescope, it has a small but easily visible round halo and a conspicuous brightening toward its center.

Under a dark rural sky, an 8-inch (20 cm) telescope reveals perhaps 30 other galaxies within the constellation, but they are all faint, and need patience to find among the throngs of faint stars.

At the southern end of Telescopium, near Alpha (α) Pavonis, is the pretty binary star **Dunlop 227**. A 4-inch (10 cm) telescope shows contrasting orange and white colors in the 6th magnitude components, which are widely separated in a background field liberally sprinkled with faint stars.

Below The constellation Telescopium is represented by three stars forming almost a right angle in the center of the image. As it is a southern constellation Telescopium was not named until the eighteenth century when Southern hemisphere discovery began.

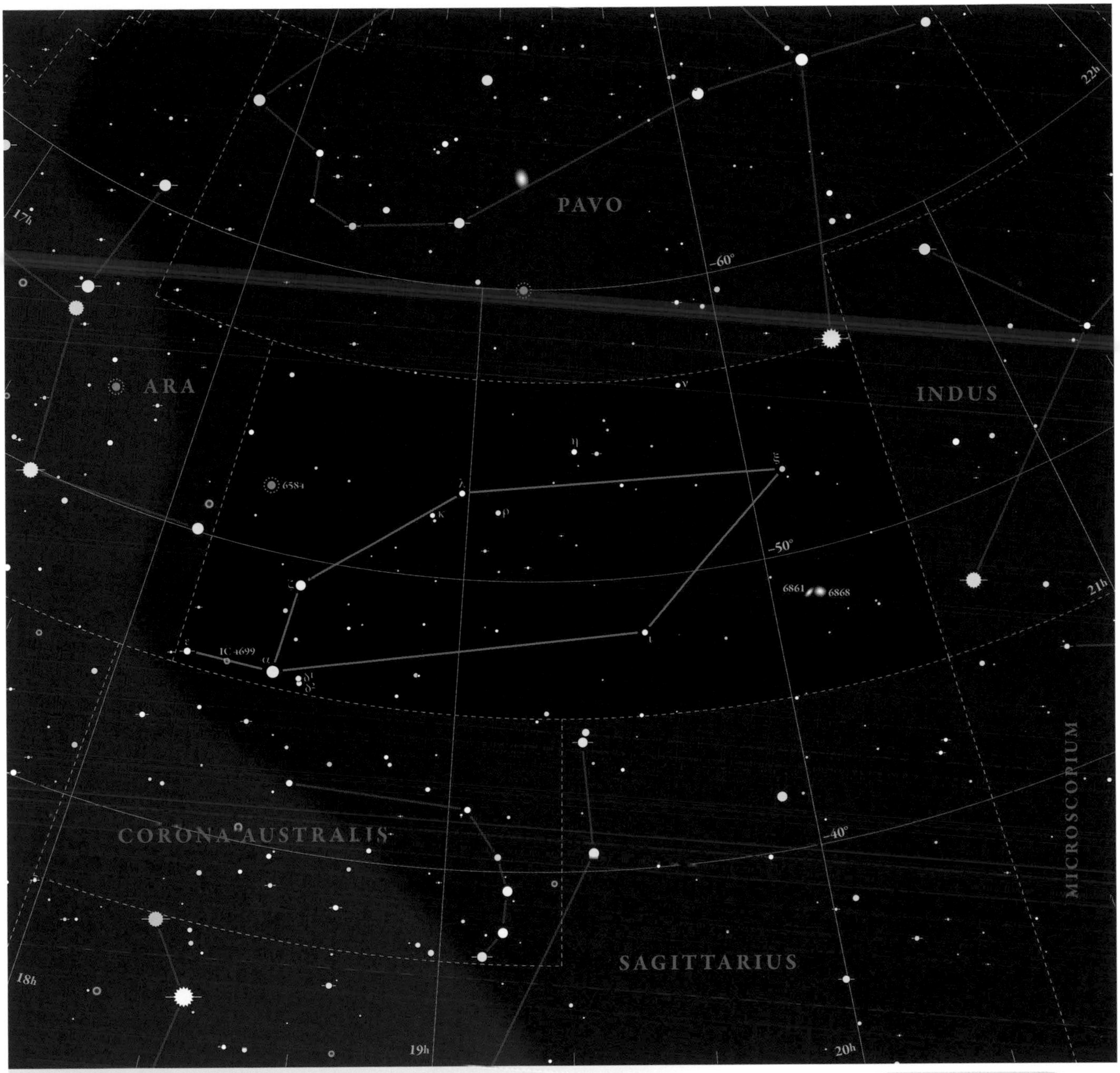

PAVO

ARA

INDUS

CORONA AUSTRALIS

MICROSCOPIUM

SAGITTARIUS

6584 IC 4699 6861 6868

−60° −50° −40°

17h 18h 19h 20h 21h 22h

FACT FILE

Telescopium
tell-eh-SKO-pee-um
The Telescope

Abbreviation
Tel

Genitive
Telescopii

Right Ascension
19 hours

Declination
−50°

Visibility
30°N to 90°S

Notable Features
NGC 6584
NGC 6868
Dunlop 227

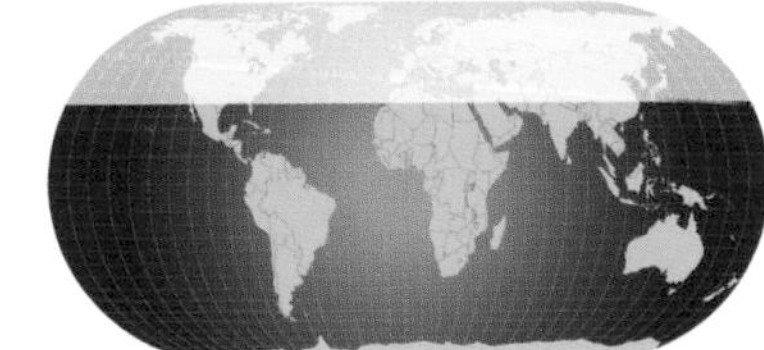

NORTHERN HEMISPHERE—as viewed from New York, USA, at 10 PM on the 15th of each month

JAN FEB MAR APR MAY JUN JUL AUG SEP OCT NOV DEC

SOUTHERN HEMISPHERE—as viewed from Sydney, Australia, at 10 PM on the 15th of each month

JAN FEB MAR APR MAY JUN JUL AUG SEP OCT NOV DEC

Triangulum

Triangulum—the Triangle—may be the eleventh smallest constellation in the sky, but it does boast M33 as its showpiece galaxy.

D FOR DELTA

Although this constellation was recorded as early as 270 BCE, it draws on little classical mythology, except possibly an association with the roughly triangular island of Sicily, which the goddess Demeter—Ceres to the Romans—begged Jupiter to place in the sky. Early Greek astronomers perceived it as their capital letter delta (Δ), and some believe that the ancients associated it with river deltas, particularly that of the Nile.

Right Galaxy Evolution Explorer image of M33, the Triangulum Galaxy. It is a perennial favorite of amateur and professional astronomers alike, due to its orientation and relative proximity to us. It is the second nearest spiral galaxy to our Milky Way Galaxy.

THE STARS OF TRIANGULUM

Three stars of 3rd and 4th magnitude form the narrow, almost isosceles triangle for which this constellation is named. Alpha (α) is also called Caput Trianguli—the head, or apex, of the triangle—and the two base angles are marked by Beta (β), the brightest of the three, and Gamma (γ), which is the dimmest.

OBJECTS OF INTEREST

Iota (ι) Trianguli is a color-contrast double for a 2-inch (6 cm) refractor. The gold and blue components are magnitudes 5.3 and 6.9, separated by 3.9 arc-seconds.

The **Triangulum Galaxy, M33 (NGC 598)** is the third largest galaxy in the Local Group, exceeded only by M31 and our own Milky Way. M33 is a large, diffuse, face-on spiral galaxy of magnitude 5.7. Beginners frequently try to view it under light-polluted suburban or moonlit skies, but its light is spread over a large area, so its surface brightness is rather low: Faint extended objects must be surrounded by darker sky in the eyepiece to be detectable by contrast.

Observe the Triangle from your astronomy club's dark-sky site, and use an optical instrument with a field of view much larger than the galaxy. M33 is obvious in binoculars at a dark transparent site. Telescope designs like Schmidt-Cassegrains, which have small fields of view, must be swept back and forth across M33's boundaries to detect the change in brightness.

In the Northern hemisphere, the basic test for an adequate club or star-party observing site is, in fact, that M33 should be visible with the unaided eye on the best nights. There are no naked-eye stars near M33's position to confuse the issue; the brightest are 8th magnitude.

An 8-inch (20 cm) Dobsonian shows the galaxy's 13th magnitude nucleus, and the two main spiral arms with several knots along the southern arm.

The main northern spiral arm ends at a bright 50 arc-second-wide knot, **NGC 604**, located 12 arc-minutes northeast of the nucleus, which even a 4-inch (10 cm) reveals. This is one of the largest emission nebulae known in any galaxy, measuring about 1,000 light-years in diameter. NGC 604 anchors a chain of seven condensations, all visible with a 16-inch (40 cm) telescope, that arc across the northern and western sides of the core.

With a detailed chart of the galaxy, an 8-inch (20 cm) telescope reveals seven cataloged emission nebulae and stellar associations within M33.

After many hours of careful star-hopping, a 16-inch (40 cm) telescope shows 31 condensations, including M33's brightest globular cluster, **C39**.

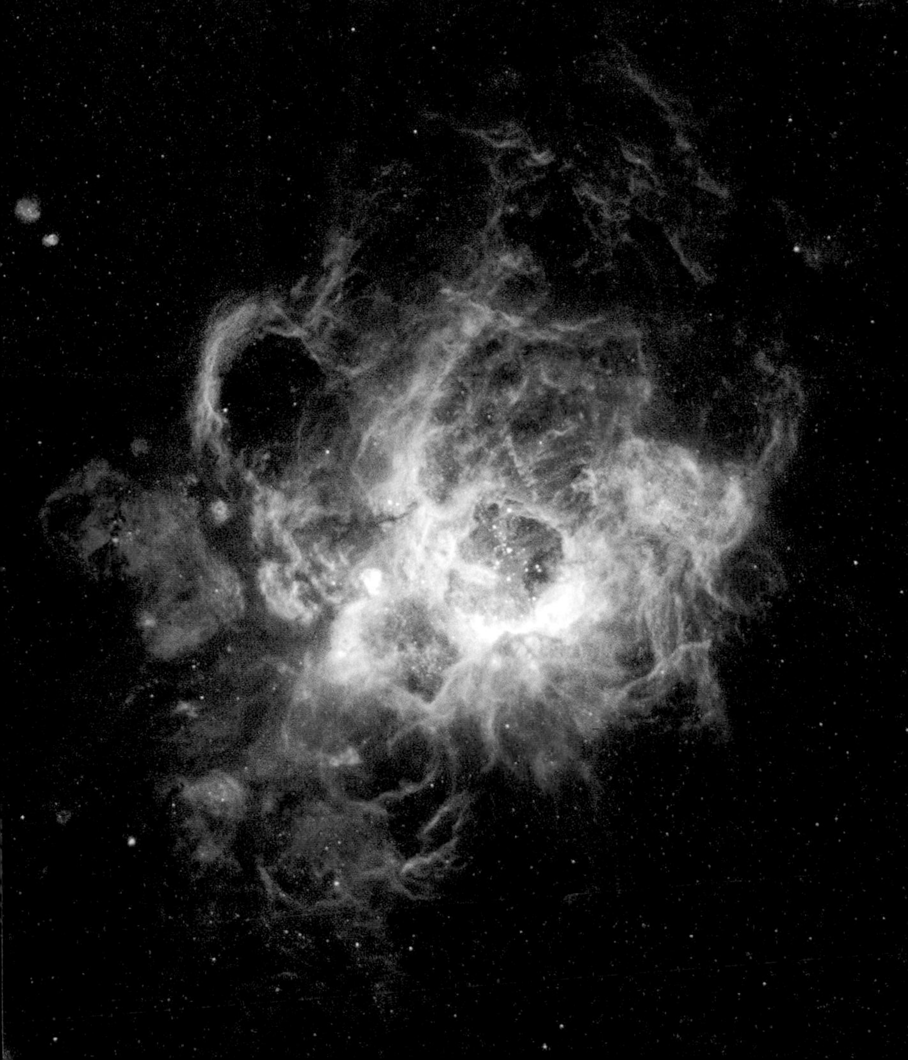

Left Hubble Space Telescope image of a vast nebula called NGC 604, which lies in the neighboring spiral galaxy M33, located 2.7 million light-years away in the constellation Triangulum.

CASSIOPEIA
+50°
PERSEUS
ANDROMEDA
+40°
925
δ β γ ε
M33
+30°
α Mothallah
784
PISCES
ARIES
+20°
4h 3h 2h 1h
50° 45° 40° 35°

FACT FILE

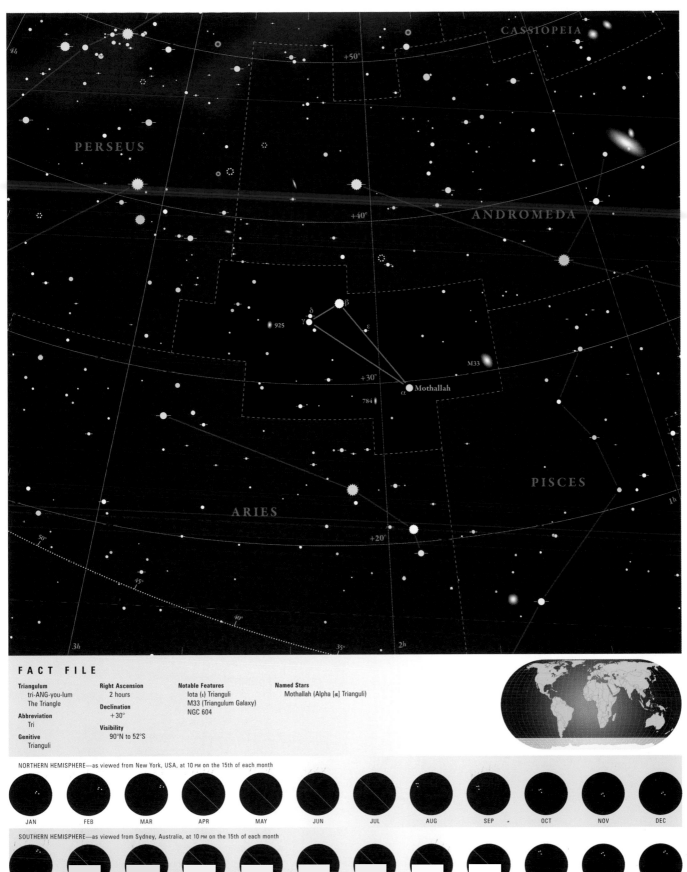

Triangulum
tri-ANG-you-lum
The Triangle

Abbreviation
Tri

Genitive
Trianguli

Right Ascension
2 hours

Declination
+30°

Visibility
90°N to 52°S

Notable Features
Iota (ι) Trianguli
M33 (Triangulum Galaxy)
NGC 604

Named Stars
Mothallah (Alpha [α] Trianguli)

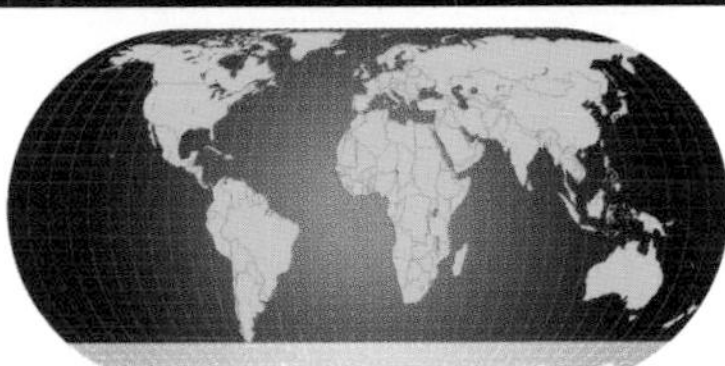

NORTHERN HEMISPHERE—as viewed from New York, USA, at 10 PM on the 15th of each month

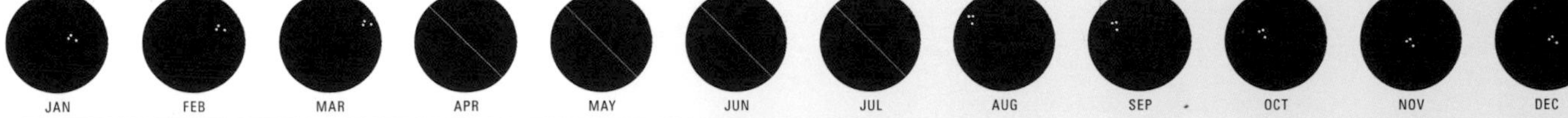

SOUTHERN HEMISPHERE—as viewed from Sydney, Australia, at 10 PM on the 15th of each month

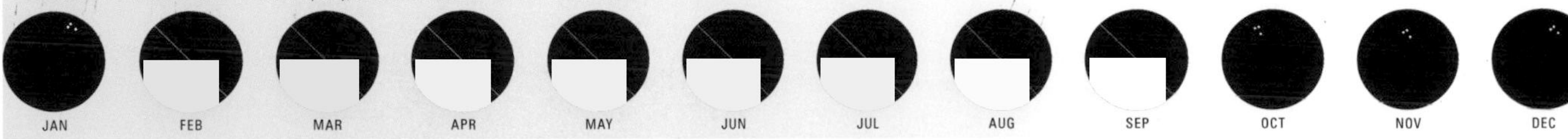

Triangulum Australe

The Southern Triangle is much brighter, larger, and easier to see than its northern counterpart, Triangulum.

SOUTHERN GEOMETRY

The stars of Triangulum Australe were too far south to be easily visible to ancient European sky-watchers. Possibly the first European to notice these stars was the Italian explorer and cartographer, Amerigo Vespucci, while he was exploring the eastern coast of South America from 1499 to 1502.

The sixteenth-century Dutch explorers and cartographers, Pieter Dirkszoon Keyser and Frederick de Houtman, invented the figure during their explorations of the East Indies.

Plancius included a southern triangle in his 1598 star-globe but it was a different group of stars than those recorded by Keyser and de Houtman. Johannes Bayer formalized the constellation in his star atlas *Uranometria* in 1603, but much later Lacaille attached a different name to the group—the Surveyor's Level. Lacaille's interpretation did not catch on, and the constellation has since regained its original name.

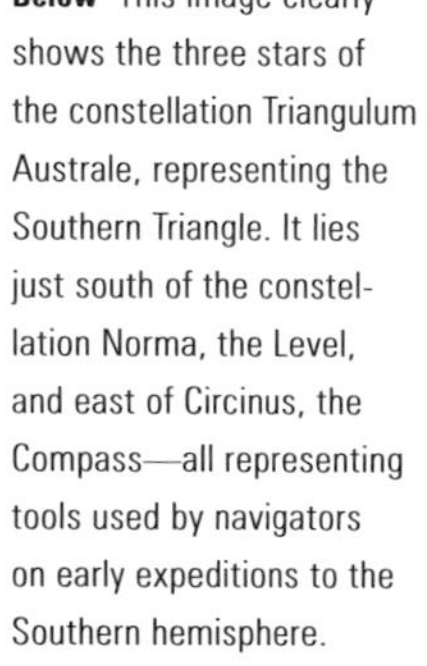

Below This image clearly shows the three stars of the constellation Triangulum Australe, representing the Southern Triangle. It lies just south of the constellation Norma, the Level, and east of Circinus, the Compass—all representing tools used by navigators on early expeditions to the Southern hemisphere.

THE STARS OF TRIANGULUM AUSTRALE

One of two heavenly triangles, Triangulum Australe is easier to spot than its Northern hemisphere counterpart. The three brightest stars of Triangulum Australe make up an appealing isosceles—nearly equilateral—triangle following a couple of hours behind the bright pointer stars to the Southern Cross, Alpha (α) and Beta (β) Centauri.

Alpha (α) Trianguli Australe, otherwise known by the unimaginative common name Atria, is the 41st brightest star in the sky and has an obvious light-orange tint. It is a bright K-class giant star about 415 light-years distant. Seven times heavier than our Sun, its total luminosity approaches 5,000 times greater. Beta (β) and Gamma (γ) have almost identical brightness to Alpha, but are white. The western end of the constellation borders on the Milky Way, and sweeping there with binoculars or a telescope at low power reveals a multitude of tightly packed faint stars.

OBJECTS OF INTEREST

Although Triangulum Australe partly borders on the Milky Way Galaxy to its west, there are relatively few star clusters of interest to observers.

NGC 6025 is the exception. Located near the border of Norma, it is a fine compact cluster of bright-white and yellowish stars that is very faintly visible to the unaided eyes in a rural sky away from lights. It is a lovely object in either binoculars or a small telescope.

Two relatively bright planetary nebulae are the other highlights of the constellation. **NGC 5979** is visible as a tiny, bright, aqua-colored disk in a 6-inch (15 cm) telescope, although care is needed to find it—the field seems to contain almost innumerable faint stars.

NGC 5844 is a larger but somewhat fainter planetary nebula. An 8-inch (20 cm) telescope shows it as a seemingly translucent, grayish, slightly oval haze. Galaxies also abound in Triangulum Australe, but they are all small and very faint objects.

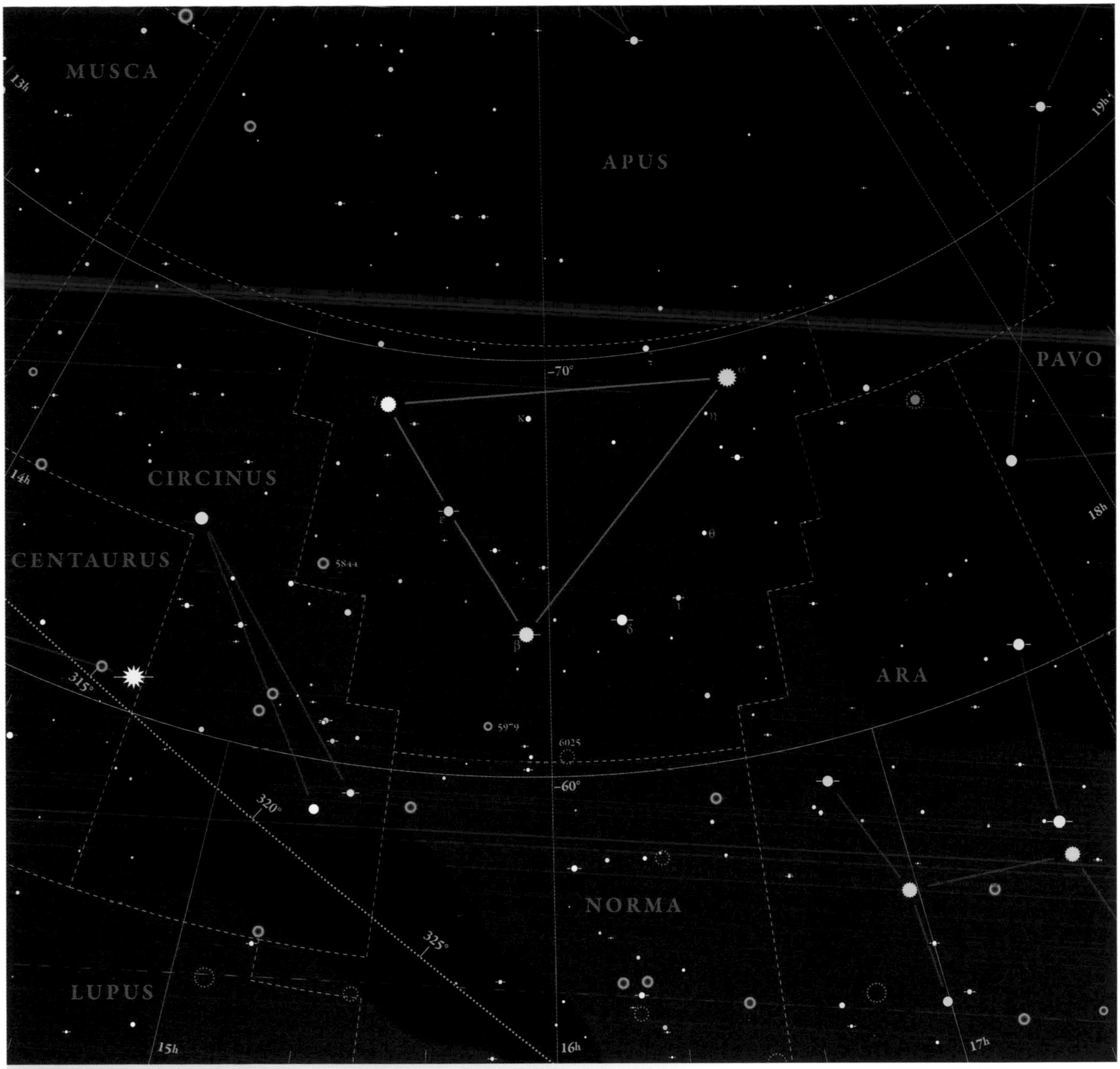

MUSCA
APUS
PAVO
CIRCINUS
CENTAURUS
ARA
NORMA
LUPUS
-70°
-60°
5844
5979
6025
315°
320°
325°
13h
14h
15h
16h
17h
18h
19h

FACT FILE

Triangulum Australe
tri-ANG-you-lum oss-TRAH-lee
The Southern Triangle

Abbreviation
TrA

Genitive
Trianguli Australe

Right Ascension
16 hours

Declination
–65°

Visibility
16°N to 90°S

Notable Features
NGC 5844
NGC 5979
NGC 6025

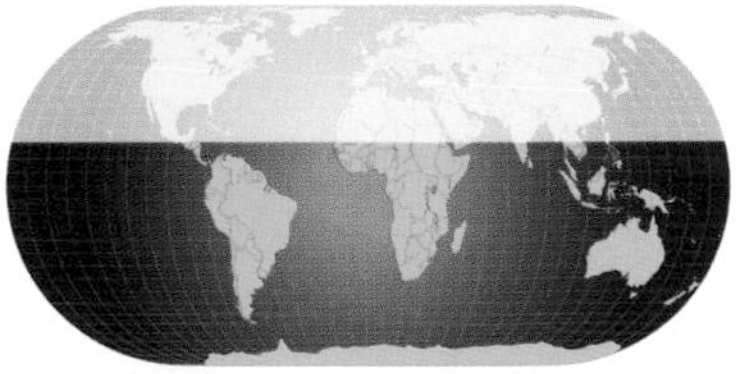

NORTHERN HEMISPHERE—as viewed from New York, USA, at 10 PM on the 15th of each month

JAN
FEB
MAR
APR
MAY
JUN
JUL
AUG
SEP
OCT
NOV
DEC

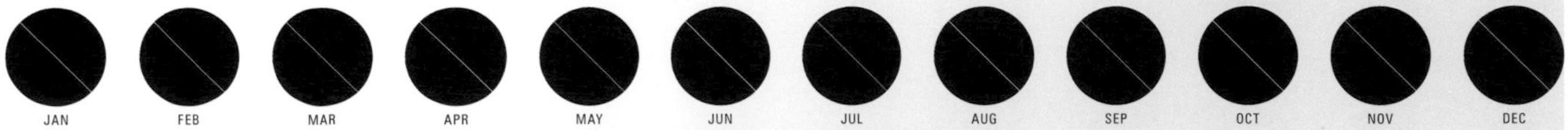

SOUTHERN HEMISPHERE—as viewed from Sydney, Australia, at 10 PM on the 15th of each month

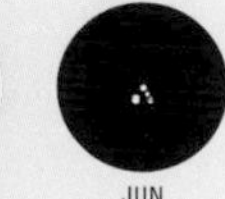

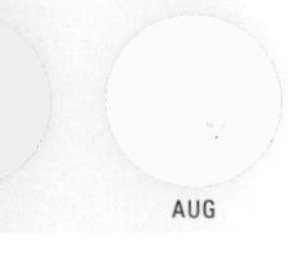

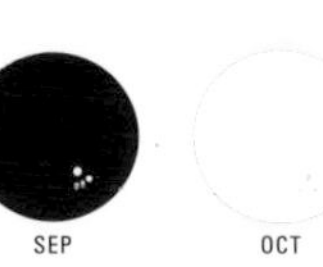

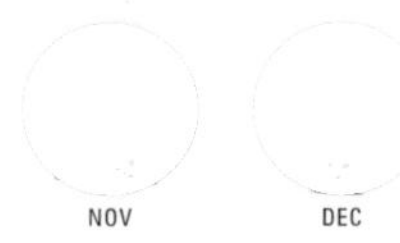

JAN
FEB
MAR
APR
MAY
JUN
JUL
AUG
SEP
OCT
NOV
DEC

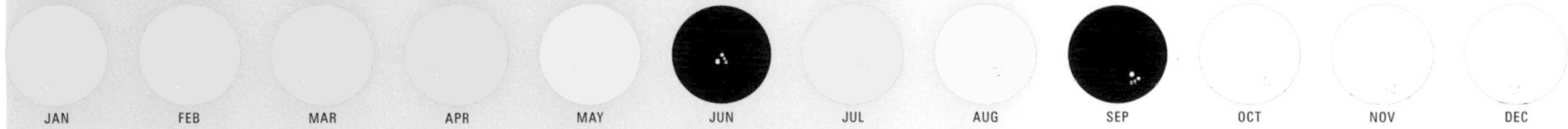

Tucana

Of the many exotic birds in the southern sky, Tucana, the Toucan, is perhaps the sky-watcher's favorite.

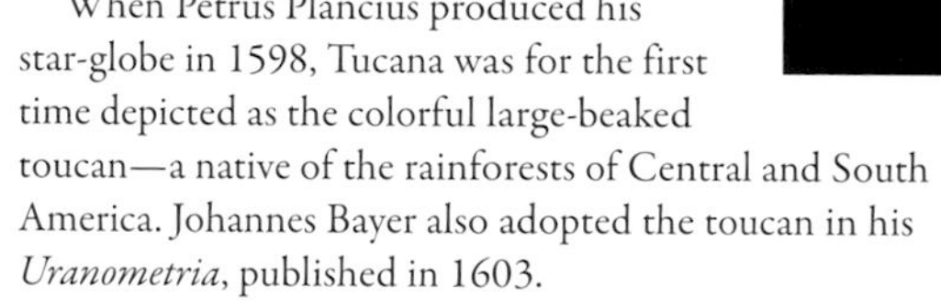

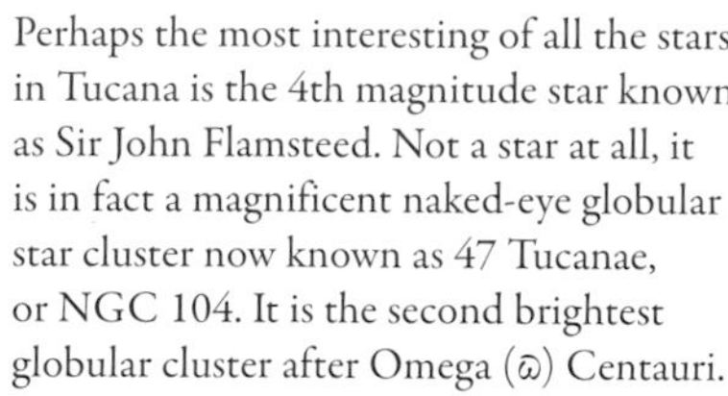

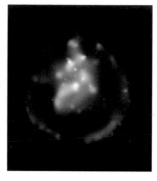

Right The supernova remnant SNR 0103-72.6 in the Small Magellanic Cloud is encircled by oxygen-rich gas heated to millions of degrees Celsius. The x-rays it emits, captured in this image, took 190,000 years to reach Earth.

BIRD'S-EYE VIEW

The star-group now recognized as Tucana is too far south to have been visible to early European observers. Johannes Kepler was possibly the first to recognize it, and he called it Anser Americanus, the American Goose. Sixteenth-century Dutch explorers Pieter Dirkszoon Keyser and Frederick de Houtman observed its stars but named it after another long-billed bird—the hornbill.

When Petrus Plancius produced his star-globe in 1598, Tucana was for the first time depicted as the colorful large-beaked toucan—a native of the rainforests of Central and South America. Johannes Bayer also adopted the toucan in his *Uranometria*, published in 1603.

THE STARS OF TUCANA

The principal stars of Tucana seem to bear little resemblance to a bird of any type, let alone a toucan, with Alpha (α), Gamma (γ), and Beta (β) forming a long isosceles triangle.

Perhaps the most interesting of all the stars in Tucana is the 4th magnitude star known as Sir John Flamsteed. Not a star at all, it is in fact a magnificent naked-eye globular star cluster now known as 47 Tucanae, or NGC 104. It is the second brightest globular cluster after Omega (ω) Centauri.

Third magnitude Alpha (α) Tucanae is a yellowish-looking K-type orange giant star located 199 light-years distant.

At the eastern end of the constellation is Beta (β) Tucanae, which looks like a close double star to the naked eye, and, to the keen-sighted, perhaps a triple. A large telescope shows it to be a complex system of six stars in three pairs, all located approximately 150 light-years away.

OBJECTS OF INTEREST

The stars of Tucana may not be bright, but the constellation is home to some of the most interesting sights in the whole sky. The **Small Magellanic Cloud**  is a satellite galaxy of the Milky Way that lies about 200,000 light-years distant. In a rural sky it appears as a small detached portion of the Milky Way, a little narrower and dimmer at its northern end. Binoculars and telescopes reveal a wealth of minute stars, stars knots, and gaseous nebulae, many of which have their own designations within various catalogs. Perhaps the best of all is just north of center; **NGC 346** has a tightly knit bunch of tiny stars wreathed in nebulosity when viewed in a 6-inch (15 cm) telescope. It is similar in brightness to a magnitude 10 star.

Next to the Small Magellanic Cloud is **47 Tucanae**, or **NGC 104** — a fuzzy star, visible to the naked eye in a suburban sky. Most southern sky-watchers agree it is the most beautiful of all the globular star clusters. In an 8-inch (20 cm) telescope it is seen to be almost the size of the full Moon and resolved into thousands of faint stars. Another similar-looking but dimmer and smaller globular cluster is **NGC 362**, which is also located near the Small Magellanic Cloud.

Left Although the infant stars in open cluster NGC 346 are three to five million years old, they are not yet burning hydrogen in their core. They reside in the center of the Small Magellanic Cloud, a satellite galaxy of the Milky Way.

MENSA
OCTANS
PAVO
HYDRUS
HOROLOGIUM
SMC
104
346
362
-80°
-70°
-60°
-50°
INDUS
ERIDANUS
GRUS
PHOENIX
19h
20h
21h
22h
23h
0h
1h
2h
3h
4h

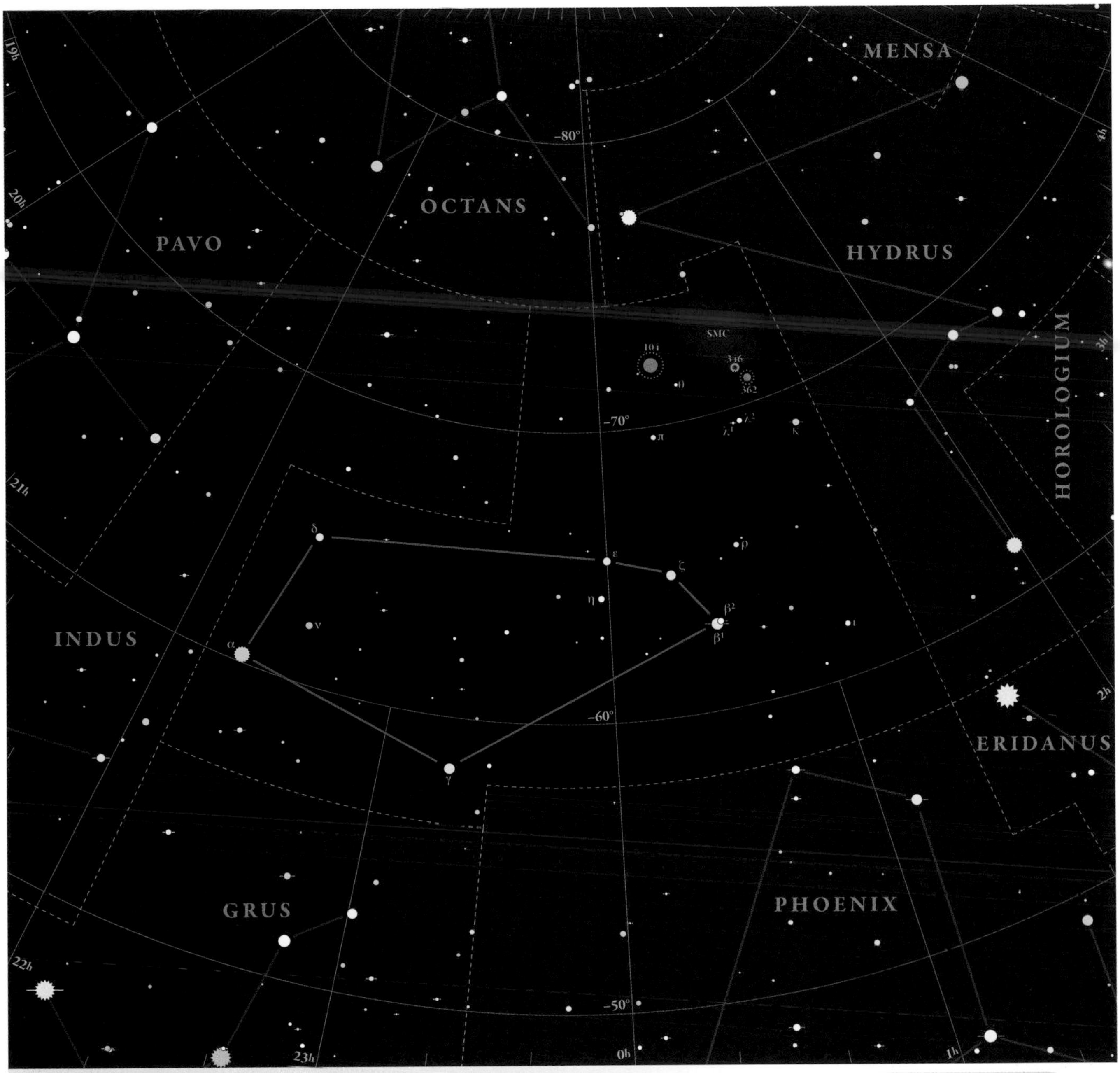

FACT FILE

Tucana
too-KAH-nah
The Toucan

Abbreviation
Tuc

Genitive
Tucanae

Right Ascension
24 hours

Declination
–65°

Visibility
20°N to 90°S

Notable Features
Small Magellanic Cloud (SMC)
NGC 104 (47 Tucanae)
NGC 346
NGC 362

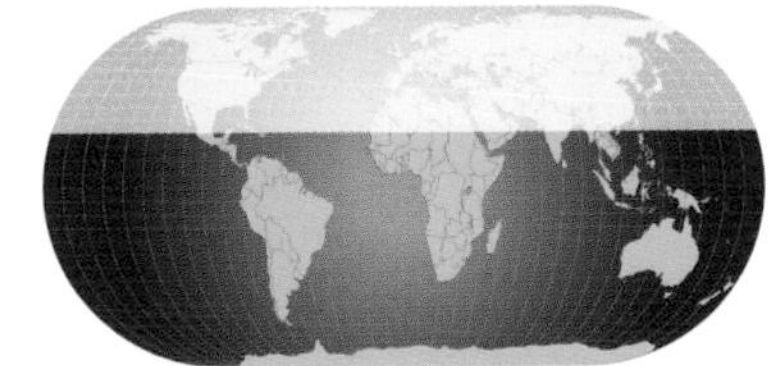

NORTHERN HEMISPHERE—as viewed from New York, USA, at 10 PM on the 15th of each month

JAN
FEB
MAR
APR
MAY
JUN
JUL
AUG
SEP
OCT
NOV
DEC

SOUTHERN HEMISPHERE—as viewed from Sydney, Australia, at 10 PM on the 15th of each month

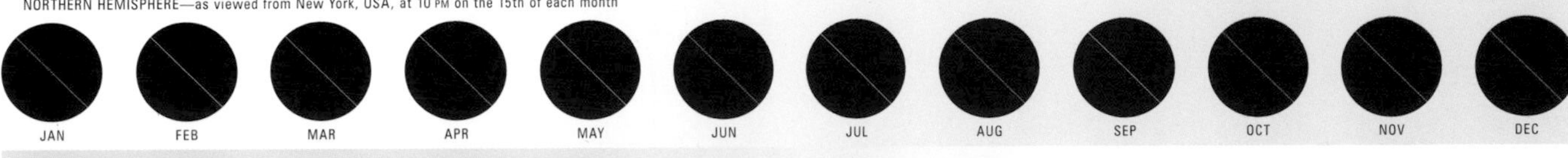

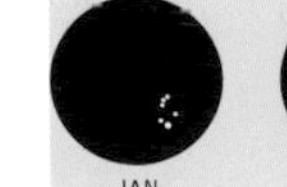

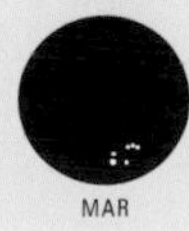

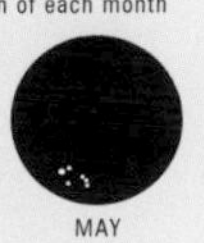

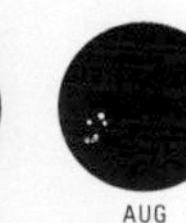

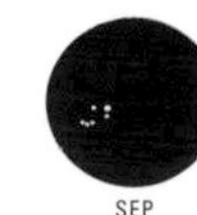

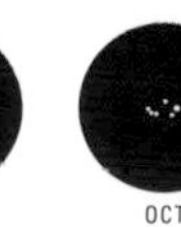

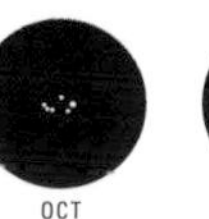

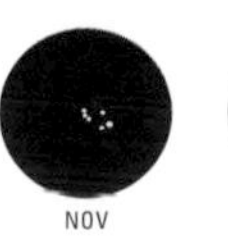

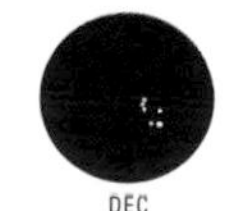

JAN
FEB
MAR
APR
MAY
JUN
JUL
AUG
SEP
OCT
NOV
DEC

Ursa Major

The Great Bear is rich in galaxies, with approximately 95 suitable for a 10-inch (25 cm) telescope. It is the third-largest constellation, at 1,280 square degrees.

GETTING THE BEARINGS

Callisto, a mortal, was one of Zeus's many conquests, and his spouse Hera was determined to punish her as soon as she delivered her son, Arcas. Hera turned Callisto into a bear. Eventually Arcas grew into a young man who was fond of hunting. One day Hera contrived to lead him to where his mother was foraging in the forest. Just as he unknowingly raised his spear, Zeus intervened and placed both Callisto and Arcas in the sky, making her the Great Bear and her son the Lesser Bear. Hera, furious that her rival had been honored with a place in the heavens, asked the God of the Sea to refuse the Bears the right to rest in the sea as the other constellations do.

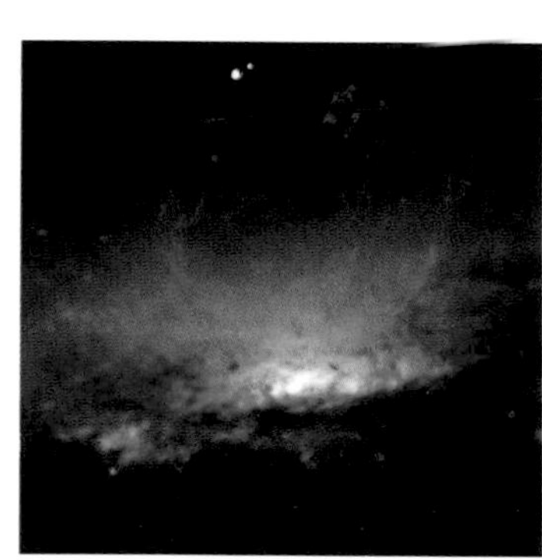

Above Hubble Space Telescope image of a lumpy bubble of hot gas rising from a cauldron of glowing matter in NGC 3079, located 50 million light-years away, in the constellation Ursa Major.

Thus the two bears forever circle the pole. While Ursa Major is no longer completely circumpolar at the latitudes of southern Greece, the myth worked perfectly thousands of years ago, when the pole star was not Polaris, but Thuban or Alpha (α) Draconis. At that time, the two bears were on opposite sides of the pole and about equidistant from it. This myth, therefore, can illustrate precession.

THE STARS OF URSA MAJOR

Seven prominent stars form a famous asterism known as the Big Dipper, or the Plow. Ursa Major makes a respectable bear, except for its strangely long tail. To find the North Star, extend the line from "The Pointers"—Beta (β) and Alpha (α) Ursae Majoris—five times.

OBJECTS OF INTEREST

Mizar, or **Zeta (ζ) Ursae Majoris**, a magnitude 2 star, and 4th magnitude **Alcor**, 12 arc-minutes apart, form a naked-eye pair known as "The Horse and Rider," considered a test of good eyesight before eyeglasses. Mizar resolves into a beautiful blue–white and greenish white binary of magnitudes 2.2 and 3.9 at 14.3 arc-seconds.

Xi (ξ) Ursae Majoris was one of the first doubles proven to be a gravitationally bound binary and to have its orbit—60 years—determined. The pale yellow pair is magnitudes 4.3 and 4.8 at 1.7 arc-seconds.

Galaxies **M81** and **M82 (NGCs 3031 and 3034)**, obvious in binoculars, are an interacting pair that are 37 arc-minutes apart. An 8-inch (20 cm) telescope adds two smaller members of M81's retinue—**NGCs 2976** and **3077**.

Oval M81 is bright (magnitude 6.9) and large (24 by 13 arc-minutes), but exhibits disappointingly little detail—just a brighter center and bright nucleus.

M82 is undergoing a burst of star formation due to a close encounter with its massive neighbor, M81. Magnitude 8.4, M82 is a 12 by 6 arc-minute edge-on irregular galaxy. An 8-inch (20 cm) reflector shows a diagonal dust lane crossing the long axis; a 25-inch (63 cm) adds a second dark band and five condensations.

A 4-inch (10 cm) telescope shows magnitude 9.9 **M97 (NGC 3587),** the **Owl Nebula**, which is a 194 arc-seconds round planetary nebula. Two indistinct darker areas within M97—the Owl Eyes—are visible with an 8-inch (20 cm) telescope, but larger apertures are recommended. The central star lies between the "eyes" and requires a 12-inch (30 cm) telescope. The mottled edge-on spiral **M108 (NGC 3556)** is in M97's field.

Magnitude 7.9, **M101 (NGC 5457)** is a 26 arc-minutes-wide, face-on, spiral galaxy of low surface brightness that challenges some beginners, although a 2¼-inch (6 cm) refractor detects it. On a superb night, an 8-inch (20 cm) telescope reveals the northern spiral arm and three knots. The knots are stellar associations and emission nebulae. Larger apertures reveal more arms and knots. An 8-inch (20 cm) also shows three small adjacent galaxies; one of them, **NGC 5474**, is a satellite of M101.

The gravitationally lensed double quasar **QSO 0957+561A/B**, 7 billion light-years away, is detectable visually with a 16-inch (40 cm) telescope. Its greatly red-shifted photons can elicit a response in our retinas after traveling across half the universe!

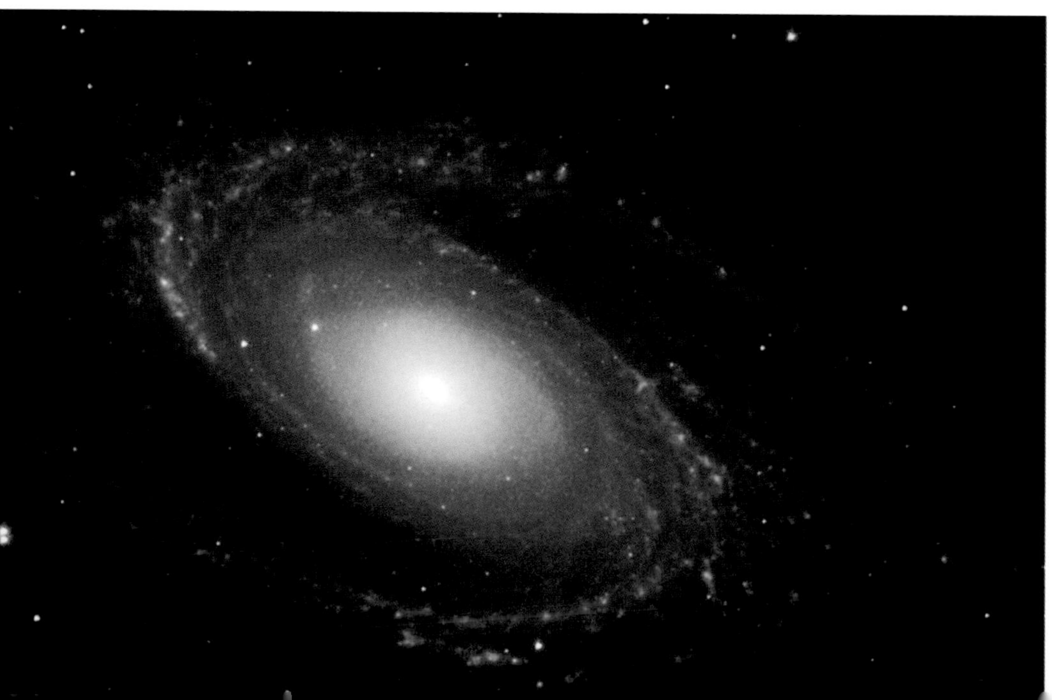

Left This Spitzer Space Telescope image highlights the magnificent spiral arms of the nearby galaxy M81. Located 12 million light-years away, in the northern constellation of Ursa Major, this galaxy is easily visible through binoculars or a small telescope.

FACT FILE

Ursa Major
ER-sah MAY-jer
The Great Bear

Abbreviation
UMa

Genitive
Ursae Majoris

Right Ascension
11 hours

Declination
+50°

Visibility
90°N to 22°S

Notable Features
Big Dipper
M81
M82
M97 (Owl Nebula)
M101
M108
NGC 2976
NGC 3034
NGC 5474
QSO 0957+561A/B
Zeta (ζ) Ursae Majoris
Xi (ξ) Ursae Majoris

Named Stars
Dubhe (Alpha [α] Ursae Majoris)
Merak (Beta [β] Ursae Majoris)
Phecda (Gamma [γ] Ursae Majoris)
Megrez (Delta [δ] Ursae Majoris)
Alioth (Epsilon [ε] Ursae Majoris)
Alcor Mizar (Zeta [ζ] Ursae Majoris)
Alkaid (Eta [η] Ursae Majoris)
Talitha (Iota [ι] Ursae Majoris)
Tania Borealis (Lambda [λ] Ursae Majoris)
Tania Australis (Mu [μ] Ursae Majoris)
Alula Borealis (Nu [ν] Ursae Majoris)
Alula Australis (Xi [ξ] Ursae Majoris)
Muscida (Omicron [ο] Ursae Majoris)

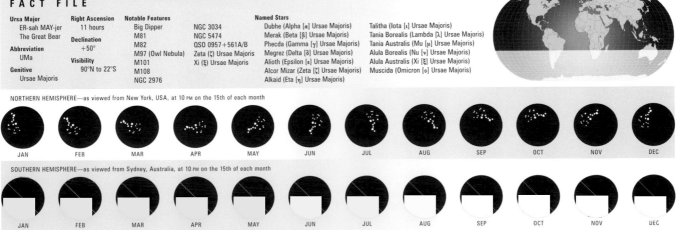

NORTHERN HEMISPHERE—as viewed from New York, USA, at 10 PM on the 15th of each month

JAN FEB MAR APR MAY JUN JUL AUG SEP OCT NOV DEC

SOUTHERN HEMISPHERE—as viewed from Sydney, Australia, at 10 PM on the 15th of each month

JAN FEB MAR APR MAY JUN JUL AUG SEP OCT NOV DEC

Ursa Minor

Ursa Minor, the Lesser Bear, houses Polaris, the North Star, which was crucial for navigation for most of history.

LITTLE TO BEAR

Arcas was the son of Callisto, who had been transformed into a bear by Hera, the wife of Zeus. When Arcas was fifteen years old, he came across a bear when hunting in the forest. Although the bear behaved strangely, looking him in the eyes, Arcas was unable to recognize his mother in her strange form, and he raised his spear to kill her.

Zeus, however, prevented him and instead transformed Arcas into a bear like his mother. The mother and son—the Greater Bear and Lesser Bear—were then taken up into the sky. Hera was angry that the pair should be given such an honor and decided to take her revenge. She convinced Poseidon to forbid them from bathing in the sea.

It is for this reason that Ursa Major and Ursa Minor are both circumpolar constellations. For thousands of years, these two constellations did not dip beneath the horizon when viewed from northern latitudes. Ursa Minor is also known as the Little Dipper, with Polaris—the North Star—marking the end of the dipper's handle.

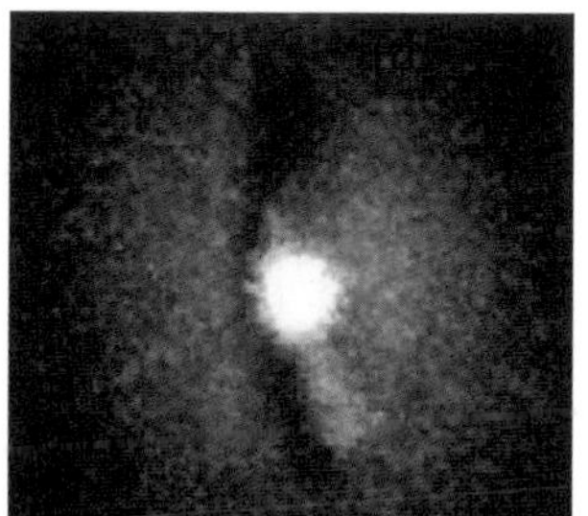

Right Hubble Space Telescope view of the center of the elliptical galaxy NGC 6251. The bright white spot at the center of the image is light from the vicinity of the black hole that is illuminating the disk. The galaxy is 300 million light-years away, in the constellation Ursa Minor.

One way that Northern hemisphere amateur astronomers can assess the sky's transparency on a given night is to determine the faintest magnitude star that is visible near Polaris with the unaided eye, since these stars are always at essentially the same altitude. Therefore any difference in what is called the "sky's limiting magnitude" must be due to a change in atmospheric transparency.

THE STARS OF URSA MINOR

These stars form the Little Dipper, an asterism. It has no resemblance to a bear since it would be mostly a tail.

Alpha (α) Ursae Minoris has at least twelve names—probably the most names of any star. Polaris is the one most commonly used today. Polaris is a Cepheid variable star, but its amplitude has become small and changeable in recent years. It is also a popular double star for small telescopes. The magnitude 2.1 yellow supergiant (the Cepheid) has a pale blue companion of magnitude 9.1 at 18.6 arc-seconds away.

OBJECTS OF INTEREST

The yellow and white double star **Pi-1 (π-1) Ursae Minoris** is magnitudes 6.6 and 7.3 at 31 arc-seconds. The magnitude 11.2 barred spiral galaxy **NGC 6217** is visible in a 6-inch (15 cm) Newtonian. It exhibits a brighter core with a nucleus.

Polarissima Borealis, NGC 3172, is the closest cataloged galaxy to the North Celestial Pole, at only 53 arc-minutes away. Amateurs observe it with 10-inch (25 cm) telescope or larger, for that reason only. It is magnitude 13.6 and is very small, at only 0.7 arc-minutes in diameter.

Left Hubble Space Telescope image of the North Star, Polaris A, a bright supergiant variable star. Just above Polaris is a small companion, Polaris Ab, which is less than two-tenths of an arc-second from Polaris.

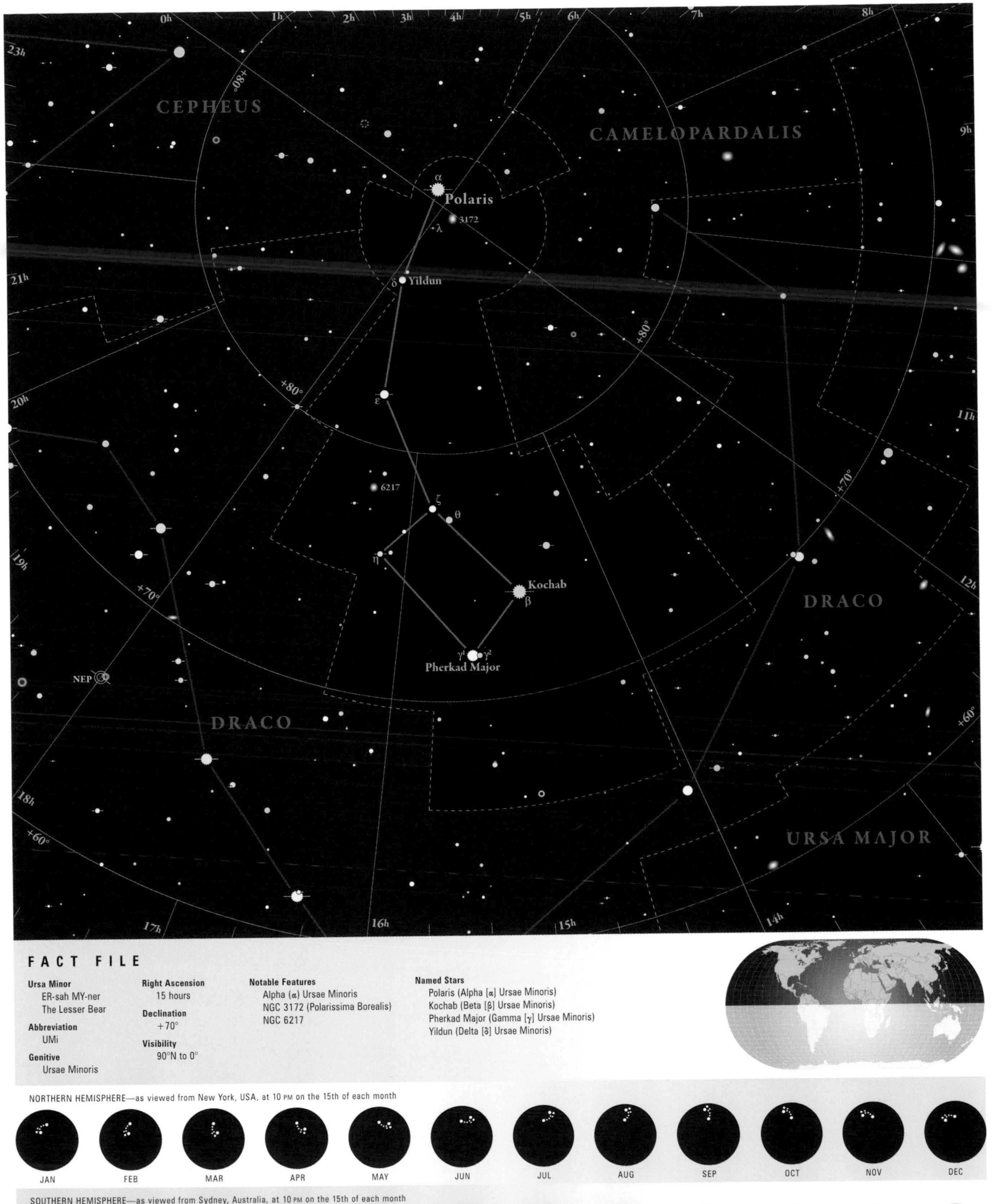

FACT FILE

Ursa Minor
ER-sah MY-ner
The Lesser Bear

Abbreviation
UMi

Genitive
Ursae Minoris

Right Ascension
15 hours

Declination
+70°

Visibility
90°N to 0°

Notable Features
Alpha (α) Ursae Minoris
NGC 3172 (Polarissima Borealis)
NGC 6217

Named Stars
Polaris (Alpha [α] Ursae Minoris)
Kochab (Beta [β] Ursae Minoris)
Pherkad Major (Gamma [γ] Ursae Minoris)
Yildun (Delta [δ] Ursae Minoris)

NORTHERN HEMISPHERE—as viewed from New York, USA, at 10 PM on the 15th of each month

JAN FEB MAR APR MAY JUN JUL AUG SEP OCT NOV DEC

SOUTHERN HEMISPHERE—as viewed from Sydney, Australia, at 10 PM on the 15th of each month

JAN FEB MAR APR MAY JUN JUL AUG SEP OCT NOV DEC

Vela

The constellation of Vela is a starry representation of the sails on Jason's ship, the *Argo*.

SET SAIL

Vela (the Sails), along with Carina (the Keel) and Puppis (the Stern), form the three sections of Jason's constellation ship, Argo Navis. (For the mythology related to Jason and the *Argo*, see the constellation description for Carina.) Abbé Nicolas Louis de Lacaille dismembered the huge constellation of Argo in the seventeenth century to make it "more manageable."

It is ironic that the mythologies surrounding the epic of the *Argo* omit any mention of *Argo* being equipped with sails. The vessel was believed to be a galley rowed by the 50 Argonauts, and probably had no sails. Despite that, virtually all the star-atlases from the sixteenth to the nineteenth centuries show it as a sailing ship with oars. When Argo was broken up, the Bayer designations of its stars remained with the individual wherever that star fell. That is why Vela has no Alpha or Beta star—its brightest is a Gamma.

THE STARS OF VELA

The stars of Vela are fairly bright, but bear little resemblance to a sail—except, vaguely, to the sail of a square-rigger. The two best known stars of Vela are Kappa (κ) and Delta (δ) Velorum, which form the northern half of the distinctive "false-cross" asterism, along with Epsilon (ε) and Iota (ι) Carinae. Delta Velorum is the brightest star in the sky that is without a common name.

Vela's lucida is the magnitude 1.8 Gamma (γ) Velorum, the thirty-third brightest star in the sky, bearing the ancient Arabic name Suhail. In more modern times, it is sometimes known as Regor. It is an extremely complex multiple star with five or six components (three are visible in small telescopes), the brightest of which is an exotic Wolf-Rayet star. Wolf-Rayet stars are high-mass, very evolved, extremely hot and luminous supergiants with powerful stellar winds. Only a few hundred are known, and of these, Suhail is the closest to us at only 830 light-years distant.

OBJECTS OF INTEREST

Very close to brilliant Gamma (γ) Velorum is the open cluster **NGC 2547**. It is a magnificent spray of moderately bright stars, about a half-moon diameter, with a multitude of faint stars in the background that shows well in a 6-inch (15 cm) telescope.

Almost next to Delta (δ) Velorum is another open cluster, **IC 2391**. It is clearly visible to the naked eye as a faint spot, and is particularly well suited to binoculars or wide-field small telescopes. At the northern end of the constellation is **NGC 3201**, a large bright globular cluster showing little compression to the center. This cluster is well resolved in an 8-inch (20 cm) telescope.

On the border with Antlia is one of the showpiece planetary nebulae of the southern sky, **NGC 3132**. Known as the **Eight-burst Nebula**, it shows in a 6-inch (15 cm) telescope as a medium-size bright blue disk with a dimmer center and a central star. That star is not actually within the shell and is just a chance alignment.

Above The globular cluster NGC 3201 in Vela consists of magnitude 13 to 16 stars and makes for interesting viewing in a 6-inch (15 cm) telescope under dark skies.

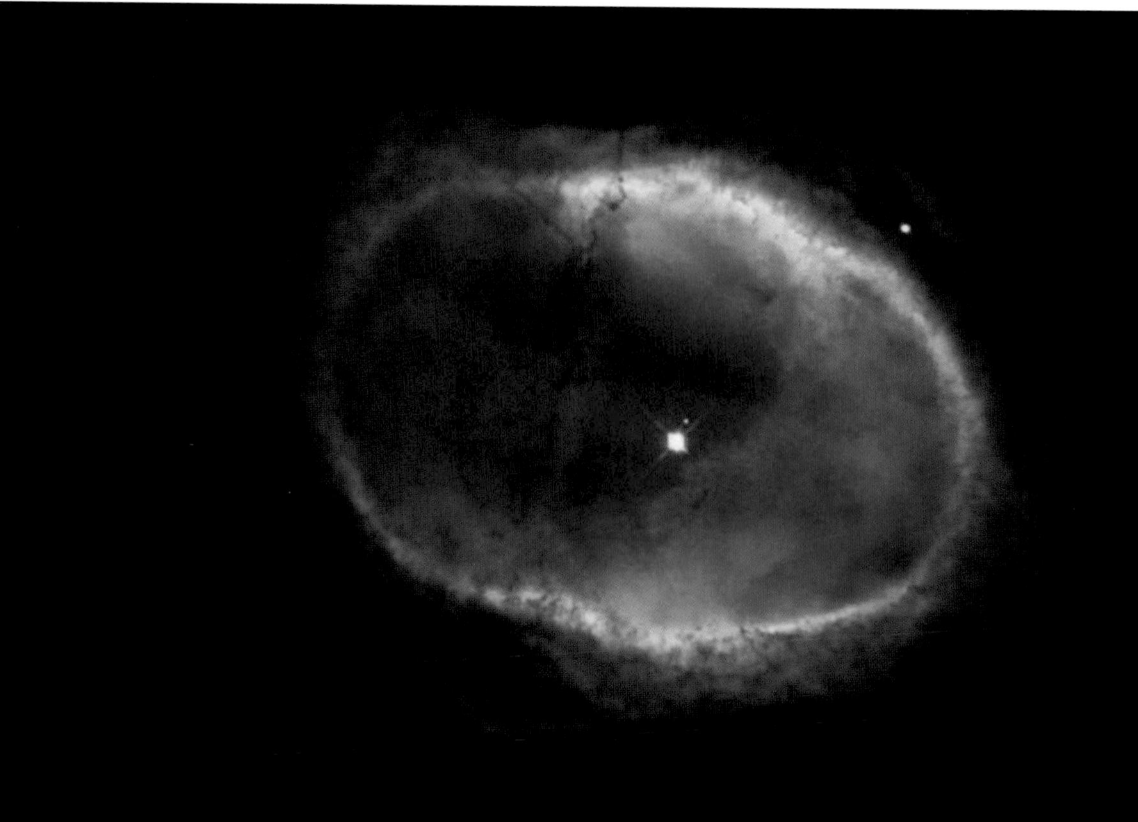

Right On the border of Antlia and Vela is the intriguing Eight-burst Nebula (NGC 3132), an oddly shaped gas cloud shrouding a dying binary system. The dim star, not the bright central one, created the nebula.

FACT FILE

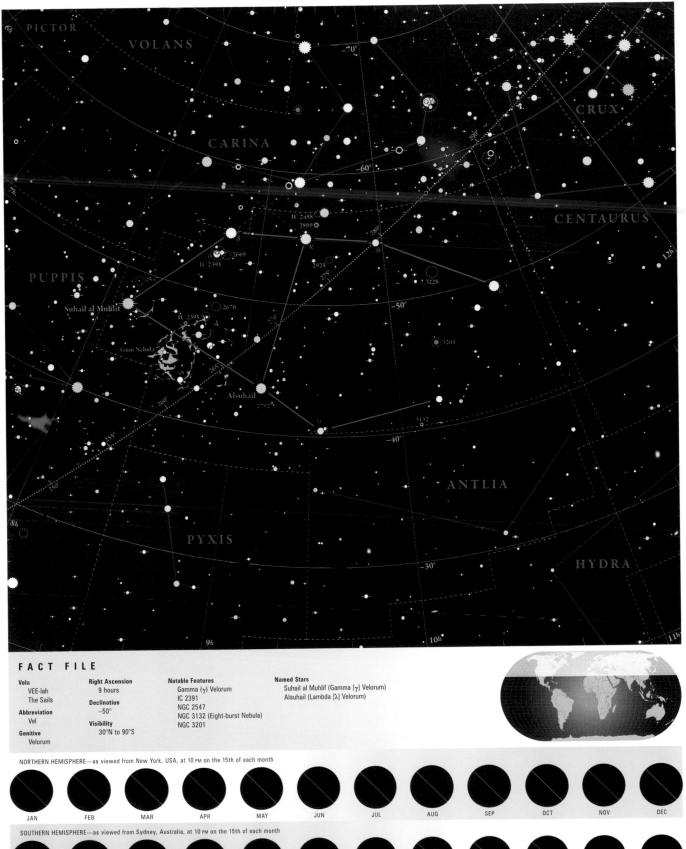

Vela
VEE-lah
The Sails

Abbreviation
Vel

Genitive
Velorum

Right Ascension
9 hours

Declination
–50°

Visibility
30°N to 90°S

Notable Features
Gamma (γ) Velorum
IC 2391
NGC 2547
NGC 3132 (Eight-burst Nebula)
NGC 3201

Named Stars
Suhail al Muhlif (Gamma [γ] Velorum)
Alsuhail (Lambda [λ] Velorum)

NORTHERN HEMISPHERE—as viewed from New York, USA, at 10 PM on the 15th of each month

JAN FEB MAR APR MAY JUN JUL AUG SEP OCT NOV DEC

SOUTHERN HEMISPHERE—as viewed from Sydney, Australia, at 10 PM on the 15th of each month

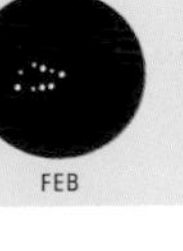

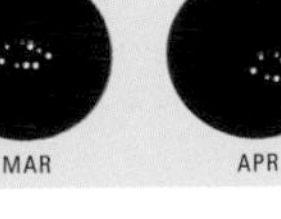
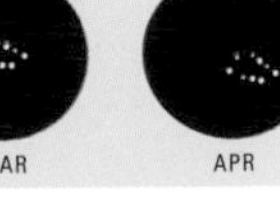
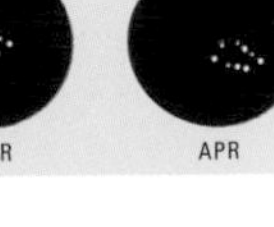

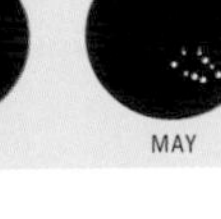

JAN FEB MAR APR MAY JUN JUL AUG SEP OCT NOV DEC

Virgo

Virgo is home to our sky's greatest number of galaxies. All of the ancient cultures perceived the stars of Virgo as a maiden or a goddess of some type.

MOTHER STAR

In Egyptian culture, the stars of Virgo represented Isis, the chief mother-goddess. The Sumerians and Chaldeans saw the constellation as Inanna, Queen of Heaven.

The Greeks seemed to have a veritable multitude of goddesses whom Virgo came to represent. The two most common identifications were with the goddess of justice, Dike, the daughter of Zeus and Themis, and with Demeter, goddess of corn, and the daughter of Cronus and Rhea. Several others also contend, including Persephone, the reluctant queen of the underworld kidnapped by Pluto, and Erigone, the daughter of Icarius and Tyche, the goddess of fortune, who is usually pictured holding a cornucopia. The Romans picked up the agricultural connection and named the constellation for Ceres, goddess of the harvest.

Right This curious straight beam, somewhat like a searchlight in space, protrudes from the center of M87 in the Virgo Cluster of Galaxies. It is thought to be superheated gas swirling around the galaxy's central, supermassive black hole.

THE STARS OF VIRGO

Located almost at right angles to the Milky Way, the stars of Virgo are generally faint, and for the best visibility they need a dark sky. The central trapezium, made up by 1st magnitude Alpha (α) along with 3rd magnitude Zeta (ζ), Gamma (γ), and Delta (δ), is the most readily recognized part of the constellation.

Alpha (α) Virginis, which is otherwise known as Spica, meaning "ear of grain," is the lucida of Virgo. The fifteenth brightest star in the sky, it consists of a very close binary of blue B-type stars separated by only 9 million miles (15 million km). Their combined luminosity approaches 2,100 times that of the Sun, and they are located 220 light-years distant.

OBJECTS OF INTEREST

Virgo is located far from the plane of the Milky Way and contains no nebulae or open clusters, but it has an incredible abundance of bright galaxies. The largest galaxy cluster within 100 million light-years is centered in northern Virgo, and perhaps 200 galaxies are visible in an 8-inch (20 cm) telescope, provided there is a truly dark sky.

The field surrounding the bright elliptical galaxies **M84** and **M86** is awash with numerous small fuzzy galaxies. A couple of low-power fields away is the enigmatic **M87**, an enormous supergiant elliptical galaxy possibly containing ten times the number of stars in our own Milky Way. In an 8-inch (20 cm) telescope, all three look like small, round, hazy patches with conspicuously brighter cores. Quite near-by to M87 are two beautiful spirals: **M90** and **M58**. M90 has an oval halo and a bright, almost stellar nucleus; M58 is seen almost face-on and has a roundish outline with a small, very bright core. **M61** is also a noteworthy face-on spiral, displaying a clear spiral structure in large amateur telescopes.

Best of all is **M104** in southern Virgo, near the Corvus border. A 4-inch (10 cm) telescope is large enough to show the thin, dramatic lane of dust that bisects its nearly edge-on form; an 8-inch (20 cm) telescope provides a view like that of a professional photograph.

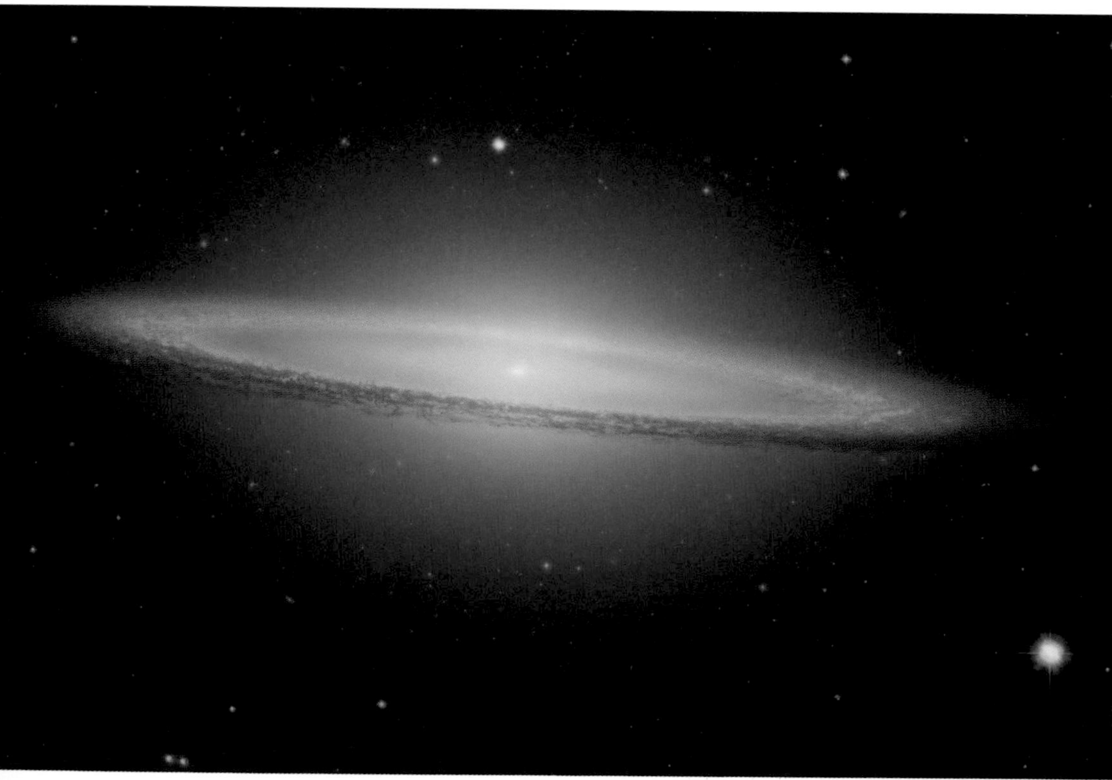

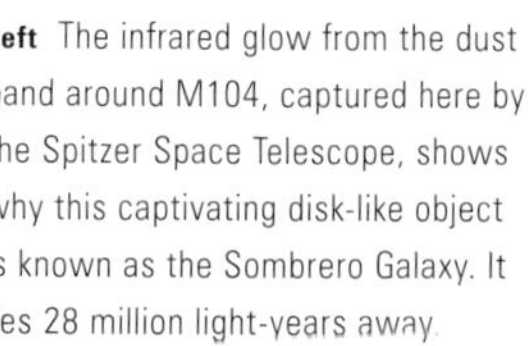

Left The infrared glow from the dust band around M104, captured here by the Spitzer Space Telescope, shows why this captivating disk-like object is known as the Sombrero Galaxy. It lies 28 million light-years away.

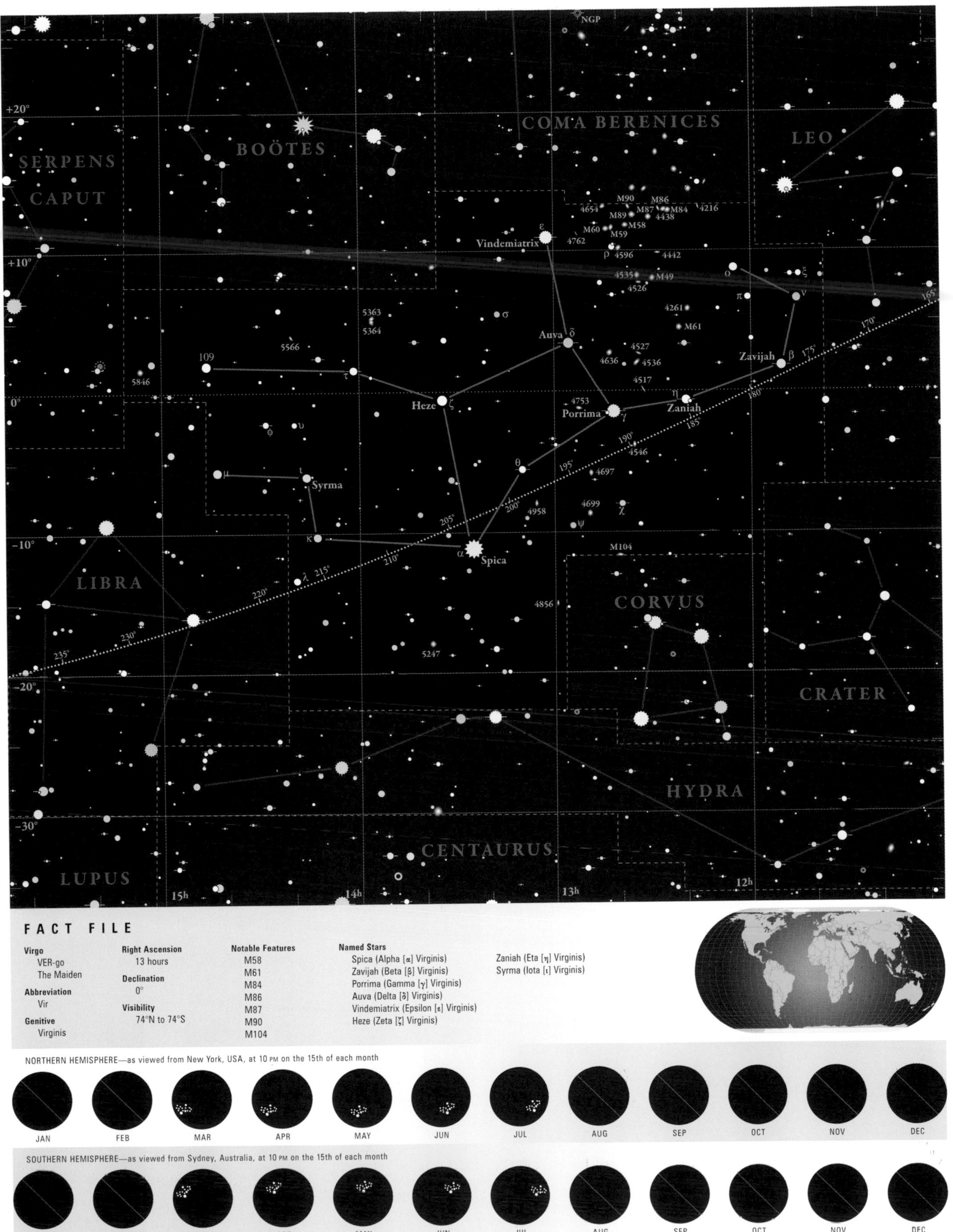

FACT FILE

Virgo
VER-go
The Maiden

Abbreviation
Vir

Genitive
Virginis

Right Ascension
13 hours

Declination
0°

Visibility
74°N to 74°S

Notable Features
M58
M61
M84
M86
M87
M90
M104

Named Stars
Spica (Alpha [α] Virginis)
Zavijah (Beta [β] Virginis)
Porrima (Gamma [γ] Virginis)
Auva (Delta [δ] Virginis)
Vindemiatrix (Epsilon [ε] Virginis)
Heze (Zeta [ζ] Virginis)
Zaniah (Eta [η] Virginis)
Syrma (Iota [ι] Virginis)

Volans

Swimming in the wash beneath the keel of the mighty ship *Argo* is the obscure group of stars known as Volans, the Flying Fish.

FISH TALE

The faint stars of Volans were too far south for ancient astronomers to recognize, and accordingly no mythology is associated with them. The sixteenth-century Dutch explorers Pieter Dirkszoon Keyser and Frederick de Houtman observed these stars, and in 1598 Petrus Plancius placed the constellation on his star-globe as Vliegendenvis—the Winged Fish.

When Bayer published his *Uranometria* in 1603, he rendered Volans as a flying fish and named it Piscis Volans. The constellation and its form have not changed since the seventeenth century, the original name Piscis Volans has become simply Volans.

THE STARS OF VOLANS

The most easily recognized part of Volans is a wide trapezium of similar-looking 4th magnitude stars, mid-way between the bright 2nd magnitude star Miaplacidus in Carina and the Large Magellanic Cloud. Alpha (α) and Beta (β) Volantis are removed from this asterism and are found near Miaplacidus. Curiously, Alpha is not the lucida of Volans, and is instead the third brightest. The beautiful double star Gamma (γ) Volantis, at only 4th magnitude, is the brightest star if both components are taken together. Beta (β) Volantis, an orange giant star, is also brighter than Alpha.

Above In the constellation of Volans, a group of 4th magnitude stars marks out the body of the flying fish, while Alpha (α) and Beta (β) Volantis distinguish the tail.

OBJECTS OF INTEREST

Volans boasts only a few objects that are worthwhile viewing with a telescope. The constellation contains no clusters, nebulae, or planetary nebulae, and most of the galaxies are faint; with **NGC 2442** being the one exception. Known as the **Meathook Galaxy**, NGC 2442 is located near the center of the trapezium of stars that make up the main configuration of the constellation. An 8-inch (20 cm) telescope shows the inner regions of the galaxy as a reasonably bright, elongated haze.

Gamma (γ) Volantis is an attractive binary star with orange and white components of magnitudes 3.8 and 5.7 that are easily resolved with a 4-inch (10 cm) telescope. The brighter star is a K-type orange giant, while the lesser is an F-type dwarf, and the pair, separated by 600 Astronomical Units (AU) with an orbit of at least 7,500 years, are 142 light-years distant.

A somewhat tighter binary is **Epsilon (ε) Volantis**—a pair of magnitude 5.4 and 6.7 bluish-white stars that are beautiful in a 6-inch (15 cm) telescope.

Visible only in the largest amateur telescopes as a patch of haze, **ESO 34-11** is a rare ring-form spiral that resulted from a head-on collision between a large spiral galaxy and a small galaxy.

Right Lying in the direction of the constellation of Volans, the galaxy seen here—AM 0644–741—is some 150,000 light-years across and larger than the Milky Way.

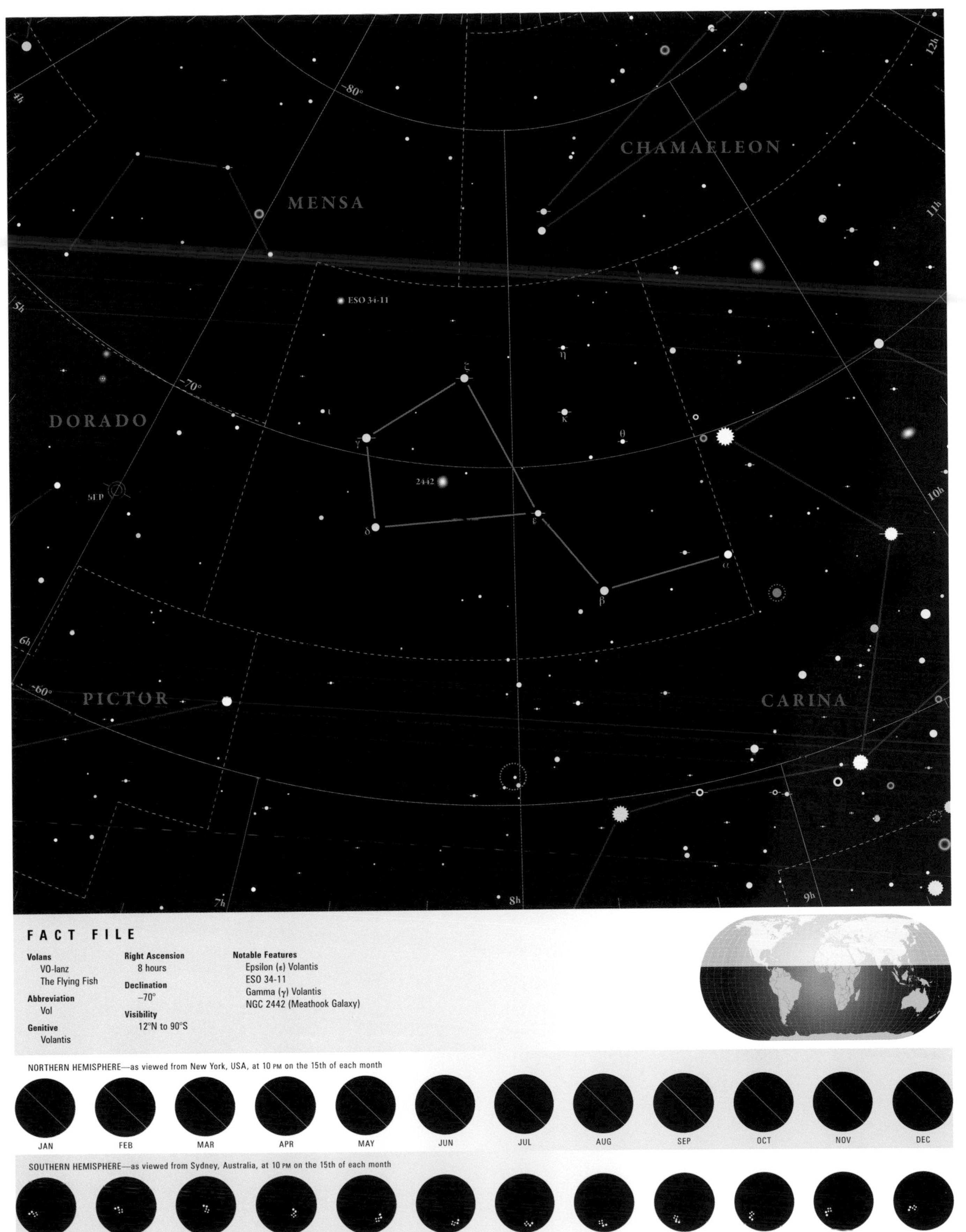

FACT FILE

Volans
VO-lanz
The Flying Fish

Abbreviation
Vol

Genitive
Volantis

Right Ascension
8 hours

Declination
–70°

Visibility
12°N to 90°S

Notable Features
Epsilon (ε) Volantis
ESO 34-11
Gamma (γ) Volantis
NGC 2442 (Meathook Galaxy)

NORTHERN HEMISPHERE—as viewed from New York, USA, at 10 PM on the 15th of each month

JAN FEB MAR APR MAY JUN JUL AUG SEP OCT NOV DEC

SOUTHERN HEMISPHERE—as viewed from Sydney, Australia, at 10 PM on the 15th of each month

JAN FEB MAR APR MAY JUN JUL AUG SEP OCT NOV DEC

Vulpecula

Vulpecula, the Little Fox, was carved out of the Cygnus Milky Way. It contains the sky's finest planetary nebula, Messier 27.

SLY SKY HUNTER

Introduced in the seventeenth century by Polish astronomer Johannes Hevelius, Vulpecula is a small constellation that he originally called Vulpecula cum Ansere, or "Little Fox with Goose." Some early star-atlases depict it as a fox carrying off a domestic goose.

THE STARS OF VULPECULA

There is no recognizable constellation figure—Vulpecula is merely thought of as that part of the Milky Way between Sagitta and Beta (β) Cygni. Its brightest star, Alpha (α) Vulpeculae, is magnitude 4.4; the next brightest are 5th magnitude, and are drowned in the background Milky Way.

Right A close-up view of the glowing knots of gas and dust in the Dumbbell Nebula, formed when a red-giant star expelled its outer layers into space. M27 was the first planetary nebula ever discovered.

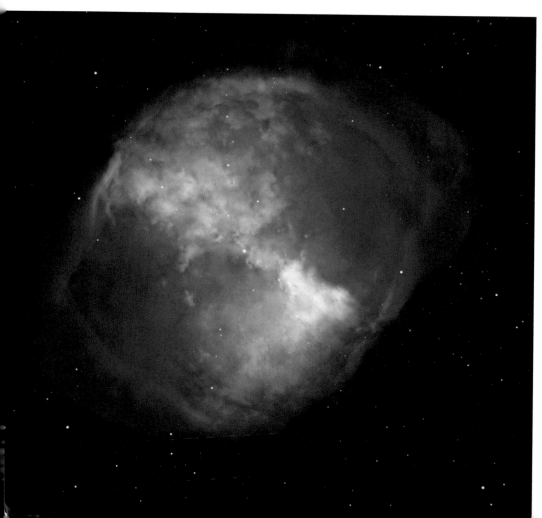

OBJECTS OF INTEREST

Collinder 399 is a naked-eye glow, two moon-diameters wide, which binoculars show to be an amusing asterism called the **Coathanger**. It was once thought to be a star cluster, but it has been proven otherwise. However, there is a real open cluster at the eastern end of the Coathanger: the 9th magnitude **NGC 6802**, which is 5 arc-minutes in diameter. An 8-inch (20 cm) Newtonian just begins to resolve the bar-shaped group.

The open cluster **NGC 6940** is magnitude 6.3 and nearly half a moon-diameter. Professionals have counted 170 stars here. It is a prominent fuzzy patch in binoculars, and is noticeable with every casual sweep of the Milky Way in a northern summer. An 8-inch (20 cm) shows its full splendor as a rich cluster in a dense Milky Way field. The cluster is arranged in a bulbous Y-shape: Each arm of the Y is a cloud of sparkles rather than just a line. An orange star marks the center.

Planetary nebula **M27 (NGC 6853)**, the **Dumbbell Nebula**, is 8.0 by 5.7 arc-seconds and magnitude 7.3. The planetary is obvious in 7 × 50 binoculars or a 2¼-inch (6 cm) refractor. The popular name refers to the pinched waist of the somewhat rectangular nebula, but Hourglass Nebula or Applecore Nebula are both better descriptions. Few people see color in this large planetary, but one famous observer calls it "pale blue" and another sees "misty green."

With an 8-inch (20 cm) telescope, faint nebulosity beyond the "applecore" fills out M27, making it oval. At a magnification of 350, the magnitude 13.8 central star becomes visible, as well as three other stars within the "applecore" portion of M27.

Star-parties give the opportunity for views of splendors through giant Dobsonians. A 36-inch (0.9 m) telescope shows that on M27's following side there is a brighter rim to the outer edge of the oval and a darker area between this rim and the "hourglass" portion. The planetary is adorned with a multitude of stars, one of which is orange. The nebulosity that fills out the oval is bright enough at this aperture to show a faint rusty tinge.

Left An ageing star's last gasp: The Dumbbell Nebula (M27, NGC 6853), a gaseous emission nebula, belches copious amounts of dust and gas as it depletes the nuclear fuel required to stoke its core. At 1,000 light-years distant, it is one of our closest planetary nebulae.

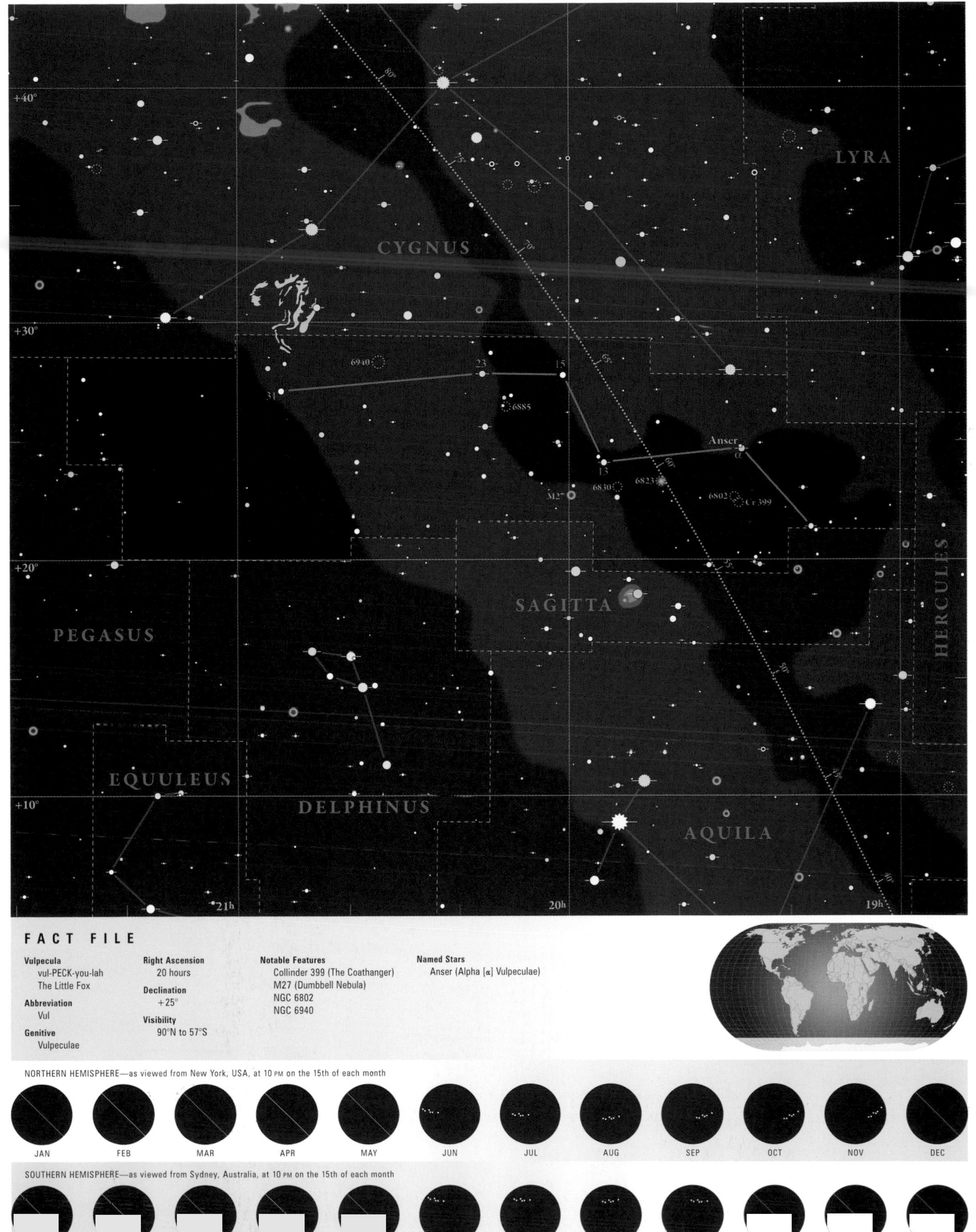

FACT FILE

Vulpecula
vul-PECK-you-lah
The Little Fox

Abbreviation
Vul

Genitive
Vulpeculae

Right Ascension
20 hours

Declination
+25°

Visibility
90°N to 57°S

Notable Features
Collinder 399 (The Coathanger)
M27 (Dumbbell Nebula)
NGC 6802
NGC 6940

Named Stars
Anser (Alpha [α] Vulpeculae)

NORTHERN HEMISPHERE—as viewed from New York, USA, at 10 PM on the 15th of each month

JAN FEB MAR APR MAY JUN JUL AUG SEP OCT NOV DEC

SOUTHERN HEMISPHERE—as viewed from Sydney, Australia, at 10 PM on the 15th of each month

JAN FEB MAR APR MAY JUN JUL AUG SEP OCT NOV DEC

Using the Monthly Sky Maps

The marvels of the night sky on a clear night are a wondrous sight, and becoming familiar with the starry markers that dot the darkness of space is part of the experience of enjoying the hobby of astronomy.

The maps on the following pages provide a month-by-month overview of the night sky. Each map presents the night sky on the 1st day of each month at approximately 9:30 P.M. in the evening. On each left-hand page is the northern polar region; on the right-hand page is the southern polar region; and at the bottom of the pages is the equatorial region. These maps show the celestial sphere, and the orientation of the constellations at the specified month and time from any place on Earth. Just as world atlas maps show east and west as if the reader is looking at the ground, so the monthly polar maps show east and west reversed, as the maps are compared to the night sky above.

As the year progresses, the constellations rotate around the north and south celestial poles, with some constellations remaining in view at all times, while others disappear from view for months at a time. Referring to these maps will easily determine the visibility of each constellation throughout the year.

Discover the constellations in the night sky that are visible in your local area, and determine the best viewing month or months for each of the 88 recognized representatives. There are some constellations that are only visible in the northern hemisphere, and likewise, there are some that are only visible to southern hemisphere observers. The majority, however, will be fully or partially visible at some time during the year for observers to track and observe the wonders of these starry regions of the night sky.

Used in conjunction with the individual constellation guides and maps, these whole sky maps introduce observers to the night sky in their hemisphere. Often minimal or even no equipment is required to view many of the constellations, though binoculars or a telescope will reveal many hidden wonders within the network of stars.

Star Magnitude

−1.5
−1
−0.5
0
0.5
1
1.5
2
2.5
3
3.5
4
4.5
5

Milky Way
ARA Constellation
Constellation locators
6h Right Ascension (hours)
+60° Declination (degrees)
NEP Poles of Ecliptic
NGP Galactic Poles
90° Galactic Equator
60° Ecliptic

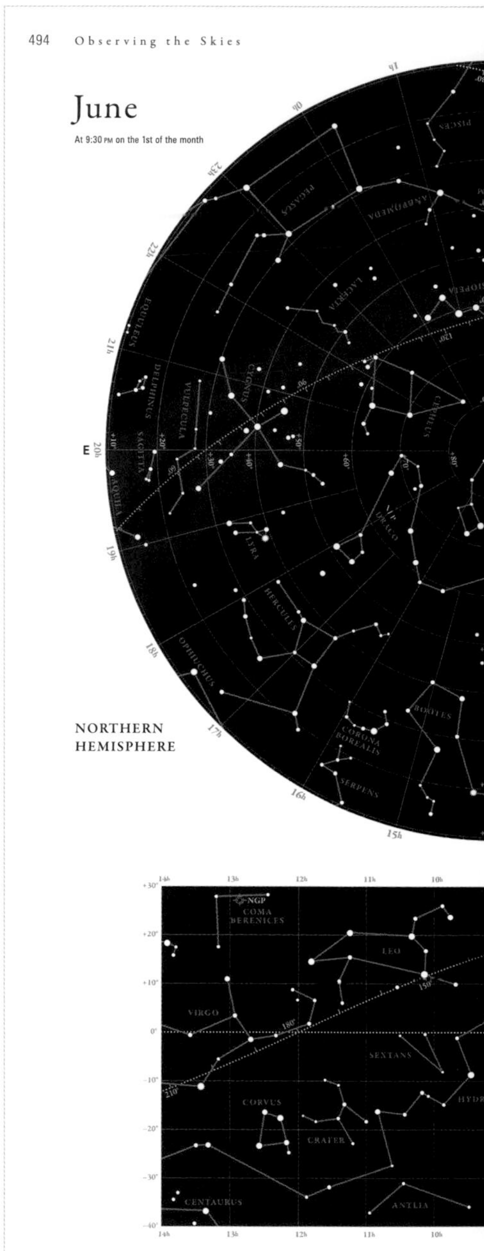

Polar Region Maps

The northern polar region and southern polar region show the night sky for each of the hemispheres. Extending from the celestial pole in each hemisphere, the map extends out 80°, overlapping with the corresponding area of the equatorial region map. At the center of the northern polar map is the north celestial pole marked by a cross, with the south celestial pole at the center of the south polar map. SGP = south galactic pole, NGP = north galactic pole, SEP = south ecliptic pole, NEP = north ecliptic pole.

Constellations

The outline of each constellation is traced out, joining the significant stars of the configuration that represent the mythological or historical figure. A guide to the magnitude of stars used on these maps is given in the legend on page 482.

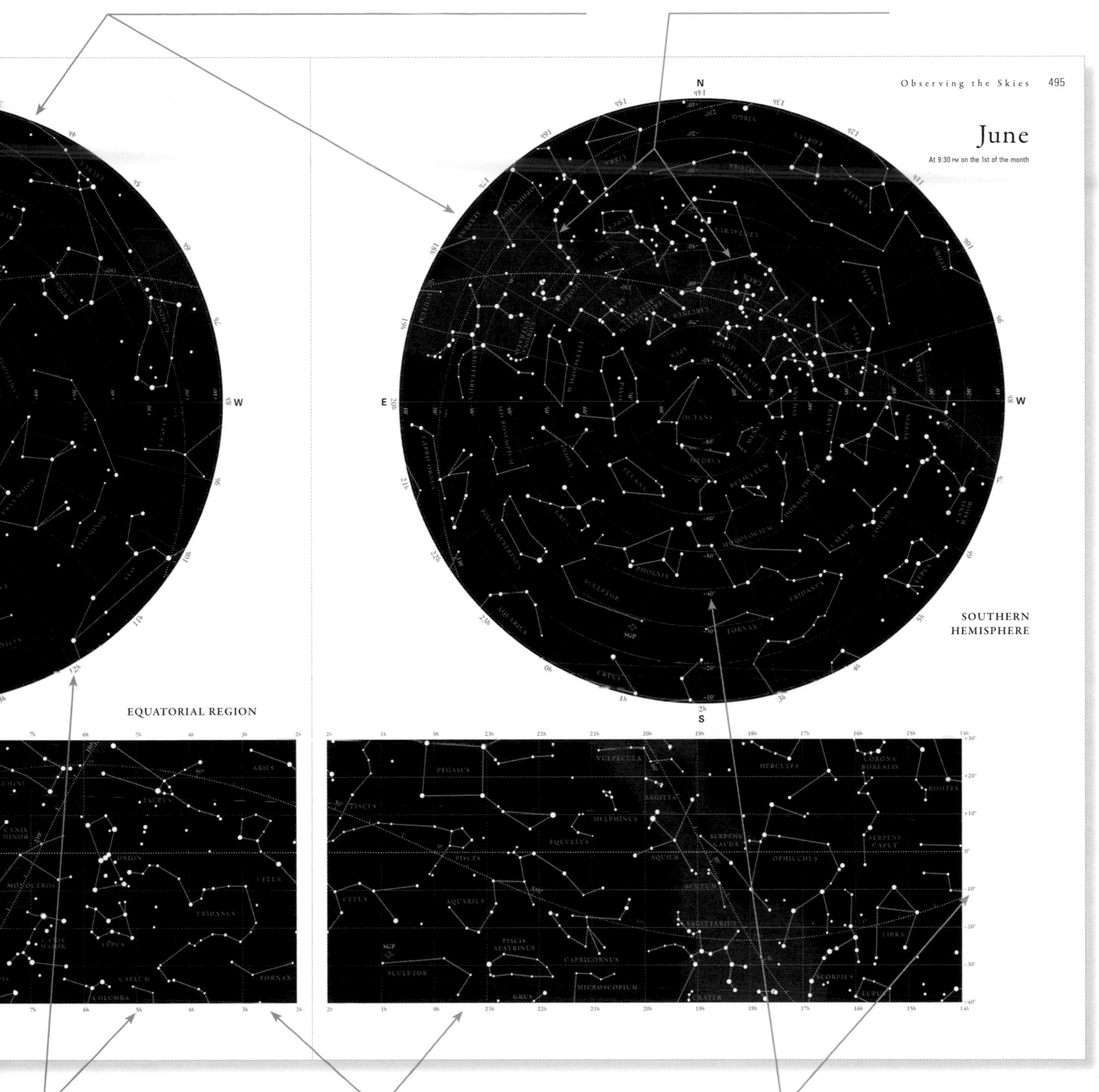

Right Ascension

Right Ascension (RA) is one of two co-ordinates used in celestial mapping to locate stars and other objects in the night sky. Right Ascension is the equivalent of longitude seen on world maps, and is measured in hours, minutes, and seconds.

Equatorial Region

This section of mapping depicts those constellations that lie in the equatorial region. The map strip details 70° of sky that links the two hemisphere maps, and overlaps with each of them. This map strip provides a clear guide and easy location of those constellations that lie in this region.

Declination

Used in conjunction with Right Ascension (RA), Declination (Dec) is used to plot the positions of stars and other objects in the night sky. Declination is the equivalent of latitude on a world map, and is measured in degrees, arc-minutes, and arc-seconds.

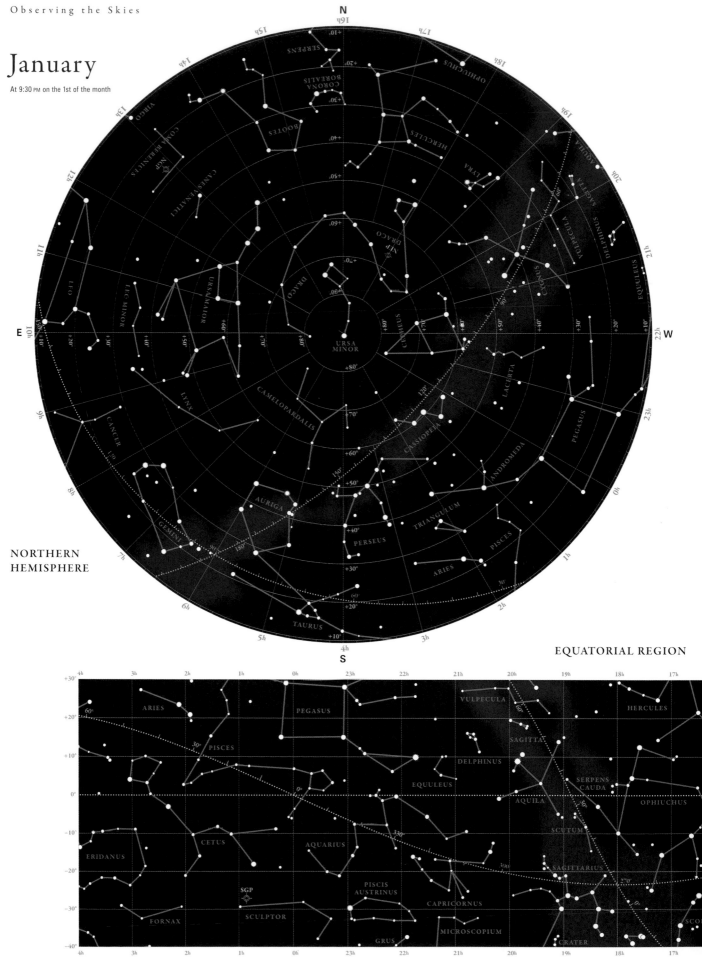

January

At 9:30 PM on the 1st of the month

NORTHERN HEMISPHERE

EQUATORIAL REGION

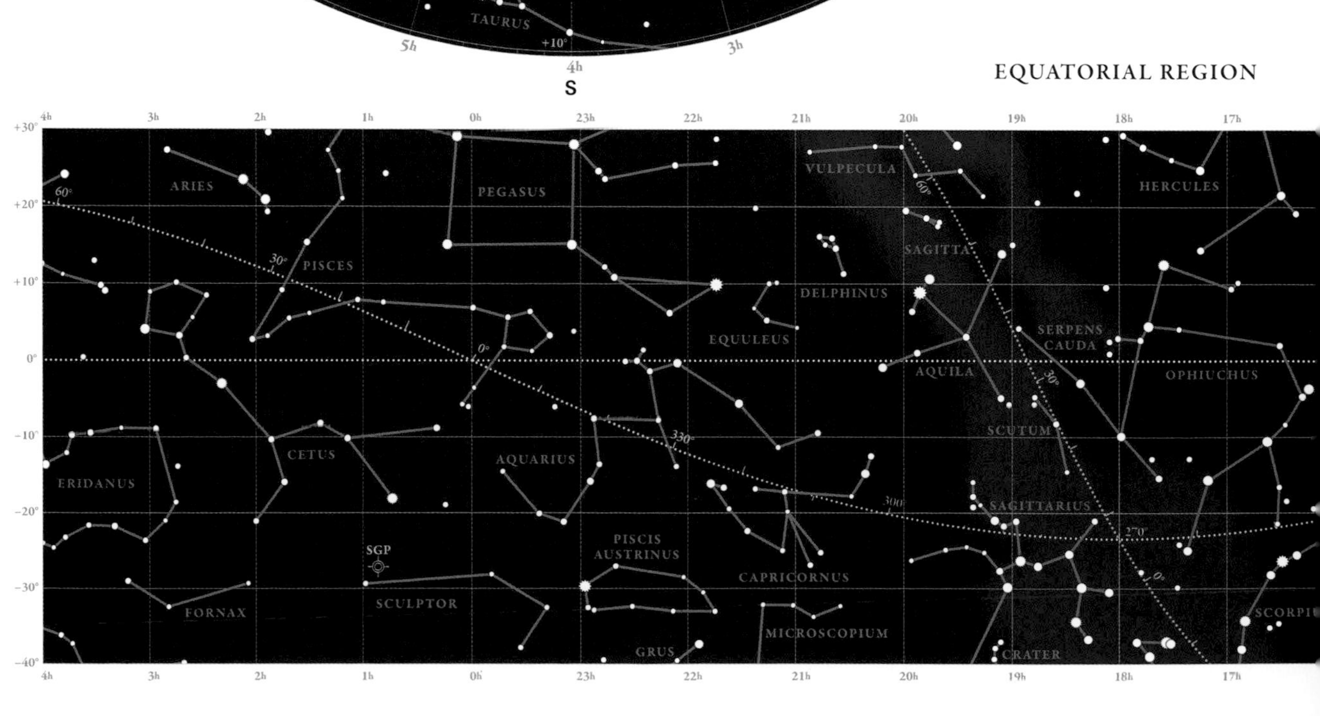

January

At 9:30 PM on the 1st of the month

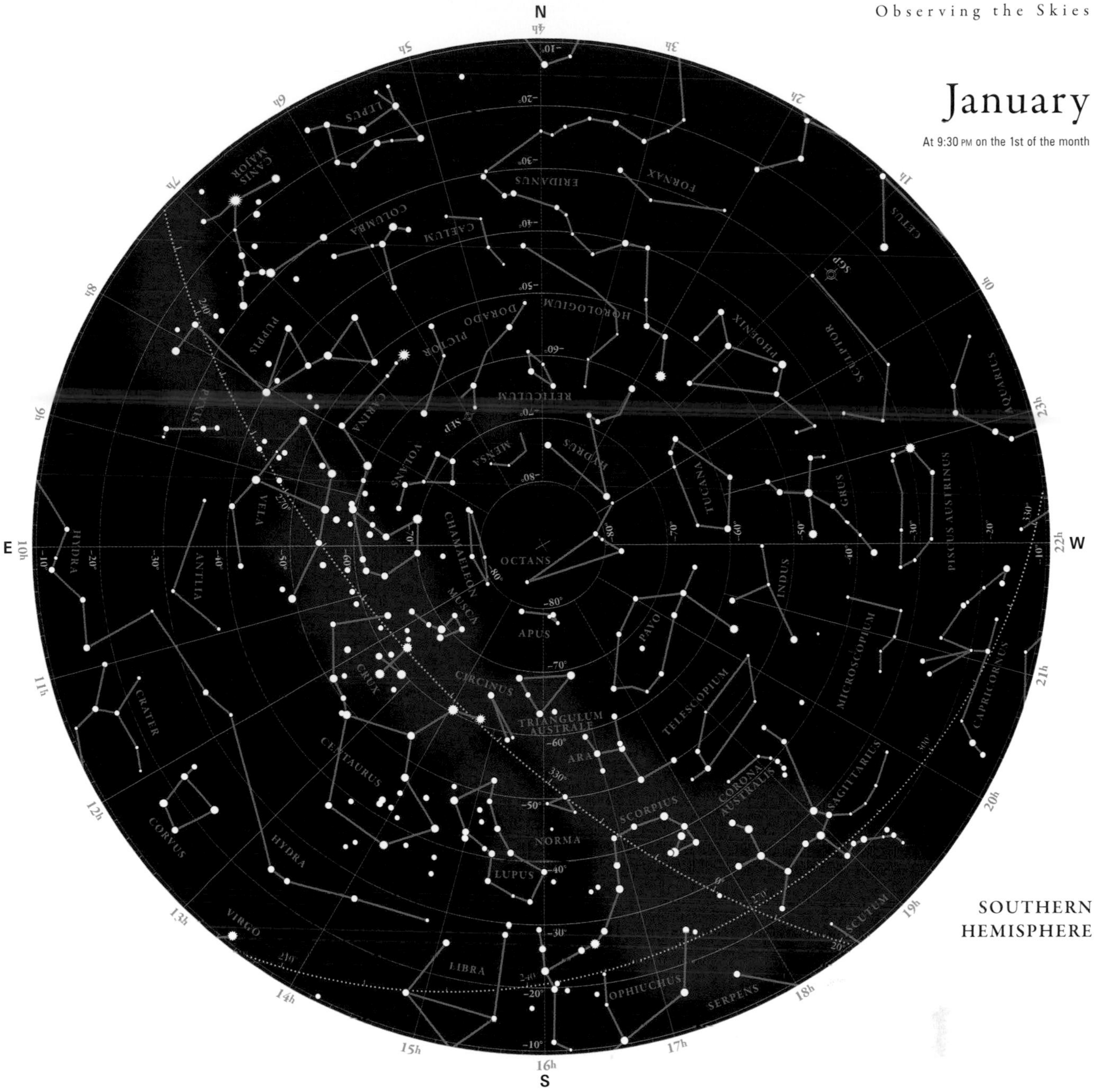

SOUTHERN HEMISPHERE

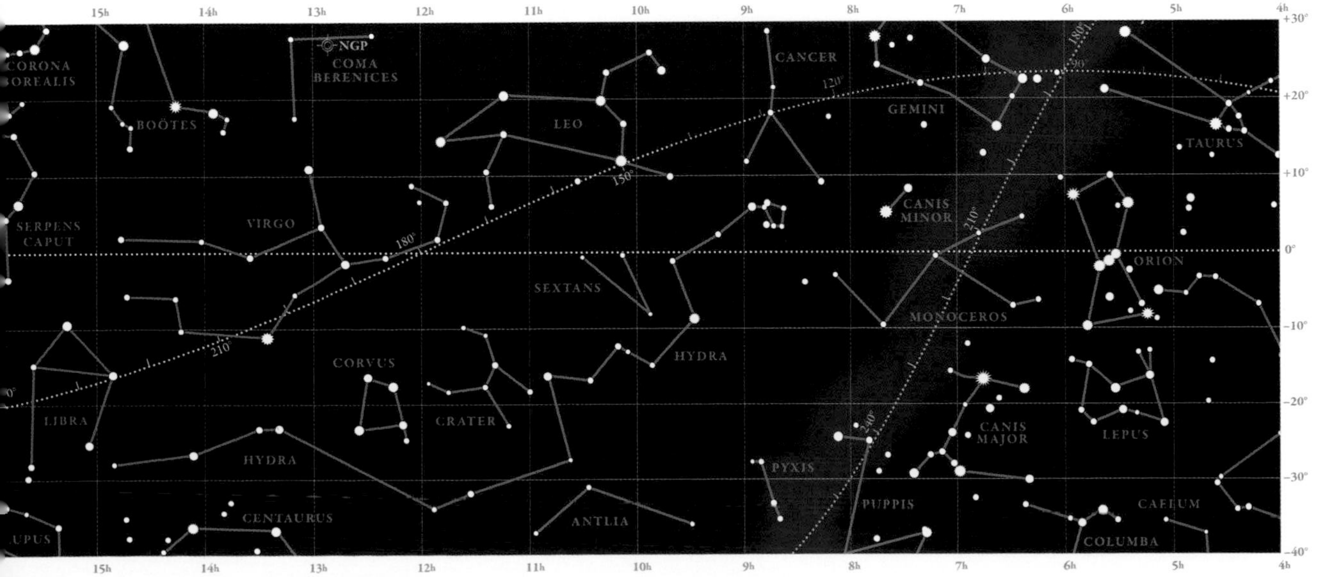

February

At 9:30 PM on the 1st of the month

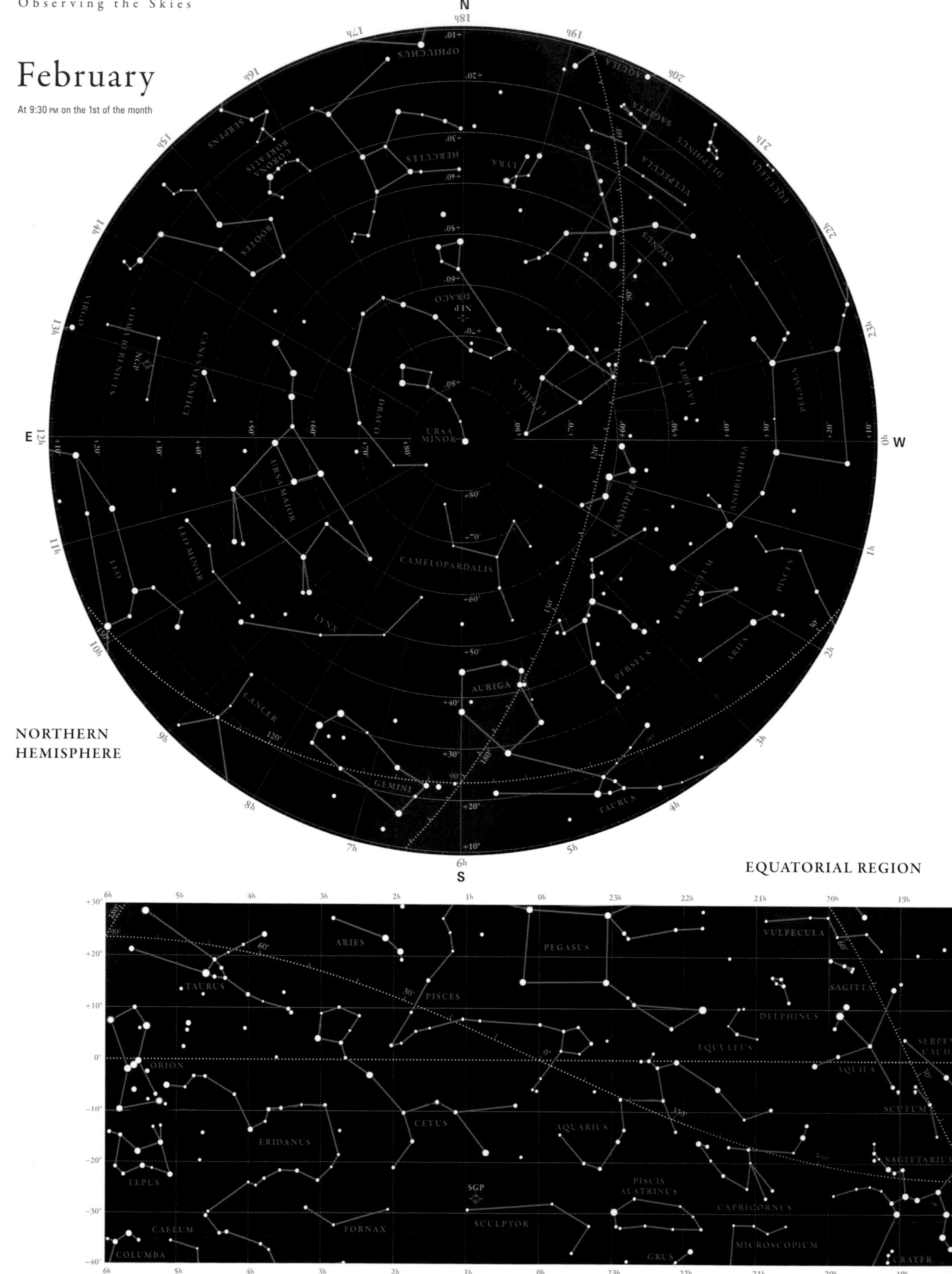

February

At 9:30 PM on the 1st of the month

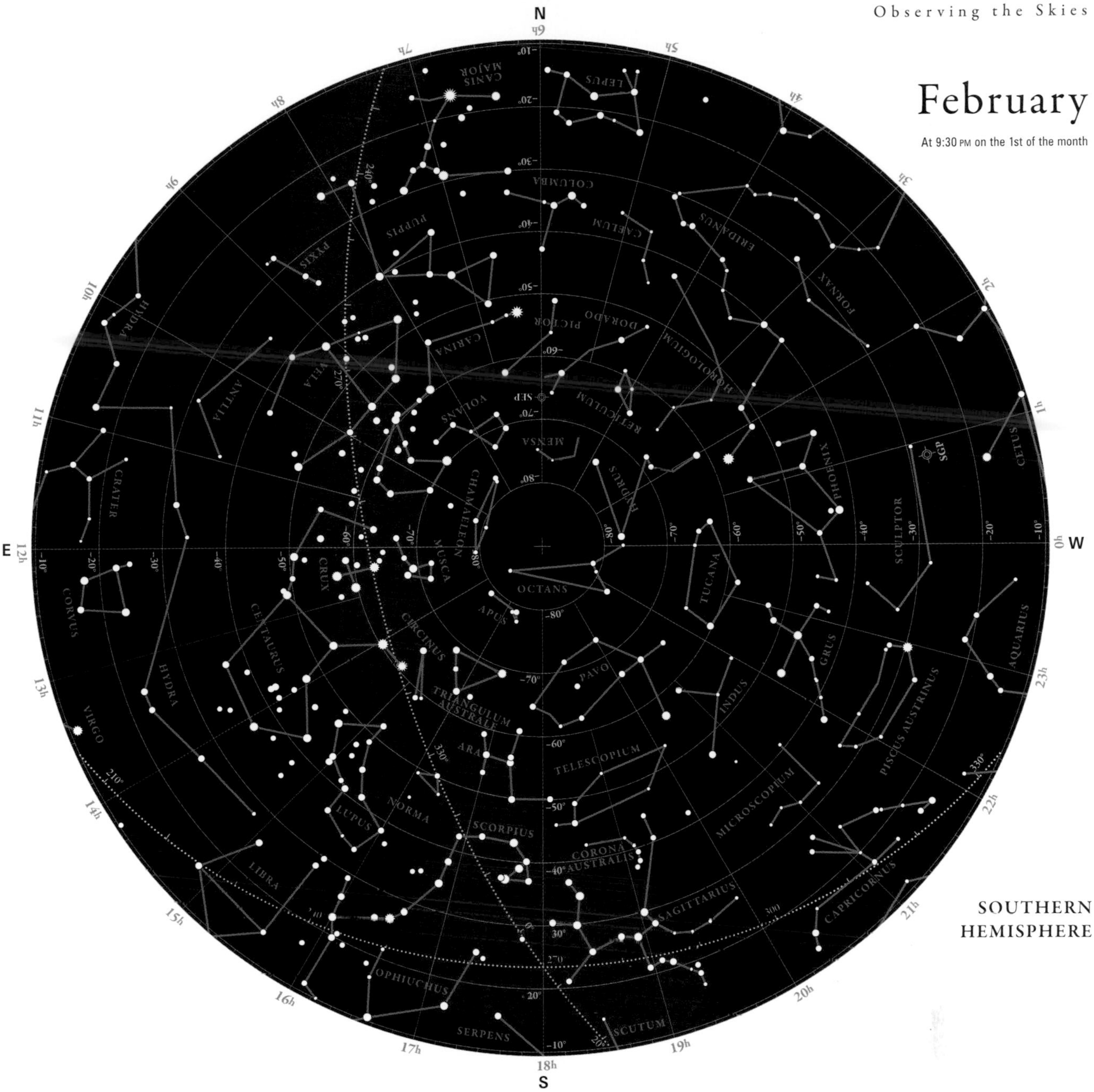

SOUTHERN HEMISPHERE

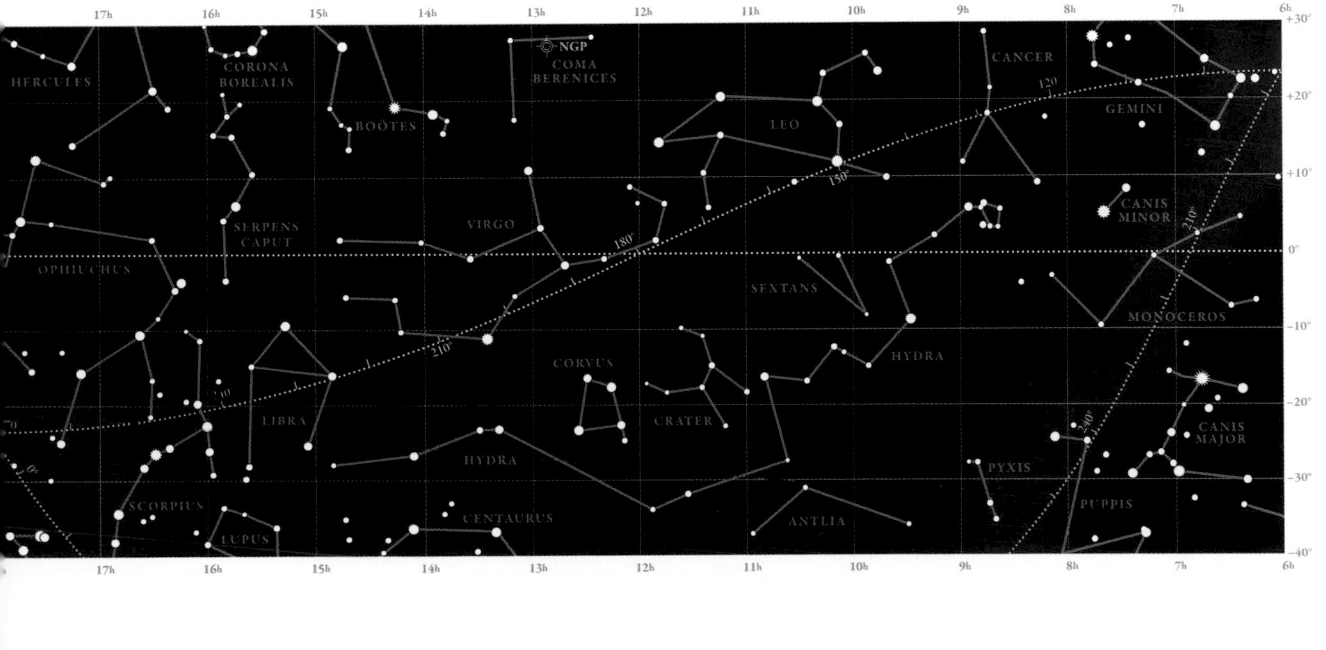

March

At 9:30 PM on the 1st of the month

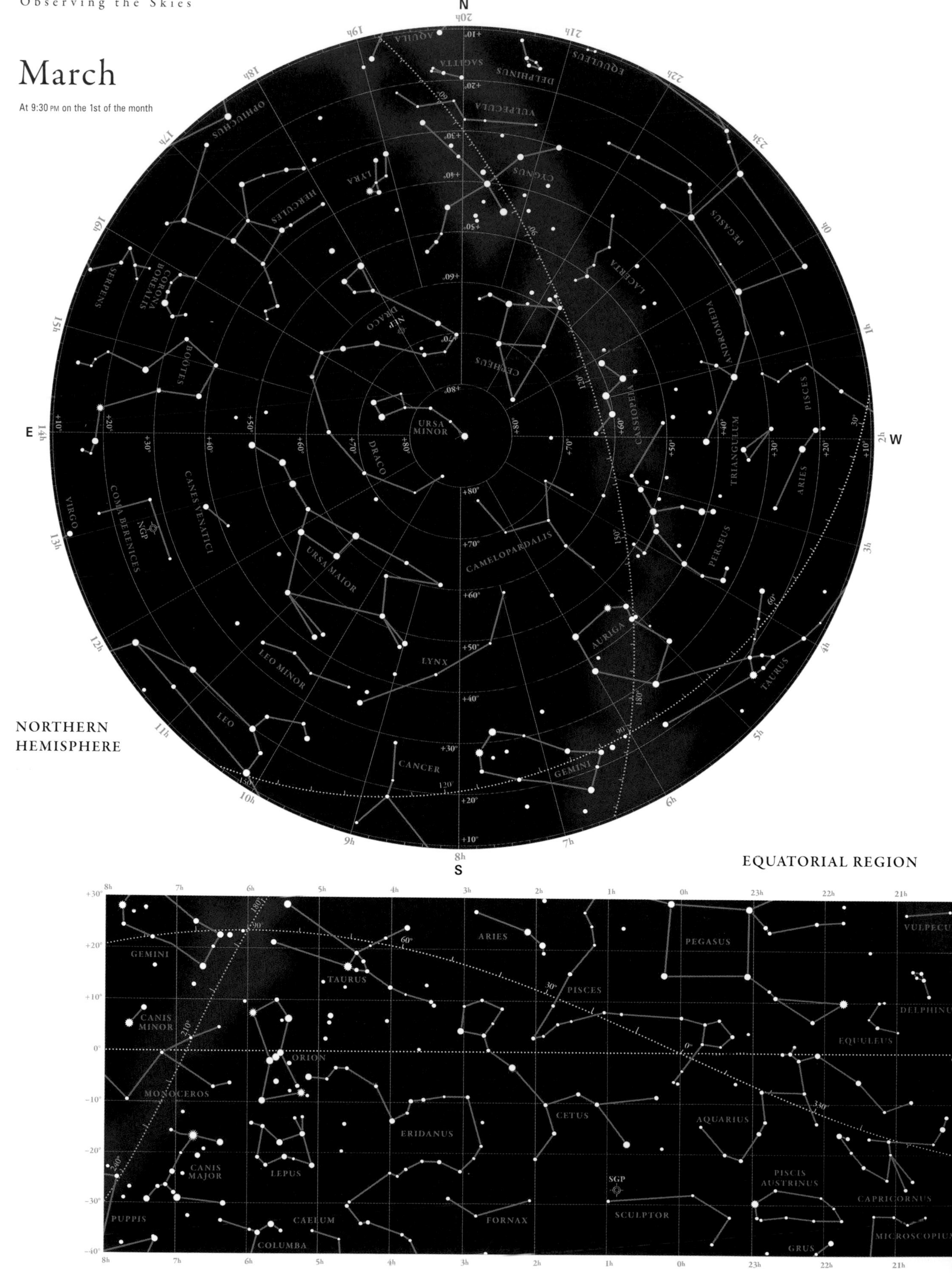

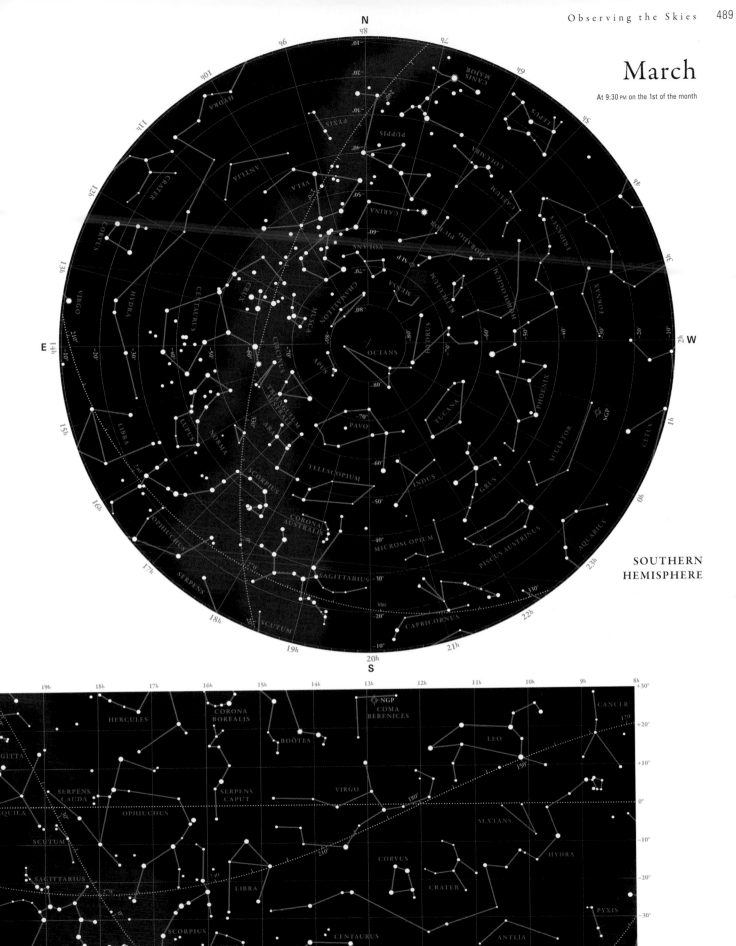

March

At 9:30 PM on the 1st of the month

SOUTHERN HEMISPHERE

April

At 9:30 PM on the 1st of the month

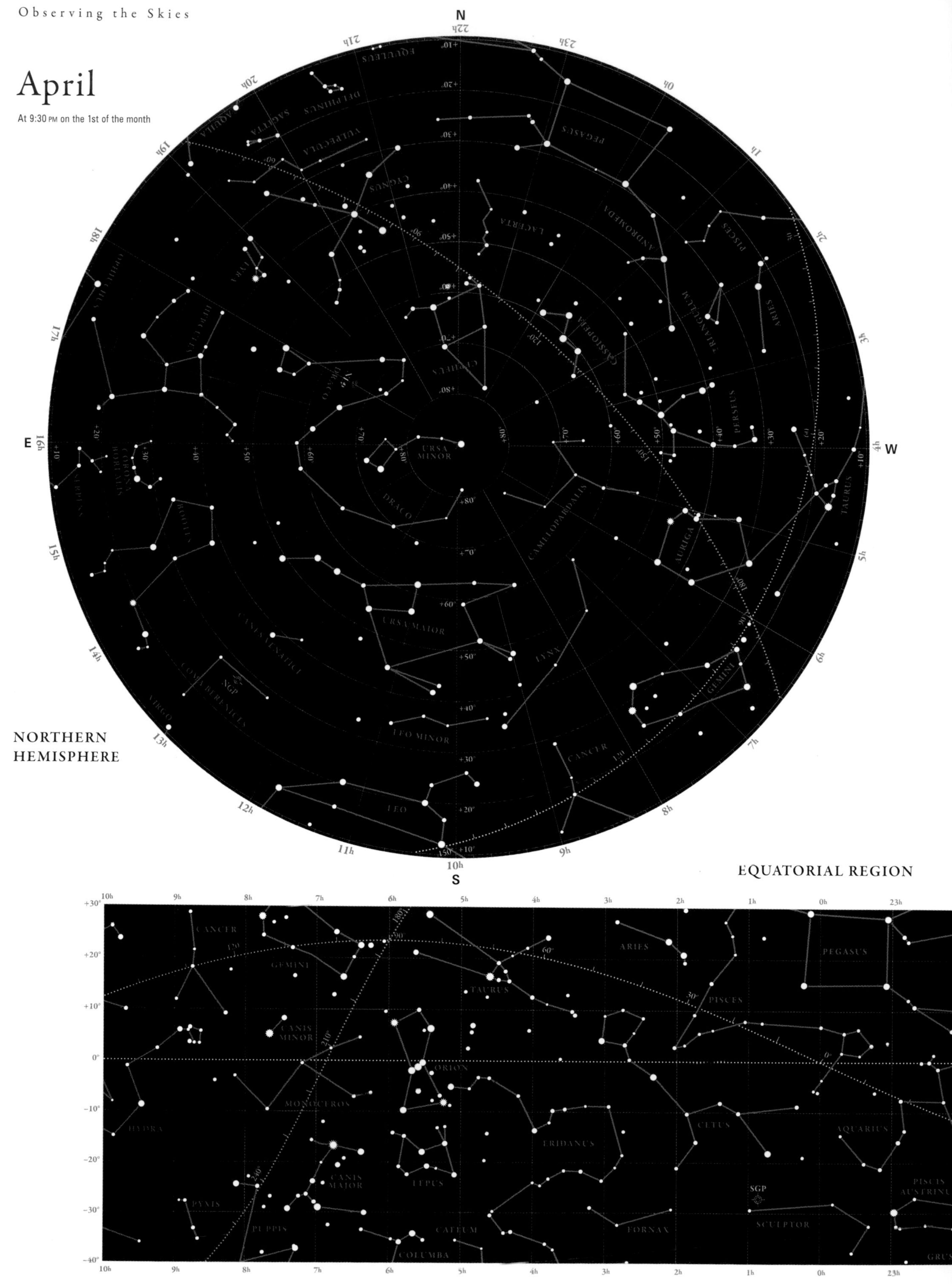

April

At 9:30 PM on the 1st of the month

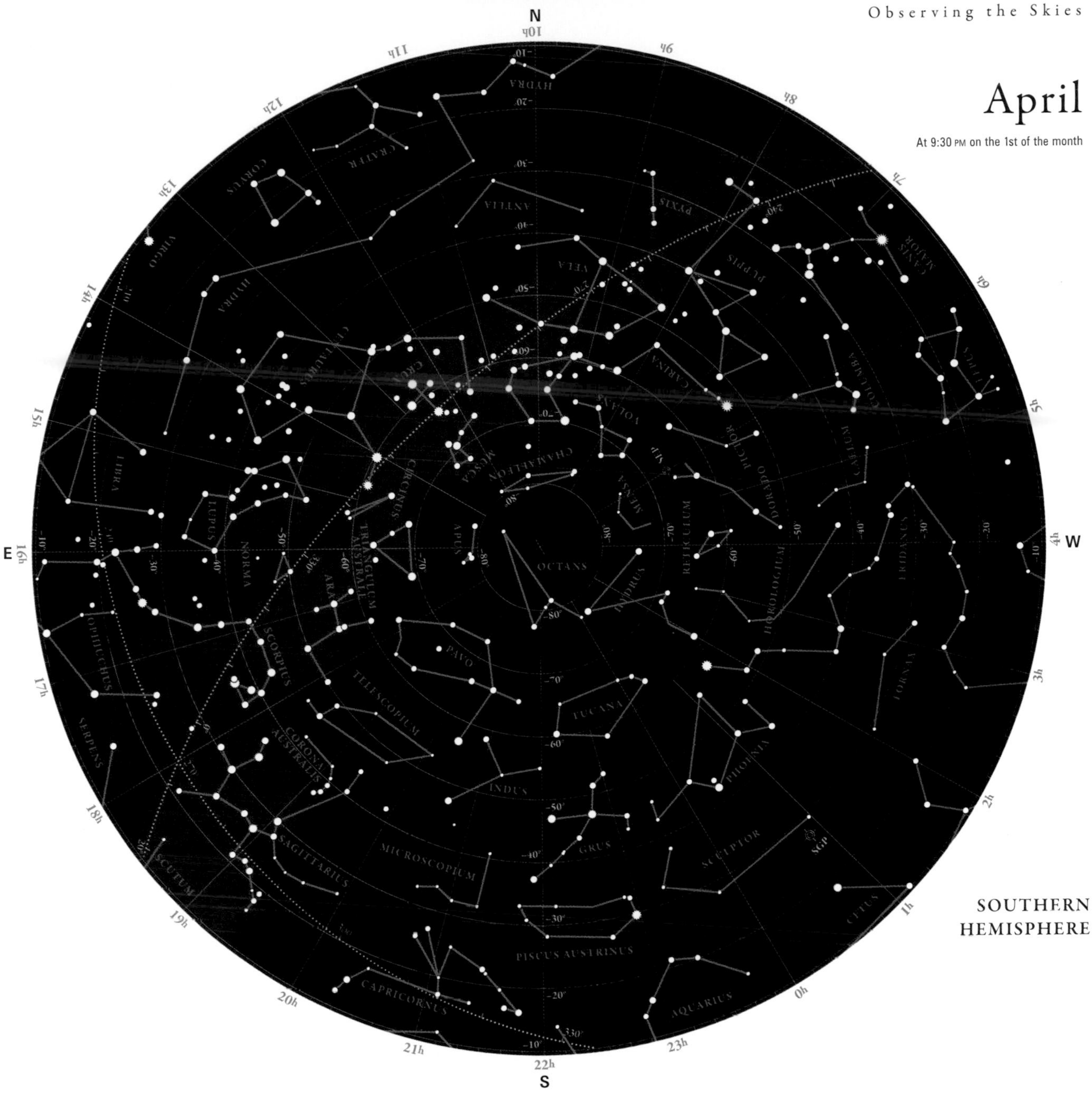

SOUTHERN HEMISPHERE

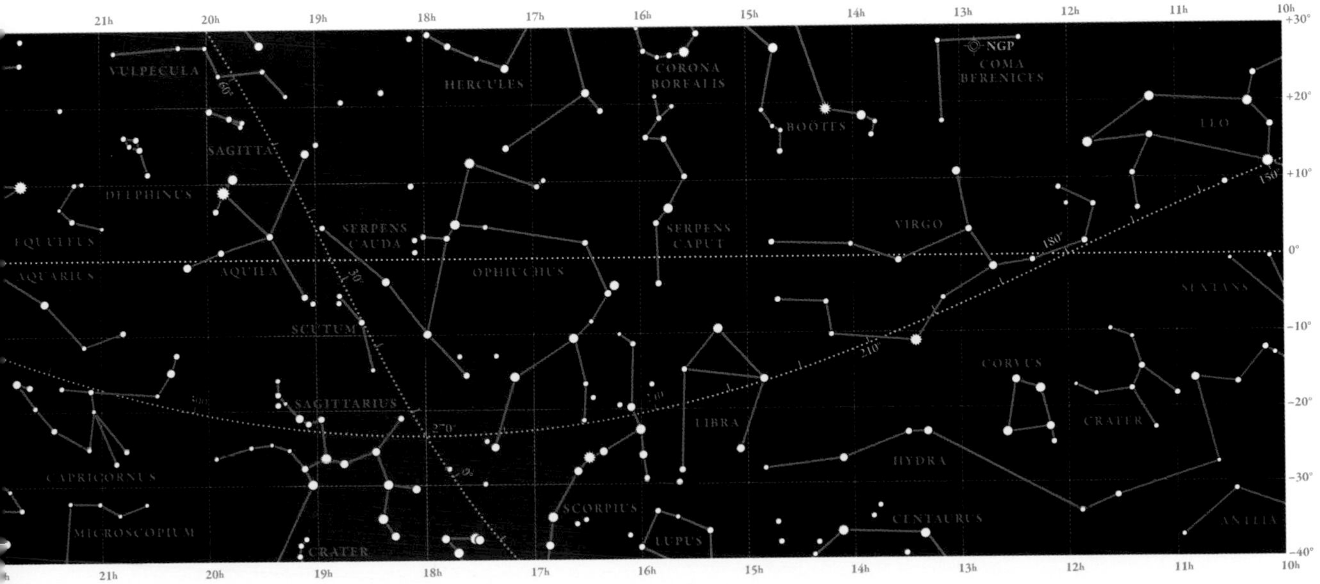

May

At 9:30 PM on the 1st of the month

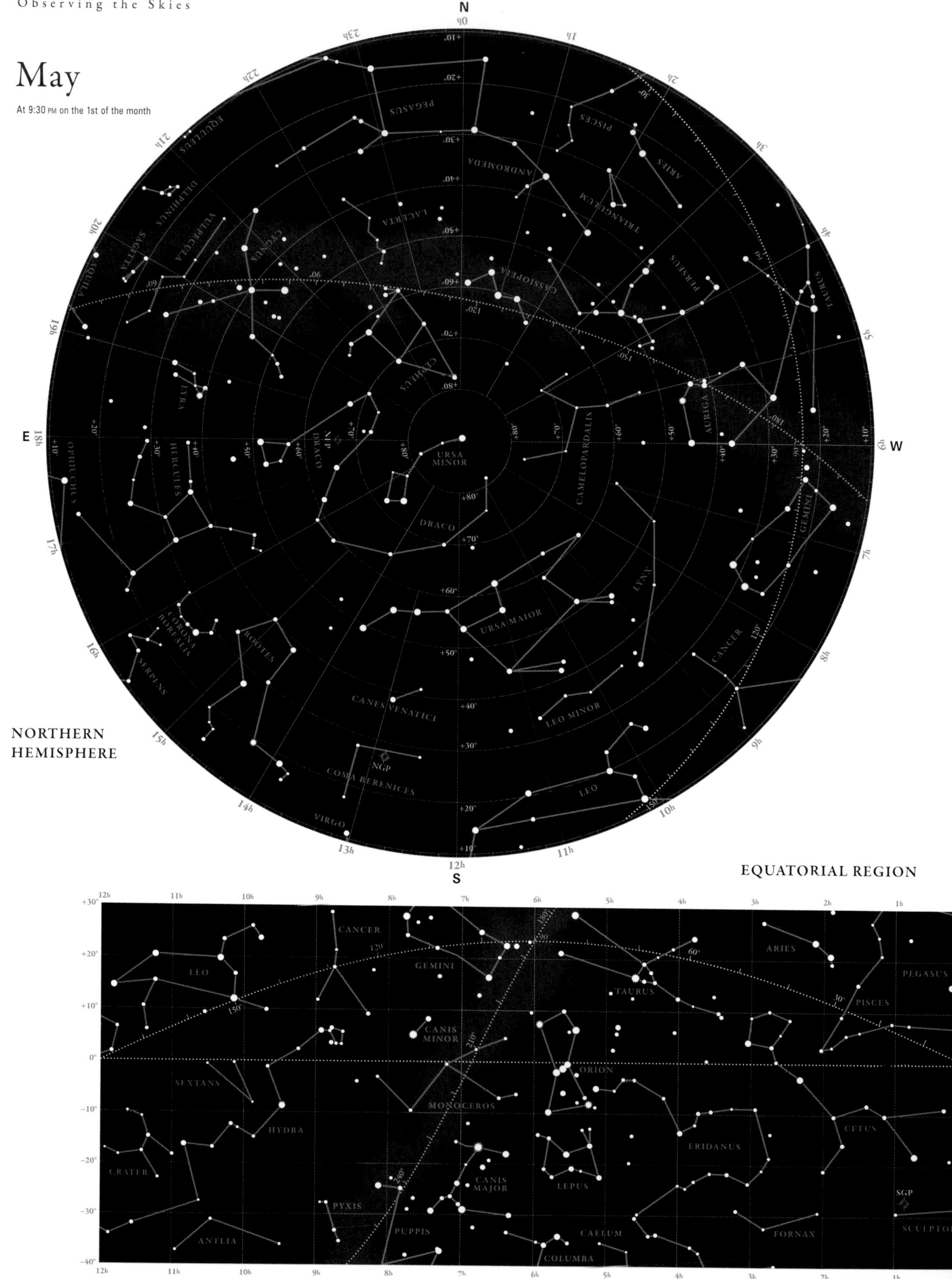

May

At 9:30 PM on the 1st of the month

SOUTHERN HEMISPHERE

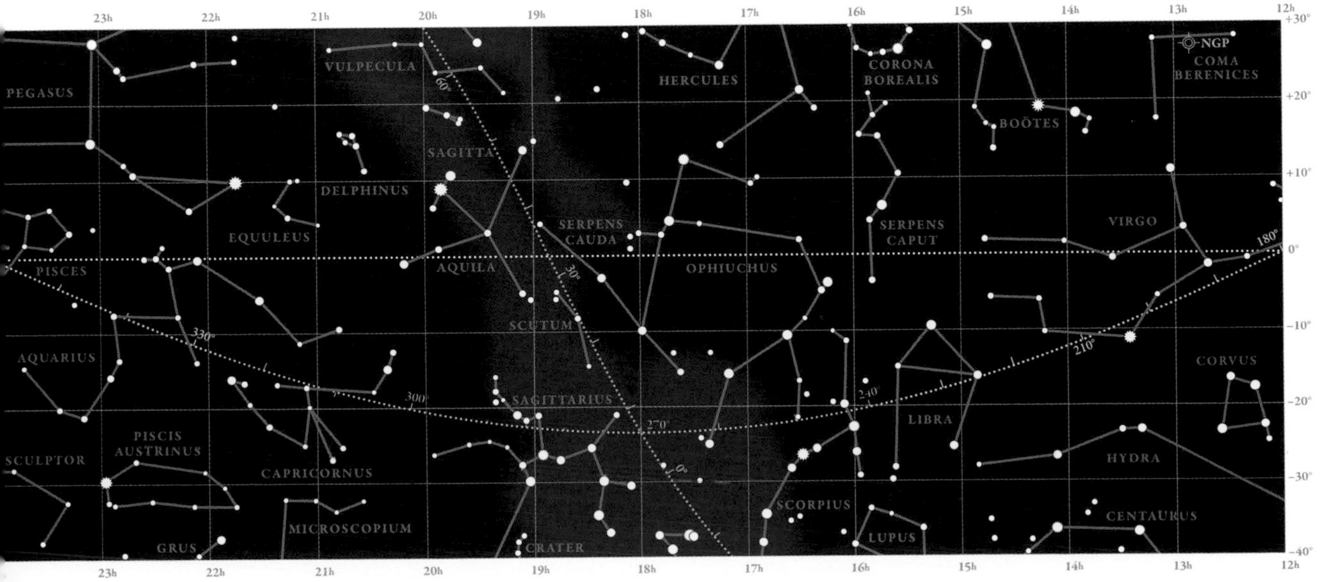

June

At 9:30 PM on the 1st of the month

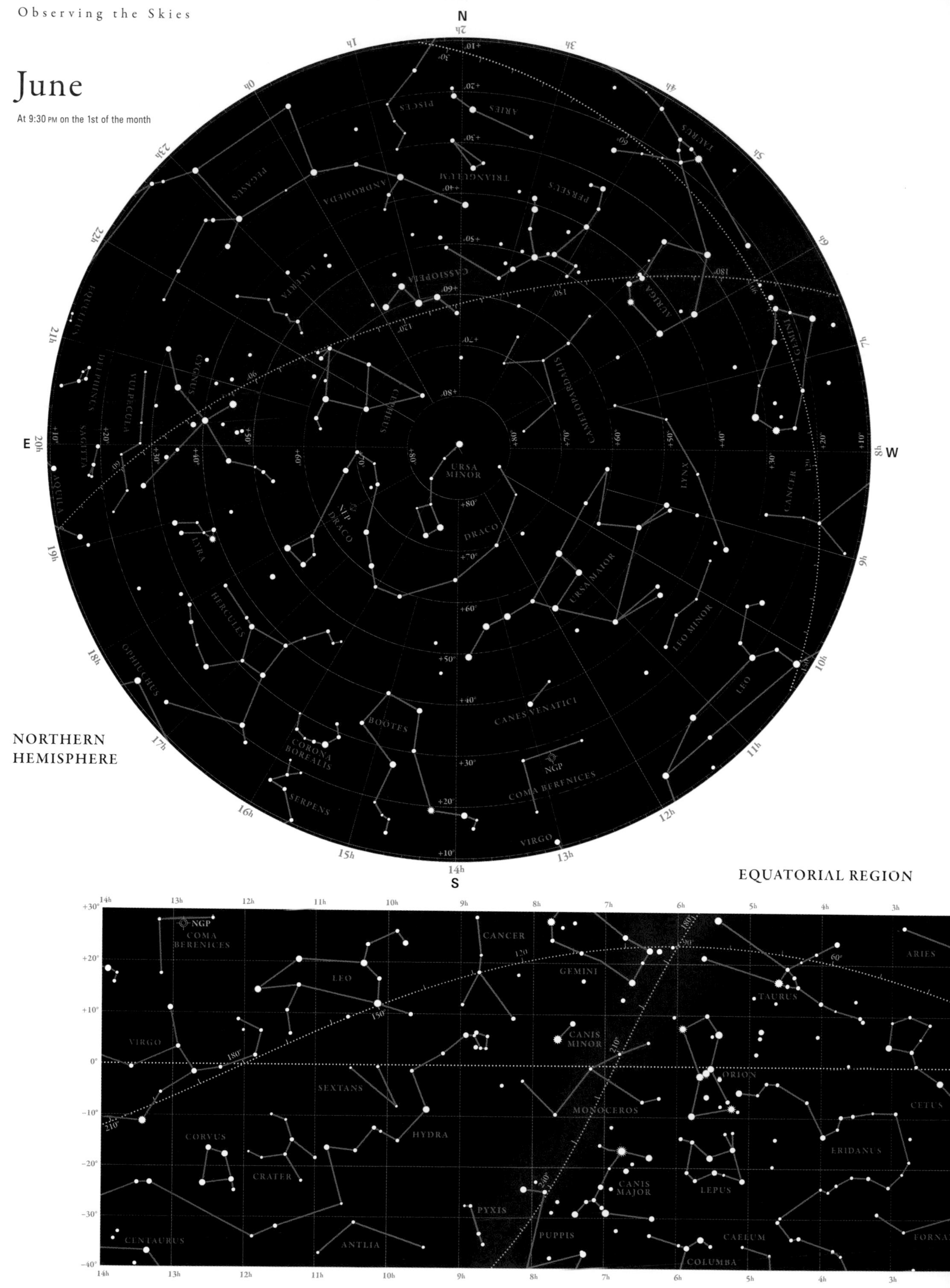

June

At 9:30 PM on the 1st of the month

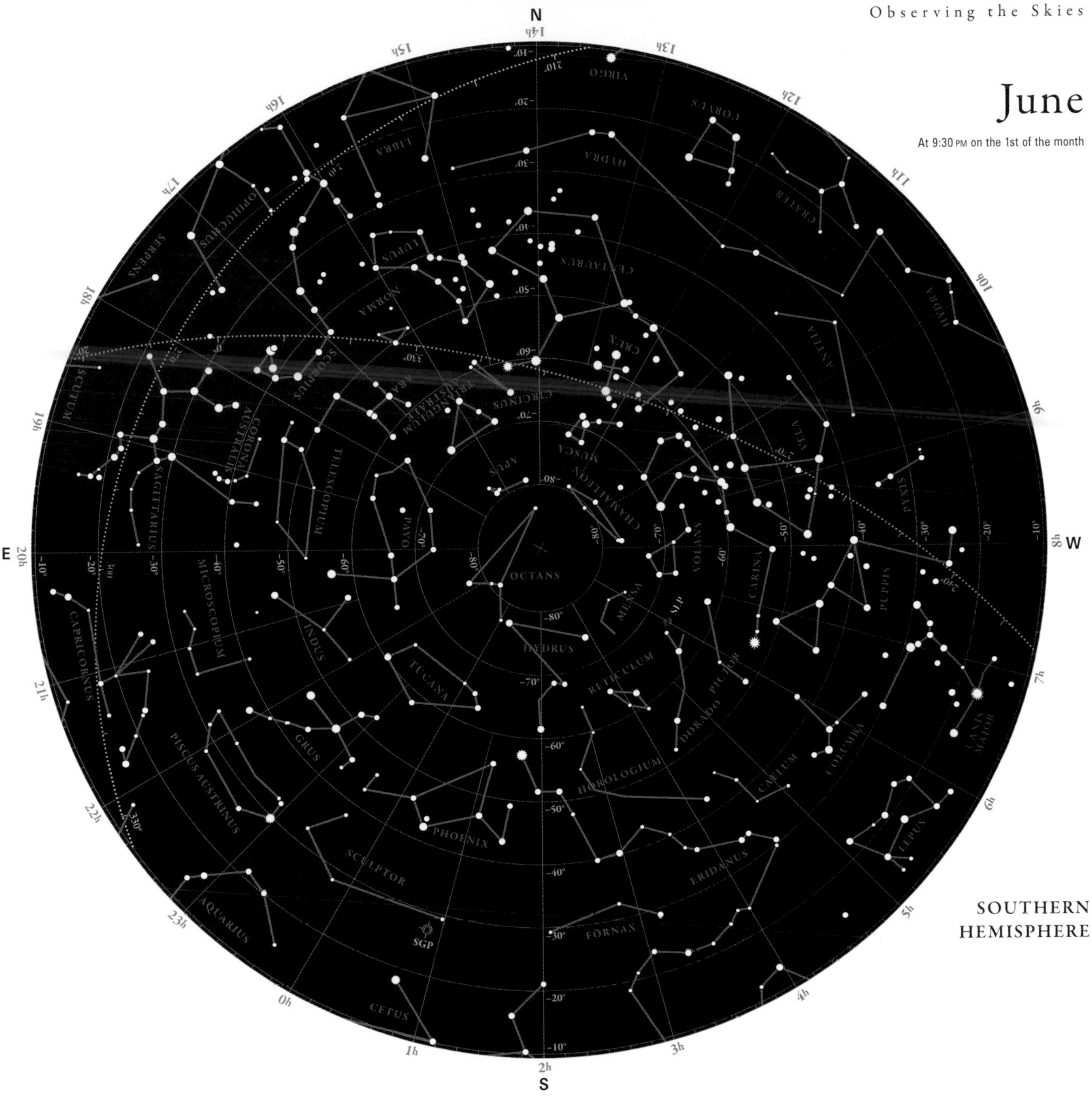

SOUTHERN HEMISPHERE

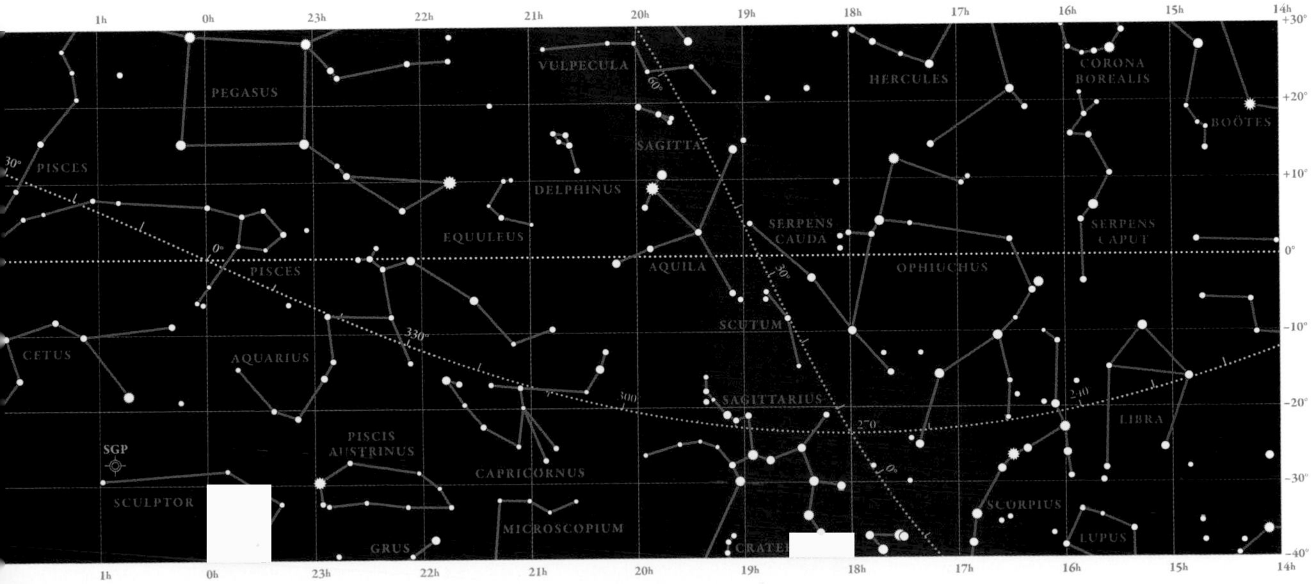

July

At 9:30 PM on the 1st of the month

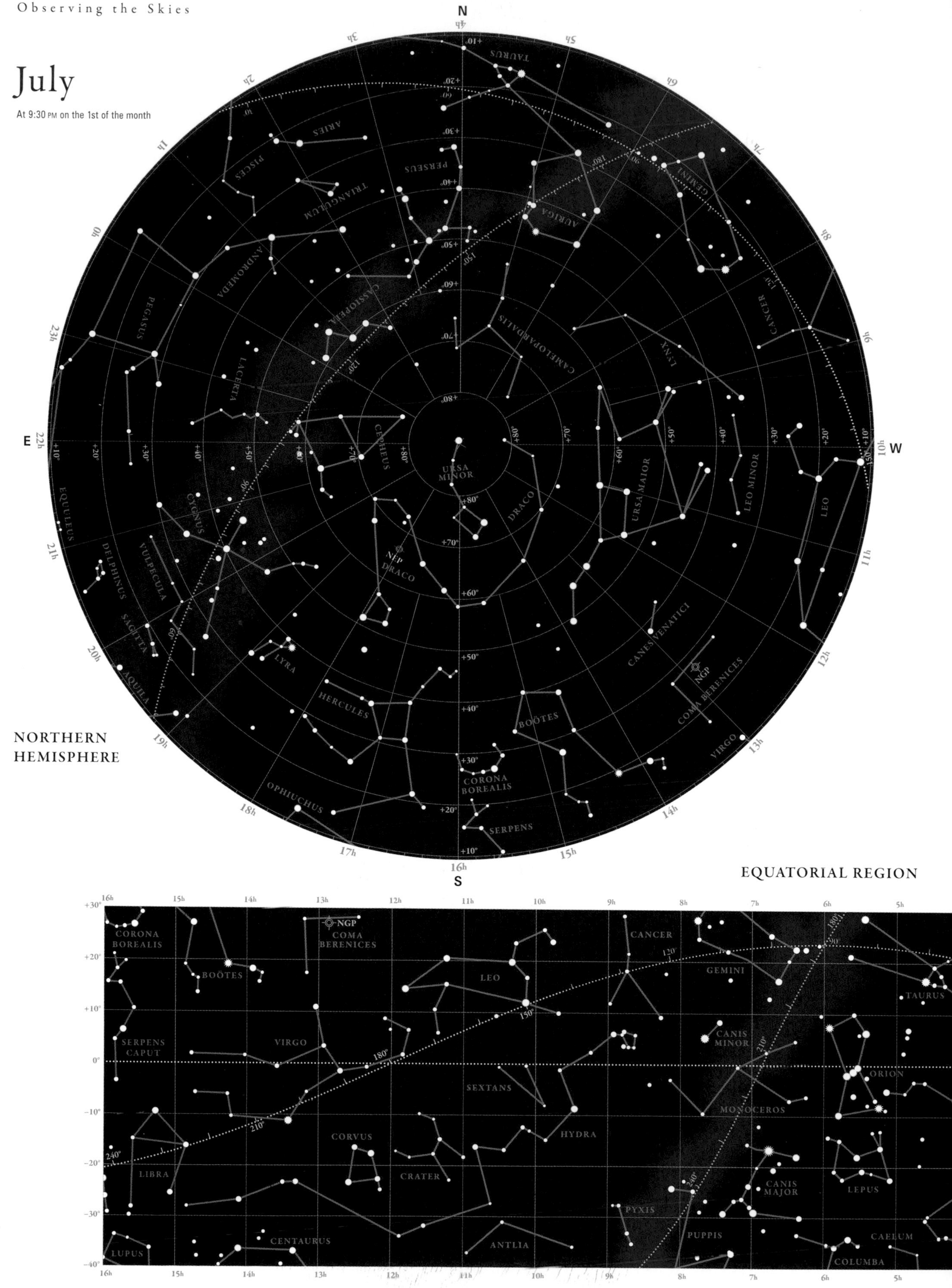

July

At 9:30 PM on the 1st of the month

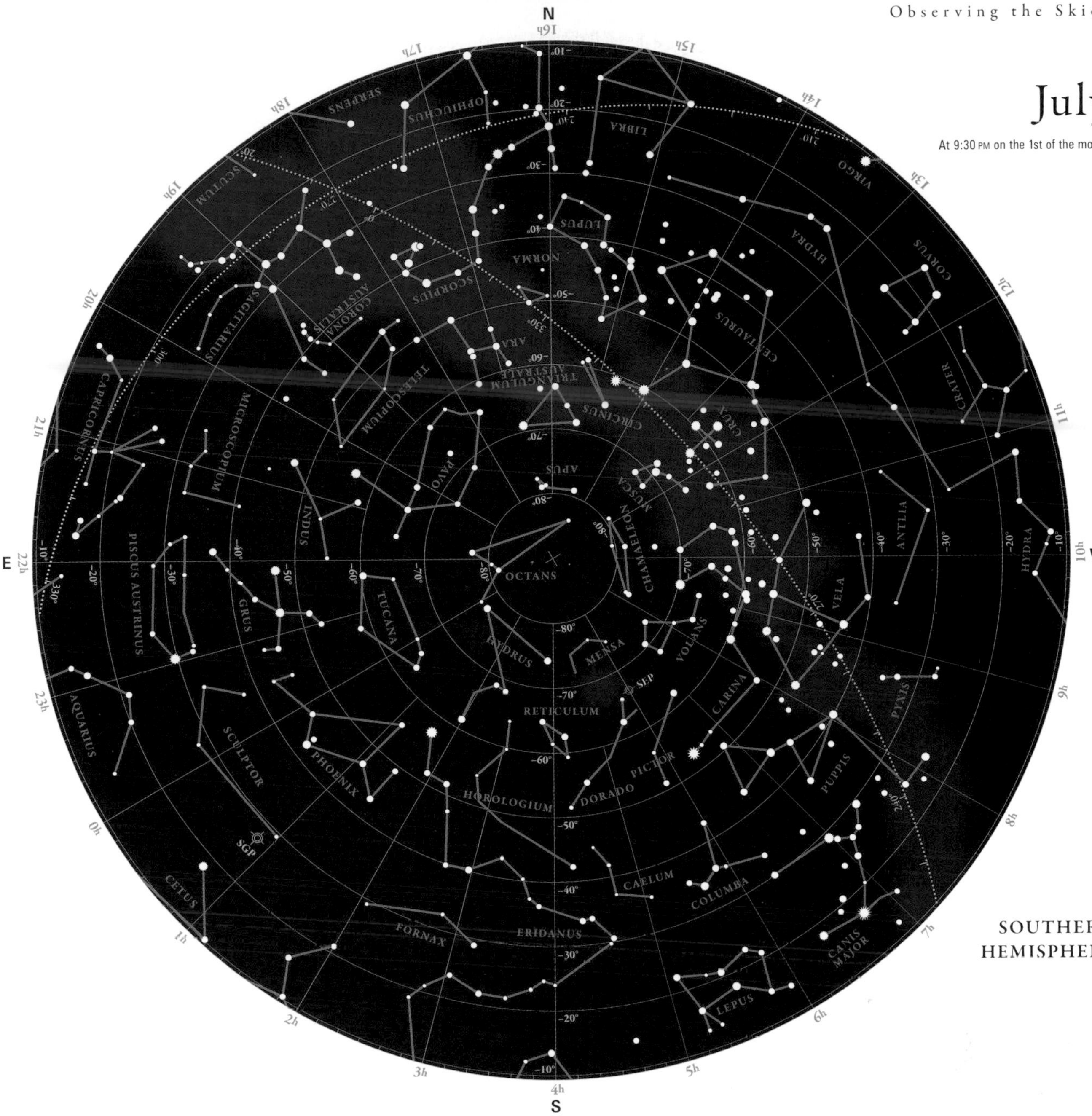

SOUTHERN HEMISPHERE

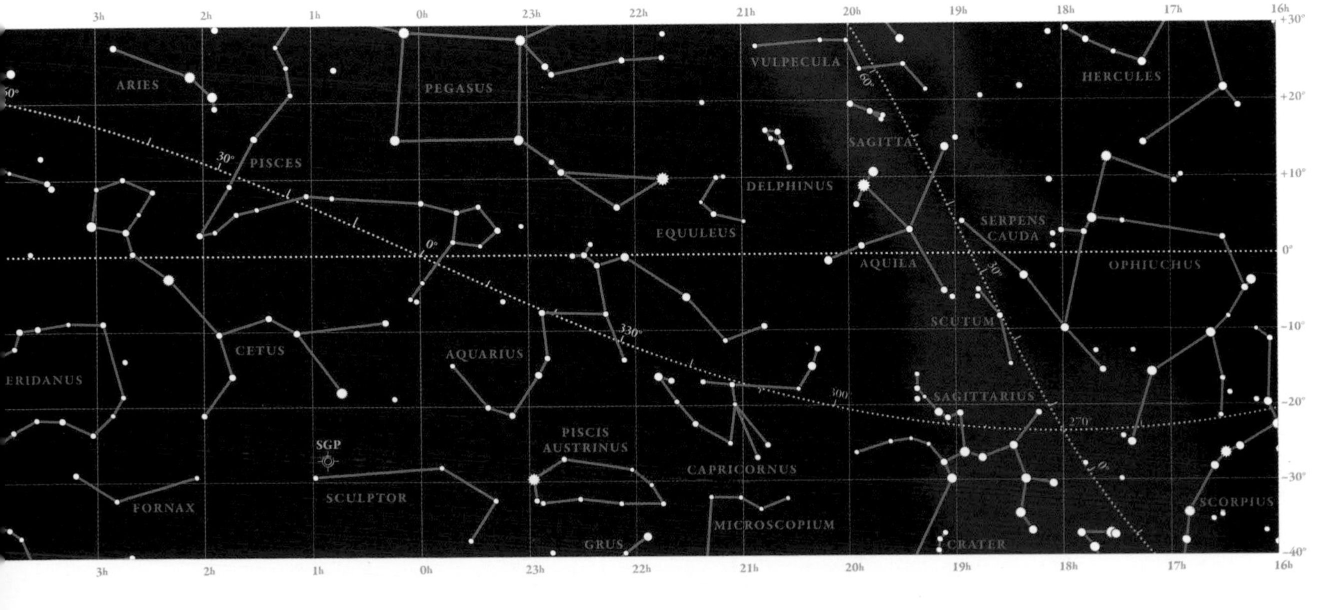

August

At 9:30 PM on the 1st of the month

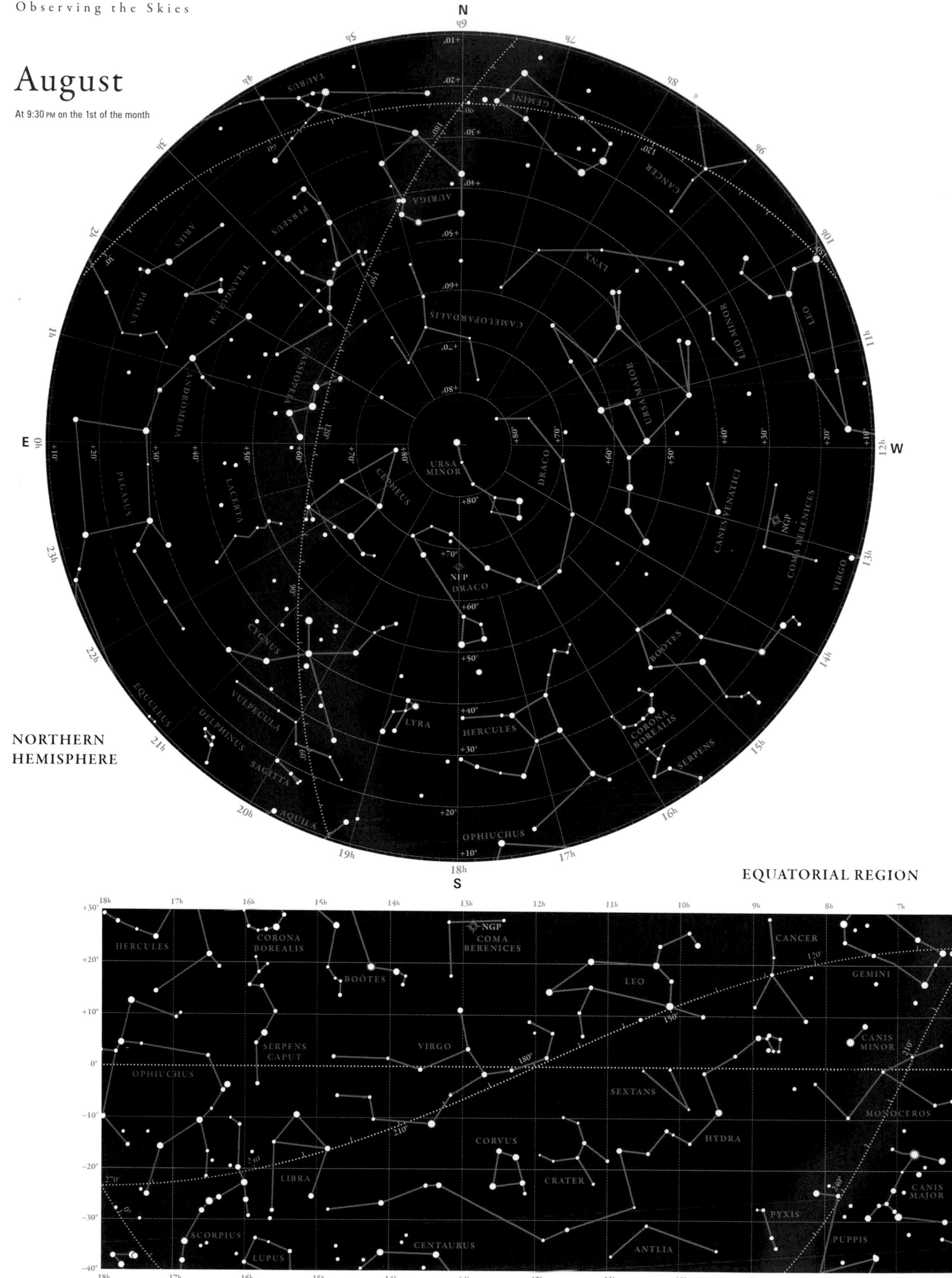

August

At 9:30 PM on the 1st of the month

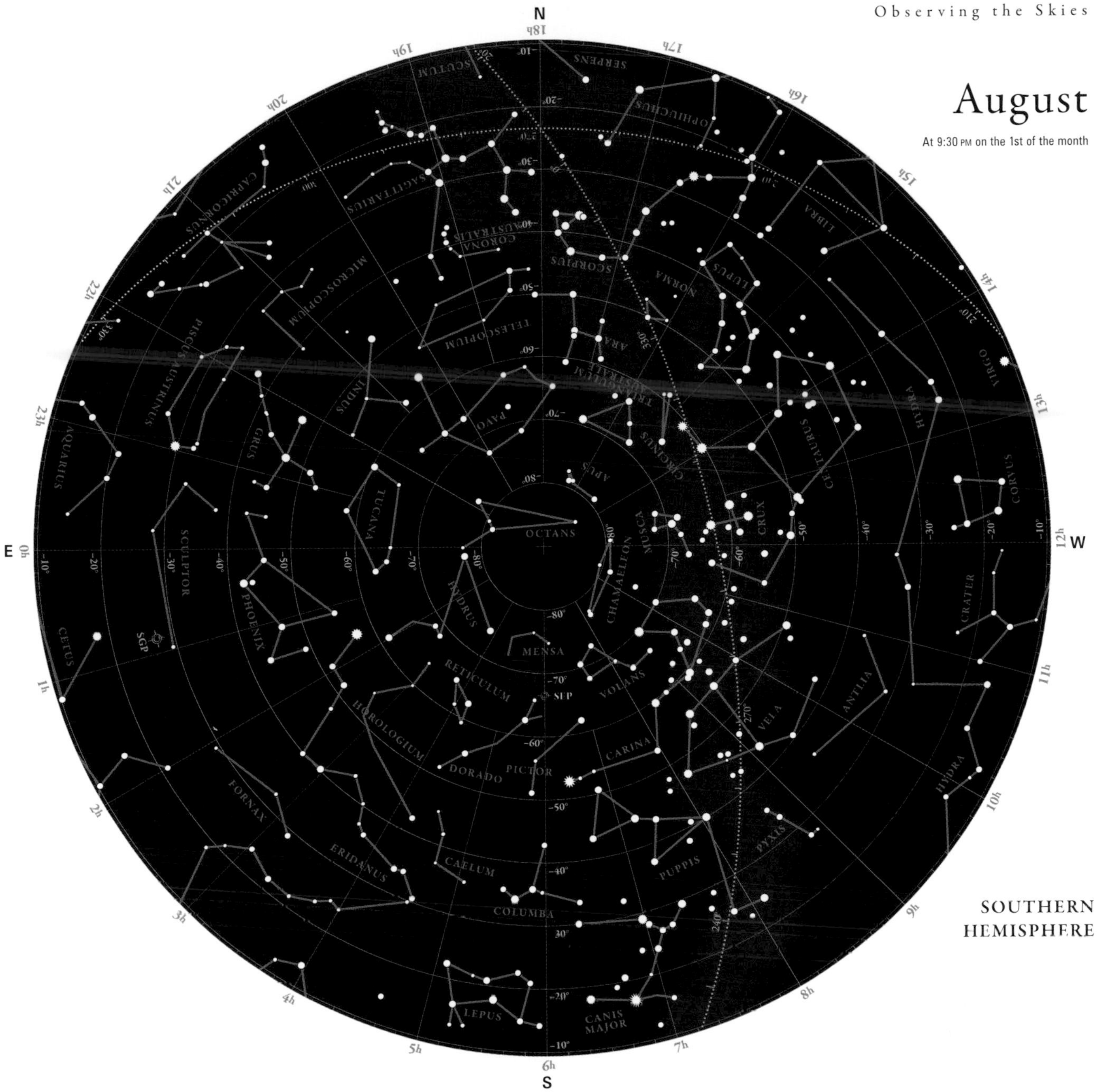

SOUTHERN HEMISPHERE

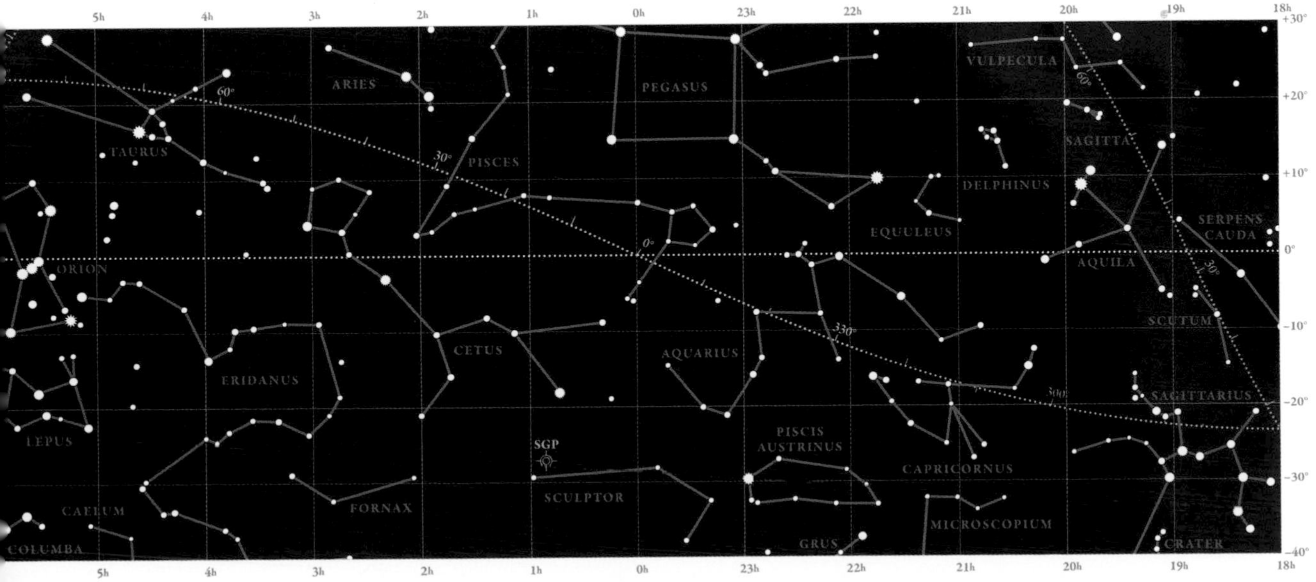

September

At 9:30 PM on the 1st of the month

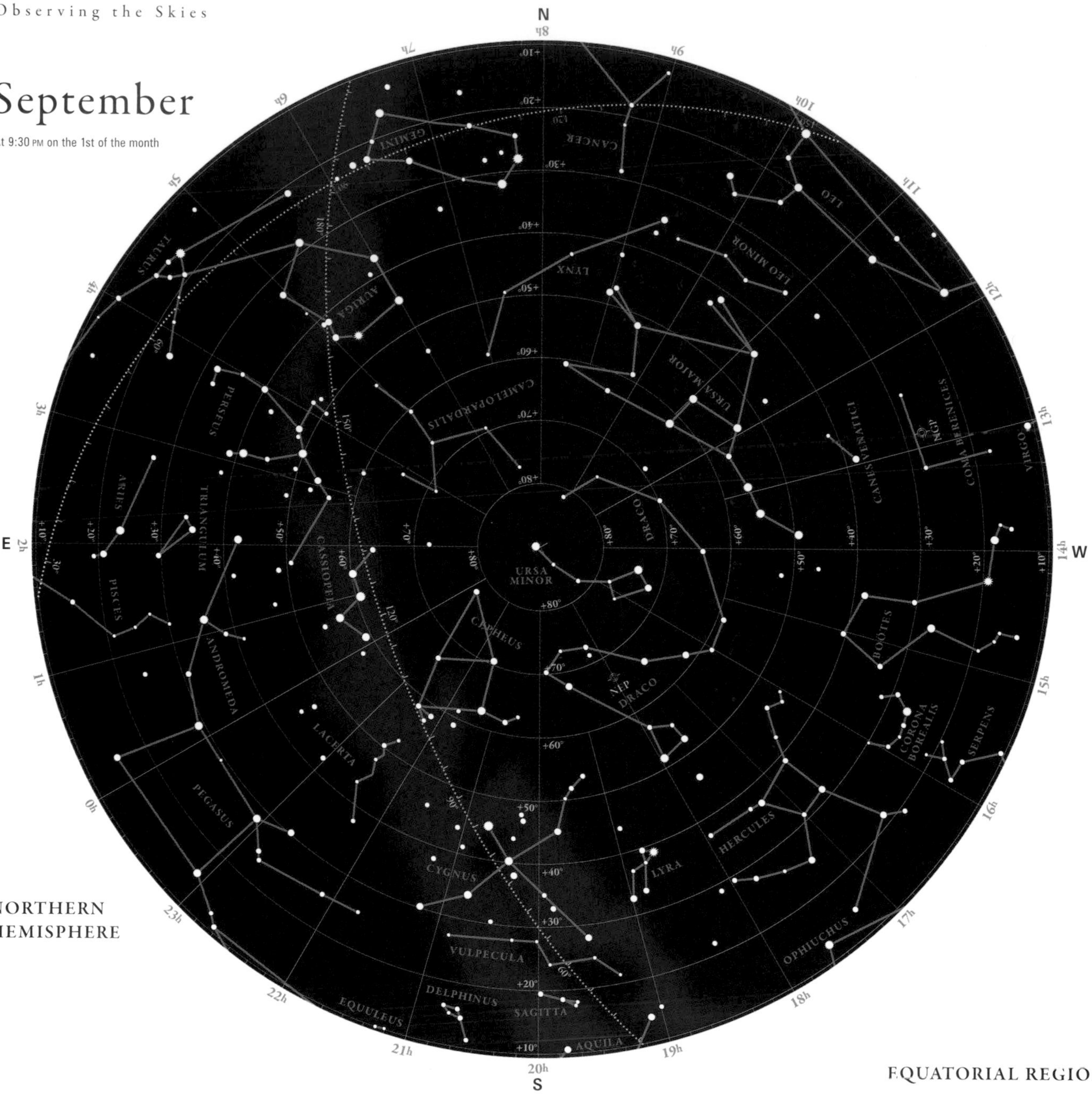

NORTHERN HEMISPHERE

EQUATORIAL REGION

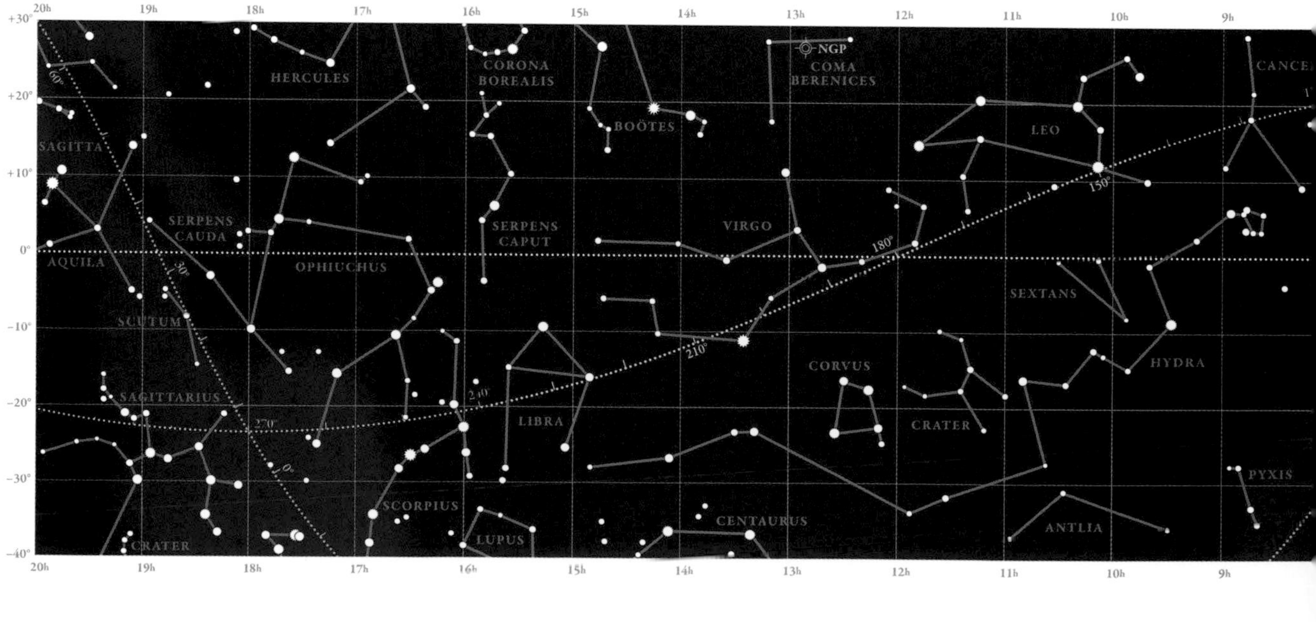

September

At 9:30 PM on the 1st of the month

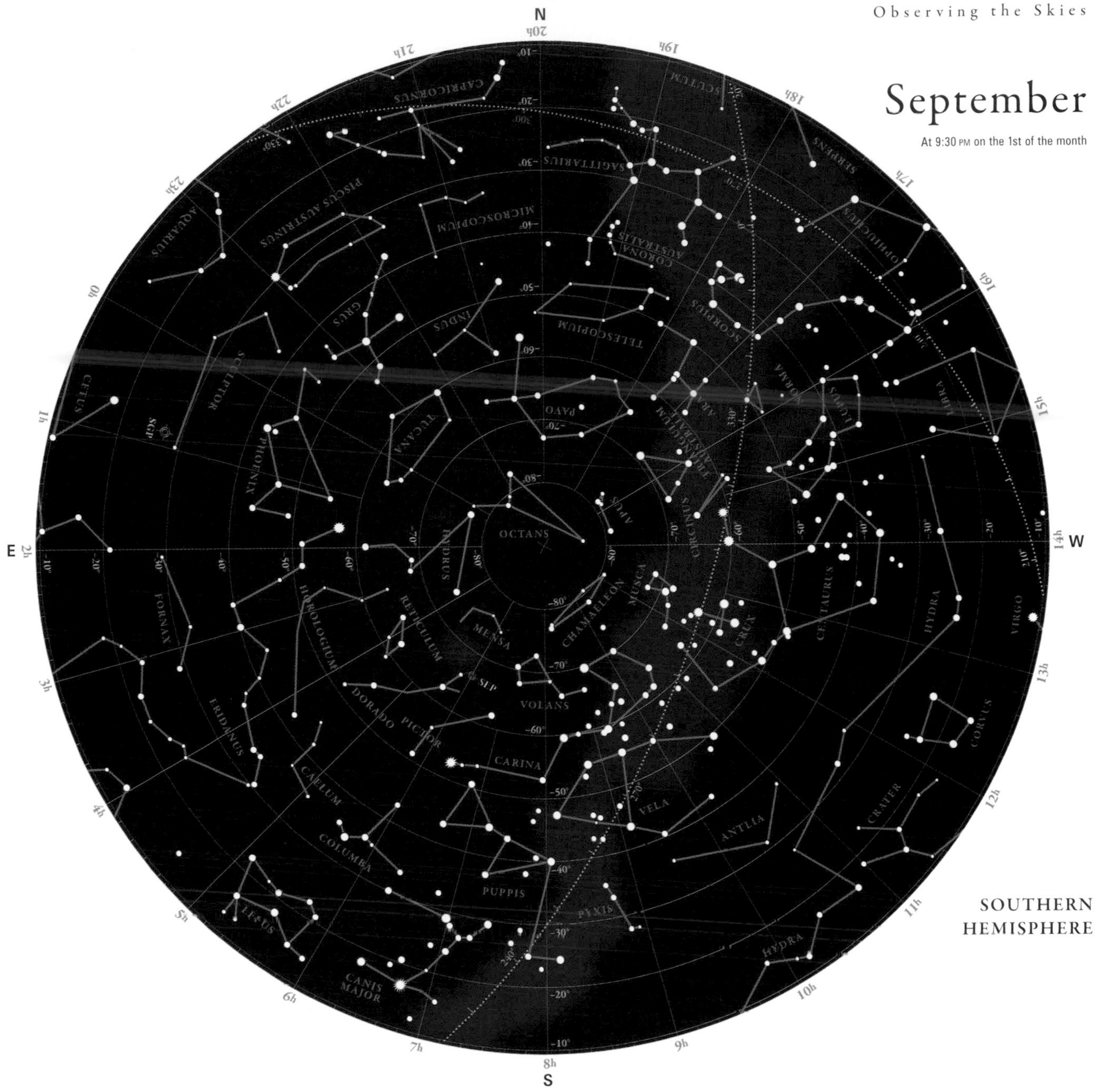

SOUTHERN HEMISPHERE

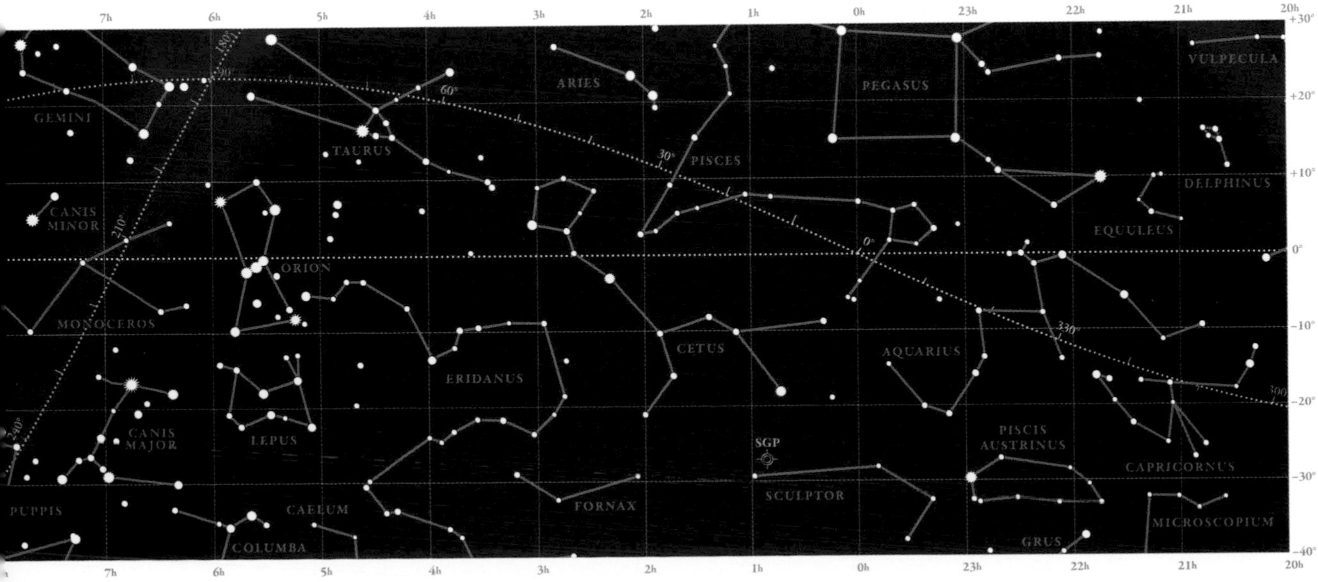

October

At 9:30 PM on the 1st of the month

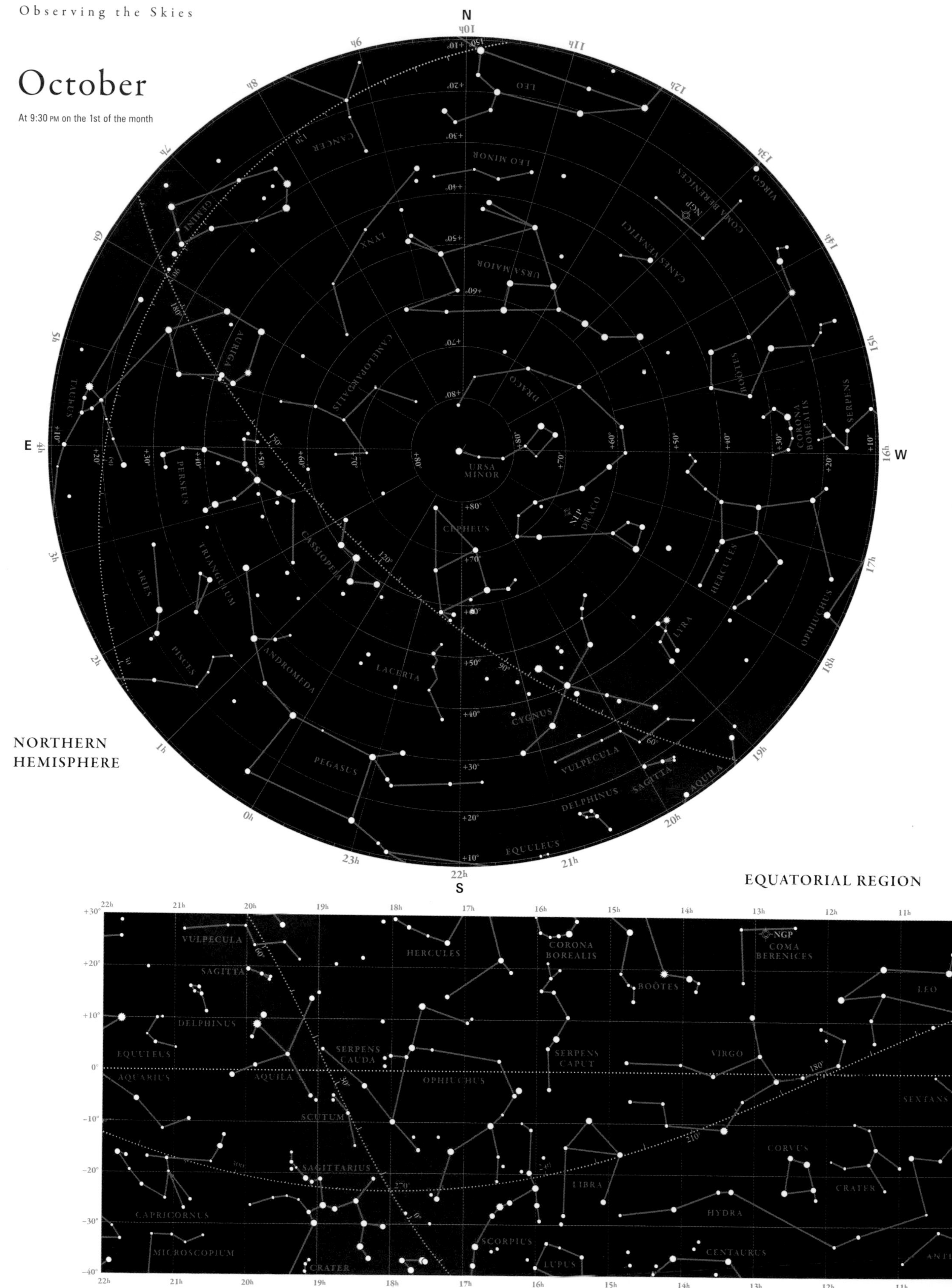

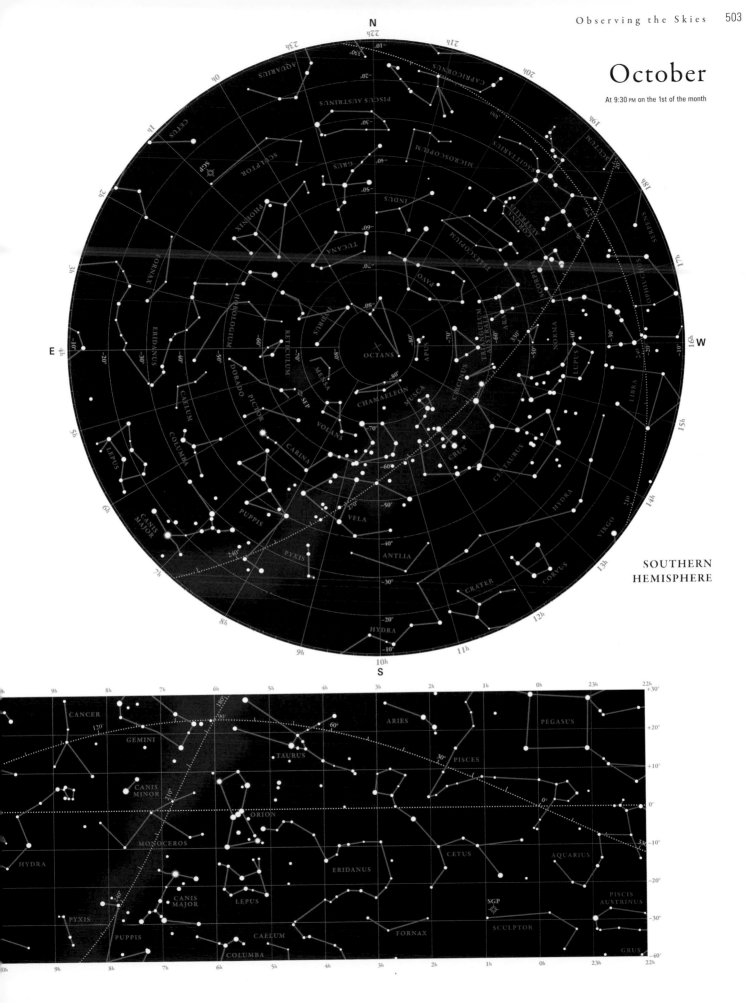

October

At 9:30 PM on the 1st of the month

SOUTHERN HEMISPHERE

November

At 9:30 PM on the 1st of the month

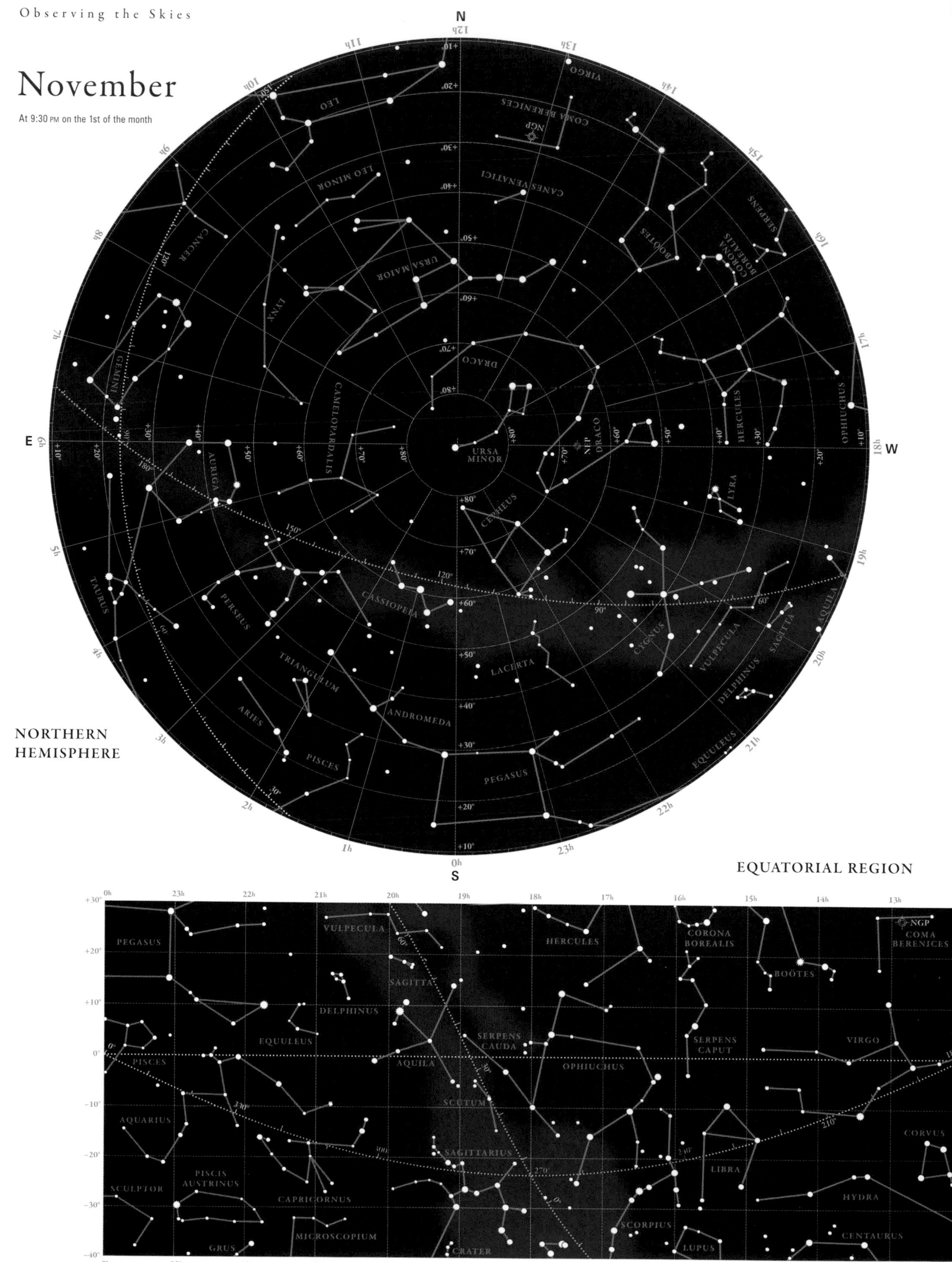

November

At 9:30 PM on the 1st of the month

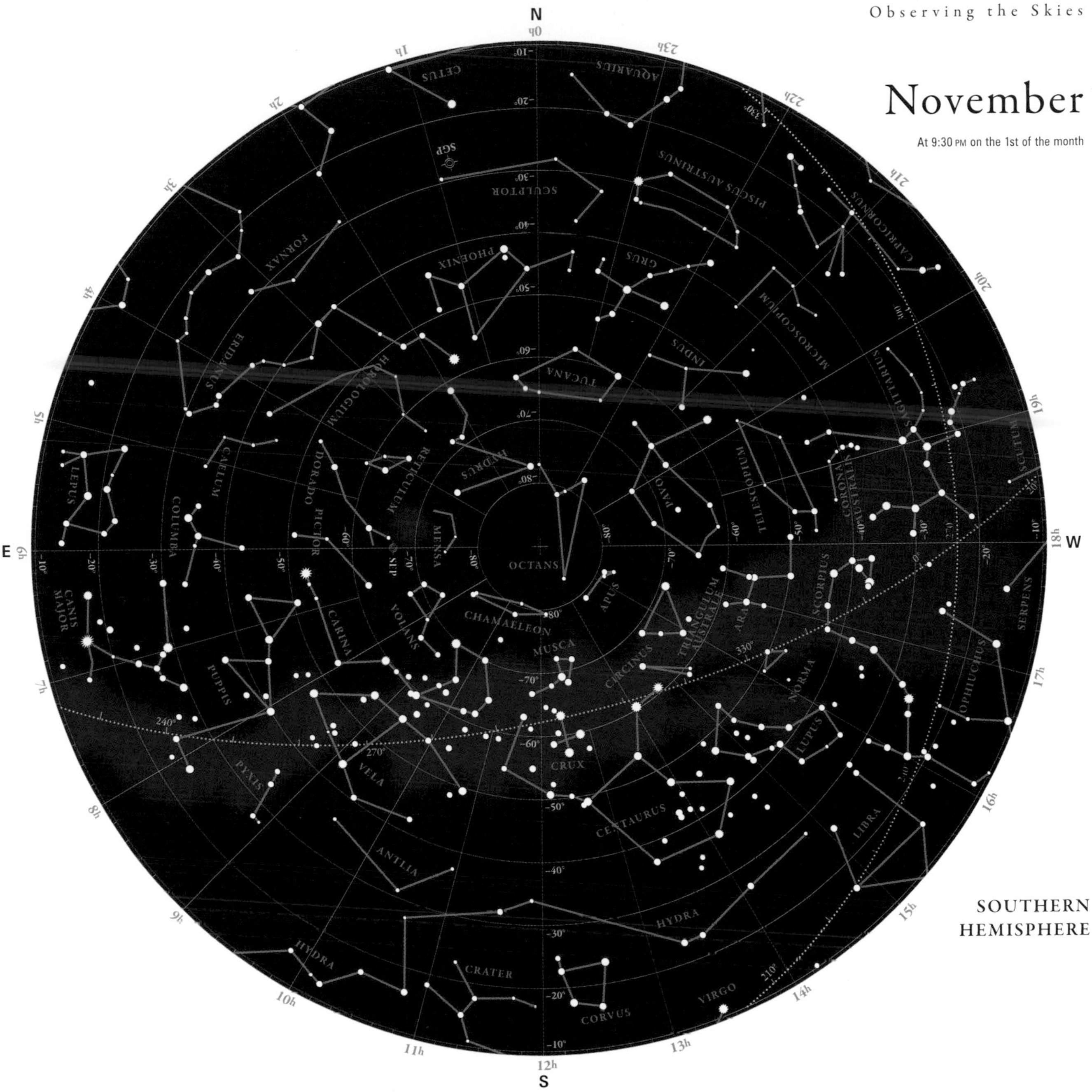

SOUTHERN HEMISPHERE

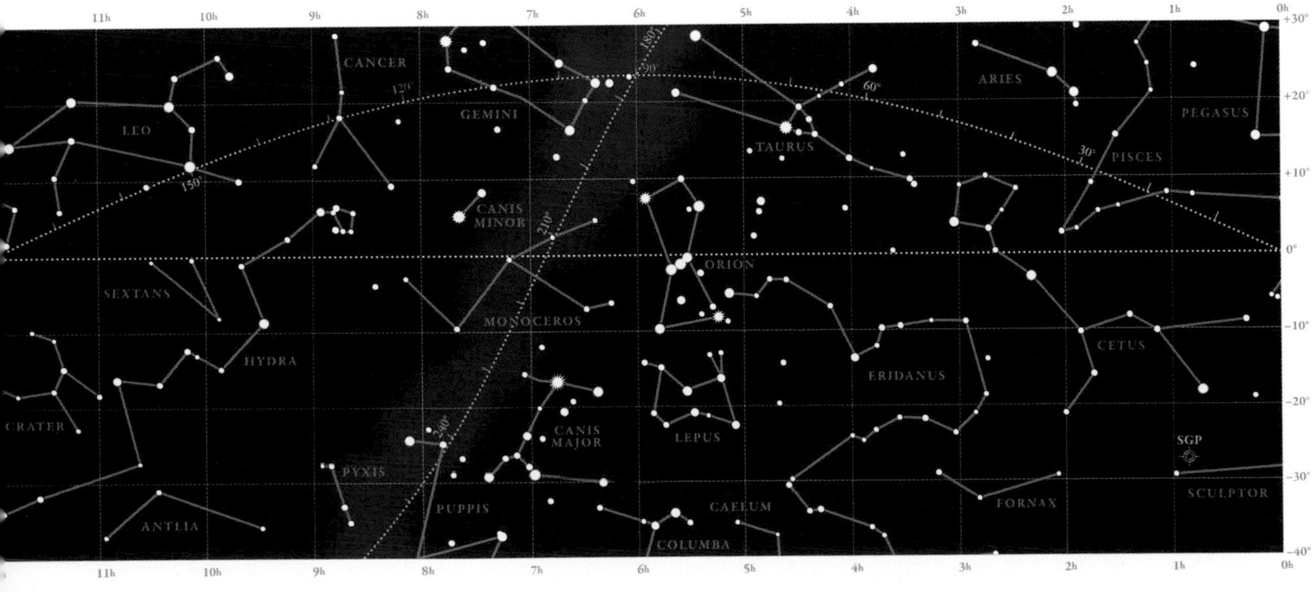

December

At 9:30 PM on the 1st of the month

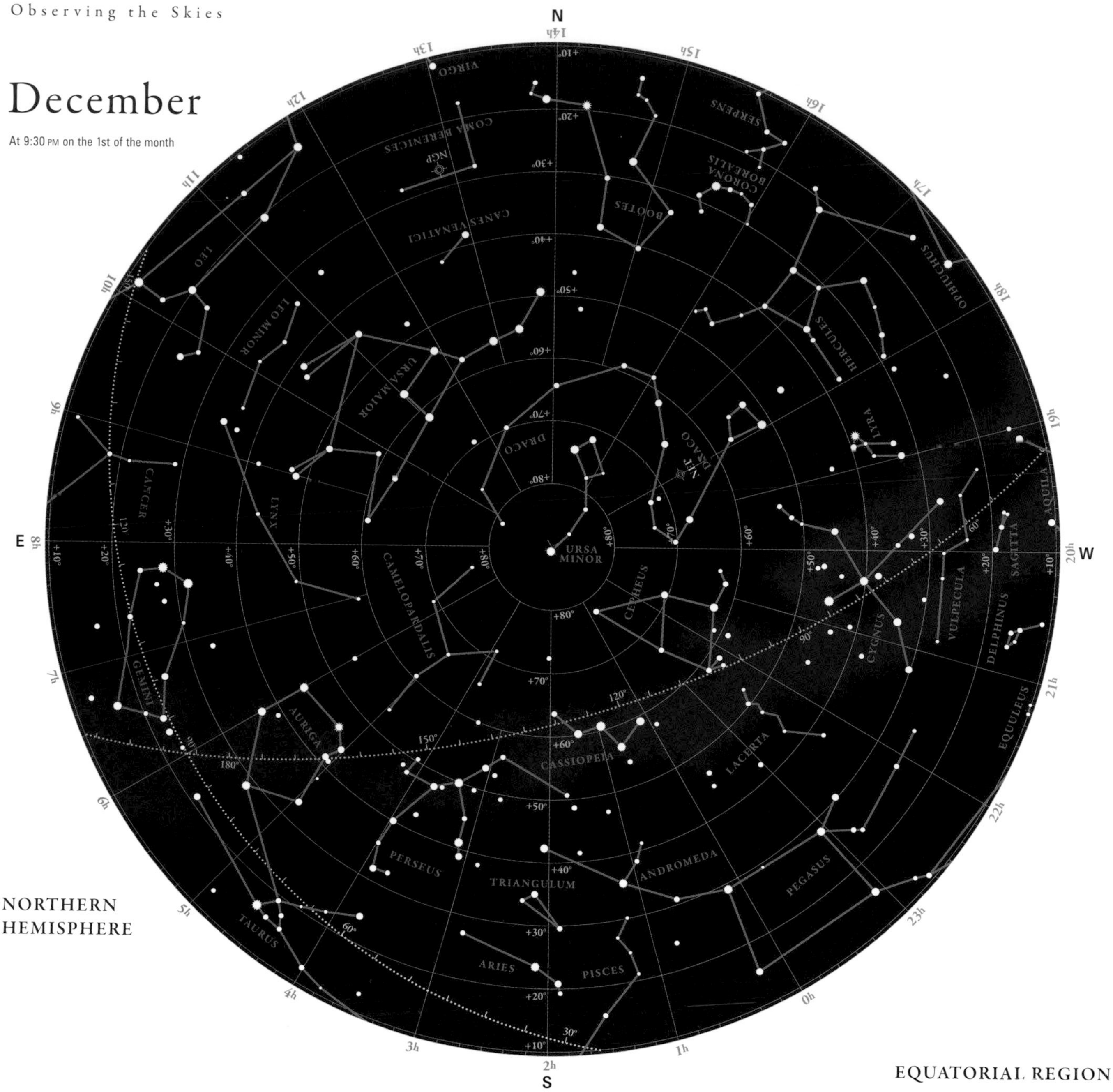

NORTHERN HEMISPHERE

EQUATORIAL REGION

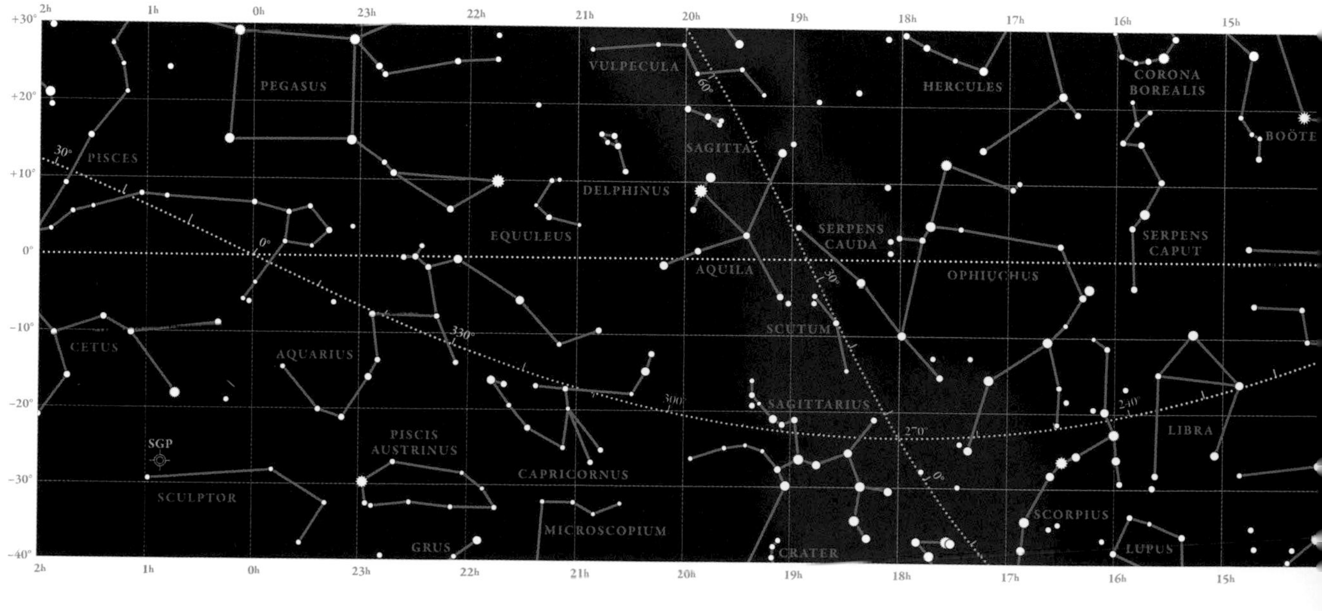

December

At 9:30 PM on the 1st of the month

SOUTHERN HEMISPHERE

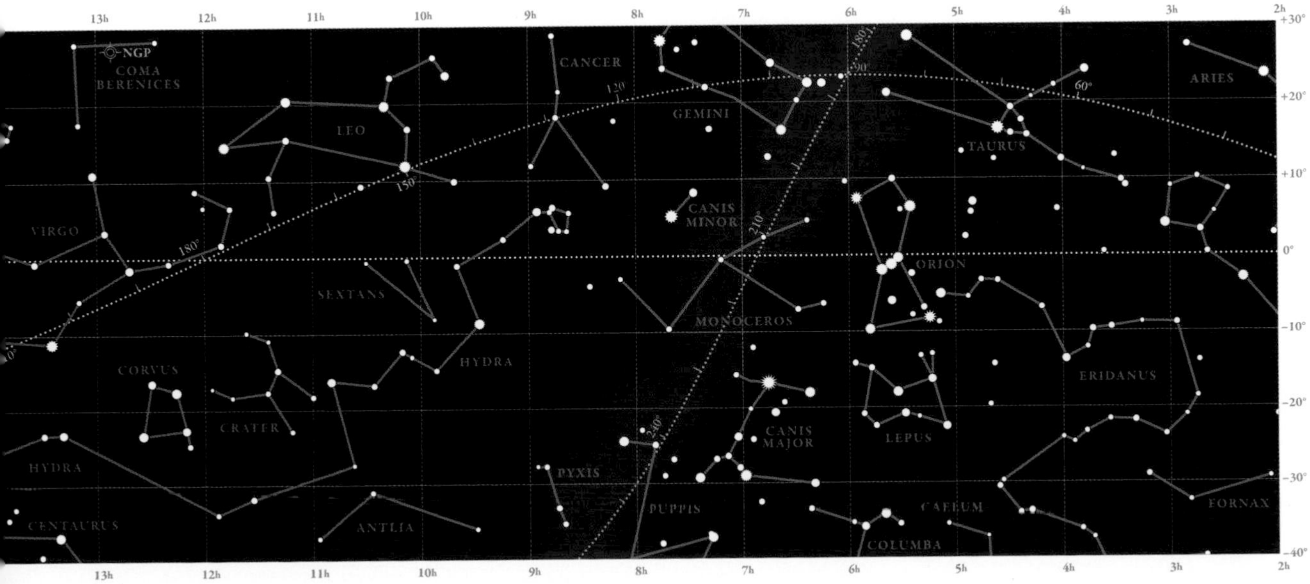

Other Celestial Phenomena

Perceiving shapes of mythical figures in the constellations that adorn the skies often requires experience and expensive equipment, not to mention a lively imagination. But anyone can observe and marvel at the dazzling displays provided by eclipses and auroras. A tablet recording a total solar eclipse in Ireland dates back as far as the fourth millennium BCE, and Chinese records reach back for four thousand years.

Above Image of a total solar eclipse showing the corona. This moment is called second contact–when the Moon covers the Sun and totality begins.

Opposite This image shows the magnificent sight of the "diamond ring," or third contact, when the first rays of sunlight reappear after an eclipse.

TOTAL SOLAR ECLIPSES

A total solar eclipse is a lifetime memory—an event not to be missed by anyone living within easy distance of the eclipse path. Tens of thousands of astronomers travel to other continents to stand in the Moon's shadow, especially when an eclipse path passes over popular travel destinations. If clouds threaten, experienced eclipse-chasers search frantically for large holes in the cloud cover as the seconds tick away to the eclipse.

On March 29, 2006, the Moon's shadow darkened the southwestern horizon a minute before the long-awaited total eclipse—the totality. Lunar mountains peeked through the Sun's rapidly dwindling crescent, breaking it up into long beads of sunlight known as Baily's Beads. For a few seconds there was a short arc of pink chromosphere before the advancing Moon covered it too, but two triangular pink prominences remained. The pearly corona, extending about 2.5° from streamer tip to streamer tip, was glorious—a classic sunspot-minimum type, with four long streamers on either side of the Moon's black disk, and short polar brushes following the curved magnetic field lines.

All types of solar eclipses begin with the Moon's first tiny dent in the Sun's disk. The excitement builds for about 75 minutes as the Moon takes an ever-larger bite out of the Sun. A few mountains and crater rims on the Moon's limb are usually visible in telescopes, silhouetted against the Sun. It is especially interesting to watch the Moon conceal any sunspots present.

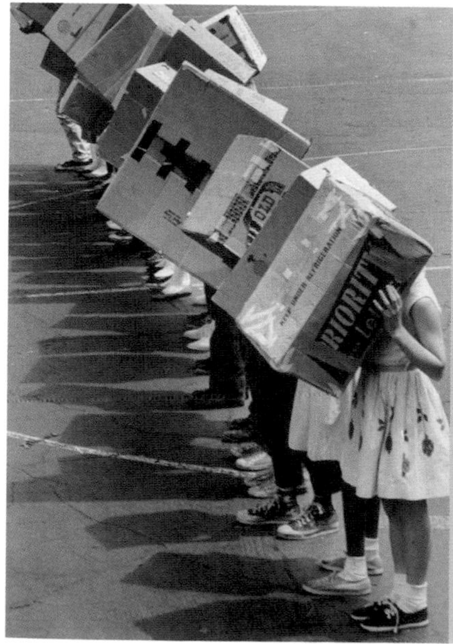

WARNING

To avoid blindness, never look at the Sun at any time, including an eclipse, without proper eye protection.

When observing the Sun with binoculars or telescopes, ensure they are fitted with a full-aperture commercial solar filter, securely attached to the front end of the tube. Aluminized mylar filters are safe for unaided-eye viewing of the partial phases, as is a shade #14 welder's glass filter.

Filters are not necessary during totality because the corona is fainter than a full moon, but when the first rays of sunlight reappear, solar filters must be used.

DO NOT USE HOMEMADE FILTERS.

These 1950s school children are facing away from the Sun to avoid eye damage.

DIAMONDS AND DAISIES

Overlapping tree leaves can form multiple pinhole images of the solar crescent on the ground. Venus becomes obvious about 10 minutes before second contact—the moment when the Moon completely covers the Sun and totality begins. Birds may settle down to roost before then, and the temperature may fall noticeably. The narrowing solar crescent generates abnormally sharp shadows and, if the atmosphere is turbulent, shadow bands—rapidly moving bands of light and shadow, like the patterns on the bottom of a swimming pool—may appear on reflective surfaces.

In the final minutes, the light becomes eerie because it is coming only from the Sun's limb, reddened by the solar atmosphere. During totality—low in the sky around the full 360° horizon—orange sunset colors emanate from surrounding areas in deep partial eclipse.

At third contact—all too soon—the first ray of sunlight appears in a deep lunar valley, and viewers shout "Diamond Ring!" The bright inner corona remains visible for a few more seconds, making the ring, while the diamond of sunlight swells, and filters are needed again for viewing. The partial phases repeat in reverse until fourth contact, but this is so anti-climatic after the strong emotions that totality arouses, that most people seldom glance at the swelling Sun; everyone is excitedly discussing the shape of the corona and prominences, and feeling elated at not being clouded out.

Sunspot-maximum eclipses usually have more prominences than at sunspot minimum. The 1991 Mexican eclipse featured a large seahorse-shaped prominence. Sunspot-maximum coronas are like a daisy, with streamers stretching out from the black Moon in all directions, and the shape and brightness of the corona also changes as active solar areas rotate and evolve. Longer eclipses may allow time to look for planets and zero-magnitude stars. At the 1998 Caribbean eclipse, magnitude –2 Jupiter and magnitude –1.5 Mercury book-ended the eclipsed Sun just beyond the corona—an exceptionally striking conjunction.

ECLIPSE GEOMETRY

Solar eclipses occur only at New Moon if the Moon's shadow falls on Earth. Lunar eclipses can happen only at Full Moon if the Moon moves through Earth's shadow. Eclipses do not happen every month because the plane of the Moon's orbit is inclined about 5° to the ecliptic plane. The two points where the orbits cross are called nodes. Eclipses can occur only near a node. Each twice-a-year opportunity lasts up to 37.5 days, and is called an eclipse season. Since the interval between new moons is 29.5 days, at least two eclipses, a solar and a lunar, and sometimes three, occur in each eclipse season. The nodes move slowly westward along the ecliptic, so the twice-yearly eclipse seasons happen 18.6 days earlier each year. If the first node-crossing occurs in mid-January, a third eclipse season will begin that year, allowing the possibility of seven eclipses in a calendar year. The Sun's angular diameter ranges from 32 minutes 32 seconds in early January, when Earth is at perihelion, to 31 minutes 28 seconds in early July, at aphelion. The Moon is largest at its monthly perigee and smallest at apogee. The Moon's perigee distance, and hence its maximum angular diameter, vary substantially.

If New Moon occurs close enough to a node, and if the Moon's angular diameter exceeds the Sun's, the Moon's shadow reaches Earth, and a total eclipse occurs along a narrow path. But at most eclipses, the Moon's disk is smaller than the Sun's, and the resulting event is called an annular eclipse because even at mid-eclipse, a ring of sunlight—*annulus* is Latin for "ring"—surrounds the Moon. Hybrid eclipses are annular on two—or rarely one—sections of the path when the Moon is near the horizon, but become short total eclipses when the Moon is near the zenith, and thus closer to Earth by one Earth radius.

Totality lasts from just seconds up to the maximum possible, 7 minutes 58 seconds, unless the observer is flying. On June 30, 1973, Concorde extended totality to 74 minutes—still a record. The longest totalities occur at a node, in late June through July when the Sun appears smallest; during an unusually close lunar perigee when the Moon appears largest; at noon when the New Moon is closer than when on the horizon; and in the tropics where the rotational speed of Earth is greatest, partially counteracting the great speed of the Moon's shadow. The periodicities involved in eclipses converge in 18 years and 11.32 days—possibly one day less, depending on leap years—called the Saros. After one Saros, an eclipse repeats with slight changes, but the one-third day means that the path moves westward one third around Earth.

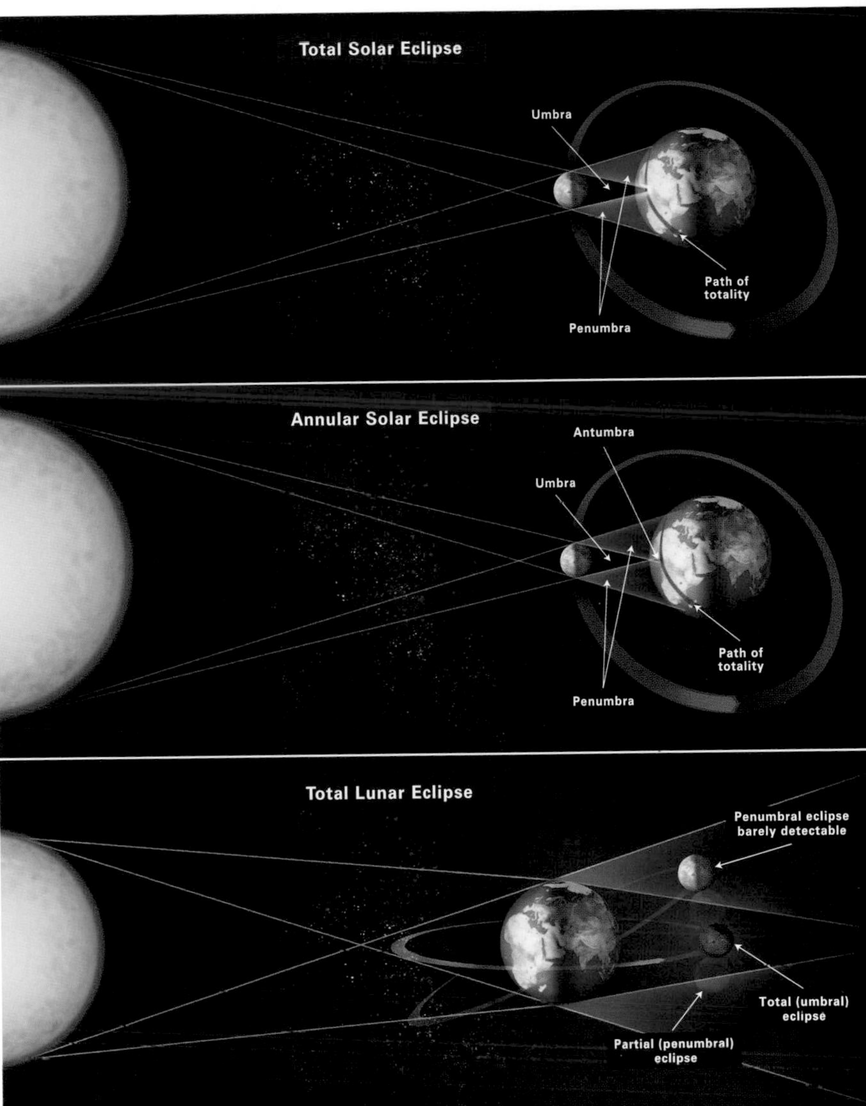

Above This illustration shows the three kinds of eclipses that we see on Earth. Top, the path of totality is the area where a total solar eclipse is seen. Middle, the path of annularity is the area where an annular solar eclipse is seen. Bottom, a lunar eclipse is where the shadow of Earth blocks the sunlight from the Moon. In all cases, a shadow's inner portion—whether the Moon's or Earth's—is called the umbra. From within the umbra the Sun is entirely hidden. Surrounding the umbra is the much larger penumbra, where only part of the Sun is hidden.

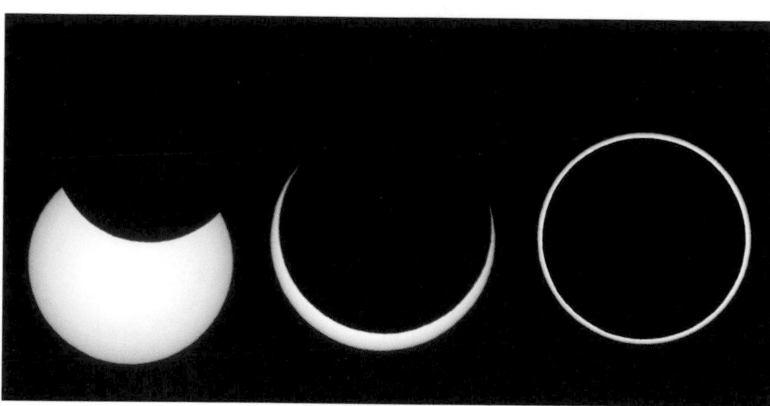

Above This is a series of images of an annular eclipse when the Sun, Moon, and Earth are exactly in line. The Moon's apparent size is smaller so the Sun appears as a bright ring, or annulus, surrounding the outline of the Moon.

Opposite This image shows the magnificent Aurora Borealis against silhouetted trees, as seen from Fairbanks, Alaska, USA.

ANNULAR AND PARTIAL SOLAR ECLIPSES

Annular eclipses are dramatic, but they are really only a special type of partial eclipse. However, annulars can produce spectacular displays of Baily's Beads lasting several minutes at second and third contact. At long total eclipses the cusps of the solar crescent are stubby, but at annular eclipses the long thin cusps are ideal for producing beads, especially at sites well south of the eclipse centerline. There the Moon's tallest mountains, those at the lunar south pole, produce spectacular displays of beads if the libration is favorable.

If the Moon's umbral shadow just misses Earth, only a partial solar eclipse occurs anywhere. Also, all annular and total eclipse paths are surrounded by a huge area that experiences a partial eclipse. When a path of totality 112 miles (180 km) wide crossed the Mediterranean Sea in 2006, a partial eclipse was visible from Kenya to Iceland. But not even watchers experiencing a 99 percent partial eclipse see the splendor that occurs only within the narrow path of totality.

LUNAR ECLIPSES

Lunar eclipses are visible wherever the eclipsed Full Moon is above the horizon. If the Moon passes only through Earth's penumbral shadow, the shading is rather slight, and penumbral eclipses can be detected only on those portions of the Moon that pass close to the umbra. If any part of the Moon enters the umbra, it is called a partial eclipse. If the Moon enters the umbra entirely, it is a total lunar eclipse. A total lunar eclipse begins as a penumbral eclipse, and then proceeds through a partial phase before becoming total. Totality can last as long as 100 minutes if the Moon passes centrally through the umbra and is moving relatively slowly in its orbit. There will also be two partial phases, each lasting about an hour, and two—mostly invisible—penumbral phases.

Earth's atmosphere refracts some reddened sunlight into its umbral shadow, so the totally eclipsed Moon is normally visible, although typically dimmed to between magnitude –2 and 1, with a color ranging from orange through brick red to ruddy. Other colors may be present, perhaps even a bluish zone at the umbra's edge. If a major volcanic eruption has recently spewed ash into the stratosphere, the Moon may be dark gray and almost invisible at mid-eclipse.

AURORAS

If powerful solar flares and the associated coronal mass ejections are facing Earth when they erupt, charged particles flow earthward much faster than the normal solar wind. Most particles are deflected by Earth's magnetosphere, but some particles spiral down at Earth's magnetic poles, collide with oxygen and nitrogen in the upper atmosphere, typically at about 70 miles (110 km) altitude, and raise their electrons to higher energy levels. When the electrons drop back down, they emit the characteristic auroral green, and, if the solar disturbance is powerful enough, occasionally reds.

Great all-sky auroras, bright enough to hide most stars, may begin with arcs, increase to moving rayed arcs, and then suddenly flame into green and pink draperies rippling up to a huge pulsating white zenithal corona—the term does not refer to the solar corona. One or two rays may be red, or even tend to violet.

Auroras occur in ovals, centered on the magnetic poles. The North Magnetic Pole has moved north 6° of latitude in over 20 years. Because the magnetic pole lies on the North American side of the geographic pole, auroras are more frequently seen from Canadian locations than at similar latitudes in Europe. During winter, both Yellowknife, Northwest Territories, Canada, and Fairbanks, Alaska, USA, attract visitors to view the auroras.

The South Magnetic Pole is on coastal Antarctica, facing Australia, near the Antarctic Circle—much further from the geographic South Pole than the North Magnetic Pole. When magnetic storms enlarge the auroral ovals, displays of the Aurora Borealis and Aurora Australis can occur at unusually low latitudes, such as Arizona, USA, and New South Wales, Australia. In March 1989, auroras were even reported from the Caribbean.

Below This combination of five images shows the various stages of the lunar eclipse seen in Sofia, Bulgaria on March 4, 2007. Total lunar eclipses occur when the Sun, Earth, and the Moon are all in alignment, and the Moon travels into the broad cone of shadow cast by Earth.

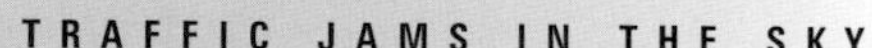

TRAFFIC JAMS IN THE SKY

A transit occurs when a planet crosses the Sun's disk, and an occultation is the concealment of one heavenly body by another. Both events are eclipses. There is no difference between a transit of Venus and a solar eclipse, which could be called a transit of the Moon. Nor is there any difference between a lunar occultation of Antares and a solar eclipse—both are eclipses of a star by the Moon. Eclipses of Jupiter's moons by its shadow and the shadows of the moons crossing Jupiter—total solar eclipses on Jupiter—can be watched several nights a week.

Captain James Cook set sail in 1768 to observe the 1769 Transit of Venus from the island of Tahiti, in the South Pacific Ocean. His success was mixed, but he sailed on to discover Terra Australis Incognita—the Unknown Land of the South.

Index

Page numbers in **bold** print refer to main entries. Page numbers in *italics* refer to photographs

O

P

Picture Credits

The Publisher would like to thank the following picture libraries and other copyright owners for permission to reproduce their images. Every attempt has been made to obtain permission for use of all images from the copyright owners, however, Millennium House would be pleased to hear from copyright owners.

KEY—(t) top of page / (b) bottom of page / (l) left side of page / (r) right side of page / (c) center of page.

NASA centers are credited as follows:
NASA-Great Images in NASA: NASA-GRIN; NASA Goddard Space Flight Center: NASA-GSFC; NASA Jet Propulsion Laboratory: NASA-JPL; NASA Johnson Space Center; NASA-JSC; NASA Kennedy Space Center: NASA-KSC; NASA Langley Research Center: NASA-LaRC; NASA Marshall Space Flight Center: NASA-MSFC

Other Organizations:
Anglo-Australian Observatory: AAO; Association of Universities for Research in Astronomy: AURA; Chandra X-ray Observatory Center: CXC; European Space Agency: ESA; European Southern Observatory: ESO; Japan Aerospace Exploration Agency: JAXA; National Optical Astronomy Observatory: NOAO; Particle Physics and Astronomy Research Council: PPARC; Solar and Heliospheric Observatory: SOHO; Space Telescope Science Institute: STScI; Wisconsin, Indiana, Yale and NOAO Observatory: WIYN

1 T. A. Rector/University of Alaska Anchorage and WIYN/AURA/NSF; 2–3 NASA-JPL-Caltech—Viking 1/digital version by Science Faction; 4–5 NASA Charles Pete Conrad; 7 Getty Images: Science Faction, NASA-ESA/digital version by Science Faction; 8–9(l–r) NASA, ESA, and The Hubble Heritage Team STScI/AURA), Acknowledgment: P. McCullough (STScI); NASA/JPL/University of Arizona; ESA/Hubble, Akira Fujii and Digitized Sky Survey; NASA, ESA and the Hubble Heritage Team STScI/AURA). Acknowledgment: J. Gallagher (University of Wisconsin), M. Mountain (STScI) and P. Puxley (NSF); A. Schaller (STScI); Getty Images: Hulton Archive; NASA-JSC; Getty Images: Science Faction, Tony Hallas; 10–11 Getty Images: Science Faction, NASA-JPL-Caltech-Cassini/digital version by Science Faction; 12(l) NASA-JPL; 12(t) NASA-JPL; 13(l) NASA-MSFC; 13(r) T.A. Rector/University of Alaska Anchorage and WIYN/AURA/NSF; 14 Getty Images: Science Faction, Peter Ginter; 15(c) NASA-JPL: 15(t) Getty Images: Science Faction, Louie Psihoyos; 18(tl) NASA and The Hubble Heritage Team (STScI/AURA), Acknowledgment: Ray A. Lucas (STScI/AURA); 18(bl) NASA/CXC/M.Markevitch et al; 19(t) NASA, ESA, S. Beckwith (STScI), and the Hubble Heritage Team; 20 NASA, ESA, M. J. Jee and H. Ford (Johns Hopkins University); 21(t) NASA, Adam Riess (STScI, Baltimore, MD)/NASA-GSFC; 21(c) Getty Images; Hulton Archive; 22–23 NASA/JPL/University of Arizona; 24(ct) The Art Archive/Biblioteca Nazionale Marciana Venice/Alfredo Dagli Orti; 24(b) NASA-GRIN; 25(tl) Getty Images: Stocktrek Images; 25(tr) ESO; 25(b) Southwest Research Institute (Dan Durda)/Johns Hopkins University Applied Physics Laboratory (Ken Moscati); 27(b) The Art Archive: Bridgeman Art Library 26; Getty Images: Dorling Kindersley; 27(bl) NASA-JPL; 28(bl) NASA/JPL-Caltech; (28br) NASA/JPL-Caltech; 28–29(t) Getty Images: Digital Vision; 29(b) NASA/Walt Feimer; 30 International Astronomical Union; 31 NASA/JPL-Caltech/R. Hurt (SSC); 32(t) The Art Archive/National Museum Beirut/Gianni Dagli Orti; 32(b) Getty Images: Purestock; 33 Getty Images: Time & Life Pictures/Getty Images; 34 NASA-ESA-SOHO/digital version by Science Faction; 35(l) Getty Images: Digital Vision; 35(r) Getty Images: Time & Life Pictures/Getty Images 35(b); 36(c) Getty Images: Time & Life Pictures/Getty Images; 37(t) NASA; 37(b) Getty Images: StockTrek; Getty Images: Photodisc Green; 38(t) Getty Images: Hulton Archive; 38(c) Getty Images: Hulton Archive; 38(b) Getty Images: Hulton Archive; 39(t) Getty Images: Science Faction; 39(b) Hinode JAXA/NASA/PPARC; 40 The Art Archive/Galleria degli Uffizi Florence/Alfredo Dagli Orti; 40 The Art Archive/Harper Collins Publishers; 41(t) NASA; 41(b) VT-2004 Programme/Heinz Kotzlowski; 42(t) NASA; 42() Getty Images: ©2006 European Space Agency; 42(b) NASA; 43 NASA-JPL; 44(l) NASA; 44(r) NASA-JPL; 44(c) NASA; 44(b) NASA; 45 Getty Images/Riser, Jim Ballard; 46(t) NASA; 46(b) NASA; 46–47(t) NASA-JPL; 47 Getty Images: Time & Life Pictures/Getty Images; 48 ESA; 49(l) The Art Archive/National Glyptothek Munich/Dagli Orti (A); 49(r) NASA-JPL: 50(t) The Art Archive/Archaeological Museum Aleppo Syria/Dagli Orti; 50(b) NASA; 51(t) NASA; 51(b) Getty Images: Time & Life Pictures/Getty Images; 52 Data provided by the Landsat 7 Team at NASA's Goddard Space Flight Center; 53 NASA; 54(t) NASA; 54(b) NASA; 54–55(b) Terra MODIS data captured by direct broadcast at Louisiana State University and processed at the University of Wisconsin-Madison by Liam Gumley, Space Science and Engineering Center; 56 NASA; 57(l) NASA; 57(r) NASA; 57(b) NASA; 58(t) NASA/Apollo 16/Erik van Meijgaarden/Kipp Teague; 58(b) Getty Images: Digital Vision; 59(b) Getty Images: Stockbyte; 60(t) The Art Archive/Museo Nazionale Palazzo Altemps Rome/Dagli Orti; 60(b) Steve Lee (University of Colorado), Jim Bell (Cornell University), Mike Wolff (Space Science Institute), and NASA/ESA; 61(t) NASA; 61(bl) Getty Images: Digital Vision; 61(br) Getty Images: Science Faction; 62(tl) NASA/JPL/UA; 62(tr) Getty Images: Science Faction; 62(br) Getty Images: NASA JPL-Caltech—Mars Global Surveyor/digital version by Science Faction; 63 Getty Images: Riser; 64(t) Getty Images: Riser; 64(b) Getty Images: Riser; 65 Getty Images: Photographer's Choice; 66(t) NASA, Jet Propulsion Laboratory; 66(b) ESA/DLR/FU Berlin (G. Neukum); 67(t) Getty Images: Photodisc Green; 67(c) Getty Images: De Agostini Picture Library; 67(bl) NASA/JPL/Malin Space Science Systems; 67(br) NASA-JPL; 68(cl) Getty Images: Purestock; 68(cr) Getty Images: Purestock; 68–69(b) IMP Team, JPL, NASA; 69(tl) Phil James (Univ. Toledo), Todd Clancy (Space Science Inst., Boulder, CO), Steve Lee (Univ. Colorado), and NASA/ESA; 69(tr) NASA/JPL/Univ. of Arizona; 70(t) Getty Images: The Bridgeman Art Library, Gustave Moreau; 70(b) NASA and the Hubble Heritage Team (STScI/AURA) Acknowledgment: NASA/ESA, John Clarke (University of Michigan); 71(tl) NASA; 71(b) NASA, ESA, I. de Pater and M. Wong (University of California, Berkeley); 72(t) Getty Images: Time Life Pictures; 72(bl) NASA/JPL/Space Science Institute; 72(br) Getty Images: Time Life Pictures; 73 NASA/JPL/Space Science Institute; 74(t) Getty Images: Time Life Pictures; 74(b) Getty Images: Time Life Pictures; 75 NASA-HQ-GRIN; 76(t) Getty Images: Time Life Pictures; 76(b) Getty Images: Time Life Pictures; 77 Getty Images: Hulton Archive; 78(t) NASA-JPL; 78–79(b) NASA-JPL; 79(t) Getty Images: Stockbyte; 79(bl) Getty Images: Science Faction; 79(br) Getty Images: Stockbyte; 80(t) The Art Archive/Collegio del Cambio Perugia/Gianni Dagli Orti; 80(b) NASA/JPL/Space Science Institute; 81(tl) NASA; 81(b) NASA-JPL; 82(t) NASA/JPL/Space Science Institute; 82(cl) NASA/JPL/University of Arizona); 82(bl) NASA/JPL/John Hopkins University); 82–83(b) NASA/JPL/Space Science Institute); 83(t) NASA/JPL/University of Arizona); 84(t) NASA/JPL/Space Science Institute); 84(c) NASA/JPL/Space Science Institute); 84(b) NASA/JPL; 85 Getty Images: Digital Vision; 86(b) NASA, ESA, and E. Karkoschka (University of Arizona); 85–86(b) NASA/JPL/Space Science Institute; 87(lt) NASA/JPL/Space Science Institute; 87(lc) NASA/JPL/Space Science Institute; 87(lb) NASA/JPL/Space Science Institute; 87(r) NASA/JPL/Space Science Institute; 87(b) ESA/NASA/JPL/University of Arizona); 88 NASA; 89(t) NASA/JPL/University of Iowa; 89(c) NASA; 89(bl) NASA/JPL/Space Science Institute; 89(br) ESA; 90 The Art Archive/Musée du Louvre Paris/Gianni Dagli Orti; 91(t) NASA; 91(cl) NASA JPL; 919(r) Getty Images: Time & Life Pictures/Getty Images; 92 Getty Images: Science Faction, William Radcliffe; 93(tl) Getty Images: Stockbyte, Jason Reed; 93(tr) NASA-JPL; 93(c) NASA-JPL; 93(b) NASA-JPL; 94(t) Getty Images: Dorling Kindersley, Demetrio Carrasco; 94(b) NASA; 95 (tl) NASA; 95(b) NASA-JPL; 96(t) Getty Images: Time & Life Pictures; 96(b) NASA; 97(t) Getty Images: NASA-JPL-Caltech—Voyager/digital version by Science Faction; 98(t) The Art Archive/Musée du Vin de Bourgogne Beaune/Gianni Dagli Orti; 98(b) NASA); 99(tl) NASA; 99(b) Getty Images: Time & Life Pictures/Getty Images; 100(c) Getty Images: The Image Bank, Antonio M. Rosario; 100(b) NASA, ESA,

H. Weaver (JHU/APL), A. Stern (SwRI), and the HST Pluto Companion Search Team; 100–101(b) NASA, ESA and G. Bacon (STScI); 101(t) NASA, ESA; 102–103(t) Johns Hopkins University Applied Physics Laboratory/Southwest Research Institute (JHUAPL/SwRI); 102(b) Gaspra, Ida: Galileo(NASA/JPL), Eros: NEAR Shoemaker (JHU/APL), Vesta and Mars: HST (NASA/STScI); 103(t) NASA; 103(c) NASA/JPL-Caltech; 104(l) NASA/JPL-Caltech/R. Hurt (SSC-Caltech); 104(r) NASA/JPL-Caltech/R. Hurt (SSC-Caltech); 105(t) Getty Images; 105(b) NASA/JPL-Caltech/R. Hurt (SSCCaltech); 106 Getty Images: Riser, Moritz Steiger; 107 NASA/JPL-Caltech/T. Pyle (SSC); 108(tl) NASA; 108(tr) NASA; 108(b) Getty Images: Stockbyte; 109(l) NASA; 109(r) A.Ikeshita/MEF/ISAS; 109(b) Don Davis, NASA; 110(t) The Art Archive/Biblioteca Nacional Madrid/Dagli Orti; 110(b) NASA, ESA, H. Weaver (APL/JHU), M. Mutchler and Z. Levay (STScI); 111(t) NASA, NOAO, NSF, T. Rector (University of Alaska, Anchorage), Z. Levay and L.Frattare (Space Telescope Science Institute); 112(t) Fredrik; 112(b) Getty Images: Taxi, Jack Zehrt; 113(t) The Art Archive/Scrovegni Chapel Padua/Dagli Orti (A); 113(r) Lowell Observatory and the National Optical Astronomy Observatories; 114 Getty Images: Photographer's Choice, Roger Ressmeyer; 115(t) NASA; 115(b) ESA; 116(t) NASA-JPL; 116(bl) ESA/MPGH/H. Uwe Keller; 116(br) NASA/JPL/UMD; 116-117(b) ESA/AOES Medialab; 118(t) Getty Images: Time & Life Pictures/Getty Images; 118(l) Getty Images: Time & Life Pictures/Getty Images; 119 Getty Images: Science Faction, Tony Hallas; 120(t) Getty Images: StockTrek; 120(c) Lorenzo Lovato; 121(tr) NASA-JPL; 121(c) Getty Images: NASA-JPL-Caltech—Mars Rover/digital version by Science Faction; 121(b) Getty Images: National Geographic, Jonathan Blair, 122–123 ESA/Hubble, Akira Fujii and Digitized Sky Survey 2; 124 NASA, ESA and Jesús Maíz Apellániz (Instituto de astrofísica de Andalucía, Spain). Acknowledgement: Davide De Martin (ESA/Hubble); 125 NASA, ESA and the Hubble Heritage Team STScI/AURA)-ESA/Hubble Collaboration; 126(l) Getty Images: Time & Life Pictures/Getty Images; 127 NASA/ESA and the Hubble Heritage Team (AURA/STScI); 128(tl) NASA, ESA and K. Luhman; 128(tr) NASA/JPL-Caltech/UT Austin); 128–129(b) Getty Images: Stocktrek Images; 129(c) Getty Images: Digital Vision; 130(t) Dick Schwartz (Univ. of Missouri-St. Louis), and NASA/ESA; 130–131(b) NASA, ESA, N. Smith (University of California, Berkeley), and The Hubble Heritage Team (STScI/AURA); 131(tl) Mohammad Heydari-Malayeri (Paris Observatory, France), NASA/ESA; 131(r) Getty Images; 132(t) SOHO (ESA & NASA)/Steele Hill; 132(b) NASA, ESA, N. Smith (University of California, Berkeley), and The Hubble Heritage Team (STScI/AURA); 133(t) NASA/ESA and The Hubble Heritage Team STScI/AURA); 133(b) NASA, ESA, and the Hubble Heritage (STScI/AURA)-ESA/Hubble Collaboration, Acknowledgment: R. Fesen (Dartmouth College) and J. Long (ESA/Hubble); 134(t) X-ray: NASA/CXC/RIT/J. Kastner et al.; Optical/IR: BD +30 & Hen 3: NASA/STScI/Univ. MD/J. P. Harrington; NGC 7027: NASA/STScI/Caltech/J. Westphal & W. Latter; Mz 3: NASA/STScI/Univ. Washington/B. Balick; 134(b) European Space Agency, NASA and Robert A. E. Fosbury (European Space Agency/Space Telescope-European Coordinating Facility, Germany); 135(t) Getty Images: NASA; 135(b) NASA/CXC/U. Amsterdam/S. Migliari et al; 136(b) Getty Images: Stocktrek Images; 137(t) NASA/ESA and The Hubble Heritage Team (AURA/STScI); 137(b) NASA/CXC/M. Weiss), 137(r) NASA and ESA; 138 S. Points, C. Smith, R. Leiton, and C. Aguilera/NOAO/AURA/NSF and Z. Levay (STScI); 139 NASA, ESA, The Hubble Heritage Team, (STScI/AURA) and A. Riess (STScI); 140(t) NASA, ESA and H. E. Bond (STScI); 141(t) NASA, ESA and H. E. Bond (STScI); 142(t) X-ray: NASA/SAO/CXC; Infrared: NASA/JPL-Caltech/A. Tappe & J. Rh; 142(b) NASA/CXC/M.Weiss; 143 ESA/Hubble; 144(t) Getty Images: Stocktrek Images; 143(b) Getty Images: Stone; 145 NASA; 146(tl) Getty Images: NASA; 146(tr) X-ray: NASA/CXC/PSU/S. Park & D. Burrows; Optical: NASA/STScI/CfA/P. Challis; 146(b) NASA/Chris Meaney; 147(t) NASA/CXC/MIT/UMass Amherst/M. D. Stage et al; 147(b) NASA-GSFC/Los Alamos National Laboratory; 148(t) The Art Archive/Musée Archéologique Naples/Alfredo Dagli Orti; 148(b) European Space Agency & NASA; 149 NASA, ESA and A. Nota (STScI/ESA); 150(t) ESA, NASA and Martino Romaniello (European Southern Observatory, Germany); 150(b) NASA, ESA and AURA/Caltech); 151(l) NASA, ESA and A. Schaller (for STScI); 151(r) NASA; plate facing page 151: NASA, ESA, M. Robberto (Space Telescope Science Institute/ESA) and the Hubble Space Telescope Orion Treasury Project Team; center plate: NASA, ESA, N. Smith (University of California, Berkeley), and The Hubble Heritage Team (STScI/AURA); plate facing page 152: NASA, ESA, and The Hubble Heritage Team STScI/AURA); 152(b) NASA, H. Ford (JHU), G. Illingworth (UCSC/LO), M. Clampin (STScI), G. Hartig (STScI), the ACS Science Team, and ESA; 152–153(t) NASA, ESA, and The Hubble Heritage Team (STScI/AURA; 153(b) NASA, ESA and J. Hester (ASU); 154(l) NASA, ESA, and The Hubble Heritage Team (STScI/AURA); 154–155(t) Hubble Heritage Team (AURA/STScI/NASA/ESA); 155(b) NASA, ESA, J. Hester and A. Loll (Arizona State University); 156(t) Getty Images. Time & Life Pictures/Getty Images; 156(b) Getty Images: Stocktrek Images; 157(t) Getty Images: Stocktrek Images: 157(b) NASA-JPL; 158(l) AFP/Getty Images; 158–159(b) ESO; 159(tl) Getty Images: Stocktrek Images; 159(tr) NASA-JPL; 160(l) Craig Attebery - NASA; 160–161(t) Getty Images: Purestock; 161(b) Getty Images; 162(t) NASA/JPL;162(b) NASA;163(t) Getty Images: Aurora/Getty Images; 163(b) NASA/JPL-Caltech; 164–165 NASA, ESA and the Hubble Heritage Team STScI/AURA). Acknowledgment: J. Gallagher (University of Wisconsin), M. Mountain (STScI) and P. Puxley (NSF); 166(b) NASA, ESA, and The Hubble Heritage Team (STScI/AURA); 167(t) NASA, ESA, and The Hubble Heritage Team STScI/AURA); 167(b) NASA/CXC, B. McNamara (University of Waterloo and Ohio University); 168(t) The Art Archive/Museo del Prado Madrid; 168(b) NASA/JPLCaltech/STScI; 169 NASA, ESA, and The Hubble Heritage Team (STScI/AURA); 170(t) NASA, the DSS-II and GSC-II Consortia (with images from the Palomar Observatory-STScI Digital Sky Survey of the northern sky, based on scans of the Second Palomar Sky Survey are copyright © 1993-1999 by the California Institute of Technology); 170(c) Image composite by Ingrid Kallick of Possible Designs, Madison, Wisconsin. The background Milky Way image is a drawing made at Lund Observatory. High-velocity clouds are from the survey done at Dwingeloo Observatory (Hulsbosch & Wakker, 1988); 171(t) NASA and The Hubble Heritage Team (STScI/AURA); 171(b) NASA-GSFC; 172(b) Palomar Observatory, Caltech and the STScI Digitized Sky Survey (AURA); 172–173(t) NASA, ESA and A. Schaller (for STScI); 173(bl) Tod R. Lauer/NASA/ESA; 173(br) Mohammad Heydari-Malayeri (Paris Observatory, France), and NASA/ESA; 174(b) NASA, ESA and the Hubble Heritage Team STScI/AURA); 175(t) Getty Images: Stocktrek; 175(b) Getty Images: Photographer's Choice, Jim Ballard; 176(l) NASA/ESA; 176(r) Ray Gallagher (University of Wisconsin-Madison), Alan Watson (Lowell Observatory, Flagstaff, AZ), and NASA/ESA; 177 NASA, ESA, and the Hubble Heritage Team STScI/AURA)-ESA/Hubble Collaboration. Acknowledgement: B. Whitmore (Space Telescope Science Institute) and James Long (ESA/Hubble); 178(t) Curt Struck and Philip Appleton (Iowa State University), Kirk Borne (Hughes STX Corporation), and Ray Lucas (STScI), and NASA/ESA; 178(b) NASA, ESA, CXC, and JPL-Caltech; 179(tl) NASA, ESA, and The Hubble Heritage Team (AURA/STScI); 179(tr) NASA, ESA and A. Zezas; 180(tl) The Art Archive; 180(tr) NASA, ESA, and The Hubble Heritage Team (STScI/AURA), Acknowledgment: M. Gregg (Univ. Calif.-Davis and Inst. for Geophysics and Planetary Physics, Lawrence Livermore Natl. Lab.); 180(b) Digitized Sky Survey, Palomar Observatory, STScI; 181(t) NASA, N. Benitez (JHU), T. Broadhurst (Racah Institute of Physics/The Hebrew University), H. Ford (JHU), M. Clampin (STScI), G. Hartig (STScI), G. Illingworth (UCO/Lick Observatory), the ACS Science Team and ESA; 181(b) NASA-JPL; 182(t) European Space Agency, A. Riera (Universitat Politecnica de Catalunya, Spain) and P. Garcia-Lario (European Space Agency ISO Data Centre, Spain); 182(b) E.J. Schreier (STScI) and NASA; 183 Dana Berry (STScI); 184(t) National Radio Astronomy Observatory/National Science Foundation; 184(b) X-ray: NASA/CXC/MIT/UCSB/P.Ogle et al.; Optical: NASA/STScI/A.Capetti et al.); 185 NASA, ESA, CXC, STScI, and B. McNamara (University of Waterloo); 186(t) NASA, Andrew S. Wilson (University of Maryland);

Patrick L. Shopbell (Caltech); Chris Simpson (Subaru Telescope); Thaisa Storchi-Bergmann and F. K. B. Barbosa (UFRGS, Brazil); and Martin J. Ward (University of Leicester, U.K.); 186(b) X-ray: NASA/CXC/Univ. of Maryland/ A.S. Wilson et al.; Optical: Pal. Obs. DSS; IR: NASA/JPL-Caltech; VLA: NRA O/AU I/NSF; 187(t) European Space Agency and Wolfram Freudling (Space Telescope-European Coordinating Facility/ European Southern Observatory, Germany); 187(b) NASA, ESA, ESO, Frédéric Courbin (Ecole Polytechnique Federale de Lausanne, Switzerland) & Pierre Magain (Universite de Liege, Belgium); 188(l) The Art Archive/Salamanca University/Dagli Orti (A); 188(r) Professor Anthony Fairall; 189 Max-Planck Institute for Astrophysics, Garching, Germany; 190–191 A. Schaller (STScI); 192(t) Getty Images: Hulton Archive; 192(b) NASA, ESA, and The Hubble Heritage Team (STScI/AURA), Acknowledgment: P. Knezek (WIYN); 193 NASA, ESA, and S. Beckwith (STScI) and the HUDF Team; 194 NASA-WMAP; 195(t) Getty Images: Time & Life Pictures/ Getty Images; 195(r) NASA/ JPLCaltech/GSFC/SDSS; 198(l) NASA/JPL-Caltech/A. Kashlinsky (GSFC); 199 Getty Images: Photographer's Choice RF, Steven Hunt; 200(l) NASA and J. Blakeslee (JHU) Science Center/DSS; 200–201(b) NASA/JPL-Caltech; 201(tl) NASA, J. English (U. Manitoba), S. Hunsberger, S. Zonak, J. Charlton, S. Gallagher (PSU), and L. Frattare (STScI) Science Credit: NASA, C. Palma, S. Zonak, S. Hunsberger, J. Charlton, S. Gallagher, P. Durrell (The Pennsylvania State University) and J. English (University of Manitoba) Heritage Team, 201(tr) Getty Images: Photodisc; 202–203 Getty Images: Hulton Archive; 204(l) The Art Archive/Bibliothèque des Arts Décoratifs Paris/Gianni Dagli Orti; 204–205(t) Getty Images: National Geographic/Getty Images; 205(b) Getty Images; 206(l) Getty Images: Time & Life Pictures/Getty Images; 207(t) Getty Images: Iconica, Macduff Everton; 207(b) The Art Archive/Biblioteca Nazionale Marciana Venice/Dagli Orti (A); 208(t) Getty Images; Hulton Archive; 208(b) The Art Archive/Gianni Dagli Orti; 209(tl) Getty Images: Hulton Archive; 209(r) The Art Archive/Biblioteca Nazionale Marciana Venice/Gianni Dagli Orti; 210(l) Getty Images: Hulton Archive; 211 The Art Archive/University Library Istanbul/Gianni Dagli Orti; 212(l) The Art Archive/Maritime Museum Kronborg Castle Denmark/ Dagli Orti; 212(r) Getty Images: Hulton Archive; 213 The Art Archive/Observatory Academy Florence Italy/Dagli Orti (A); 214(t) The Art Archive/Royal Society/Eileen Tweedy; 214(b) Getty Images: Hulton Archive; 215 Getty Images: Time & Life Pictures/ Getty Images; 216 Getty Images: Time & Life Pictures/Getty Images; 217(tr) Menahem Kahana/AFP/ Getty Images; 217(tr) Getty Images: Science Faction, Peter Ginter; 217(c) Getty Images: Hulton Archive; 218(t) Getty Images: Time & Life Pictures/Getty Images; 218(b) Getty Images: The Bridgeman Art Library, Polish School; 219(t) Getty Images: Riser, Steve Allen; 219(r) Swedish Solar Telescope, Royal Swedish Academy of Sciences, Institute for Solar Physics; 220(tl) Getty Images: Gallo Images ROOTS RF collection, Shaen Adey; 220–221(t) Getty Images; 222(l) Getty Images: Time & Life Pictures/Getty Images; 222(r) NASA/JPL; 2223 NASA/ JPL/Space Science Institute; 224(tl) NASA/Kepler Mission; 224(c) NASA-JPL; 224–225(tr) NASA/ JPL-Caltech/Laboratorio de Astrofísica Espacial y Física Fundamental; 225(b) Getty Images: 1999 Joe McNally; 226–227 NASA-JSC; 228–229(tc) NASA-MSFC; 228(cl) NASA-MSFC; 28(b) NASA-MSFC; 229 Getty Images: Hulton Archive; 229(b) The Art Archive/Christies/ Eileen Tweedy; 230(t) Asif Siddiqi; 230(cl) NASA-MSFC; 230(c) NASA-GFSC; 230–231(c) Getty Images: Hulton Archive; 231(t) Asif Siddiqi; 231(b) Getty Images: Hulton Archive; 232(l) Getty Images: Time & Life Pictures/Getty Images; 232(r) Getty Images: Hulton Archive; 232(b) Asif Siddiqi; 233 Getty Images: Hulton Archive; 234(l) Getty Images: Time & Life Pictures/Getty Images; 234(r) Getty Images: Hulton Archive; 235(t) The Art Archive: Paramount/The Kobal Collection 235(bl) Getty Images: Time & Life Pictures, Howard Sochurek; 235(c) Getty Images: Time & Life Pictures/Getty Images; 236(t) Getty Images: Time & Life Pictures, Michael Rougier; 236(b) Getty Images: Hulton Archive; 236–237(t) NASA-MSFC; 237(tr) NASA (GRIN): 237(b) Getty Images: Hulton Archive; 238(l) NASA; 238(t) Getty Images: Hulton Archive; 238(br) Getty Images: Hulton Archive; 239(tl) Getty Images: Hulton Archive; 239(r) NASA; 240(l) Getty Images: Hulton Archive; 240(c) Getty Images: Hulton Archive; 240–241(b) Getty Images: AFP/ Getty Images; 241(r) NASA-KSC; 241(b) NASA (GRIN); 242(l) NASA (GRIN); 242(bl) Peter Gorin; 242(br) Getty Images: Hulton Archive; 243 Getty Images: NASA—Gemini/digital version by Science Faction; 244(tl) NASA (GRIN); 244(tr) Getty Images, Hulton Archive; 244(b) NASA (GRIN); 245(l) NASA (GRIN); 245(r) Getty Images: Time & Life Pictures/ Getty Images; 246(tl) Getty Images: Time & Life Pictures/Getty Images; 246(r) Getty Images: Hulton Archive; 246(b) Getty Images: Time & Life Pictures/Getty Images; 247(t) Getty Images: Hulton Archive; 247(b) Getty Images: Time & Life Pictures/Getty Images; 247(l) NASA (GRIN); 247(r) Getty Images: Hulton Archive; 249(t) NASA-LaRC; 249(b) NASA (GRIN); 250(t) Getty Images; 250(b) Getty Images; 251(t) NASA-(GRIN); 251(b) Getty Images, Hulton Archive; 252 NASA (GRIN); 253 (l) Getty Images; 253(c) Getty Images: Hulton Archive; 253(t) NASA; 254(l) Getty Images: NASA-Apollo/digital version by Science Faction; 254(b) NASA David R. Scott; 255 NASA-HQ-GRIN; 256 Getty Images: Stan Wayman; 257 © Mark Wade; 257(tr) Michel Koivisto; 257(c) Asif Siddiqi; 257 (b) Asif Siddiqi; 258(l) NASA; 258(r) NASA (GRIN); 259(t) NASA (GRIN); 259(bl) Getty Images: Time & Life Pictures/Getty Images; 259(br) NASA-MSFC; 260(l) NASA-MSFC; 260(r) ©Mark Wade; 260(b) Dietrich Haeseler; 261 RIA Novosti; 262 Getty Images: Photodisc, StockTrek; 263(t) Getty Images: Time & Life Pictures/Getty Images; 263(c) NASA; 264(t) NASA-MSFC; 264(l) Getty Images: Time & Life Pictures/Getty Images; 266–267 Getty Images: Hulton Archive; 267(tl) NASA; 267(tr) Getty Images: Time & Life Pictures/Getty Images; 267(b) Getty Images: Hulton Archive; 267 Getty Images: Digital Vision; 268 NASA-MSFC; 269(t) NASA-HQ-GRIN; 269(b) NASA; 270(t) Getty Images; 270(b) Getty Images; 271(t) Stringer/AFP/ Getty Image; 271(b) Getty Images; 272(l) Getty Images; 272(r) Getty Images: Time & Life Pictures/Getty Images; 273 Getty Images: Taxi, Space Frontiers; 274(l) NASA; 274(b) NASA; 274(r) Getty Images: Photodisc, PhotoLink; 275 Getty Images: Stone, World Perspectives; 276(l) Getty Images: Stocktrek Images; 276–277 Getty Images: Photographer's Choice, Joe Drivas; 278(tl) Getty Images: Digital Vision, Steve Taylor; 278(tr) NASA-(GRIN); 278(b) Getty Images; Digital Vision; 279 NASA (GRIN); 280 NASA (GRIN); 281(tl) NASA; 281(tr) NASA; 281(bl) NASA; 281(cr) Getty Images; 281(lb) Getty Images; 282(l) ESA/CNES/Arianespace-S. Corvaja via Getty Images); 282(r) Getty Images; 282–283(b) Getty Images; 283(tr) Getty Images: AFP/Getty Images; 283(cr) Getty Images: J. Huart/AFP/Getty Images; 284(l) STR/AFP/Getty Images; 284(b) Getty Images: China Photos; 284–285(t) AFP/Getty Images; 285(r) Getty Images: Chinafotopress; 286–287(t) NASA-MSFC; 286(b) Roberto Schmidt/AFP/ Getty Images; 287(b) STRDEL/ AFP/Getty Images; 287(r) Prakash Singh/AFP/Getty Images; 288 NASA-KSC; 289(t) NASA; 289(c) Getty Images: Time & Life Pictures/ Getty Images; 289(b) Getty Images: Time & Life Pictures/Getty Images; 290(l) NASA; 290(r) Getty Images: Time & Life Pictures/Getty Images; 290(b) NASA (GRIN); 291(t) NASA; 291(b) NASA; 292(l) NASA; 292(r) NASA/ESA/JHU/ R. Sankrit & W. Blair; 293(t) NASA-JSC; 293(b) NASA-MSFC; 294(t) Getty Images: Iconica, Don Klumpp; 294(b) NASA-GSFC; 295(t) Getty Images: Stone, Erik Simonsen; 295(b) European Space Agency DLR/FU Berlin (G. Neukum); 296–297 Getty Images: Science Faction, Tony Hallas; 298(t) NASA-JPL; 298(b) ©2007 Paul Downing; 299 NASA (GRIN); 300–301 iStockphoto: ©Monte Wilson; 302(l) iStockphoto: ©Thomas Tuchan; 302(t) T.A.Rector (NOAO/AURA/NSF) and Hubble Heritage Team (STScI/AURA/ NASA; 303(b) iStock: Bryan Delodder; 303(t) iStock: James Benet; 302(c) iStock: Tomasz Resiak; 303(b) Getty Images: Digital Vision, Nick Dolding; 306(t) NASA/Hubblesource; 306 ESA/© 2002 R. Gendler, Photo by R. Gendler; 308(t) Image courtesy History of Science Collections, University of Oklahoma Libraries; copyright the Board of Regents of the University of Oklahoma; 308(b) ESO; 310(t) Image courtesy History of Science Collections, University of Oklahoma Libraries; copyright the Board of Regents of the University of Oklahoma; 310(b) Getty Images: Purestock/Getty Images; 312(tl) NASA/Hubblesource; 312(b) NASA-JPL; 314(tl) Image courtesy History of Science Collections, University of Oklahoma Libraries; copyright the Board of Regents of the University of Oklahoma; 314(c) NASA-HQ-GRIN; 314(b) David Talent/NOAO/AURA/NSF; 316(b) NASA/ESA and the Hubble Heritage Team (AURA/STScI); 318(tl) Image courtesy History of Science Collections, University of Oklahoma Libraries; copyright the Board of Regents of the University of Oklahoma; 318(b) Image provided by Jim Misti; 320(tl) NASA/Hubblesource; 320(b) T.A.Rector and B.A.Wolpa/NOAO/AURA/NSF; 322(tl) NASA/Hubblesource; 322(l) B.Jannuzi, A.Dey, NDWFS team/NOAO/AURA/NSF; 322(b) NOAO/Bob Rickert; 324(tl) Image courtesy History of Science Collections, University of Oklahoma Libraries; copyright the Board of Regents of the University of Oklahoma; 324(b) ESO; 326(tl) Image courtesy History of Science Collections, University of Oklahoma Libraries; copyright the Board of Regents of the University of Oklahoma; 326(l) © 2007 Paul Downing; 326(r) T. A. Rector/University of Alaska Anchorage, H. Schweiker/WIYN and NOAO/AURA/NSF; 328(tl) NASA/Hubblesource; 328(c) NOAO/AURA/NSF; 328(b) NOAO/AURA/NSF; 330(tl) Image courtesy History of Science Collections, University of Oklahoma Libraries; copyright the Board of Regents of the University of Oklahoma; 330(b) NASA-GSFC; 332(tl) NASA/Hubblesource; 332(l) NASA, ESA and R. Humphreys (University of Minnesota); 332(r) Akira Fujii; 334(t) Image courtesy History of Science Collections, University of Oklahoma Libraries; copyright the Board of Regents of the University of Oklahoma; 334(t) REU program/NOAO/AURA/NSF; 334(b) ©2007 Paul Downing; 336(t) Image courtesy History of Science Collections, University of Oklahoma Libraries; copyright the Board of Regents of the University of Oklahoma; 336(c) NASA; 336(b) NASA, The Hubble Heritage Team (AURA/STScI); 338 Nathan Smith, University of Minnesota/NOAO/AURA/NSF; 339(t) NOAO/AURA/NSF; 339(b) NASA, The Hubble Heritage Team (AURA/STScI), Acknowledgment: S. Casertano STScI); 340(tl) NASA/Hubblesource; 340(bl) T.A. Rector/University of Alaska Anchorage and WIYN/AURA/NSF; 342(tl) Image courtesy History of Science Collections, University of Oklahoma Libraries; copyright the Board of Regents of the University of Oklahoma; 342(c) SSRO/PROMPT and NOAO/AURA/NSF; 342(b) Eric Peng (JHU), Holland Ford (JHU/STScI), Ken Freeman (ANU), Rick White (STScI), NOAO/AURA/NSF; 344(tl) Image courtesy History of Science Collections, University of Oklahoma Libraries; copyright the Board of Regents of the University of Oklahoma; 344(c) Gemini Observatory/Travis Rector, University of Alaska Anchorage; 344(b) WIYN/NOAO/NSF; 346(tl) Image courtesy History of Science Collections, University of Oklahoma Libraries; copyright the Board of Regents of the University of Oklahoma; 346(l) NOAO/AURA/NSF; 346(r) NOAO/AURA/NSF; 348(tl) Image courtesy History of Science Collections, University of Oklahoma Libraries; copyright the Board of Regents of the University of Oklahoma; 348(c) NASA/JPL-Caltech/J. Bally (Univ. of Colo.); 348(b) FORS Team, 8.2-meter VLT Antu, ESO; 350(tl) Image courtesy History of Science Collections, University of Oklahoma Libraries; copyright the Board of Regents of the University of Oklahoma; 350(l) X-ray: NASA/Penn State/F. Bauer et al; Optical: NASA/A.Wilson et al); 350(r) Andrew S. Wilson (University of Maryland), Patrick L. Shopbell (Caltech), Chris Simpson (Subaru Telescope), Thaisa Storch; 352(tl) Image Courtesy US Naval Observatory Library; 352(l) NASA/JPL-Caltech/SSC/Ricardo Schiavo (UVA); 352(r) ESO; 354(tl) Image courtesy History of Science Collections, University of Oklahoma Libraries; copyright the Board of Regents of the University of Oklahoma; 354(l) NASA and The Hubble Heritage Team (AURA/STScI); 354(r) © 2007 Paul Downing; 356(tl) Image courtesy History of Science Collections, University of Oklahoma Libraries; copyright the Board of Regents of the University of Oklahoma; 356(t) UH88/Nedachi et al; 356(b) David Malin, UK Schmidt Telescope, AAO; 358(tl) NASA/Hubblesource; 358(c) Anja von der Linden: Image taken at Hoher List Observatory; 358(b) ©Paul Howell; 360(tl) NASA/Hubblesource; 360(tr) Image Courtesy US Naval Observatory Library; 360(l) NASA, ESA, Z. Levay (STScI) and A. Fujii; 360(r) NASA, ESA, and the Hubble Heritage Team (STScI/AURA)-ESA/Hubble Collaboration;362(tl) Image courtesy History of Science Collections, University of Oklahoma Libraries; copyright the Board of Regents of the University of Oklahoma; 362(b) NASA; 364(tl) Image Courtesy US Naval Observatory Library; 364(cl) Andrew Butler, Calvin Observatory; 364(b) T. A. Rector (U. Alaska), WIYN, NOAO, AURA, NSF; 366(tl) Image Courtesy US Naval Observatory Library; 366(b) NOAO/Hubble; 368(tr) B. Brandl (Cornell & Leiden) et al, JPL, Caltech, NASA; 368(bl) Hubble Heritage Team (STScI/AURA), Y. Chu (UIUC) et al, NASA; 370(tl) Image Courtesy US Naval Observatory Library; 370(bl) NASA, ESA, and The Hubble Heritage Team (STScI/AURA) Acknowledgment: W. Keel (University of Alabama, Tuscaloosa); 370(br) NASA J. P. Harrington and K. J. Borkowski University of Maryland; 372(tl) Image courtesy History of Science Collections, University of Oklahoma Libraries; copyright the Board of Regents of the University of Oklahoma; 372(bl) NASA and The Hubble Heritage Team (STScI/AURA); 374(tl) Image courtesy History of Science Collections, University of Oklahoma Libraries; copyright the Board of Regents of the University of Oklahoma; 374(cl) Gemini Observatory/Travis Rector, University of Alaska Anchorage; 374(r) NASA, ESA, and The Hubble Heritage Team (STScI/AURA); 376(tl) Image courtesy History of Science Collections, University of Oklahoma Libraries; copyright the Board of Regents of the University of Oklahoma; 376(cl) NASA, ESA, and The Hubble Heritage Team (STScI/AURA); 376(b) NASA and John Trauger (JPL); 378(tl) NASA/Hubblesource; 378(c) N. A. Sharp/NOAO/AURA/NSF; 378(b) (NASA-GSFC); 380(t) Image courtesy History of Science Collections, University of Oklahoma Libraries; copyright the Board of Regents of the University of Oklahoma; 380(b) ESO; 382(tl) NASA/Hubblesource; 382(c) Robert Rubin and Christopher Ortiz (NASA Ames Research Center), Patrick Harrington and Nancy Jo Lame (University of Maryland), Reginald Dufour (Rice University), and NASA; 382(b) Bernard Hubl; 834(tl) Image courtesy History of Science Collections, University of Oklahoma Libraries; copyright the Board of Regents of the University of Oklahoma; 384(b) NASA, ESA, and D. Maoz (Tel-Aviv University and Columbia University); 386(tl) Image courtesy History of Science Collections, University of Oklahoma Libraries; copyright the Board of Regents of the University of Oklahoma; 386(l) The Two Micron All Sky Survey (2MASS) at IPAC; 386(b) NASA and The Hubble Heritage Team (STScI/AURA), Acknowledgment: C. Conselice (U. Wisconsin/STScI); 388(tl) Image courtesy History of Science Collections, University of Oklahoma Libraries; copyright the Board of Regents of the University of Oklahoma; 388(b) Christopher J. Picking; 390(t) Image courtesy History of Science Collections, University of Oklahoma Libraries; copyright the Board of Regents of the University of Oklahoma; 390(b) Christopher J. Picking; 392(tl) Image Courtesy US Naval Observatory Library; 392(c) European Space Agency and the Digitized Sky Survey; 392(b) Charles Shahar; 394(tl) NASA/Hubblesource; 394(tr) Image courtesy History of Science Collections, University of Oklahoma Libraries; copyright the Board of Regents of the University of Oklahoma; 394(cl) NASA-JPL; 394(b) © 2007 Paul Downing; 396(tl) Image courtesy History of Science Collections, University of Oklahoma Libraries; copyright the Board of Regents of the University of Oklahoma; 396(r) NOAO/AURA/NSF; 396(b) NASA and The Hubble Heritage Team (STScI/AURA, Acknowledgment: Dr. Raghvendra Sahai (JPL) and Dr. Arsen R. Hajian (USNO); 398(tl) Image courtesy History of Science Collections, University of Oklahoma Libraries; copyright the Board of Regents of the University of Oklahoma; 398(b) Allan Sandage, Carnegie Observatories; 400(tl) Image courtesy History of Science Collections, University of Oklahoma Libraries; copyright the Board of Regents of the University of Oklahoma; 400(b) NASA-GSFC; 402(tl) Image courtesy History of Science Collections, University of Oklahoma Libraries; copyright the Board of Regents of the University of Oklahoma; 402(c) Shardha Jogee (Yale)/WIYN/NOAO/NSF; 402(bl) European Space Agency, NASA and Robert A.E. Fosbury (European Space Agency/Space Telescope-European Coordinating Facility, Germany); 404(tl) NASA/Hubblesource; 404(c) NASA-JPL; 404(b) C.F.Claver/WIYN/NOAO/NSF; 406(l) NASA, The Hubble Heritage Team, STScI, AURA; 406(l) NASA-JPL; 408(t) Image courtesy History of Science Collections, University of Oklahoma Libraries; copyright the Board of Regents of the University of Oklahoma; 408(b) NASA/ESA/G. Bacon (STScI); 410(tl) Image courtesy History of Science Collections, University of Oklahoma Libraries; copyright the Board of

Regents of the University of Oklahoma; 410(c) NASA-GRIN; 410(b) NASA-JPL; 412(tl) Image Courtesy US Naval Observatory Library; 412(cl) Christopher J Picking; 412(b) R. Sahai and J. Trauger (JPL), WFPC2, HST, NASA; 414(tl) Image courtesy History of Science Collections, University of Oklahoma Libraries; copyright the Board of Regents of the University of Oklahoma; 414(c) NASA-JPL; 414(b) Gemini Observatory, AURA, NSF; 416(t) NASA/Hubblesource; 416(b) Chris and Dawn Schur; 418(tl) NASA/Hubblesource; 418(l) NASA-JPL; 418(b) NASA/ESA and The Hubble Heritage Team STScI/AURA); 420(tl) NASA/Hubblesource; 420(cl) Andrea Dupree (Harvard-Smithsonian CfA), Ronald Gilliland (STScI), NASA and ESA; 420(b) K.L. Luhman (Harvard-Smithsonian Center for Astrophysics, Cambridge, Mass.); and G. Schneider, E. Young, G. Rieke, A. Cotera, H. Chen, M. Rieke, R. Thompson (Steward Observatory, University of Arizona, Tucson, Ariz.) and NASA/ESA; 422(tl) Image courtesy History of Science Collections, University of Oklahoma Libraries; copyright the Board of Regents of the University of Oklahoma; 422(cl) NASA and The Hubble Heritage Team (STScI/AURA), Acknowledgment: R. Windhorst (ASU); 422(b) Eberly College of Science; 424(tl) NASA/Hubblesource; 424(c) NASA/ESA, Jayanne English (University of Manitoba), Sally Hunsberger (Pennsylvania State University), Zolt Levay (Space Telescope Science Institute), Sarah Gallagher (Pennsylvania State University), and Jane Charlton (Pennsylvania State University); 424(b) Edward Conway, Nathan Horleston and Steve Maddox (University of Nottingham) and Nik Szymanek (University of Hertfordshire); 426(tl) NASA/Hubblesource; 426(cl) © Paul Downing; 426(b) NASA/JPL-Caltech/R. A. Gutermuth (Harvard-Smithsonian CfA); 428(t) Image courtesy History of Science Collections, University of Oklahoma Libraries; copyright the Board of Regents of the University of Oklahoma; 428(b) Knut Olsen (CTIO) & Phillip Massey (Lowell Observatory)(NOAO/CTIO/KPNO); 430(tl) Millennium House Pty Ltd; 430(r) NASA, ESA, and The Hubble Heritage Team (STScI/AURA); 430(b) NASA-GSFC; 432(tl) Image Courtesy US Naval Observatory Library; 432(cl) NASA-GSFC, 432(b) © 2007 Paul Downing; 434(tl) Image courtesy History of Science Collections, University of Oklahoma Libraries; copyright the Board of Regents of the University of Oklahoma; 434(cl) NASA-JPL; 434(b) NASA, ESA, P. Kalas and J. Graham (University of California, Berkeley), and M. Clampin (NASA-GSFC); 436(tl) Image courtesy History of Science Collections, University of Oklahoma Libraries; copyright the Board of Regents of the University of Oklahoma; 436(cl) NASA-GSFC; 436(b) Gemini Observatory/Travis Rector, University of Alaska Anchorage; 438(t) Image courtesy History of Science Collections, University of Oklahoma Libraries; copyright the Board of Regents of the University of Oklahoma; 438(b) NASA; 440(t) Image courtesy History of Science Collections, University of Oklahoma Libraries; copyright the Board of Regents of the University of Oklahoma; 440(b) Christopher J Picking; 442(tl) Image courtesy History of Science Collections, University of Oklahoma Libraries; copyright the Board of Regents of the University of Oklahoma; 442(b) NOAO; 444(t) NASA/Hubblesource; 444(b) Lawrence Johnson and Kathy Ortega/Adam Block/NOAO/AURA/NSF; 446 NOAO; 447(t) NASA, H. Ford (JHU), G. Illingworth (USCS/LO), M. Clampin (STScI),G. Hartig (STScI), the ACS Science Team, and ESA; 447(cr) N. A. Sharp, REU program/NOAO/AURA/NSF; 448(t) NASA/Hubblesource; 448(b) NOAO/AURA/NSF; 450(l) © Paul Downing; 450–451(t) N. A. Sharp, REU program/NOAO/AURA/NSF; 452(tl) Image courtesy History of Science Collections, University of Oklahoma Libraries; copyright the Board of Regents of the University of Oklahoma; 452(c) © Paul Downing; 452(b) NASA/JPL; 454(tl) Image courtesy History of Science Collections, University of Oklahoma Libraries; copyright the Board of Regents of the University of Oklahoma; 454(cl) N. A. Sharp, REU program/NOAO/AURA/NSF; 454(b) NASA/JPL-Caltech, D. Figer (STScI/Rochester Institute of Technology), E. Churchwell (University of Wisconsin, Madison) and the GLIMPSE Legacy Team; 456(tl) NASA/Hubblesource; 456(cl) Jeff Hester and Paul Scowen (Arizona State University), and NASA/ESA; 456(b) NASA/JPL; 458(tl) Image Courtesy US Naval Observatory Library; 458(bl) S. D. Van Dyk (IPAC/Caltech) et al; 458(br) Romano Corradi (ING); 460(tl) NASA/Hubblesource; 460(cl) NASA-JPL; 460(r) NASA-JPL; 462(tl) Image courtesy History of Science Collections, University of Oklahoma Libraries; copyright the Board of Regents of the University of Oklahoma; 462(b) Christopher J. Picking; 464(tl) NASA/Hubblesource; 464(c) NASA/JPL; 464(b) NASA-HQ-GRIN; 466(tl) Image courtesy History of Science Collections, University of Oklahoma Libraries; copyright the Board of Regents of the University of Oklahoma; 466(b) Christopher J Picking; 468(tl) Image courtesy History of Science Collections, University of Oklahoma Libraries; copyright the Board of Regents of the University of Oklahoma; 468(c) NASA/CXC/Penn Stat/S.Park et al.; 468(b) A. Nota (ESA/STScI) et al, ESA, NASA; 470(tl) NASA/Hubblesource; 470(c) NASA-MSFC; 470(b) NASA-JPL; 472(tl) Image courtesy History of Science Collections, University of Oklahoma Libraries; copyright the Board of Regents of the University of Oklahoma; 472(c) Philippe Crane (European Southern Observatory), and NASA; 472(b) NASA, ESA, G. Bacon (STScI); 474(tl) Image courtesy History of Science Collections, University of Oklahoma Libraries; copyright the Board of Regents of the University of Oklahoma; 474(cl) ESO; 474(b) NASA-GSFC; 476(tl) NASA/Hubblesource; 476(c) NASA-GSFC; 476(b) Infrared NASA/JPL-Caltech/R. Kennicutt (University of Arizone) and the SINGS Team Visible: Hubble Space Telescope/Hubble Heritage Team; 478(tl) Image courtesy History of Science Collections, University of Oklahoma Libraries; copyright the Board of Regents of the University of Oklahoma; 478(cl) Christopher J. Picking; 478(b) Image Credit: NASA, ESA, and The Hubble Heritage Team (AURA/STScI), Acknowledgment: J. Higdon (Cornell U.) and I. Jordan (STScI); 480(tl) Image Courtesy US Naval Observatory Library; 480(c) NASA/JPL; 480(b) Michael Pierce, Robert Berrington (Indiana University), Nigel Sharp, Mark Hanna (NOAO)/WIYN/NSF; 508(tl) Getty Images: Photonica, Image Acquisition; 508(bl) Getty Images: Time & Life Pictures, Francis Miller; 509(tr) Millennium House Pty Ltd; 509(b) Getty Images: Science Faction, Jay Pasachoff; 510(tl) Getty Images: AFP/Getty Images; 510(b) Getty Images: AFP/Getty Images; 511Getty Images: Photographer's Choice, Kevin Schafer; 512–513 NASA/JPL-Caltech/Cornell: 514 Getty Images: NASA-ESA-SOHO/digital version by Science Faction; 515 Getty Images: Science Faction, William Radcliffe; 516 NASA-JPL; 517 NASA and NASA Goddard Space Flight Center Image by Reto Stöckli (land surface, shallow water, clouds). Enhancements by Robert Simmon (ocean color, compositing, 3D globes, animation). Data and technical support: MODIS Land Group; MODIS Science Data Support Team; MODIS Atmosphere Group; MODIS Ocean Group Additional data: USGS EROS Data Center (topography); USGS Terrestrial Remote Sensing Flagstaff Field Center (Antarctica); Defense Meteorological Satellite Program (city lights); 518 Getty Images: Photographer's Choice, World Perspectives; 519 NASA, ESA, I. de Pater and M. Wong (University of California, Berkeley); 520 NASA-JPL; 521 Getty Images: Science Faction, William Radcliffe; 522 NASA; 523 NASA, ESA and G. Bacon (STScI); 539 NASA, ESA and A. Nota (ESA/STScI, STScI/AURA); 540(tl) Photo courtesy AstroInfo Services, London; 540(b) NASA; 540(tr) NASA; 540(cr) Photo courtesy AstroInfo Services, London; 541(tl) US Army; 541(tr) A-4/V-2 Resource Site; 541(b) US Army; 542(bl) Chris Dubbs collection; 542(cb) NASA; 542(tr) Photo courtesy AstroInfo Services, London; 542(cr) NASA; 543(ct) NASA; 543(tr) NASA; 543(lb) NASA; 543(br) Photo courtesy AstroInfo Services, London; 544(tl) NASA; 544(tr) NASA; 544(b) NASA; 545(c) NASA; 545(tl) NASA; 545(bl) NASA; 545(br) NASA; 546(tl) NASA; 546(bl) NASA; 546(br) NASA; 547(tl) NASA; 547(ct) NASA; 547(br) NASA; 548(tl) NASA; 548(cl) NASA-JPL; 548(cb) NASA-JPL; 548(br) NASA-JPL; 549(tr) NASA-JPL; 550–551: Getty Images: Stockbyte, John Foxx; 562–563: NASA, ESA

Chapter Opener images:

pages 22–23 Jupiter and Io.
pages 122–123 Star forming region in N66.
pages: 164–165 Messier 82 (M82).
pages 190–191 Stellar "Fireworks Finale" came first in the young universe.
pages 202–203 Map of the celestial hemispheres *c.* 1680.
pages 226–227 *Atlantis* docked with *Mir.*
pages 296–297 The Pleiades star cluster.

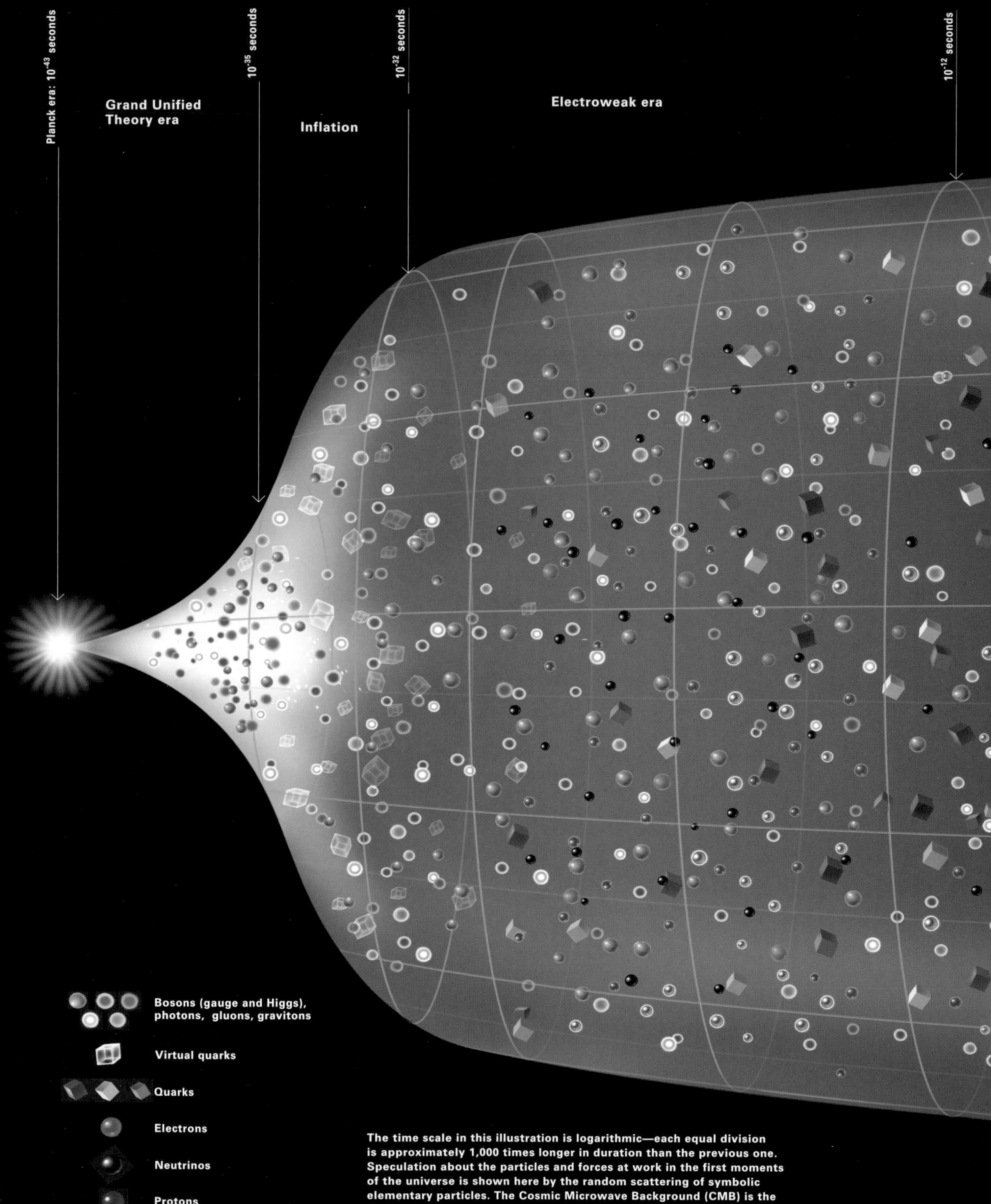

The time scale in this illustration is logarithmic—each equal division is approximately 1,000 times longer in duration than the previous one. Speculation about the particles and forces at work in the first moments of the universe is shown here by the random scattering of symbolic elementary particles. The Cosmic Microwave Background (CMB) is the first real evidence of what the early universe looked like. The first stars and galaxies appeared some time later, although cosmologists are still looking for evidence of which came first.